AF323685

Frontiers of Research with the Chinese Academy of Sciences — Vol. I

LANDAU–LIFSHITZ EQUATIONS

Boling Guo

*Beijing Institute of Applied Physics and
Computational Mathematics, China*

Shijin Ding

South China Normal University, China

World Scientific

NEW JERSEY · LONDON · SINGAPORE · BEIJING · SHANGHAI · HONG KONG · TAIPEI · CHENNAI

Published by

World Scientific Publishing Co. Pte. Ltd.

5 Toh Tuck Link, Singapore 596224

USA office: 27 Warren Street, Suite 401-402, Hackensack, NJ 07601

UK office: 57 Shelton Street, Covent Garden, London WC2H 9HE

Library of Congress Cataloging-in-Publication Data
Guo, Boling.
 Landau-Lifshitz equations / by Boling Guo & Shijin Ding.
  p. cm. -- (Frontiers of Research with the Chinese Academy of Sciences ; v. 1)
 Includes bibliographical references.
 ISBN-13: 978-981-277-875-8 (hardcover : alk. paper)
 ISBN-10: 981-277-875-6 (hardcover : alk. paper)
 1. Differential equations, Partial--Numerical solutions. 2. Maxwell equations--Numerical
solutions. 3. Geometry. 4. Mathematical physics. I. Ding, Shijin, 1959– II. Title.
 QC776.G86 2008
 518'.64--dc22

                 2007044801

British Library Cataloguing-in-Publication Data
A catalogue record for this book is available from the British Library.

Printed in Singapore by World Scientific Printers (S) Pte Ltd

Preface

In studying the dispersive theory of magnetization of ferromagnets in 1935, Landau–Lifshitz [102] proposed the equations of ferromagnetic spin chain which are important magnetization equations, called Landau–Lifshitz equations now. Later on, such equations were also found in the condensed matter physics. In the 1960s, Soviet physicists A. Z. Akhiezer, V. G. Beryahltar, S. V. Peletninskii studied spin wave, the equations of ferromagnetic chain and the traveling wave solutions in detail in their book "Spin Waves"[3]. In 1974, K. Nakamura, T. Sasada [122] first observed that there is a soliton solution to the one-dimensional Landau–Lifshitz equations without Gilbert damping. Then, many mathematicians and physicists studied the soliton theory of Landau–Lifshitz equations using the approaches including inverse scattering method, infinite many conservation laws, geometry expression method and gauge equivalence of nonlinear Schrödinger equations and so on. Early in 1957, Suhl [131] had studied the infinite dimensional dynamic system of the Landau–Lifshitz equations with Gilbert damping term. A series of further studies on the theory of dynamics and numerical results have appeared since then. In recent years, the ferromagnetic materials have been widely applied in the video and recording apparatus. This is one of the applications for Landau–Lifshitz equations.

From 1982, mathematicians began their studies on the well-posedness for Landau–Lifshitz equations. In China, a group headed by Yulin Zhou and Boling Guo proved the existence of the global weak solutions to the initial value problems and initial boundary value problems for Landau–Lifshitz equations from one dimension to multi-dimensions [150–157]. Alouges and Soyeur [4] proved similar results by penalty method in 1992. We also refer the readers to the result by P L. Sulem, C. Sulem and C. Bardos [132]. Since then, many other results on the global existence were obtained [20–22]. However, the regularity and the uniqueness were unsolved in the 1980s due to the complexity of Landau–Lifshitz equations.

However, in 1991, Zhou, Guo and Tan [158] obtained the existence and uniqueness of global smooth solution to one-dimensional Landau–Lifshitz equations with or without Gilbert damping by using a mobile frame on S^2 and some fine *a priori* estimates.

In 1993, Guo and Hong began the studies on two-dimensional Landau–Lifshitz equations. They established in [77] the relations between two-dimensional Landau–Lifshitz equations and harmonic maps and applied the approaches studying harmonic maps to get the global existence and uniqueness of partially regular weak solution.

This conclusion has been cited by many others up to now and gives rise to many successive works (see, for example, [47–49, 52, 89, 113, 115, 143]). Later on, in 1998, Chen, Ding and Guo [29] further proved that all the weak solutions with finite energy must be the Chen–Struwe solutions [34]. The uniqueness was also given. This says that the weak solution with finite energy is globally smooth with exception of finitely many singular points at most.

From 1998 to 2001, Guo and Ding discussed many other Landau–Lifshitz equations such as inhomogeneous equations, unsaturated equations and compressible equations [50, 51, 75, 108].

From the beginning of the new century, more and more mathematicians are interested in the researches of Landau–Lifshitz equations. We refer the readers to the works by Guo, Su, Carbou and Harpes, *et al.* [20–23, 80–84, 89] on Landau–Lifshitz equations and Landau–Lifshitz–Maxwell equations.

A natural question is the regularity of weak solutions to the higher dimensional Landau–Lifshitz equations. In this aspect, in 2004, Liu [109] proved that the "stationary" weak solutions of higher dimensional Landau–Lifshitz equations are partially regular. The Hausdorff dimensions and the Hausdorff measures of the singular set were estimated. These extend the results on harmonic map heat flow by Feldman [58] to Landau–Lifshitz equations. At the same time, Moser [115] obtained the similar results for lower dimensional Landau–Lifshitz equations by different methods.

We know that the "stationary" conditions are hard to verify. So, in 2005, Melcher [113] proved the partial regularity for the weak solutions to the initial value problems of Landau–Lifshitz equations. However, as stated by Melcher, his method does not fit the other dimensional problems and, the partial regularity of weak solutions to the boundary value problems are still unsolved.

This attracted the attention of Changyou Wang at the University of Kentucky. Wang [143], using the method of [142], proved the partial regularity for the weak solutions of the initial value problems and initial boundary value problems on three- and four-dimensional manifolds.

However, all the results on the partial regularity only answered the questions on the singular set such as how many points there are in the set or how large the set is, provided that the singularity does exist. But, the existence of finite time singularity of weak solutions is not answered. For the harmonic map heat flow, the similar questions were answered by Chen and Ding [30] in 1990 ($n \geq 3$) and by Chang, Ding and Ye [26] ($n = 2$) in 1992 (see also [39] and many others).

Does the weak solution of Landau–Lifshitz equations really blow-up at finite time? Pistella and Valente [123] in 2002, Bartels, Ko and Prohl [13] in 2005, gave positive answers respectively by numerical analysis.

In 2007, Ding and Wang [52] rigorously proved that in three and four dimensions, some Dirichlet problems and Neumann problems for Landau–Lifshitz equations indeed admit finite time blow-up solutions. Comparing with the similar proofs for harmonic map heat flows, we do not have the monotonicity inequality and the Bochner identity.

Unfortunately, our method does not apply to two-dimensional problems and higher dimensional problems. We do not know the types of the singularities either.

In recent years, there are many papers discussing the other problems such as domain wall, energy concentrations and vortices. We refer to [44–46, 98–100, 116, 117] and references therein.

The aim of this book is to introduce the readers the key works of the group headed by Yulin Zhou and Boling Guo in China from 1980s. There is a bibliography comment at the end of every chapter to introduce other mathematicians' works in this field and briefly state the development in recent years. However, it is not possible to include all the works and achievements throughout the world in such a short comment behind every chapter.

The authors are deeply grateful to Professor Yulin Zhou for his support to this work. We also want to thank Professors Yongqian Han, Jianqing Chen for their help in the preparation of this book. This work was partially supported by the projects of Natural Science Foundation of China (Grant No. 10471050), the National 973 Program of China (Grant No. 2006CB805902), Natural Science Foundation of Guangdong Province (Grant No. 7005795) and University Special Research Fund for Ph.D. Program of China (Grant No. 20060574002).

The Authors
March 2007

Contents

LANDAU–LIFSHITZ
EQUATIONS

Chapter 1

Spin Waves and Equations of Ferromagnetic Spin Chain

1.1 Physics Background for the Equations of Ferromagnetic Spin Chain

1.1.1 Motion Equations for Magnetization

Studying the dispersive theory of magnetization of ferromagnets, Landau–Lifshitz proposed the following motion equation of magnetization

$$\vec{S}_t = \lambda_1 \vec{S} \times \vec{H}^e - \lambda_2 \vec{S} \times (\vec{S} \times \vec{H}^e), \qquad (1.1.1)$$

where $\vec{S} = (S_1, S_2, S_3)$ is the vector of magnetization, $\vec{H}^e$ is the effective magnetic field applied to magnetic moment,

$$\vec{H}^e = \frac{\partial}{\partial \vec{S}} e_{\mathrm{mag}}(\vec{S}). \qquad (1.1.2)$$

Here $e_{\mathrm{mag}}(\vec{S})$ denotes the density of the total magnetic energy, $\vec{H}^e$ is also related to Maxwell equations, λ_1, λ_2 are constants, $\lambda_2 > 0$.

If a bounded multiply connected domain $\Omega \subset R^3$ is occupied by a ferromagnet under the constant temperature below Curie temperature, and if mechanics effects are not taking into account, one can consider the following magnetic energy functional:

1. *Anisotropic energy*

Anisotropic energy reads as

$$\mathcal{E}_{\mathrm{an}} = \int_\omega \Phi(\vec{S}) dx, \qquad (1.1.3)$$

in which $\Phi : R^3 \to R^+$ is a convex function and depends on the crystal structure of the materials.

Near the Curie temperature, taking first order approximation, one has

$$\Phi(\vec{S}) := \sum_{lm} b_{lm} S_l S_m, \qquad (1.1.4)$$

where $\{b_{lm}\}$ is a symmetric, positively definite tensor. $\Phi(\vec{S}) = k|\vec{S}|^2(\sin\theta)^2$ for example for the uniaxial crystal, where θ is the angle between $\vec{S}$ and the direction of "easy" magnetization, k is a positive constant.

2. *Exchange energy*

The behavior of ferromagnets is that the quantum force makes the magnetic field of molecules arrange in order. the most important quantity is the exchange energy

$$\mathcal{E}_{\text{ex}} := \frac{1}{2}\sum_{l,m} a_{lm} \int_\Omega \frac{\partial\vec{S}}{\partial x_l}\frac{\partial\vec{S}}{\partial x_m}dx, \qquad (1.1.5)$$

in which $\{a_{lm}\}$ is a symmetric, positive definite tensor.

3. *The energy to the magnetic field $\vec{H}$*

The energy of magnetic field $\vec{H}$ is:

$$\mathcal{E}_{\text{H}}(\vec{S}) := \frac{1}{8\pi}\int_{R^3} H^2 dx, \qquad (1.1.6)$$

where $\vec{H}$ and $\vec{S}$ being given by Maxwell's equations. Note that here the integral is extended to the whole space R^3 since $\vec{H}$ does not vanish outside the domain Ω.

The total magnetic energy has the form

$$\mathcal{E}_{\text{mag}}(\vec{S}) := \mathcal{E}_{\text{an}}(\vec{S}) + \mathcal{E}_{\text{ex}}(\vec{S}) + \mathcal{E}_{\text{H}}(\vec{S}) \qquad (1.1.7)$$

and at equilibrium state $\mathcal{E}_{\text{mag}}$ attains an absolute minimum. The magnetostatic Maxwell equation is:

$$\nabla\cdot(\vec{H} + 4\pi\vec{S}) = 0, \quad \text{in}\ \ R^3, \qquad (1.1.8)$$

$$\nabla\times\vec{H} = 0, \quad \text{in}\ \ R^3, \qquad (1.1.9)$$

where $\nabla\cdot = \text{div}$, $\nabla\times = \text{curl}$, "$\cdot$" denotes the inner product and "$\times$" denotes the vector product. $\vec{S}$ satisfies the non-convex condition:

$$|\vec{S}(x)| = S_0, \quad \text{in}\ \ \Omega. \qquad (1.1.10)$$

1.1.2 Landau–Lifshitz Equations

One model of dynamical system is the Landau–Lifshitz equation:

$$\frac{\partial\vec{S}}{\partial t} = \lambda_1\vec{S}\times\vec{H}^e - \lambda_2\vec{S}\times(\vec{S}\times\vec{H}^e), \qquad \text{in}\ \ \Omega\times(0,T), \qquad (1.1.11)$$

$$\vec{H}^e := -\frac{\partial\Phi(\vec{S})}{\partial\vec{S}} + \sum_{l,m} a_{lm}\frac{\partial^2\vec{S}}{\partial x_l\partial x_m} + \vec{H}, \quad \text{in}\ \ \Omega\times(0,T), \qquad (1.1.12)$$

in which λ_1, λ_2 are constants in physics, $\lambda_2 > 0$. The first term on the right-hand side of (1.1.11) which is not dissipative but resulted from the motion of $\vec{S}$ around $\vec{H}^e$, has a constant angle; the second term expresses the ordered arrangement of $\vec{S}$ according to $\vec{H}^e$ and it is due to the viscosity and then dissipative.

The initial condition is as follows

$$\vec{S}(x,0) = \vec{S}_0(x), \quad x \in \Omega. \tag{1.1.13}$$

It follows from (1.1.11) that

$$\frac{\partial}{\partial t}|\vec{S}|^2 = 2\vec{S}\frac{\partial\vec{S}}{\partial t} = 0. \tag{1.1.14}$$

Then, if $\vec{S}_0(x)$ satisfies (1.1.10), one has $|\vec{S}(x,t)| = S_0$, $x \in \Omega$, $t \geq 0$; if $\vec{S} \times \vec{H}^e \neq 0$, $\vec{S} \times \vec{H}^e$ and $\vec{S} \times (\vec{S} \times \vec{H}^e)$ is an orthogonal base which are on the tangential plane of the sphere $|\vec{S}| = S_0$. Hence, (1.1.11) is a dissipative nonlinear evolutionary equation of $\vec{S}$ on the surface of sphere.

The fields $\vec{H}$ and $\vec{S}$ solve the Maxwell equation

$$\nabla \times \vec{H} = \frac{\varepsilon}{c}\frac{\partial\vec{E}}{\partial t} + \frac{4\pi}{c}\vec{J}, \qquad \text{in } R^3 \times [0,T], \tag{1.1.15}$$

$$\nabla \times \vec{E} = -\frac{1}{c}\frac{\partial}{\partial t}(\mu_0\vec{H} + 4\pi\vec{S}), \quad \text{in } R^3 \times [0,T]. \tag{1.1.16}$$

It follows from Ohm law that:

$$\vec{J} = \sigma(\vec{E} + \vec{f}), \tag{1.1.17}$$

where $\vec{E}$ represents the electric field, $\vec{J}$ the density of current, σ the conductivity, ε magnetization rate of the electric medium, c the speed of light and $\vec{f}$ is the given non-induction electric force. Usually there holds: $\vec{B} := \mu_0\vec{H} + 4\pi\vec{S}$.

The initial data are

$$\vec{E}(x,0) = \vec{E}_0(x), \tag{1.1.18}$$

$$\vec{B}(x,0) = \vec{B}_0(x), \quad \nabla \cdot \vec{B}_0 = 0. \tag{1.1.19}$$

1.2 A Simple Derivation of Landau–Lifshitz Equation

1.2.1 Magnetically Ordered Crystals

Many crystals have an ordered magnetic structure. This means that in the absence of an external magnetic field, the mean magnetic moment of at least one of the atoms in each unit of cell of the crystal is non-zero. In the simplest type of magnetically ordered crystals, i.e. ferromagnets such as Fe, Ni, Co and Dy, the mean magnetic moments of all the atoms have the same orientation provided that the temperature of the ferromagnet does not exceed a critical value, i.e. the Curie temperature. For this reason ferromagnets have spontaneous magnetic moments, i.e. non-zero macroscopic magnetic moments, even in the absence of an external magnetic field.

In antiferromagnets, these include carbonates, anhydrous sulphates, oxides and fluorides of transition metals Mn, Ni, Co and Fe, the mean atomic magnetic moments compensate each other within each unit cell (in zero external magnetic field). In other words, an antiferromagnet consists of a set of sublattices (called magnetic sublattices), each of which has a non-zero mean moment. This type of magnetic order occurs if the temperature of the antiferromagnets is less than a critical temperature, known as the Neel temperature.

Finally, there is one further type of magnetically ordered crystal — that of the ferrites — which consists of a number of magnetic sublattices whose magnetic moments are uncompensated (in contrast to antiferromagnets); thus ferrites exhibit spontaneous magnetic moments. Examples of this type are compounds of transition metals such as the salts $MnO.Fe_2O_3, 3Y_2O_3.5Fe_2O_3$.

1.2.2   The Wave Function and Spin Operator for the System of Two Electrons

Let us consider a simple molecule model. Assume the molecule of Argon has two electrons and two protons between which there is no interaction with each other since the mass of the protons is much larger. The interaction of this system is as in Figure 1.2.1 in which a, b denote protons, 1, 2 represent electrons provided that there is Coulomb force between them.

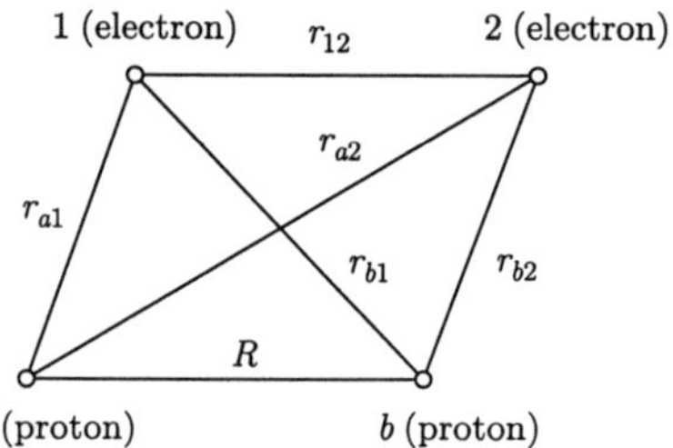

Figure 1.2.1.   *Interaction of oxygen molecular system.*

1. *Wave function of electrons*
Consider two-body problem:

$$\left(-\frac{\hbar^2}{2m}\nabla_1^2 - \frac{e^2}{r_{a_1}}\right)\varphi(r_{a_1}) = E_0\varphi(r_{a_1}), \tag{1.2.1}$$

$$\left(-\frac{\hbar^2}{2m}\nabla_2^2 - \frac{e^2}{r_{b_2}}\right)\varphi(r_{b_2}) = E_0\varphi(r_{b_2}), \tag{1.2.2}$$

where $\varphi(r_{a_1})$ is the wave function of electron "1", $\varphi(r_{b_2})$ is the wave function of electron "2", $\hbar$ is the Planck constant, m is the mass of the electron and e is the charge. The Hamiltonian for this system is $\hat{\mathcal{H}} = \hat{\mathcal{H}}_0 + \hat{\mathcal{H}}_1$ where:

$$\hat{\mathcal{H}}_0 = -\frac{\hbar^2}{2m}(\nabla_1^2 + \nabla_2^2) - \frac{e^2}{r_{a_1}} - \frac{e^2}{r_{b_1}}, \tag{1.2.3}$$

$$\hat{\mathcal{H}}_1 - \frac{e^2}{R} + \frac{e^2}{r} - \frac{e^2}{r_{a_2}} - \frac{e^2}{r_{b_2}}, \tag{1.2.4}$$

$\hat{\mathcal{H}}_1$ is a small perturbation. To find the eigenvalue E such that $\hat{\mathcal{H}}\Psi = E\Psi$, we take single electron approximation:

$$\varphi_1 = \varphi(r_{a_1})\varphi(r_{b_2}), \quad \varphi_2 = \varphi(r_{a_2})\varphi(r_{b_1}), \tag{1.2.5}$$

which is the wave function of the system. Let

$$\psi = C_1\varphi_1 + C_2\varphi_2, \tag{1.2.6}$$

and assume that φ_1, φ_2 are normalized. It follows from the identical principle that electrons "1" and "2" are identical or are invariant after two exchanges. Thus we have

$$\begin{aligned}
\int \varphi_1^* \hat{\mathcal{H}} \psi dx &= \int E\varphi_1^* \psi dx \\
&= E \int \varphi_1^*(C_1\varphi_1 + C_2\varphi_2)dx \\
&= C_1 E_0 + C_2 E_0 \nu^2 ,
\end{aligned} \tag{1.2.7}$$

where

$$\nu^2 = \int \varphi^*(r_{b_2})\varphi^*(r_{a_1})\varphi(r_{a_2})\varphi(r_{b_1})dr_1 dr_2 .$$

The left-hand side of (1.2.7) is

$$\int \varphi_1^* \hat{\mathcal{H}} \psi dx = C_1 \int \varphi_1^* \hat{\mathcal{H}} \varphi_1 dx + C_2 \int \varphi_1^* \hat{\mathcal{H}} \varphi_2 dx = C_1 \alpha_{11} + C_2 \alpha_{12}, \tag{1.2.8}$$

with

$$\begin{aligned}
\alpha_{11} &= \int \varphi^*(r_{b_2})\varphi^*(r_{a_1})\hat{\mathcal{H}}_0 \varphi(r_{a_1})\varphi(r_{b_2})dr_1 dr_2 \\
&\quad + \int \varphi^*(r_{b_2})\varphi^*(r_{a_1})\frac{e^2}{R}\varphi(r_{a_1})\varphi(r_{b_2})dr_1 dr_2 \\
&\quad + \int \varphi^*(r_{b_2})\varphi^*(r_{a_1})\left(\frac{e^2}{r} - \frac{e^2}{r_{a_2}} - \frac{e^2}{r_{b_1}}\right)\varphi(r_{a_1})\varphi(r_{b_2})dr_1 dr_2 \\
&= 2E_0 + \frac{e^2}{R} + A(r),
\end{aligned} \tag{1.2.9}$$

here

$$A(r) = \int \varphi^*(r_{b_2})\varphi^*(r_{a_1})\left(\frac{e^2}{r} - \frac{e^2}{r_{a_2}} - \frac{e^2}{r_{b_1}}\right)\varphi(r_{a_1})\varphi(r_{b_2})dr_1 dr_2, \tag{1.2.10}$$

and

$$\begin{aligned}
\alpha_{12} = &\int \varphi^*(r_{b_2})\varphi^*(r_{a_1})\hat{\mathcal{H}}_0\varphi(r_{a_2})\varphi(r_{b_1})dr_1 dr_2 \\
&+ \frac{e^2}{R}\int \varphi^*(r_{b_2})\varphi^*(r_{a_2})\varphi(r_{a_2})\varphi(r_{b_1})dr_1 dr_2 \\
&+ \int \varphi^*(r_{b_2})\varphi^*(r_{a_1})\left(\frac{e^2}{r} - \frac{e^2}{r_{a_2}} - \frac{e^2}{r_{b_1}}\right)\varphi(r_{b_1})\varphi(r_{b_2})dr_1 dr_2 \\
= &\left(2E_0 + \frac{e^2}{R}\right)\nu^2 + B(r),
\end{aligned} \tag{1.2.11}$$

here

$$B(r) = \int \varphi^*(r_{b_2})\varphi^*(r_{a_1})\left(\frac{e^2}{r} - \frac{e^2}{r_{a_2}} - \frac{e^2}{r_{b_1}}\right)\varphi(r_{b_1})\varphi(r_{b_2})dr_1 dr_2. \tag{1.2.12}$$

Hence we have from (1.2.7) and (1.2.8) that

$$C_1\alpha_{11} + C_2\alpha_{12} = C_1 E_0 + C_2 E_0 \nu^2. \tag{1.2.13}$$

Similarly it follows from $\varphi_2^*\hat{\mathcal{H}}\psi = E_0\varphi_2^*\psi$ that

$$C_1\alpha_{21} + C_2\alpha_{22} = C_1 E_0 \nu^2 + C_2 E_0, \tag{1.2.14}$$

where

$$\begin{cases} \alpha_{11} = (2E_0 + \frac{e^2}{R})\nu^2 + B(r), \\ \alpha_{22} = 2E_0 + \frac{e^2}{R} + A(r). \end{cases} \tag{1.2.15}$$

We have from (1.2.13) and (1.2.14) that

$$C_1(E_0 - \alpha_{11}) + C_2(E_0\nu^2 - \alpha_{12}) = 0,$$
$$C_1\left\{\left[E_0 - \left(2E_0 + \frac{e^2}{R}\right)\right]\nu^2 + B(r)\right\} + C_2\left\{\left[E_0 - \left(2E_0 + \frac{e^2}{R}\right)\right] + A(r)\right\} = 0.$$

If $C_1, C_2 \not\equiv 0$, one has

$$\begin{vmatrix} E_0 - \alpha_{11} & E_0\nu^2 - \alpha_{12} \\ E_0\nu^2 - \alpha_{21} & E_0 - \alpha_{22} \end{vmatrix} = 0 \tag{1.2.16}$$

and the corresponding eigenfunctions are

$$(1) \qquad C_1 = -C_2 = C_a,$$
$$\psi_a = C_a[\varphi(r_{a_1})\varphi(r_{b_2}) - \varphi(r_{a_2})\varphi(r_{b_1})]. \tag{1.2.17}$$
$$(2) \qquad C_1 = C_2 = C_s,$$
$$\psi_s = C_s[\varphi(r_{a_1})\varphi(r_{b_2}) + \varphi(r_{a_2})\varphi(r_{b_1})]. \tag{1.2.18}$$

Since $\int \psi_a \psi_a^* = \int \psi_s \psi_s^* = 1$, we have

$$C_a = \frac{1}{\sqrt{2(1-\gamma^2)}}, \quad C_a = \frac{1}{\sqrt{2(1+\gamma^2)}}, \tag{1.2.19}$$

where

$$\gamma = \int \varphi(r_{a_1})\varphi(r_{b_1})dr_1. \tag{1.2.20}$$

2. *Spin wave function*

The wave function ψ for the system of two electrons can be written as a product of the space and the spin wave functions

$$\psi(r_1\sigma_1, r_2\sigma_2) = \varphi(r_1, r_2)\chi(\sigma_1, \sigma_2), \tag{1.2.21}$$

where σ_1, σ_2 are the projections of the electron spins along a given axis. In accordance with the Pauli principle, the wave function ψ must be antisymmetric with respect to the simultaneous interchange of the coordinates and of the spin variables of electrons. This means that an antisymmetric space function will be associated with symmetric spin function, and conversely, a symmetric space function will be associated with an antisymmetric spin function.

The function χ will be symmetric if the resultant spin S of the two electrons is equal to unity ($S = 1$) and antisymmetric if $S = 0$. Therefore the space wave function will be antisymmetric for $S = 1$ and symmetric for $S = 0$. We shall denote these wave functions by φ_a (for $S = 1$) and φ_s (for $S = 0$):

$$\psi_a = C_a[\varphi(r_{a_1})\varphi(r_{b_2}) - \varphi(r_{a_2})\varphi(r_{b_1})], \quad S = 1. \tag{1.2.22}$$
$$\psi_s = C_s[\varphi(r_{a_1})\varphi(r_{b_2}) + \varphi(r_{a_2})\varphi(r_{b_1})], \quad S = 0. \tag{1.2.23}$$

The energies of the molecules in states corresponding to $S = 1$ and $S = 0$ are related to the functions φ_a and φ_s by

$$E_{\uparrow\uparrow}(r) = \int \varphi_a(r_1, r_2)\hat{\mathcal{H}}\varphi_a(r_1, r_2)dr_1 dr_2, \tag{1.2.24}$$

$$E_{\uparrow\downarrow}(r) = \int \varphi_s(r_1, r_2)\hat{\mathcal{H}}_1\varphi_s(r_1, r_2)dr_1 dr_2 \tag{1.2.25}$$

in which we have omitted the symbol indicating complex conjugate in the integrals, since φ_a and φ_s are real functions (the atoms are assumed to be in the ground states). Substituting the expressions of ψ_a and ψ_s into (1.2.24) and (1.2.25), we find that

$$E_{\uparrow\uparrow}(r) = 2E_0 + \frac{e^2}{R} + \frac{A(r) - B(r)}{1 - \gamma^2}, \quad S = 1, \tag{1.2.26}$$

$$E_{\uparrow\downarrow}(r) = 2E_0 + \frac{e^2}{R} + \frac{A(r) + B(r)}{1 + \gamma^2}, \quad S = 0. \tag{1.2.27}$$

Choosing wave functions such that $\gamma = 0$, we have

$$E_{\uparrow\uparrow}(r) = 2E_0 + \frac{e^2}{R} + (A(r) - B(r)), \quad S = 1, \tag{1.2.28}$$

$$E_{\uparrow\downarrow}(r) = 2E_0 + \frac{e^2}{R} + (A(r) + B(r)), \quad S = 0. \tag{1.2.29}$$

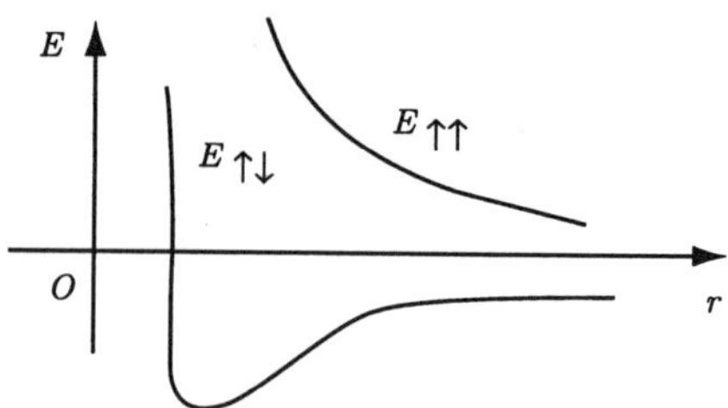

Figure 1.2.2. *Total energy of a system of two electrons.*

As shown in Figure 1.2.2, the total energy of the system can be represented by

$$E = \hat{\mathcal{H}} = 2E_0 + \frac{e^2}{R} + A - \frac{B}{2} - 2B\vec{S}_1 \cdot \vec{S}_2$$
$$= \hat{\mathcal{H}}_0 + \hat{\mathcal{H}}_{ex}, \tag{1.2.30}$$

where $\hat{\mathcal{H}}_{ex} = -2B\vec{S}_1 \cdot \vec{S}_2$ is called exchange Hamiltonian which was first obtained by Dirac in 1929. $\vec{S}_1$ and $\vec{S}_2$ are electron spin operators. We want to prove that (1.2.30) is identical to (1.2.28) and (1.2.29).

In fact,

$$\vec{S}_1 \cdot \vec{S}_2 = \frac{1}{2}\vec{S}^2 - \frac{1}{2}(\vec{S}_1^2 + \vec{S}_2^2)$$
$$= \frac{1}{2}S(S+1) - \frac{3}{4}$$
$$= \begin{cases} -\frac{3}{4}, & S = 0 \\ \frac{1}{4}, & S = 1, \end{cases} \tag{1.2.31}$$

therefore we get $E_{\uparrow\uparrow}$ ($S = 1$) and $E_{\uparrow\downarrow}$ ($S = 0$). We see that the term $-2B\vec{S}_1 \cdot \vec{S}_2$ denote the multibody effect.

For any operator $\hat{F}$ we can obtain the following motion equation:

$$\frac{d\hat{F}}{dt} = \frac{\partial \hat{F}}{\partial t} + \frac{1}{h}[\hat{\mathcal{H}}, \hat{\mathcal{H}}], \tag{1.2.32}$$

where $\hat{\mathcal{H}}$ is the Hamiltonian operator, $[\hat{\mathcal{H}}, \hat{F}] = \hat{\mathcal{H}}\hat{F} - \hat{F}\hat{\mathcal{H}}$. For the spin operator $\vec{S}_1$, we have

$$\frac{d\vec{S}_1}{dt} = \frac{\partial \vec{S}_1}{\partial t} + \frac{1}{ih}[\hat{\mathcal{H}}, \vec{S}_1] = [\hat{\mathcal{H}}, \vec{S}_1]$$
$$= \frac{1}{ih}\left[-\frac{h^2}{2m}(\nabla_1^2 + \nabla_2^2) - A - \frac{1}{2}B - 2B\vec{S}_1 \cdot \vec{S}_2, \vec{S}_1 \right]$$
$$= \frac{1}{ih}[-2B\vec{S}_1 \cdot \vec{S}_2, \vec{S}_1] = \frac{-2B}{ih}[\vec{S}_1 \cdot \vec{S}_2, \vec{S}_1],$$
$$[\vec{S}_1 \cdot \vec{S}_2, \vec{S}_1] = [\vec{S}_1 \cdot \vec{S}_2, \vec{S}_{1x}]\vec{i} + [\vec{S}_1 \cdot \vec{S}_2, \vec{S}_{1y}]\vec{j} + [\vec{S}_1 \cdot \vec{S}_2, \vec{S}_{1z}]\vec{k},$$

where

$$\begin{aligned}
[\vec{S}_1 \cdot \vec{S}_2, \vec{S}_{1x}] &= [\vec{S}_{1x} \cdot \vec{S}_{2x} + \vec{S}_{1y} \cdot \vec{S}_{2y} + \vec{S}_{1z} \cdot \vec{S}_{2z}, \vec{S}_{1x}] \\
&= [\vec{S}_{1y} \cdot \vec{S}_{2y} + \vec{S}_{1z} \cdot \vec{S}_{2z}, \vec{S}_{1x}] \\
&= [\vec{S}_{1y}, \vec{S}_{1x}]\vec{S}_{2y} + [\vec{S}_{1z}, \vec{S}_{1x}]\vec{S}_{2z} \\
&= ih(\vec{S}_{1y}\vec{S}_{2z} - \vec{S}_{1z}\vec{S}_{2y})
\end{aligned}$$

and the other terms can be derived in the similar manner. Hence we finally obtain

$$\frac{d\vec{S}_1}{dt} = \vec{S}_1 \times (-2B\vec{S}_2) = \vec{S}_1 \times \vec{H}_{\text{eff}}, \tag{1.2.33}$$

where

$$\hat{\mathcal{H}}_e = -2B\vec{S}_1 \cdot \vec{S}_2 = \vec{S}_1(-2B\vec{S}_2) = \vec{S}_1 \cdot \vec{H}_{\text{eff}}. \tag{1.2.34}$$

1.2.3 Multi-electron Wave Function and Spin Operator

1. *Equation for the isotropic ferromagnetic chain*

Now we consider the spin operator for multi-electron system, i.e. one-dimensional homogeneous Heisenberg model (see Figure 1.2.3): Assume that the ferromagnets are isotropic.

$$\hat{H}_e = -2B\vec{S}_i \cdot \sum_{j=1}^{k} \vec{S}_j = -2B\vec{S}_i(\vec{S}_{i-1} + \vec{S}_{i+1}). \tag{1.2.35}$$

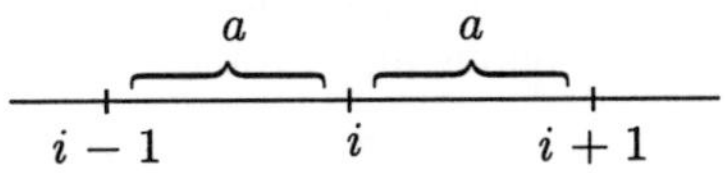

Figure 1.2.3. *One-dimensional homogeneous Heisenberg model.*

$$\vec{S}_{i+1} = \vec{S}_i + \frac{\partial \vec{S}_i}{\partial x}a + \frac{\partial^2 \vec{S}_i}{\partial x^2}a^2,$$

$$\vec{S}_{i-1} = \vec{S}_i - \frac{\partial \vec{S}_i}{\partial x}a + \frac{\partial^2 \vec{S}_i}{\partial x^2}a^2,$$

therefore

$$\begin{aligned}
\frac{d\vec{S}_i}{dt} &= \vec{S}_i \times \hat{H}_e = \vec{S}_i \times (-2B)\left(2\vec{S}_i + a^2\frac{\partial^2 \vec{S}_i}{\partial x^2}\right) \\
&= -2Ba^2\vec{S}_i \cdot \vec{S}_i \times \frac{\partial^2 \vec{S}_i}{\partial x^2}. \tag{1.2.36}
\end{aligned}$$

Passing to the continuous case: $\vec{S}_i \rightarrow \vec{S}(x,t)$ and setting the total energy of the system to be

$$\mathcal{H} = \frac{1}{2}\int \left(\frac{\partial \vec{S}}{\partial x}\right)^2 dx, \tag{1.2.37}$$

we get the magnetization spin motion equation as follows:

$$\frac{\partial \vec{S}}{\partial t} = \vec{S} \times H_{\text{eff}}, \tag{1.2.38}$$

$$\vec{H}_{\text{eff}} = -\frac{\partial \mathcal{H}}{\partial S}. \tag{1.2.39}$$

For the nonhomogeneous isotropic Heisenberg chain, the nonhomogeneous isotropic Heisenberg exchange Hamiltonian is

$$\mathcal{H} = -J \sum_{i=1}^{N-1} f_i \vec{S}_i \cdot \vec{S}_{i+1}, \tag{1.2.40}$$

The motion equation of $\vec{S}_i$ is

$$\frac{\partial \vec{S}_i}{\partial t} = J f_i (\vec{S}_i \times \vec{S}_{i+1}) + J f_{i-1} (\vec{S}_i \times \vec{S}_{i-1}). \tag{1.2.41}$$

Considering the continuous situation: $\vec{S}_i \to \vec{S}(x,t)$, $f_i \to f(x,t)$ and $\vec{S}_i$, f_i varies slowly in the same lattices (with length a), taking Taylor expansions for $\vec{S}(x+a,t)$, $f(x-a,t)$, we have from (1.2.41) that $\vec{S}(x,t)$ meets the motion equation

$$\vec{S}_t(x,t) = f(x)(\vec{S} \times \vec{S}_{xx}) + f_x(x)(\vec{S} \times \vec{S}_x), \tag{1.2.42}$$

where the variable of times has the scaling factor Ja^2.

2. *Anisotropic ferromagnetic chain equations*

Now let us consider the continuous anisotropic chain equation for the nonhomogeneous ferromagnets:

$$\mathcal{H} = \frac{1}{2} \int \left[\left(\frac{\partial \vec{S}}{\partial x} \right)^2 - J_1 S_1^2 - J_2 S_2^2 - J_3 S_3^2 \right]. \tag{1.2.43}$$

It follows from this that

$$\frac{\partial \vec{S}}{\partial t} = \vec{S} \times \vec{H}_{\text{eff}}, \tag{1.2.44}$$

in which

$$\vec{H}_{\text{eff}} = -\frac{\partial \mathcal{H}}{\partial \vec{S}} = \begin{pmatrix} J_1 & 0 & 0 \\ 0 & J_2 & 0 \\ 0 & 0 & J_3 \end{pmatrix} \cdot \begin{pmatrix} S_1 \\ S_2 \\ S_3 \end{pmatrix}. \tag{1.2.45}$$

1.3   Equations for the Antiferromagnets

1.3.1   Antiferromagnetic Moments and Magnetic Energy

Some ferromagnets exhibit the so-called antiferromagnetic property if placed under the temperature much less than the Curie temperature. These ferromagnets are called

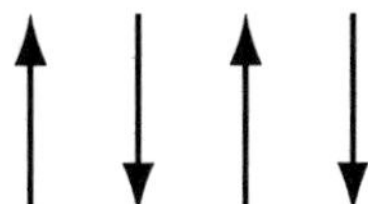

Figure 1.3.1   *Orientation of electron spin.*

antiferromagnets. Their electron spins are in different directions and compensate with each other as shown in Figure 1.3.1.

We can divide antiferromagnets into two systems: moment densities $\mathcal{M}_1(r,t)$ and $\mathcal{M}_2(r,t)$, and $|\mathcal{M}_1(r,t)| = |\mathcal{M}_2(r,t)|$. The energy of magnetic dipole interaction W_m and the energy W_H of the ferromagnet in the external magnetic field are given by the following formulae

$$W_m = \frac{-1}{2} \int_V (\mathcal{M}_1 + \mathcal{M}_2) H^{(m)} d\vec{r}, \tag{1.3.1}$$

$$W_H = - \int_V (\mathcal{M}_1 + \mathcal{M}_2) H_0^{(e)} d\vec{r}, \tag{1.3.2}$$

where $H^{(m)}$ represents the magnetic field due to the magnetic moment of the atoms in the antiferromagnets and the integrals are evaluated over the volume of the ferromagnet.

1.3.2   Equations for the Antiferromagnets

The density of the exchange energy W_e of an antiferromagnet is therefore of the form

$$W_e = f(\mathcal{M}_1, \mathcal{M}_2, \mathcal{M}_1^2, \mathcal{M}_2^2) + \frac{1}{2}\alpha_{ik}\left(\frac{\partial \mathcal{M}_1}{\partial x_i}\frac{\partial \mathcal{M}_1}{\partial x_k} + \frac{\partial \mathcal{M}_2}{\partial x_i}\frac{\partial \mathcal{M}_2}{\partial x_k}\right)$$

$$+ \alpha'_{ik}\frac{\partial \mathcal{M}_1}{\partial x_i}\frac{\partial \mathcal{M}_2}{\partial x_k}, \tag{1.3.3}$$

where f is a symmetric function of the magnetic moments $\mathcal{M}_1, \mathcal{M}_2$ and $\alpha_{ik}, \alpha'_{ik}$ are tensors. The first term in this expression represents the exchange energy density of uniformly magnetized sublattices, and the second and the third terms represent the exchange energy density connected with the non-uniformity of the magnetic moments. At the same time, the second term describes the exchange interaction in each of the sublattices, and the third term the exchange interaction between the sublattices. At sufficiently low temperatures the squares of the magnetic moment densities are practically constant and the function f can be regarded as depending only on the single variable $\mathcal{M}_1 \cdot \mathcal{M}_2$. In the simplest model of an antiferromagnet it is assumed that

$$f(\mathcal{M}_1, \mathcal{M}_2, \mathcal{M}_1^2, \mathcal{M}_2^2) = \delta \mathcal{M}.\mathcal{M}_2, \quad \delta > 0. \tag{1.3.4}$$

The total exchange energy of an antiferromagnet is therefore of the form

$$
\begin{aligned}
W_e &= \int_V w_e dr \\
&= \int_V \left\{ \delta\mathcal{M} \cdot \mathcal{M}_2 + \frac{1}{2}\alpha_{ik}\left(\frac{\partial\mathcal{M}_1}{\partial x_i}\frac{\partial\mathcal{M}_1}{\partial x_k} + \frac{\partial\mathcal{M}_2}{\partial x_i}\frac{\partial\mathcal{M}_2}{\partial x_k}\right) + \alpha'_{ik}\frac{\partial\mathcal{M}_1}{\partial x_i}\frac{\partial\mathcal{M}_2}{\partial x_k} \right\} dr,
\end{aligned}
$$

$$(1.3.5)$$

and the total exchange energy of an isotropic antiferromagnet is

$$
\begin{aligned}
\mathcal{H} = \int_V d^3x\Bigg[&\frac{1}{2}\alpha_{ik}\left(\frac{\partial\mathcal{M}_1}{\partial x_i}\frac{\partial\mathcal{M}_1}{\partial x_k} + \frac{\partial\mathcal{M}_2}{\partial x_i}\frac{\partial\mathcal{M}_2}{\partial x_k}\right) + \alpha'_{ik}\frac{\partial\mathcal{M}_1}{\partial x_i}\frac{\partial\mathcal{M}_2}{\partial x_k} \\
&+ \delta\mathcal{M}_1 \cdot \mathcal{M}_2 - \frac{1}{2}(\mathcal{M}_1 + \mathcal{M}_2)H^{(m)} - (\mathcal{M}_1 + \mathcal{M}_2)H_0^{(e)}\Bigg],
\end{aligned}
$$

$$(1.3.6)$$

In some simplified situations, we have

$$
\mathcal{H} = \int_V d^3x[k_1|\nabla\mathcal{M}_1|^2 + k_{12}|\nabla\mathcal{M}_2|^2 + k_{12}\nabla\mathcal{M}_1 \cdot \nabla\mathcal{M}_2].
$$

$$(1.3.7)$$

In this case the motion equation is of the form

$$
\begin{cases}
\dfrac{\partial\mathcal{M}_1}{\partial t} = \mathcal{M}_1 \times [2k_1\nabla^2\mathcal{M}_1 + k_{12}\nabla^2\mathcal{M}_2], \\[2mm]
\dfrac{\partial\mathcal{M}_2}{\partial t} = \mathcal{M}_2 \times [2k_2\nabla^2\mathcal{M}_2 + k_{12}\nabla^2\mathcal{M}_1].
\end{cases}
$$

$$(1.3.8)$$

1.4   Spin Waves in Ferromagnets

1.4.1   Equilibrium State of Ferromagnets

1. *Equilibrium state conditions of ferromagnets*

Consider a uniaxial ferromagnet. Assume that

$$
\mathcal{M}(r,t) = \mathcal{M}_0(r,t) + \vec{m}(r,t), \tag{1.4.1}
$$

$$
\vec{H}^{(i)}(r,t) = \vec{H}_0^{(i)} + \vec{h}(r,t), \tag{1.4.2}
$$

where $\vec{m}, \vec{h}$ are small derivations from $\mathcal{M}_0$ and $\vec{H}_0^{(i)}$, $\mathcal{M}_0$ is the equilibrium magnetization, $\vec{H}_0^{(i)}$ denotes the magnetic field inside the ferromagnet. According to the equilibrium conditions one has

$$
\vec{H}_0^{(i)} + \beta\vec{n}(\mathcal{M}_0 \cdot \vec{n}) - 2\mathcal{M}_0 f'(\mathcal{M}_0^2) = 0, \tag{1.4.3}
$$

or

$$
\vec{H}_0^{(e)} + \beta\vec{n}(\mathcal{M}_0 \cdot \vec{n}) - 4\pi\hat{N} \cdot \mathcal{M}_0 - 2\mathcal{M}_0 f'(\mathcal{M}_0^2) = 0, \tag{1.4.4}
$$

where β is a constant, $\vec{n}$ is a unit vector along the anisotropy axis, $\hat{N} = \hat{N}(r)$ is the demagnetization tensor with elements

$$N_{ik}(r) = \frac{1}{4\pi}\frac{\partial^2}{\partial x_i \partial x_k}\int_V \frac{dr'}{|r-r'|}. \qquad (1.4.5)$$

When the ferromagnet is an ellipsoid, $\mathcal{M} = $ constant, and then

$$\int_V \frac{dr'}{|r-r'|} = \pi abc \int_0^\infty \frac{ds}{R_s}\left(1 - \frac{x^2}{a^2+s} - \frac{y^2}{b^2+s} - \frac{z^2}{c^2+s}\right), \qquad (1.4.6)$$

where $R_s = \sqrt{(a^2+s)(-b^2+s)(c^2+s)}$, a,b,c are the semi-axes of the ellipsoid, x,y,z are the projection of the radial vector $\vec{r}$ in any point of the ellipsoid onto the main axses. The elements on the diagonal of the tensor $\vec{N}$ are

$$\begin{cases} N_1 = \dfrac{1}{2}abc\displaystyle\int_0^\infty \frac{ds}{(a^2+s)R_s}, \\[2ex] N_2 = \dfrac{1}{2}abc\displaystyle\int_0^\infty \frac{ds}{(b^2+s)R_s}, \\[2ex] N_3 = \dfrac{1}{2}abc\displaystyle\int_0^\infty \frac{ds}{(c^2+s)R_s}. \end{cases} \qquad (1.4.7)$$

It is clear that $N_1 + N_2 + N_3 = 1$. If the ferromagnet occupies a sphere, then $N_1 = N_2 = N_3 = 1/3$. If it occupies a cylinder ($a = \infty, b = c$), $N_1 = 0$, $N_2 = N_3 = 1/2$. What we are concerned are the equilibrium states when (1.4.7) admits solution:

$$N_3 > N_2 > N_1, \text{ if } \beta > 0; \quad N_3 < N_2 < N_1, \text{ if } \beta < 0,$$

see Figures 1.4.1 and 1.4.2.

2. *The equation of motion for the magnetization*

The expression of effective magnetic field related to (1.4.1) and (1.4.2) is

$$\begin{aligned} \mathcal{H} &= \vec{h} + \alpha_{ik}\frac{\partial^2 \vec{m}}{\partial x_i \partial x_k} - \frac{1}{\mathcal{M}_0^2}\{\mathcal{M}_0 \cdot H_0^{(i)} + \beta(\mathcal{M}_0 \cdot \vec{n})^2\}\vec{m} \\ &\quad + \beta\vec{n}(\vec{m}\cdot\vec{n}) - 4\mathcal{M}_0 f''(\mathcal{M}_0^2)(\mathcal{M}_0 \cdot \vec{m}). \end{aligned} \qquad (1.4.8)$$

Therefore the equation of motion for the magnetization in the case of small departures from the equilibrium value, which we shall refer to as the linearized equation of motion, will be of the form

$$\frac{\partial \vec{m}}{\partial t} = g\left[\mathcal{M}_0, \vec{h} + \alpha_{ik}\frac{\partial^2 \vec{m}}{\partial x_i \partial x_k} + \beta\vec{n}(\vec{m}\cdot\vec{n}) - \frac{1}{\mathcal{M}_0^2}\{\mathcal{M}_0 \cdot H_0^{(i)} + \beta(\mathcal{M}_0 \cdot \vec{n})^2\}\vec{m}\right].$$

If the ferromagnet exhibits magnetic anisotropy of the "easy axis" type ($\beta > 0$) and $H_0^{(i)}$ is parallel to $\vec{n}$, the linearized effective magnetic field is given by

$$\mathcal{H} = \vec{h} - \left(\beta + \frac{H_0^{(i)}}{\mathcal{M}_0}\right)\vec{m} + \alpha_{ik}\frac{\partial^2 \vec{m}}{\partial x_i \partial x_k} + (\beta - 4\mathcal{M}_0 f''(\mathcal{M}_0^2))(\vec{m}\cdot\hat{n})n. \qquad (1.4.9)$$

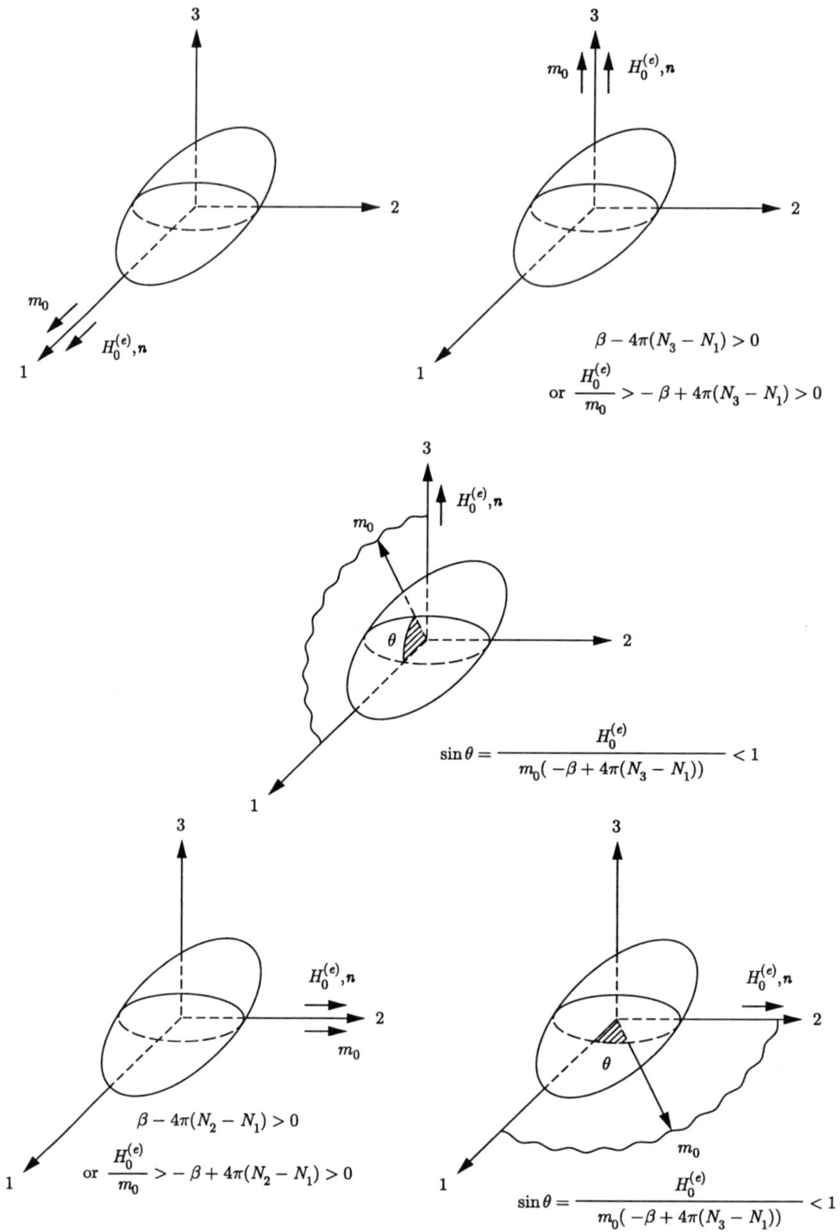

Figure 1.4.1. *Equilibrium states of a ferromagnet-1 $N_3 > N_2 > N_1$, if $\beta > 0$.*

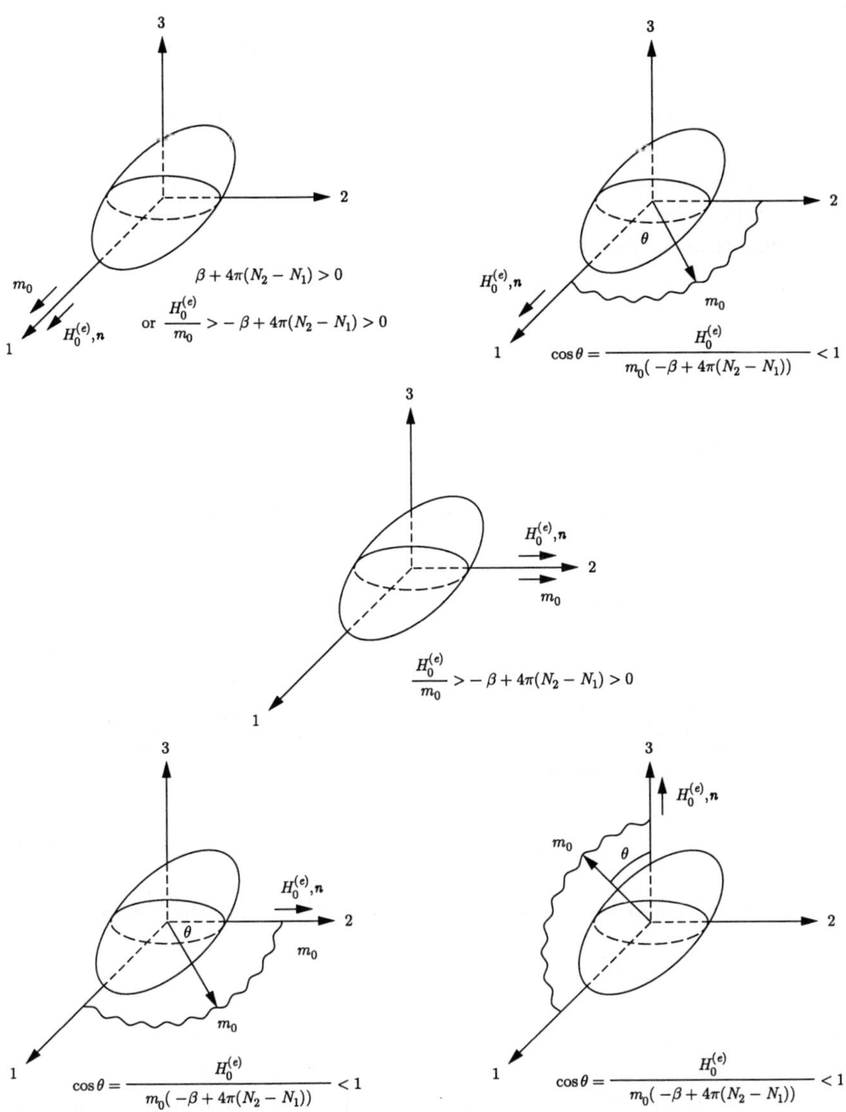

Figure 1.4.2. *Equilibrium states of a ferromagnet-2 $N_3 < N_2 < N_1$, if $\beta < 0$.*

If the ferromagnet exhibits magnetic anisotropy of the "easy plane" type ($\beta < 0$) and $H_0^{(i)}$ is perpendicular to $\vec{n}$, the linearized effective magnetic field is given by

$$\mathcal{H} = \left(\vec{h} - \frac{H_0^{(i)}}{\mathcal{M}_0}\right)\vec{m} + \alpha_{ik}\frac{\partial^2 \vec{m}}{\partial x_i \partial x_k} + \beta\vec{n}(\vec{m}\cdot\vec{n}) - 4\mathcal{M}_0 f''(\mathcal{M}_0^2)(\mathcal{M}_0\cdot\vec{m}). \quad (1.4.10)$$

The linearized equation of motion for the magnetization (1.4.8) must be augmented by the boundary conditions for the magnetization. The general boundary condition for the magnetization was formulated as

$$\left.\frac{\partial\mathcal{F}}{\partial(\partial\mathcal{M}/\partial x_k)}\nu_k\right|_S = 0, \quad (1.4.11)$$

where $\mathcal{F}$ is the energy density of the ferromagnet:

$$\mathcal{F}\left(\mathcal{M}, \frac{\partial\mathcal{M}_i}{\partial x_k}\right) = \frac{1}{2}\alpha_{ik,lm}(\mathcal{M})\frac{\partial\mathcal{M}_i}{\partial x_k}\frac{\partial\mathcal{M}_k}{\partial x_m} + \gamma_{ik}(\mathcal{M})\frac{\partial\mathcal{M}_i}{\partial x_k}$$
$$+ \frac{1}{2}\beta_{ik}\mathcal{M}_i\cdot\mathcal{M}_k + f(\mathcal{M}^2), \quad (1.4.12)$$

where ν is the unit vector along the outward normal to the surface of the ferromagnet.

Since we are interested in small departures of the magnetic moment from the equilibrium value and small magnetization gradients, we have

$$\frac{\partial\mathcal{F}}{\partial(\partial\mathcal{M}_i/\partial x_k)} = \alpha_{kj}\frac{\partial\vec{m}_i}{\partial x_j} + \gamma_{ik,l}\vec{m}_l, \qquad \gamma_{ik,l} = \left.\left(\frac{\partial\gamma_{ik}}{\partial\mathcal{M}_l}\right)\right|_{\mathcal{M}=\mathcal{M}_0},$$

and the boundary condition assumes the form

$$\left.\left(\alpha_{kj}\frac{\partial m_i}{\partial x_j} + \gamma_{ik,l}m_l\right)\nu_k\right|_S = 0. \quad (1.4.13)$$

The linearized equation of motion for the magnetization is then of the form

$$\frac{\partial\vec{m}}{\partial t} = g[\mathcal{M}_0, \mathcal{H}], \quad (1.4.14)$$

where

$$\mathcal{H} = \vec{h} - \beta(z)\vec{m} + \alpha\Delta\vec{m}. \quad (1.4.15)$$

Assuming that the function $\beta(z)$ increases rapidly in the thin layer of thickness δ in which $\vec{m}$ and $\vec{h}$ are practically constant, we obtain after integrating this equation with respect to z between 0 and δ,

$$\delta\frac{\partial\vec{m}}{\partial t} = g\left[\mathcal{M}_0, \vec{h}\cdot\delta - \vec{m}\int_0^\delta \beta(z)dz + \left.\alpha\frac{\partial\vec{m}}{\partial z}\right|_{z=\delta}\right.$$
$$\left. + \alpha\left(\frac{\partial^2\vec{m}}{\partial x^2} + \frac{\partial^2\vec{m}}{\partial y^2}\right)\delta\right], \quad (1.4.16)$$

where we have taken into account the fact that

$$\frac{\partial \vec{m}}{\partial z}\bigg|_{z=0} = 0. \tag{1.4.17}$$

Assuming further that as $\delta \to 0$, the integral

$$\int_0^\delta \beta(z)dz < \infty, \tag{1.4.18}$$

we obtain from the last equation the following effective boundary condition:

$$d\frac{\partial \vec{m}}{\partial z} - \vec{m}|_{z=0} = 0, \tag{1.4.19}$$

where

$$d = \frac{\alpha}{\int_0^\delta \beta(z)dz}. \tag{1.4.20}$$

In order to be able to neglect in (1.4.16) terms proportional to δ, it is clearly necessary that the frequency ω of the variation in the magnetization, the wavelength λ and the quantity $\beta(0)$ must satisfy the conditions:

$$\omega \ll g\mathcal{M}_0\beta(0), \quad \lambda \gg \sqrt{\frac{\alpha}{\beta(0)}}, \quad \beta(0) \gg 1. \tag{1.4.21}$$

Then

$$\alpha\frac{\partial^2 \vec{m}}{\partial z^2} - \beta(0)\vec{m} = 0,$$

it can readily be concluded that

$$\delta \ll \sqrt{\frac{\alpha}{\beta(0)}}$$

or

$$\delta \ll d.$$

If the wavelength λ satisfies the inequalities $\sqrt{\delta}d \ll \lambda \ll d$, the effective boundary condition assumes the form

$$\frac{\partial \vec{m}}{\partial z}\bigg|_{z=0} = 0, \quad \sqrt{\delta}d \ll \lambda \ll d. \tag{1.4.22}$$

If on the other hand $\lambda \gg d$, then

$$\vec{m}|_{z=0} = 0, \quad \lambda \gg d. \tag{1.4.23}$$

1.4.2   Spin Waves in Ferromagnets

1. *Motion of spin waves in ferromagnets*

Now we consider the propagations of spin waves in ferromagnets. Using the Fourier representations

$$\vec{m}(r,t) = \int \vec{m}(\vec{k},\omega)\exp[i(\vec{k}\cdot\vec{r}-\omega t)]d\vec{k}d\omega,$$

$$\vec{h}(r,t) = \int \vec{h}(\vec{k},\omega)\exp[i(\vec{k}\cdot\vec{r}-\omega t)]d\vec{k}d\omega,$$

$$(1.4.24)$$

we have from (1.4.8) that

$$-i\omega\vec{m}(\vec{k},\omega) = g\Bigg[\mathcal{M}_0, h(\vec{k},\omega) - \bigg\{\alpha_{ij}k_ik_j + \frac{\mathcal{M}_0\cdot H_0^{(i)}}{\mathcal{M}_0^2} + \beta\frac{(\mathcal{M}_0\cdot\vec{n})^2}{\mathcal{M}_0^2}\bigg\}\vec{m}(\vec{k},\omega)$$

$$+ \beta\vec{n}(\vec{n}\cdot\vec{m}(\vec{k},\omega))\Bigg].$$

$$(1.4.25)$$

This equation gives the relationship between the Fourier components $\vec{m}(\vec{k},\omega)$ and $h(\vec{k},\omega)$ which we shall write in the form

$$m_i(\vec{k},\omega) = \chi_{ij}(\vec{k},\omega)h_j(\vec{k},\omega),$$

$$(1.4.26)$$

where

$$\chi_{ij}(\vec{k},\omega) = \begin{pmatrix} \chi_{xx} & \chi_{xy} & 0 \\ \chi_{yx} & \chi_{yy} & 0 \\ 0 & 0 & 0 \end{pmatrix},$$

$$(1.4.27)$$

$$\chi_{xx} = \frac{g\mathcal{M}_0\Omega_1}{\Omega_1\Omega_2 - \omega^2}, \quad \chi_{yy} = \frac{g\mathcal{M}_0\Omega_2}{\Omega_1\Omega_2 - \omega^2}, \quad \chi_{xy} = -\chi_{yx} = \frac{i\omega g\mathcal{M}_0}{\Omega_1\Omega_2 - \omega^2},$$

$$\Omega_1 = g\mathcal{M}_0\left(\alpha_{ij}k_ik_j + \frac{\mathcal{M}_0\cdot H_0^{(i)}}{\mathcal{M}_0^2} + \beta\cos^2\psi\right),$$

$$\Omega_2 = g\mathcal{M}_0\left(\alpha_{ij}k_ik_j + \frac{m_0\cdot H_0^{(i)}}{\mathcal{M}_0^2} + \beta\cos 2\psi\right),$$

$$(1.4.28)$$

and ψ is the angle between the anisotropy axis $\vec{n}$ and the vector $\mathcal{M}_0$; the z-axis lies along $\mathcal{M}_0$ and the x-axis lies in the plane containing the vector $\vec{n}$ and $\mathcal{M}_0$. The quantities $\chi_{ij}(\vec{k},\omega)$ form a tensor called the high-frequency magnetic susceptibility tensor of the ferromagnet.

If the ferromagnet exhibits anisotropy of the "easy axis" type ($\beta > 0$) and m_0, $\vec{n}$, $H_0^{(i)}$ are parallel with each other, then

$$\Omega_1 = \Omega_2 = \Omega,$$

$$(1.4.29)$$

where

$$\Omega = g\mathcal{M}_0\left(\alpha_{ij}k_ik_j + \frac{H_0^{(i)}}{\mathcal{M}_0} + \beta\right). \tag{1.4.30}$$

The formulae corresponding to this case are valid also for ferromagnets with cubic symmetry. All that is required is to replace β by $2\beta'\mathcal{M}_0^2$, if the easy magnetization axis lies along the edge of the cube, and by $\frac{4}{3}|\beta'|\mathcal{M}_0^2$, if the easy magnetization axis lies along the space diagonal of the cube, where β' is the anisotropy constant.

If the ferromagnet exhibits magnetic anisotropy of the "easy plane" type ($\beta < 0$) and $\mathcal{M}_0 \perp \vec{n}$, $\mathcal{M}_0$ is parallel to $H_0^{(i)}$,

$$\begin{cases} \Omega_1 = g\mathcal{M}_0\left(\alpha_{ij}k_ik_j + \frac{H_0^{(i)}}{\mathcal{M}_0}\right), \\ \Omega_2 = g\mathcal{M}_0\left(\alpha_{ij}k_ik_j + \frac{H_0^{(i)}}{\mathcal{M}_0} + |\beta|\right). \end{cases} \tag{1.4.31}$$

2. *Dispersion of spin waves*

We must now establish the dependence of the frequency ω of the spin wave on its wave vector $\vec{k}$. This requires the use of both the equation of motion for the magnetic moment and the Maxwell equations. However, spin waves are low-frequency magnetic waves, so that the electric field can be neglected and the magnetic field may be assumed to be irrotational. In other words, spin waves can be treated in terms of the magnetostatic approximation, i.e. we can assume that $\vec{m}(r,t)$ and $\vec{h}(r,t)$ satisfy the equations

$$\begin{cases} \mathrm{rot}\,\vec{h}(r,t) = 0, \\ \mathrm{div}\,\vec{h}(r,t) = 4\pi\,\mathrm{div}\,\vec{m}(r,t). \end{cases} \tag{1.4.32}$$

Transforming to the Fourier components in these equations, we obtain

$$\begin{cases} [\vec{k}, \vec{h}(\vec{k},\omega)] = 0, \\ \vec{k}\cdot\vec{h}(\vec{k},\omega) = 4\pi\vec{k}\cdot\vec{m}(\vec{k},\omega). \end{cases} \tag{1.4.33}$$

From the first equation it follows that the magnetic field $\vec{h}(\vec{k},w)$ is parallel to the wave vector $\vec{k}$:

$$\vec{h}(\vec{k},\omega) = -i\vec{k}\cdot\varphi(\vec{k},\omega),$$

where $\varphi(\vec{k},\omega)$ is the Fourier component of the magnetic potential. Since

$$\vec{m}(\vec{k},\omega) = \chi(\vec{k},\omega)\vec{h}(\vec{k},\omega),$$

we can write the second equation in (1.4.33) in the form

$$(k^2 + 4\pi k_ik_j\chi_{ij}(\vec{k},\omega))\varphi(\vec{k},\omega) = 0$$

and hence

$$k^2 + 4\pi k_ik_j\chi_{ij}(\vec{k},\omega) = 0. \tag{1.4.34}$$

This equation relates ω and the wave vector $\vec{k}$ of the spin wave (it is called the dispersion relation). Using (1.4.27) we can reduce the dispersion relation (1.4.34) to the form

$$1 + \frac{4\pi g \mathcal{M}_0 \Omega_1}{\Omega_1 \Omega_2 - \omega^2 k^2} \frac{k_x^2}{k^2} + \frac{4\pi g \mathcal{M}_0 \Omega_2}{\Omega_1 \Omega_2 - \omega^2 k^2} \frac{k_y^2}{k^2} = 0$$

and hence

$$\omega_s(\vec{k}) = \sqrt{\Omega_1 \Omega_2 + 4\pi g \mathcal{M}_0 (\Omega_1 \cos^2 \varphi_k + \Omega_2 \sin^2 \varphi_k) \sin^2 \theta_k}, \qquad (1.4.35)$$

where θ_k and φ_k are the polar and azimuthal angles of the wave vector $\vec{k}$. We recall that the z-axis lies along the vector $\mathcal{M}_0$ and the x-axis lies in the plane containing $\mathcal{M}_0$ and $\vec{n}$. For wave vectors with $\alpha k^2 \gg 1$, the equation for the frequency of the spin wave becomes much simpler:

$$\omega_S(\vec{k}) = g \mathcal{M}_0 \alpha_{ij} k_i k_j. \qquad (1.4.36)$$

In the isotropic case this formula assumes the form

$$\omega_S(\vec{k}) = \frac{\theta_C}{\hbar} (ak)^2, \qquad (1.4.37)$$

where $\theta_C = \hbar g \mathcal{M}_0 \alpha / a^2$ (θ_C is of the order of the Curie temperature). It follows that for large wave vectors ($\alpha k^2 \gg 1$), the spin wave frequency is proportional to the square of the wave vector.

1.4.3   Damping of Spin Waves

1. *Expression of the damping of spin waves*

Now we consider the damping of spin waves. Damping is due to the interaction of spin waves with each other and also with lattice vibrations and conduction electrons. The phenomenological description can be obtained from the equation of motion for the magnetization, containing the relaxation term R:

$$R = \frac{1}{\tau_2} \mathcal{H} - \frac{1}{\tau_1} [\vec{n}, [\vec{n}, \mathcal{H}]], \qquad (1.4.38)$$

where $\vec{n} = \frac{\mathcal{M}_0}{|\mathcal{M}_0|}$ and τ_1, τ_2 are constants which have the dimensions of time, ($\frac{1}{\tau_2} > 0$, $\frac{1}{\tau_1} + \frac{1}{\tau_2} > 0$). Then the equation for $\mathcal{M}$ will be the form

$$\frac{\partial \mathcal{M}}{\partial t} = g[\mathcal{M}, \mathcal{H}] + \frac{1}{\tau_2} \mathcal{H} - \frac{1}{\tau_1} [\vec{n}, [\vec{n}, \mathcal{H}]], \qquad (1.4.39)$$

the effective magnetic field is

$$\mathcal{H} = \vec{h} - \left(\beta + \frac{H_0^{(i)}}{\mathcal{M}_0} \right) \vec{m} + \alpha_{ik} \frac{\partial^2 \vec{m}}{\partial x_i \partial x_k} + (\beta - 4\mathcal{M}_0 f''(\mathcal{M}_0^2))(\vec{m} \cdot \vec{n})\vec{n}, \quad (1.4.40)$$

here we mainly focus on the uniaxial ferromagnet and assume that the field $H_0^{(i)}$ is parallel to the easy magnetization axis. The quantity $f''(\mathcal{M}_0^2)$ in the expression for $\mathcal{H}$ can readily be related to the static susceptibility of the ferromagnet, $\chi_{zz}^0 = \partial \mathcal{M}_0 / \partial H_0^{(i)}$. We have seen that in the state of equilibrium

$$H_0^{(i)} + \beta \mathcal{M}_0 - 2\mathcal{M}_0 f'(\mathcal{M}_0^2) = 0 \tag{1.4.41}$$

and hence

$$(2\mathcal{M}_0)^2 f'(\mathcal{M}_0^2) = \frac{1}{\chi_{zz}^0} - \frac{H_0^{(i)}}{\mathcal{M}_0}. \tag{1.4.42}$$

From these formulae we obtain the following expression for the high-frequency susceptibility tensor χ:

$$\chi(\vec{k}, \omega) = \begin{pmatrix} \chi_{xx} & \chi_{xy} & 0 \\ \chi_{yx} & \chi_{yy} & 0 \\ 0 & 0 & \chi_{zz} \end{pmatrix}, \tag{1.4.43}$$

in which

$$\chi_{xx} = \chi_{yy} = \frac{g\mathcal{M}_0\Omega - (i\omega/\tau) + (\Omega/g\mathcal{M}_0\tau^2)}{\Omega^2 - (\omega + i\Omega/g\mathcal{M}_0\tau)^2},$$

$$\chi_{zz} = \frac{\chi_{zz}^0}{1 + \chi_{zz}^0(\alpha_{ij}k_ik_j - i\omega\tau_2)},$$

$$\chi_{xy} = -\chi_{yx} = \frac{i\omega g\mathcal{M}_0}{\Omega^2 - (\omega + i\Omega/g\mathcal{M}_0\tau)^2}, \tag{1.4.44}$$

$$\frac{1}{\tau} = \frac{1}{\tau_1} + \frac{1}{\tau_2}, \quad \Omega = g\mathcal{M}_0\left(\alpha_{ij}k_ik_j + \frac{H_0^{(i)}}{\mathcal{M}_0} + \beta\right).$$

We note that $1/g\mathcal{M}_0\tau \ll 1$. We also note that the component χ_{zz} on the tensor $\chi(\vec{k}, \omega)$ is not zero, whereas it does vanish when $\tau_2 = \infty$. Substituting (1.4.44) into (1.4.34) we obtain the equation for the frequency of the spin wave as a function of its wave vector. When $R \neq 0$, this equation has complex roots whose real part determines the frequencies of the spin waves and whose imaginary part determines the damping rate.

2. *Damping rate*

Since $\Omega/g\mathcal{M}_0\tau \ll |\omega|$, the damping rate $\gamma_S(\vec{k})$ is given by

$$\gamma_S(\vec{k}) = \frac{1}{g\mathcal{M}_0\tau}(\Omega + 2\mathcal{M}_0\sin^2\theta_k). \tag{1.4.45}$$

If $\alpha k^2 \ll 1$, then

$$\gamma_S(\vec{k}) = \frac{\beta^2}{24\pi}g\mathcal{M}_0\frac{\mu_0\mathcal{M}_0}{\theta_C}\left(\frac{T}{\theta_C}\right)^2; \tag{1.4.46}$$

if, on the other hand $\alpha k^2 \gg 1$, then

$$\gamma_S(\vec{k}) \approx \frac{\theta_C}{\hbar}(ak)^3\left(\frac{T}{\theta_C}\right), \quad \hbar\omega_S(\vec{k}) \gg T. \tag{1.4.47}$$

Substitute

$$\mathcal{M} = \mathcal{M}_0 + \vec{m}$$

into (1.4.39), here $\vec{m}$ is a small addition to the equilibrium magnetization $\mathcal{M}_0$, which we shall assume that $\mathcal{M}_0 = \mathcal{M}_0(t)$ is independent of r. The effective magnetic field is given by, to within terms linear in $\vec{m}$,

$$\mathcal{H} = -\left(\frac{4\pi}{3} + \beta + \frac{H_0^{(e)}}{\mathcal{M}_0}\right)\vec{m} + \left(\beta - \frac{4\pi}{3} + \frac{H_0^{(e)}}{\mathcal{M}_0} - \frac{1}{\chi_{zz}^0}\right)(\vec{m}\cdot\vec{n})\vec{n},$$

and the solution of (1.4.39) is of the form

$$\begin{cases} m_z = m_{z0}\exp(-\gamma_z t), \\ m_x + im_y = (m_{x0} + im_{y0})\exp(-\gamma_\perp t)\exp(i\omega_0 t), \end{cases} \tag{1.4.48}$$

where

$$\gamma_z = \frac{1}{\tau_2}\left(\frac{4\pi}{3} + \frac{1}{\chi_{zz}^0}\right), \quad \gamma_\perp = \left(\frac{1}{\tau_1} + \frac{1}{\tau_2}\right)\left(\beta + \frac{H_0^{(e)}}{\mathcal{M}_0}\right), \tag{1.4.49}$$

and m_{z0} and $m_{x0} + im_{y0}$ are the initial values of the longitudinal and transverse (relative to $\mathcal{M}_0$) components of the deviation of the magnetization $\vec{m}$ from the equilibrium value $\mathcal{M}_0$.

The quantities

$$\tau_z = \frac{1}{\gamma_z}, \quad \tau_\perp = \frac{1}{\gamma_\perp}, \tag{1.4.50}$$

are the relaxation times for the longitudinal and transverse components of the magnetic moment. Since $\mathcal{M}_0$ is only slightly dependent on $H_0^{(i)}$, it follows that $\chi_{zz}^0 \ll 1$ and consequently $\gamma_z \approx 1/\tau_2\chi_{zz}^0$. If $\beta + H_0^{(e)}/\mathcal{M}_0 \gg 1$, we have

$$\gamma_z = \frac{1}{\tau_2\chi_{zz}^0}, \quad \gamma_\perp = \left(\frac{1}{\tau_1} + \frac{1}{\tau_2}\right)\left(\beta + \frac{H_0^{(e)}}{\mathcal{M}_0}\right). \tag{1.4.51}$$

1.5   Spin Waves in Antiferromagnets

1.5.1   Equilibrium States of Antiferromagnets

1. *Motion equations of magnetic moments*

If we do not take dissipation of energy into account, the equation of motion for the magnetizations $\mathcal{M}_1(r,t)$ and $\mathcal{M}_2(r,t)$ of the two magnetic sublattices will be of the form

$$
\begin{cases}
\dfrac{\partial \mathcal{M}_1}{\partial t} = g[\mathcal{M}_1, \mathcal{H}_1], \\[2mm]
\dfrac{\partial \mathcal{M}_2}{\partial t} = g[\mathcal{M}_2, \mathcal{H}_2],
\end{cases}
\tag{1.5.1}
$$

where g is the gyromagnetic ratio and $\mathcal{H}_1$ and $\mathcal{H}_2$ are the effective magnetic fields acting on the moments $\mathcal{M}_1(r,t)$ and $\mathcal{M}_2(r,t)$

$$
\mathcal{H}_1 = \frac{\delta W}{\delta \mathcal{M}_1}, \quad \mathcal{H}_2 = \frac{\delta W}{\delta \mathcal{M}_2},
\tag{1.5.2}
$$

W is the energy of the antiferromagnets. Using the expression of W, we obtain

$$
\begin{cases}
\mathcal{H}_1 = H^{(i)} - \dfrac{\partial \mathcal{F}}{\partial \mathcal{M}_1} + \dfrac{\partial}{\partial x_k} \dfrac{\partial \mathcal{F}}{\partial(\partial \mathcal{M}_1/\partial x_k)}, \\[3mm]
\mathcal{H}_2 = H^{(i)} - \dfrac{\partial \mathcal{F}}{\partial \mathcal{M}_2} + \dfrac{\partial}{\partial x_k} \dfrac{\partial \mathcal{F}}{\partial(\partial \mathcal{M}_2/\partial x_k)},
\end{cases}
\tag{1.5.3}
$$

where $H^{(i)}$ is the magnetic field inside the antiferromagnets.

We note that the equations of motion for the magnetizations are consistent with the conservation of energy and lead to the following expression for the energy flux density in a ferromagnet

$$
\begin{aligned}
\pi_i = {} & \frac{c}{4\pi}[\vec{E}, H^{(m)}]_i - \alpha_{ij}\left(\frac{\partial \mathcal{M}_1}{\partial x_j}\frac{\partial \mathcal{M}_1}{\partial t} + \frac{\partial \mathcal{M}_2}{\partial x_j}\frac{\partial \mathcal{M}_2}{\partial t}\right) \\
& - \alpha'_{ij}\left(\frac{\partial \mathcal{M}_1}{\partial x_j}\frac{\partial \mathcal{M}_2}{\partial t} + \frac{\partial \mathcal{M}_2}{\partial x_j}\frac{\partial \mathcal{M}_1}{\partial t}\right).
\end{aligned}
\tag{1.5.4}
$$

2. Equilibrium states of antiferromagnets

Consider the equilibrium values of the magnetizations $\mathcal{M}_{10}, \mathcal{M}_{20}$ of the two sublattices about which the oscillations of the magnetizations $\mathcal{M}_1$ and $\mathcal{M}_2$ take place. To do this we must equate the effective magnetic fields to zero. The equilibrium state of the antiferromagnets corresponding to the direction of magnetic moments of sublattices for which the energy density of the antiferromagnets

$$
\begin{aligned}
w = {} & \delta \mathcal{M}_1 \cdot \mathcal{M}_2 - \frac{1}{2}\beta[(\mathcal{M}_1 \cdot \vec{n})^2 + (\mathcal{M}_2 \cdot \vec{n})^2] \\
& - \beta'(\mathcal{M}_1 \cdot \vec{n}) \cdot (\mathcal{M}_2 \cdot \vec{n}) - H_0^{(e)}(\mathcal{M}_1 + \mathcal{M}_2)
\end{aligned}
$$

reaches a minimum. We have neglected in this expression the energy w_d which is responsible for the appearance of weak ferromagnetism and the energy

$$
2\pi(\mathcal{M}_1 - \mathcal{M}_2)\hat{N}(\mathcal{M}_1 + \mathcal{M}_2)
$$

which depends on the shape of the body. Let us assume that the external magnetic field is zero. If at the same time, $\beta - \beta' > 0$, it is readily to verify that the minimum of w is reached when the magnetic moments of the sublattices lie along the anisotropy axis and $\mathcal{M}_{10} + \mathcal{M}_{20} = 0$. Such antiferromagnets are said to have magnetic anisotropy of the "easy axis" type.

When $\beta - \beta' < 0$, the minimum of w is reached when the magnetic moments of the sublattices are perpendicular to the anisotropy axis and $\mathcal{M}_{10} + \mathcal{M}_{20} = 0$. In this case it is said that the antiferromagnets has a magnetic anisotropy of "easy plane" type. Examples of the "easy axis" anisotropy are $CuCl_2.2H_2O$, Cr_2O_3 and $FeCO_3$; antiferromagnets with magnetic anisotropy of the "easy plane" type are hematite in its low temperature phase, the carbonates and fluorides of transition metals.

The equilibrium directions of $\mathcal{M}_{10}$ and $\mathcal{M}_{20}$ are readily determined even in the presence of an external magnetic field $H_0^{(e)}$. The results of the corresponding calculations are given in Figures 1.5.1 and 1.5.2.

1.5.2   Spin Waves in Antiferromagnets

1. *Motion of spin waves in antiferromagnets*

Consider the spin waves in antiferromagnets. Substitute

$$\begin{cases} \mathcal{M}_1(r,t) = \mathcal{M}_{10} + \vec{m}_1(r,t), \\ \mathcal{M}_2(r,t) = \mathcal{M}_{20} + \vec{m}_2(r,t), \\ H^{(i)}(r,t) = H_0^{(i)} + \vec{h}(r,t). \end{cases} \qquad (1.5.5)$$

into (1.5.1) where $\vec{m}_1, \vec{m}_2$ and $\vec{h}$ are small deviations from the equilibrium values $\mathcal{M}_{10}$, $\mathcal{M}_{20}$ and $H_0^{(i)}$, and then linearize these equations. This results in a set of two linear differential equations for $\vec{m}_1$ and $\vec{m}_2$. We can now express the Fourier component of the deviations of the magnetic moments by $\vec{m}_1(\vec{k},\omega)$, $\vec{m}_2(\vec{k},\omega)$ and $\vec{h}(\vec{k},\omega)$

$$\vec{m}(\vec{k},\omega) = \vec{m}_1(\vec{k},\omega) + \vec{m}_2(\vec{k},\omega) = \chi(\vec{k},\omega)\vec{h}(\vec{k},\omega), \qquad (1.5.6)$$

where $\chi(\vec{k},\omega)$ is a tensor which depends on $\vec{k}$ and ω and on the quantities characterizing the equilibrium state of the antiferromagnets. If the magnetic field $H_0^{(e)}$ is parallel to the anisotropy axis and $H_0^{(e)} < H_1$, $H_1 = \mathcal{M}_0\sqrt{2\delta(\beta - \beta')}$, then the linearized equations for $\vec{m}_1(\vec{k},\omega)$ and $\vec{m}_2(\vec{k},\omega)$ are of the form

$$\begin{cases} -i\omega\vec{m}_1(\vec{k},\omega) = g\left\{\mathcal{M}_{10}, \vec{h}(\vec{k},\omega) - \left(\delta + \dfrac{H_0^{(e)}}{\mathcal{M}_0} + \beta - \beta' + \alpha_{ij}k_ik_j\right)\cdot\vec{m}_1(\vec{k},\omega)\right. \\ \qquad\qquad \left. - (\delta + \alpha'_{ij}k_ik_j)\vec{m}_2(\vec{k},\omega)\right\}, \\[2mm] -i\omega\vec{m}_2(\vec{k},\omega) = g\left\{\mathcal{M}_{20}, \vec{h}(\vec{k},\omega) - \left(\delta + \dfrac{H_0^{(e)}}{\mathcal{M}_0} + \beta - \beta' + \alpha_{ij}k_ik_j\right)\cdot\vec{m}_2(\vec{k},\omega)\right. \\ \qquad\qquad \left. - (\delta + \alpha'_{ij}k_ik_j)\vec{m}_1(\vec{k},\omega)\right\}, \end{cases}$$

$$(1.5.7)$$

$$\beta - \beta' > 0$$

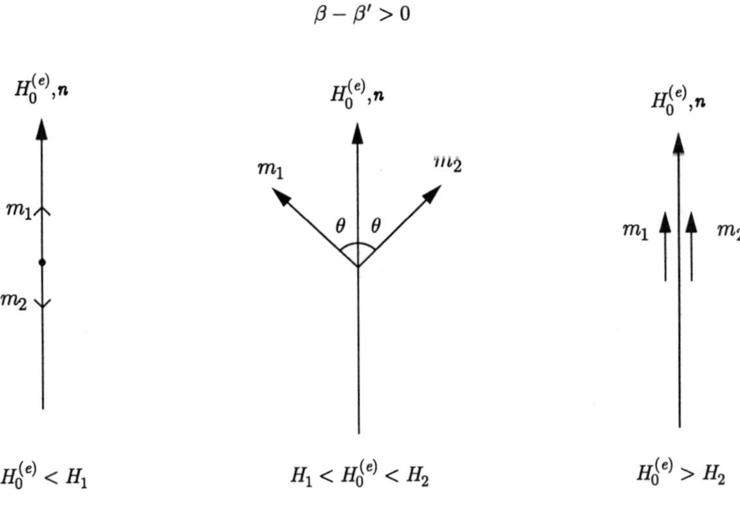

$$H_1 = m_0 \sqrt{2\delta(\beta - \beta')}, \quad H_2 = 2\delta m_0, \quad \cos\theta = \frac{H_0^{(e)}}{H_2}$$

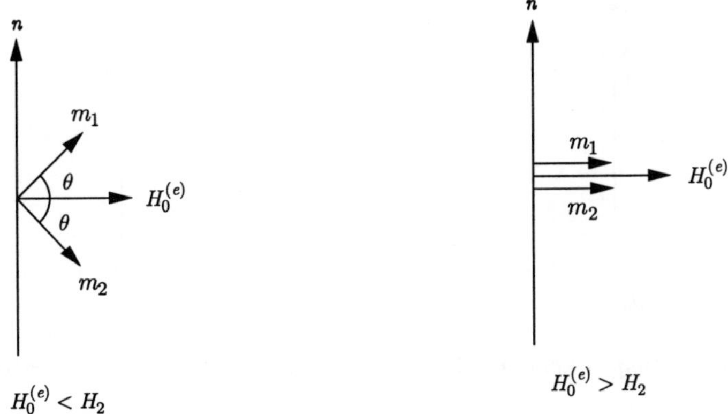

Figure 1.5.1. *Numerical results for the equilibrium states of an antiferromagnet (1).*

$$\beta - \beta' < 0$$

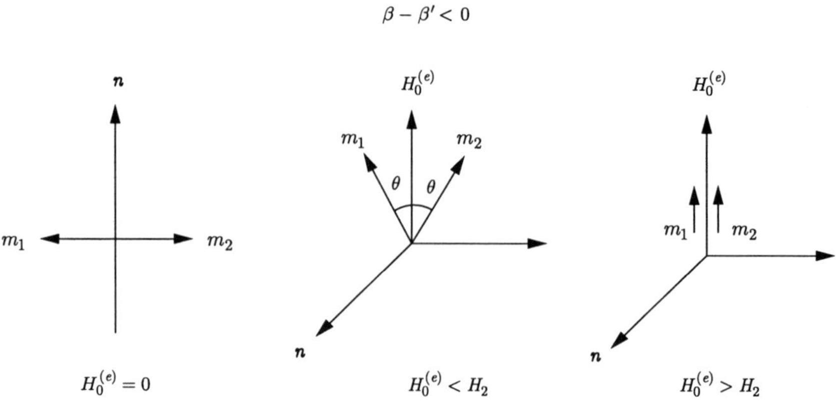

$$H_1 = m_0 \sqrt{2\delta(\beta - \beta')}, \quad H_2 = 2\delta m_0, \quad \cos\theta = \frac{H_0^{(e)}}{H_2}$$

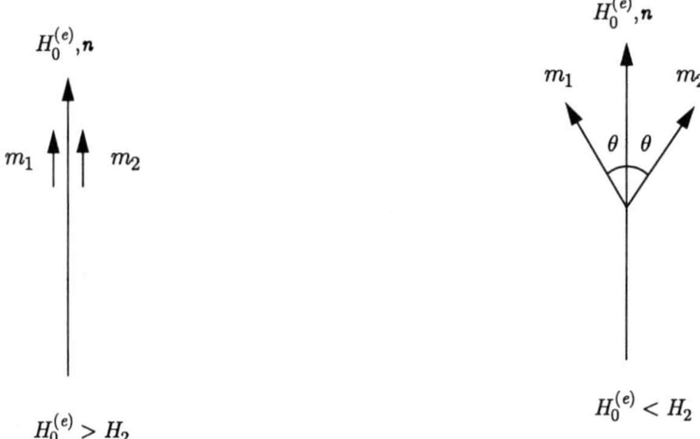

Figure 1.5.2. *Computational results for the equilibrium states of an antiferromagnet* (2).

$$\hat{\chi}(\vec{k},\omega) = \begin{pmatrix} \chi_{xx} & \chi_{xy} & 0 \\ \chi_{yx} & \chi_{yy} & 0 \\ 0 & 0 & 0 \end{pmatrix}, \tag{1.5.8}$$

where

$$\begin{cases} \chi_{xx} = \chi_{yy} = \dfrac{1}{2}\chi_0 \left(\dfrac{\Omega_+(\Omega_+ - gH_0^{(e)})}{\Omega_+^2 - \omega^2} + \dfrac{\Omega_-(\Omega_- + gH_0^{(e)})}{\Omega_-^2 - \omega^2} \right), \\[4mm] \chi_{xy} = -\chi_{yx} = i\omega\chi_0 \left(\dfrac{(\Omega_+ - gH_0^{(e)})}{\Omega_+^2 - \omega^2} + \dfrac{(\Omega_- + gH_0^{(e)})}{\Omega_-^2 - \omega^2} \right), \\[4mm] \Omega_\pm = g\mathcal{M}_0\sqrt{2\delta(\alpha_{ij} - \alpha_{ij'})k_i k_j + (H_1/\mathcal{M}_0)^2} \pm gH_0^{(e)}, \end{cases} \tag{1.5.9}$$

in which $\chi_0 = \delta^{-1}$.

Proceeding in an analogous fashion, we can determine the high-frequency magnetic susceptibility tensor of an antiferromagnet when the magnetic field $H_0^{(i)}$ is at right angles to the anisotropy axis:

$$\hat{\chi}(\vec{k},\omega) = \begin{pmatrix} \chi_{xx} & 0 & 0 \\ 0 & \chi_{yy} & \chi_{yz} \\ 0 & \chi_{zy} & \chi_{zz} \end{pmatrix}, \tag{1.5.10}$$

where

$$\begin{cases} \chi_{xx} = \chi_0\dfrac{\Omega_2^2}{\Omega_2^2 - \omega^2}, \quad \chi_{yy} = \chi_0\dfrac{\Omega_1^2}{\Omega_1^2 - \omega^2}, \\[4mm] \chi_{zz} = \chi_0\dfrac{(gH_0^{(e)})^2}{\Omega_1^2 - \omega^2}, \quad \chi_{yz} = -\chi_{zy} = i\omega\chi_0\dfrac{gH_0^{(e)}}{\Omega_1^2 - \omega^2}, \\[4mm] \Omega_1 = g\mathcal{M}_0\sqrt{2\delta(\alpha_{ij} - \alpha_{ij'})k_i k_j + (H_1/\mathcal{M}_0)^2 + (H_0^{(e)}/\mathcal{M}_0)^2}, \\[4mm] \Omega_2 = g\mathcal{M}_0\sqrt{2\delta(\alpha_{ij} - \alpha_{ij'})k_i k_j + (H_1/\mathcal{M}_0)^2}. \end{cases} \tag{1.5.11}$$

It is assumed that $H_0^{(e)} \ll H_2$ and $H_2 = 2\delta\mathcal{M}_0$ (the z-axis lies along the anisotropy axis and the x-axis lies along $H_0^{(e)}$).

Consider now an antiferromagnet with magnetic anisotropy of the "easy plane" type. If the field $H_0^{(e)}$ is perpendicular to the anisotropy axis and $H_0^{(e)} \ll H_2$, the high-frequency magnetic susceptibility tensor is given by

$$\hat{\chi}(\vec{k},\omega) = \begin{pmatrix} \chi_{xx} & 0 & 0 \\ 0 & \chi_{yy} & \chi_{yz} \\ 0 & \chi_{zy} & \chi_{zz} \end{pmatrix}, \tag{1.5.12}$$

where

$$
\begin{cases}
\chi_{xx} = \chi_0 \dfrac{\Omega_2'^2}{\Omega_2'^2 - \omega^2}, \quad \chi_{yy} = \chi_0 \dfrac{(gH_0^{(e)})^2}{\Omega_1'^2 - \omega^2}, \\[2ex]
\chi_{zz} = \chi_0 \dfrac{\Omega_1'^2}{\Omega_1'^2 - \omega^2}, \quad \chi_{yz} = -\chi_{zy} = -i\omega\chi_0 \dfrac{gH_0^{(e)}}{\Omega_1'^2 - \omega^2}, \\[2ex]
\Omega_1' = g\mathcal{M}_0\sqrt{2\delta(\alpha_{ij} - \alpha_{ij'})k_ik_j + (H_0^{(e)}/\mathcal{M}_0)^2}, \\[2ex]
\Omega_2' = g\mathcal{M}_0\sqrt{2\delta(\alpha_{ij} - \alpha_{ij'})k_ik_j + (H_1/\mathcal{M}_0)^2}
\end{cases}
\tag{1.5.13}
$$

(the z-axis lies along the anisotropy axis and the x-axis lies along the magnetic field $H_0^{(e)}$).

If the magnetic field $H_0^{(e)}$ is parallel to the anisotropy axis, then

$$
\hat{\chi}(\vec{k}, \omega) = \begin{pmatrix} \chi_{xx} & \chi_{xy} & 0 \\ \chi_{yx} & \chi_{yy} & 0 \\ 0 & 0 & \chi_{zz} \end{pmatrix},
\tag{1.5.14}
$$

where

$$
\begin{cases}
\chi_{xx} = \chi_0 \dfrac{(gH_0^{(e)})^2}{\Omega_1''^2 - \omega^2}, \quad \chi_{yy} = \chi_0 \dfrac{\Omega_1''^2}{\Omega_1''^2 - \omega^2}, \\[2ex]
\chi_{zz} = \chi_0 \dfrac{\Omega_2''^2}{\Omega_2''^2 - \omega^2}, \quad \chi_{yz} = -\chi_{zy} = i\omega\chi_0 \dfrac{gH_0^{(e)}}{\Omega_1''^2 - \omega^2}, \\[2ex]
\Omega_1'' = g\mathcal{M}_0\sqrt{2\delta(\alpha_{ij} - \alpha_{ij'})k_ik_j + (H_1/\mathcal{M}_0)^2 + (H_0^{(e)}/\mathcal{M}_0)^2}, \\[2ex]
\Omega_2'' = g\mathcal{M}_0\sqrt{2\delta(\alpha_{ij} - \alpha_{ij'})k_ik_j(1 - (H_1/\mathcal{M}_0)^2)}.
\end{cases}
\tag{1.5.15}
$$

Here the z-axis lies along the anisotropy axis and the x-axis lies in the plane of the magnetic moments of the sublattices.

2. *The spin-wave spectrum in antiferromagnets*

If we know the high-frequency magnetic susceptibility tensor of an antiferromagnet, we can readily find its spin-wave spectrum. To do this we must use the general dispersion relation (1.4.42) which determines the spin-wave spectrum in the magnetostatic approximation of both ferromagnets and antiferromagnets,

$$
k^2 + 4\pi k_i k_j \chi_{ij}(\vec{k}, \omega) = 0.
\tag{1.5.16}
$$

However, in the case of an antiferromagnet, this equation needs not be solved since the components of the tensor $\chi_{ij}(\vec{k}, \omega)$ for an antiferromagnet are proportional to a small parameter χ_0 and therefore to within terms of the order of $g\mathcal{M}_0\chi_0$, the spin-wave frequencies must coincide with the poles of the tensor $\hat{\chi}(\vec{k}, \omega)$. Hence it follows, for example, that in the case of antiferromagnets with magnetic anisotropy of the "easy axis" type, the spin-wave frequencies are given by

$$\begin{cases} \omega_{S1} \equiv \Omega_+ = g\mathcal{M}_0\sqrt{2\delta(\alpha_{ij} - \alpha'_{ij})k_i k_j + (H_1/\mathcal{M}_0)^2} + gH_0^{(e)}, \\ \omega_{S2} = \Omega_- = g\mathcal{M}_0\sqrt{2\delta(\alpha_{ij} - \alpha'_{ij})k_i k_j + (H_1/\mathcal{M}_0)^2} - gH_0^{(e)}, \end{cases} \quad (1.5.17)$$

if the field $H_0^{(e)}$ is parallel to the anisotropy axis and $H_0^{(e)} < H_1$, and by

$$\begin{cases} \omega_{S1} \equiv \Omega_1 = g\mathcal{M}_0\sqrt{2\delta(\alpha_{ij} - \alpha'_{ij})k_i k_j + (H_1/\mathcal{M}_0)^2 + (H_0^{(e)}/\mathcal{M}_0)^2}, \\ \omega_{S2} \equiv \Omega_2 = g\mathcal{M}_0\sqrt{2\delta(\alpha_{ij} - \alpha'_{ij})k_i k_j + (H_1/\mathcal{M}_0)^2}, \end{cases} \quad (1.5.18)$$

if the field $H_0^{(e)}$ is perpendicular to the anisotropy axis.

In the case of antiferromagnets with magnetic anisotropy of the "easy plane" type, the spin-wave frequencies are given by

$$\begin{cases} \omega_{S1} \equiv \Omega_1'' = g\mathcal{M}_0\sqrt{2\delta(\alpha_{ij} - \alpha'_{ij})k_i k_j + (H_1/\mathcal{M}_0)^2 + (H_0^{(e)}/\mathcal{M}_0)^2}, \\ \omega_{S2} \equiv \Omega_2'' = g\mathcal{M}_0\sqrt{2\delta(\alpha_{ij} - \alpha'_{ij})k_i k_j (1 - (H_1/\mathcal{M}_0)^2)}, \end{cases} \quad (1.5.19)$$

if the magnetic field $H_0^{(e)}$ is parallel to the anisotropy axis, and by

$$\omega_{S1} \equiv \Omega_1' = g\mathcal{M}_0\sqrt{2\delta(\alpha_{ij} - \alpha'_{ij})k_i k_j + (H_0^{(e)}/\mathcal{M}_0)^2},$$

$$\omega_{S2} \equiv \Omega_2' = g\mathcal{M}_0\sqrt{2\delta(\alpha_{ij} - \alpha'_{ij})k_i k_j + (H_1/\mathcal{M}_0)^2}$$

if the magnetic field $H_0^{(e)}$ is perpendicular to the anisotropy axis.

1.5.3 Electromagnetic Waves in Magnetically Ordered Crystals

1. *Dispersion relation for electromagnetic waves*

First of all we discuss the dispersion relation for electromagnetic waves. Assume that the magnetic moments have magnetostatic oscillations. This leads to

$$\vec{m}(\vec{k}, \omega) = \hat{\chi}(\vec{k}, \omega) \cdot \vec{h}(\vec{k}, \omega), \quad (1.5.20)$$

where $\vec{m}(\vec{k}, \omega)$ and $\vec{h}(\vec{k}, \omega)$ are the amplitudes of the oscillating components of the magnetization and the the magnetic field, and $\hat{\chi}(\vec{k}, \omega)$ is the high-frequency magnetic susceptibility tensor of a ferromagnet or an antiferromagnets.

Maxwell's equations for plane waves with allowance for (1.5.20) are of the form

$$[k, e] = \frac{\omega}{c}b, \quad [k, h] = -\frac{\omega}{c}d, \quad (1.5.21)$$

where $b = \hat{\mu}\vec{h}$ and $d = \hat{\epsilon}e$ are the amplitudes of the oscillating components of the magnetic and electric induction, $\hat{\mu}(\vec{k},\omega) = 1 + 4\pi\hat{\chi}(\vec{k},\omega)$ is the magnetic permeability tensor and $\hat{\epsilon}$ is the permittivity tensor. We shall assume for simplicity that $\epsilon_{ij} = \epsilon\delta_{ij}$ and that ϵ is independent of $\vec{k}$ and ω. Equating the determinant of (1.5.21) to zero, we obtain the dispersion relation connecting the frequency and the wave vectors of the electromagnetic waves in magnetically ordered crystals

$$D(\vec{k},\omega) = A(\vec{k},\omega)n^4 + B(\vec{k},\omega)n^2 + C(\vec{k},\omega), \qquad (1.5.22)$$

where $n = ck/\omega\sqrt{\epsilon}$ is the refractive index,

$$A(\vec{k},\omega) = 1 + \frac{4\pi}{k^2}k_i k_j \chi_{ij}(\vec{k},\omega),$$

$$B(\vec{k},\omega) = \left(\frac{k_i k_j}{k^2} - \delta_{ij}\right)\Delta_{ij}(\vec{k},\omega), \qquad (1.5.23)$$

$$C(\vec{k},\omega) = \det\mu_{ij}(\vec{k},\omega),$$

and $\Delta_{ij}(\vec{k},\omega)$ are the minors of the determinant $|\mu_{ij}(\vec{k},\omega)|$. This dispersion relation will in general define not one but several frequencies for a given wave vector $\vec{k}$. For a ferromagnet there are three such frequencies, whereas for an antiferromagnet there are four. Different frequencies corresponding to the same $\vec{k}$ define different branches of the oscillations. In fact, dividing (1.5.22) by n^4, and allowing n to tend to infinity, we obtain

$$A(\vec{k},\omega) = 0, \qquad (1.5.24)$$

which is the same as the dispersion relation

$$1 + \frac{4\pi}{k^2}k_i k_j \chi_{ij}(\vec{k},\omega) \qquad (1.5.25)$$

for the spin waves.

Since the spin wave frequency $\omega_S(\vec{k})$ is of the order of $g\mathcal{M}_0$, it may be said that spin waves correspond to wave vector $\vec{k}$ satisfying the inequality

$$k \gg \frac{g\mathcal{M}_0}{c} \qquad (1.5.26)$$

or in terms of wave length

$$\lambda \ll \frac{c}{g\mathcal{M}_0}.$$

When $k \gg \frac{g\mathcal{M}_0}{c}$, in addition to the spin wave there are also two proper electromagnetic waves characterized by the dispersion relation

$$\omega = \frac{ck}{\sqrt{\epsilon}}. \qquad (1.5.27)$$

Our problem now is to determine the properties of the branches of electromagnetic oscillations for $k \ll \frac{g\mathcal{M}_0}{c}$.

Since $\frac{1}{\sqrt{\alpha}} \gg \frac{g\mathcal{M}_0}{c}$, it follows that in this region of wave vectors $\alpha k \ll 1$ and, consequently, the spatial dispersion of the high-frequency magnetic susceptibility tensor is unimportant. This means that the coefficients A, B and C in the dispersion relation (1.5.22) will depend only on the frequency in the direction of the wave vector when $k \ll \frac{g\mathcal{M}_0}{c}$, but not on its magnitude.

Under these conditions the dispersion relation can conveniently be looked upon as the equation for n or, what amounts to the same thing, as an equation for the modulus of the wave vector $\vec{k}$ for given frequency and directions of propagation.

The solution of this equation is of the form

$$n_{1,2}^2 = \left(\frac{ck_{1,2}}{\omega\sqrt{\epsilon}}\right)^2 = \frac{-B(\kappa,\omega) \pm \sqrt{B^2(\kappa,\omega) - 4A(\kappa,\omega)C(\kappa,\omega)}}{2A(\kappa,\omega)}, \qquad (1.5.28)$$

where $\kappa = \frac{\vec{k}}{k}$.

Real refractive indices n_i correspond to waves propagating with phase velocity

$$v_i(\kappa,\omega) = \frac{c}{\sqrt{\epsilon}\,n_i(\kappa,\omega)}, \qquad (1.5.29)$$

which is a function of κ and ω.

2. *Interaction between proper electromagnetic waves and spin waves*

Further study of the branches of electromagnetic oscillations for $k \leq \frac{g\mathcal{M}_0}{c}$ requires detailed knowledge of the high-frequency magnetic susceptibility tensor. Here we shall confine our attention to a uniaxial ferromagnet and will assume that the external magnetic field lies along the anisotropy axis. The tensor $\hat{\mu}(\vec{k},\omega)$ is then of the form

$$\hat{\mu} = \left\{ \begin{array}{ccc} \mu & i\mu' & 0 \\ -i\mu' & \mu & 0 \\ 0 & 0 & 1 \end{array} \right\}, \qquad (1.5.30)$$

where

$$\begin{cases} \mu(\omega) = 1 + \dfrac{4\pi g\mathcal{M}_0\Omega_0}{\Omega_0 - \omega^2}, \\[2mm] \mu'(\omega) = 1 + \dfrac{4\pi g\mathcal{M}_0\omega}{\Omega_0 - \omega^2}, \\[2mm] \Omega_0 = g\mathcal{M}_0\left(\beta + \dfrac{H_0^{(i)}}{\mathcal{M}_0}\right), \end{cases} \qquad (1.5.31)$$

and the z-axis lies along the anisotropy axis.

Using the unit vectors

$$\vec{J}_1 = \frac{[\vec{n},\vec{k}]}{|[\vec{n},\vec{k}]|}, \quad \vec{J}_2 = \frac{[\vec{k},[\vec{n},\vec{k}]]}{|[\vec{k},[\vec{n},\vec{k}]]|}, \quad \vec{J}_3 = \frac{\vec{k}}{k}, \qquad (1.5.32)$$

where $\vec{n}$ is a unit vector along the anisotropy axis, we can write the vectors $\vec{b}, \vec{h}$ and $\vec{e}$ in the form

$$\begin{cases} \vec{b} = b_1 \vec{J}_1 + b_2 \vec{J}_2, \\ \vec{h} = h_1 \vec{J}_1 + h_2 \vec{J}_2 + h_3 \vec{J}_3, \\ \vec{e} = e_1 \vec{J}_1 + e_2 \vec{J}_2. \end{cases} \tag{1.5.33}$$

Eliminating the vector $\vec{e}$ from (1.5.21) we obtain

$$\vec{k}(\vec{k} \cdot \vec{h}) - k^2 \vec{h} = -\frac{\omega^2 \epsilon}{c^2} \vec{b}, \tag{1.5.34}$$

and hence

$$b_1 = n^2 h_1, \quad b_2 = n^2 h_2. \tag{1.5.35}$$

Since $\vec{b} = \hat{\mu}\vec{h}$, it follows that

$$\begin{cases} \left\{ n^2 - \dfrac{\mu \cos^2 \theta_k + (\mu^2 - \mu'^2) \sin^2 \theta_k}{\mu \sin^2 \theta_k + \cos^2 \theta_k} \right\} b_1 - i \dfrac{\mu' \cos \theta_k}{\mu \sin^2 \theta_k + \cos^2 \theta_k} b_2 = 0, \\[4mm] i \dfrac{\mu' \cos \theta_k}{\mu \sin^2 \theta_k + \cos^2 \theta_k} b_1 + \left\{ n^2 - \dfrac{\mu}{\mu \sin^2 \theta_k + \cos^2 \theta_k} \right\} b_2 = 0, \end{cases} \tag{1.5.36}$$

where θ_k is the angle between the wave vector $\vec{k}$ and the anisotropy axis.

If we eliminate the amplitudes b_1 and b_2 from these equations, we obtain the following values for the refractive index

$$n_{1,2}^2 = \frac{1}{2}(\mu \sin^2 \theta_k + \cos^2 \theta_k)^{-1}\{\mu(1 + \cos^2 \theta_k) + (\mu^2 - \mu'^2) \sin^2 \theta_k$$

$$\pm \sqrt{(\mu^2 - \mu'^2 - \mu) \sin^4 \theta_k + 4\mu'^2 \cos^2 \theta_k}\}. \tag{1.5.37}$$

These formulae are a special case of (1.4.79) for a uniaxial crystal. The value of b_1 and b_2 for waves with refractive index n_j are related by

$$\frac{b_1^{(j)}}{b_2^{(j)}} = i\rho_j, \tag{1.5.38}$$

where

$$\rho_j = \frac{n_j^2 \cos^2 \theta_k - \mu(1 - n_j^2 \sin^2 \theta_k)}{\mu' \cos \theta_k}. \tag{1.5.39}$$

We note that

$$\rho_1 \rho_2 = -1 \tag{1.5.40}$$

and consequently

$$\frac{b_1^{(1)}}{b_2^{(1)}} = -\frac{b_2^{(2)}}{b_1^{(2)}} = i\rho_1. \tag{1.5.41}$$

It follows from (1.5.36) that

$$\frac{h_1^{(j)}}{h_2^{(j)}} = i\rho_j, \quad \frac{h_3^{(j)}}{h_2^{(j)}} = \frac{\mu' \sin\theta_k + (\mu - 1)\sin\theta_k \cos\theta_k)}{\mu \sin^2\theta_k + \cos^2\theta_k},$$

where as before the subscript j represents waves with refractive index n_j.

Let us now return to the expressions given by (1.5.37) for the refractive indices and find the values of ω for which the wave vector is zero. Since the right-hand sides of (1.5.37) have finite limits, which are equal to $\mu(0)$ and

$$\frac{\mu(0)}{\cos^2\theta_k + \mu(0)\sin^2\theta_k}, \tag{1.5.42}$$

it follows that k will be zero together with ω, and when $\omega \ll g\mathcal{M}_0$,

$$k_1 = \omega \frac{\sqrt{\epsilon\mu(0)}}{c}, \quad k_2 = \omega \frac{\sqrt{\epsilon\mu(0)}}{c} \frac{1}{\sqrt{\cos^2\theta_k + \mu(0)\sin^2\theta_k}}. \tag{1.5.43}$$

Moreover, the wave vector will vanish for a certain value of ω of the order of $g\mathcal{M}_0$ and, in particular, for

$$\omega = \omega_0 = g(H_0^{(i)} + 4\pi\mathcal{M}_0 + \beta\mathcal{M}_0). \tag{1.5.44}$$

3. *Properties of the branches of electromagnetic oscillations in ferromagnets*

The above results can be used to obtain a schematic representation of the properties of the branches of electromagnetic oscillations in uniaxial ferromagnets. In Figure 1.5.3 spin waves correspond to the broken curve $\omega = \omega_S(\vec{k})$, whilst the proper electromagnetic waves correspond to the broken curve $\omega = ck/\sqrt{\epsilon}$. This curve tends asymptotically to a part of branch I for $k \gg g\mathcal{M}_0/c$, and the straight line is the common asymptote of branches II and III (also for $k \gg g\mathcal{M}_0/c$). It is readily shown that the deviation of these curves from the straight line $\omega = ck/\sqrt{\epsilon}$ for $k \gg g\mathcal{M}_0/c$ is given by

$$\omega_{\mathrm{II,III}} = \frac{ck}{\sqrt{\epsilon}}\left(1 \pm \frac{2\pi g\mathcal{M}_0}{ck}\sqrt{\epsilon}\cos\theta_k\right). \tag{1.5.45}$$

Consider now the polarization properties of the above branches of electromagnetic oscillations. It is readily shown that as $k \to 0$, the induction vector $\vec{b}$ lies along the vector $\vec{J}_1$ for oscillations in branch I, and along the vector $\vec{J}_2$ for oscillations in branch II. Branch III is characterized by elliptical polarization for $k \to 0$, and

$$\frac{b_1^{\mathrm{III}}}{b_2^{\mathrm{III}}} = \frac{1}{\cos\theta_k}.$$

For large $k \gg g\mathcal{M}_0/$, branches II and II have circular polarizations.

For large k the magnetic field for branches II and III is transverse and for branch I longitudinal. For small k, on the other hand, there are both transverse and longitudinal magnetic field components, and both are of the same order of magnitude.

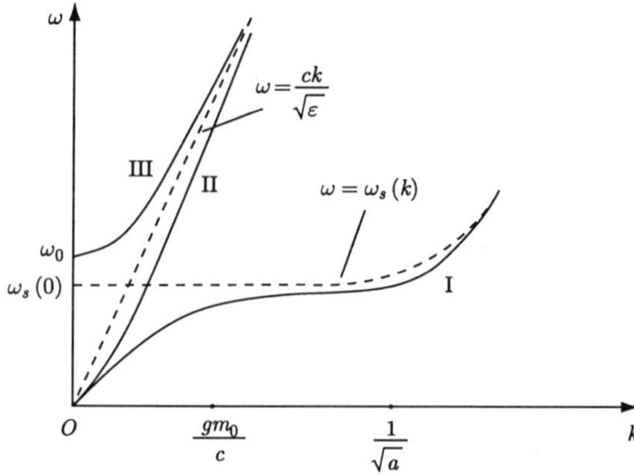

Figure 1.5.3. *Electromagnetic oscillating bifurcation property of a uni-axial ferromagnet.*

1.6 Bibliography Comments

In this chapter, we give the physics backgrounds and the derivations for the Landau–Lifshitz equations. There are two different approaches to derive Landau–Lifshitz equations. One is to consider the macroscopic motion of the ferromagnets and evaluate the total energy, then derive Landau–Lifshitz equations from the Hamiltonian (see the original paper by Landau and Lifshitz [102], [140]). The other is to begin with the microscopic aspects. Applying the quantum-mechanical spin theory, one can make out the Hamiltonian for the Heisenberg chain, and then get the Landau–Lifshitz equations and at the same time the equations of ferromagnets. For the ferromagnetic equations for multi-media and the propagations and interactions of spin waves in ferromagnets, we refer to the book "Spin waves" by Akhiezer *et al.* [3] and references therein.

Chapter 2

Integrability of Heisenberg Chain

This chapter is concerned with the integrability of the Heisenberg chain without Gilbert damping including spin waves, solitary waves, the geometry representation for the Landau–Lifshitz equations, nonhomogeneous Heisenberg chain and the spherical (cylindrical) symmetric Heisenberg chain equations.

2.1 Spin Waves and Solitary Waves

2.1.1 Spin Waves

In this section, we shall consider the spin waves and solitary waves, the approximate solutions and the equivalence to Schrödinger equations of the following Landau–Lifshitz equation without Gilbert damping

$$\frac{\partial \vec{S}}{\partial t} = \vec{S} \times \frac{\partial^2 \vec{S}}{\partial x^2}.$$ (2.1.1)

Suppose that equation (2.1.1) admits a solution of the form

$$\vec{S}(x,t) = \vec{A}\cos\alpha + \{\vec{B}\cos(kx - \omega t) + \vec{C}\sin(kx - \omega t)\}\sin\alpha,$$ (2.1.2)

where $(\vec{A}, \vec{B}, \vec{C})$ forms a unit right-hand orthogonal system, α, k are any real numbers, $\omega = k^2 \sin\alpha$. The solutions of such form are called spin waves. Their energy density ε and density flow j are constants of the form

$$\varepsilon = \frac{1}{2}k^2\sin^2\alpha, \quad j = k^3\sin^2\alpha\cos\alpha = v\varepsilon,$$ (2.1.3)

in which $v = 2k\cos\alpha = \frac{d\omega}{dk}$ is the group velocity of the wave. The solution (2.1.2) of spin wave is in fact the form $\vec{S}(x,t) = \vec{S}(u)$ with $u = x - ct$. In fact, substituting this form into (2.1.1), we have

$$-c\vec{S}' = \vec{S} \times \vec{S}'',$$ (2.1.4)

where "$'$" denotes the differential with respect to u. Noting that $\vec{S}' \times \vec{S}' = 0$ one has from (2.1.4) that

$$\frac{d}{du}(\vec{S} \times \vec{S}' + c\vec{S}) = 0, \tag{2.1.5}$$

and integrating this equality one further has

$$\vec{S} \times \vec{S}' + c\vec{S} = \vec{S}_0, \tag{2.1.6}$$

where $\vec{S}_0$ is a constant vector. Taking inner product of (2.1.6) with $\vec{S}$, we have by noting $|\vec{S}| = 1$ that:

$$\vec{S} \cdot \vec{S}_0 = c. \tag{2.1.7}$$

Let $\vec{A}$ be the unit vector along the direction $\vec{S}_0$ and construct the other two vectors $\vec{B}, \vec{C}$ such that they are orthogonal to $\vec{A}$ and orthogonal with each other in the right-hand system. $\vec{S}(u)$ can be written as

$$\vec{S}(u) = \vec{A}\cos\alpha + \{\vec{B}\cos\varphi(u) + \vec{C}\sin\varphi(u)\}\sin\alpha, \tag{2.1.8}$$

where α is the constant angle between $\vec{S}$ and $\vec{S}_0$, $\varphi(u)$ is the angle between the projection of vector $\vec{S}$ onto $(\vec{B}, \vec{C})$-plane and vector $\vec{B}$. Substituting (2.1.8) into (2.1.4), we get

$$\varphi'(u) = \frac{c}{\cos\alpha}. \tag{2.1.9}$$

This implies that $\varphi(u) = kx - \omega t$, where $k = c/\cos\alpha$, $\omega = k^2\cos\alpha$. This is the spin wave (2.1.2).

2.1.2   Solitary Waves

In order to study equation (2.1.1) in detail, we introduce the polar coordinates as follows:

$$\begin{cases} u = \sin\theta(x,t)\cos\varphi(x,t), \\ v = \sin\theta(x,t)\sin\varphi(x,t), \\ w = \cos\theta(x,t) \end{cases} \tag{2.1.10}$$

where $\vec{S}(x,t) = (u(x,t), v(x,t), w(x,t))$. It follows from (2.1.1) that

$$\begin{cases} \dfrac{d\theta}{dt} = -2\theta_x\varphi_x\cos\theta - \varphi_{xx}\sin\theta, \tag{2.1.11} \\[2mm] \varphi_t\sin\theta = \theta_{xx} - (\varphi_x)^2\sin\theta\cos\theta. \tag{2.1.12} \end{cases}$$

The energy density is of the form

$$\varepsilon(x,t) = \frac{1}{2}(\theta_x^2 + \chi^2), \quad \chi = \varphi_x\sin\theta. \tag{2.1.13}$$

First we assume that θ and φ are both the functions of $u = x - ct$ which corresponds to the case of spin wave. Then we assume that $\theta(x,t)$ be of the function of u but φ take the form

$$\varphi(x,t) = \overline{\varphi}(u) + \Omega t. \tag{2.1.14}$$

Define
$$x(u) = \overline{\varphi}'(u) \sin \theta(u). \tag{2.1.15}$$

It follows from (2.1.11) and (2.1.12) that
$$\begin{cases} x' = \theta'(c - \chi \cos \theta), & \tag{2.1.16} \\ \theta'' = -x(c - \chi \cos \theta) + \Omega \sin \theta, & \tag{2.1.17} \end{cases}$$

where "$'$" denote the differential with respect to u. It follows from this equation that
$$(\theta')^2 + x^2 + 2\Omega \cos \theta = 2\alpha, \tag{2.1.18}$$

where α is a quantity independent of u. Now the energy density (2.1.13) reads as follows:
$$\varepsilon(u) = -\Omega \cos \theta(u) + \alpha. \tag{2.1.19}$$

If we define
$$\begin{cases} Z(u) = \cos \theta(u), \\ g(u) = \overline{\varphi}'(u) \sin^2 \theta(u), \end{cases} \tag{2.1.20}$$

then it follows from (2.1.16) that
$$y' = -cz' \tag{2.1.21}$$

and
$$Z'' - cy + \Omega(1 - Z^2) + \frac{Z(Z' + y^2)}{1 - Z^2} = 0. \tag{2.1.22}$$

From (2.1.18), we have
$$Z' + y^2 = 2(\alpha - \Omega Z)(1 - Z^2). \tag{2.1.23}$$

And, (2.1.22) becomes
$$Z'' - cy - 3\Omega Z^2 + 2\alpha Z + \Omega = 0. \tag{2.1.24}$$

It is not difficult to get the general solutions to (2.1.21) and (2.1.22). The solitary solution is then of the form
$$\begin{cases} Z(u) = (\tanh \tfrac{1}{2}cu)^2, \\ y(u) = c\,\mathrm{sech}^2 \tfrac{1}{2}cu, \end{cases} \tag{2.1.25}$$

where
$$\alpha = \Omega = \frac{1}{2}c^2. \tag{2.1.26}$$

It follows from (2.1.20) that
$$\frac{d\overline{\varphi}}{du} = \frac{c}{1 + \tanh^2 \tfrac{1}{2}cu}. \tag{2.1.27}$$

Then we have by integrating this equation that

$$\overline{\varphi} = \tan^{-1}\left[\tanh\frac{1}{2}cu\right] + \frac{1}{2}cu. \qquad (2.1.28)$$

It also follows from (2.1.14) and (2.1.26) that

$$\varphi = \tan^{-1}\left[\tanh\frac{1}{2}c(x-ct)\right] + \frac{1}{2}cx, \qquad (2.1.29)$$

and the energy density and the flow density are

$$\varepsilon(u) = \frac{c^2}{2\cosh^2\frac{1}{2}cu}, \quad j(u) = c\varepsilon(u). \qquad (2.1.30)$$

2.1.3   Approximate Solutions

Let $\vec{S}(x,t) = \vec{S}(\eta)$, $\eta = xt^\alpha$ with $\alpha = -\frac{1}{2}$, i.e.

$$\eta = xt^{-\frac{1}{2}}. \qquad (2.1.31)$$

Substituting this into (2.1.1) we have

$$\eta\vec{S}' = -2\vec{S}\times\vec{S}''. \qquad (2.1.32)$$

For large η, the approximate solution to (2.1.32) is of the form

$$\vec{S}(\eta) = \left(\frac{2q}{\eta}\sin\frac{1}{4}\eta^2, \ -\frac{2q}{\eta}\cos\frac{1}{4}\eta^2, \ 1\right) + O(\eta^{-2}), \qquad (2.1.33)$$

with $|\vec{S}'(\eta)| = q$. In this case, the energy density and the flow density are of the form

$$\varepsilon(x,t) = \frac{1}{2t}|\vec{S}'(\eta)|^2,$$
$$j(x,t) = \frac{1}{t^{3/2}}\vec{S}\cdot(\vec{S}'\times\vec{S}''), \qquad (2.1.34)$$

or, from (2.1.32)

$$j(x,t) = \frac{\eta}{2t^{3/2}}|\vec{S}'(\eta)|^2. \qquad (2.1.35)$$

Since it follows from (2.1.32) that $\vec{S}'\cdot\vec{S}'' = 0$, we get

$$|\vec{S}'(\eta)|^2 = q^2 = \text{constant}.$$

Hence

$$\varepsilon(x,t) = \frac{q^2}{2t}, \quad j(x,t) = \frac{q^2x}{2t^2} = \frac{x}{t}\varepsilon(x,t). \qquad (2.1.36)$$

2.1.4 Equivalence to Nonlinear Schrödinger Equations

1. *Vector representation for the L–L equations without damping*

Denote the unit vector $\vec{S}(x,t)$ by $\vec{e}_1(x,t)$ to rewrite equation (2.1.1) as

$$\vec{e}_{1t} = \vec{e}_1 \times \vec{e}\,'', \tag{2.1.37}$$

In order to have an equation independent of the coordinate system, we let $\vec{e}_1$ be a tangential vector of a curve with curvature k and torsion τ:

$$k(x,t) = (\vec{e}_1' \cdot \vec{e}_1')^{\frac{1}{2}}, \tag{2.1.38}$$

$$\tau(x,t) = k^{-2}\vec{e}_1 \cdot (\vec{e}_1' \times \vec{e}_1''). \tag{2.1.39}$$

Now establish an orthogonal three-side-body $\vec{e}_1, \vec{e}_2, \vec{e}_3$ where $\vec{e}_2$ lies along the direction of $\vec{e}_1'$, $\vec{e}_3 = \vec{e}_1 \times \vec{e}_2$. For such vectors we have from the Serret–Frenet formula that

$$\begin{cases} \vec{e}_1' = k\vec{e}_2, \\ \vec{e}_2' = -k\vec{e}_1 + \tau\vec{e}_3, \\ \vec{e}_3' = -\tau\vec{e}_3. \end{cases} \tag{2.1.40}$$

It follows from these equations and (2.1.37) that

$$\begin{cases} \vec{e}_1' = -k\tau\vec{e}_2 + k'\vec{e}_3, \\ \vec{e}_2' = k\tau\vec{e}_1 + \left(\dfrac{k''}{k} - \tau^2\right)\vec{e}_3, \\ \vec{e}_3' = -k'\vec{e}_1 + \left(\tau^2 - \dfrac{k''}{k}\right)\vec{e}_3, \end{cases} \tag{2.1.41}$$

with compatible conditions

$$\frac{d}{dt}\dot{\vec{e}}_j = \frac{\partial}{\partial x}\dot{\vec{e}}_j, \quad j = 1, 2, 3. \tag{2.1.42}$$

These conditions lead to the following equations of k and τ:

$$\begin{cases} k_t = -2k_x\tau - k\tau_x, \\ \tau_t = \left(\dfrac{k_{xx}}{k} - \tau^2\right)_x + kk'. \end{cases} \tag{2.1.43}$$

Equation (2.1.43) is equivalent to the continuum equation

$$\frac{\partial\varepsilon}{\partial t} + \frac{\partial j}{\partial x} = 0, \tag{2.1.44}$$

where $\varepsilon = \frac{1}{2}|\frac{\partial\vec{S}}{\partial x}|^2$, $j(x,t) = \vec{S} \cdot (\vec{S}_x \times \vec{S}_{xx})$ and

$$\varepsilon(x,t) = \frac{1}{2}x^2, \quad j(x,t) = k^2\tau. \tag{2.1.45}$$

2. Equivalence to nonlinear Schrödinger equations

Using the complex change of variables

$$\Psi(x,t) = k(x,t) \exp\left\{ i \int \tau(x,t)dx \right\}, \tag{2.1.46}$$

we may change system (2.1.43) into the nonlinear Schrödinger equations

$$i\Psi_t + \Psi_{xx} + \frac{1}{2}|\Psi|^2\Psi = 0 \tag{2.1.47}$$

which admits solitary solution of the form:

$$\Psi(x,t) = 4\eta \exp\{-4(\xi^2 - \eta^2)t - 2i\xi x\}\operatorname{sech}[2\eta(x - x_0) + 8\eta\xi t], \tag{2.1.48}$$

where $\zeta = \xi + i\eta$ is the characteristic parameter, x_0 is a constant. It follows from this that the energy density is

$$\varepsilon(x,t) = \frac{1}{2}k^2(x,t) = 8\eta^2\operatorname{sech}^2[2\eta(x - x_0) + 8\eta\xi t], \tag{2.1.49}$$

and the density of momentum is

$$j(x,t) = k^2(x,t)\tau(x,t) = 32\xi\eta^2\operatorname{sech}^2[2\eta(x - x_0) + 8\eta\xi t]. \tag{2.1.50}$$

Take $\xi = \eta$, $\tau = -2\xi = \frac{c}{2} = $ const. to give

$$\varepsilon(x,t) = \frac{c^2}{2}\operatorname{sech}^2\left[\frac{c}{2}(x - x_0 - ct)\right]. \tag{2.1.51}$$

2.2   Geometric Representation for the Landau–Lifshitz Equations

2.2.1   Establishing the Natural Coordinate System

Consider the following Landau–Lifshitz equation:

$$\frac{\partial \vec{S}}{\partial t} = \vec{S} \times \vec{S}_{xx} + \lambda[\vec{S}_{xx} - (\vec{S} \cdot \vec{S}_{xx})\vec{S}], \tag{2.2.1}$$

$$\vec{S}^2(x,t) = 1, \quad \vec{S} = (S_1, S_2, S_3), \tag{2.2.2}$$

where $\lambda \geq 0$ is the Gilbert damping constant. In order to get the geometric representation of this system, we let $[\vec{e}_1, \vec{e}_2, \vec{e}_3]$ be the natural coordinate consisting of the usual unit tangential vector, unit normal vector and unit conormal vector and

$$\vec{e}_{1t} = \vec{e}_1 \times \vec{e}_{1xx} + \lambda[\vec{e}_{1xx} - (\vec{e}_1 \cdot \vec{e}_{1xx})\vec{e}_1]. \tag{2.2.3}$$

It follows from Serret–Frenet formula that

$$\vec{e}_{ix} = \vec{d} \times \vec{e}_i,$$
$$\vec{d} = k\vec{e}_3 + \tau\vec{e}_1, \tag{2.2.4}$$

where k, τ are the curvature and torsion of the space curve with tangential vector $\vec{e}_1$, and

$$\vec{e}_{it} = \vec{\omega} \times \vec{e}_i,$$
$$\vec{\omega} = \omega_1\vec{e}_1 + \omega_2\vec{e}_2 + \omega_3\vec{e}_3, \tag{2.2.5}$$

in which

$$\omega_1 = \left[\frac{k_{xx}}{k} - \tau^2\right] + \frac{\lambda}{k}(2k_x\tau + k\tau_x), \tag{2.2.6}$$

$$\omega_2 = -(k_x + \lambda k\tau), \tag{2.2.7}$$

$$\omega_3 = -k\tau + \lambda k_x. \tag{2.2.8}$$

It follows from (2.2.4) and (2.2.5) and the compatible condition $(\vec{e}_{it})_x = (\vec{e}_{ix})_t$ that

$$k_t = -2k_x\tau - k\tau_x + \lambda(k_{xx} - k\tau^2), \tag{2.2.9}$$

$$\tau_t = \left[\frac{k_{xx}}{k} - \tau^2\right]_x + kk_x + \lambda\left[\left(\frac{1}{k^2}(k^2\tau)_x\right)_x + k^2\tau\right], \tag{2.2.10}$$

or by the densities of energy and flow:

$$\varepsilon_t = -j_x + \lambda\left[\varepsilon_{xx} - \frac{\varepsilon_x^2}{2\varepsilon} - \frac{j^2}{2\varepsilon}\right], \tag{2.2.11}$$

$$j_t = \varepsilon_{xx} - \frac{2\varepsilon_x}{\varepsilon}\left[\varepsilon_{xx} - \frac{\varepsilon_x^2}{2\varepsilon}\right] + \left[\varepsilon^2 - \frac{j^2}{\varepsilon}\right]_x + 2\varepsilon j$$
$$+ \lambda\left[\frac{j\varepsilon_{xx}}{\varepsilon} - \frac{j\varepsilon_x^2}{2\varepsilon^2} - \frac{j^3}{2\varepsilon^2} + j_{xx} - \frac{j_x\varepsilon_x}{\varepsilon}\right]. \tag{2.2.12}$$

Hence the total energy of the ferromegnet is

$$E(t) = \int_{-\infty}^{+\infty} \varepsilon(x,t)dx = \frac{1}{2}\int_{-\infty}^{+\infty}\left|\frac{\partial\vec{S}}{\partial x}\right|^2 dx = \frac{1}{2}\int_{-\infty}^{+\infty} k^2(x,t)dx. \tag{2.2.13}$$

2.2.2 Geometric Representation of Landau–Lifshitz Equation

It follows from (2.2.9)–(2.2.12) that

$$\frac{dE(t)}{dt} = \frac{d}{dt}\int_{-\infty}^{+\infty}\frac{1}{2}k^2 dx = -\lambda\int_{-\infty}^{+\infty}[(k_x)^2 + k^2\tau^2]dx \le 0.$$

Making complex change for the unknowns

$$\Psi(x,t) = k(x,t)\exp\left[i\int_{-\infty}^{+\infty}\tau(x,t)dx\right]$$

$$= [2E(x,t)]^{1/2}\exp\left[i\int_{-\infty}^{+\infty}\frac{j(x,t)}{2\varepsilon(x,t)}dx\right], \tag{2.2.14}$$

we can rewrite (2.2.9) and (2.2.10) into the Schrödinger with damping

$$i\Psi_t + (1-i\lambda)\Psi_{xx} + \frac{1}{2}|\Psi|^2\Psi$$

$$+\frac{i\lambda}{2}\Psi\int_{-\infty}^{+\infty}(\Psi\Psi_x^* - \Psi^*\Psi_x)dx = 0. \tag{2.2.15}$$

It follows from (2.2.15) and its complex conjugate that

$$i\frac{d(\Psi\Psi^*)}{dt} + (1-i\lambda)\Psi^*\Psi_{xx} - (1+i\lambda)\Psi_{xx}^*\Psi = 0. \tag{2.2.16}$$

Integrate (2.2.14) and (2.2.16) to give

$$i\tau\frac{dE}{dt} = (1+i\lambda)\int_{-\infty}^{+\infty}\Psi_{xx}^*\Psi dx - (1-i\lambda)\int_{-\infty}^{+\infty}\Psi^*\Psi_{xx}dx. \tag{2.2.17}$$

This equation is of the same form as (2.2.13). We know that the single solitary solution to (2.2.15) without damping ($\lambda = 0$) is of the form

$$\Psi(x,t) = c\left[\operatorname{sech}\frac{c}{2}x\right]\left[\exp i\frac{c}{2}x\right], \tag{2.2.18}$$

then

$$k(x,t) = c\operatorname{sech}\frac{c}{2}x, \quad \tau(x,t) = \frac{c}{2}, \tag{2.2.19}$$

where c is a constant. In the case of $\lambda > 0$, c should be regarded as a function of t, i.e. $c = c(t)$. Then it follows from (2.2.7) and (2.2.18) that

$$\frac{dc}{dt} = -\frac{2}{3}\lambda c^3, \tag{2.2.20}$$

and therefore

$$c = c(t) = \frac{\sqrt{3}}{2}(\lambda t + \delta)^{-1/2}, \tag{2.2.21}$$

where δ is the integral constant. Setting

$$E_l = \frac{e_{1l} + ie_{2l}}{1 - e_{3l}}, \quad e_{1l}^2 + e_{2l}^2 + e_{3l}^2 = 1,$$

where e_{1l} is the lth component of $\vec{e}_1$, one get the following two Riccati equations

$$Z_{lx} = -ikZ_l + \frac{1}{2}i\tau(Z_i^2 - 1), \tag{2.2.22}$$

$$\begin{aligned}
Z_{lt} = &-i\left[\left(\frac{k_{xx}}{k} - \tau^2\right) + \frac{\lambda}{k}(2k_x\tau + k\tau_x)\right]Z_i \\
&- \frac{1}{2}[(k\tau - \lambda k_x) + i(k_x + \lambda\tau k)]Z_i^2 \\
&- \frac{1}{2}[(k\tau - \lambda k_x) - i(k_x + \lambda\tau k)].
\end{aligned} \tag{2.2.23}$$

If we know the curvature k and the torsion τ, then we can solve the spin equation by virtue of (2.2.22) and (2.2.23). In fact, we have

$$\begin{cases}
\varepsilon(x,t) = \dfrac{1}{2}k^2 = \dfrac{3}{8}(\lambda t + \delta)^{-1}\operatorname{sech}^2\left(\dfrac{\sqrt{3}x}{4}\right)(\lambda t + \delta)^{-1/2}, \\[2mm]
j(x,t) = k^2\tau = \dfrac{3\sqrt{3}}{16}(\lambda t + \delta)^{-3/2}\operatorname{sech}^2\left(\dfrac{\sqrt{3}x}{4}\right)\dfrac{1}{2}(\lambda t + \delta)^{-1/2}.
\end{cases} \tag{2.2.24}$$

And, we also have

$$\begin{cases}
S^x = \operatorname{sech}\left[\left(\dfrac{\sqrt{3}x}{4}\right)(\lambda t + \delta)^{-1/2}\right]\left\{\tan h\left[\left(\dfrac{\sqrt{3}x}{4}\right)(\lambda t + \delta)^{-1/2}\right]\right. \\[2mm]
\qquad \left. \sin\left[\left(\dfrac{\sqrt{3}x}{4}\right)(\lambda t + \delta)^{-1/2}\right] - \cos\left[\left(\dfrac{\sqrt{3}x}{4}\right)(\lambda t + \delta)^{-1/2}\right]\right\} \\[2mm]
S^y = -\operatorname{sech}\left[\left(\dfrac{\sqrt{3}x}{4}\right)(\lambda t + \delta)^{-1/2}\right]\left\{\tan h\left[\left(\dfrac{\sqrt{3}x}{4}\right)(\lambda t + \delta)^{-1/2}\right]\right. \\[2mm]
\qquad \left. \cos\left[\left(\dfrac{\sqrt{3}x}{4}\right)(\lambda t + \delta)^{-1/2}\right] - \sin\left(\dfrac{\sqrt{3}x}{4}\right)(\lambda t + \delta)^{-1/2}\right]\right\} \\[2mm]
S^z = \left\{\tan h^2\left[\left(\dfrac{\sqrt{3}x}{4}\right)(\lambda t + \delta)^{-1/2}\right].
\end{cases} \tag{2.2.25}$$

2.3 Inhomogeneous Heisenberg Chain

2.3.1 Inhomogeneous Ferromagnetic Equations

Consider the nonhomogeneous ferromagnetic chain equation

$$\vec{S}_t(x,t) = f(x)\vec{S} \times \vec{S}_{xx} + f_x(x)(\vec{S} \times \vec{S}_x). \tag{2.3.1}$$

Let $\vec{e}_1(x,t), \vec{e}_2(x,t), \vec{e}_3(x,t)$ be the tangential vector, normal vector and co-normal vector of a moving space curve and form a natural coordinate system. From the Serret–Frenet formula

$$\begin{cases}
\vec{e}_{1x} = k\vec{e}_2, \\
\vec{e}_{2x} = -k\vec{e}_1 + \tau\vec{e}_3, \\
\vec{e}_{3x} = -\tau\vec{e}_3,
\end{cases} \tag{2.3.2}$$

where the curvature $k = (\vec{e}_{1x} \cdot \vec{e}_{1x})^{1/2}$ and the torsion $\tau = k^{-2}\vec{e}_1 \cdot (\vec{e}_{1x} \times \vec{e}_{1xx})$. Set

$$\begin{cases} N = (\vec{e}_2 + i\vec{e}_3) \exp\left[i \int_{-\infty}^{+\infty} \tau(x,t)dx\right], \\ g = \dfrac{k}{2} \exp\left[i \int_{-\infty}^{+\infty} \tau(x,t)dx\right], \end{cases} \tag{2.3.3}$$

then $N_x = -2q\vec{e}_1$. Let

$$\begin{cases} \vec{N}_t = \alpha\vec{N} + \beta\vec{N}^* + \gamma\vec{e}_1, \\ \vec{e}_{1t} = \lambda\vec{N} + \mu\vec{N}^* + \nu\vec{e}_1, \end{cases} \tag{2.3.4}$$

we have $\mu = -\frac{1}{2} = \lambda^*$, $\beta = 0$, $\alpha = iR$, R is real number. Hence

$$\begin{cases} \vec{N}_t = iR\vec{N} + \gamma\vec{e}_1, \\ \vec{e}_{1t} = -\frac{1}{2}(\gamma^*\vec{N}^* + \gamma\vec{N}), \end{cases} \tag{2.3.5}$$

and it follows from $N_{xt} = N_{tx}$ that

$$\begin{cases} q_t + \frac{1}{2}\gamma_x - iRq = 0, \\ R_x = i(\gamma q^* - \gamma^* q). \end{cases} \tag{2.3.6}$$

Note that $|\vec{S}|^2 = 1$. We let the vector $\vec{S}$ be $\vec{e}_1$. Then equation (2.3.1) becomes

$$\vec{e}_{1t} = -k\tau f\vec{e}_2 + (kf)_x\vec{e}_3 = i[(qf)_x\vec{N}^* - (q^*f)_x\vec{N}]. \tag{2.3.7}$$

Comparing this equation with (2.3.5), we may take, if f be a real function,

$$\gamma = -2i(gf)_x. \tag{2.3.8}$$

Substituting (2.3.8) into (2.3.6), replacing q by $q\exp(i(ax + bt))$ and eliminating the integral constant, one may get the Schrödinger equation with coefficients depending on x:

$$iq_t + fq_{xx} + 2f|q|^2q + 2q\int_{-\infty}^{+\infty} f_x|q|^2dx + qf_{xx} + 2f_xq_x = 0. \tag{2.3.9}$$

From the Hamiltonian change

$$\mathcal{H} = -J\sum_{i=1}^{N-1} f_i\vec{S}_i \cdot \vec{S}_{i+1}, \tag{2.3.10}$$

we get the energy density $E(x,t) = \frac{1}{2}f\vec{S}_x^2$ and the continuum equation

$$E_t + P_x = 0, \tag{2.3.11}$$

where $P(x,t) = f^2\vec{S} \cdot (\vec{S}_x \times \vec{S}_{xx})$ or by k, τ

$$E = \frac{1}{2}fk^2, \quad P = f^2k^2\tau.$$

For equation (2.3.9), we have

$$E = \frac{1}{2}f|q|^2, \quad P = 4f^2|q|^2(\arg q)_x.$$

2.3.2 Inhomogeneous Heisenberg Chain

Now consider another ferromagnetic chain equation:

$$\begin{cases} \vec{S}_t(x,t) = (\gamma_2 + \mu_2 x)\vec{S} \times \vec{S}_{xx} + \mu_2(\vec{S} \times \vec{S}_x) - (\gamma_1 + \mu_1 x)\vec{S}_x, \\ \vec{S} = (S_1, S_2, S_3), \qquad |\vec{S}|^2 = 1. \end{cases} \tag{2.3.12}$$

Using the Serret–Frenet formula:

$$\vec{e}_{ix} = d \wedge \vec{e}_i, \quad d = -k\vec{e}_3 + \tau\vec{e}_1, \tag{2.3.13}$$

and taking $\vec{e}_1 = \vec{S}$ in (2.3.12), we have

$$\vec{e}_{1t} = (\gamma_2 + \mu_2 x)\vec{e}_1 \times \vec{e}_{1xx} + \mu_2(\vec{e}_1 \times \vec{e}_{1x}) - (\gamma_1 + \mu_1 x)\vec{e}_{1x}.$$

It follows from (2.3.13) that

$$\vec{e}_{1t} = -[(\gamma_2 + \mu_2 x)k\tau + (\gamma_2 + \mu_1 x)k]\vec{e}_2 + [(\gamma_2 + \mu_2 x)k]_x\vec{e}_3.$$

The equations for $\vec{e}_{2t}$, $\vec{e}_{3t}$ are similar as above. If we take solid body motion, we have

$$\begin{aligned} \vec{e}_{it} &= \vec{\omega} \wedge \vec{e}_i, \\ \vec{\omega} &= \omega_1\vec{e}_1 + \omega_2\vec{e}_2 + \omega_3\vec{e}_3, \end{aligned} \tag{2.3.14}$$

where

$$\begin{cases} \omega_1 = k^{-1}[(\gamma_2 + \mu_2 x)k]_{xx} - (\gamma_2 + \mu_2 x)\tau^2, -(\gamma_1 + \mu_1 x)\vec{e}_{1x} - (\gamma_1 + \mu_1 x)\tau, \\ \omega_2 = -[(\gamma_2 + \mu_2 x)k]_x, \\ \omega_3 = -[(\gamma_2 + \mu_2 x)k\tau + (\gamma_1 + \mu_1 x)k]. \end{cases} \tag{2.3.15}$$

Since

$$R_x = \vec{e}_1, \quad R_t = (\gamma_1 + \mu_2 x)\vec{e}_1 \wedge \vec{e}_{1x} - (\gamma_1 + \mu_1 x)\vec{e}_1 + \mu_1 R,$$

we have from (2.3.13), (2.3.14) and the compatible condition $(\vec{e}_{it})_x = (\vec{e}_{ix})_t$, $i = 1, 2, 3$ that

$$\begin{cases} k_t = -2\mu_2 k\tau - (\gamma_2 + \mu_2 x)[k\tau_x + 2k_x\tau] - (\gamma_1 + \mu_1 x)k_x - \mu_1 k, \\ \tau_t = \mu_2[k^{-1}k_{xx} - \tau^2 + \frac{1}{2}k^2] \\ \qquad + (\gamma_2 + \mu_2 x)(k^{-1}k_{xx} - \tau^2 + \frac{1}{2}k^2)x + 2\mu_2(k^{-1}k_x)_x \\ \qquad - \mu_1\tau - (\gamma_1 + \mu_1 x)\tau_x + \frac{1}{2}\mu_2 k^2. \end{cases} \tag{2.3.16}$$

Setting $q = \frac{1}{2}k \exp(i \int_{-\infty}^{+\infty} \tau dx)$, we have

$$iq_t + i\mu_1 q + i(\gamma_1 + \mu_1 x)q_x + (\gamma_2 + \mu_2 x)(q_{xx} + 2q|q|^2)$$
$$+ 2\mu_2\left(q_x + q\int_{-\infty}^{+\infty} |q(x',t)|^2 dx'\right) = 0. \tag{2.3.17}$$

2.4 Spherical (Cylindrical) Symmetric Heisenberg Equations of Ferromagnetic Spin Chain

2.4.1 Radial Symmetric Equations

Consider n-dimensional Landau–Lifshitz equation

$$\begin{cases} \vec{S}_t(\vec{r}, t) = \vec{S} \times \Delta \vec{S}, \quad \vec{r} = (r_1, r_2, \ldots, r_n), \\ \vec{S}^2 = 1, \quad \vec{S} = (S^x, S^y, S^z), \quad \Delta = \dfrac{\partial^2}{\partial r_1^2} + \cdots + \dfrac{\partial^2}{\partial r_n^2}. \end{cases} \tag{2.4.1}$$

The radial symmetric equation is of the form

$$\begin{cases} \vec{S}_t = \vec{S} \times \vec{S}_{rr} + \dfrac{n-1}{r} \vec{S} \times \vec{S}_r, \\ r = (r_1^2 + \cdots + r_n^2)^{1/2}, \quad 0 < r < \infty. \end{cases} \tag{2.4.2}$$

In equation (2.4.2), the curvature of the space curve is

$$k(r, t) = [\vec{S}_r \cdot \vec{S}_r]^{1/2}, \tag{2.4.3}$$

and the torsion is

$$\tau(x, t) = k^{-2}[\vec{S} \cdot (\vec{S}_r \times \vec{S}_{rr})]. \tag{2.4.4}$$

Using Serret–Frenet relation, equation (2.4.2), in the natural coordinate system, exhibits in the form

$$\vec{e}_{ir} = \vec{D} \wedge \vec{e}_i, \quad \vec{e}_{it} = \vec{e}_i \wedge \Omega, \quad i = 1, 2, 3, \tag{2.4.5}$$

where

$$\vec{D} = k\vec{e}_3 + \tau\vec{e}_1, \quad \Omega = \omega_1\vec{e}_1 + \omega_2\vec{e}_2 + \omega_3\vec{e}_3, \tag{2.4.6}$$

$$\omega_1 = \dfrac{k_{rr}}{k} - \tau^2 + \left(\dfrac{n-1}{r}\right)\left(\dfrac{k_r}{r} - \dfrac{1}{r}\right), \tag{2.4.7}$$

$$\omega_2 = -k_r - [(n-1)/r]k, \tag{2.4.8}$$

$$\omega_3 = -k\tau. \tag{2.4.9}$$

It follows from the compatible condition $(\vec{e}_{ir})_t = (\vec{e}_{it})_r$, $i = 1, 2, 3$ and complex change of unknowns that

$$q(r, t) = \dfrac{k(r, t)}{2} \exp\left[i \int_0^r \tau(r', t)dr'\right], \tag{2.4.10}$$

and then the following radial symmetric nonlinear Schrödinger equation is derived

$$iq_t + q_{rr} + 2q|q|^2 + \dfrac{n-1}{r}q_r - \dfrac{n-1}{r^2}q$$

$$+ 4(n-1)q\int_0^r \dfrac{|q|^2}{r'}dr' = 0, \quad \text{if } n = 1. \tag{2.4.11}$$

This explains the equivalence of one-dimensional Heisenberg chain equation to nonlinear Schrödinger equation. If $n \neq 1$, we need to study the integrability of system (2.4.2) and equation (2.4.11).

2.4.2 Painleve Property of the Radial Symmetric Nonlinear Schrödinger Equation

First of all, we give the Painleve analysis for equation (2.4.11). For this reason we rewrite (2.4.2) as

$$iq_t + q_{rr} + \frac{n-1}{r}q_r + 2kq - \frac{n-1}{r^2}q = 0, \tag{2.4.12}$$

$$R_r - (|q|^2)_r - 2\frac{n-1}{r}|q|^2 = 0. \tag{2.4.13}$$

Let $q = a + ib$, where a, b are real numbers. It follows from (2.4.12), (2.4.13) that

$$a_t + b_{rr} + \frac{n-1}{r}b_r + 2kb - \frac{n-1}{r^2}b = 0, \tag{2.4.14}$$

$$-b_t + a_{rr} + \frac{n-1}{r}a_r + 2ka - \frac{n-1}{r^2}a = 0, \tag{2.4.15}$$

$$R_r - (a^2 + b^2)_r - 2\frac{n-1}{r}(a^2 + b^2) = 0. \tag{2.4.16}$$

Suppose that the first order solution of (2.4.14)–(2.4.16) are

$$a \sim a_0\varphi^\alpha, \quad b \sim b_0\varphi^\beta, \quad R \sim R_0\varphi^\gamma, \tag{2.4.17}$$

where a_0, b_0, R_0 are analytical functions of (r, t), α, β, γ are determined by (2.4.14)–(2.4.16):

$$\alpha = \beta = -1, \quad \gamma = -2, \tag{2.4.18}$$

and

$$a_0^2 + b_0^2 = -\varphi_r^2, \quad k_0 = -\varphi_r^2. \tag{2.4.19}$$

It is clear that one of the functions is arbitrary, a_0 or b_0. In order to derive the resonance conditions, we take Laurant expansion

$$a = \sum_j a_j\varphi^{j-1}, \quad b = \sum_j b_j\varphi^{j-1}, \quad R = \sum_j \varphi^{j-1}. \tag{2.4.20}$$

Reserving the terms with highest order in equation (2.4.14)–(2.4.16), we get the matrix equation for the coefficients of the terms with lower order as

$$\begin{bmatrix} (j^2 - 3j)\varphi_r^2 & 0 & 2a_0 \\ 0 & (j^2 - 3j)\varphi_r^2 & 2b_0 \\ -2a_0(j-2) & -2b_0(j-2) & j-2 \end{bmatrix} \begin{bmatrix} a_j \\ b_j \\ R_j \end{bmatrix} = 0. \tag{2.4.21}$$

It follows from (2.4.19) that the resonance conditions are

$$j = -1, 0, 2, 3, 4. \tag{2.4.22}$$

Obviously, resonance $j = -1$ implies the arbitrariness of the singular flow $\varphi(x, t) = 0$, resonance $j = 0$ indicates the arbitrariness of the functions a_0 or b_0. In order to verify the existence of sufficiently many functions with respect to the resonance $j = 2, 3, 4$, we substitute the expansions (2.4.20) into (2.4.14)–(2.4.16) and first combine the coefficients of $(\varphi^{-2}, \varphi^{-2}, \varphi^{-2})$ to get the equation

$$
\begin{bmatrix} 2R_0 & 0 & 2a_0 \\ 0 & 2R_0 & 2b_0 \\ 2a_0 & 2b_0 & 1 \end{bmatrix} \begin{bmatrix} a_1 \\ b_1 \\ R_1 \end{bmatrix} = \begin{bmatrix} -b_0\varphi_t + \dfrac{n-1}{r}a_0\varphi_r + 2a_0\varphi_r + a_0\varphi_{rr} \\[2ex] a_0\varphi_t + \dfrac{n-1}{r}b_0\varphi_r + 2b_0\varphi_r + b_0\varphi_{rr} \\[2ex] -2\dfrac{n-1}{r}\varphi_r \end{bmatrix}. \tag{2.4.23}
$$

It follows from (2.4.23) that

$$
a_1 = \frac{1}{2\varphi_r^2}\left(b_0\varphi_t - 2a_{0r}\varphi_{rr} + a_0\varphi_{rr} + \frac{n-1}{r}\varphi_{rr}\right), \tag{2.4.24}
$$

$$
b_1 = \frac{1}{2\varphi_r^2}\left(b_0\varphi_{rr} + \frac{n-1}{r}b_0\varphi_r - a_0\varphi_t - 2b_{0r}\varphi_r\right), \tag{2.4.25}
$$

$$
R_1 = \frac{n-1}{r}\varphi_r + \varphi_{rr}. \tag{2.4.26}
$$

Similarly we have from the coefficients of $(\varphi^{-1}, \varphi^{-1}, \varphi^{-1})$ that

$$
\begin{aligned}
2\varphi_r^2[R_2 + a_0a_2 + b_0b_2] &= b_0a_{0t} - a_0b_{0t} - \varphi_r\varphi_{rr}/r \\
&\quad + b_0b_{0rr} + a_0a_{0rr} + \varphi_{rr}^2,
\end{aligned} \tag{2.4.27}
$$

$$
\begin{aligned}
2\varphi_r^2(a_0b_2 + b_0a_2) &= -\varphi_r\varphi_{rr} + \frac{1}{r}(a_0b_{0r} - b_0a_{0r}) \\
&\quad + (a_0b_{0rr} - b_0a_{0rr}) + 2R_1(a_0b_1 - b_0a_1),
\end{aligned} \tag{2.4.28}
$$

$$
R_{1r} - 2(a_0a_1 + b_0b_1)x - \frac{4(n-1)}{r}(a_0a_1 + b_0b_1) = 0. \tag{2.4.29}
$$

It follows from (2.4.27) and (2.4.28) that one of the functions a_2, b_2, R_2 is arbitrary which corresponds to the resonance $j = 2$. On the other hand, It also follows from (2.4.29), (2.4.19) and (2.4.24)–(2.4.26) that

$$
(n - 1) = (n - 1)^2. \tag{2.4.30}
$$

We should introduce exponential movable singular manifold except for $n = 1, 2$. Similarly, it follows from the coefficients of $(\varphi_0, \varphi_0, \varphi_0)$ that

$$
\begin{aligned}
a_{1t} + a_2\varphi_t &+ \frac{1}{r}(b_{1r} + b_2\varphi_r) + b_{1rr} + 2b_{2r}\varphi_r + b_2\varphi_{rr} \\
&+ 2(R_1b_2 + R_2b_1 + R_3b_0) - b_1/r^3 = 0,
\end{aligned} \tag{2.4.31}
$$

$$-b_{1t} - b_2\varphi_t + \frac{1}{r}(a_{1r} + a_2\varphi_r) + a_{1rr} + 2a_{2r}\varphi_r + a_2\varphi_{rr}$$

$$+ 2(R_1 a_2 + R_2 a_1 + R_3 a_0) - a_1/r^3 = 0, \tag{2.4.32}$$

$$R_3\varphi_r - 2(a_0 a_3 + b_0 b_3)\varphi_r + R_{2r} - 2(a_0 a_2 + +a_1^2/2 + b_0 b_2 + b_1^2/2)$$

$$- 2(a_1 a_2 + b_1 b_2)\varphi_r = 0. \tag{2.4.33}$$

It can be inferred from (2.4.31) and (2.4.32) that R_3 is identical ($n = 1, 2$), and then it follows from (2.4.33) that one of the functions a_3, b_3 is arbitrary. Finally, it can be verified from the coefficients of $(\varphi^1, \varphi^1, \varphi^1)$ that a_4 or b_4 is arbitrary. Therefore, the general solutions $(a(r,t), b(r,t), R(r,t))$ of (2.4.14)–(2.4.16) admits the required number of arbitrary functions. For $n = 1, 2$, the elements lead to the movable critical manifolds. Hence, the Painleve property is satisfied for $n = 1$ or $n = 2$. The conclusion is obvious for $n = 1$. If $n = 2$, (2.4.11) is of the form

$$iq_t + q_{rr} + 2|q|^2 q + \frac{q_r}{r} - \frac{q}{r^2} + 4\int_0^r \frac{|q|^2}{r'}dr' = 0. \tag{2.4.34}$$

2.4.3 Normal Change for Radial Symmetric Equation and Matrix Form for the Radial Symmetric Heisenberg Equations

Now consider the normal change. Let $a_j = b_j = 0$, $j \geq 2$, $R_j = 0$, $j \geq 3$ in (2.4.20). Then

$$\begin{cases} a = a_0\varphi^{-1} + a_1, \quad b = b_0\varphi^{-1} + b_1, \\ R = R_0\varphi^{-2} + R_1\varphi^{-1} + R_2. \end{cases} \tag{2.4.35}$$

Substituting (2.4.35) into (2.4.34), we get the differential system of a_1, b_1, R_2. Define new variables

$$\begin{aligned} \vec{A}_0 = a_0 + ib_0, \quad & q = \vec{A} = a + ib, \\ \vec{B}_0 = a_0 - ib_0, \quad & q^* = \vec{B} = a - ib. \end{aligned} \tag{2.4.36}$$

It follows from simple computations that

$$\left[\frac{\vec{A}_{0r}}{r\vec{A}_0} + \frac{2\vec{A}\varphi_r}{r\vec{A}_0}\right]_r = \left[\frac{\vec{B}_{0r}}{r\vec{B}_0} + \frac{2\vec{B}\varphi_r}{r\vec{B}_0}\right]_r = 0. \tag{2.4.37}$$

Let $i\lambda$ be the integral constant, we have

$$\vec{A}_{0r} + i\lambda\vec{A}_0 = -2q\varphi_r, \quad \vec{B}_{0r} - i\lambda\vec{B}_0 = 2\vec{B}\varphi_r. \tag{2.4.38}$$

Define square characteristic functions as follows:

$$\vec{A}_0 = i\Psi_1^2, \quad \vec{B}_0 = i\Psi_2^2, \quad \varphi_r = -i\Psi_1\Psi_2. \tag{2.4.39}$$

We obtain the dispersion problem

$$\Psi_r = V_1\Psi, \quad \Psi = (\Psi_1, \Psi_2)^T, \tag{2.4.40}$$

where

$$V_1 = \begin{bmatrix} -i\lambda r/2 & q \\ -q^* & i\lambda r/2 \end{bmatrix}. \tag{2.4.41}$$

Similarly one can also get the evolutionary part of the dispersion problem from (2.4.24) and (2.4.25) as follows:

$$\Psi_t = V_1\Psi = \begin{bmatrix} A & B \\ -B^* & -A \end{bmatrix}, \tag{2.4.42}$$

where

$$A = -\frac{i\lambda^2 r^2}{2} + 2i\left[\int_0^t \frac{|q|^2}{r'}dr' + \frac{|q|^2}{2}\right], \tag{2.4.43}$$

$$B = \lambda rq + i\left(q_r + \frac{q}{r}\right), \tag{2.4.44}$$

and the compatible condition is

$$U_{1t} - V_{1r} + [U_1, V_1] = 0.$$

These lead to the non-identical spectrum representation for equation (2.4.34):

$$\lambda_t = 2\lambda^2. \tag{2.4.45}$$

Introduce normal change:

$$\hat{\Psi} = g^{-1}\Psi, \quad g(r,t) \in \mathrm{GL}(2, C). \tag{2.4.46}$$

Substitute (2.4.40) into (2.4.42) to give new eigenvalue equation

$$\hat{\Psi}_r = \hat{U}_r\hat{\Psi}, \quad \hat{\Psi}_t = \hat{V}_1\hat{\Psi}, \tag{2.4.47}$$

where

$$\begin{cases} \hat{U}_1 = g^{-1}U_1 g - g^{-1}g_r, \quad \hat{V}_1 = g^{-1}V_1 g - g^{-1}g_t, \\ g_r = U_1(\lambda = 0)g, \quad g_t = V_1(\lambda = 0)g. \end{cases} \tag{2.4.48}$$

Define spin matrix of the form

$$\vec{S} = -g^{-1}\sigma_3 g = \vec{S} \cdot \vec{\sigma},$$

$$\vec{\sigma} = (\sigma_1, \sigma_2, \sigma_3) \text{ is the Pauli matrix.}$$

We have

$$\hat{U}_1 = \frac{i\lambda_r}{2}\vec{S}, \quad \hat{V}_1 = \frac{i\lambda^2 r^2}{2}\vec{S} + \frac{\lambda r}{4}[\vec{S}, \vec{S}_r]. \tag{2.4.49}$$

The compatible condition

$$\hat{U}_{1t} - \hat{V}_{1t} + [\hat{U}_1, \hat{V}_1] = 0,$$

and $\lambda_1 = 2\lambda^2$ lead to

$$\vec{S}_t = \frac{1}{\tau_i}[\vec{S}, \vec{S}_r] + \frac{1}{\tau_i r}[\vec{S}, \vec{S}_r].$$

This is the matrix form of radial symmetric Heisenberg equation (2.4.2).

2.4.4 Bäcklund Change for Radial Symmetric Nonlinear Schrödinger Equation and Solitary Solutions

Now we consider the Bäcklund change for equation (2.4.34) and its solitary solution. Let

$$\Gamma = \Psi_1/\Psi_2. \tag{2.4.50}$$

From (2.4.40) and (2.4.42) we can get the following Riccati equation of Γ

$$\begin{cases} \Gamma_r = -i\lambda r\Gamma + q/2 + (q^*/2)\Gamma^2, \\ \Gamma_t = B + 2A\Gamma + B^*\Gamma^2. \end{cases} \tag{2.4.51}$$

We can also construct Γ' which meets the same equations (2.4.51) and has potential $q'(r)$, and

$$q'(r,t) = q(r,t) + f(\Gamma,\lambda). \tag{2.4.52}$$

On the other hand, eliminating Γ from (2.4.51), one gets the Bäcklund change for (2.4.34). Take, for example, Γ' and q' as follows:

$$\begin{cases} \Gamma' = 1/\Gamma^*, \\ q' + q = -4\beta r\Gamma/(1 + |\Gamma|^2). \end{cases} \tag{2.4.53}$$

Then Γ' satisfies (2.4.51) with potential $q'(r,t)$ where $\lambda(t) = \alpha(t) + i\beta(t)$, $\beta > 0$. Solve Γ from (2.4.53) as

$$\Gamma = -[2\beta\gamma + 4\beta^2 r^2 - |q' + q|^{1/2}]/(q'^* + q^*). \tag{2.4.54}$$

Substitute (2.4.54) into (2.4.51) to give the $B\Gamma$ of (2.4.34):

$$\begin{aligned} q_r + q'_r = {}& (q + q')/r \\ & + \frac{1}{2}(q' + q)(4\beta^2 r^2 - |q' + q|^2)^{1/2} - i\alpha r(q' + q), \\ q_t + iq'_t = {}& 4\alpha(q + q') + \left(iq_r + \frac{iq}{r} - \alpha rq'\right)(4\beta^2 r^2 - |q' + q|^2)^{1/2} \\ & + \frac{iq}{4}|q' + q|^2 + i(q' + q)\int_0^r r\frac{|q|^2}{r'}dr' - ir^2 q(\alpha^2 - \beta^2) \\ & - ir^2 q(\alpha^2 + \beta^2) - \frac{i}{4}(q'^2 q^r + |q|^2 q). \end{aligned} \tag{2.4.55}$$

Now we may construct the solitary solution of (2.4.34). Assuming $q(0) = 0$, we get from (2.4.40), (2.4.42) and (2.4.53) that

$$q(1) = -2\beta\gamma\, \mathrm{sech}\left[\frac{\beta r^2}{2} + \delta_1\right]\exp\left[-\frac{i\alpha r^2}{2} + \delta_2\right], \tag{2.4.56}$$

where δ_1, δ_2 are phase constants. It follows from

$$\alpha_t = 2(\alpha^2 - \beta^2), \quad \beta_t = 4\alpha\beta, \tag{2.4.57}$$

that

$$\alpha(t) = -\frac{\alpha(0) + 2t}{(\alpha(0) + 2t)^2 + \beta(0)^2},$$
$$\beta(t) = -\frac{\beta(0)}{(\alpha(0) + 2t)^2 + \beta(0)^2}.$$

$$(2.4.58)$$

we can then construct $q(2)$ from (2.4.53) and $\Psi(1)$, and so on. In this way we obtain a series of solitary solutions.

2.5   Bibliography Comments

This chapter is mainly concerned with the theory of integrability of Heisenberg chain without Gilbert damping. Since Nakamura and Sasada [122] found the solitary solution to L–L equations in 1974, there have been a series studies on the solitary solution to L–L equation, scattering method, geometric representation, infinite conservation law and the normal equivalence to nonlinear Schrödinger equations, for the details we refer to references [140], [71]. At the same time, we refer to references [12, 59, 110] and [124] for the discussions on the one-dimensional inhomogeneous and compressible flow Heisenberg chain, the Painleve properties for the solitary waves of spherical symmetric Heisenberg ferromagnetic systems.

Chapter 3

One-Dimensional Landau–Lifshitz Equations

In this chapter, we discuss the boundary value problems of one-dimensional Landau–Lifshitz equations. We state and prove some results on the existence and uniqueness of weak solutions and smooth solution.

3.1 Initial Boundary Value Problem of One-dimensional Ferromagnetic Spin Chain Equations

This section deals with the ferromagnetic spin chain system of the form

$$\vec{Z}_t = \vec{Z} \times \vec{Z}_{xx} + \vec{f}(x, t, \vec{Z}) \tag{3.1.1}$$

with initial-boundary condition of the form

$$\begin{cases} \vec{Z}(0, t) = \vec{Z}(l, t) = 0, \\ \vec{Z}(x, 0) = \vec{Z}_0(x), \quad 0 \le x \le l, \end{cases} \tag{3.1.2}$$

where $\vec{Z}(x, t) = (u(x, t), v(x, t), w(x, t))$ is a three-dimensional unknown, $\vec{f}(x, t, \vec{Z})$ is a known three-dimensional vector of variables $x, t, \vec{Z}$, $\times$ denotes the cross product of two three-dimensional vectors. In the classical studies, isotropic Heisenberg chain equations and Landau–Lifshitz equations are the special cases of (3.1.1).

3.1.1 Initial Boundary Value Problem of Ferromagnetic Spin Chain Equations

System (3.1.1) is a strongly coupling and strongly degenerate equation. In fact, (3.1.1) can be rewritten as

$$\vec{Z}_t = A(\vec{Z})\vec{Z}_{xx} + \vec{f}(x, t, \vec{Z}), \tag{3.1.3}$$

53

where

$$\vec{Z} = \begin{pmatrix} u \\ v \\ w \end{pmatrix}, \qquad A(\vec{Z}) = \begin{bmatrix} 0 & -w & v \\ w & 0 & -u \\ -v & u & 0 \end{bmatrix}. \tag{3.1.4}$$

It is not difficult to verify that

(1) The matrix $A(\vec{Z})$ is zero definite, i.e.

$$\xi \cdot A(\vec{Z})\xi = 0, \quad \forall \, \xi, \vec{Z} \in R^3.$$

(2) $A(\vec{Z})$ is singular, i.e.

$$\det A(\vec{Z}) = 0, \quad \forall \, \vec{Z} \in R^3.$$

This implies that (3.1.1) is strongly degenerate.

3.1.2   Theory on Quasilinear Parabolic Equations

1. *Initial boundary value problem of linear parabolic systems*
We consider the following linear parabolic systems

$$\vec{u}_t - A(x,t)\vec{u}_{xx} + B(x,t)\vec{u}_x + C(x,t)\vec{u} = \vec{f}(x,t) \tag{3.1.5}$$

with initial boundary condition

$$\begin{cases} \vec{u}|_{x=0} = \vec{u}|_{x=l} = 0, \\ \vec{u}|_{t=0} = u_0(x), \end{cases} \tag{3.1.6}$$

where $\vec{u}(x,t)$ and $\vec{u}_0(x)$ are two N-dimensional vector-valued functions.

Lemma 3.1.1 *If* (3.1.5) *and* $\vec{u}_0(x)$ *satisfy the following conditions*
    (i) $A(x,t)$ *is a positively definite* $n \times n$ *matrix over* $Q_T = [0,l] \times [0,T]$;
    (ii) $A(x,t), B(x,t)$ *and* $C(x,t)$ *are measurable bounded* $N \times N$ *matrix*;
    (iii) $\vec{f}(x,t)$ *is quadratically integrable* N-*dimensional vector-valued function over* Q_T;
    (iv) $\vec{u}_0(x) \in H_0^1(x)$, *then problem* (3.1.5)–(3.1.6) *admits a unique solution*

$$\vec{u}(x,t) \in L^\infty(0,T; H_0^1(0,l)) \bigcap W_2^{(2,1)}(Q_T), \tag{3.1.7}$$

and the following estimate holds

$$\sup_{0 \le t \le T} \|\vec{u}(\cdot,t)\|_{H_0^1(0,l)} + \|\vec{u}_t\|_{L^2(Q_T)} + \|\vec{u}_{xx}\|_{L^2(Q_T)}$$

$$\le K(\|\vec{u}_0\|_{H_0^1(0,l)} + \|\vec{f}\|_{L^2(Q_T)}), \tag{3.1.8}$$

where K depends on $A(x,t), B(x,t), C(x,t)$, but is independent of $\vec{u}(x,t)$.

Proof. Taking inner products of (3.1.5) with vectors $\vec{u}$ and $\vec{u}_{xx}$ respectively, we obtain

$$(\vec{u}, \vec{u}_t) - (\vec{u}, A(x,t)\vec{u}_{xx}) + (\vec{u}, B(x,t)\vec{u}_x) + (\vec{u}, C(x,t)\vec{u}) = (\vec{u}, \vec{f}(x,t));$$
$$(\vec{u}_{xx}, \vec{u}_t) - (\vec{u}_{xx}, A(x,t)\vec{u}_{xx}) + (\vec{u}_{xx}, B(x,t)\vec{u}_x) + (\vec{u}_{xx}, C(x,t)\vec{u}) = (\vec{u}_{xx}, \vec{f}(x,t)).$$

Subtracting one from the other in the above two equations, we get

$$\int_0^l (\vec{u}_{xx}(x,t), \vec{u}_t(x,t))dx = (\vec{u}_x(x,t), \vec{u}_t(x,t))|_{x=0}^{x=l}$$
$$- \int_0^l (\vec{u}_x(x,t), \vec{u}_{xt}(x,t))dx.$$

Using boundary condition (3.1.6), we see that the first term on the right-hand side of the above equality is zero. Therefore we have

$$\int_0^l (\vec{u}_{xx}(x,t), \vec{u}_t(x,t))dx = -\frac{1}{2}\frac{d}{dt}\|\vec{u}_x(\cdot, t)\|^2_{L^2(0,l)}.$$

In the rectangular domain $Q_\tau = \{0 \le x \le l, 0 \le t \le \tau\}$, $(0 \le \tau \le T)$, making subtraction of these two equalities leads to

$$\|\vec{u}(\cdot, t)\|^2_{H_0^1(0,l)} - \|\vec{u}_0\|^2_{H_0^1(0,l)} + 2\int\int_{Q_\tau} (\vec{u}_{xx}, A\vec{u}_{xx})dxdt$$
$$= 2\int\int_{Q_\tau} (\vec{u}_{xx}, B\vec{u}_x + C\vec{u} - \vec{f})dxdt$$
$$+ 2\int\int_{Q_\tau} (\vec{u}, A\vec{u}_{xx} - B\vec{u}_x - C\vec{u} + \vec{f})dxdt.$$

It is easy to derive (3.1.8) from the above equality by the conditions of this lemma. And, using the parameter extension method, one gets the existence of the solution to the above problem. The uniqueness can be derived from (3.1.8). Then, the proof is complete. $\qquad\square$

2. *Initial boundary value problem of spin systems*

Now we turn to equation (3.1.1) by viscosity method, that is, we consider the following system with small diffusion term $\varepsilon\vec{Z}_{xx}$

$$\vec{Z}_t = \varepsilon\vec{Z}_{xx} + \vec{Z} \times \vec{Z}_{xx} + \vec{f}(x,t,\vec{Z}) \tag{3.1.9}$$

which is called spin system. It is clear that this is a parabolic system whose initial boundary condition is (3.1.6).

Take three-dimensional vector-valued function space $B = L^\infty(Q_T)$ as the base space of fixed point argument. We define a functional mapping with parameter $0 \le \lambda \le 1$ from the base space onto itself by $T_\lambda : B \to B$: For every $\vec{u} \in B$, the image $\vec{Z} = T_\lambda(\vec{u})$ is the solution of the linear parabolic system

$$\vec{Z}_t = \varepsilon\vec{Z}_{xx} + \vec{u} \times \vec{Z}_{xx} + \vec{f}(x,t,\vec{u}) \tag{3.1.10}$$

subject to the condition (3.1.6), where $0 \leq \lambda \leq 1$. It follows from Lemma 3.1.1 that $\vec{Z} = T_\lambda \vec{u}$ is the unique solution of (3.1.10) under condition (3.1.6). This unique solution is in the space G where $G = L^\infty(0, T; H_0^1(0, l)) \cap W_2^{(2,1)}(Q_T)$.

It is easy to see that for every λ the operator T_λ is completely continuous and for every bounded set M in B the operator T_λ is uniformly continuous in $\lambda : 0 \leq \lambda \leq 1$.

In order to prove the existence of weak solution to spin equation (3.1.9) under (3.1.6) by fixed point theorem, we should make out the *a priori* uniform estimate in $\lambda \in (0, 1)$ for all possible fixed point of the mapping T_λ.

To this aim, we assume

(1) Vector $\vec{f}(x, t, \vec{Z})$ is continuously differentiable in $x, \vec{Z}$, and, 3×3 Jacobi matrix $\vec{f}_{\vec{Z}}(x, t, \vec{Z})$ is semi-bounded, i.e., there is a constant $b > 0$ such that for any $\xi \in R^3$ there holds

$$\xi \cdot \vec{f}_{\vec{Z}}(x, t, \vec{Z}) \cdot \xi \leq b|\xi|^2, \tag{3.1.11}$$

for any $(x, t, \vec{Z}) \in Q_T \times R^3$. Moreover, $\vec{f}_0(x, t) \equiv \vec{f}(x, t, 0) \in L^2(Q_T)$.

(2) For $(x, t, \vec{Z}) \in Q_T \times R^3$, there holds

$$|\vec{f}_x(x, t, \vec{Z})| \leq c(x, t)|\vec{Z}|^3 + d(x, t), \tag{3.1.12}$$

where $c(x, t) \in L^\infty(Q_T), d(x, t) \in L^2(Q_T)$.

(3) $\vec{Z}_0(x) \in H_0^1(0, l)$.

Construct the following vector system

$$\vec{Z}_t = \varepsilon \vec{Z}_{xx} + \lambda \vec{Z} \times \vec{Z}_{xx} + \lambda \vec{f}(x, t, \vec{Z}) \tag{3.1.13}$$

and take inner product of this system with $\vec{Z}(x, t)$ to give

$$\vec{Z} \cdot \vec{Z}_t = \varepsilon \vec{Z}_{xx} \cdot \vec{Z} + \lambda \vec{Z} \cdot (\vec{Z} \times \vec{Z}_{xx}) + \lambda \vec{Z} \cdot \vec{f}(x, t, \vec{Z}). \tag{3.1.14}$$

Simple calculations give that

$$\vec{Z} \cdot \vec{f}(x, t, \vec{Z}) = \vec{Z} \cdot \left(\int_0^1 \vec{f}_{\vec{Z}}(x, t, \tau\vec{Z})d\tau \right) \vec{Z} + \vec{Z} \cdot \vec{f}_0(x, t)$$

$$\leq (b + \delta)|\vec{Z}|^2 + \frac{1}{4\delta}|\vec{f}_0(x, t)|^2,$$

where $\delta > 0$. Integrate (3.1.14) over the rectangular domain Q_τ $(0 \leq \tau \leq T)$ to get

$$\|\vec{Z}(\cdot, \tau)\|_{L^2(0,l)}^2 - \|\vec{Z}_0\|_{L^2(0,l)}^2$$

$$\leq 2\lambda(b + \delta) \int_0^\tau \|\vec{Z}(\cdot, t)\|_{L^2(0,l)}^2 dt + \frac{\lambda}{2\delta}\|\vec{f}_0\|_{L^2(Q_T)}^2$$

and hence there holds

$$\|\vec{Z}(\cdot, t)\|_{L^2(0,l)}^2 \leq \left(\|\vec{Z}_0\|_{L^2(0,l)}^2 + \frac{\lambda}{2\delta}\|\vec{f}_0\|_{L^2(0,l)}^2 \right) e^{2\lambda(b+\delta)t}, \tag{3.1.15}$$

where $0 \le \lambda \le 1$, $0 \le t \le T$, $\delta > 0$. Therefore we have

$$\sup_{0 \le t \le T} \|\vec{Z}(\cdot, t)\|^2_{L^2(0,l)} \le K_1, \tag{3.1.16}$$

where K_1 is a constant independent of ε, λ, l.

Now taking the inner product of (3.1.13) with $\vec{Z}_{xx}$, we have

$$\vec{Z}_{xx} \cdot \vec{Z}_t = \varepsilon \vec{Z}_{xx} \cdot \vec{Z}_{xx} + \lambda \vec{Z}_{xx} \cdot (\vec{Z} \times \vec{Z}_{xx}) + \lambda \vec{Z}_{xx} \cdot \vec{f}(x, t, \vec{Z}). \tag{3.1.17}$$

For the left-hand side of (3.1.17), we have

$$\int_0^l (\vec{Z}_{xx} \cdot \vec{Z}_t) dx = -\frac{1}{2} \frac{d}{dt} \|\vec{Z}_x(\cdot, t)\|^2_{L^2(0,l)}. \tag{3.1.18}$$

For the last term on the right of (3.1.17), we have

$$\int_0^l (\vec{Z}_{xx} \cdot \vec{f}(x, t, \vec{Z})) dx = \vec{Z}_x(l, t) \cdot \vec{f}(l, t, \vec{Z}(l, t))$$
$$- \vec{Z}_x(0, t) \cdot \vec{f}(0, t, \vec{Z}(0, t)) - \int_0^l (\vec{Z}_x \cdot D_x \vec{f}) dx. \tag{3.1.19}$$

It follows from assumption (1) that

$$\int \int_{Q_\tau} (\vec{Z}_x \cdot D_x \vec{f}) dx dt = \int \int_{Q_\tau} (\vec{Z}_x \cdot \vec{f}_x(x, t, \vec{Z}) \vec{Z}_x) dx dt$$
$$+ \int \int_{Q_\tau} (\vec{Z}_x \cdot \vec{f}_x(x, t, \vec{Z})) dx dt$$
$$\le \left(b + \frac{1}{2} \right) \int_0^l \|\vec{Z}_x(\cdot, t)\|^2_{L^2(0,l)} dt$$
$$+ \frac{1}{2} \int \int_{Q_\tau} |\vec{f}_x(x, t, \vec{Z})|^2 dx dt.$$

It follows from assumption (2) and interpolation formula that

$$\frac{1}{2} \int \int_{Q_\tau} |\vec{f}_x|^2 dx dt \le \|c(x, t)\|^2_{L^\infty(Q_T)} \int \int_{Q_\tau} |\vec{Z}|^6 dx dt + \|d\|^2_{L^2(Q_T)}$$

and

$$\int_0^l |\vec{Z}(x, t)|^6 dx \le C_1 \|\vec{Z}(\cdot, t)\|^4_{L^2(0,l)} \cdot \|\vec{Z}(\cdot, t)\|^2_{H_0^1(0,l)}.$$

Therefore (3.1.19) can be rewritten in the following inequality:

$$\int \int_{Q_\tau} (\vec{Z}_{xx} \cdot \vec{f}(x, t, \vec{Z})) dx$$
$$\le \int_0^\tau [\vec{Z}_x(l, t) \cdot \vec{f}(l, t, \vec{Z}(l, t)) - \vec{Z}_x(0, t) \cdot \vec{f}(0, t, \vec{Z}(0, t))] dt$$
$$+ C_2 \int_0^\tau \|\vec{Z}_x(\cdot, t)\|^2_{L^2(0,l)} dt + C_3,$$

where $0 \leq \tau \leq T$, and

$$C_2 = \left(b + \frac{1}{2}\right) + C_1 \|c\|_{L^\infty(Q_T)}^2 K_1^4, \quad C_3 = \|d\|_{L^2(Q_T)}^2.$$

When system (3.1.9) is homogeneous, i.e. $\vec{f}(x,t,0) \equiv 0$, we can obtain

$$\int\int_{Q_\tau} (\vec{Z}_{xx} \cdot \vec{f}(x,t,\vec{Z}))dx \leq C_2 \int_0^\tau \|\vec{Z}_x(\cdot,t)\|_{L^2(0,l)}^2 dt + C_3. \tag{3.1.20}$$

Integrating (3.1.17) over $Q_\tau, 0 \leq \tau \leq T$, we finally get

$$\sup_{0 \leq t \leq T} \|\vec{Z}_x(\cdot,t)\|_{L^2(0,l)} \leq K_2, \tag{3.1.21}$$

where K_2 is independent of λ and ε, and also of l if $\|c\|_{L^\infty(Q_\tau)}$ and $\|d\|_{L^2(Q_\tau)}$ are independent of the width l, then K_2 is also independent of l. Therefore we have

Theorem 3.1.1 *Let assumptions (1)–(3) hold, (3.1.9) is homogeneous, i.e.*
$\vec{f}(x,t,0) \equiv 0$. *Then the spin system (3.1.9) with (3.1.6) admits unique global weak solution*

$$\vec{Z}(x,t) \in L^\infty(0,T; H_0^1(0,l)) \bigcap W_2^{(2,1)}(Q_T).$$

If $b < 0$, we take $\delta > 0$ such that $b + \delta < 0$. If $T = \infty$, denote (1)–(3) by (1_∞)–(3_∞), respectively; hence, when $b < 0$, the norm $\|\vec{Z}(\cdot,t)\|_{L^2(0,l)}$ of the global weak solution $\vec{Z}(x,t)$ of the spin system (3.1.9) (with (3.1.6)) tend to zero when $t \to \infty$, i.e. $\lim_{t\to\infty} \|\vec{Z}(\cdot,t)\|_{L^2(0,l)} = 0$.

3.1.3   Approximate Solution to the Initial Boundary Problem of the System of Ferromagnetic Spin Chain

Now we begin to consider the initial-boundary problem (3.1.1)–(3.1.2). In the proof of the above theorem, we see that the *a priori* estimates for the approximate solutions are independent of λ. Denote the approximate solution obtained in the above subsection by $\vec{Z}_\varepsilon = \vec{Z}_\varepsilon(x,t)$.

    1. *Approximate solutions to initial-boundary value problem*

Lemma 3.1.2 *The solution $\vec{Z}_\varepsilon(x,t)$ obtained in Theorem 3.1.1 meets the estimates*

$$\|\vec{Z}_\varepsilon(\cdot,t)\|_{L^2(0,l)} \leq K_3(\|\vec{Z}_0\|_{L^2(0,l)} + \|\vec{f}_0\|_{L^2(Q_T)})e^{(b+\delta)t}, \tag{3.1.22}$$

$$\sup_{0 \leq t \leq T} \|\vec{Z}_\varepsilon(\cdot,t)\|_{H_0^1(0,l)} \leq K_4, \tag{3.1.23}$$

where $0 \leq t \leq T$, $\delta > 0$, K_3, K_4 are independent of ε and of l if $\|\vec{Z}_0\|_{H_0^1(0,l)}$, $\|\vec{f}_0\|_{L^2(Q_T)}$, $\|c\|_{L^\infty(Q_T)}$ and $\|d\|_{L^2(Q_T)}$ are independent of l.

Lemma 3.1.3 *For the solution $\vec{Z}_\varepsilon(x,t)$ obtained in Theorem 3.1.1, there holds*

$$\sup_{0\le t\le T} \|\vec{Z}_{\varepsilon t}\|_{H^{-1}(0,l)} \le K_5, \tag{3.1.24}$$

where K_5 is independent of ε and of l if $\|\vec{Z}_0\|_{H_0^1(0,l)}$, $\|\vec{f}_0\|_{L^2(Q_T)}$, $\|c\|_{L^\infty(Q_T)}$ and $\|d\|_{L^2(Q_T)}$ are independent of l

Proof. For any $\psi(x) \in H_0^1(0,l)$ we have

$$\int_0^l \psi(x)\vec{Z}_{\varepsilon t}dx = -\varepsilon \int_0^l \psi_x(x)\vec{Z}_{\varepsilon x}(x,t)dx$$

$$- \int_0^l \psi_x(x)[\vec{Z}_\varepsilon(x,t) \times \vec{Z}_{\varepsilon x}(x,t)]dx$$

$$+ \int_0^l \psi(x)\vec{f}(x,t,\vec{Z}_\varepsilon(x,t))dx.$$

It follows from (3.1.23) that

$$\left| \int_0^l \psi(x)\vec{Z}_{\varepsilon t}dx \right| \le C_4 \|\psi\|_{H_0^1(0,l)},$$

where C_4 is independent of ε and l. Lemma 3.1.3 follows from the definition of Sobolev space of negative order.

2. *Estimates for approximate solutions*

Lemma 3.1.4 *For the above approximate solutions there holds the estimate*

$$\|\vec{Z}_\varepsilon(x,t)\|_{C^{\frac{1}{2},\frac{1}{4}}(Q_T)} \le (1+l)K_6, \tag{3.1.25}$$

with K_6 independent of ε and if $\|\vec{Z}_0\|_{H_0^1(0,l)}$, $\|\vec{f}_0\|_{L^2(Q_T)}$, $\|c\|_{L^\infty(Q_T)}$ and $\|d\|_{L^2(Q_T)}$ are independent of l, then K_6 is also independent of l.

Proof. Let $\vec{w}_\varepsilon(x,t) = \int_0^x \vec{Z}_\varepsilon(\xi,t)d\xi$. Then $\vec{w}_{\varepsilon t}(x,t) = \int_0^x \vec{Z}_{\varepsilon t}(\xi,t)d\xi$, $\vec{w}_{\varepsilon x t}(x,t) = \vec{Z}_{\varepsilon t}(x,t)$, $\vec{w}_{\varepsilon x}(x,t) = \vec{Z}_\varepsilon(x,t)$ and $\vec{w}_{\varepsilon x x}(x,t) = \vec{Z}_{\varepsilon x}(x,t)$. For any $\psi \in H_0^1(0,l)$, the following holds

$$\left| \int_0^l \psi'(x)\vec{w}_{\varepsilon t}(x,t)dx \right| = \left| \int_0^l \psi(x)\vec{Z}_{\varepsilon t}(x,t)dx \right|$$

$$\le C_4 \|\psi\|_{H_0^1(0,l)} \le C_4(1+l)\|\psi'\|_{L^2(0,l)},$$

this implies

$$\sup_{0\le t\le T} \|\vec{w}_{\varepsilon t}(\cdot,t)\|_{L^2(0,l)} \le (1+l)K_4,$$

and then for any given $l > 0$, $\{\vec{w}_\varepsilon(x,t)\}$ is uniformly bounded in ε in the space $L^\infty(0,T;H^2(0,l)) \cap W_\infty^{(1)}(0,T;L^2(0,l))$ so that

$$|\vec{Z}_\varepsilon(x,t_2) - \vec{Z}_\varepsilon(x,t_1)| = |\vec{w}_{\varepsilon x}(x,t_2) - \vec{w}_{\varepsilon x}(x,t_1)|$$

$$\le C_5 \|\vec{w}_\varepsilon(\cdot,t_2) - \vec{w}_\varepsilon(\cdot,t_1)\|_{L^2(0,l)}^{1/2} \|\vec{w}_\varepsilon(\cdot,t_2) - \vec{w}_\varepsilon(\cdot,t_1)\|_{H^2(0,l)}^{3/4}$$

$$\le C_6 |t_2 - t_1|^{1/2} \sup_{0\le t\le T} \|\vec{w}_{\varepsilon t}(\cdot,t)\|_{L^2(0,l)}^{1/4} \sup_{0\le t\le T} \|\vec{w}_\varepsilon(\cdot,t)\|_{H^2(0,l)}^{3/4}.$$

This completes the proof of Lemma 3.1.4.

3.1.4   Weak Solution to the Ferromagnetic Spin Chain Equation

1. *Weak solution to the initial-boundary value problem of ferromagnetic spin chain system*

In this subsection we discuss the limit of the approximate solutions when $\varepsilon \to 0$. From this procedure, we can get the weak solution to the initial-boundary value problem (3.1.1) and (3.1.2).

Definition 3.1.1 *A three-dimensional vector-valued function* $\vec{Z}(x,t) \in L^2(0,T; H^1(0,l)) \cap C(Q_T)$ *is called a weak solution of* (3.1.1) *and* (3.1.2) *if for any test function* $\varphi(x,t) \in \Phi := \{\varphi : \varphi \in C^1(Q_T), \varphi(x,T) \equiv 0\}$ *there holds*

$$\int\int_{Q_T}[\varphi_t\vec{Z} - \varphi_x(\vec{Z} \times \vec{Z}_x) + \varphi\vec{f}(x,t,\vec{Z})]dxdt + \int_0^l \varphi(x,0)\vec{Z}_0(x)dx = 0, \qquad (3.1.26)$$

in which $\varphi(0,t) = 0$ *(or* $\varphi(l,t) = 0$*) if the condition on* $x = 0$ *(or* $x = l$*) is* $\vec{Z}(0,t) = 0$ *(or* $\vec{Z}(l,t) = 0$*).*

For the approximate solution $\vec{Z}_\varepsilon(x,t)$ and any test function $\varphi \in \Phi$, it is clear that the following integral identity holds

$$\int\int_{Q_T}[\varphi_t\vec{Z}_\varepsilon - \varepsilon\varphi_x\vec{Z}_{\varepsilon x} - \varphi_x(\vec{Z}_\varepsilon \times \vec{Z}_{\varepsilon x})$$

$$+ \varphi\vec{f}(x,t,\vec{Z}_\varepsilon)]dxdt + \int_0^l \varphi(x,0)\vec{Z}_0(x)dx = 0. \qquad (3.1.27)$$

It follows from Lemmas 3.1.2–3.1.4 that the set of approximate solutions $\{\vec{Z}_\varepsilon(x,t)\}$ to problem (3.1.1) and (3.1.2) is uniformly bounded in ε in the space $L^\infty(0,T; H^1(0,l)) \cap C^{(\frac{1}{2},\frac{1}{4})}(Q_T)$. It follows from the *a priori* estimates that there is a subsequence $\{\vec{Z}_{\varepsilon_i}\}$ of $\{\vec{Z}_\varepsilon\}$ such that it uniformly converges on Q_T to a vector-valued function $\vec{Z}(x,t)$. Hence $\{\vec{f}(x,t,\vec{Z}_{\varepsilon_i})\}$ uniformly converges on Q_T to $\vec{f}(x,t,\vec{Z}(x,t))$. Moreover, $\{\vec{Z}_{\varepsilon_i x}(x,t)\}$ converges weakly to $\vec{Z}_x(x,t)$. This means that $\vec{Z}(x,t) \in L^\infty(0,T; H^1(0,l)) \cap C^{(\frac{1}{2},\frac{1}{4})}(Q_T)$.

In order to verify the convergence of the third term of (3.1.27), we estimate as follows:

$$\int\int_{Q_T} \varphi_x(\vec{Z}_{\varepsilon_i} \times \vec{Z}_{\varepsilon_i x})dxdt - \int\int_{Q_T} \varphi_x(\vec{Z} \times \vec{Z}_x)dxdt$$

$$= \int\int_{Q_T} \varphi_x[(\vec{Z}_{\varepsilon_i} - \vec{Z}) \times \vec{Z}_{\varepsilon_i x}]dxdt + \int\int_{Q_T} \varphi_x[\vec{Z} \times (\vec{Z}_{\varepsilon_i x} - \vec{Z}_x)]dxdt.$$

Since $\{\vec{Z}_{\varepsilon_i x}(x,t)\}$ converges weakly to $\vec{Z}_x(x,t)$, the second integral on the right tends to zero, and for the first integral, when $\varepsilon_i \to 0$, we have

$$\left| \int\int_{Q_T} \varphi_x[(\vec{Z}_{\varepsilon_i} - \vec{Z}) \times \vec{Z}_{\varepsilon_i x}]dxdt \right|$$

$$\leq \|\varphi_x\|_{L^2(Q_T)}\|\vec{Z}_{\varepsilon_i x}\|_{L^2(Q_T)}\|\vec{Z}_{\varepsilon_i} - \vec{Z}\|_{L^\infty(Q_T)} \to 0.$$

Hence, as $\varepsilon_i \to 0$, the integral identity (3.1.27) tends to (3.1.26). This means that the limit $\vec{Z}(x,t)$ is a weak solution of (3.1.1) and (3.1.2).

Theorem 3.1.2 *Suppose that the homogeneous ferromagnetic spin chain system (3.1.1) (i.e. $\vec{f}(x,t,0) \equiv 0$) and the initial data $\vec{Z}_0(x)$ meet the conditions (1)–(3). Then problem (3.1.1)–(3.1.2) admits at least one global weak solution:*

$$\vec{Z}(x,t) \in L^{\infty}(0,T; H^1(0,l)) \bigcap C^{(\frac{1}{2},\frac{1}{4})}(Q_T).$$

From estimates (3.1.22), we have

Theorem 3.1.3 *If $b < 0$ and $T = \infty$, then the L^2-norm (in x) of the solution $\vec{Z}(x,t)$ obtained in Theorem 3.1.2 tends to 0 when $t \to \infty$, i.e.*

$$\lim_{t \to \infty} \|\vec{Z}(\cdot,t)\|_{L^2(0,l)} = 0.$$

Now we consider the first problem of spin equation (3.1.9) over the domain $Q_T^* = \{x \in R^+, 0 \le t \le T\}$ as follows

$$\vec{Z}(0,t) = 0, \quad \vec{Z}(x,0) = \vec{Z}_0(x), \tag{3.1.28}$$

where $\vec{Z}_0(x)$ is the initial data defined on the infinite interval $R^+ = [0,\infty)$. Denote the assumptions (1), (2), (3) for $l = \infty$ by (1*), (2*), (3*).

2. *Weak solutions to spin equations*

Theorem 3.1.4 *Assume that $\vec{f}(x,t,\vec{Z})$ and $\vec{Z}_0(x)$ meet conditions (1*), (2*), (3*). Then the homogeneous spin equation (3.1.9) with the first boundary condition (3.1.28) admits unique global weak solution*

$$\vec{Z}(x,t) \in L^{\infty}(0,T; H^1(R^+)) \bigcap W_2^{(2,1)}(Q_T^*).$$

Proof. Select a subsequence l_s such that when $s \to \infty$, $l_s \to \infty$ and for every s, construct a three-dimensional vector-valued function $\vec{Z}_0^{(s)}(x)$ defined on $[0, l_s]$ satisfying the boundary condition at $x = 0$ and $x = l_s$ such that for $x \in [0, l_{s-1}]$, $\vec{Z}_0^{(s)}(x) \equiv \vec{Z}_0(x)$. At the same time, the $H^1(0, l_s)$-norm of $\vec{Z}_0^{(s)}(x)$ is bounded uniformly in s. Therefore we can first consider the first boundary problem of (3.1.9) on the rectangular domain $Q_T^{(s)} = \{0 \le x \le l_s, 0 \le t \le T\}$ whose unique global weak solution on $Q_T^{(s)}$ will be denoted by $\vec{Z}^{(s)}(x,t)$. It is clear that $\vec{Z}^{(s)}(x,t) \in L^{\infty}(0,T; H^1(0,l_s)) \bigcap W_2^{(2,1)}(Q_T^{(s)})$. Therefore we have

$$\sup_{0 \le t \le T} \|\vec{Z}^{(s)}(\cdot,t)\|_{H_0^1(0,l_s)} + \|\vec{Z}_t^{(s)}\|_{L^2(Q_T^{(s)})} + \|\vec{Z}_{xx}^{(s)}\|_{L^2(Q_T^{(s)})} \le K_7, \tag{3.1.29}$$

where K_7 is independent of l_s but depends on ε.

Therefore we can choose a subsequence $\{\vec{Z}^{(s_i)}(x,t)\}$ from $\{\vec{Z}^{(s)}(x,t)\}$ such that $\vec{Z}^{(s_i)}(x,t)$ and $\vec{Z}_x^{(s_i)}(x,t)$ uniformly converge to $\vec{Z}(x,t)$ and $\vec{Z}_x(x,t)$ on any rectangular domain $\tilde{Q}_T = \{0 \le x \le l, 0 \le t \le T\}$ $(l > 0)$, and, at the same time, $\vec{Z}_{xx}^{(s_i)}(x,t)$ and $\vec{Z}_t^{(s_i)}(x,t)$ weakly converge to $\vec{Z}_{xx}(x,t)$ and $\vec{Z}_t(x,t)$. It is easy to see that this vector $\vec{Z}(x,t)$ is the global solution of (3.1.9) and (3.1.28). This finishes the proof of the theorem.

3. *Estimates of approximate solutions and the solvability*

Theorem 3.1.5 *If $b < 0$ and $T = \infty$, then for the solution obtained in Theorem 3.1.4, there holds*

$$\lim_{t \to \infty} \|\vec{Z}(\cdot, t)\|_{L^2(0, +\infty)} = 0. \tag{3.1.30}$$

Now we intend to give the uniform estimates for the approximate solutions of (3.1.9) and (3.1.28) on the unbounded domain $Q_T^* = \{x \in R^+, 0 \le t \le T\}$.

Lemma 3.1.5 *Suppose that (1^*), (2^*), (3^*) hold. Then for the solution of (3.1.9) and (3.1.28), there holds*

$$\sup_{0 \le t \le T} \|\vec{Z}_\varepsilon(\cdot, t)\|_{H^1(R^+)} \le K_8, \tag{3.1.31}$$

and hence

$$\|\vec{Z}_\varepsilon(x, t)\|_{L^\infty(R^+)} \le K_9, \tag{3.1.32}$$

where K_8, K_9 are independent of ε.

Corollary 3.1.1 *Under the conditions of Lemma 3.1.5 and that $\vec{f}(x, t, 0) \in L^n(Q_T^*)$, $\vec{Z}_0(x) \in L^n(R^+)$, there holds*

$$\sup_{0 \le t \le T} \|\vec{Z}_\varepsilon(\cdot, t)\|_{L^n(R^+)} \le K_{10}(n), \tag{3.1.33}$$

where K_{10} is independent of ε, but depends on n (≥ 2).

Lemma 3.1.6 *Suppose that (1^*), (2^*), (3^*) and one of the following conditions hold.*
(4_1^*). For any $(x, t, \vec{Z}) \in Q_T^* \times R^3$:

$$|\vec{f}(x, t, \vec{Z})| \le \bar{c}(x, t)F(\vec{Z}) + \bar{d}(x, t), \tag{3.1.34}$$

where $F(\vec{Z})$ is a continuous function in $\vec{Z} \in R^3$, $\bar{c}(x, t), \bar{d}(x, t) \in L^\infty(0, T; L^2(R^+))$;
(4_2^*). For any $(x, t, \vec{Z}) \in Q_T^* \times R^3$:

$$|\vec{f}(x, t, \vec{Z})| \le c(x, t)|\vec{Z}|^l + \bar{d}(x, t), \tag{3.1.35}$$

where $l \ge 0$, $c(x, t) \in L^\infty(Q_T^*)$, $\bar{d}(x, t) \in L^\infty(0, T; L^s(R^+))$, $1 < s \le 2$.
Then for the solution $\vec{Z}_\varepsilon(x, t)$ of (3.1.9) and (3.1.28), there holds

$$\sup_{0 \le t \le T} \|\vec{Z}_{\varepsilon t}(\cdot, t)\|_{H^{-1}(R^+)} \le K_{11}, \tag{3.1.36}$$

where K_{11} is independent of ε.

Proof. Testing the equation (3.1.9) by $\psi \in H_0^1(R^+)$, we have

$$\int_0^\infty \psi(x)\vec{Z}_{\varepsilon t}(x,t)dx = -\varepsilon \int_0^\infty \psi_x(x)\vec{Z}_{\varepsilon x}(x,t)dx$$
$$- \int_0^\infty \psi_x(x)[\vec{Z}_\varepsilon(x,t) \times \vec{Z}_{\varepsilon x}(x,t)]dx$$
$$+ \int_0^\infty \psi(x)\vec{f}(x,t,\vec{Z}_\varepsilon(x,t))dx.$$

For the first two terms on the right-hand side of the above equality, we have

$$\left| \int_0^\infty \psi_x(x)\vec{Z}_{\varepsilon x}(x,t)dx \right| \leq K_7 \|\psi_x\|_{L^2(R^+)}$$

and

$$\left| \int_0^\infty \psi_x(x)[\vec{Z}_\varepsilon(x,t) \times \vec{Z}_{\varepsilon x}(x,t)]dx \right| \leq K_7 K_8 \|\psi_x\|_{L^2(R^+)}.$$

Under condition (4_1^*), the third term on the right-hand side can be estimated as follows

$$\left| \int_0^\infty \psi(x)\vec{f}(x,t,\vec{Z}_\varepsilon(x,t))dx \right| \leq \|\psi\|_{L^2(R^+)} \|\vec{f}(x,t,\vec{Z}_\varepsilon(x,t))\|_{L^2(R^+)}.$$

It follows from (3.1.34) that

$$|\vec{f}(x,t,\vec{Z}_\varepsilon)| \leq M\bar{c}(x,t) + \bar{d}(x,t), \tag{3.1.37}$$

where $M = \sup_{|\vec{Z}| \leq K_8} |F(\vec{Z})|$. This means that $\vec{f}(x,t,\vec{Z}_\varepsilon(x,t))$ is uniformly (in ε) bounded in the space $L^\infty(0,T; L^2(R^+))$.

Under the condition (4_2^*), it follows from Corollary 3.1.1 that $\{\vec{Z}_\varepsilon(x,t)\}$ is uniformly bounded (in ε) in $L^\infty(0,T; L^{sl}(R^+))$. Then

$$\left| \int_0^\infty \psi(x)\vec{f}(x,t,\vec{Z}_\varepsilon(x,t))dx \right| \leq C_5 \|\psi\|_{L^r(R^+)} \|\vec{f}(x,t,\vec{Z}_\varepsilon(x,t))\|_{L^s(R^+)},$$

where $\frac{1}{r} + \frac{1}{s}$, $1 < s \leq 2$, $2 \leq r < \infty$, C_5 is independent of ε and $t \in [0,T]$. Using (3.1.35), we have that

$$\|\vec{f}(x,t,\vec{Z}_\varepsilon(x,t))\|_{L^s(R^+)} \leq C_6 \|c\|_{L^\infty(Q_T^*)} [K_{10}(ls)]^l$$
$$+ \|\bar{d}\|_{L^\infty(0,T;L^s(R^+))},$$

where C_6 is independent of ε and $t \in [0,T]$.

Therefore in both cases, we have

$$\left| \int_0^{+\infty} \varphi(x)\vec{Z}_{\varepsilon t}(x,t)dx \right| \leq C_7 \|\psi\|_{H^1(R^+)}$$

where C_7 is independent of ε and $t \in [0,T]$.

Lemma 3.1.6 follows.

Now consider the solvability of problem (3.1.1) and (3.1.28) over the infinite rectangular domain $Q_T^* = \{x \in R^+, 0 \le t \le T\}$.

Definition 3.1.2 *A vector $\vec{Z}(x,t) \in L^\infty(0,T; H^1(R^+)) \cap C(Q_T^*)$ is called a weak solution of (3.1.1) and (3.1.28) if for any test function $\varphi(x,t) \in \Phi^* = \{\varphi : \varphi \in C(Q_T^*), \varphi(x,T) = 0, \mathrm{supp}\,\varphi < \infty\}$, $\vec{Z}(x,t)$ meets*

$$\int\!\!\int_{Q_T^*} [\varphi_t \vec{Z} - \varphi_x(\vec{Z} \times \vec{Z}_x) + \varphi \vec{f}(x,t,\vec{Z}(x,t))]dxdt$$

$$+ \int_0^\infty \varphi(x,t)\vec{Z}_0(x)dx = 0, \tag{3.1.38}$$

where $\varphi(0,t) = 0$.

4. *Global weak solutions*

Now we prove that when $\varepsilon \to 0$, the approximate solutions $\vec{Z}_\varepsilon(x,t)$ of spin equation (3.1.9) and (3.1.28) tend to the global weak solution of (3.1.1) and (3.1.28). First we note that for the approximate solutions $\vec{Z}_\varepsilon(x,t)$ of spin equation (3.1.9) and (3.1.8), there hold

$$\int\!\!\int_{Q_T^*} [\varphi_t \vec{Z}_\varepsilon - \varepsilon\varphi_x \vec{Z}_{\varepsilon x} - \varphi_x(\vec{Z}_\varepsilon \times \vec{Z}_{\varepsilon x}) + \varphi \vec{f}(x,t,\vec{Z}_\varepsilon(x,t))]dxdt$$

$$+ \int_0^\infty \varphi(x,t)\vec{Z}_0(x)dx = 0 \tag{3.1.39}$$

for all $\varphi \in \Phi^*$ with $\varphi(0,t) = 0$.

Lemmas 3.1.1 and 3.1.6 imply that the set of approximate solutions $\vec{Z}_\varepsilon(x,t)$ is a uniformly (in ε) bounded set in the space $L^\infty(0,T; H^1(R^+)) \cap W_\infty^{(1)}(0,T; H^{-1}(R^+))$. Hence, in any rectangular domain $\overline{Q}_T$ with width L: $\overline{Q}_T = \{0 \le x \le L, 0 \le t \le T\}$, the set $\{\vec{Z}_\varepsilon(x,t)\}$ is also unformly bounded in $L^\infty(0,T; H^1(0,L)) \cap W_\infty^{(1)}(0,T; H^{-1}(0,L))$. It follows from the similar proof as that of Lemma 3.1.5 that for any rectangular domain $\overline{Q}_T$, there holds

$$\|\vec{Z}_\varepsilon\|_{C^{(\frac{1}{2},\frac{1}{4})}(\overline{Q}_T)} \le (1+L)K_{12}, \tag{3.1.40}$$

where K_{12} is independent of ε and L.

Choose a suitable subsequence $\{\vec{Z}_{\varepsilon_i}(x,t)\}$ from $\{\vec{Z}_\varepsilon(x,t)\}$ such that for some vector-valued function $\vec{Z}(x,t)$ there holds $\vec{Z}_{\varepsilon_i} \to \vec{Z}$ uniformly in any $\overline{Q}_T$ for any width L and $\vec{f}(x,t,\vec{Z}_{\varepsilon_i}) \to \vec{f}(x,t,\vec{Z})$ uniformly in any $\overline{Q}_T$ too. Moreover, we may have that $\vec{Z}_{\varepsilon_i x}$ weakly converges to $\vec{Z}_x$ and verify that $\vec{Z}(x,t) \in L^\infty(0,T; H^1(R^+)) \cap C_{\mathrm{loc}}^{(1/2,1/4)}(Q_T^*)$. By the similar method we can see that the integrale relation (3.1.38) is the limit of the relation (3.1.39). These imply that $\vec{Z}(x,t)$ is a global weak solution of (3.1.1) and (3.1.28).

Therefore we have the following:

Theorem 3.1.6 *Suppose that $\vec{f}(x,t,\vec{Z})$ and $\vec{Z}_0(x)$ meet conditions (1^*), (2^*), (3^*) and one of (4_1^*) and (4_2^*) and assume that (3.1.1) is homogeneous. Then the first* *initial value problem (3.1.1) and (3.1.28) admits at least one global weak solution*

$$L^\infty(0, T; H^1(R^+)) \bigcap C_{\text{loc}}^{(\frac{1}{2}, \frac{1}{4})}(Q_T^*).$$

Theorem 3.1.7 *Suppose that conditions of Theorem 3.1.6 hold and $b < 0$. Then for the weak solution $\vec{Z}(x, t)$ to the first initial boundary value problem (3.1.1) and (3.1.28), there holds*

$$\lim_{t \to \infty} \|\vec{Z}(\cdot, t)\|_{L^2(R^+)} = 0.$$

Corollary 3.1.2 *For the system of ferromagnetic spin chain*

$$\vec{Z}_t = \vec{Z} \times \vec{Z}_{xx} + \vec{f}(x, t, \vec{Z}), \tag{3.1.41}$$

we can prove by similar method that there exists at least one global weak solution with the same properties as above under one of the following conditions:

(1) the second boundary value problem:

$$\vec{Z}_x(0, t) = \vec{Z}_x(l, t) = 0,$$
$$\vec{Z}|_{t=0} = \vec{Z}_0(x), \quad 0 \le x \le l; \tag{3.1.42}$$

(2) the mixed condition:

$$\vec{Z}(0, t) = \vec{Z}_x(l, t) = 0, \quad or \quad \vec{Z}_x(0, t) = \vec{Z}(l, t) = 0,$$
$$\vec{Z}|_{t=0} = \vec{Z}_0(x), \quad 0 \le x \le l; \tag{3.1.43}$$

(3) the periodic boundary condition

$$\vec{Z}(x, t) = \vec{Z}(x + 2D, t) = 0, \quad \vec{Z}|_{t=0} = \vec{Z}_0(x), \quad D > 0; \tag{3.1.44}$$

(4) the initial condition

$$\vec{Z}|_{t=0} = \vec{Z}_0(x), \quad x \in R. \tag{3.1.45}$$

Remark *Under the above assumption (2), the condition (3.1.12) is not essential. In fact, the growth condition of $\vec{f}(x, t, \vec{Z})$ on $\vec{Z}$ can be eliminated. For example we may make use of the cutoff method to obtain the proof.*

3.2 Nonlinear Initial-boundary Value Problem for the System of Ferromagnetic Spin Chain

We continue to consider the equation in the above section, but with nonlinear boundary conditions. This time, we will use the difference method to prove the existence of global weak solutions.

3.2.1   Nonlinear Initial-boundary Value Problem for the System of Ferromagnetic Spin Chain

1. *Equations, boundary conditions and initial conditions*

Consider the following system of ferromagnetic spin chain:

$$\vec{Z}_t = \vec{Z} \times \vec{Z}_{xx} + \vec{f}(x, t, \vec{Z}) \tag{3.2.1}$$

and the relative spin equation:

$$\vec{Z}_t = \varepsilon \vec{Z}_{xx} + \vec{Z} \times \vec{Z}_{xx} + \vec{f}(x, t, \vec{Z}) \tag{3.2.2}$$

with the nonlinear boundary condition

$$\begin{cases} \vec{Z}_x(0, t) = \operatorname{grad} \psi_0(t, \vec{Z}(0, t)), \\ -\vec{Z}_x(l, t) = \operatorname{grad} \psi_1(t, \vec{Z}(l, t)) \end{cases} \tag{3.2.3}$$

and initial condition

$$\vec{Z}(x, 0) = \varphi(x), \tag{3.2.4}$$

where $\psi_0(t, \vec{Z})$ and $\psi_1(t, \vec{Z})$ are vector-valued functions of variables $t \in [0, T]$ and $\vec{Z} \in R^3$, "grad" denotes the gradient operator with respect to $\vec{Z}$. On the other hand we also consider, in this section, the global solution for problem (3.2.1) and (3.2.2) over the semi-infinite domain $Q_T^* = \{x \in R^+, 0 \le t \le T\}$ with the nonlinear boundary condition:

$$\begin{cases} \vec{Z}_x(0, t) = \operatorname{grad} \psi_0(t, \vec{Z}(0, t)), \\ \vec{Z}_x(x, 0) = \varphi(x). \end{cases} \tag{3.2.5}$$

To this aim, we will use the difference (in spatial variable) method to prove the existence of global weak solution. We suppose:

(1) $\psi_0(t, \vec{Z})$ and $\psi_1(t, \vec{Z})$ are two scalar functions which are continuously first-order differentiable in $t \in [0, T]$ and continuously second-order differentiable in $t, \vec{Z}$ mixed. 3×3 Hessian matrix $H_0(t, \vec{Z}) = \nabla^2 \psi_0(t, \vec{Z})$ and $H_1(t, \vec{Z}) = \nabla^2 \psi_1(t, \vec{Z})$ are non-negatively definite and $\operatorname{grad} \psi_0(t, 0) = \operatorname{grad} \psi_1(t, 0) = 0$.

(2) $\vec{f}(x, t, \vec{Z})$ is a three-dimensional vector-valued function which is continuously differentiable with respect to $(x, t) \in Q_T$ and $\vec{Z} \in R^3$. Moreover, condition (3.1.11) holds, that is

$$\xi \cdot \vec{f}_{\vec{Z}}(x, t, \vec{Z}) \cdot \xi \le b|\xi|^2, \quad \forall \, \xi \in R^3, \tag{3.2.6}$$

(3) $\varphi(x) \in H^2(0, l)$ is the three-dimensional vector-valued initial data which, on the end points of the interval $[0, l]$, is compatible with the nonlinear condition (3.2.3).

Now divide the interval $[0, l]$ into J small intervals by points $x_j = jh$ ($j = 0, 1, \ldots, J$) where $Jh = l$, J is a positive integer, h is the length of the step. Vector $\vec{Z}_h(t) = \{\vec{Z}_j(t) : j = 0, 1, \ldots, J\}$ is function defined on the points x_j ($j = 0, 1, \ldots, J$).

2. *Construct the nonlinear ordinary differential systems*

We establish the following difference-differential equation:

$$\frac{d\vec{Z}_j}{dt} = \varepsilon\frac{\Delta_+\Delta_-\vec{Z}_j}{h^2} + \vec{Z}_j \times \frac{\Delta_+\Delta_-\vec{Z}_j}{h^2}$$
$$+ \vec{f}(x_j, l, z_j), \quad j - 1, 2, \ldots, J \quad 1, \tag{3.2.7}$$

and the nonlinear boundary condition relative to (3.2.3):

$$\frac{\Delta_+\vec{Z}_0(t)}{h} = \operatorname{grad}\psi_0(t, \vec{Z}_0(t)), \tag{3.2.8}$$

$$-\frac{\Delta_-\vec{Z}_J(t)}{h} = \operatorname{grad}\psi_1(t, \vec{Z}_J(t))$$

$$-\sum_{j=0}^{J-1}\int_0^1 h\left(\frac{\Delta_+\vec{Z}_j}{h} \cdot \vec{f}_z(x_j, t, \tau\vec{Z}_{j+1} + (1-\tau)\vec{Z}_j)\frac{\Delta_+\vec{Z}_j}{h}\right)d\tau$$

$$- \vec{f}(0, t, \vec{Z}_0)\operatorname{grad}\psi_0(t, \vec{Z}_0) - \vec{f}(l, t, \vec{Z}_J) \cdot \operatorname{grad}\psi_1(t, \vec{Z}_J). \tag{3.2.9}$$

Multiplying (3.2.7) by $\vec{Z}_j h$ and summing from $j = 1$ to J and integrating over $[0, t]$, we have

$$\|\delta\vec{Z}_h\|_2^2 + 2\varepsilon\int_0^t \|\delta^2\vec{Z}_h(t)\|_2^2 dt \le (2b + 1)\int_0^t \|\delta\vec{Z}_h\|_2^2 dt + 2G,$$

where

$$G = \frac{1}{2}\int_0^t \sum_{j=0}^{J-1}\left|\int_0^1 \vec{f}_x(x_{j+2}, t, \vec{Z}_{j+1})d\tau\right|^2 h dt + \frac{1}{2}\|\delta\psi_h\|_2^2$$

$$+ \int_0^t [\vec{f}(0, t, \vec{Z}_0) \cdot \operatorname{grad}\psi_0(t, \vec{Z}_0) + \vec{f}(l, t, \vec{Z}_J) \cdot \operatorname{grad}\psi_1(t, \vec{Z}_J)]dt$$

$$+ \int_0^t [\psi_0(t, \vec{Z}_0) + \psi_1(t, \vec{Z}_J)]dt$$

$$- [\psi_0(t, \vec{Z}_0(t)) + \psi_1(t, \vec{Z}_J(t))] + [\psi_0(0, \overline{\varphi}_0(t)) + \psi_1(l, \overline{\varphi}_J(t))]$$

which is clearly bounded.

Since

$$\|\delta\vec{Z}_h(t)\|_2^2 = \sum_{j=1}^{J-1}\left|\frac{\Delta_+\vec{Z}_j}{h}\right|^2 h,$$

$$\|\delta^2\vec{Z}_h(t)\|_2^2 = \sum_{j=1}^{J-1}\left|\frac{\Delta_+\Delta_-\vec{Z}_j}{h^2}\right|^2 h.$$

Therefore, we have

$$\|\delta\vec{Z}_h\|_2^2 + 2\varepsilon\int_0^t \|\delta^2\vec{Z}_h(t)\|_2^2 dt \le (2b + 1)\int_0^t \|\delta\vec{Z}_h\|_2^2 dt + C_2,$$

where C_2 is independent of $h, \varepsilon \ge 0$.

3.2.2 Discrete Solution of Nonlinear Ordinary Differential Systems

1. *Estimates for discrete solutions*

It follows from the above discussions that we have

Lemma 3.2.1 *Let assumption* (1)–(3) *hold. For the solution* $\vec{Z}_j$ *of the nonlinear ordinary differential system* (3.2.7) *with* (3.2.8) *and* (3.2.9), *there holds*

$$
\sup_{0 \le t \le T} \|\delta \vec{Z}_h\|_\infty + \sup_{0 \le t \le T} \|\delta \vec{Z}_h\|_2 + \sqrt{\varepsilon}\left(\int_0^T \|\delta^2 \vec{Z}_h(t)\|_2^2 dt \right)^{1/2}
$$
$$
+ \sqrt{\varepsilon}\left(\int_0^T \|\delta \vec{Z}_h'\|_2^2 dt \right)^{1/2} \le K_1, \tag{3.2.10}
$$

where K_1 *is a constant independent of* h, $\varepsilon > 0$.

Now we estimate $y(t) = \vec{Z}_h'(t) = \{\vec{Z}_j'(t) : j = 0, 1, \ldots, J\}$. Differentiating (3.2.7) with respect to t, we have

$$
y_j' = \varepsilon \frac{\Delta_+ \Delta_- y_j}{h^2} + y_j \times \frac{\Delta_+ \Delta_- y_j}{h^2} + \vec{Z}_j \times \frac{\Delta_+ \Delta_- y_j}{h^2}
$$
$$
+ \vec{f}_t(x_j, t, \vec{Z}_j) + \vec{f}_z(x_j, t, \vec{Z}_j)y_j, \qquad j = 1, 2, \ldots, J - 1. \tag{3.2.11}
$$

The relative boundary conditions are

$$
\begin{cases}
\dfrac{\Delta_+ y_0}{h} = \operatorname{grad} \psi_{0t}(t, \vec{Z}_0) + H_0(t, \vec{Z}_0)y_0, \\[2mm]
-\dfrac{\Delta_- y_J}{h} = \operatorname{grad} \psi_{1t}(t, \vec{Z}_J) + H_1(t, \vec{Z}_J)y_J,
\end{cases} \tag{3.2.12}
$$

and

$$
\begin{cases}
y_j(0) = \varepsilon \dfrac{\Delta_+ \Delta_- \overline{\varphi}_j}{h^2} + \overline{\varphi}_j \times \dfrac{\Delta_+ \Delta_- \overline{\varphi}_j}{h^2} + \vec{f}(x_j, 0, \overline{\varphi}_j), \\[2mm]
(E + hH_0(0, \overline{\varphi}_0))y_0(0) = y_1(0) - h \operatorname{grad} \psi_{0t}(0, \vec{\psi}_0), \\[2mm]
(E + hH_1(0, \overline{\varphi}_J))y_J(0) = y_{J-1}(0) - h \operatorname{grad} \psi_{1t}(0, \vec{\psi}_J),
\end{cases} \tag{3.2.13}
$$

where E is a 3×3 unit matrix.

Taking inner product of system (3.2.11) with $y_j h$ and summing from $j = 1$ to $J - 1$, we have

$$
\sum_{j=1}^{J-1} y_j y_j' h = \varepsilon \sum_{j=1}^{J-1} y_j \frac{\Delta_+ \Delta_- y_j}{h^2} h + \sum_{j=1}^{J-1} y_j \left(\vec{Z}_j \times \frac{\Delta_+ \Delta_- y_j}{h^2} \right) h
$$
$$
+ \sum_{j=1}^{J-1} y_j \vec{f}_t(x_j, t, \vec{Z}_j)h + \sum_{j=1}^{J-1} y_j \vec{f}_z(x_j, t, \vec{Z}_j)y_j h. \tag{3.2.14}
$$

It is obvious that

$$\sum_{j=1}^{J-1} y_j y_j' h = \frac{1}{2}\frac{d}{dt}\left(\sum_{j=1}^{J-1} |y_j|^2 h\right),$$

$$\left|\sum_{j=1}^{J-1} y_j \vec{f}_t(x_j, t, \vec{Z}_j) h\right| \le \frac{1}{2}\sum_{j=1}^{J-1} |y_j|^2 h + \frac{1}{2}\sum_{j=1}^{J-1} |\vec{f}_t(x_j, t, \vec{Z}_j)|^2 h,$$

$$\sum_{j=1}^{J-1} y_j \vec{f}_z(x_j, t, \vec{Z}_j) y_j h \le b\sum_{j=1}^{J-1} |y_j|^2 h.$$

For the first term on the right-hand side of (3.2.14), we have

$$\sum_{j=1}^{J-1} y_j \frac{\Delta_+ \Delta_- y_j}{h^2} h = -\|\delta y_h(t)\|_2^2 - y_0(\operatorname{grad}\psi_{0t}(t, \vec{Z}_0) + H_0(t, \vec{Z}_0)y_0)$$
$$- y_J(\operatorname{grad}\psi_{1t}(t, \vec{Z}_J) + H_1(t, \vec{Z}_J)y_J),$$

where

$$|y_0 \operatorname{grad}\psi_{0t}(t, \vec{Z}_0)| \le C_3\|y_h\|_\infty \le \frac{1}{8}\|\delta y_h\|_2^2 + C_4\|y_h(t)\|_2^2 + C_5,$$

$$|y_J \operatorname{grad}\psi_{1t}(t, \vec{Z}_J)| \le \frac{1}{8}\|\delta y_h\|_2^2 + C_4\|y_h(t)\|_2^2 + C_5.$$

It follows from (3.2.12) that

$$|y_0| \le C_6|y_1| + C_7,$$
$$|y_J| \le C_6|y_{J-1}| + C_7.$$

Hence

$$\|y_h(t)\|_2^2 \le C_8 \sum_{j=1}^{J-1} |y_j|^2 h + C_9. \tag{3.2.15}$$

For the second term on the right-hand side of (3.2.14), direct computations give

$$\sum_{j=1}^{J-1} y_j \left(\vec{Z}_j \times \frac{\Delta_+ \Delta_- y_j}{h^2}\right) h = -\sum_{j=1}^{J-1} y_j \left(\frac{\Delta_+ \vec{Z}_j}{h} \times \frac{\Delta_+ y_j}{h}\right) h$$
$$- y_0 \cdot \left(\vec{Z}_0 \times \frac{\Delta_+ y_0}{h}\right) + y_J \cdot \left(\vec{Z}_J \times \frac{\Delta_- y_J}{h}\right), \tag{3.2.16}$$

where

$$\left|y_0 \cdot \left(\vec{Z}_0 \times \frac{\Delta_+ y_0}{h}\right)\right| = |y_0 \cdot (\vec{Z}_0 \times (\operatorname{grad}\psi_{0t}(t, \vec{Z}_0) + H_0(t, \vec{Z}_0)y_0))|$$
$$\le C_{10}\|y_h(t)\|_\infty^2 + C_{11}$$
$$\le \frac{\varepsilon}{8}\|\delta y_h(t)\|_2^2 + C_{12}\|y_h(t)\|_2^2 + C_{13},$$

$$\left|y_J \cdot \left(\vec{Z}_J \times \frac{\Delta_- y_J}{h}\right)\right| \le \frac{\varepsilon}{8}\|\delta y_h(t)\|_2^2 + C_{12}\|y_h(t)\|_2^2 + C_{13}.$$

For the first term on the right-hand side of (3.2.16), we have

$$\left| \int_0^t \left(\sum_{j=1}^{J-1} y_j \frac{\Delta_+ \vec{Z}_j}{h} \times \frac{\Delta_+ y_j}{h} \right) h\,dt \right|$$

$$\leq \int_0^t \|y_h(t)\|_\infty \|\delta \vec{Z}_h(t)\|_2 \|\delta y_h(t)\|_2 dt$$

$$\leq C_{14} K_1 \int_0^t \left(\|y_h(t)\|_2^{1/2} \|\delta y_h(t)\|_2^{3/2} + \|y_h(t)\|_2 \|\delta y_h(t)\|_2 \right) dt$$

$$\leq \frac{\varepsilon}{4} \int_0^t \|\delta y_h(t)\|_2^2 dt + C_{15} \int_0^t \|y_h(t)\|_2^2 dt,$$

where C_{15} is a constant independent of h but depends on $\varepsilon > 0$.

Therefore we have from (3.2.14) that

$$\|y_h(t)\|_2^2 - \|y_h(0)\|_2^2 + \frac{\varepsilon}{2} \int_0^t \|\delta y_h(t)\|_2^2 dt \leq C_{16} \int_0^t \|y_h(t)\|_2^2 dt + C_{17},$$

where C_{16}, C_{17} are positive constants independent of h but depending on ε. This implies that $\sup_{0 \leq t \leq T} \|\vec{Z}_h'(t)\|_2$ and $\int_0^t \|\delta \vec{Z}_h'(t)\|_2 dt$ are uniformly bounded in h.

Lemma 3.2.2 *Let* (1)–(3) *hold. For the solution* $\vec{Z}_h(t)$ *of* (3.2.7)–(3.2.9), *there holds*

$$\sup_{0 \leq t \leq T} \|\vec{Z}_h'(t)\|_2 + \left(\int_0^t \|\delta \vec{Z}_h'(t)\|_2^2 dt \right)^{1/2} \leq K_2(\varepsilon). \tag{3.2.17}$$

Since the coefficient matrix of $\frac{\Delta_+ \Delta_- \vec{Z}_i}{h^2}$ in (3.2.7) is positively definite, $\sup_{0 \leq t \leq T} \|\delta^2 \vec{Z}_h(t)\|_2$ is uniformly bounded. Similarly, for the differential-difference equation, we have

$$\frac{\Delta_+ \vec{Z}_j'(t)}{h} = \varepsilon \frac{\Delta_+^2 \Delta_- \vec{Z}_j(t)}{h^3} + \vec{Z}_{j+1}(t) \times \frac{\Delta_+^2 \Delta_- \vec{Z}_j(t)}{h^3}$$

$$+ \frac{\Delta_+ \vec{Z}_j(t)}{h} \times \frac{\Delta_+ \Delta_- \vec{Z}_j(t)}{h^2} + \int_0^1 \vec{f}_x(x_{j+1}, t, \vec{Z}_{j+1}) dz$$

$$+ \int_0^1 \vec{f}_z(x_j, t, \tau \vec{Z}_{j+1} + (1-\tau)\vec{Z}_j) d\tau \frac{\Delta_+ \vec{Z}_j(t)}{h}.$$

This implies that $\int_0^T \|\delta^2 \vec{Z}_h(t)\|_2^2 dt$ is uniformly bounded in h. Hence, we have

Lemma 3.2.3 *Let* (1)–(3) *hold. For the solution* $\vec{Z}_h(t)$ *of* (3.2.7)–(3.2.9), *there holds*

$$\sup_{0 \leq t \leq T} \|\delta^2 \vec{Z}_h(t)\|_2 + \left(\int_0^T \|\delta^3 \vec{Z}_h(t)\|_2^2 dt \right)^{1/2} \leq K_3(\varepsilon), \tag{3.2.18}$$

where $K_3(\varepsilon)$ is a constant independent of h.

2. *Further estimates for the discrete solutions*

Using the discrete interpolation formula, we get from the above estimates the following lemma.

Lemma 3.2.4 *Let* (1)–(3) *hold. For the solution* $\vec{Z}_h(t)$ *of* (3.2.7)–(3.2.9), *there holds for* $t \in [0, T - \Delta t]$

$$\sup_{0 \le t \le T} |\Delta_+ \vec{Z}_j(t)| \le K_4 h, \quad j = 0, 1, \ldots, J - 1, \tag{3.2.19}$$

$$\sup_{0 \le t \le T} |\Delta_+^2 \vec{Z}_j(t)| \le K_5 h^{1/2}, \quad j = 0, 1, \ldots, J - 1, \tag{3.2.20}$$

$$\max_{j=0,1,\ldots,J} |\vec{Z}_j(t + \Delta t) - \vec{Z}_j(t)| \le K_6 \Delta t^{1/2}, \tag{3.2.21}$$

$$\max_{j=0,1,\ldots,J-1} |\Delta_+ \vec{Z}_j(t + \Delta t) - \Delta_+ \vec{Z}_j(t)| \le K_7 h \Delta t^{1/4}, \tag{3.2.22}$$

where K_j, $(j = 4, 5, 6, 7)$ *are constants independent of* h *but depending on* ε.

3.2.3   Global Weak Solution for the Spin System

1. *Construct three-dimensional vector-valued functions*

In order to establish the existence of global weak solution of (3.2.2) with nonlinear boundary condition (3.2.3) and initial condition (3.2.4), we approximate it by the discrete solution of (3.2.7)–(3.2.9). For this reason, we construct a vector-valued function $\vec{Z}_h^*(x, t)$ defined on Q_T as follows: on $Q_T^j = \{x_j \le x \le x_{j+1}, 0 \le t \le T\}$, $j = 0, 1, \ldots, J - 1$:

$$\vec{Z}_h^*(x, t) = \frac{x - x_j}{h} \vec{Z}_{j+1}(t) + \frac{x_{j+1} - x}{h} \vec{Z}_j(t).$$

It follows from (3.2.10), (3.2.19) and (3.2.21) that $\{\vec{Z}_h^*(x, t)\}$ is uniformly bounded and equi-continuous over $Q_T^j = \{x_j \le x \le x_{j+1}, 0 \le t \le T\}$ $(j = 0, 1, \ldots, J - 1)$, $h > 0$. We may choose a sequence $\{h_j\}$ such that $\{\vec{Z}_{h_j}^*(x, t)\}$ uniformly converge over Q_T to a vector-valued function $\vec{Z}(x, t)$ as $h_j \to 0$. Similarly, construct on Q_T^j

$$\overline{\vec{Z}}_h^*(x, t) = \frac{x - x_j}{h} \frac{\Delta_+ \vec{Z}_{j+1}(t)}{h} + \frac{x_{j+1} - x}{h} \frac{\Delta_+ \vec{Z}_j(t)}{h}, \quad j = 0, 1, \ldots, J - 2,$$

and

$$\overline{\vec{Z}}_h^*(x, t) = \frac{\Delta_- \vec{Z}_{J-1}}{h}, \quad \text{on } Q_T^{J-1}.$$

It follows from (3.2.10), (3.2.20) and (3.2.22) that we may choose a subsequence of $\{h_j\}$, still denoted by $\{h_j\}$, such that $\{\overline{\vec{Z}}_{h_j}^*(x, t)\}$ uniformly converges over Q_T to a vector-valued function $\vec{Z}_x(x, t)$ as $h_j \to 0$.

For the discrete vector-valued function $\vec{Z}_h(t) = \{\vec{Z}_j(t) : j = 0, 1, \ldots, J\}$ let

$$\vec{Z}_h(x,t) = \vec{Z}_j(t), \qquad \vec{Z}_h^{(1)}(x,t) = \frac{\Delta_+ \vec{Z}_j(t)}{h}, \quad \text{on} \ \ Q_T^{(j)}, \ \ j = 0, 1, \ldots, J-1$$

$$\tilde{\vec{Z}}_h(x,t) = \vec{Z}_j'(t), \qquad \tilde{\vec{Z}}_h^{(1)}(x,t) = \frac{\Delta_+ \vec{Z}_j'(t)}{h}, \quad \text{on} \ \ Q_T^{(j)}, \ \ j = 0, 1, \ldots, J-1$$

$$\vec{Z}_h^{(2)}(x,t) = \frac{\Delta_+ \Delta_- \vec{Z}_j(t)}{h^2}, \quad \text{on} \ \ Q_T^{(j)}, \ \ j = 1, 2, \ldots, J-1,$$

$$\vec{Z}_h^{(2)}(x,t) = \frac{\Delta_+ \Delta_- \vec{Z}(t)}{h^2}, \quad \text{on} \ \ Q_T^{(0)},$$

$$\vec{Z}_h^{(3)}(x,t) = \frac{\Delta_+^2 \Delta_- \vec{Z}_j(t)}{h^3}, \quad \text{on} \ \ Q_T^{(j)}, \ \ j = 1, 2, \ldots, J-2,$$

$$\vec{Z}_h^{(3)}(x,t) = \frac{\Delta_+^2 \Delta_- \vec{Z}_1(t)}{h^3}, \quad \text{on} \ \ Q_T^{(0)},$$

$$\vec{Z}_h^{(3)}(x,t) = \frac{\Delta_+^2 \Delta_- \vec{Z}_{J-2}(t)}{h^3}, \quad \text{on} \ \ Q_T^{(J-2)}.$$

These functions are piecewise constant in direction of the variable x on Q_T. It is clear that

$$|\vec{Z}_h^*(x,t) - \vec{Z}_h(x,t)| \leq K_4 h,$$
$$|\tilde{\vec{Z}}_h^{\,*}(x,t) - \vec{Z}_h^{(1)}(x,t)| \leq K_5 h^{1/2}.$$

On the other hand, it follows from Lemmas 3.2.2 and 3.2.3 that

$$\|\vec{Z}_h\|_{L^\infty(Q_T)} + \|\vec{Z}_h^{(1)}\|_{L^\infty(Q_T)} + \sup_{0 \leq t \leq T} \|\vec{Z}_h^{(2)}(\cdot,t)\|_{L^2(0,l)}$$
$$+ \sup_{0 \leq t \leq T} \|\tilde{\vec{Z}}_h(\cdot,t)\|_{L^2(0,l)} + \|\tilde{\vec{Z}}_h^{(1)}\|_{L^2(Q_T)} + \|\vec{Z}_h^{(3)}\|_{L^2(Q_T)}$$
$$\leq K_8,$$

where K_8 is a constant independent of h but depending on ε.

We may choose $\{h_j\}$ such that when $h_j \to 0$, $\{\vec{Z}_{h_j}\}$ and $\{\vec{Z}_{h_j}^{(1)}\}$ uniformly converge to $\vec{Z}(x,t)$ and $\vec{Z}_x(x,t)$ in Q_T respectively; and $\{\vec{Z}_{h_j}^{(2)}\}$ and $\{\vec{Z}_{h_j}^{(1)}\}$ weakly converge in $L^p(0,T; L^2(0,l))$ $(2 \leq p < \infty)$ to $\vec{Z}^{(2)}(x,t)$ and $\tilde{\vec{Z}}(x,t)$ respectively; and, $\{\vec{Z}_{h_j}^{(3)}\}$ and $\{\tilde{\vec{Z}}_{h_j}^{(1)}\}$ weakly converge in $L^2(Q_T)$ to $\vec{Z}^{(3)}(x,t)$ and $\tilde{\vec{Z}}^{(1)}(x,t)$ respectively. The norms of $\vec{Z}^{(2)}(x,t)$ and $\tilde{\vec{Z}}(x,t)$ in $L^p(0,T; L^2(0,l))$ are uniformly bounded in p. This implies that $\vec{Z}^{(2)}$ and $\tilde{\vec{Z}}(x,t) \in L^\infty(0,T; L^2(0,l))$.

2. *The characteristics of vector-valued functions*

By the standard method we can prove that $\vec{Z}^{(2)} = \vec{Z}_{xx}$, $\vec{Z}^{(3)} = \vec{Z}_{xxx}$, $\tilde{\vec{Z}} = \vec{Z}_t$ and $\tilde{\vec{Z}}^{(1)} = \vec{Z}_{xt}$. That is, the limit functions $\vec{Z}(x,t)$ is weakly differentiable and $\vec{Z}_x, \vec{Z}_{xx}, \vec{Z}_t \in L^\infty(0,T; L^2(0,l))$, $\vec{Z}_{xt}, \vec{Z}_{xxx} \in L^2(Q_T)$.

Let $\Phi(x,t)$ be any test function, and $\Phi_h(x,t) = \Phi_j(t) = \Phi(x_j,t)$ in $Q_T^{(0)}$, $j = 0,1,\ldots,J-1$. Similarly we can define $F_h(x,t) = \vec{f}(x_j,t,\vec{Z}_j(t))$ in $Q_T^{(j)}$, $j = 0,1,\ldots,J-1$. It follows from (3.2.7) that

$$\int_0^T \sum_{j=0}^{J-1} \Phi_j(t)\left[\vec{Z}_j'(t) - \varepsilon\frac{\Delta_+\Delta_-\vec{Z}_j(t)}{h^2}\right.$$
$$\left. - \vec{Z}_j(t) \times \frac{\Delta_+\Delta_-\vec{Z}_j(t)}{h^2} - \vec{f}(x_j,t,\vec{Z}_j(t))\right]hdt = 0.$$

This is equivalent to the following integral identity:

$$\int_0^T \int_0^l \Phi_h(x,t)[\tilde{\vec{Z}}_h(x,t) - \varepsilon\vec{Z}_h^{(2)}(x,t)$$
$$- \vec{Z}_h(x,t) \times \vec{Z}_h^{(2)}(x,t) - F_h(x,t)]dxdt = 0.$$

Sending $h_i \to 0$, we have that $\Phi_h(x,t), \vec{Z}_h(x,t), F_h(x,t)$ uniformly converge on Q_T to $\Phi(x,t), \vec{Z}(x,t), \vec{f}(x,t,\vec{Z}(x,t))$; $\tilde{\vec{Z}}_h(x,t)$ and $\vec{Z}_h^{(2)}(x,t)$ weakly converge in $L^2(Q_T)$ to $\vec{Z}_t(x,t)$ and $\vec{Z}_{xx}$, respectively. Therefore by sending $h_i \to 0$, we get

$$\int_0^T \int_0^l \Phi(x,t)[\vec{Z}_t(x,t) - \varepsilon\vec{Z}_{xx}(x,t)$$
$$-\vec{Z}(x,t) \times \vec{Z}_{xx}(x,t) - \vec{f}(x,t,\vec{Z}(x,t))]dxdt = 0.$$

3. *Nonlinear weak solution to the spin equations*

It follows from above discussion that $\vec{Z}(x,t)$ solves the spin equation (3.2.2) in the sense of distribution.

Since $\{\vec{Z}_{h_i}(x,t)\}$ uniformly converges on Q_T to $\vec{Z}(x,t)$, and $\vec{Z}_{h_i}^{(1)}(x,t)$ uniformly converges on Q_T to $\vec{Z}_x(x,t)$ when $h_i \to 0$, the nonlinear boundary condition (3.2.3) and initial condition (3.2.4) are satisfied by $\vec{Z}(x,t)$ in the classical sense.

Theorem 3.2.1 *Under conditions (1)–(3), problem (3.2.2)–(3.2.4) admits at least one global weak solution $\vec{Z}(x,t)$ which has classical first order differential $\vec{Z}_x(x,t)$ and weak derivatives $\vec{Z}_{xx}, \vec{Z}_t \in L^\infty(0,T;L^2(0,l))$ and $\vec{Z}_{xt}(x,t), \vec{Z}_{xxx}(x,t) \in L^2(Q_T)$. Moreover, the equation is met in the sense of distribution and the boundary condition (3.2.3) and initial condition (3.2.4) in the classical sense.*

Theorem 3.2.2 *Under conditions (1)–(3), the solution to problem (3.2.2)–(3.2.4) is unique.*

Proof. If there are two solutions $\vec{Z}(x,t)$ and $\overline{\vec{Z}}(x,t)$ to problem (3.2.2)–(3.2.4), then $w(x,t) = \vec{Z}(x,t) - \overline{\vec{Z}}(x,t)$ solves the following equation

$$\int\int_{Q_T} \Phi[w_t - \varepsilon w_{xx} - w \times \vec{Z}_{xx} - \overline{\vec{Z}} \times w_{xx}$$
$$- [\vec{f}(x,t,\vec{Z}(x,t)) - \vec{f}(x,t,\overline{\vec{Z}}(x,t))]]dxdt = 0,$$

$$w_x(0,t) = H_0(t, \tilde{\vec{Z}}(0,t))w(0,t),$$
$$-w_x(l,t) = H_1(t, \tilde{\vec{Z}}(l,t))w(l,t),$$
$$w(x,0) = 0,$$

where $\Phi(x,t) \in L^2(Q_T)$ is any test function and

$$H_0(t, \tilde{\vec{Z}}(0,t)) = \int_0^1 H_0(t, \tau\tilde{\vec{Z}}(0,t) + (1-\tau)\overline{\vec{Z}}(0,t))d\tau,$$
$$H_1(t, \tilde{\vec{Z}}(l,t)) = \int_0^1 H_1(t, \tau\tilde{\vec{Z}}(l,t) + (1-\tau)\overline{\vec{Z}}(l,t))d\tau.$$

Take test function $\Phi = w(x,t)$ to give

$$\|w(\cdot,t)\|^2_{L^2(0,l)} + \varepsilon\|w_x\|^2_{L^2(Q_T)}$$
$$\leq \varepsilon \int_0^t [w(l,t)w_x(l,t) - w(0,t)w_x(0,t)]dt$$
$$+ \int_0^l [w(l,t)\overline{\vec{Z}}(l,t) \times w_x(l,t) - w(0,t)\overline{\vec{Z}}(0,t) \times w_x(0,t)]dt$$
$$- \int\int_{Q_T} w\overline{\vec{Z}}_x \times w_x dxdt + b\|w\|^2_{L^2(Q_T)}.$$

Since

$$w(l,t)w_x(l,t) - w(0,t)w_x(0,t)$$
$$= -w(l,t) \cdot H_1(t, \tilde{\vec{Z}}(l,t))w(l,t) - w(0,t) \cdot H_0(t, \tilde{\vec{Z}}(0,t))w(0,t) \leq 0;$$
$$|w(l,t)\overline{\vec{Z}}(l,t) \times w_x(l,t) - w(0,t)\overline{\vec{Z}}(0,t) \times w_x(0,t)|$$
$$= |w(l,t) \cdot \overline{\vec{Z}}(l,t) \times H_1(t, \tilde{\vec{Z}}(l,t))w(l,t)$$
$$+ w(0,t) \cdot \overline{\vec{Z}}(0,t) \times H_0(t, \tilde{\vec{Z}}(0,t))w(0,t)|$$
$$\leq C_{18}\|w(\cdot,t)\|^2_{L^\infty(0,l)}$$
$$\leq \frac{\varepsilon}{2}\|w_x(\cdot,t)\|^2_{L^2(0,l)} + C_{19}(\varepsilon)\|w(\cdot,t)\|^2_{L^2(0,l)}$$

and

$$\left|\int\int_{Q_T} w \cdot \vec{Z}_x \times w_x dxdt\right| \leq \frac{\varepsilon}{2}\|w_x\|^2_{L^2(Q_T)} + C_{20}(\varepsilon)\|w\|^2_{L^2(Q_T)}.$$

Hence

$$\|w(\cdot,t)\|^2_{L^2(0,l)} \leq C_{21}(\varepsilon) \int_0^t \|w(\cdot,t)\|^2_{L^2(0,l)}dt.$$

This implies $\|w(\cdot,t)\|_{L^2(0,l)} = 0$, $\forall\, t \in [0,T]$. The proof of the theorem is complete.

3.2.4 Global Weak Solution to the Equations of Ferromagnetic Spin Chain

1. Global weak solutions to spin equations

Now we prove the existence of global weak solution to equation of ferromagnetic spin chain (3.2.1) with nonlinear boundary condition (3.2.3) and initial condition (3.2.4). We have obtained the existence of global weak solution of (3.2.2)–(3.2.4) for $\varepsilon > 0$ and have proved some uniform *a priori* estimates for such solutions.

Lemma 3.2.5 *Let condition* (1), (2) *and* (3) *hold. Then the following estimate for the solution* $\vec{Z}_\varepsilon(x,t)$ $(\varepsilon > 0)$ *to problem* (3.2.2)–(3.2.4) *holds*

$$\|\vec{Z}_\varepsilon\|_{L^\infty(Q_T)} + \sup_{0 \leq t \leq T} \|\vec{Z}_{\varepsilon x}\|_{L^2(0,l)} \leq K_9, \tag{3.2.23}$$

where K_9 is independent of $\varepsilon > 0$.

Let $g(x,t) \in H^1(0,l)$ be any test function. Simple computations give

$$\int_0^l g(x)\vec{Z}_{\varepsilon t}(x,t) = -\varepsilon \int_0^l g_x(x)\vec{Z}_{\varepsilon x}(x,t)dx$$
$$- \int_0^l g_x(x)(\vec{Z}_\varepsilon(x,t) \times \vec{Z}_{\varepsilon x}))dx$$
$$+ \int_0^l g(x)\vec{f}(x,t,\vec{Z}_\varepsilon(x,t))dx,$$

where $g(0) = g(l) = 0$. It follows from (3.2.23) that there is a constant C_{22} independent of $t \in [0,T]$ and $\varepsilon > 0$ such that

$$\left| \int_0^l g(x)\vec{Z}_{\varepsilon x}(x,t)dx \right| \leq C_{22}\|g\|_{H^1(0,l)}.$$

This means that $\{\vec{Z}_{\varepsilon t}(x,t)\}$ is bounded in $L^\infty(0,T;H^{-1}(0,l))$ uniform in ε. Then we have the following lemma.

Lemma 3.2.6 *Let* (1)–(3) *hold. Then for the global weak solution of* (3.2.2)–(3.2.4) *there holds*

$$\sup_{0 \leq t \leq T} \|\vec{Z}_{\varepsilon t}(\cdot,t)\|_{H^{-1}(0,l)} \leq K_{10}, \tag{3.2.24}$$

where K_{10} is independent of ε.

Lemma 3.2.7 *Let* (1)–(3) *hold. Then for the global weak solution of* (3.2.2)–(3.2.4), *there holds*

$$\begin{cases} |\vec{Z}_\varepsilon(x_1,t) - \vec{Z}_\varepsilon(x_2,t)| \leq K_{11}|x_1 - x_2|^{1/2}, \\ |\vec{Z}_\varepsilon(x,t_1) - \vec{Z}_\varepsilon(x,t_2)| \leq K_{12}|t_1 - t_2|^{1/4}, \end{cases} \tag{3.2.25}$$

where K_{11}, K_{12} are independent of ε, $x_1, x_2 \in [0,l]$ and $t_1, t_2 \in [0,T]$.

Proof. Let $S_\varepsilon(x,t) = \int_0^x \vec{Z}_\varepsilon(x,t)dx$, then $S_{\varepsilon t}(x,t) = \int_0^x \vec{Z}_{\varepsilon t}(x,t)dx$, $S_{\varepsilon x}(x,t) = \vec{Z}_\varepsilon(x,t)$, $S_{\varepsilon xx}(x,t) = \vec{Z}_{\varepsilon x}(x,t)$. For any $g(x) \in H_0^1(0,l)$ there holds

$$\left| \int_0^l g'(x)S_{\varepsilon t}(x,t) \right| = \left| \int_0^l g(x)\vec{Z}_{\varepsilon t}(x,t)dx \right|$$
$$\leq K_{10}\|g\|_{H_0^1(0,l)} \leq C_{23}\|g\|_{L^2(0,l)},$$

where $g(0) = g(l) = 0$, C_{23} is independent of $t \in [0,T]$ and $\varepsilon > 0$. This implies

$$\sup_{0 \leq t \leq T} \|S_{\varepsilon t}(\cdot,t)\|_{L^2(0,l)} \leq C_{24}.$$

Then it is clear that

$$\sup_{0 \leq t \leq T} \|S_\varepsilon(\cdot,t)\|_{H^2(0,l)} \leq C_{25}$$

with C_{25} independent of $\varepsilon > 0$.

It follows from the interpolation inequality that

$$|S_{\varepsilon x}(x,t_1) - S_{\varepsilon x}(x,t_2)|$$
$$\leq C_{26}\|S_\varepsilon(x,t_1) - S_\varepsilon(x,t_2)\|_{L^2}^{1/4}\|S_\varepsilon(x,t_1) - S_\varepsilon(x,t_2)\|_{H^2}^{3/4},$$
$$\leq C_{26}\left\| \int_{t_1}^{t_2} S_{\varepsilon t}(x,t)dt \right\|_{L^2}^{1/4}\|S_\varepsilon(x,t_1) - S_\varepsilon(x,t_2)\|_{H^2}^{3/4}$$
$$\leq C_{27}|t_1 - t_2|^{1/4} \cdot \sup_{0 \leq t \leq T} \|S_{\varepsilon t}(x,t)\|_{L^2}^{1/4}\|S_\varepsilon(\cdot,t)\|_{H^2}^{3/4}, \quad \forall\, x \in [0,l].$$

Similarly, we have

$$|S_{\varepsilon x}(x_1,t) - S_{\varepsilon x}(x_2,t)| \leq C_{28}|x_1 - x_2|^{1/2}\|S_{\varepsilon xx}(\cdot,t)\|_{L^2(0,l)}, \quad \forall\, x \in [0,l].$$

The constants C_{26}, C_{27}, C_{28} are independent of ε. Therefore we get the first estimate of (3.2.25). The second one can be proved in the similar manner.

2. *Global weak solution*

Definition 3.2.1 *Vector* $\vec{Z}(x,t) \in L^2(0,T;H_0^1(0,l)) \bigcap C(Q_T)$ *is called a weak solution of problem* (3.2.1) *and* (3.2.3)–(3.2.4), *if for any test function* $g(x,t) \in H^1(Q_T)$, $g(x,T) = 0$, *there holds*

$$\int\int_{Q_T} [g_t(x,t)\vec{Z}(x,t) - g_x(x,t)(\vec{Z}(x,t) \times \vec{Z}_x(x,t))$$
$$+ g(x,t)\vec{f}(x,t,\vec{Z}(x,t))]dxdt + \int_0^l g(x,0)\varphi(x)dx$$
$$- \int_0^T [g(l,t)(\vec{Z}(l,t) \times \operatorname{grad}\psi_1(t,\vec{Z}(l,t)))$$
$$+ g(0,t)(\vec{Z}(0,t) \times \operatorname{grad}\psi_0(t,\vec{Z}(0,t)))]dt = 0. \qquad (3.2.26)$$

Theorem 3.2.3 *Let (1)–(3) hold. Problem (3.2.1) and (3.2.3)–(3.2.4) admits at least one weak solution $\vec{Z}(x,t) \in L^\infty(0,T; H_0^1(0,l)) \cap C^{(\frac{1}{2},\frac{1}{4})}(Q_T)$.*

Proof. Let $\vec{Z}_\varepsilon(x,t)$ be the unique weak solution of problem (3.2.2)–(3.2.4) as above. Since the estimates (3.2.23) and (3.2.25) is uniform in ε, $\{\vec{Z}_\varepsilon(x,t)\}$ is uniformly bounded and equi-continuous. We may choose a subsequence $\{\vec{Z}_{\varepsilon_i}(x,t)\}$ from $\{\vec{Z}_\varepsilon(x,t)\}$ such that $\{\vec{Z}_{\varepsilon_i}(x,t)\}$ uniformly converges on Q_T to some vector $\vec{Z}(x,t)$. Moreover, $\vec{Z}(x,t) \in C^{(\frac{1}{2},\frac{1}{4})}(Q_T)$ since $\{\vec{Z}_\varepsilon(x,t)\}$ is uniformly bounded in $C^{(\frac{1}{2},\frac{1}{4})}(Q_T)$. On the other hand, we may assume that $\{\vec{Z}_{\varepsilon_i x}(x,t)\}$ weakly converges in $L^p(0,T; L^2(0,l))$ to $\vec{Z}_x(x,t)$ for any $2 \le p \le \infty$. Since $\{\vec{Z}_{\varepsilon_i x}(x,t)\}$ is bounded in $L^p(0,T; L^2(0,l))$ uniformly in p, the norm of $\vec{Z}_x(x,t)$ in $L^p(0,T; L^2(0,l))$ is bounded uniformly in p. Therefore, $\vec{Z}_x(x,t) \in L^\infty(0,T; L^2(0,l))$.

Furthermore, for $\vec{Z}_\varepsilon(x,t)$ there holds

$$\int\int_{Q_T} [g_t(x,t)\vec{Z}_\varepsilon(x,t) - \varepsilon g_x(x,t)\vec{Z}_{\varepsilon x}(x,t)$$
$$- g_x(x,t)(\vec{Z}_\varepsilon(x,t) \times \vec{Z}_{\varepsilon x}(x,t))$$
$$+ g(x,t)\vec{f}(x,t,\vec{Z}_\varepsilon(x,t))]dxdt + \int_0^l g(x,0)\varphi(x)dx$$
$$- \varepsilon \int_0^T [g(l,t)\mathrm{grad}\,\psi_1(t,\vec{Z}_\varepsilon(l,t))$$
$$+ g(0,t)\mathrm{grad}\,\psi_0(t,\vec{Z}_\varepsilon(0,t))]dt$$
$$- \int_0^T [g(l,t)(\vec{Z}_\varepsilon(l,t) \times \mathrm{grad}\,\psi_1(t,\vec{Z}_\varepsilon(l,t)))$$
$$+ g(0,t)(\vec{Z}_\varepsilon(0,t) \times \mathrm{grad}\,\psi_0(t,\vec{Z}_\varepsilon(0,t)))]dt = 0,$$

where $g(x,t) \in H^1(Q_T)$, $g(x,T) = 0$.

Since $\{\vec{Z}_{\varepsilon_i}(x,t)\}$ uniformly converges on Q_T to $\vec{Z}(x,t)$, $\vec{f}(x,t,\vec{Z}_{\varepsilon_i}(x,t))$ uniformly converges to $\vec{f}(x,t,\vec{Z}(x,t))$. Similarly, $\{\vec{Z}_{\varepsilon_i}(0,t)\}$, $\{\vec{Z}_{\varepsilon_i}(l,t)\}$, $\{\mathrm{grad}\,\psi_0(t,\vec{Z}_{\varepsilon_i}(0,t))\}$ and $\{\mathrm{grad}\,\psi_1(t,\vec{Z}_{\varepsilon_i}(l,t))\}$ uniformly converge on Q_T to $\vec{Z}(0,t)$, $\vec{Z}(l,t)$, $\mathrm{grad}\,\psi_0(t,\vec{Z}(0,t))$ and $\mathrm{grad}\,\psi_1(t,\vec{Z}(l,t))$ respectively. The corresponding integrals are also bounded. The diffusion term with small parameter ε tends to zero.

Therefore by sending $\varepsilon_i \to 0$, we see that the above integral equality tends to the integral equality (3.2.26). This indicates that $\vec{Z}(x,t)$ is a global weak solution of problem (3.2.1) and (3.2.3)–(3.2.4).

3.2.5　Mixed Boundary Value Problem

Now consider the mixed boundary value problem: (3.2.1) with

$$\vec{Z}_x(0,t) = \mathrm{grad}\,\psi_0(t,\vec{Z}(0,t)), \quad \vec{Z}(l,t) = 0, \tag{3.2.27}$$
$$\vec{Z}(x,0) = \varphi(x), \quad x \in [0,l]. \tag{3.2.28}$$

Let conditions (2) and (3) hold and replace (1) by

(1'). $\psi_0(t, \vec{Z})$ is scalar function with continuous derivative with respect to $t \in [0, T]$, and the mixed derivative with respect to $t \in [0, T]$ and $\vec{Z} \in R^3$, and the second derivative with respect to $\vec{Z} \in R^3$ are continuous. The Hessian matrix $H_0(t, \vec{Z})$ of $\psi_0(t, \vec{Z})$ is assumed to be non-negative definite and $\mathrm{grad}\, \psi_0(t, 0) = 0$, $\varphi(l) = 0$.

We first consider the existence of the solution $\vec{Z}_\varepsilon(x, t)$ to problem (3.2.2), (3.2.27) and (3.2.28).

Theorem 3.2.4 *Let* (1'), (2) *and* (3) *hold. Then the mixed boundary value problem* (3.2.2), (3.2.27) *and* (3.2.28) *admits unique weak solution*

$$\vec{Z}_\varepsilon(x, t) \in Z_l = L^\infty(0, T; H^2(0, l)) \bigcap W_\infty^{(1)}(0, T; L^2(0, l)) \bigcap L^2(0, T; H^3(0, l)).$$

Proof. The proof is similar to that of Theorems 3.2.1 and 3.2.2.

We construct the nonlinear boundary condition for the ordinary differential equation (3.2.7)

$$\frac{\Delta_+ \vec{Z}_0}{h} = \mathrm{grad}\, \psi_0(t, \vec{Z}_0), \qquad \vec{Z}_J = 0, \tag{3.2.29}$$

and initial condition

$$\vec{Z}_j(0) = \overline{\varphi}_j, \quad j = 0, 1, \ldots, J, \tag{3.2.30}$$

whose solution (approximate solution) is denoted by $\vec{Z}_j(t)$, where $\overline{\varphi}_j = \varphi_j = \varphi_j(x_j)$, $(j = 1, 2, \ldots, J - 1)$, $\overline{\varphi}_J = \varphi_J = \varphi(l) = 0$, $\overline{\varphi}_0$ is subject to:

$$\varphi_1 = \overline{\varphi}_0 + h\, \mathrm{grad}\, \psi_0(0, \overline{\varphi}_0).$$

Before applying fixed point theorem, we must prove that problem (3.2.7), (3.2.29) and (3.2.30) admits at least one solution $\vec{Z}_h = \{\vec{Z}_j(t) : j = 0, 1, \ldots, J\}$ with $\vec{Z}_j(t) \in C^{(2)}([0, T])$ $(j = 1, 2, \ldots, J)$ and $\vec{Z}_0(t) \in C^{(1)}([0, T])$. In order to get solution $\vec{Z}_\varepsilon(t)$ of (3.2.2), (3.2.27) and (3.2.28), we first give the *a priori* estimates for solution $\vec{Z}_h(t)$ of (3.2.7), (3.2.29) and (3.2.30) and then send h to zero.

Multiplying (3.2.7) by $|\vec{Z}_j|^{p-2} \vec{Z}_j h$, $2 \le p < \infty$, and summing from $j = 1$ to $j = J$, and integrating over $t \in [0, t]$, $0 \le t \le T$, we have

$$\sum_{j=1}^{J-1} |\vec{Z}_j|^p h - \sum_{j=1}^{J-1} |\overline{\varphi}_j| h$$

$$= -\frac{p\varepsilon}{2} \int_0^t \sum_{j=0}^{J-1} (|\vec{Z}_{j+1}|^{p-2} + |\vec{Z}_j|^{p-2}) \left| \frac{\Delta_+ \vec{Z}_j}{h} \right|^2 h\, dt$$

$$- \frac{p\varepsilon}{2} \int_0^t \sum_{j=0}^{J-1} (|\vec{Z}_{j+1}|^{p-2} - |\vec{Z}_j|^{p-2})(|\vec{Z}_{j+1}|^2 - |\vec{Z}_j|^2) \frac{dt}{h}$$

$$- p\varepsilon \int_0^t |\vec{Z}_0|^{p-2} \vec{Z}_0\, \mathrm{grad}\, \psi_0(t, \vec{Z}_0)\, dt$$

$$+ p \int_0^t \sum_{j=1}^{J-1} |\vec{Z}_j|^{p-2} \vec{Z}_j \cdot \vec{f}(x_j, t, \vec{Z}_j) h\, dt,$$

where $\vec{Z}_J(t) = 0$. As before we may conclude that $z_h(t) = \{\vec{Z}_j(t) : j = 0, 1, \ldots, J\}$ is uniform bounded in $j = 0, 1, \ldots, J$ and $t \in [0, T]$.

Then we test (3.2.7) by $\frac{\Delta_+\Delta_-\vec{Z}_j}{h^2}h$, and sum from $j = 1$ to $j = J - 1$, and integrate over $t \in [0, t]$, $0 \le t \le T$ to give

$$\sum_{j=0}^{J-1} \left|\frac{\Delta_+\vec{Z}_j(t)}{h}\right|^2 h + 2\int_0^t \sum_{j=1}^{J-1} \left|\frac{\Delta_+\Delta_-\vec{Z}_j}{h^2}\right|^2 hdt + 2\psi_0(t, \vec{Z}_0(t))$$

$$= \sum_{j=0}^{J-1} \left|\frac{\Delta_+\overline{\varphi}_j}{h}\right|^2 h + 2\psi_0(0, \overline{\varphi}_0)$$

$$+ 2\int_0^t \psi_{0t}(t, \vec{Z}_0(t))dt$$

$$+ 2\int_0^t \sum_{j=0}^{J-1} \frac{\Delta_+\vec{Z}_j}{h} \int_0^1 \vec{f}_x(x_{j+2}, t, \vec{Z}_{j+1})hd\tau dt$$

$$+ 2\int_0^t \sum_{j=0}^{J-1} \left(\frac{\Delta_+\vec{Z}_j}{h} \int_0^1 \vec{f}_z(x_j, t, \tau\vec{Z}_{j+1} + (1-\tau)\vec{Z}_j)\right)\frac{\Delta_+\vec{Z}_j}{h}hdtd\tau$$

$$+ 2\int_0^t \vec{f}(0, t, \vec{Z}_0) \cdot \operatorname{grad}\psi_0(t, \vec{Z}_0)dt - 2\int_0^t \vec{f}(l, t, 0)\frac{\Delta_-\vec{Z}_J}{h}dt,$$

where $\vec{Z}_J'(t) = 0$. All terms except for the last term on the right-hand side can be controlled by $A\int_0^t \|\delta\vec{Z}_h\|_{L^2}^2 dt + B$, where A, B are independent of $h > 0$, $\varepsilon \ge 0$. For the last term, we have

$$\left|2\int_0^t \vec{f}(l, t, 0)\frac{\Delta_-\vec{Z}_J}{h}dt\right| \le C_{28}\int_0^t \|\delta\vec{Z}_h(t)\|_\infty dt + \varepsilon\int_0^t \|\delta^2\vec{Z}_h(t)\|_{L^2}^2 dt + \frac{1}{\varepsilon}C_{29}.$$

The above estimates are independent of h but depending on ε. If $\vec{f}(l, t, 0) = 0$, then these estimates are independent of both h and ε.

Next we estimate $y_h(t) = \vec{Z}_h'(t) = \{\vec{Z}_j'(t) : j = 0, 1, \ldots, J\}$. We have

$$\frac{\Delta_+y_0}{h} = \operatorname{grad}\psi_{0t}(t, \vec{Z}_0) + H_0(t, \vec{Z}_0)y_0, \quad y_J = 0;$$

$$y_j(0) = \varepsilon\frac{\Delta_+\Delta_-\overline{\varphi}_j}{h^2} + \overline{\varphi}_j \times \frac{\Delta_+\Delta_-\overline{\varphi}_j}{h^2} + \vec{f}(x_j, t, \overline{\varphi}_j), \quad j = 1, 2, \ldots, J-1;$$

$$(E + hH_0(0, \overline{\varphi}_0))y_0(0) = y_1(0) - h\operatorname{grad}\psi_0(0, \overline{\varphi}_0), \quad y_J(0) = 0,$$

where E is a 3×3 unit matrix. From this we can get the estimates of $\sup_{0\le t\le T} \|\vec{Z}_j'(t)\|_2^2$ and $\frac{\Delta_+\Delta_-\vec{Z}_j}{h^2}$, $\frac{\Delta_+\vec{Z}_h'}{h}$ and $\frac{\Delta_+^2\Delta_-\vec{Z}_h}{h^3}$. we have

$$\sup_{0\le t\le T} \|\vec{Z}_h(t)\|_2 + \sup_{0\le t\le T} \|\delta\vec{Z}_h(t)\|_2 + \sup_{0\le t\le T} \|\delta^2\vec{Z}_h(t)\|_2$$

$$+ \sup_{0\le t\le T} \|\vec{Z}_h'(t)\|_2 + \left(\int_0^t \|\delta^2\vec{Z}_h'(t)\|_2^2 dt\right)^{1/2} + \left(\int_0^t \|\delta^3\vec{Z}_h(t)\|_2^2 dt\right)^{1/2}$$

$$\le K_{13}(\varepsilon),$$

where $K_{13}(\varepsilon)$ is independent of h.

Now, as before, we may send $h \to 0$ (for some subsequence if necessary) to get a limit vector $\vec{Z}_\varepsilon(x,t)$ which has weak derivatives $\vec{Z}_{\varepsilon x}, \vec{Z}_{\varepsilon x}x$ and $\vec{Z}_{\varepsilon t} \in L^\infty(0,T; L^2(0,l))$ and $\vec{Z}_{\varepsilon x}t, \vec{Z}_{\varepsilon x}xx \in L^2(Q_T)$. Moreover, $\vec{Z}_\varepsilon(x,t)$ solves (3.2.2) in the sense of distribution and satisfies (3.2.27) and (3.2.28) in the classical sense. The uniqueness of solution $\vec{Z}_\varepsilon(x,t)$ to the problem (3.2.2) and (3.2.27)–(3.2.28) can be given by standard argument. The theorem is proved.

2. *Approximate solution of the mixed initial-boundary problem of the spin equation*

In order to prove the existence of equation (3.2.1) with nonlinear boundary condition (3.2.27) and initial condition (3.2.28), we must give the uniform (in ε) *a priori* estimates for the approximate solution $\vec{Z}_\varepsilon(x,t)$ obtained above.

Lemma 3.2.8 *Let (1′), (2), (3) hold and $\vec{f}(l,t,0) = 0$. Then for the approximate solution $\vec{Z}_\varepsilon(x,t)$ obtained above, we have*

$$\|\vec{Z}_\varepsilon\|_{L^\infty} + \sup_{0\leq t\leq T} \|\vec{Z}_{\varepsilon x}(\cdot,t)\|_{L^2} + \sup_{0\leq t\leq T} \|\vec{Z}_{\varepsilon t}(\cdot,t)\|_{H^{-1}(0,l)} \leq K_{14}, \qquad (3.2.31)$$

where K_{14} is a constant independent of $\varepsilon > 0$.

Lemma 3.2.9 *Let (1′), (2), (3) hold and $\vec{f}(l,t,0) = 0$. Then for the approximate solution $\vec{Z}_\varepsilon(x,t)$ of problem (3.2.2) and (3.2.27)–(3.2.28), we have*

$$|\vec{Z}_\varepsilon(x_1,t) - \vec{Z}_\varepsilon(x_2,t)| \leq K_{15}|x_1 - x_2|^{1/2}, \qquad (3.2.32)$$
$$|\vec{Z}_\varepsilon(x,t_1) - \vec{Z}_\varepsilon(x,t_2)| \leq K_{16}|t_1 - t_2|^{1/4}, \qquad (3.2.33)$$

where K_{15}, K_{16} are constants independent of $\varepsilon > 0$.

3.2.6   The Mixed Boundary Problem of Equations of Ferromagnetic Spin Chain

1. *Weak solution for the mixed boundary problem of equations of ferromagnetic spin chain*

Definition 3.2.2 *Vector $\vec{Z}(x,t) \in L^2(0,T; H^1(0,l)) \cap C(Q_T)$ with $\vec{Z}(l,t) = 0$ is called a weak solution of equation (3.2.1) subject to mixed nonlinear boundary condition (3.2.27) and initial condition (3.2.28), if for any test function $g(x,t) \in H^1(Q_T)$, $g(l,t) = g(x,T) = 0$ there holds*

$$\int\int_{Q_T} [g_t(x,t)\vec{Z}(x,t) - g_x(x,t)(\vec{Z}(x,t) \times \vec{Z}_x(x,t))$$

$$+ g(x,t)\vec{f}(x,t,\vec{Z}(x,t))]dxdt + \int_0^l g(x,0)\varphi(x)dx$$

$$- \int_0^T g(0,t)(\vec{Z}(0,t) \times \operatorname{grad} \psi_0(t,\vec{Z}_0(0,t)))dt = 0. \qquad (3.2.34)$$

Theorem 3.2.5 *Let* (1'), (2), (3) *hold and* $\vec{f}(l, t, 0) = 0$. *Then equation* (3.2.1) *with the mixed nonlinear boundary condition* (3.2.27) *and initial condition* (3.2.28) *admits at least one weak solution:*

$$\vec{Z}(x, t) \in L^\infty(0, T; H^1(0, l)) \bigcap C^{(\frac{1}{2}, \frac{1}{4})}(Q_T).$$

Proof. The proof is similar to that of Theorem 3.2.3.

In the following we consider another problem.

Let $Q_T^* = \{x \in R^+, 0 \le t \le T\}$. Consider the nonlinear boundary condition

$$\vec{Z}_x(0, t) = \operatorname{grad} \psi_0(t, \vec{Z}(0, t)), \tag{3.2.35}$$

$$\vec{Z}(x, 0) = \varphi(x). \tag{3.2.36}$$

(2*) $\vec{f}(x, t, \vec{Z})$ is a vector-valued function and continuously differentiable with respect to $(x, t) \in Q_T^*$ and $\vec{Z} \in R^3$ and the Jacobi matrix $\vec{f}_{\vec{Z}}(x, t, \vec{Z})$ is semi-bounded, that is, there is a constant $b > 0$ such that

$$\xi \cdot \vec{f}_{\vec{Z}}(x, t, \vec{Z}) \cdot \xi \le b|\xi|^2, \quad \xi \in R^3, (x, t, \xi) \in Q_T^* \times R^3;$$

(3*) $\varphi(x) \in H^1(R^+)$, $\varphi'(x) = \operatorname{grad} \psi_0(0, \varphi)$.

(4*) $\vec{f}(x, t, \vec{Z})$, $\vec{f}_x(x, t, \vec{Z})$, $\vec{f}_t(x, t, \vec{Z})$ meet one of the following conditions:

$$|\vec{f}(x, t, \vec{Z})|, |\vec{f}_x(x, t, \vec{Z})|, |\vec{f}_t(x, t, \vec{Z})|$$
$$\le a(x, t)F(\vec{Z}) + b(x, t); \tag{3.2.37}$$

or

$$|\vec{f}(x, t, \vec{Z})|, |\vec{f}_x(x, t, \vec{Z})|, |\vec{f}_t(x, t, \vec{Z})|$$
$$\le c(x, t)|\vec{Z}|^k + d(x, t), \tag{3.2.38}$$

where $a(x, t), b(x, t), d(x, t) \in L^\infty(0, T; L^2(R^+))$, $c(x, t) \in L^\infty(Q_T^*)$, $k \ge 0$ is a constant, $F(\vec{Z})$ is a continuous scalar function.

2. *Weak solution to mixed boundary value problem of system of ferromagnetic spin chain*

Theorem 3.2.6 *Let* (1'), (2*), (3*), (4*) *hold. Then the mixed boundary value problem of system* (3.2.2) *defined on* Q_T^* *with* (3.2.35) *and* (3.2.36) *admits unique weak solution:*

$$\vec{Z}_\varepsilon(x, t) \in Z_\infty = L^\infty(0, T; H^2(R^+)) \bigcap W_\infty^{(1)}(0, T; L^2(R^+)) \bigcap L^2(0, T; H^3(R^+)).$$

Proof. For any $l > 0$, construct a continuously differentiable vector-valued function $\vec{f}^{(l)}(x, t, \vec{Z})$, for $(x, t) \in Q_T^{(l)} = \{0 \le x \le l, 0 \le t \le T\}$ and $\vec{Z} \in R^3$ such that

(i) For $x \in [0, l - 1], t \in [0, T], \vec{Z} \in R^3, \vec{f}^{(l)}(x, t, \vec{Z}) \equiv \vec{f}(x, t, \vec{Z}), \vec{f}(x, t, 0) \equiv 0.$

(ii) Jacobi matrix $\vec{f}_{\vec{Z}}^{(l)}(x,t,\vec{Z})$ is semi-bounded, that is, there is a constant $\bar{b} > 0$ such that

$$\xi \cdot \vec{f}_{\vec{Z}}^{(l)}(x,t,\vec{Z}) \cdot \xi \le \bar{b}|\xi|^2, \quad l > 0, \tag{3.2.39}$$

where $\bar{b}$ is independent of $l > 0$.

(iii) $\vec{f}^{(l)}(x,t,\vec{Z})$, $\vec{f}_x^{(l)}(x,t,\vec{Z})$, $\vec{f}_t^{(l)}(x,t,\vec{Z})$ meet one of the following conditions:

$$|\vec{f}^{(l)}(x,t,\vec{Z})|, |\vec{f}_x^{(l)}(x,t,\vec{Z})|, |\vec{f}_t^{(l)}(x,t,\vec{Z})|$$
$$\le \bar{a}(x,t)\overline{F}(\vec{Z}) + \bar{b}(x,t); \tag{3.2.40}$$

or

$$|\vec{f}^{(l)}(x,t,\vec{Z})|, |\vec{f}_x^{(l)}(x,t,\vec{Z})|, |\vec{f}_t^{(l)}(x,t,\vec{Z})|$$
$$\le \bar{c}(x,t)|\vec{Z}|^{\bar{k}} + \bar{d}(x,t), \tag{3.2.41}$$

where the norms of $\bar{a}(x,t), \bar{b}(x,t), \bar{d}(x,t)$ in $L^\infty(0,T;L^2(0,l))$, the norm of $c(x,t)$ in $L^\infty(Q_T^{(l)})$, and $\bar{k}$, $\overline{F}$ are all independent of $l > 0$.

We also construct for any $l > 0$ the vector $\varphi^{(l)}(x) \in H^1(0,l)$ such that $\varphi^{(l)}(x) = \varphi(x)$ for $x \in [0, l-1]$, $\varphi^{(l)}(l) = 0$ and $\|\varphi^{(l)}\|_{H^1(0,l)}$ is uniform bounded in $l > 0$.

For any $l > 0$, consider mixed nonlinear boundary condition (3.2.27) and the initial condition:

$$\vec{Z}(x,0) = \varphi^{(l)}(x). \tag{3.2.42}$$

Over the rectangular domain $Q_T^{(l)}$, we consider the spin equation:

$$\vec{Z}_t = \varepsilon\vec{Z}_{xx} + \vec{Z} \times \vec{Z}_{xx} + \vec{f}^{(l)}(x,t,\vec{Z}). \tag{3.2.43}$$

It follows from Theorem 3.2.4 that problem (3.2.43), (3.2.27), (3.2.42) admits unique weak solution:

$$\vec{Z}^{(l)}(x,t) \in Z_l = L^\infty(0,T;H^2(0,l)) \bigcap W_\infty^{(1)}(0,T;L^2(0,l))$$
$$\bigcap L^2(0,T;H^3(0,l)).$$

Multiplying (3.2.43) by $|\vec{Z}|^{p-2}\vec{Z}$, $(2 \le p \le \infty)$ and integrating it over $Q_t^{(l)}$, we have

$$\|\vec{Z}(\cdot,t)\|_{L^p(0,l)}^p - \|\varphi^{(l)}\|_{L^p(0,l)}$$
$$= -p\varepsilon \int\!\!\int_{Q_t^{(l)}} |\vec{Z}|^{p-2}(\vec{Z}_x \cdot \vec{Z}_x)dxdt$$
$$- p(p-2)\varepsilon \int\!\!\int_{Q_t^{(l)}} |\vec{Z}|^{p-2}\left(\frac{\vec{Z}}{|\vec{Z}|} \cdot \vec{Z}_x\right)^2 dxdt$$
$$- p\varepsilon \int_0^t |\vec{Z}(0,t)|^{p-2}\vec{Z}(0,t) \cdot \operatorname{grad}\psi_0(t,\vec{Z}(0,t))dt$$
$$+ p \int\!\!\int_{Q_t^{(l)}} |\vec{Z}|^{p-2}\vec{Z} \cdot \vec{f}(x,t,\vec{Z})dxdt.$$

This yields

$$\sup_{0 \le t \le T} \|\vec{Z}(\cdot, t)\|_{L^p(0,l)} \le K_{17} e^{(\bar{b}+\delta)T} [\|\varphi^{(l)}\|_{L^p(0,l)} + \|\vec{f}_0^{(l)}\|_{L^p(Q_T^{(l)})}], \tag{3.2.44}$$

where $\vec{f}_0^{(l)} = \vec{f}^{(l)}(x, l, 0)$, $\delta > 0$, and constant K_7 depends on $\delta > 0$. Therefore,

$$\|\varphi^{(l)}\|_{L^p(0,l)} \le C_{30} \|\varphi^{(l)}\|_{L^2(0,l)}^{1/2+1/p} \|\varphi^{(l)}\|_{H^1(0,l)}^{1/2-1/p},$$

$$\|\vec{f}_0^{(l)}\|_{L^p(Q_T^{(l)})} \le C_{31} \big[\sup_{0 \le t \le T} \|\vec{f}_x^{(l)}(\cdot, t, 0)\|_{L^2(0,l)}$$

$$+ \sup_{0 \le t \le T} \|\vec{f}_x^{(l)}(\cdot, t, 0)\|_{L^p(0,l)} \big]. \tag{3.2.45}$$

It follows from (3.2.40) and (3.2.41) that the right-hand side of (3.2.45) is uniformly bounded in $l > 0$. Then

$$\sup_{0 \le t \le T} \|\vec{Z}^{(l)}(\cdot, t)\|_{L^p(0,l)} \le C_{32},$$

where C_{32} is uniform in both $2 \le p < \infty$ and $l > 0$, hence

$$\sup_{0 \le t \le T} \|\vec{Z}^{(l)}(\cdot, t)\|_{L^\infty(0,l)} \le C_{32}.$$

Multiplying equation (3.2.43) by $\vec{Z}_{xx}$ and then integrating it over $Q_T^{(l)}$, we have

$$\|\vec{Z}_x(\cdot, t)\|_{L^2(0,l)}^2 - \|\varphi_x^{(l)}\|_{L^2(0,l)}^2 + 2\varepsilon \|\vec{Z}_{xx}\|_{L^2(0,l)}^2$$
$$+ \psi_0(t, \vec{Z}(0, t)) - \psi_0(0, \varphi^{(l)}(0))$$
$$= 2 \int_0^l \psi_{0t}(t, \vec{Z}(0, t)) dt$$
$$+ 2 \int_0^t \operatorname{grad} \psi_0(t, \vec{Z}(0, t)) \cdot \vec{f}^{(l)}(0, t, \vec{Z}(0, t)) dt$$
$$+ 2 \int \int_{Q_t^{(l)}} \vec{Z}_x \cdot \vec{f}_x^{(l)}(x, t, \vec{Z}) dx dt$$
$$+ 2 \int \int_{Q_T^{(l)}} \vec{Z}_x \cdot \vec{f}_x^{(l)}(x, t, \vec{Z}) dx dt,$$

where $\vec{Z}(l, t) = \vec{Z}_t(l, t) \equiv 0$, $\vec{f}^{(l)}(x, t, 0) \equiv 0$. It follows from the above that

$$\|\vec{Z}_x^{(l)}(\cdot, t)\|_{L^2(0,l)}^2 + 2\varepsilon \|\vec{Z}_{xx}^{(l)}\|_{L^2(Q_t^{(l)})}^2$$
$$\le (2\bar{b} + 1) \|\vec{Z}_x^{(l)}(\cdot, t)\|_{L^2(Q_t^{(l)})}^2$$
$$+ 2 \int \int_{Q_t^{(l)}} |\vec{f}_x^{(l)}(x, t, \vec{Z}^{(l)})|^2 dx dt + C_{33}, \tag{3.2.46}$$

where the constant C_{33} is independent of $l > 0$.

If $\vec{f}_x(x, t, \vec{Z})$ meets condition (3.2.37), $\vec{f}_x^{(l)}(x, t, \vec{Z})$ satisfies condition (3.2.40). We have

$$\int\int_{Q_t^{(l)}} |\vec{f}_x^{(l)}(x, t, \vec{Z}^{(l)}(x, t))|^2 dxdt$$

$$\leq 2 \sup_{|\vec{Z}| \leq C_{32}} |\overline{F}(\vec{Z})|^2 \|\overline{a}\|^2_{L^2(Q_t^{(l)})} + 2\|\overline{b}\|^2_{L^2(Q_t^{(l)})}.$$

If $\vec{f}_x(x, t, \vec{Z})$ meets condition (3.2.38), $\vec{f}_x^{(l)}(x, t, \vec{Z})$ satisfies condition (3.2.41). We have

$$\int\int_{Q_t^{(l)}} |\vec{f}_x^{(l)}(x, t, \vec{Z}^{(l)}(x, t))|^2 dxdt$$

$$\leq 2\|\overline{c}\|_{L^\infty(Q_t^{(l)})} \sup_{0 \leq t \leq T} \|\vec{Z}^{(l)}(x, t)\|^{2\overline{k}}_{L^{2\overline{k}}(0,l)} + 2\|\overline{d}\|^2_{L^2(Q_t^{(l)})},$$

whose right-hand side is independent of $l > 0$. Combining these inequalities, we have

$$\sup_{0 \leq t \leq T} \|\vec{Z}_x^{(l)}(\cdot, t)\|_{L^2(0,l)} + \sqrt{\varepsilon}\|\vec{Z}_{xx}^{(l)}\|_{L^2(Q_T^{(l)})} \leq C_{34},$$

where C_{34} is independent of $l > 0$.

In order to get the estimate for $y^{(l)}(x, t) = \vec{Z}_t^{(l)}(x, t)$ in $L^\infty(0, T; L^2(0, l))$, we differentiate (3.2.43) with respect to t to obtain

$$y_t = \varepsilon y_{xx} + y \times \vec{Z}_{xx} + \vec{Z} \times y_{xx} + \vec{f}_t^{(l)}(x, t, \vec{Z}) + \vec{f}_{\vec{Z}}^{(l)}(x, t, \vec{Z})y. \tag{3.2.47}$$

Assumption (4^*) implies that $\int\int_{Q_t^{(l)}} |\vec{f}_t(x, t, \vec{Z}^{(l)}(x, t))|^2 dxdt$ is bounded uniform in l. By the similar method as above we have

$$\sup_{0 \leq t \leq T} \|\vec{Z}_t^{(l)}(\cdot, t)\|_{L^2(0,l)} + \sqrt{\varepsilon}\|\vec{Z}_{xx}^{(l)}\|_{L^2(Q_T^{(l)})} \leq C_{35}, \tag{3.2.48}$$

where C_{35} is independent of $l > 0$.

Differentiating (3.2.43) with respect to x, we obtain

$$\vec{Z}_{xt} = \varepsilon \vec{Z}_{xxx} + \vec{Z} \times \vec{Z}_{xxx} + \vec{Z}_x \times \vec{Z}_{xx} + \vec{f}_x^{(l)}(x, t, \vec{Z}) + \vec{f}_{\vec{Z}}^{(l)}(x, t, \vec{Z})\vec{Z}_x. \tag{3.2.49}$$

Then the uniform boundedness in l of $\vec{Z}_{xxx}^{(l)}(x, t)$ in $L^2(Q_T^{(l)})$ can be derived from the uniform boundedness in l of $\vec{Z}_{xt}^{(l)}(x, t)$ in $L^2(Q_T^{(l)})$ since the coefficient matrix of $\vec{Z}_{xxx}$ is positively definite and non-singular.

Hence, we may choose $\{l_j\}$ such that when $l_j \to \infty$, $\{\vec{Z}^{(l_j)}\}$ and $\{\vec{Z}_x^{(l_j)}\}$ uniformly converge to $\vec{Z}(x, t)$ and $\vec{Z}_x(x, t)$ on any rectangular domain $\{0 \leq x \leq l, 0 \leq t \leq T\}$ respectively; and, $\{\vec{Z}_{xx}^{(l_j)}\}$ and $\{\vec{Z}_t^{(l_j)}\}$ weakly converge in $L^\infty(0, T; L^2(0, l))$ to $\vec{Z}_{xx}(x, t)$ and $\vec{Z}_t(x, t)$ respectively; and, $\{\vec{Z}_{xxx}^{(l_j)}\}$ and $\{\vec{Z}_{xt}^{(l_j)}\}$ weakly converge in $L^2(Q_T^*)$ to $\vec{Z}_{xxx}(x, t)$ and $\vec{Z}_{xt}(x, t)$ ($\varepsilon > 0$) respectively. This means that the limit $\vec{Z}(x, t)$ is a weak solution of (3.2.2) and (3.2.35) and (3.2.36), and it satisfies the conditions (3.2.35) and (3.2.36) in the classical sense. By the same method as above, we can prove the uniqueness. Let $l \to \infty$, we have

Lemma 3.2.10 *Under the condition* (1^*), (2^*), (3^*) *and* (4^*), *for the weak solution* $\vec{Z}_\varepsilon(x,t)$ *of the problem* (3.2.2) *and* (3.2.35) *and* (3.2.36) $(\varepsilon > 0)$, *there holds*

$$
\begin{aligned}
\sup_{0 \le t \le T} \|\vec{Z}_\varepsilon(\cdot,t)\|_{L^2(R^+)} \\
+ \sup_{0 \le t \le T} \|\vec{Z}_{\varepsilon T}(\cdot,t)\|_{L^2(R^+)} + \sup_{0 \le t \le T} \|\vec{Z}_{\varepsilon t}(\cdot,t)\|_{H^{-1}(R^+)} \\
\le K_{18}.
\end{aligned}
\tag{3.2.50}
$$

The constant K_{18} *is independent of* $\varepsilon > 0$.

Proof. Let $g(x) \in H_0^1(R^+)$ be any test function. We have

$$
\begin{aligned}
\int_0^\infty g(x)\vec{Z}_{\varepsilon t}(x,t)dx = -\varepsilon \int_0^\infty g_x(x)\vec{Z}_{\varepsilon x}(x,t)dx \\
- \int_0^\infty g_x(x)(\vec{Z}_\varepsilon(x,t) \times \vec{Z}_{\varepsilon x}(x,t))dx \\
+ \int_0^\infty g(x)\vec{f}(x,t,\vec{Z}_\varepsilon(x,t))dx.
\end{aligned}
$$

And we have

$$
\left| \int_0^\infty g(x)\vec{f}(x,t,\vec{Z}_\varepsilon(x,t))dx \right|
$$

$$
\le \|g\|_{L^2(R^+)} \left(\int_0^\infty |\vec{f}(x,t,\vec{Z}_\varepsilon(x,t))|^2 dx \right)^{1/2}.
$$

In the case of (3.2.37) or (3.2.38), we have

$$
\int_0^\infty |\vec{f}(x,t,\vec{Z}_{\varepsilon t}(x,t))|^2 dx
$$

$$
\le 2 \left(\sup_{|\vec{Z}| \le K_{18}} |F(\vec{Z})|^2 \sup_{0 \le t \le T} \|a(\cdot,t)\|_{L^2(R^+)}^2 + \sup_{0 \le t \le T} \|b(\cdot,t)\|_{L^2(R^+)}^2 \right)
$$

and

$$
\int_0^\infty |\vec{f}(x,t,\vec{Z}_\varepsilon(x,t))|^2 dx
$$

$$
\le 2 \left(\sup_{0 \le t \le T} \|\vec{Z}_\varepsilon(\cdot,t)\|_{L^{2k}(R^+)}^{2k} \|c\|_{L^\infty(Q_T^*)} + \sup_{0 \le t \le T} \|d(\cdot,t)\|_{L^2(R^+)}^2 \right).
$$

This implies that (3.2.50) is true.

Lemma 3.2.11 *Under the condition* $(1')$, (2^*), (3^*) *and* (4^*), *for the weak solution* $\vec{Z}_\varepsilon(x,t)$ *of the problem* (3.2.2) *and* (3.2.38) *and* (3.2.39) $(\varepsilon > 0)$, *there holds*

$$
|\vec{Z}_\varepsilon(x,t_1) - \vec{Z}_\varepsilon(x,t_2)| \le K_{19}(x)|t_1 - t_2|^{1/4}, \tag{3.2.51}
$$

$$
|\vec{Z}_\varepsilon(x_1,t) - \vec{Z}_\varepsilon(x_2,t)| \le K_{20}|x_1 - x_2|^{1/2}, \tag{3.2.52}
$$

where $K_{19}(x)$, K_{20} *are independent of* ε, $K_{19}(x)$ *depends on* x.

It follows from the discussion of Lemma 3.2.10 that the uniform boundedness in $\varepsilon > 0$ of the vector $S_\varepsilon(x,t) = \int_0^x \vec{Z}_\varepsilon(x,t)dx$ and its derivative $S_{\varepsilon t}(x,t) = \int_0^x \vec{Z}_{\varepsilon t}(x,t)dx$ is only true on the bounded domain, therefore the estimate (3.2.51) holds only for finite x.

Now we are in the position to consider the existence of the weak solution to problem (3.2.1) and (3.2.35) and (3.2.36).

3. *Existence of weak solution to the nonlinear boundary problem of the equations of ferromagnetic spin chain*

Definition 3.2.3 *A vector $\vec{Z}(x,t) \in L^2(0,T;H^1(R^+)) \bigcap C(Q_T^*)$ is called a weak solution to problem (3.2.1) and (3.2.35) and (3.2.36), if for any test function $g(x,t) \in H^1(Q_T^*)$, $g(x,T) = 0$, $\operatorname{supp} g(x,t) < \infty$ there holds*

$$\int\int_{Q_T^*} [g_t(x,t)\vec{Z}(x,t) - g_x(x,t)(\vec{Z}(x,t) \times \vec{Z}_x(x,t))$$

$$+ g(x,t)\vec{f}(x,t,\vec{Z}(x,t))]dxdt + \int_0^\infty g(x,0)\varphi(x)dx$$

$$- \int_0^T g(0,t)(\vec{Z}(0,t) \times \operatorname{grad}\psi_0(t,\vec{Z}_0(0,t)))dt = 0.$$

Theorem 3.2.7 *Let $(1')$, (2^*), (3^*) and (4^*) hold. Then there is at least one weak solution $\vec{Z}(x,t)$ of the problem (3.2.1) and (3.2.35) and (3.2.36) on Q_T^*:*

$$\vec{Z}(x,t) \in L^\infty(0,T;H^1(R^+)) \bigcap C_{\text{loc}}^{(\frac{1}{2},\frac{1}{4})}(Q_T^*).$$

4. *Properties of weak solution*

In the following we discuss the "blow-up" problem and the asymptotic properties for the solution.

Theorem 3.2.8 *Let (1)–(3) hold for any $T > 0$ and $\vec{f}(x,t,0) \in L^p(Q_\infty)$ for any $2 \le p < \infty$. If in (3.2.6), $b < 0$, then the solution to spin equation (3.2.2) with (3.2.3) and (3.2.4) satisfies*

$$\lim_{t \to \infty} \|\vec{Z}_\varepsilon(\cdot,t)\|_{L^p(0,l)} = 0.$$

Proof. For any fixed $2 \le p < \infty$, take $\delta > 0$ such that $b + \delta < 0$ to give from the L^p-estimates of $\vec{Z}_\varepsilon(x,t)$ that

$$\|\vec{Z}_\varepsilon(\cdot,t)\|_{L^p(0,l)}$$

$$\le e^{(b+\delta)t}\left\{\|\varphi\|_{L^p(0,l)} + (\delta(p-1))^{1/p}\left(\frac{p-1}{2p}\right)\|\vec{f}_0\|_{L^p(Q_\infty)}\right\}.$$

This implies the conclusion.

Similarly we have

Theorem 3.2.9 *Under the same conditions as in Theorem 3.2.8, the solution to equation (3.2.1) with (3.2.3) and (3.2.4) satisfies*

$$\lim_{t \to \infty} \|\vec{Z}(\cdot,t)\|_{L^p(0,l)} = 0, \quad 2 \le p < \infty.$$

Theorem 3.2.10 *Let* $(1')$, (2) *and* (3) *hold and* $\vec{f}(x,t,0) \in L^p(Q_\infty)$ *for any* $2 \leq p < \infty$, $b < 0$. *Then the solution to equation* $(3.2.1)$ *with* $(3.2.35)$ *and* $(3.2.36)$ *satisfies*

$$\lim_{t \to \infty} \|\vec{Z}(\cdot,t)\|_{L^p(0,l)} = 0, \quad 2 \leq p < \infty.$$

Theorem 3.2.11 *Let* $(1')$, (2^*), (3^*) *and* (4^*) *hold and* $\vec{f}(x,t,0) \in L^p(Q_\infty^*)$ *for any* $2 \leq p < \infty$, $b < 0$. *Then the solution to equation* $(3.2.1)$ *and* $(3.2.2)$ *with* $(3.2.35)$ *and* $(3.2.36)$ *satisfies*

$$\lim_{t \to \infty} \|\vec{Z}(\cdot,t)\|_{L^p(0,l)} = 0, \quad 2 \leq p < \infty.$$

Now consider the "blow-up" problem. If there holds

$$\vec{Z} \cdot \vec{f}(x,t,\vec{Z}) \geq C_0|\vec{Z}|^{2+\delta}, \quad (x,t) \in Q_T, \ \vec{Z} \in R^3 \tag{3.2.53}$$

where $C_0 > 0, \delta > 0$, we have by multiplying $(3.2.1)$ by $|\vec{Z}|^{p-2}\vec{Z}$ and integrating over $[0,l]$ with respect to x that

$$\frac{1}{p}\frac{d}{dt}\|\vec{Z}(\cdot,t)\|_{L^p(0,l)}^p = \int_0^l |\vec{Z}(x,t)|^{p-2}\vec{Z}(x,t) \cdot \vec{f}(x,t,\vec{Z})dx.$$

This combined with $(3.2.53)$ yields

$$\frac{d}{dt}\|\vec{Z}(\cdot,t)\|_{L^p(0,l)}^p \geq C_0 l^{-\frac{\delta}{p}}\|\vec{Z}(\cdot,t)\|_{L^p(0,l)}^{1+\delta},$$

and hence

$$\|\vec{Z}(\cdot,t)\|_{L^p(0,l)} \geq (\|\varphi\|_{L^p(0,l)}^{-\delta} - C_0 t\delta l^{-\frac{\delta}{p}})^{-\frac{1}{\delta}}, \quad 2 \leq p < \infty. \tag{3.2.54}$$

It follows from $(3.2.54)$ that

Theorem 3.2.12 *If* $\vec{f}(x,t,\vec{Z})$ *satisfies* $(3.2.53)$ *and* $\|\varphi\|_{L^p(0,l)} \neq 0, 2 \leq p < \infty$, *then the solution to* $(3.2.1)$ *blows up at finite time, i.e.*

$$\lim_{t \to \infty} \|\vec{Z}(\cdot,t)\|_{L^p(0,l)} = +\infty.$$

3.3 Smooth Solution for the Ferromagnetic Spin Chain Systems

In the above sections, we have obtained the existence and uniqueness of the global weak solution for one-dimensional problems. In this section, we shall prove the existence and uniqueness of smooth solution for the one-dimensional periodic initial-boundary value problem and initial value problem. Our method is to use the mobile framework on S^2 in the procedure of viscosity vanishing to get the uniform estimate.

3.3.1   Smooth Solution to the Nonlinear Systems with Periodic Initial Boundary Conditions

1. *The problem*

In order to establish the existence and uniqueness of smooth solution to the equation of ferromagnetic spin chain

$$\vec{Z}_t = \vec{Z} \times \vec{Z}_{xx} \tag{3.3.1}$$

with periodic initial-boundary conditions

$$\vec{Z}(x,0) = \vec{Z}_0(x), \quad \vec{Z}(x+D,t) = \vec{Z}(x-D,t), \quad |\vec{Z}_0(x)| \equiv 1, \quad x \in R^1, \tag{3.3.2}$$

or initial condition

$$\vec{Z}(x,0) = \vec{Z}_0(x), \quad |\vec{Z}_0(x)| \equiv 1, \quad x \in R^1, \tag{3.3.3}$$

we consider the following nonlinear equations with small parameter $\varepsilon > 0$

$$\vec{Z}_t = -\varepsilon \vec{Z} \times (\vec{Z} \times \vec{Z}_{xx}) + \vec{Z} \times \vec{Z}_{xx}, \tag{3.3.4}$$

with (3.3.2) to approximate the desired solution. It can be proved that:

$$\begin{cases} \vec{Z}_t = \varepsilon \vec{Z}_{xx} + \vec{Z} \times \vec{Z}_{xx} + \varepsilon |\vec{Z}_x|^2 \vec{Z}, \\ \vec{Z}(x,0) = \vec{Z}_0(x), \\ \vec{Z}(x+D,t) = \vec{Z}(x-D,t) \end{cases} \tag{3.3.5}$$

is equivalent to (3.3.4) with (3.3.2) in the classical sense. Here we assume that at the initial time the ferromagnet is saturated, that is, $|\vec{Z}_0(x)| = C_0$ (positive constant, $C_0 = 1$, for example).

We use the difference method to prove the existence of local smooth solution of periodic initial value problem (3.3.5). This also means the existence of local smooth solution of (3.3.4) and (3.3.2) ($\varepsilon > 0$). Then we use the uniform estimates in ε and in D for solutions of (3.3.4) and (3.3.2) to get the existence of solution to (3.3.1) and (3.3.2) by sending $\varepsilon \to 0$, and next to obtain the existence of solution to (3.3.1) and (3.3.3) by sending $D \to \infty$. The uniqueness of smooth solution can be easily given.

2. *Smooth solution to the difference-differential system*

Using difference in the spatial direction, we prove that (3.3.5) admits at least one local smooth solution. For simplicity, we let $\varepsilon = 1$ and establish the following difference-differential equation:

$$\begin{cases} \dfrac{d\vec{Z}_j}{dt} = \dfrac{\Delta_+\Delta_-\vec{Z}_j}{h^2} + \vec{Z}_j \times \dfrac{\Delta_+\Delta_-\vec{Z}_j}{h^2} + \left|\dfrac{\Delta_+\vec{Z}_j}{h}\right|^2 \vec{Z}_j, \\ \vec{Z}_j|_{t=0} = \vec{Z}_{0j} = \vec{Z}_0(jh), \\ \vec{Z}_{j+J} = \vec{Z}_j, \quad j = 0, \pm 1, \ldots, \pm J, \end{cases} \tag{3.3.6}$$

where $\vec{Z}_j = \vec{Z}(x_j, t)$, $x_j = jh, j = 0, \pm 1, \ldots, \pm J$, $h = 2D/J$, $J > 0$, Δ_+, Δ_- denote the forward and backward difference respectively.

It is not difficult to know that the initial value problem (3.3.6) admits a local smooth solution. For such solution, we shall give some estimates uniformly in h. In this section we always denote the smooth solution of (3.3.6) by $\vec{Z}_j$.

Lemma 3.3.1 *If $\vec{Z}_0(x) \in H^1(\Omega)$ ($\Omega = (-D, D)$), then there are constants $T_0 > 0$, $C > 0$ independent of h such that*

$$\sup_{0 \le t \le T_0} \|\vec{Z}_h(t)\|_2 \le C, \qquad \sup_{0 \le t \le T_0} \|\delta \vec{Z}_h(t)\|_2 \le C, \tag{3.3.7}$$

where $u_h = \{u_j = u(x_j) \mid j = 0, 1, 2, \ldots, J\}$, $x_j = jh$, $h = 2D/J$, and

$$\|\cdot\|_p = \|\cdot\|_{L^p(\Omega)}, \qquad \|\delta^k u_h\|_p = \left(\sum_{i=0}^{J-k} \left|\frac{\Delta_+^k u_i}{h^k}\right|^p h\right)^{\frac{1}{p}}.$$

Proof. Multiplying (3.3.6) by $\vec{Z}_j h$ and summing from $j = 1$ to J, we have

$$\frac{1}{2}\frac{d}{dt}\sum_{j=1}^{J} |\vec{Z}_j|^2 h = \sum_{j=1}^{J} \vec{Z}_j \frac{\Delta_+ \Delta_- \vec{Z}_j}{h^2} h + \sum_{j=1}^{J} \left|\frac{\Delta_+ \vec{Z}_j}{h}\right|^2 |\vec{Z}_j|^2 h$$

$$= -\sum_{j=0}^{J-1} \left|\frac{\Delta_+ \vec{Z}_j}{h}\right|^2 h + \sum_{j=1}^{J} \left|\frac{\Delta_+ \vec{Z}_j}{h}\right|^2 |\vec{Z}_j|^2 h.$$

Since

$$\|\vec{Z}_h\|_\infty \le C\|\vec{Z}_h\|_2^{\frac{1}{2}} \left(\|\delta \vec{Z}_h\|_2 + \frac{1}{2D}\|\vec{Z}_h\|_2\right)^{\frac{1}{2}},$$

we have

$$\frac{d}{dt}\|\vec{Z}_h\|_2^2 + \|\delta \vec{Z}_h\|_2^2 \le C(\|\vec{Z}_h\|_2^4 + \|\delta \vec{Z}_h\|_2^4). \tag{3.3.8}$$

Moreover, multiplying (3.3.6) by $\frac{\Delta_+ \Delta_- \vec{Z}_j}{h}$ and summing from $j = 1$ to J, we get

$$\sum_{j=1}^{J} \vec{Z}_{jt} \frac{\Delta_+ \Delta_- \vec{Z}_j}{h} = \sum_{j=1}^{J} \left|\frac{\Delta_+ \Delta_- \vec{Z}_j}{h^2}\right|^2 h + \sum_{j=1}^{J} \frac{\Delta_+ \Delta_- \vec{Z}_j}{h} \left|\frac{\Delta_+ \vec{Z}_j}{h}\right|^2 \vec{Z}_j.$$

Therefore, one gets

$$\frac{1}{2}\frac{d}{dt}\sum_{j=0}^{J-1} \left|\frac{\Delta_+ \vec{Z}_j}{h}\right|^2 h + \sum_{j=1}^{J} \left|\frac{\Delta_+ \Delta_- \vec{Z}_j}{h^2}\right|^2 h$$

$$\le \frac{1}{4}\sum_{j=1}^{J} \left|\frac{\Delta_+ \Delta_- \vec{Z}_j}{h^2}\right|^2 h + C \max_{1 \le j \le J} |\vec{Z}_j|^2 \sum_{j=1}^{J} \left|\frac{\Delta_+ \vec{Z}_j}{h}\right|^4 h.$$

Applying interpolation inequalities

$$\|\vec{Z}_h\|_\infty \le C\|\vec{Z}_h\|_2^{\frac{3}{4}}(\|\delta^2\vec{Z}_h\|_2 + \|\vec{Z}_h\|_2)^{\frac{1}{4}},$$

$$\|\delta\vec{Z}_h\|_4 \le C\|\delta\vec{Z}_h\|_2^{\frac{3}{4}}(\|\delta^2\vec{Z}_h\|_2 + \|\delta\vec{Z}_h\|_2)^{\frac{1}{4}}$$

and Hölder inequality, we have

$$\frac{d}{dt}\|\delta\vec{Z}_h\|_2^2 + \|\delta^2\vec{Z}_h\|_2^2 \le C(1 + \|\vec{Z}_h\|_2^8 + \|\delta\vec{Z}_h\|_2^8). \tag{3.3.9}$$

Combining (3.3.8) with (3.3.9), we have

$$\frac{d}{dt}(\|\vec{Z}_h\|_2^2 + \|\delta\vec{Z}_h\|_2^2) + \|\delta^2\vec{Z}_h|_2^2 \le C + C(\|\vec{Z}_h\|_2^2 + \|\delta\vec{Z}_h\|_2^2)^4.$$

This inequality combined with Gronwall inequality implies that there exist constants $T_0, C > 0$ independent of h such that

$$\|\vec{Z}_h(t)\|_2 + \|\delta\vec{Z}_h(t)\|_2 \le C, \quad \forall\, t \in [0, T_0],$$

and

$$\int_0^{T_0} \|\delta^2\vec{Z}_h(t)|_2^2 \le C.$$

The lemma is proved.

Corollary 3.3.1 *Under the conditions in Lemma* 3.3.1, *we have, for some constant* C *independent of* h,

$$\sup_{0\le t\le T_0; 0\le j\le J} |\vec{Z}_j| \le C. \tag{3.3.10}$$

Lemma 3.3.2 *If* $\vec{Z}_0(x) \in H^2(\Omega)$, *there are constants* $T_0 > 0$, $C > 0$ *independent of* h *such that*

$$\sup_{0\le t\le T_0} \|\delta^2\vec{Z}_h(t)\|_2 \le C, \quad \int_0^{T_0} \|\delta^3\vec{Z}_h(t)\|_2 \le C. \tag{3.3.11}$$

Proof. It follows from (3.3.6) that

$$\frac{d}{dt}\Delta_+\vec{Z}_j = \frac{\Delta_+^2\Delta_-\vec{Z}_j}{h^2} + \Delta_+\left(\vec{Z}_j \times \frac{\Delta_+\Delta_-\vec{Z}_j}{h^2} + \left|\frac{\Delta_+\vec{Z}_j}{h}\right|^2 \vec{Z}_j\right).$$

Multiplying this equality by $\frac{\Delta_+^2\Delta_-\vec{Z}_j}{h^3}$, summing it from $j = 1$ to J and noting that

$$\sum_{j=1}^{J} \frac{\Delta_+^2\Delta_-\vec{Z}_j}{h^3} \cdot \Delta_+\vec{Z}_{jt} = -\frac{1}{2}\frac{d}{dt}\sum_{j=0}^{J-1}\left|\frac{\Delta_+^2\vec{Z}_j}{h^2}\right|^2 h,$$

$$\sum_{j=1}^{J} \frac{\Delta_+^2\Delta_-\vec{Z}_j}{h^3} \cdot \Delta_+\left(\vec{Z}_j \times \frac{\Delta_+\Delta_-\vec{Z}_j}{h^2}\right) = \sum_{j=1}^{J} \frac{\Delta_+^2\Delta_-\vec{Z}_j}{h^3} \cdot \left(\Delta_+\vec{Z}_j \times \frac{\Delta_+\Delta_-\vec{Z}_j}{h^2}\right)$$

$$\leq \frac{1}{4}\sum_{j=1}^{J}\left|\frac{\Delta_+^2\Delta_-\vec{Z}_j}{h^3}\right|^2 h + \max_{1\leq j\leq J}\left|\frac{\Delta_+\vec{Z}_j}{h}\right|^2 \sum_{j=1}^{J}\left|\frac{\Delta_+\Delta_-\vec{Z}_j}{h^2}\right|^2 h,$$

$$\sum_{j=1}^{J}\frac{\Delta_+^2\Delta_-\vec{Z}_j}{h^3}\cdot\Delta_+\left(\left|\frac{\Delta_+\vec{Z}_j}{h}\right|^2\vec{Z}_j\right) \leq \frac{1}{4}\sum_{j=1}^{J}\left|\frac{\Delta_+^2\Delta_-\vec{Z}_j}{h^3}\right|^2 h$$

$$+ C\left(\max_{1\leq j\leq J}\left|\frac{\Delta_+\vec{Z}_j}{h}\right|^2 \sum_{j=1}^{J}\left|\frac{\Delta_+^2\vec{Z}_j}{h^2}\right|^2 h + \sum_{j=1}^{J}\left|\frac{\Delta_+\vec{Z}_j}{h}\right|^6 h\right),$$

as well as the following inequalities

$$\|\delta\vec{Z}_h\|_\infty \leq C\|\delta\vec{Z}_h\|_2^{1/2}(\|\delta^2\vec{Z}_h\|_2 + \|\delta\vec{Z}_h\|_2)^{1/2},$$
$$\|\delta\vec{Z}_h\|_6 \leq C\|\delta\vec{Z}_h\|_2^{2/3}(\|\delta^2\vec{Z}_h\|_2 + \|\delta\vec{Z}_h\|_2)^{1/3},$$

we finally get

$$\frac{d}{dt}\|\delta^2\vec{Z}_h\|_2^2 + \|\delta^3\vec{Z}_h\|_2^2 \leq C + C\|\delta^2\vec{Z}_h\|_2^3.$$

Lemma 3.3.2 follows from the Gronwall's inequality.

Corollary 3.3.2 *Under the conditions in Lemma 3.3.2, we have, for some C independent of h,*

$$\begin{cases} \displaystyle\sup_{0\leq t\leq T_0; 1\leq j\leq J}\left|\frac{\Delta_+\vec{Z}_j(t)}{h}\right| \leq C, \\ \displaystyle\int_0^{T_0}\|\delta\vec{Z}_{ht}(t)\|_2^2 dt \leq C. \end{cases} \tag{3.3.12}$$

3. *Local smooth solution to the nonlinear equation with periodic initial data*

By the similar method as in the proof of Lemmas 3.3.1 and 3.3.2 and using induction argument, we have

Lemma 3.3.3 *If $\vec{Z}_0(x) \in H^k(\Omega)$, then there are constants $T_0 > 0, C > 0$ independent of h such that*

$$\sup_{0\leq t\leq T_0}\|\delta^k\vec{Z}_h\|_2 \leq C,$$

$$\sup_{0\leq t\leq T_0}\|\delta^{k-2}\vec{Z}_{ht}\|_2 \leq C, \quad k\geq 2, \tag{3.3.13}$$

$$\sup_{0\leq t\leq T_0}\|\delta^{k-4}\vec{Z}_{htt}\|_2 \leq C, \quad k\geq 4.$$

From the above uniform estimates in h of $\vec{Z}_j(t)$ and from the standard method, we can prove that the solutions of (3.3.6) approximates the solution of (3.3.5), and then we get the existence of local smooth solution to (3.3.5). That is, we have

Theorem 3.3.1 *Let $\varepsilon > 0$, $\vec{Z}_0(x) \in H^k(\Omega)$ and $\vec{Z}_0(x - D) = \vec{Z}_0(x + D)$. Then problem (3.3.5) admits at least one local smooth solution $\vec{Z}(x,t)$ in*

$$\vec{Z}(x,t) \in G(T_0)$$

$$= \left(\bigcap_{s=0}^{[\frac{k}{2}]} W_\infty^s(0, T_0; H^{k-2s}(\Omega)) \right) \bigcap \left(\bigcap_{s=0}^{[\frac{k+1}{2}]} H^s(0, T_0; H^{k+1-2s}(\Omega)) \right), \quad (3.3.14)$$

where $T_0 > 0$ is independent of k, s and k are non-negative integers with $k - 2s \geq 0$.

Theorem 3.3.2 *Let conditions in Theorem 3.3.1 hold and $|\vec{Z}_0(x)| = 1$, $x \in R^1$. Then problem (3.3.5) is equivalent to the following problem:*

$$\begin{cases} \vec{Z}_t = -\varepsilon\vec{Z} \times (\vec{Z} \times \vec{Z}_{xx}) + \vec{Z} \times \vec{Z}_{xx}, \\ \vec{Z}(x,0) = \vec{Z}_0(x), \\ \vec{Z}(x + D, t) = \vec{Z}(x - D, t). \end{cases} \quad (3.3.15)$$

Proof. First we let $\vec{Z}(x,t)$ be a classical solution of (3.3.15) ($\varepsilon > 0$) and prove that it is also a solution of (3.3.5).

In fact it is easy to get that $|\vec{Z}(x,t)| \equiv 1$. Then we have

$$-\varepsilon\vec{Z} \times (\vec{Z} \times \vec{Z}_{xx}) = \varepsilon|\vec{Z}|^2\vec{Z}_{xx} - \varepsilon(\vec{Z} \cdot \vec{Z}_{xx})\vec{Z}$$

$$= \varepsilon\vec{Z}_{xx} + \varepsilon|\vec{Z}_x|^2\vec{Z},$$

this implies that $\vec{Z}$ solves (3.3.5) in the classical sense.

On the other hand, if $\vec{Z}$ is a classical solution of (3.3.5), we want to prove $|\vec{Z}(x,t)| \equiv 1$ in Q_T. In fact, denoting $u(x,t) = |\vec{Z}(x,t)|^2$, we have from (3.3.5)

$$u_t = \varepsilon u_{xx} + 2\varepsilon|\vec{Z}_x|^2(u - 1),$$

$$u(x,0) = 1,$$

$$u(x - D, t) = u(x + D, t).$$

It is clear that $\bar{u} = 1$ also solves this problem. Let $w = u - \bar{u} = |\vec{Z}(x,t)|^2 - 1$. Then

$$w_t = \varepsilon w_{xx} + 2\varepsilon|\vec{Z}_x|^2w, \quad (3.3.16)$$

$$w(x,0) = 0, \quad (3.3.17)$$

$$w(x - D, t) = w(x + D, t). \quad (3.3.18)$$

Multiplying (3.3.16) by w and integrating over $[-D, D]$, we have

$$\frac{1}{2}\int_{-D}^{D}|w|^2dx + \varepsilon\int_{-D}^{D}|w_x|^2dx \leq 2\varepsilon \max_{x,t}|\vec{Z}_x|^2 \int_{-D}^{D}|w|^2dx.$$

This combined with Gronwall inequality yields the conclusion.

4. *Global smooth solution to nonlinear system with periodic initial data*

It follows from Theorems 3.3.1 and 3.3.2 that the problem (3.3.5) also admits a local smooth solution. In order to establish the global existence of smooth solution for problem (3.3.15). We must derive the *a priori* estimates for the solutions of (3.3.5).

Lemma 3.3.4 *Let $\vec{Z}(x,t)$ be a smooth solution of (3.3.5) $|\vec{Z}_0(x)| = 1$. Then we have*

$$\begin{cases} |\vec{Z}(x,t)| = 1, \quad \forall\, (x,t) \in R^1 \times [0,T], \\ \sup_{0 \le t \le T} \|\vec{Z}_x(\cdot,t)\|_2 \le C, \end{cases} \tag{3.3.19}$$

where C is independent of T and D.

Proof. Since (3.3.15) is equivalent to (3.3.5), the lemma can be proved by multiplying (3.3.5) by $\vec{Z}(x,t)$ and $\vec{Z}_{xx}$ and integrating by parts.

Lemma 3.3.5 *Let $\vec{Z}(x,t)$ be the solution of (3.3.5). Then under the conditions of Lemma 3.3.4 there hold*

$$\begin{cases} \sup_{0 \le t \le T_2} \|\vec{Z}_{xx}(\cdot,t)\|_{L^2(\Omega)} \le C, \\ \|\vec{Z}_{xxx}(x,t)\|_{L^2(Q_T)} \le C, \\ \|\vec{Z}_{xt}(\cdot,t)\|_{L^2(Q_T)} \le C. \end{cases} \tag{3.3.20}$$

Proof. It follows from (3.3.5) (let $\varepsilon = 1$ for simplicity):

$$\vec{Z}_{xxt} \cdot \vec{Z}_{xx} = \vec{Z}_{xxxx} \cdot \vec{Z}_{xx} + (\vec{Z} \times \vec{Z}_{xx})_{xx} + (|\vec{Z}_x|^2 \vec{Z})_{xx} \cdot \vec{Z}_{xx}.$$

Integrating this equality over Ω, we get

$$\frac{1}{2}\frac{d}{dt}\int_{-D}^{D} |\vec{Z}_{xx}|^2 dx + \int_{-D}^{D} |\vec{Z}_{xxx}|^2 dx$$

$$\le \int_{-D}^{D} |\vec{Z}_x||\vec{Z}_{xx}||\vec{Z}_{xxx}|dx + \int_{-D}^{D} |\vec{Z}_x|^3|\vec{Z}_{xxx}|dx$$

$$+ 2\int_{-D}^{D} |\vec{Z}||\vec{Z}_x||\vec{Z}_{xx}||\vec{Z}_{xxx}|dx$$

$$\le 3\Big(\sup_{x} |\vec{Z}_x|\Big)\|\vec{Z}_{xx}\|_{L^2(\Omega)}\|\vec{Z}_{xxx}\|_{L^2(\Omega)}$$

$$+ \|\vec{Z}_x\|_{L^6(\Omega)}^3\|\vec{Z}_{xxx}\|_{L^2(\Omega)}. \tag{3.3.21}$$

From the interpolation inequalities

$$\|\vec{Z}_x\|_{L^\infty(\Omega)} \le C\|\vec{Z}_x\|_{L^2(\Omega)}^{3/4}\|\vec{Z}_{xxx}\|_{L^2(\Omega)}^{1/4},$$

$$\|\vec{Z}_x\|_{L^6(\Omega)} \le C\|\vec{Z}_x\|_{L^2(\Omega)}^{5/6}\|\vec{Z}_{xxx}\|_{L^2(\Omega)}^{1/6},$$

$$\|\vec{Z}_{xx}\|_{L^2(\Omega)} \le C\|\vec{Z}_x\|_{L^2(\Omega)}^{1/2}\|\vec{Z}_{xxx}\|_{L^2(\Omega)}^{1/2},$$

we have from (3.3.21) and Lemma 3.3.4 and Hölder inequality that

$$\frac{d}{dt}\int_{-D}^{D} |\vec{Z}_{xx}|^2 dx + \frac{1}{2}\int_{-D}^{D} |\vec{Z}_{xxx}|^2 dx \le C.$$

This inequality and Gronwall inequality yields the desired conclusion.

Corollary 3.3.3 *Under the conditions of Lemma 3.3.5, there holds*

$$\sup_{Q_T} |\vec{Z}_x| \leq C, \tag{3.3.22}$$

where C is independent of D and T.

By the similar method in proving Lemma 3.3.5, we have from induction argument

Lemma 3.3.6 *Let $\vec{Z}(x,t)$ be a classical solution of* (3.3.5). *Under the conditions of Theorem 3.3.1 and $|\vec{Z}_0(x)| = 1$, there holds*

$$\begin{cases} \sup_{0 \leq t \leq T} \|\vec{Z}_{x^{k-2s}t^s}(\cdot, t)\|_{L^2(\Omega)} \leq C, & k - 2s > 0, \\ \|\vec{Z}_{x^{k+1-2s}t^s}\|_{L^2(Q_T)} \leq C, & k + 1 - 2s \geq 0, \end{cases} \tag{3.3.23}$$

where k, s are non-negative integers and C is independent of D.

It follows from Lemma 3.3.6 and extension method that problem (3.3.5) and (3.3.15) admits a global smooth solution. That is

Theorem 3.3.3 *Let $\vec{Z}_0(x) \in H^k(\Omega)$ $(k \geq 4)$, $|\vec{Z}_0(x)| = 1$, $\vec{Z}_0(x - D) = \vec{Z}_0(x + D)$. Problem* (3.3.15) *$(\varepsilon > 0)$ admits a global smooth solution $\vec{Z}(x,t) \in G(T)$, where T is any positive number.*

3.3.2   Smooth Solution to the Equations of Ferromagnetic Spin Chain

1. *Estimates of smooth solution*

In order to get the existence of global smooth solution to the periodic initial value problem (3.3.1) and (3.3.2), we must have the *a priori* estimates uniform in ε for the smooth solutions of problem (3.3.15) with $\varepsilon > 0$.

Lemma 3.3.7 *Let $\vec{Z}_\varepsilon(x,t)$ be a classical solution of* (3.3.5). *Under the conditions of Theorem 3.3.3, there holds*

$$\begin{cases} |\vec{Z}_\varepsilon(x,t)| = 1, & \forall \ (x,t) \in R \times [0,T], \\ \sup_{0 \leq t \leq T} \|\vec{Z}_{\varepsilon x}(\cdot, t)\|_{L^2(\Omega)} \leq K_1, \end{cases} \tag{3.3.24}$$

where K_1 is independent of D, ε.

Lemma 3.3.8 *Under the conditions of Theorem 3.3.3 $(k \geq 3)$, for the classical solutions of* (3.3.5), *there holds*

$$\begin{cases} \sup_{0 \leq t \leq T} \|\vec{Z}_{\varepsilon x}x(\cdot, t)\|_{L^2(\Omega)} \leq K_2, \\ \sup_{0 \leq t \leq T} \|\vec{Z}_{\varepsilon t}(\cdot, t)\|_{L^2(\Omega)} \leq K_2, \end{cases} \tag{3.3.25}$$

where K_2 is independent of D, ε.

Proof. Drop the subscript ε for $\vec{Z}_\varepsilon$ for simplicity. Since $|\vec{Z}(x,t)| \equiv 1$, the vectors $\vec{Z}, \vec{Z}_x, \vec{Z} \times \vec{Z}_x$ form a unit orthogonal basis of R^3. Let $\vec{Z}_{xx} = \alpha\vec{Z} + \beta\vec{Z}_x + \gamma\vec{Z} \times \vec{Z}_x$. Then $\alpha = -|\vec{Z}_x|^2$, $\beta = \vec{Z}_x \cdot \vec{Z}_{xx}/|\vec{Z}_x|^2$, $\gamma = (\vec{Z} \times \vec{Z}_x) \cdot \vec{Z}_{xx}/|\vec{Z}_x|^2$. Since

$$\frac{1}{2}\frac{d}{dt}\int_{-D}^{D}|\vec{Z}_{xx}|^2 dx = \int_{-D}^{D}\vec{Z}_{xx} \cdot (\varepsilon\vec{Z}_{xx} + \vec{Z} \times \vec{Z}_{xx} + \varepsilon|\vec{Z}_{xx}|^2\vec{Z})_{xx}dx$$

$$= -\varepsilon\int_{-D}^{D}|\vec{Z}_{xxx}|^2 dx - \int_{-D}^{D}(\vec{Z}_x \times \vec{Z}_{xx}) \cdot \vec{Z}_{xxx}dx$$

$$- \varepsilon\int_{-D}^{D}\vec{Z}_{xxx} \cdot (|\vec{Z}_x|^2\vec{Z}_x + 2\vec{Z}(\vec{Z}_x \cdot \vec{Z}_{xx}))dx,$$

where

$$\int_{-D}^{D}\vec{Z}_{xxx} \cdot |\vec{Z}_x|^2\vec{Z}_x dx = -\int_{-D}^{D}|\vec{Z}_x|^2|\vec{Z}_{xx}|^2 dx - 2\int_{-D}^{D}|\vec{Z}_x \cdot \vec{Z}_{xx}|^2 dx,$$

$$\int_{-D}^{D}\vec{Z}_{xxx} \cdot (\vec{Z}(\vec{Z}_x \cdot \vec{Z}_{xx}))dx = -3\int_{-D}^{D}|\vec{Z}_x \cdot \vec{Z}_{xx}|^2 dx,$$

$$\int_{-D}^{D}(\vec{Z}_x \times \vec{Z}_{xx}) \cdot \vec{Z}_{xxx}dx$$

$$= \int_{-D}^{D}[\vec{Z}_x \times (-|\vec{Z}_x|^2\vec{Z} + \frac{(\vec{Z} \times \vec{Z}_x) \cdot \vec{Z}_{xx}}{|\vec{Z}_x|^2}(\vec{Z} \times \vec{Z}_x))] \cdot \vec{Z}_{xxx}dx$$

$$= \int_{-D}^{D}|\vec{Z}_x|^2(\vec{Z} \times \vec{Z}_x) \cdot \vec{Z}_{xxx}dx$$

$$+ \int_{-D}^{D}\frac{(\vec{Z} \times \vec{Z}_x) \cdot \vec{Z}_{xx}}{|\vec{Z}_x|^2}|\vec{Z}_x|^2\vec{Z} \cdot \vec{Z}_{xxx}dx$$

$$= \frac{5}{2}\int_{-D}^{D}|\vec{Z}_x|^2(\vec{Z} \times \vec{Z}_x) \cdot \vec{Z}_{xxx}dx.$$

In the above relations we have used

$$\vec{Z} \cdot \vec{Z}_{xxx} = -\frac{3}{2}(|\vec{Z}_x|^2)_x.$$

In summary we have

$$\frac{1}{2}\frac{d}{dt}\int_{-D}^{D}|\vec{Z}_{xx}|^2 dx + \varepsilon\int_{-D}^{D}|\vec{Z}_{xxx}|^2 dx$$

$$= \varepsilon\int_{-D}^{D}|\vec{Z}_x|^2|\vec{Z}_{xx}|^2 dx + 8\varepsilon\int_{-D}^{D}|\vec{Z}_x \cdot \vec{Z}_{xx}|^2 dx$$

$$- \frac{5}{2}\int_{-D}^{D}|\vec{Z}_x| \cdot (\vec{Z} \times \vec{Z}_x) \cdot \vec{Z}_{xxx}dx. \tag{3.3.26}$$

On the other hand,

$$\frac{1}{4}\frac{d}{dt}\int_{-D}^{D}|\vec{Z}_x|^4 dx$$

$$= \int_{-D}^{D}|\vec{Z}_x|^2\vec{Z}_x\cdot(\varepsilon\vec{Z}_{xx}+\vec{Z}\times\vec{Z}_{xx}+\varepsilon|\vec{Z}_x|^2\vec{Z})_x dx$$

$$= \int_{-D}^{D}|\vec{Z}_x|^2\vec{Z}_x\cdot(\varepsilon\vec{Z}_{xxx}+\vec{Z}_x\times\vec{Z}_{xx}+\vec{Z}\times\vec{Z}_{xxx}$$

$$+\varepsilon|\vec{Z}_x|^2\vec{Z}_x+2\varepsilon\vec{Z}(\vec{Z}_x\cdot\vec{Z}_{xx}))dx$$

$$= -\varepsilon\int_{-D}^{D}|\vec{Z}_x|^2|\vec{Z}_{xx}|^2 dx - 2\varepsilon\int_{-D}^{D}|\vec{Z}_x\cdot\vec{Z}_{xx}|^2 dx$$

$$+\varepsilon\int_{-D}^{D}|\vec{Z}_x|^6 dx - \int_{-D}^{D}|\vec{Z}_x|^2(\vec{Z}\times\vec{Z}_x)\cdot\vec{Z}_{xxx}dx,$$

and hence

$$\frac{1}{4}\frac{d}{dt}\int_{-D}^{D}|\vec{Z}_x|^4 dx + \varepsilon\int_{-D}^{D}|\vec{Z}_x|^2|\vec{Z}_{xx}|^2 + 2\varepsilon\int_{-D}^{D}|\vec{Z}_x\cdot\vec{Z}_{xx}|^2 dx$$

$$= \varepsilon\int_{-D}^{D}|\vec{Z}_x|^6 dx - \int_{-D}^{D}|\vec{Z}_x|^2(\vec{Z}\times\vec{Z}_x)\cdot\vec{Z}_{xxx}dx. \tag{3.3.27}$$

It follows from (3.3.26) and (3.3.27) that

$$4\frac{d}{dt}\int_{-D}^{D}|\vec{Z}_x|^2 dx + 8\varepsilon\int_{-D}^{D}|\vec{Z}_{xxx}|^2 dx + 20\varepsilon\int_{-D}^{D}|\vec{Z}_x|^6 dx$$

$$= 5\frac{d}{dt}\int_{-D}^{D}|\vec{Z}_x|^4 dx + 28\varepsilon\int_{-D}^{D}|\vec{Z}_x|^2|\vec{Z}_{xx}|^2 dx$$

$$+104\varepsilon\int_{-D}^{D}|\vec{Z}_x\cdot\vec{Z}_{xx}|^2 dx.$$

It follows from Lemma 3.3.7 that

$$28\int_{-D}^{D}|\vec{Z}_x|^2|\vec{Z}_{xx}|^2 dx + 104\int_{-D}^{D}|\vec{Z}_x\cdot\vec{Z}_{xx}|^2 dx$$

$$\le 132\int_{-D}^{D}|\vec{Z}_x|^2|\vec{Z}_{xx}|^2 dx$$

$$\le \delta\int_{-D}^{D}|\vec{Z}_x|^6 dx + C(\delta)\int_{-D}^{D}|\vec{Z}_{xx}|^3 dx$$

$$\le \delta\int_{-D}^{D}|\vec{Z}_x|^6 dx + C(\delta)\|\vec{Z}_x\|_{L^2(\Omega)}^{5/4}\|\vec{Z}_{xxx}\|_{L(\Omega)}^{7/4}$$

$$\le \delta\int_{-D}^{D}|\vec{Z}_x|^6 dx + C(\delta)\|\vec{Z}_{xxx}\|_{L(\Omega)}^2 + C'(\delta).$$

Taking $\delta = \frac{1}{4}$, we have

$$4\frac{d}{dt}\int_{-D}^{D}|\vec{Z}_{xx}|^4 dx + 4\varepsilon\int_{-D}^{D}|\vec{Z}_{xxx}|^2 dx + 16\varepsilon\int_{-D}^{D}|\vec{Z}_x|^6 dx$$

$$\le 5\frac{d}{dt}\int_{-D}^{D}|\vec{Z}_x|^4 dx + C,$$

where C is independent of ε. Integrating this inequality over $t \in [0, T]$ and using the following imbedding inequality

$$\|\vec{Z}_x\|_{L^4(\Omega)} \leq C\|\vec{Z}_x\|_{L^2(\Omega)}^{3/4}\|\vec{Z}_{xx}\|_{L^2(\Omega)}^{1/4},$$

we can get the estimates on $\|\vec{Z}_{xx}\|_{L^2(\Omega)}$ immediately. Using equation (3.3.5), we also obtain the estimates for $\|\vec{Z}_t\|_{L^2(\Omega)}$.

Corollary 3.3.4 *Let $\vec{Z}_\varepsilon(x, t)$ be a smooth solution to problem (3.3.5). We have*

$$\sup_{x,t} |\vec{Z}_{\varepsilon x}(x, t)| \leq K_3, \tag{3.3.28}$$

where K_3 is independent of ε, D.

2. Uniform boundedness

Lemma 3.3.9 *Under the conditions of Theorem 3.3.2 ($k \geq 4$), for the smooth solution of (3.3.5) $\vec{Z}_\varepsilon(x, t)$, there holds*

$$\begin{cases} \displaystyle\sup_{0 \leq t \leq T} \|\vec{Z}_{\varepsilon x}xx(\cdot, t)\|_{L^2(\Omega)} \leq K_4, \\ \displaystyle\sup_{0 \leq t \leq T} \|\vec{Z}_{\varepsilon x}t(\cdot, t)\|_{L^2(\Omega)} \leq K_4, \end{cases} \tag{3.3.29}$$

where K_4 is independent of ε, D.

Proof. Differentiating equation (3.3.5) three times with respect to x, and multiplying it by $\vec{Z}_{xxx}$, then integrating in x, we have

$$\frac{1}{2}\frac{d}{dt}\int_{-D}^{D} |\vec{Z}_{xxx}|^2 dx + \varepsilon \int_{-D}^{D} |\vec{Z}_{xxxx}|^2 dx$$

$$= -2\int_{-D}^{D} (\vec{Z}_x \times \vec{Z}_{xxx}) \cdot \vec{Z}_{xxxx} dx - \varepsilon \int_{-D}^{D} (|\vec{Z}_x|^2\vec{Z})_{xx} \cdot \vec{Z}_{xxxx} dx.$$

Since

$$\left|\int_{-D}^{D} (|\vec{Z}_x|^2\vec{Z})_{xx} \cdot \vec{Z}_{xxxx} dx\right|$$

$$\leq C + \frac{1}{4}\left|\int_{-D}^{D} |\vec{Z}_{xxxx}|^2 dx + C\int_{-D}^{D} |\vec{Z}_{xxx}|^2 dx + C\int_{-D}^{D} |\vec{Z}_{xx}|^4 dx\right|$$

and the following imbedding inequality

$$\|\vec{Z}_{xx}\|_4 \leq C\|\vec{Z}_{xx}\|_2^{7/8}\|\vec{Z}_{xxxx}\|_2^{1/8},$$

we have from Hölder inequality that

$$\frac{1}{2}\frac{d}{dt}\int_{-D}^{D} |\vec{Z}_{xxx}|^2 dx + \frac{\varepsilon}{2}\int_{-D}^{D} |\vec{Z}_{xxxx}|^2 dx$$

$$\leq C + C\int_{-D}^{D} |\vec{Z}_{xxx}|^2 dx - 2\int_{-D}^{D} (\vec{Z}_x \times \vec{Z}_{xxx}) \cdot \vec{Z}_{xxxx} dx. \tag{3.3.30}$$

Since $\vec{Z} \cdot \vec{Z}_x = 0$, $\vec{Z} \cdot \vec{Z}_{xxx} = -3|\vec{Z}_{xx}|^2 - 4\vec{Z}_x \cdot \vec{Z}_{xxx}$, assuming $\vec{Z}_{xxx} = \alpha'\vec{Z} + \beta'\vec{Z}_x + \gamma'\vec{Z} \times \vec{Z}_x$, where $\alpha' = \vec{Z} \cdot \vec{Z}_{xxx} = -3\vec{Z} \cdot \vec{Z}_{xx}$, $\beta' = \vec{Z}_x \cdot \vec{Z}_{xxx}/|\vec{Z}_x|^2$, $\gamma' = \frac{(\vec{Z}\times\vec{Z}_x)\cdot\vec{Z}_{xxx}}{|\vec{Z}_x|^2}$, we have

$$\int_{-D}^{D} (\vec{Z}_x \times \vec{Z}_{xxx}) \cdot \vec{Z}_{xxxx}$$

$$= \int_{-D}^{D} \left[-3(\vec{Z}_x \cdot \vec{Z}_{xx}) \cdot (\vec{Z}_x \times \vec{Z}) + \frac{(\vec{Z} \times \vec{Z}_x) \cdot \vec{Z}_{xxx}}{|\vec{Z}_x|^2}|\vec{Z}_x|^2 \cdot \vec{Z}\right] \cdot \vec{Z}_{xxxx}dx$$

$$= -3\int_{-D}^{D} [(\vec{Z}_x \cdot \vec{Z}_{xx}) \cdot (\vec{Z} \times \vec{Z}_x)]_x \cdot \vec{Z}_{xxx}dx$$

$$- 4\int_{-D}^{D} ((\vec{Z}_x \times \vec{Z}_x) \cdot \vec{Z}_{xxx})\vec{Z}_x \cdot \vec{Z}_{xxx}dx$$

$$- 3\int_{-D}^{D} ((\vec{Z} \times \vec{Z}_x) \cdot \vec{Z}_{xxx})|\vec{Z}_{xx}|^2dx$$

$$\leq C\left(\int_{-D}^{D} |\vec{Z}_{xxx}|^2dx + \int_{-D}^{D} |\vec{Z}_{xx}|^4dx \right).$$

Here we have used the following Sobolev inequality

$$\|\vec{Z}_{xx}\|_{L^4(\Omega)} \leq C\|\vec{Z}_{xx}\|_{L^2(\Omega)}^{3/4}\|\vec{Z}_{xxx}\|_{L^2(\Omega)}^{1/4}$$

and the Hölder inequality. From (3.3.30), we finally get

$$\frac{1}{2}\frac{d}{dt}\int_{-D}^{D} |\vec{Z}_{xxx}|^2dx + \frac{\varepsilon}{2}\int_{-D}^{D} |\vec{Z}_{xxxx}|^2dx \leq C + C\int_{-D}^{D} |\vec{Z}_{xxx}|^2dx.$$

The lemma then follows from Gronwall inequality.

Corollary 3.3.5 *Under the conditions of Lemma 3.3.9, for the smooth solution of* (3.3.5) $\vec{Z}_\varepsilon(x,t)$, *there holds*

$$\begin{cases} \sup\limits_{x,t} |\vec{Z}_{\varepsilon x}x(x,t)| \leq K_5, \\ \sup\limits_{x,t} \|\vec{Z}_{\varepsilon t}(x,t)| \leq K_5, \end{cases} \tag{3.3.31}$$

where K_5 is independent of ε, D.

By induction, we also have

Lemma 3.3.10 *Under the conditions of Theorem 3.3.3, for the smooth solution of* (3.3.5) $\vec{Z}_\varepsilon(x,t)$, *there holds*

$$\sup\limits_{0\leq t\leq T} \|\vec{Z}_{\varepsilon x^{k-2s}t^s}(\cdot,t)| \leq K_6, \tag{3.3.32}$$

where K_6 is independent of ε, D, k, s are any integers such that $k - 2s > 0$, $s \geq 0$, $k \geq 3$.

3. *Smooth solution to the equations of ferromagnetic spin chain*

From the above estimates uniform in ε and choosing subsequence by the standard method, sending $\varepsilon \to 0$ in (3.3.5), we obtain the global smooth solution to problem (3.3.1) and (3.3.2). That is we have

Theorem 3.3.4 *Let* $\vec{Z}_0(x) \in H^k(\Omega)$, $k > 4$ *and* $|\vec{Z}_0(x)| \equiv 1$, $\vec{Z}_0(x - D) = \vec{Z}_0(x + D)$, $D > 0$. *Then problem* (3.3.1) *and* (3.3.2) *admits a global smooth solution* $\vec{Z}(x, t) \in$
$$\bigcap_{s=0}^{[k/2]} W_\infty^s(0, T; H^{k-2s}(\Omega)).$$

Similarly, sending $\varepsilon \to 0$ and $D \to \infty$, we know that the initial value problem (3.3.1) and (3.3.3) admits a global smooth solution.

By the standard method, one can prove the uniqueness of such smooth solution.

3.4 Smooth Solution for the 1D Inhomogeneous Heisenberg Chain Equations

3.4.1 Inhomogeneous Heisenberg Chain Equations

In this section, we are concerned with the existence and uniqueness of smooth solution to the one-dimensional periodic initial value problem of the inhomogeneous non-automorphic Landau–Lifshitz equation

$$\vec{Z}_t = f(x, t)\vec{Z} \times \vec{Z}_{xx} + \frac{\partial f(x,t)}{\partial x}\vec{Z} \times \vec{Z}_x, \tag{3.4.1}$$

$$\vec{Z}(x, 0) = \vec{Z}_0(x), \quad \vec{Z}(x + D, t) = \vec{Z}(x - D, t), \quad |\vec{Z}_0(x)| \equiv 1, \quad x \in R^1, \tag{3.4.2}$$

where $D > 0$ is a constant, $f(x, t)$ and $\vec{Z}_0(x)$ are smooth functions and $f(x, t) \geq f_0 > 0$ for some constant f_0, $\vec{Z}(x, t) = (Z_1(x, t), Z_2(x, t), Z_3(x, t))$. We assume that $f(x, t)$ and $\vec{Z}_0(x)$ are periodic functions with period 2D.

Equation (3.4.1) is related to the generalized model of inhomogeneous ferromagnetisms and the simplified compressible ferromagnetisms which reads as

$$\vec{Z}_t = (G(\vec{Z}_x)\vec{Z} \times \vec{Z}_x)_x, \tag{3.4.3}$$

in which $G(\xi) = A + B|\xi|^2$. Equation (3.4.3) with $B = 0$, $A = f(x)$ corresponds to the inhomogeneous Heisenberg chain equation; $B = 0$, $A = f(x, t)$ is just equation (3.4.1).

In order to prove the existence of smooth solution to problem (3.4.1)–(3.4.2), we use the viscosity vanishing method as in the above section.

First, by difference method, we establish the existence and uniqueness of smooth solution to the following problem

$$\vec{Z}_t = \varepsilon\vec{Z}_{xx} + \varepsilon|\vec{Z}_x|^2\vec{Z} + f(x, t)\vec{Z} \times \vec{Z}_{xx} + \frac{\partial f(x,t)}{\partial x}\vec{Z} \times \vec{Z}_x, \quad \varepsilon > 0, \tag{3.4.4}$$

$$\vec{Z}(x, 0) = \vec{Z}_0(x), \quad \vec{Z}(x + D, t) = \vec{Z}(x - D, t), \quad |\vec{Z}_0(x)| \equiv 1, \quad x \in R^1. \tag{3.4.5}$$

Then we prove some *a priori* estimates for the solution of (3.4.4)–(3.4.5) uniform in ε by constructing some new energy laws, then we send ε to zero to obtain the existence of smooth solution to problem (3.4.1)–(3.4.2). In this procedure, as in the above section, we use the orthogonal base $\{\vec{Z}, \vec{Z}_x, \vec{Z} \times \vec{Z}_x\}$ in our proof.

In the classical sense, equation (3.4.4) is equivalent to

$$\vec{Z}_t = -\varepsilon \vec{Z} \times (\vec{Z} \times \vec{Z}_{xx}) + f(x,t)\vec{Z} \times \vec{Z}_{xx} + \frac{\partial f(x,t)}{\partial x}\vec{Z} \times \vec{Z}_x, \qquad (3.4.6)$$

since $\vec{A} \times (\vec{B} \times \vec{C}) = (\vec{A} \cdot \vec{C})\vec{B} - (\vec{A} \cdot \vec{B})\vec{C}$ and $|\vec{Z}(x,t)| \equiv 1$.

3.4.2　$\varepsilon > 0$: Local Smooth Solution

Lemma 3.4.1 *Let q, r be real numbers and j, m be integers such that $1 \leq q, r \leq \infty$, $0 \leq j < m$. If $u \in W^{m,r}(\Omega) \cap L^q(\Omega)$, then*

$$\|D^j u\|_p \leq C\|u\|_q^{1-\alpha}\|D^m u\|_r^\alpha,$$

where $\|\cdot\|_p = \|\cdot\|_{L^p(\Omega)}$, $p \geq 1$, $\frac{j}{m} \leq \alpha \leq 1$ and

$$\frac{1}{p} - j = \frac{1-\alpha}{q} + \alpha\left(\frac{1}{r} - m\right), \quad \Omega \subset R^1.$$

Lemma 3.4.2 *Let p be real number and j, m be integers such that $2 \leq p \leq \infty$, $0 \leq j < m$. Then*

$$\|\delta^j u_h\|_p \leq C\|u_h\|_2^{1-\alpha}\left(\|\delta^m u_h\|_2 + \frac{\|u_h\|_2}{(2D)^m}\right)^\alpha,$$

where $u_h = \{u_j = u(x_j) \mid j = 0, 1, 2, \ldots, J\}$, $x_j = jh$, $h = 2D/J$, $\alpha = \frac{1}{m}(j + \frac{1}{2} - \frac{1}{p})$,

$$\|\delta^k u_h\|_p = \left(\sum_{i=0}^{J-k}\left|\frac{\Delta_+^k u_i}{h^k}\right|^p h\right)^{\frac{1}{p}}.$$

Lemma 3.4.3 *Let $u_h = \{u_j \mid j = 0, \pm 1, \pm 2, \ldots, \pm J, \ldots\}$, $v_h = \{v_j \mid j = 0, \pm 1, \pm 2, \ldots, \pm J, \ldots\}$ such that $u_{j+J} = u_j$, $v_{j+J} = v_j$. We have*

$$\text{(i)} \qquad \sum_{j=0}^{J-1} u_j \Delta_+ v_j = -\sum_{j=1}^{J} v_j \Delta_- u_j,$$

$$\text{(ii)} \qquad \sum_{j=1}^{J} u_j \Delta_+ \Delta_- v_j = -\sum_{j=0}^{J-1} \Delta_+ u_j \Delta_+ v_j,$$

$$\text{(iii)} \qquad \Delta_+(u_j v_j) = u_{j+1}\Delta_+ v_j + v_j \Delta_+ u_j,$$

where Δ_+, Δ_- denote the forward and backward difference respectively.

We use the differential-difference method to prove the local existence of smooth solution of (3.4.4)–(3.4.5). We establish the following difference-differential equation:

$$\frac{d\vec{Z}_j}{dt} = \frac{\Delta_+\Delta_-\vec{Z}_j}{h^2} + \left|\frac{\Delta_+\vec{Z}_j}{h}\right|^2 \vec{Z}_j + f_j\vec{Z}_j \times \frac{\Delta_+\Delta_-\vec{Z}_j}{h^2} + \frac{\Delta_+ f_j}{h}\vec{Z}_j \times \frac{\Delta_+\vec{Z}_j}{h}, \quad (3.4.7)$$

$$\vec{Z}_j|_{t=0} = \vec{Z}_{0j} = \vec{Z}_0(jh), \quad (3.4.8)$$

$$\vec{Z}_{j+J} = \vec{Z}_j, \quad (3.4.9)$$

where $j = 0, \pm 1, \ldots, \pm J, \ldots$, $h = \frac{2D}{J}$, $J > 0$.

It is clear that the initial value problem for ordinary differential equation (3.4.7)–(3.4.9) admits a local smooth solution. For such solution, we shall give some estimates uniformly in h and then get a local smooth solution to problem (3.4.4)–(3.4.5).

In this section we always denote a smooth solution of (3.4.7)–(3.4.9) by $\vec{Z}_j$.

Lemma 3.4.4 *If $\vec{Z}_0(x) \in H^1(\Omega)$, $\frac{\partial f}{\partial x} \in L^\infty(Q)$ then there are constants $T_0 > 0$, $C > 0$ independent of h such that*

$$\sup_{0 \leq t \leq T_0} \|\vec{Z}_h(t)\|_2 \leq C, \quad (3.4.10)$$

$$\sup_{0 \leq t \leq T_0} \|\delta\vec{Z}_h(t)\|_2 \leq C. \quad (3.4.11)$$

Proof. Multiplying (3.4.7) by $\vec{Z}_j h$ and summing from $j = 1$ to J, we have

$$\frac{1}{2}\frac{d}{dt}\sum_{j=1}^{J}|\vec{Z}_j|^2 h = \sum_{j=1}^{J}\vec{Z}_j \cdot \frac{\Delta_+\Delta_-\vec{Z}_j}{h^2}h + \sum_{j=1}^{J}\left|\frac{\Delta_+\vec{Z}_j}{h}\right|^2|\vec{Z}_j|^2 h$$

$$= -\sum_{j=0}^{J-1}\left|\frac{\Delta_+\vec{Z}_j}{h}\right|^2 h + \sum_{j=1}^{J}\left|\frac{\Delta_+\vec{Z}_j}{h^2}\right|^2|\vec{Z}_j|^2 h.$$

It follows from Lemma 3.4.2 that

$$\|\vec{Z}_h\|_\infty \leq C\|\vec{Z}_h\|_2^{\frac{1}{2}}\left(\|\delta\vec{Z}_h\|_2 + \frac{1}{2D}\|\vec{Z}_h\|_2\right)^{\frac{1}{2}}.$$

Therefore we have

$$\frac{d}{dt}\|\vec{Z}_h\|_2^2 + \|\delta\vec{Z}_h\|_2^2 \leq C(\|\vec{Z}_h\|_2^4 + \|\delta\vec{Z}_h\|_2^4). \quad (3.4.12)$$

Moreover, multiplying (3.4.7) by $\frac{\Delta_+\Delta_-\vec{Z}_j}{h}$ and summing from $j = 1$ to J, we get

$$\sum_{j=1}^{J}\frac{\Delta_+\Delta_-\vec{Z}_j}{h}\cdot\vec{Z}_{jt} = \sum_{j=1}^{J}\left|\frac{\Delta_+\Delta_-\vec{Z}_j}{h^2}\right|^2 h + \sum_{j=1}^{J}\frac{\Delta_+\Delta_-\vec{Z}_j}{h}\cdot\left(\left|\frac{\Delta_+\vec{Z}_j}{h}\right|^2\vec{Z}_j\right)$$

$$+ \sum_{j=1}^{J}\frac{\Delta_+ f_j}{h}\left(\vec{Z}_j \times \frac{\Delta_+\vec{Z}_j}{h}\right)\cdot\frac{\Delta_+\Delta_-\vec{Z}_j}{h^2}h.$$

Therefore, one gets

$$\frac{1}{2}\frac{d}{dt}\sum_{j=0}^{J-1}\left|\frac{\Delta_+\vec{Z}_j}{h}\right|^2 h + \sum_{j=1}^{J}\left|\frac{\Delta_+\Delta_-\vec{Z}_j}{h^2}\right|^2 h$$

$$\leq \frac{1}{4}\sum_{j=1}^{J}\left|\frac{\Delta_+\Delta_-\vec{Z}_j}{h^2}\right|^2 h + C\max_{1\leq j\leq J}|\vec{Z}_j|^2\sum_{j=1}^{J}\left|\frac{\Delta_+\vec{Z}_j}{h}\right|^4 h + C\max_{1\leq j\leq J}|\vec{Z}_j|^2\sum_{j=1}^{J}\left|\frac{\Delta_+\vec{Z}_j}{h}\right|^2 h.$$

Applying Lemma 3.4.2, we have

$$\|\vec{Z}_h\|_\infty \leq C\|\vec{Z}_h\|_2^{\frac{3}{4}}(\|\delta^2\vec{Z}_h\|_2 + \|\vec{Z}_h\|)^{\frac{1}{4}},$$
$$\|\delta\vec{Z}_h\|_4 \leq C\|\delta\vec{Z}_h\|^{\frac{3}{4}}(\|\delta^2\vec{Z}_h\|_2 + \|\delta\vec{Z}_h\|_2)^{\frac{1}{4}}.$$

Then it follows from the Young's inequality that

$$\frac{d}{dt}\left(\|\delta\vec{Z}_h\|_2^2\right) + \|\delta^2\vec{Z}_h\|_2^2 \leq C\left(1 + \|\vec{Z}_h\|_2^{18} + \|\delta\vec{Z}_h\|_2^{18}\right). \tag{3.4.13}$$

Hence, putting (3.4.12) and (3.4.13) together, we have

$$\frac{d}{dt}\left(\|\delta\vec{Z}_h\|_2^2 + \|\delta\vec{Z}_h\|_2^2\right) \leq C\left(1 + \|\vec{Z}_h\|_2^2 + \|\delta\vec{Z}_h\|_2^2\right)^9.$$

This inequality, combined with Gronwall's inequality, implies that there are constants $T_0, C > 0$, independent of h such that

$$\|\vec{Z}_h(t)\|_2 + \|\delta\vec{Z}_h(t)\|_2 \leq C, \quad \forall\, t \in [0, T_0],$$
$$\int_0^{T_0}\|\delta^2\vec{Z}_h(t)\|_2^2 dt \leq C.$$

Lemma 3.4.4 is proved.

Corollary 3.4.1 *Under the conditions in Lemma 3.4.4 and $f(x,t) \in L^\infty(Q)$, we have, for some constant C independent of h,*

$$\sup_{0\leq t\leq T_0; 1\leq j\leq J}|\vec{Z}_j| \leq C. \tag{3.4.14}$$

Lemma 3.4.5 *If $\vec{Z}_0(x) \in H^2(\Omega)$, $f(x,t)$, $\frac{\partial f}{\partial x} \in L^\infty(Q)$, $\frac{\partial^2 f}{\partial x^2} \in L^\infty(Q)$, then there are constants $T_0 > 0$, $C > 0$ independent of h such that*

$$\sup_{0\leq t\leq T_0}\|\delta^2\vec{Z}_h(t)\|_2 \leq C, \tag{3.4.15}$$

$$\int_0^{T_0}\|\delta^3\vec{Z}_h(t)\|_2 \leq C. \tag{3.4.16}$$

Proof. It follows from (3.4.7) that

$$
\frac{d}{dt}\Delta_+\vec{Z}_j = \frac{\Delta_+^2\Delta_-\vec{Z}_j}{h^2} + \Delta_+\left(\left|\frac{\Delta_+\vec{Z}_j}{h}\right|^2\vec{Z}_j\right) + f_{j+1}\Delta_+\left(\vec{Z}_j\times\frac{\Delta_+\Delta_-\vec{Z}_j}{h^2}\right)
$$
$$
+ \Delta_+f_j\left(\vec{Z}_j\times\frac{\Delta_+\Delta_-\vec{Z}_j}{h^2}\right) + \frac{\Delta_+f_{j+1}}{h}\Delta_+\left(\vec{Z}_j\times\frac{\Delta_+\vec{Z}_j}{h}\right)
$$
$$
+ \frac{\Delta_+^2 f_j}{h}\left(\vec{Z}_j\times\frac{\Delta_+\vec{Z}_j}{h}\right).
$$

Multiplying this equality by $\frac{\Delta_+^2\Delta_-\vec{Z}_j}{h^3}$, summing it from $j=1$ to J and noting that

$$
\sum_{j=1}^{J}\frac{\Delta_+^2\Delta_-\vec{Z}_j}{h^3}\cdot\Delta_+\vec{Z}_{jt} = -\frac{1}{2}\frac{d}{dt}\sum_{j=0}^{J-1}\left|\frac{\Delta_+^2\vec{Z}_j}{h^2}\right|^2 h,
$$

$$
\sum_{j=1}^{J}\frac{\Delta_+^2\Delta_-\vec{Z}_j}{h^3}\cdot\Delta_+f_j\left(\vec{Z}_j\times\frac{\Delta_+\Delta_-\vec{Z}_j}{h^2}\right)
$$
$$
\leq \frac{1}{10}\sum_{j=1}^{J}\left|\frac{\Delta_+^2\Delta_-\vec{Z}_j}{h^3}\right|^2 h + C\max_{1\leq j\leq J}|\vec{Z}_j|^2\sum_{j=1}^{J}\left|\frac{\Delta_+\Delta_-\vec{Z}_j}{h^2}\right|^2 h
$$

and

$$
\sum_{j=1}^{J}\frac{\Delta_+^2\Delta_-\vec{Z}_j}{h^3}\cdot f_{j+1}\Delta_+\left(\vec{Z}_j\times\frac{\Delta_+\Delta_-\vec{Z}_j}{h^2}\right)
$$
$$
= \sum_{j=1}^{J}\frac{\Delta_+^2\Delta_-\vec{Z}_j}{h^3}\cdot f_{j+1}\left(\Delta_+\vec{Z}_j\times\frac{\Delta_+\Delta_-\vec{Z}_j}{h^2}\right)
$$
$$
\leq \frac{1}{10}\sum_{j=1}^{J}\left|\frac{\Delta_+^2\Delta_-\vec{Z}_j}{h^3}\right|^2 h + \max_{1\leq j\leq J}\left|\frac{\Delta_+\vec{Z}_j}{h}\right|^2\sum_{j=1}^{J}\left|\frac{\Delta_+\Delta_-\vec{Z}_j}{h^2}\right|^2 h,
$$

as well as

$$
\|\delta\vec{Z}_h\|_\infty \leq C\|\delta\vec{Z}_h\|_2^{\frac{1}{2}}(\|\delta^2\vec{Z}_h\|_2 + \|\delta\vec{Z}_h\|_2)^{\frac{1}{2}},
$$
$$
\|\delta\vec{Z}_h\|_6 \leq C\|\delta\vec{Z}_h\|_2^{\frac{2}{3}}(\|\delta^2\vec{Z}_h\|_2 + \|\delta\vec{Z}_h\|_2)^{\frac{1}{3}}
$$

we finally get

$$
\frac{d}{dt}\sum_{j=0}^{J-1}\left|\frac{\Delta_+^2\vec{Z}_j}{h^2}\right|^2 h + \sum_{j=0}^{J-1}\left|\frac{\Delta_+^2\Delta_-\vec{Z}_j}{h^3}\right|^2 h \leq C + C\left(\sum_{j=0}^{J-1}\left|\frac{\Delta_+^2\vec{Z}_j}{h^2}\right|^2 h\right)^3.
$$

Then, Lemma 3.4.5 follows from the Gronwall's inequality.

Corollary 3.4.2 *Under the conditions in Lemma 3.4.5, we have, for some C independent of h,*

$$\sup_{0\leq t\leq T_0; 1\leq j\leq J}\left|\frac{\Delta_+\vec{Z}_j(t)}{h}\right|\leq C, \tag{3.4.17}$$

$$\int_0^{T_0}\|\delta\vec{Z}_{ht}(t)\|_2^2 dt \leq C. \tag{3.4.18}$$

Lemma 3.4.6 *If $\vec{Z}_0(x)\in H^3(\Omega)$, $f(x,t)$, $\frac{\partial^j f}{\partial x^j}\in L^\infty(Q)$, $j=1,2,3$, then there are constants $T_0>0$, $C>0$ independent of h such that*

$$\sup_{0\leq t\leq T_0}\|\delta^3\vec{Z}_h(t)\|_2\leq C, \tag{3.4.19}$$

$$\int_0^{T_0}\|\delta^4\vec{Z}_h(t)\|_2\leq C. \tag{3.4.20}$$

Proof. From (3.4.7) we have

$$\sum_{j=1}^{J}\frac{\Delta_+^3\Delta_-\vec{Z}_j}{h^4}\cdot\frac{\Delta_+^2\vec{Z}_{jt}}{h^2}h$$

$$=\sum_{j=1}^{J}\left|\frac{\Delta_+^3\Delta_-\vec{Z}_j}{h^4}\right|^2 h+\sum_{j=1}^{J}\left|\frac{\Delta_+\vec{Z}_{j+2}}{h}\right|^2\frac{\Delta_+\vec{Z}_j}{h^2}\cdot\frac{\Delta_+^3\Delta_-\vec{Z}_j}{h^4}h$$

$$+\sum_{j=1}^{J}\left(\frac{\Delta_+^2\vec{Z}_{j+1}}{h^2}\cdot\frac{\Delta_+\vec{Z}_{j+2}}{h}\right)\frac{\Delta_+\vec{Z}_j}{h}\cdot\frac{\Delta_+^3\Delta_-\vec{Z}_j}{h^4}h$$

$$+\sum_{j=1}^{J}\left(\frac{\Delta_+\vec{Z}_{j+2}}{h}\cdot\frac{\Delta_+^2\vec{Z}_{j+1}}{h^2}+\frac{\Delta_+^2\vec{Z}_{j+1}}{h^2}\cdot\frac{\Delta_+\vec{Z}_{j+1}}{h}\right)\frac{\Delta_+\vec{Z}_j}{h}\cdot\frac{\Delta_+^3\Delta_-\vec{Z}_j}{h^4}h$$

$$+\sum_{j=1}^{J}\left(\frac{\Delta_+\vec{Z}_{j+2}}{h}\cdot\frac{\Delta_+^3\vec{Z}_j}{h^3}+\frac{\Delta_+^2\vec{Z}_{j+1}}{h^2}\cdot\frac{\Delta_+^2\vec{Z}_j}{h^2}\right)\vec{Z}_j\cdot\frac{\Delta_+^3\Delta_-\vec{Z}_j}{h^4}h$$

$$+\sum_{j=1}^{J}\left(\frac{\Delta_+\vec{Z}_{j+1}}{h}\cdot\frac{\Delta_+^2\vec{Z}_{j+1}}{h^2}\right)\left(\frac{\Delta_+\vec{Z}_j}{h}\cdot\frac{\Delta_+^3\Delta_-\vec{Z}_j}{h^3}\right)$$

$$+\sum_{j=1}^{J}\left(\frac{\Delta_+\vec{Z}_{j+1}}{h}\cdot\frac{\Delta_+^3\vec{Z}_j}{h^3}+\left|\frac{\Delta_+^2\vec{Z}_j}{h^2}\right|^2\right)\left(\vec{Z}_j\cdot\frac{\Delta_+^3\Delta_-\vec{Z}_j}{h^4}\right)h$$

$$+2\sum_{j=1}^{J}f_{j+2}\left(\frac{\Delta_+\vec{Z}_{j+1}}{h}\times\frac{\Delta_+^2\Delta_-\vec{Z}_j}{h^3}\right)\cdot\frac{\Delta_+^3\Delta_-\vec{Z}_j}{h^4}h$$

$$+\sum_{j=1}^{J}f_{j+2}\left(\frac{\Delta_+^2\vec{Z}_j}{h^2}\times\frac{\Delta_+\Delta_-\vec{Z}_j}{h^2}\right)\cdot\frac{\Delta_+^3\Delta_-\vec{Z}_j}{h^4}h$$

$$+2\sum_{j=1}^{J}\frac{\Delta_+f_{j+1}}{h}\left(\frac{\Delta_+\vec{Z}_j}{h}\times\frac{\Delta_+\Delta_-\vec{Z}_j}{h^3}\right)\cdot\frac{\Delta_+^3\Delta_-\vec{Z}_j}{h^4}h$$

$$+ 2 \sum_{j=1}^{J} \frac{\Delta_+ f_{j+1}}{h} \left(\vec{Z}_{j+1} \times \frac{\Delta_+^2 \Delta_- \vec{Z}_j}{h^3} \right) \cdot \frac{\Delta_+^3 \Delta_- \vec{Z}_j}{h^4} h$$

$$+ \sum_{j=1}^{J} \frac{\Delta_+^2 f_j}{h^2} \left(\vec{Z}_j \times \frac{\Delta_+ \Delta_- \vec{Z}_j}{h^2} \right) \cdot \frac{\Delta_+^3 \Delta_- \vec{Z}_j}{h^4} h$$

$$+ \sum_{j=1}^{J} \frac{\Delta_+ f_{j+2}}{h} \left(\vec{Z}_{j+2} \times \frac{\Delta_+^3 \vec{Z}_j}{h^3} \right) \cdot \frac{\Delta_+^3 \Delta_- \vec{Z}_j}{h^4} h$$

$$+ \sum_{j=1}^{J} \frac{\Delta_+ f_{j+2}}{h} \left(\frac{\Delta_+ \vec{Z}_{j+1}}{h} \times \frac{\Delta_+^2 \vec{Z}_j}{h^2} \right) \cdot \frac{\Delta_+^3 \Delta_- \vec{Z}_j}{h^4} h$$

$$+ \sum_{j=1}^{J} \frac{\Delta_+^2 f_{j+1}}{h^2} \left(\vec{Z}_j \times \frac{\Delta_+ \vec{Z}_j}{h} \right) \cdot \frac{\Delta_+^3 \Delta_- \vec{Z}_j}{h^4} h$$

$$+ \sum_{j=1}^{J} \frac{\Delta_+^2 f_{j+1}}{h^2} \left(\vec{Z}_{j+1} \times \frac{\Delta_+^2 \vec{Z}_{j+1}}{h^2} \right) \cdot \frac{\Delta_+^3 \Delta_- \vec{Z}_j}{h^4} h$$

$$+ \sum_{j=1}^{J} \frac{\Delta_+^3 f_j}{h^3} \left(\vec{Z}_j \times \frac{\Delta_+ \vec{Z}_j}{h} \right) \cdot \frac{\Delta_+^3 \Delta_- \vec{Z}_j}{h^4} h.$$

From Lemma 3.4.2, we have

$$\| \delta^2 \vec{Z}_h \|_4 \leq C \| \vec{Z}_h \|_2^{\frac{1}{4}} \left(\| \delta^3 \vec{Z}_h \|_2 + \| \vec{Z}_h \|_2 \right)^{\frac{3}{4}}.$$

By Lemmas 3.4.3–3.4.5, one gets

$$\sum_{j=1}^{J} \frac{\Delta_+^3 \Delta_- \vec{Z}_j}{h^4} \cdot \frac{\Delta_+^2 \vec{Z}_{jt}}{h^2} h = -\frac{1}{2} \frac{d}{dt} \sum_{j=0}^{J-1} \left| \frac{\Delta_+^3 \vec{Z}_j}{h^3} \right|^2 h,$$

$$\sum_{j=1}^{J} \left(\frac{\Delta_+^2 \vec{Z}_{j+1}}{h^2} \cdot \frac{\Delta_+ \vec{Z}_{j+2}}{h} \right) \frac{\Delta_+ \vec{Z}_j}{h} \cdot \frac{\Delta_+^3 \Delta_- \vec{Z}_j}{h^4} h$$

$$\leq \frac{1}{32} \sum_{j=1}^{J} \left| \frac{\Delta_+^3 \Delta_- \vec{Z}_j}{h^4} \right|^2 h + C \sum_{j=1}^{J} \left| \frac{\Delta_+^3 \vec{Z}_j}{h^3} \right|^2 h + C.$$

The others are given in the same way.

Therefore we have

$$\frac{1}{2} \frac{d}{dt} \sum_{j=0}^{J-1} \left| \frac{\Delta_+^3 \vec{Z}_j}{h^3} \right|^2 h + \frac{1}{2} \sum_{j=1}^{J} \left| \frac{\Delta_+^3 \Delta_- \vec{Z}_j}{h^4} \right|^2 h \leq C \sum_{j=0}^{J-1} \left| \frac{\Delta_+^3 \vec{Z}_j}{h^3} \right|^2 h + C.$$

Lemma 3.4.6 follows from the Gronwall's inequality.

By induction method we have

Lemma 3.4.7 *If $\vec{Z}_0(x) \in H^k(\Omega, S^2)$, $f(x,t) \in L^\infty(Q)$, $\frac{\partial^j f}{\partial x^j} \in L^\infty(Q)$, $j = 1, 2, \ldots, k$, then there are constants $T_0 > 0$, $C > 0$ independent of h such that*

$$\sup_{0 \leq t \leq T_0} \|\delta^k \vec{Z}_h\|_2 \leq C, \tag{3.4.21}$$

$$\sup_{0 \leq t \leq T_0} \|\delta^{k-2} \vec{Z}_{ht}\|_2 \leq C, \quad k \geq 2, \tag{3.4.22}$$

$$\sup_{0 \leq t \leq T_0} \|\delta^{k-4} \vec{Z}_{htt}\|_2 \leq C, \quad k \geq 4. \tag{3.4.23}$$

From Lemma 3.4.7 and the *a priori* estimates for solutions to the differential-difference equation (3.4.7)–(3.4.9), we conclude that there exists a constant $T_0 > 0$ such that problem (3.4.4)–(3.4.5) admits a smooth solution in $\Omega \times [0, T_0]$. This result is stated as follows.

Theorem 3.4.1 *Let $\epsilon > 0$, $\vec{Z}_0(x) \in H^k(\Omega, S^2)$, $f(x,t) \in L^\infty(Q)$, $\frac{\partial^j f}{\partial x^j} \in L^\infty(Q)$, $j = 1, 2, \ldots, k$, $f(x,t) \geq f_0 > 0$. Then (3.4.4)–(3.4.5) admits a local smooth solution $\vec{Z}(x,t)$ in $[0, T_0]$ with T_0 depending on k:*

$$\vec{Z}(x,t) \in \left(\bigcap_{s=0}^{[\frac{k}{2}]} W_\infty^s(0, T_0; H^{k-2s}(\Omega)) \right) \bigcap \left(\bigcap_{s=0}^{[\frac{k+1}{2}]} H^s(0, T_0; H^{k+1-2s}(\Omega)) \right).$$

3.4.3 $\varepsilon > 0$: Global Solution and Uniform Estimates

In the above subsection, we have obtained a local smooth solution for (3.4.4)–(3.4.5) when $\varepsilon > 0$ is fixed. In this subsection, we intend to prove the global existence of smooth solution to problem (3.4.4)–(3.4.5) for fixed $\varepsilon > 0$ by deriving the global (in time) estimates. To meet the need of sending ε to zero, these estimates have to be equal both global in time and uniform in ε.

Lemma 3.4.8 *If $\vec{Z}_0(x) \in H^1(\Omega, S^2)$ and $\vec{Z}(x,t)$ is a global smooth solution of problem (3.4.4)–(3.4.5), then we have*

$$|\vec{Z}(x,t)| = 1, \quad \forall \, (x,t) \in R^1 \times [0, +\infty). \tag{3.4.24}$$

Lemma 3.4.9 *Under the same conditions of Lemma 3.4.8 and $\frac{\partial f}{\partial t}, \frac{\partial f}{\partial x} \in L^\infty(Q)$, we have, for any given $T > 0$, there exists $C > 0$ independent of ε and D,*

$$\sup_{0 \leq t \leq T} \|\vec{Z}_x(\cdot, t)\|_2 \leq C. \tag{3.4.25}$$

Proof. From (3.4.6), we have

$$\vec{Z}_{xt} = -\varepsilon \left(\vec{Z} \times \left(\vec{Z} \times \vec{Z}_{xx} \right) \right)_x + \frac{\partial^2 f}{\partial x^2} \vec{Z} \times \vec{Z}_x + 2\frac{\partial f}{\partial x} \vec{Z} \times \vec{Z}_{xx} + f \vec{Z}_x \times \vec{Z}_{xx} + f \vec{Z} \times \vec{Z}_{xxx}.$$

By $(\vec{A} \times (\vec{B} \times \vec{C})) \cdot \vec{B} = (\vec{A} \times \vec{B}) \cdot (\vec{C} \times \vec{B})$, we get

$$
\int_\Omega f \vec{Z}_x \cdot \vec{Z}_{xt} = \varepsilon \int_\Omega \frac{\partial f}{\partial x} \left(\vec{Z} \times (\vec{Z} \times \vec{Z}_{xx}) \right) \cdot \vec{Z}_x - \varepsilon \int_\Omega f \left(\vec{Z} \times (\vec{Z}_{xx} \times \vec{Z}) \right) \cdot \vec{Z}_{xx}
$$
$$
+ 2 \int_\Omega f \frac{\partial f}{\partial x} \left(\vec{Z} \times \vec{Z}_{xx} \right) \cdot \vec{Z}_x + \int_\Omega f^2 \left(\vec{Z} \times \vec{Z}_{xx} \right)_x \cdot \vec{Z}_x
$$
$$
= \varepsilon \int_\Omega \frac{\partial f}{\partial x} \left(\vec{Z} \times (\vec{Z} \times \vec{Z}_{xx}) \right) \cdot \vec{Z}_x - \varepsilon \int_\Omega f \left| \vec{Z} \times \vec{Z}_{xx} \right|^2 .
$$

Therefore, one gets

$$
\frac{1}{2} \frac{d}{dt} \int_\Omega f \left| \vec{Z}_x \right|^2 + \varepsilon \int_\Omega f \left| \vec{Z} \times \vec{Z}_{xx} \right|^2 = \frac{1}{2} \int_\Omega \frac{\partial f}{\partial t} \left| \vec{Z}_x \right|^2 + \varepsilon \int_\Omega \frac{\partial f}{\partial x} \left(\vec{Z} \times (\vec{Z} \times \vec{Z}_{xx}) \right) \cdot \vec{Z}_x
$$
$$
\leq C \int_\Omega f \left| \vec{Z}_x \right|^2 dx + \frac{\varepsilon}{2} \int_\Omega f \left| \vec{Z} \times \vec{Z}_{xx} \right|^2 .
$$

It is easy to see that

$$
\frac{1}{2} \frac{d}{dt} \int_\Omega f \left| \vec{Z}_x \right|^2 dx \leq C \int_\Omega f \left| \vec{Z}_x \right|^2 dx.
$$

Then lemma 3.4.9 follows. $\qquad\square$

Lemma 3.4.10 *Let* $\vec{Z}_0(x) \in H^2(\Omega, S^2)$, $f(x,t)$, $\frac{\partial f}{\partial t}$, $\frac{\partial^j f}{\partial x^j} \in L^\infty(Q)$, $j = 1,2,3$, *then for any given* $T > 0$, *there are constants* $C > 0$ *independent of* ε, D, *such that*

$$
\sup_{0 \leq t \leq T} \| \vec{Z}_{xx}(\cdot, t) \|_2 \leq C, \tag{3.4.26}
$$

$$
\sup_{0 \leq t \leq T} \| \vec{Z}_t(\cdot, t) \|_2 \leq C. \tag{3.4.27}
$$

Proof. In the proof, we shall use the following identities which follow from the fact $|\vec{Z}(x,t)| = 1$:

$$
\vec{Z} \cdot \vec{Z}_t = 0, \quad \vec{Z} \cdot \vec{Z}_x = 0, \quad \vec{Z} \cdot \vec{Z}_{xx} = -|\vec{Z}_x|^2, \quad \vec{Z} \cdot \vec{Z}_{xxx} = -3\vec{Z}_x \cdot \vec{Z}_{xx}. \tag{3.4.28}
$$

It follows from (3.4.4) that

$$
\vec{Z}_{xxt} = \varepsilon \vec{Z}_{xxxx} + \varepsilon \left| \vec{Z}_x \right|^2 \vec{Z}_{xx} + 2\varepsilon \left(\vec{Z}_x \cdot \vec{Z}_{xx} \right) \vec{Z}_x
$$
$$
+ 2\varepsilon \left[\left(\vec{Z}_x \cdot \vec{Z}_{xx} \right) \vec{Z} \right]_x + \frac{\partial^3 f}{\partial x^3} \vec{Z} \times \vec{Z}_x + 3\frac{\partial^2 f}{\partial x^2} \vec{Z} \times \vec{Z}_{xx} + 3\frac{\partial f}{\partial x} \vec{Z}_x \times \vec{Z}_{xx}
$$
$$
+ 3\frac{\partial f}{\partial x} \vec{Z} \times \vec{Z}_{xxx} + 2f \vec{Z}_x \times \vec{Z}_{xxx} + f \vec{Z} \times \vec{Z}_{xxxx}. \tag{3.4.29}
$$

From (3.4.29) we get

$$\frac{1}{2}\frac{d}{dt}\int_\Omega f^2\left|\vec{Z}_{xx}\right|^2 dx$$

$$= \int_\Omega f\frac{\partial f}{\partial t}\left|\vec{Z}_{xx}\right|^2 dx + \int_\Omega f^2 \vec{Z}_{xx}\cdot\vec{Z}_{xxt}dx$$

$$= -2\varepsilon\int_\Omega f\frac{\partial f}{\partial x}\vec{Z}_{xx}\cdot\vec{Z}_{xxx}dx - \varepsilon\int_\Omega f^2\left|\vec{Z}_{xxx}\right|^2 dx + \varepsilon\int_\Omega f^2\left|\vec{Z}_{x}\right|^2\left|\vec{Z}_{xx}\right|^2 dx$$

$$+ 8\varepsilon\int_\Omega f^2\left|\vec{Z}_{x}\cdot\vec{Z}_{xx}\right|^2 dx - 4\varepsilon\int_\Omega f\frac{\partial f}{\partial x}\left(\vec{Z}_{x}\cdot\vec{Z}_{xx}\right)\left(\vec{Z}\cdot\vec{Z}_{xx}\right)dx + \int_\Omega f\frac{\partial f}{\partial t}\left|\vec{Z}_{xx}\right|^2 dx$$

$$+ \int_\Omega f^2\frac{\partial^3 f}{\partial x^3}\left(\vec{Z}\times\vec{Z}_{x}\right)\cdot\vec{Z}_{xx}dx - \int_\Omega f^3\left(\vec{Z}_{x}\times\vec{Z}_{xx}\right)\cdot\vec{Z}_{xxx}dx. \qquad (3.4.30)$$

Note that, if $|\vec{Z}_x|\neq 0$, then the vectors $\vec{Z}$, $\vec{Z}_x$, $\vec{Z}\times\vec{Z}_x$ form an orthogonal basis of R^3. Let

$$\vec{Z}_{xx} = \alpha\vec{Z} + \beta\vec{Z}_x + \gamma\vec{Z}\times\vec{Z}_x,$$

and it is easy to see that

$$\alpha = -|\vec{Z}_x|^2, \quad \beta = \frac{\vec{Z}_x\cdot\vec{Z}_{xx}}{|\vec{Z}_x|^2}, \quad \gamma = \frac{(\vec{Z}\times\vec{Z}_x)\cdot\vec{Z}_{xx}}{|\vec{Z}_x|^2}.$$

Therefore we have

$$-\int_\Omega f^3\left(\vec{Z}_x\times\vec{Z}_{xx}\right)\cdot\vec{Z}_{xxx}$$

$$= -\int_\Omega f^3\left\{\vec{Z}_x\times\left[-|\vec{Z}_x|^2\vec{Z} + \frac{(\vec{Z}\times\vec{Z}_x)\cdot\vec{Z}_{xx}}{|\vec{Z}_x|^2}(\vec{Z}\times\vec{Z}_x)\right]\right\}\cdot\vec{Z}_{xxx}$$

$$= -\int_\Omega f^3\left|\vec{Z}_x\right|^2\left(\vec{Z}\times\vec{Z}_x\right)\cdot\vec{Z}_{xxx} - \int_\Omega f^3\left[\left(\vec{Z}\times\vec{Z}_x\right)\cdot\vec{Z}_{xx}\right]\left(-\frac{3}{2}|\vec{Z}_x|^2\right)_x$$

$$= -\frac{5}{2}\int_\Omega f^3\left|\vec{Z}_x\right|^2\left(\vec{Z}\times\vec{Z}_x\right)\cdot\vec{Z}_{xxx} - \frac{9}{2}\int_\Omega f^2\frac{\partial f}{\partial x}\left|\vec{Z}_x\right|^2\left(\vec{Z}\times\vec{Z}_x\right)\cdot\vec{Z}_{xx}, \qquad (3.4.31)$$

where we have used the fact that $\vec{A}\times(\vec{B}\times\vec{C}) = (\vec{A}\cdot\vec{C})\vec{B} - (\vec{A}\cdot\vec{B})\vec{C}$.

Combining (3.4.30) with (3.4.31), we get

$$2\frac{d}{dt}\int_\Omega f^2\left|\vec{Z}_{xx}\right|^2 + 4\varepsilon\int_\Omega f^2\left|\vec{Z}_{xxx}\right|^2$$

$$= -8\varepsilon\int_\Omega f\frac{\partial f}{\partial x}\vec{Z}_{xx}\cdot\vec{Z}_{xxx} + 4\varepsilon\int_\Omega f^2\left|\vec{Z}_x\right|^2\left|\vec{Z}_{xx}\right|^2 + 32\varepsilon\int_\Omega f^2\left|\vec{Z}_x\cdot\vec{Z}_{xx}\right|^2$$

$$- 16\varepsilon\int_\Omega f\frac{\partial f}{\partial x}\left(\vec{Z}_x\cdot\vec{Z}_{xx}\right)\left(\vec{Z}\cdot\vec{Z}_{xx}\right) + 4\int_\Omega f\frac{\partial f}{\partial t}\left|\vec{Z}_{xx}\right|^2$$

$$+ 4\int_\Omega f^2\frac{\partial^3 f}{\partial x^3}\left(\vec{Z}\times\vec{Z}_x\right)\cdot\vec{Z}_{xx} - 10\int_\Omega f^3\left|\vec{Z}_x\right|^2\left(\vec{Z}\times\vec{Z}_x\right)\cdot\vec{Z}_{xxx}$$

$$- 18\int_\Omega f^2\frac{\partial f}{\partial x}\left|\vec{Z}_x\right|^2\left(\vec{Z}\times\vec{Z}_x\right)\cdot\vec{Z}_{xx}. \qquad (3.4.32)$$

On the other hand, from (3.4.4) we get that

$$\frac{5}{2}\frac{d}{dt}\int_\Omega f^2\left|\vec{Z}_x\right|^4 dx = 5\int_\Omega f\frac{\partial f}{\partial t}\left|\vec{Z}_x\right|^4 dx + 10\int_\Omega f^2|\vec{Z}_x|^2\vec{Z}_x\cdot\vec{Z}_{xt}dx$$

$$= 5\int_\Omega f\frac{\partial f}{\partial t}\left|\vec{Z}_x\right|^4 dx + 10\varepsilon\int_\Omega f^2\left|\vec{Z}_x\right|^2\vec{Z}_x\cdot\vec{Z}_{xxx}dx$$

$$+ 10\varepsilon\int_\Omega f^2\left|\vec{Z}_x\right|^6 dx - 20\int_\Omega f^2\frac{\partial f}{\partial x}\left|\vec{Z}_x\right|^2\left(\vec{Z}\times\vec{Z}_x\right)\cdot\vec{Z}_{xx}dx$$

$$- 10\int_\Omega f^3\left|\vec{Z}_x\right|^2\left(\vec{Z}\times\vec{Z}_x\right)\cdot\vec{Z}_{xxx}dx. \tag{3.4.33}$$

Putting (3.4.32) and (3.4.33) together, we have

$$2\frac{d}{dt}\int_\Omega f^2\left|\vec{Z}_{xx}\right|^2 dx + 4\varepsilon\int_\Omega f^2\left|\vec{Z}_{xxx}\right|^2 dx + 10\varepsilon\int_\Omega f\left|\vec{Z}_x\right|^6 dx - \frac{5}{2}\frac{d}{dt}\int_\Omega f^2\left|\vec{Z}_x\right|^4 dx$$

$$= -5\int_\Omega f\frac{\partial f}{\partial t}\left|\vec{Z}_x\right|^4 dx + 4\int_\Omega f\frac{\partial f}{\partial t}\left|\vec{Z}_{xx}\right|^2 dx - 8\varepsilon\int_\Omega f\frac{\partial f}{\partial x}\vec{Z}_{xx}\cdot\vec{Z}_{xxx}dx$$

$$+ 4\varepsilon\int_\Omega f^2\left|\vec{Z}_x\right|^2\left|\vec{Z}_{xx}\right|^2 dx + 32\varepsilon\int_\Omega f^2\left|\vec{Z}_x\cdot\vec{Z}_{xx}\right|^2 dx + 16\varepsilon\int_\Omega f\frac{\partial f}{\partial x}\left|\vec{Z}_x\right|^2\vec{Z}_x\cdot\vec{Z}_{xx}dx$$

$$- 10\varepsilon\int_\Omega f^2\left|\vec{Z}_x\right|^2\vec{Z}_x\cdot\vec{Z}_{xxx}dx + 2\int_\Omega f^3\frac{\partial f}{\partial x}\left|\vec{Z}_x\right|^2\left(\vec{Z}\times\vec{Z}_x\right)\cdot\vec{Z}_{xx}dx$$

$$+ 4\int_\Omega f^2\frac{\partial^3 f}{\partial x^3}\left(\vec{Z}\times\vec{Z}_x\right)\cdot\vec{Z}_{xx}dx. \tag{3.4.34}$$

By Lemmas 3.4.1 and 3.4.8, we have

$$4\varepsilon\int_\Omega f^2\left|\vec{Z}_x\right|^2\left|\vec{Z}_{xx}\right|^2 dx + 32\varepsilon\int_\Omega f^2\left|\vec{Z}_x\cdot\vec{Z}_{xx}\right|^2 dx$$

$$\leq C\delta_1\varepsilon\int_\Omega f^2\left|\vec{Z}_x\right|^6 dx + C\varepsilon\|\vec{Z}_x\|_2^{\frac{5}{4}}\|\vec{Z}_{xxx}\|_2^{\frac{7}{4}}$$

$$\leq C\delta_1\varepsilon\int_\Omega f^2\left|\vec{Z}_x\right|^6 dx + C\varepsilon\delta_2\int_\Omega f^2\left|\vec{Z}_{xxx}\right|^2 dx + C.$$

Letting $\delta_1 = \frac{2}{C}, \delta_2 = \frac{1}{C}$, we get

$$2\frac{d}{dt}\int_\Omega f^2\left|\vec{Z}_{xx}\right|^2 + 3\varepsilon\int_\Omega f^2\left|\vec{Z}_{xxx}\right|^2 + 8\varepsilon\int_\Omega f^2\left|\vec{Z}_x\right|^6 - \frac{5}{2}\frac{d}{dt}\int_\Omega f^2\left|\vec{Z}_x\right|^4$$

$$\leq 4\int_\Omega f\frac{\partial f}{\partial t}\left|\vec{Z}_{xx}\right|^2 - 5\int_\Omega f\frac{\partial f}{\partial t}\left|\vec{Z}_x\right|^4 - 8\varepsilon\int_\Omega f\frac{\partial f}{\partial x}\vec{Z}_{xx}\cdot\vec{Z}_{xxx}$$

$$- 10\varepsilon\int_\Omega f^2\left|\vec{Z}_x\right|^2\vec{Z}_x\cdot\vec{Z}_{xxx} + 16\varepsilon\int_\Omega f\frac{\partial f}{\partial x}\left|\vec{Z}_x\right|^2\vec{Z}_x\cdot\vec{Z}_{xx}$$

$$+ 2\int_\Omega f^3\frac{\partial f}{\partial x}\left|\vec{Z}_x\right|^2\left(\vec{Z}\times\vec{Z}_x\right)\cdot\vec{Z}_{xx} + 4\int_\Omega f^2\frac{\partial^3 f}{\partial x^3}\left(\vec{Z}\times\vec{Z}_x\right)\cdot\vec{Z}_{xx} + C$$

$$\leq C\int_\Omega f^2\left|\vec{Z}_{xx}\right|^2 + 6\varepsilon\int_\Omega f^2\left|\vec{Z}_x\right|^6 + \varepsilon\int_\Omega f^2\left|\vec{Z}_{xxx}\right|^2 + C, \tag{3.4.35}$$

where we have used the fact that $\|\vec{Z}_x\|_6 \leq \|\vec{Z}_x\|_2^{\frac{2}{3}}\|\vec{Z}_{xx}\|_2^{\frac{1}{3}}$.

Therefore one gets

$$\frac{d}{dt}\int_\Omega f^2\left|\vec{Z}_{xx}\right|^2 dx \leq C\frac{d}{dt}\int_\Omega f^2\left|\vec{Z}_x\right|^4 dx + C\int_\Omega f^2\left|\vec{Z}_{xx}\right|^2 dx + C. \qquad (3.4.36)$$

Integrating (3.4.36), by Lemma 3.4.9 and $\|\vec{Z}_x\|_4 \leq C\|\vec{Z}_x\|_2^{\frac{3}{4}}\|\vec{Z}_{xx}\|_2^{\frac{1}{4}}$, we have

$$\int_\Omega f^2\left|\vec{Z}_{xx}\right|^2 dx \leq C\int_0^t\int_\Omega f^2\left|\vec{Z}_{xx}\right|^2 dxdt + C.$$

The conclusion of Lemma 3.4.10 follows from this inequality and Gronwall's inequality.

Similarly, we have the following lemma:

Lemma 3.4.11 *Let* $\vec{Z}_0(x) \in H^3(\Omega)$, $f(x,t), \frac{\partial f}{\partial t}, \frac{\partial^j f}{\partial x^j} \in L^\infty(Q)$, $j = 1, 2, 3, 4$. *For any given* $T > 0$, *there are* $C > 0$ *independent of* ε, D *such that*

$$\sup_{0\leq t\leq T}\|\vec{Z}_{xxx}(\cdot,t)\|_2 \leq C, \qquad (3.4.37)$$

$$\sup_{0\leq t\leq T}\|\vec{Z}_{xt}(\cdot,t)\|_2 \leq C. \qquad (3.4.38)$$

Proof. It follows from (3.4.29) that

$$\frac{1}{2}\frac{d}{dt}\int_\Omega f^3|\vec{Z}_{xxx}|^2 dx$$

$$= \frac{3}{2}\int_\Omega f\frac{\partial f}{\partial t}|\vec{Z}_{xxx}|^2 dx + \int_\Omega f^3\vec{Z}_{xxx}\cdot\vec{Z}_{xxxt}dx$$

$$= \frac{3}{2}\int_\Omega f\frac{\partial f}{\partial t}|\vec{Z}_{xxx}|^2 dx - \varepsilon\int_\Omega f^3|\vec{Z}_{xxxx}|^2 dx - 3\varepsilon\int_\Omega f^2\frac{\partial f}{\partial x}\vec{Z}_{xxx}\cdot\vec{Z}_{xxxx}dx$$

$$+ \varepsilon\int_\Omega f^3|\vec{Z}_x|^2|\vec{Z}_{xxx}|^2 dx + 4\varepsilon\int_\Omega f^3\left(\vec{Z}_x\cdot\vec{Z}_{xx}\right)\left(\vec{Z}_{xx}\cdot\vec{Z}_{xxx}\right)dx$$

$$+ 2\varepsilon\int_\Omega f^3|\vec{Z}_{xx}|^2\left(\vec{Z}_x\cdot\vec{Z}_{xxx}\right)dx + 2\varepsilon\int_\Omega f^3\left|\vec{Z}_x\cdot\vec{Z}_{xxx}\right|^2 dx$$

$$- 2\varepsilon\int_\Omega f^3\left[|\vec{Z}_{xx}|^2\vec{Z} + \left(\vec{Z}_x\cdot\vec{Z}_{xxx}\right)\vec{Z} + \left(\vec{Z}_x\cdot\vec{Z}_{xx}\right)\vec{Z}_x\right]\cdot\vec{Z}_{xxxx}dx$$

$$- 6\varepsilon\int_\Omega f^2\frac{\partial f}{\partial x}\left[|\vec{Z}_{xx}|^2\vec{Z} + \left(\vec{Z}_x\cdot\vec{Z}_{xxx}\right)\vec{Z} + \left(\vec{Z}_x\cdot\vec{Z}_{xx}\right)\vec{Z}_x\right]\cdot\vec{Z}_{xxx}dx$$

$$+ \int_\Omega f^3\frac{\partial^4 f}{\partial x^4}\left(\vec{Z}\times\vec{Z}_x\right)\cdot\vec{Z}_{xxx}dx + 4\int_\Omega f^3\frac{\partial^3 f}{\partial x^3}\left(\vec{Z}\times\vec{Z}_{xx}\right)\cdot\vec{Z}_{xxx}dx$$

$$+ 6\int_\Omega f^3\frac{\partial^2 f}{\partial x^2}\left(\vec{Z}_x\times\vec{Z}_{xx}\right)\cdot\vec{Z}_{xxx}dx - 2\int_\Omega f^4\left(\vec{Z}_x\times\vec{Z}_{xxx}\right)\cdot\vec{Z}_{xxxx}dx. \qquad (3.4.39)$$

Then if $|\vec{Z}_x| \neq 0$, let $\vec{Z}_{xxx} = \alpha'\vec{Z} + \beta'\vec{Z}_x + \gamma'\vec{Z}\times\vec{Z}_x$.
It is easy to see that

$$\alpha' = -3\vec{Z}_x\cdot\vec{Z}_{xx}, \quad \beta' = \frac{\vec{Z}_x\cdot\vec{Z}_{xxx}}{|\vec{Z}_x|^2}, \quad \gamma' = \frac{(\vec{Z}\times\vec{Z}_x)\cdot\vec{Z}_{xxx}}{|\vec{Z}_x|^2}.$$

Thus

$$-2 \int_\Omega f^4 \left(\vec{Z}_x \times \vec{Z}_{xxx} \right) \cdot \vec{Z}_{xxxx} dx$$

$$= -2 \int_\Omega f^4 \left\{ \vec{Z}_x \times \left[\left(-3\vec{Z}_x \cdot \vec{Z}_{xx} \right) \vec{Z} + \frac{(\vec{Z} \times \vec{Z}_x) \cdot \vec{Z}_{xxx}}{|\vec{Z}_x|^2} \left(\vec{Z} \times \vec{Z}_x \right) \right] \right\} \cdot \vec{Z}_{xxxx} dx$$

$$- 24 \int_\Omega f^3 \frac{\partial f}{\partial x} \left(\vec{Z}_x \cdot \vec{Z}_{xx} \right) \left(\vec{Z} \times \vec{Z}_x \right) \cdot \vec{Z}_{xxx} dx + 12 \int_\Omega f^4 \left| \vec{Z}_{xx} \right|^2 \left(\vec{Z} \times \vec{Z}_x \right) \cdot \vec{Z}_{xxx} dx$$

$$+ 6 \int_\Omega f^4 \left(\vec{Z}_x \cdot \vec{Z}_{xx} \right) \left(\vec{Z} \times \vec{Z}_{xx} \right) \cdot \vec{Z}_{xxx} dx + 6 \int_\Omega f^4 \left(\vec{Z}_x \cdot \vec{Z}_{xxx} \right) \left(\vec{Z} \times \vec{Z}_x \right) \cdot \vec{Z}_{xxx} dx$$

$$+ 8 \int_\Omega f^4 \left[\left(\vec{Z} \times \vec{Z}_x \right) \cdot \vec{Z}_{xxx} \right] \left(\vec{Z}_x \cdot \vec{Z}_{xxx} \right) dx$$

$$\leq C \int_\Omega f^3 \left| \vec{Z}_{xxx} \right|^2 dx + C, \tag{3.4.40}$$

where we have used the fact that $\vec{Z} \cdot \vec{Z}_{xxxx} = -3|\vec{Z}_{xx}|^2 - 4\vec{Z}_x \cdot \vec{Z}_{xxx}$.

Putting (3.4.39) and (3.4.40) together, we have

$$\frac{1}{2}\frac{d}{dt} \int_\Omega f^3 \left| \vec{Z}_{xxx} \right|^2 + \varepsilon \int_\Omega f^3 |\vec{Z}_{xxxx}|^2 \leq C \int_\Omega f^3 \left| \vec{Z}_{xxx} \right|^2 + \frac{\varepsilon}{2} \int_\Omega f^3 \left| \vec{Z}_{xxxx} \right|^2 + C.$$

Employing Gronwall's inequality, we obtain (3.4.37). Using

$$\vec{Z}_{xt} = \varepsilon \vec{Z}_{xxx} + \varepsilon|\vec{Z}_x|^2\vec{Z}_x + 2\varepsilon(\vec{Z}_x \cdot \vec{Z}_{xx})\vec{Z} + \frac{\partial^2 f}{\partial x^2}\vec{Z} \times \vec{Z}_x$$

$$+ 2\frac{\partial f}{\partial x}\vec{Z} \times \vec{Z}_{xx} + f\vec{Z}_x \times \vec{Z}_{xx} + f\vec{Z} \times \vec{Z}_{xxx},$$

we get (3.4.38).

By induction, we have

Lemma 3.4.12 *Let* $\vec{Z}_0(x) \in H^k(\Omega)$, $f(x,t)$, $\frac{\partial f}{\partial t}$, $\frac{\partial^j f}{\partial x^j} \in L^\infty(Q)$, $j = 1, 2, \ldots, k+1$. *For any given* $T > 0$, *there is* $C > 0$ *independent of* ε, D *such that*

$$\sup_{0 \leq t \leq T} \|\partial_t^s \partial_x^{k-2s} \vec{Z}(\cdot, t)\|_2 \leq C, \quad 0 \leq s \leq [k/2]. \tag{3.4.41}$$

Combining the local existence obtained above and the global in time estimates in Lemmas 3.4.8–3.4.12, we can get the global existence of smooth solution to problem (3.4.4)–(3.4.5) for fixed $\varepsilon > 0$ in the following sense:

Theorem 3.4.2 *Let* $\vec{Z}_0(x) \in H^k(\Omega)$, $f(x,t)$, $\frac{\partial f}{\partial t}$, $\frac{\partial^j f}{\partial x^j} \in L^\infty(Q)$, $j = 1, 2, \ldots, k+1$, $f(x,t) \geq f_0 > 0$. *Then, for any given* $T > 0$, *problem* (3.4.4)–(3.4.5) *admits at least one smooth solution* $\vec{Z}(x,t)$ *in* $[0,T]$:

$$\vec{Z}(x,t) \in \bigcap_{s=0}^{[\frac{k}{2}]} W_\infty^s(0, T; H^{k-2s}(\Omega)). \tag{3.4.42}$$

All the *a priori* estimates in Lemmas 3.4.8–3.4.12 are uniform in ε.

3.4.4 $\varepsilon = 0$: Global Solution and Uniqueness

In the above section, we have obtained a global smooth solution for (3.4.4)–(3.4.5) for fixed $\varepsilon > 0$, and the estimates are all uniform in ε. These uniform estimates allow us to pass to the limit $\varepsilon \to 0$ in equation (3.4.4) and then get the global smooth solution of problem (3.4.1)–(3.4.2). Therefore, we have the following theorem:

Theorem 3.4.3 *Let $\vec{Z}_0(x) \in H^k(\Omega)$, $f(x,t)$, $\frac{\partial f}{\partial t}$, $\frac{\partial^j f}{\partial x^j} \in L^\infty(Q)$, $j = 1, 2, \ldots, k+1$, $f(x,t) \geq f_0 > 0$. Then (1.1), (1.2) admits a global smooth solution $\vec{Z}(x,t)$:*

$$\vec{Z}(x,t) \in \bigcap_{s=0}^{[\frac{k}{2}]} W_\infty^s(0, \infty; H^{k-2s}(\Omega)).$$

On the other hand, all the estimates in Sec. 3 are independent of D. Thus, by sending D to ∞, we get the global existence of smooth solution to the Cauchy problem of (3.4.1).

Theorem 3.4.4 *Let $\vec{Z}_0(x) \in H^k(\Omega)$, $f(x,t)$, $\frac{\partial f}{\partial t}$, $\frac{\partial^j f}{\partial x^j} \in L^\infty(Q)$, $j = 1, 2, \ldots, k+1$, $f(x,t) \geq f_0 > 0$. Then the Cauchy problem of (3.4.1) admits a global smooth solution $\vec{Z}(x,t)$:*

$$\vec{Z}(x,t) \in \bigcap_{s=0}^{[\frac{k}{2}]} W_\infty^s(0, \infty; H^{k-2s}(\Omega)).$$

Remark 3.4.1 For any given $\varepsilon > 0$, it is not difficult to see that the problem (3.4.4)–(3.4.5) (or the Cauchy problem) admits at least one global smooth solution.

Now we turn to prove that the global smooth solution of (3.4.1) and (3.4.2) obtained in Theorem 3.4.3 is unique. That is, we prove

Theorem 3.4.5 *The global smooth solution of (3.4.1)–(3.4.2) obtained in Theorem 3.4.3 is unique.*

Proof. First of all, we prove that the smooth solution to periodic problem (3.4.4)–(3.4.5) is unique. Let $\vec{Z}_1(x,t)$ and $\vec{Z}_2(x,t)$ be smooth solutions of (3.4.4) and (3.4.5). Let $\vec{W}(x,t) = \vec{Z}_1 - \vec{Z}_2$.

Then, we have

$$\vec{W}_t = \varepsilon \vec{W}_{xx} + \varepsilon \vec{Z}_2 \left[\left(\vec{Z}_{1x} + \vec{Z}_{2x} \right) \cdot \vec{W} \right] + \varepsilon \left| \vec{Z}_{1x} \right|^2 \vec{W}$$

$$+ f\vec{Z}_2 \times \vec{W}_{xx} + f\vec{W} \times \vec{Z}_{1xx} + \frac{\partial f}{\partial x} \vec{Z}_2 \times \vec{W}_x + \frac{\partial f}{\partial x} \vec{W} \times \vec{Z}_{1x}, \quad (3.4.43)$$

with homogeneous initial boundary conditions. Multiplying (3.4.43) by $\vec{W}$, using Hölder inequality and noting that $\vec{Z}_1, \vec{Z}_2$ are smooth, we have

$$\frac{1}{2} \frac{d}{dt} \int_{-D}^{D} \left| \vec{W} \right|^2 dx + \varepsilon \int_{-D}^{D} \left| \vec{W}_x \right|^2 dx$$

$$\leq \frac{\varepsilon}{4} \int_{-D}^{D} \left| \vec{W}_{xx} \right|^2 dx + C \int_{-D}^{D} \left| \vec{W}_x \right|^2 dx + C \int_{-D}^{D} \left| \vec{W} \right|^2 dx. \quad (3.4.44)$$

On the other hand, multiplying (3.4.43) by $\vec{W}_{xx}$, we have

$$\frac{1}{2}\frac{d}{dt}\int_{-D}^{D}\left|\vec{W}_x\right|^2 dx + \varepsilon\int_{-D}^{D}\left|\vec{W}_{xx}\right|^2 dx$$

$$\leq \frac{\varepsilon}{4}\int_{-D}^{D}\left|\vec{W}_{xx}\right|^2 dx + C\int_{-D}^{D}\left|\vec{W}_x\right|^2 dx + C\int_{-D}^{D}\left|\vec{W}\right|^2 dx. \qquad (3.4.45)$$

Combining (3.4.44) and (3.4.45), using Gronwall's inequality and noting that $\vec{W}(x,0) = 0, W_x(x,0) = 0$, we can get the uniqueness when $\varepsilon > 0$.

In the second step, we let $\varepsilon = 0$ in (3.4.43) to get

$$\vec{W}_t = f(x,t)\vec{Z}_2 \times \vec{W}_{xx} + f(x,t)\vec{W} \times \vec{Z}_{1xx} + \frac{\partial f}{\partial x}\vec{Z}_2 \times \vec{W}_x + \frac{\partial f}{\partial x}\vec{W} \times \vec{Z}_{1x}. \qquad (3.4.46)$$

Therefore we have,

$$\frac{1}{2}\frac{d}{dt}\int_{-D}^{D}|\vec{W}|^2 dx = \int_{-D}^{D} f(x,t)\left(\vec{Z}_2 \times \vec{W}_{xx}\right)\cdot\vec{W}dx + \int_{-D}^{D}\frac{\partial f}{\partial x}\left(\vec{Z}_2 \times \vec{W}_x\right)\cdot\vec{W}dx$$

$$= \int_{-D}^{D}\frac{\partial f}{\partial x}\left(\vec{Z}_2 \times \vec{W}\right)\cdot\vec{W}_x dx + \int_{-D}^{D} f\left(\vec{Z}_{2x} \times \vec{W}\right)\cdot\vec{W}_x dx$$

$$- \int_{-D}^{D}\frac{\partial f}{\partial x}\left(\vec{Z}_2 \times \vec{W}\right)\cdot\vec{W}_x dx$$

$$\leq C\int_{-D}^{D}\left|\vec{W}\right|^2 dx + C\int_{-D}^{D} f\left|\vec{W}_x\right|^2 dx. \qquad (3.4.47)$$

On the other hand, from (3.4.46) we have

$$\vec{W}_{tx} = 2\frac{\partial f}{\partial x}\vec{Z}_2 \times \vec{W}_{xx} + f\vec{Z}_{2x} \times \vec{W}_{xx} + f\vec{Z}_2 \times \vec{W}_{xxx} + 2\frac{\partial f}{\partial x}\vec{W} \times \vec{Z}_{1xx}$$

$$+ f\vec{W}_x \times \vec{Z}_{1xx} + f\vec{W} \times \vec{Z}_{1xxx} + \frac{\partial^2 f}{\partial x^2}\vec{Z}_2 \times \vec{W}_x + \frac{\partial f}{\partial x}\vec{Z}_{2x} \times \vec{W}_x$$

$$+ \frac{\partial^2 f}{\partial x^2}\vec{W} \times \vec{Z}_{1x} + \frac{\partial f}{\partial x}\vec{W}_x \times \vec{Z}_{1x}.$$

Therefore using Hölder inequality and noting that $\vec{Z}_1, \vec{Z}_2$ are smooth, we have

$$\frac{1}{2}\frac{d}{dt}\int_{D}^{D} f\left|\vec{W}_x\right|^2 dx = \frac{1}{2}\int_{-D}^{D}\frac{\partial f}{\partial t}\left|\vec{W}_x\right|^2 dx + \int_{-D}^{D} f\vec{W}_x \cdot \vec{W}_{xt}dx$$

$$= \frac{1}{2}\int_{-D}^{D}\frac{\partial f}{\partial t}\left|\vec{W}_x\right|^2 dx + \int_{-D}^{D}\left[f^2\left(\vec{Z}_2 \times \vec{W}_{xx}\right)\cdot\vec{W}_x\right]_x dx$$

$$+ 2\int_{-D}^{D} f\frac{\partial f}{\partial x}\left(\vec{W} \times \vec{Z}_{1xx}\right)\cdot\vec{W}_x dx$$

$$+ \int_{-D}^{D} f^2\left(\vec{W} \times \vec{Z}_{1xxx}\right)\cdot\vec{W}_x dx + \int_{-D}^{D} f\frac{\partial^2 f}{\partial x^2}\left(\vec{W} \times \vec{Z}_{1x}\right)\cdot\vec{W}_x dx$$

$$\leq C\int_{-D}^{D}\left|\vec{W}\right|^2 dx + C\int_{-D}^{D} f\left|\vec{W}_x\right|^2 dx. \qquad (3.4.48)$$

Putting (3.4.47) and (3.4.48) together, we get

$$\frac{d}{dt}\left(\int_{-D}^{D}\left|\vec{W}\right|^{2}dx + \int_{-D}^{D}f\left|\vec{W}_{x}\right|dx\right) \leq C\left(\int_{-D}^{D}\left|\vec{W}\right|^{2}dx + \int_{-D}^{D}f\left|\vec{W}_{x}\right|^{2}dx\right).$$

Using Gronwall's inequality and noting that $\vec{W}(x,0) = 0$, $W_{x}(x,0) = 0$, we can get the conclusion of Theorem 3.4.5.

3.5   Measure-Valued Solution to the Strongly Degenerate Compressible Heisenberg Chain Equations

3.5.1   Compressible Heisenberg Chain Model and Compressible Heisenberg Chain System

In this section, we intend to establish the existence of measure-valued solution to the following periodic initial value problem:

$$\vec{Z}_{t} = (G(\vec{Z}_{x})\vec{Z} \times \vec{Z}_{x})_{x}, \quad x \in R^{1},\ t \in R_{+}, \tag{3.5.1}$$

$$\vec{Z}(x,0) = \vec{Z}_{0}(x),\ \vec{Z}(x+D,t) = \vec{Z}(x-D,t),\ |\vec{Z}_{0}(x)| \equiv 1,\ x \in R^{1}, \tag{3.5.2}$$

where $G(\xi) = A + B|\xi|^{2}$ and A, B, $D > 0$ are constants by considering the viscosity problem

$$\vec{Z}_{t} = \varepsilon(G(\vec{Z}_{x})\vec{Z}_{x})_{x} + (G(\vec{Z}_{x})\vec{Z} \times \vec{Z}_{x})_{x}, \quad x \in R^{1},\ t \in R_{+}, \tag{3.5.3}$$

$$\vec{Z}(x,0) = \vec{Z}_{0}(x),\ \vec{Z}(x+D,t) = \vec{Z}(x-D,t),\ |\vec{Z}_{0}(x)| \equiv 1,\ x \in R^{1}. \tag{3.5.4}$$

We first prove that problem (3.5.3)–(3.5.4) admits at least one global weak solution, and then give the *a priori* estimates uniform in ε for such solutions to obtain the existence of the measure-valued solution to (3.5.1)–(3.5.2) by letting $\varepsilon \to 0$.

To get the existence of local solution of (3.5.3)–(3.5.4), we apply the difference-differential method. In the sequel, we denote $\Omega = (-D, D)$. For simplicity, we let $\varepsilon = 1$ in this subsection.

We establish the following difference-differential equation:

$$\frac{d\vec{Z}_{j}}{dt} = \frac{\Delta_{-}\left(G\left(\frac{\Delta_{+}\vec{Z}_{j}}{h}\right)\frac{\Delta_{+}\vec{Z}_{j}}{h}\right)}{h} + \frac{\Delta_{-}\left(\vec{Z}_{j} \times G\left(\frac{\Delta_{+}\vec{Z}_{j}}{h}\right)\frac{\Delta_{+}\vec{Z}_{j}}{h}\right)}{h}, \tag{3.5.5}$$

$$\vec{Z}_{j}|_{t=0} = \vec{Z}_{0j} = \vec{Z}_{0}(jh), \tag{3.5.6}$$

$$\vec{Z}_{j+J} = \vec{Z}_{j}, \tag{3.5.7}$$

where $j = 0, \pm 1, \ldots, \pm J, \ldots, h = 2D/J,\ J > 0$.

It is clear that the initial value problem of ordinary differential equation (3.5.5)–(3.5.7) admits a local smooth solution. We shall give some estimates uniformly in h

for such solution, and then get the local solution to problem (3.5.3)–(3.5.4). In this section we always denote the solution of (3.5.5)–(3.5.7) by $\vec{Z}_j$.

Lemma 3.5.1 *If $\vec{Z}_0(x) \in W^{1,4}(\Omega)$, there are constants $T_0 > 0$, $C > 0$ independent of h such that*

$$\sup_{0 \leq t \leq T_0} \left(\|\vec{Z}_h(t)\|_2 + \|\delta\vec{Z}_h(t)\|_2 + \|\delta\vec{Z}_h(t)\|_4 \right) \leq C, \tag{3.5.8}$$

$$\int_0^t \int_\Omega \|\delta^2 \vec{Z}_h(t)\|_2 \leq C. \tag{3.5.9}$$

Proof. Multiplying (3.5.5) by $\vec{Z}_j h$ and summing from $j = 1$ to J, we have

$$\frac{1}{2}\frac{d}{dt}\sum_{j=1}^{J}|\vec{Z}_j|^2 h = -\sum_{j=0}^{J-1}G\left(\frac{\Delta_+\vec{Z}_j}{h}\right)\left|\frac{\Delta_+\vec{Z}_j}{h}\right|^2 h$$

$$+ \sum_{j=0}^{J-1}\left(\vec{Z}_j \times G\left(\frac{\Delta_+\vec{Z}_j}{h}\right)\frac{\Delta_+\vec{Z}_j}{h}\right) \cdot \frac{\Delta_+\vec{Z}_j}{h} h$$

$$= -\sum_{j=0}^{J-1}\left(A + B\left|\frac{\Delta_+\vec{Z}_j}{h}\right|^2\right)\left|\frac{\Delta_+\vec{Z}_j}{h}\right|^2 h.$$

Therefore, we have

$$\frac{1}{2}\frac{d}{dt}\|\vec{Z}_h\|_2^2 + A\|\delta\vec{Z}_h\|_2^2 + B\|\delta\vec{Z}_h\|_4^4 = 0. \tag{3.5.10}$$

It follows from (3.5.5) that

$$\frac{d\Delta_+\vec{Z}_j}{dt} = \frac{\Delta_+\Delta_-\left(G\left(\frac{\Delta_+\vec{Z}_j}{h}\right)\frac{\Delta_+\vec{Z}_j}{h}\right)}{h} + \frac{\Delta_+\Delta_-\left(\vec{Z}_j \times G\left(\frac{\Delta_+\vec{Z}_j}{h}\right)\frac{\Delta_+\vec{Z}_j}{h}\right)}{h}.$$

Multiplying this equation by $G\left(\frac{\Delta_+\vec{Z}_j}{h}\right)\frac{\Delta_+\vec{Z}_j}{h}$ and summing from $j = 0$ to $j = J-1$ to give

$$\sum_{j=0}^{J-1}G\left(\frac{\Delta_+\vec{Z}_j}{h}\right)\frac{\Delta_+\vec{Z}_j}{h}\frac{d\Delta_+\vec{Z}_j}{dt} = \sum_{j=0}^{J-1}\frac{\Delta_+\Delta_-\left(G\left(\frac{\Delta_+\vec{Z}_j}{h}\right)\frac{\Delta_+\vec{Z}_j}{h}\right)}{h}G\left(\frac{\Delta_+\vec{Z}_j}{h}\right)\frac{\Delta_+\vec{Z}_j}{h}$$

$$+ \sum_{j=0}^{J-1}\frac{\Delta_+\Delta_-\left(\vec{Z}_j \times G\left(\frac{\Delta_+\vec{Z}_j}{h}\right)\frac{\Delta_+\vec{Z}_j}{h}\right)}{h}G\left(\frac{\Delta_+\vec{Z}_j}{h}\right)\frac{\Delta_+\vec{Z}_j}{h}$$

$$= -\sum_{j=1}^{J}\left|\frac{\Delta_-\left(G\left(\frac{\Delta_+\vec{Z}_j}{h}\right)\frac{\Delta_+\vec{Z}_j}{h}\right)}{h}\right|^2 h$$

$$-\sum_{j=1}^{J}\frac{\Delta_-\left(\vec{Z}_j \times G\left(\frac{\Delta_+\vec{Z}_j}{h}\right)\frac{\Delta_+\vec{Z}_j}{h}\right)}{h}\frac{\Delta_-\left(G\left(\frac{\Delta_+\vec{Z}_j}{h}\right)\frac{\Delta_+\vec{Z}_j}{h}\right)}{h}h. \tag{3.5.11}$$

We claim

$$\sum_{j=1}^{J} \frac{\Delta_-\left(\vec{Z}_j \times G\left(\frac{\Delta_+\vec{Z}_j}{h}\right)\frac{\Delta_+\vec{Z}_j}{h}\right)}{h} \frac{\Delta_-\left(G\left(\frac{\Delta_+\vec{Z}_j}{h}\right)\frac{\Delta_+\vec{Z}_j}{h}\right)}{h} h = 0. \qquad (3.5.12)$$

In fact, since

$$\sum_{j=1}^{J} \frac{\Delta_-\left(\vec{Z}_j \times G\left(\frac{\Delta_+\vec{Z}_j}{h}\right)\frac{\Delta_+\vec{Z}_j}{h}\right)}{h} \frac{\Delta_-\left(G\left(\frac{\Delta_+\vec{Z}_j}{h}\right)\frac{\Delta_+\vec{Z}_j}{h}\right)}{h} h$$

$$= \sum_{j=1}^{J} \left(\frac{\Delta_-\vec{Z}_j}{h} \times G\left(\frac{\Delta_+\vec{Z}_{j-1}}{h}\right)\frac{\Delta_+\vec{Z}_{j-1}}{h}\right) \frac{\Delta_-\left(G\left(\frac{\Delta_+\vec{Z}_j}{h}\right)\frac{\Delta_+\vec{Z}_j}{h}\right)}{h} h$$

$$= \sum_{j=1}^{J} \left(\frac{\Delta_+\vec{Z}_{j-1}}{h} \times G\left(\frac{\Delta_+\vec{Z}_{j-1}}{h}\right)\frac{\Delta_+\vec{Z}_{j-1}}{h}\right) \frac{\Delta_-\left(G\left(\frac{\Delta_+\vec{Z}_j}{h}\right)\frac{\Delta_+\vec{Z}_j}{h}\right)}{h} h$$

$$= 0,$$

where we have used $\vec{a} \times \vec{a} = 0$, $(\vec{a} \times \vec{b}) \cdot \vec{b} = 0$ and $\Delta_-\vec{Z}_j = \Delta_+\vec{Z}_{j-1}$, the claim is proved.

Therefore we have

$$A\frac{d}{dt}\sum_{j=0}^{J-1}\left|\frac{\Delta_+\vec{Z}_j}{h}\right|^2 h + \frac{1}{2}B\frac{d}{dt}\sum_{j=0}^{J-1}\left|\frac{\Delta_+\vec{Z}_j}{h}\right|^4 h + A^2\sum_{j=1}^{J}\left|\frac{\Delta_+\Delta_-\vec{Z}_j}{h^2}\right|^2 h$$

$$+ AB\sum_{j=1}^{J}\left|\frac{\Delta_+\vec{Z}_j}{h}\right|^2\left|\frac{\Delta_+\Delta_-\vec{Z}_j}{h^2}\right|^2 h + B\sum_{j=1}^{J}G\left(\frac{\Delta_+\vec{Z}_j}{h}\right)\left|\frac{\Delta_+\Delta_-\vec{Z}_j}{h^2}\frac{\Delta_+\vec{Z}_{j-1}}{h}\right|^2 h$$

$$+ B^2\sum_{j=1}^{J}\left|\frac{\Delta_+\vec{Z}_{j-1}}{h}\right|^2\left|\frac{\Delta_+\vec{Z}_j}{h}\frac{\Delta_+\Delta_-\vec{Z}_j}{h^2} + \frac{\Delta_+\vec{Z}_{j-1}}{h}\frac{\Delta_+\Delta_-\vec{Z}_j}{h^2}\right|^2 h \leq 0. \qquad (3.5.13)$$

This inequality combined with (3.5.10) leads to

$$\frac{d}{dt}\left(\|\vec{Z}_h\|_2^2 + \|\delta\vec{Z}_h\|_2^2 + \|\delta\vec{Z}_h\|_4^4\right) + \|\vec{Z}_h\|_2^2 + \|\delta\vec{Z}_h\|_2^2 + \|\delta\vec{Z}_h\|_4^4 + \|\delta^2\vec{Z}_h\|_2^2 \leq 0. \qquad (3.5.14)$$

Inequality (3.5.14) combined with Gronwall inequality implies that there exist constants $T_0, C > 0$ independent of h such that

$$\|\vec{Z}_h(t)\|_2 + \|\delta\vec{Z}_h(t)\|_2 + \|\delta\vec{Z}_h(t)\|_4^4 \leq C, \quad \forall\, t \in [0, T_0],$$

$$\int_0^{T_0} \|\delta^2\vec{Z}_h(t)\|_2^2 \leq C.$$

The lemma is proved.

Corollary 3.5.1 *Under the conditions in Lemma 3.5.1, we have, for some constant C independent of h,*

$$\sup_{0\le t\le T_0;1\le j\le J}|\vec{Z}_j|\le C. \tag{3.5.15}$$

Sending $h\to 0$, we obtain the local existence of the solution to (3.5.3)–(3.5.4). For the global solution to the viscosity problem, we prove

Theorem 3.5.1 *Let $\varepsilon=1$, $\vec{Z}_0(x)\in W^{1,4}(\Omega)$. Then (3.5.3)–(3.5.4) admits a local solution $\vec{Z}(x,t)$ in $[0,T_0]$ in the space*

$$\vec{Z}(x,t)\in L^\infty(0,T_0;W^{1,4}(\Omega))\bigcap L^2(0,T_0;H^2(\Omega)) \tag{3.5.16}$$

and the following estimates hold

$$\sup_{0\le t\le T_0;x\in\Omega}|\vec{Z}|\le C, \tag{3.5.17}$$

$$\|\vec{Z}(t)\|_2+\|\vec{Z}_x(t)\|_2+\|\vec{Z}_x\|_4^4\le C,\quad \forall\, t\in[0,T_0], \tag{3.5.18}$$

$$\int_0^{T_0}\|\vec{Z}_{xx}(t)\|_2^2\le C. \tag{3.5.19}$$

In order to prove the global existence of weak solution to (3.5.3)-(3.5.4), we need the following *a priori* estimates for the solution of (3.5.3)-(3.5.4).

Lemma 3.5.2 *Let $\varepsilon=1$, $\vec{Z}_0(x)\in W^{1,4}(\Omega)$, $T>0$ and $\vec{Z}(x,t)\in L^\infty(0,T;W^{1,4}(\Omega))\cap L^2(0,T;H^2(\Omega))$ is a solution of (3.5.3)–(3.5.4). Then the following estimates hold*

$$\sup_{0\le t\le T;1\le j\le J}|\vec{Z}|\le C, \tag{3.5.20}$$

$$\|\vec{Z}(t)\|_2+\|\vec{Z}_x(t)\|_2+\|\vec{Z}_x\|_4^4\le C,\quad \forall\, t\in[0,T], \tag{3.5.21}$$

$$\int_0^T\|\vec{Z}_{xx}(t)\|_2^2\le C. \tag{3.5.22}$$

Proof. Multiplying (3.5.3) by $\vec{Z}(x,t)$ and integrating it over Ω, we get

$$\frac{1}{2}\frac{d}{dt}\|\vec{Z}(\cdot,t)\|_2^2+A\|\vec{Z}_x\|_2^2+B\|\vec{Z}_x\|_4^4=0. \tag{3.5.23}$$

Differentiating (3.5.3) with respect to x and then testing it by $G(\vec{Z}_x)\vec{Z}_x$, one has

$$G(\vec{Z}_x)\vec{Z}_x\vec{Z}_{xt}=(G(\vec{Z}_x)\vec{Z}_x)_{xx}G(\vec{Z}_x)\vec{Z}_x+(G(\vec{Z}_x)\vec{Z}\times\vec{Z}_x)_{xx}G(\vec{Z}_x)\vec{Z}_x.$$

Integrating this equation by parts, we have

$$\frac{A}{2}\frac{d}{dt}\|\vec{Z}_x\|_2^2+\frac{B}{4}\frac{d}{dt}\|\vec{Z}_x\|_4^4+\int_\Omega|(G(\vec{Z}_x)\vec{Z}_x)_x|^2=-\int_\Omega(G(\vec{Z}_x)\vec{Z}\times\vec{Z}_x)_x(G(\vec{Z}_x)\vec{Z}_x)_x$$

which implies

$$\frac{1}{2}A\frac{d}{dt}\|\vec{Z}_x\|_2^2 + \frac{1}{4}B\frac{d}{dt}\|\vec{Z}_x\|_4^4 + \int_\Omega |(G(\vec{Z}_x)\vec{Z}_x)_x|^2 = 0. \tag{3.5.24}$$

Since

$$\begin{aligned}
|(G(\vec{Z}_x)\vec{Z}_x)_x|^2 &= |G(\vec{Z}_x)\vec{Z}_{xx} + 2B(\vec{Z}_x \cdot \vec{Z}_{xx})\vec{Z}_x|^2 \\
&= G^2(\vec{Z}_x)|\vec{Z}_{xx}|^2 + 4B^2|(\vec{Z}_x \cdot \vec{Z}_{xx})\vec{Z}_x|^2 + 4BG(\vec{Z}_x)|\vec{Z}_x \cdot \vec{Z}_{xx}|^2 \\
&\geq A^2|\vec{Z}_{xx}|^2,
\end{aligned}$$

where C depends only on $\|\vec{Z}_0\|_{W^{1,4}(\Omega)}$, it follows from (3.5.24) that

$$\frac{1}{2}A\frac{d}{dt}\|\vec{Z}_x\|_2^2 + \frac{1}{4}B\frac{d}{dt}\|\vec{Z}_x\|_4^4 + A^2\int_\Omega |\vec{Z}_{xx}|^2 \leq 0.$$

Putting this inequality and (3.5.23) together, we get from the Gronwall's inequality that (3.5.21) and (3.5.22) hold. (3.5.20) can be derived from these inequalities. The lemma is proved.

Now, we can use the extension method to give

Theorem 3.5.2 *Let $\varepsilon > 0$ be fixed and $\vec{Z}_0 \in W^{1,4}(\Omega)$. Then problem (3.5.7)–(3.5.8) admits a global solution $\vec{Z}_\varepsilon(x,t)$ in the space*

$$\vec{Z}_\varepsilon(x,t) \in L^\infty(0,\infty; W^{1,4}(\Omega)) \bigcap L^2(0,\infty; H^2(\Omega)) \tag{3.5.25}$$

and the following estimates hold

$$\sup_{0 \leq t < \infty; x \in \Omega} |\vec{Z}_\varepsilon(x,t)| \leq C_1, \tag{3.5.26}$$

$$\|\vec{Z}_\varepsilon(t)\|_2 + \|\vec{Z}_{\varepsilon x}(t)\|_2 + \|\vec{Z}_{\varepsilon x}(t)\|_4^4 \leq C_1, \quad \forall\, t \in [0,\infty), \tag{3.5.27}$$

$$\|\vec{Z}_{\varepsilon t}\|_{L^{4/3}(0,\infty; W^{-1,4/3}(\Omega))} \leq C_1, \tag{3.5.28}$$

$$\int_0^\infty \|\vec{Z}_{\varepsilon xx}(t)\|_2^2 \leq C_\varepsilon, \tag{3.5.29}$$

where C_1 depends only on $\|\vec{Z}_0\|_{W^{1,4}(\Omega)}$.

3.5.2   Measure-Valued Solution to the Strongly Degenerate Equations

The *a priori* estimates we have obtained are not enough for us to obtain the weak solution for (3.5.1)–(3.5.2), we apply the notion of measure-valued solution.

Denote $M = C_1$ where C_1 is given in Theorem 3.5.2 which depends only on $\|\vec{Z}_0\|_{W^{1,4}(\Omega)}$. Let $\mathcal{Z}^\varepsilon = (\vec{Z}_\varepsilon, \vec{Z}_{\varepsilon x})$, $\tau(\xi) = B|(\xi_4,\xi_5,\xi_6)|^2(\xi_1,\xi_2,\xi_3) \times (\xi_4,\xi_5,\xi_6) : (R^3 \cap \{|(\xi_1,\xi_2,\xi_3)| \leq M\}) \times R^3 \to R^3$. Then $\mathcal{Z}^\varepsilon$ is uniformly bounded in $L^4(Q)^6$ where $Q = \Omega \times (0,T) \in R^2$ and

$$|\tau(\xi)| \leq CM(1 + |\xi|)^3, \quad \forall\, \xi \in (R^3 \cap \{|(\xi_1,\xi_2,\xi_3)| \leq M\}) \times R^3.$$

Lemma 3.5.3 *Let $Q \subset R^2$ be a bounded open set. Let $\mathcal{Z}^{\varepsilon_n}$ be uniformly bounded in $L^4(Q)^6$. Then there exists a subsequence, still denoted by $\mathcal{Z}^{\varepsilon_n}$, and a measure-valued function ν such that for all $\tau : (R^3 \cap \{|(\xi_1, \xi_2, \xi_3)| \leq M\}) \times R^3 \to R^3$ satisfying for some $q > 0$ the growth condition*

$$|\tau(\xi)| \leq C(1 + |\xi|)^3, \quad \forall\, \xi \in (R^3 \cap \{|(\xi_1, \xi_2, \xi_3)| \leq M\}) \times R^3,$$

we have

$$\tau(\mathcal{Z}^{\varepsilon_n}) \rightharpoonup \bar{\tau} \quad \text{weakly in } L^r(Q),$$
$$\bar{\tau}(y) = \langle \nu_y, \tau \rangle \quad \text{a.e. } y \in Q,$$

provided that

$$1 < r \leq \frac{4}{3}.$$

Proof. It suffices to verify the condition (2.7) of Theorem 4.2.1 of [34, Chap. 4]. Taking the Young function $\Psi(u) = u^r$, we have

$$\int_Q \Psi(\tau|\mathcal{Z}^{\varepsilon_n}|) = \int_Q |\mathcal{Z}^{\varepsilon_n}|^r \leq C^r \int_Q (1 + |\mathcal{Z}^{\varepsilon_n}|)^{3r}$$

and the last term is uniformly bounded (with respect to n) if $3r \leq 4$. The lower bound $r > 1$ follows from the properties of Orlicz functions, namely from $\lim_{s \to \infty} \Psi(s)/s = \infty$.

Definition 3.5.1 *A pair $(\vec{Z}, \nu)$ is called a measure-valued solution of (3.5.1)–(3.5.2) if*

$$\vec{Z} \in L^\infty(0, \infty; W^{1,4}(\Omega)), \tag{3.5.30}$$
$$\nu \in L^\infty_\omega(Q; \mathrm{Prob}(R^6)), \tag{3.5.31}$$

and if for any $\varphi \in \mathcal{D}(-\infty, T; C^\infty_{\mathrm{per}}(\Omega))$ there holds

$$\int_Q \vec{Z}_0 \varphi = \int_Q \vec{Z}\varphi_t - A \int_Q \varphi_x \vec{Z} \times \vec{Z}_x - \int_Q \varphi_x \int_{R^6} \tau(\lambda) d\nu_{t,x}(\lambda) dx dt \tag{3.5.32}$$

where τ is defined as above, $Q = \Omega \times (0, T)$.

Theorem 3.5.3 *Let $\vec{Z}_0 \in W^{1,4}(\Omega)$. Then problem (3.5.1)–(3.5.2) admits a measure-valued solution.*

Proof. It follows from Lemma 3.5.3 that there exists a subsequence $\mathcal{Z}^{\varepsilon_n}$ and a measure-valued function ν such that

$$\tau(\mathcal{Z}^{\varepsilon_n}) \rightharpoonup \bar{\tau} \quad \text{weakly in } L^r(Q), \ 1 < r \leq \frac{4}{3}, \tag{3.5.33}$$
$$\bar{\tau}(x, t) = \langle \nu_{t,x}, \tau \rangle \quad \text{a.e. } (x, t) \in Q. \tag{3.5.34}$$

To complete the proof, we only need to prove, for some subsequence $\vec{Z}_{\varepsilon_n}$, that

$$\int_Q \vec{Z}_{\varepsilon_n t}\varphi \to -\int_Q \vec{Z}_t\varphi + \int_\Omega \vec{Z}_0\varphi(x,0), \tag{3.5.35}$$

$$\int_Q (\vec{Z}_{\varepsilon_n} \times \vec{Z}_{\varepsilon_n x})\varphi_x \to \int_Q (\vec{Z} \times \vec{Z}_x)\varphi_x, \tag{3.5.36}$$

$$Z_x^i = \int_{\mathrm{R}^6} \lambda_{i+3} d\nu_{t,x}(\lambda), \ \text{a.e. } (x,t) \in Q, \ i = 1,2,3. \tag{3.5.37}$$

In view of (3.5.27), we have for some subsequence $\vec{Z}_{\varepsilon_n}$ that

$$\vec{Z}_{\varepsilon_n} \rightharpoonup \vec{Z} \text{ weakly in } L^r(0,T;L^2(\Omega)), \forall \, r > 1, \tag{3.5.38}$$

$$\vec{Z}_{\varepsilon_n x} \rightharpoonup \vec{Z}_x \text{ weakly in } L^r(0,T;L^4(\Omega)), \forall \, r > 1. \tag{3.5.39}$$

To prove (3.5.35), we take $\varphi \in \mathcal{D}(-\infty,T;C_{\mathrm{per}}^\infty(\Omega))$ to give

$$\int_0^T \int_\Omega \vec{Z}_{\varepsilon_n t}\varphi = -\int_0^T \int_\Omega \vec{Z}_{\varepsilon_n}\varphi_t + \int_\Omega \vec{Z}_0(x)\varphi(x,0)$$
$$\longrightarrow -\int_0^T \int_\Omega \vec{Z}\varphi_t + \int_\Omega \vec{Z}_0(x)\varphi(x,0),$$

this proves (3.5.35).

Now we prove (3.5.36) and (3.5.37).

Since $\vec{Z}_{\varepsilon_n}$ is uniformly bounded in the space

$$\{v : \ v \in L^r(0,T;W_{\mathrm{per}}^{1,4}(\Omega)), v_t \in L^{4/3}(0,T;(W^{1,4}(\Omega))^*)\} \tag{3.5.40}$$

for any $r > 1$ and since $W_{\mathrm{per}}^{1,4}(\Omega) \hookrightarrow\hookrightarrow L^r(\Omega) \hookrightarrow (W^{1,4}(\Omega))^*$, applying Aubin–Lions Lemma [8, Lemma 1.2.48] with $X_0 = W_{\mathrm{per}}^{1,4}(\Omega)$, $X = L^r(\Omega)$, $X_1 = (W^{1,4}(\Omega))^*$, $\alpha = r$ $(r > 1)$, $\beta = 4/3$, we know that the space defined in (3.5.40) is compactly imbedded into $L^r(0,T;L^r(\Omega))$, that is

$$\vec{Z}_{\varepsilon_n} \to \vec{Z} \text{ strongly in } L^r(0,T;L^r(\Omega)). \tag{3.5.41}$$

Since

$$\int_Q \left((\vec{Z}_{\varepsilon_n} \times \vec{Z}_{\varepsilon_n x})\varphi_x - (\vec{Z} \times \vec{Z}_x)\varphi_x\right)$$
$$= \int_Q ((\vec{Z}_{\varepsilon_n} - \vec{Z}) \times \vec{Z}_{\varepsilon_n x})\varphi_x + \int_Q (\vec{Z} \times (\vec{Z}_{\varepsilon_n x} - \vec{Z}_x))\varphi_x$$
$$= I_1 + I_2,$$

it follows from (3.5.26) and (3.5.39) that

$$I_2 \to 0 \text{ as } n \to \infty,$$

and it follows from (3.5.27) and (3.5.41) that

$$|I_1| \le \left(\int_0^T \int_\Omega |\vec{Z}_{\varepsilon_n} - \vec{Z}|^4 \right)^{1/4} \left(\int_0^T \int_\Omega |\varphi_x|^4 \right)^{1/4} \left(\int_0^T \int_\Omega |\vec{Z}_{\varepsilon_n x}|^2 \right)^{1/2}$$

$$\le C \left(\int_0^T \int_\Omega |\vec{Z}_{\varepsilon_n} - \vec{Z}|^4 \right)^{1/4}$$

$$\to 0 \text{ as } n \to \infty.$$

The proof of (3.5.36) is complete.

Since Lemma 3.5.3 is true for all τ, if we let $\tau = \mathrm{Id}$, then for $r = 4$, $q = 1$, $\forall \, \psi \in L^{4/3}(Q)$, we have for $\mathcal{Z}^{\varepsilon_n} = (\vec{Z}_{\varepsilon_n}, \vec{Z}_{\varepsilon_n x})$ that

$$\int_Q \mathcal{Z}^{\varepsilon_n} \psi \, dx dt \to \int_Q \psi \int_{\mathrm{R}^6} \lambda \, d\nu_{t,x}(\lambda) dx dt.$$

However, $\vec{Z}_{\varepsilon_n} \to \vec{Z}$ strongly in $L^r(Q)$ and $\vec{Z}_{\varepsilon_n x} \rightharpoonup \vec{Z}_x$ in $L^4(Q)$, we know

$$Z_x^i = \int_{\mathrm{R}^6} \lambda_{i+3} \nu_{t,x}(\lambda), \text{ a.e. } (x,t) \in Q, \ i = 1,2,3.$$

This verifies (3.5.37).

Corollary 3.5.2 *Since the former estimates are independent of D, we get by letting $D \to \infty$ that the Cauchy problem of (3.5.1)–(3.5.2) admits a solution $\vec{Z}(x,t)$:*

$$\vec{Z}(x,t) \in L^\infty(R_+; W^{1,4}(R^1)) \bigcap L^2(R_+; H^2(R^1)).$$

3.6 Bibliography Comments

In this chapter we have introduced the mathematical theory of the global solutions to the equations of ferromagnetic spin chain with or without Gilbert damping term in one dimension. Applying Leray–Schauder principle, we proved the global existence of weak solution to one-dimensional ferromagnetic spin chain equation. By the differential-difference method we get the existence of global weak solution to the one-dimensional nonlinear initial boundary value problem for the ferromagnetic spin chain. To get the existence and uniqueness of smooth solution, Zhou, Guo and Tan [158] created a particular mobile frame on S^2, this method was first used by them and enabled one to get the existence and uniqueness of the smooth solution by the special estimates for the higher derivatives, which is a challenging problem for a long time. Applying the similar method, Ding, Guo and Su [51] obtained the smooth solution for the system of inhomogeneous Heisenberg chain. However, the frame on S^2 was not used for the inhomogeneous problem in [51] until the work by Lin and Ding [108] in 2006. Introducing the notion of measure-valued solution, Ding, Guo and Su [50] discussed the compressible Heisenberg chain equations.

Chapter 4

Landau–Lifshitz Equations and Harmonic Maps

In Chapter 3, we have proved some global existence and uniqueness results for the one-dimensional Landau–Lifshitz equations. It is natural to ask how about the higher-dimensional problems.

We will find, in general, that there exist global weak solutions for the initial-boundary value problems. However, in higher dimensions, there exists global smooth solution to the Landau Lifshitz equations with the Gilbert damping term only for small initial data. If the initial data is large, one can only prove the short time existence of smooth solution.

Generally speaking, higher-dimensional Landau–Lifshitz equations with large initial data may develop singularities. In this aspect, the Landau–Lifshitz equations behave like the heat flow of harmonic maps.

4.1 Weak Solution to Multidimensional Ferromagnetic Spin Chain Equations

4.1.1 Existence of Weak Solution to Multidimensional Ferromagnetic Spin Chain Equations

1. *Initial boundary value problem for multidimensional ferromagnetic spin chain equations*

Let $\Omega \in R^m$ be a bounded domain with boundary $\partial\Omega \in C^2$. We study the ferromagnetic spin chain system with m variables of the form

$$\vec{Z}_t = \vec{Z} \times \Delta\vec{Z} + \vec{f}(x, t, \vec{Z}), \tag{4.1.1}$$

where $\vec{Z}(x, t) = (u(x, t), v(x, t), w(x, t))$ is a three-dimensional unknown and $\vec{f}(x, t)$ is a known three dimensional vector of variables $x \in R^m$, $t \in R^+$, $\vec{Z} \in R^3$, $\Delta = \sum_{i=1}^{m} \frac{\partial^2}{\partial x_i^2}$. In the cylinder $Q_T = \{x \in \Omega, 0 \le t \le T\}$, we give the homogeneous boundary

123

condition of the form

$$\vec{Z}(x,t) = 0, \quad x \in \partial\Omega, \quad 0 \le t \le T, \tag{4.1.2}$$

and initial condition

$$\vec{Z}(x,0) = \varphi(x), \quad x \in \overline{\Omega}, \tag{4.1.3}$$

where $\varphi(x)$ is a given vector.

We also consider the spin system:

$$\vec{Z}_t = \varepsilon\Delta\vec{Z} + \vec{Z} \times \Delta\vec{Z} + \vec{f}(x,t,\vec{Z}). \tag{4.1.4}$$

We establish the existence of a global weak solution to initial boundary value problem (4.1.1)–(4.1.3) and problem (4.1.4), (4.1.2) and (4.1.3) by the Galerkin method.

2. *Approximate solution to the initial boundary value problem of spin systems*

In order to prove the existence of global weak solution to the spin system, we use the Galerkin method. To this aim we construct the Galerkin approximate solution.

Let $\{w_n\}_{n=1}^{\infty}$ be the normal eigenfunctions of the equation

$$\Delta u + \lambda u = 0,$$
$$u|_{\partial\Omega} = 0,$$

corresponding to eigenvalue $\{\lambda_n\}_{n=1}^{\infty}$. It is well known that $\{w_n\}$ forms a normal orthogonal basis of $H_0^1(\Omega)$. We denote the approximate solution of (4.1.4), (4.1.2), and (4.1.3) by

$$\vec{Z}_N(x,t) = \sum_{n=1}^{N} \alpha_{nN}(t)w_n(x). \tag{4.1.5}$$

According to the Galerkin method, coefficients α_{nN} solve the first order ordinary differential system as follows:

$$\int_{\Omega} \vec{Z}_{Nt}(x,t)w_s(x)dx = \int_{\Omega}(\vec{Z}_N(x,t) \times \Delta\vec{Z}_N(x,t))w_s(x)dx$$
$$+ \varepsilon\int_{\Omega} \Delta\vec{Z}_N(x,t)w_s(x)dx + \int_{\Omega} \vec{f}(x,t,\vec{Z}_N)w_s(x)dx,$$
$$s = 1,2,\ldots,N \tag{4.1.6}$$

with the initial condition

$$\int_{\Omega} \vec{Z}_N(x,0)w_s(x)dx = \int_{\Omega} \varphi(x)w_s(x)dx, \quad s = 1,2,\ldots,N. \tag{4.1.7}$$

It is clear that

$$\begin{cases} \int_{\Omega} \vec{Z}_{Nt}(x,t)w_s(x)dx = \alpha'_{sN}(t), \\ \int_{\Omega} \vec{Z}_N(x,0)w_s(x)dx = \alpha_{sN}(0), \end{cases} \tag{4.1.8}$$

and

$$\varphi_s = \int_\Omega \varphi(x) w_s(x) dx, \quad s = 1, 2, \ldots, N$$

are the coefficients of expansion $\varphi(x) = \sum_{s=1}^N \varphi_s w_s(x)$.

3. *Estimates for the approximate solutions*

In order to prove the existence of a global weak solution, we assume the following:

(1) The 3×3 Jacobi matrix $\vec{f}_{\vec{Z}}(x, t, \vec{Z})$ is semi-bounded, i.e., there is a constant $b > 0$ such that for any $\xi \in R^3$

$$\xi \cdot \vec{f}_{\vec{Z}} \cdot \xi \le b|\xi|^2 \tag{4.1.9}$$

holds, where "$\cdot$" denotes the three-dimensional inner product.

(2) (4.1.1) and (4.1.4) are homogeneous, that is, $\vec{f}(x, t, 0) \equiv 0$. Moreover, it is assumed that for some constants A and B:

$$\begin{cases} |\vec{f}(x, t, \vec{Z})| \le A|\vec{Z}|^l + B, \quad 2 \le l \le 2 + \frac{4}{m-2}, \ m \ge 2, \\ |\nabla_x \vec{f}(x, t, \vec{Z})| \le A|\vec{Z}|^{1+\frac{2}{m}} + B. \end{cases} \tag{4.1.10}$$

(3) Vector $\varphi(x) \in H_0^1(\Omega)$.

Lemma 4.1.1 *Let condition (1) hold, $\varphi(x) \in L^2(\Omega)$ and $\vec{f}(x, t, 0) \in L^2(Q_T)$. Then the approximate solution $\vec{Z}_N(x, t)$ meets the estimate*

$$\sup_{0 \le t \le T} \|\vec{Z}_N(\cdot, t)\|_{L^2(\Omega)} \le K_1, \tag{4.1.11}$$

where K_1 is independent of ε and N.

Proof. Multiplying (4.1.6) by $\alpha_{nN}(t)$ and summing from $s = 1$ to $s = N$, we have

$$\int_\Omega \vec{Z}_{Nt} \cdot \vec{Z}_N dx = \int_\Omega (\vec{Z}_N \times \Delta \vec{Z}_N) \cdot \vec{Z}_N dx$$
$$+ \varepsilon \int_\Omega \Delta \vec{Z}_N \cdot \vec{Z}_N dx + \int_\Omega \vec{f}(x, t, \vec{Z}_N) \cdot \vec{Z}_N dx, \tag{4.1.12}$$

or simply write as

$$\int_\Omega \vec{Z}_{Nt} \cdot \vec{Z}_N dx = \int_\Omega (\vec{Z}_N \times \Delta \vec{Z}_N) \cdot \vec{Z}_N dx$$
$$+ \varepsilon \int_\Omega \Delta \vec{Z}_N \cdot \vec{Z}_N dx + \int_\Omega \vec{f}(\vec{Z}_N) \cdot \vec{Z}_N dx. \tag{4.1.13}$$

We have

$$\int_\Omega \vec{Z}_{Nt} \cdot \vec{Z}_N dx = -\frac{1}{2} \frac{d}{dt} \|\vec{Z}_N(\cdot, t)\|_{L^2(\Omega)}^2,$$
$$\int_\Omega (\vec{Z}_N \times \Delta \vec{Z}_N) \cdot \vec{Z}_N dx = 0,$$
$$\int_\Omega \Delta \vec{Z}_N \cdot \vec{Z}_N = \int_{\partial\Omega} (\nabla \vec{Z}_N \cdot \vec{Z}_N) * \nu_m - \int_\Omega \nabla \vec{Z}_N * \nabla \vec{Z}_N,$$

where ν_m is the unit outer normal vector to $\partial\Omega$, $\nabla\vec{Z}_N$ is an $m \times 3$ tensor and $*$ is the inner product in an m-dimensional space. It follows from the homogeneous condition that

$$\int_\Omega \Delta\vec{Z}_N \cdot \vec{Z}_N = -\|\nabla\vec{Z}_N\|^2_{L^2(\Omega)}.$$

For the last term on the right-hand side of (4.1.13), we have

$$\int_\Omega \vec{f}(\vec{Z}_N) \cdot \vec{Z}_N = \int_0^1 d\tau \int_\Omega \left(\frac{\partial\vec{f}(x,t,\tau\vec{Z}_N(x,t))}{\partial\vec{Z}} \cdot \vec{Z}_N(x,t) \right) \cdot \vec{Z}_N(x,t)dx$$

$$+ \int_\Omega \vec{f}(x,t,0) \cdot \vec{Z}_N(x,t)dx,$$

and hence

$$\int_\Omega \vec{f}(\vec{Z}_N) \cdot \vec{Z}_N \leq \left(b + \frac{1}{2} \right) \|\vec{Z}_N(\cdot,t)\|^2_{L^2(\Omega)}$$

$$+ \frac{1}{2}\|\vec{f}(\cdot,t,0)\|^2_{L^2(\Omega)}.$$

Therefore it follows from (4.1.13) that

$$\frac{d}{dt}\|\vec{Z}_N(\cdot,t)\|^2_{L^2(\Omega)} + 2\varepsilon\|\nabla\vec{Z}_N(\cdot,t)\|^2_{L^2(\Omega)}$$

$$\leq (2b+1)\|\vec{Z}_N(\cdot,t)\|^2_{L^2(\Omega)} + \|\vec{f}(\cdot,t,0)\|^2_{L^2(\Omega)}. \tag{4.1.14}$$

This proves the lemma.

Lemma 4.1.2 *Let conditions* (1)–(3) *hold. For the approximate solution* $\vec{Z}_N(x,t)$,

$$\sup_{0 \leq t \leq T} \|\nabla\vec{Z}_N\|_{L^2(\Omega)} \leq K_2 \tag{4.1.15}$$

holds, where K_2 *is independent of* ε *and* N.

Proof. Multiplying (4.1.6) by $-\lambda_s\alpha_{sN}(t)$ and summing from $s = 1$ to $s = N$, we have

$$\int_\Omega \vec{Z}_{Nt} \cdot \Delta\vec{Z}_N dx = \int_\Omega (\vec{Z}_N \times \Delta\vec{Z}_N) \cdot \Delta\vec{Z}_N dx$$

$$+ \varepsilon\int_\Omega \Delta\vec{Z}_N \cdot \Delta\vec{Z}_N dx + \int_\Omega \vec{f}(\vec{Z}_N) \cdot \Delta\vec{Z}_N dx. \tag{4.1.16}$$

The first term in the above equality is

$$\int_\Omega \vec{Z}_{Nt} \cdot \Delta\vec{Z}_N = -\frac{1}{2}\frac{d}{dt}\|\nabla\vec{Z}_N(\cdot,t)\|^2_{L^2(\Omega)}. \tag{4.1.17}$$

The last term on the right-hand side of (4.1.16) is

$$\int_\Omega \vec{f}(\vec{Z}_N) \cdot \Delta\vec{Z}_N = \int_{\partial\Omega} (\vec{f}(\vec{Z}_N) * \nabla\vec{Z}_N) * \nu_m - \int_\Omega D\vec{f}(\vec{Z}_N) * \Delta\vec{Z}_N$$

$$= -\int_\Omega \nabla_x\vec{f} * \nabla\vec{Z}_N - \int_\Omega \left(\frac{\partial\vec{f}}{\partial\vec{Z}_N} \cdot \nabla\vec{Z}_N \right) * \nabla\vec{Z}_N. \tag{4.1.18}$$

Since $\frac{\partial \vec{f}}{\partial \vec{Z}}$ is semi-bounded, one has

$$\int_\Omega \left(\frac{\partial \vec{f}}{\partial \vec{Z}_N} \cdot \nabla \vec{Z}_N \right) * \nabla \vec{Z}_N \leq b \|\nabla \vec{Z}_N(\cdot, t)\|_{L^2(\Omega)}^2.$$

The first term on the right of (4.1.18) can be estimated by

$$\left| \int_\Omega \nabla \vec{f} * \nabla \vec{Z}_N \right| \leq \frac{1}{2} \|\nabla \vec{Z}_N(\cdot, t)\|_{L^2(\Omega)}^2 + \frac{1}{2} \|\nabla \vec{f}(\cdot, t, \vec{Z}_N(\cdot, t))\|_{L^2(\Omega)}^2.$$

It follows from (4.1.10) that

$$\|\nabla \vec{f}(\cdot, t, \vec{Z}_N(\cdot, t))\|_{L^2(\Omega)}^2 \leq C_1 \int_\Omega |\vec{Z}_N(x, t)|^{2 + \frac{4}{m}} dx + C_2,$$

where C_1, C_2 are constants. From the Sobolev inequality, we have

$$\int_\Omega |\vec{Z}_N(x, t)|^{2 + \frac{4}{m}} dx \leq C_3 \|\vec{Z}_N(\cdot, t)\|_{L^2(\Omega)}^{4/m} \|\nabla \vec{Z}_N(\cdot, t))\|_{L^2(\Omega)}^2$$
$$+ C_4 \|\vec{Z}_N(\cdot, t))\|_{L^2(\Omega)}^{2 + \frac{4}{m}},$$

where C_3, C_4 are constants.

Finally, we have from (4.1.16) that

$$\frac{d}{dt} \|\nabla \vec{Z}_N(\cdot, t)\|_{L^2(\Omega)}^2 + 2\varepsilon \|\Delta \vec{Z}_N(\cdot, t)\|_{L^2(\Omega)}^2$$
$$\leq (2b + C_5) \|\nabla \vec{Z}_N(\cdot, t)\|_{L^2(\Omega)}^2 + C_6,$$

where C_5, C_6 are constants depending only on $\sup_{0 \leq t \leq T} \|\vec{Z}_N(\cdot, t)\|_{L^2(\Omega)}$ but independent of N. Applying the Gronwall inequality to the last inequality, we can finish the proof of this lemma.

Lemma 4.1.3 *Let* (1)–(3) *hold. For above approximate solutions,*

$$\sup_{0 \leq t \leq T} \|\vec{Z}_{Nt}(\cdot, t)\|_{H^{-(2 + [\frac{m}{2}])}(\Omega)} \leq K_3 \qquad (4.1.19)$$

holds with K_3 independent of ε and N.

Proof. Let $\nu(x) \in H_0^{2 + [\frac{m}{2}]}(\Omega)$ be a test function. We have

$$\nu(x) = \nu_N(x) + \overline{\nu}_N(x),$$
$$\nu_N(x) = \sum_{n=1}^N \beta_n w_n(x),$$
$$\overline{\nu}_N(x) = \sum_{n=N+1}^\infty \beta_n w_n(x),$$

where $\beta_n = \int_\Omega \nu(x) w_n(x) dx$, $n = 1, 2, \ldots$. If $s \geq N + 1$, then

$$\int_\Omega \vec{Z}_{Nt} w_s = 0.$$

If $1 \leq s \leq N$, one has

$$\int_\Omega \vec{Z}_{Nt} w_s = \int_{\partial\Omega} [(\vec{Z}_N \times \nabla \vec{Z}_N) * \nu_m] w_s - \int_\Omega (\vec{Z}_N \times \nabla \vec{Z}_N) * \nabla w_s$$
$$+ \varepsilon \int_{\partial\Omega} (\vec{Z}_N * \nu_m) w_s - \varepsilon \int_\Omega \nabla \vec{Z}_N * \nabla w_s$$
$$+ \int_\Omega \vec{f}(x, t, \vec{Z}_N(x, t)) w_s(x) dx. \tag{4.1.20}$$

We estimate every term on the right-hand side of (4.1.20). The first and the third term are equal to zero. For the second term, we have

$$\left| \int_\Omega (\vec{Z}_N \times \nabla \vec{Z}_N) * \nabla w_s \right| \leq C_7 \|\nabla w_s\|_{L^\infty(\Omega)} \|\vec{Z}_N(\cdot, t)\|_{L^2(\Omega)} \|\nabla \vec{Z}_N(\cdot, t)\|_{L^2(\Omega)}.$$

It follows from the Sobolev inequality that

$$\|\nabla w_s\|_{L^\infty(\Omega)} \leq C_8 \|w_s\|_{H^{2 + [\frac{m}{2}]}(\Omega)}.$$

Therefore, we have

$$\left| \int_\Omega (\vec{Z}_N \times \nabla \vec{Z}_N) * \nabla w_s \right| \leq C_9 \|w_s\|_{H^{2 + [\frac{m}{2}]}(\Omega)},$$

where C_9 depends only on $\sup_{0 \leq t \leq T} \|\vec{Z}_N(\cdot, t)\|_{H^1(\Omega)}$ but is independent of N.

For the fourth term on the right-hand side of (4.1.20), we have

$$\left| \int_\Omega \nabla \vec{Z}_N * \nabla w_s \right| \leq C_{10} \|\nabla \vec{Z}_N(\cdot, t)\|_{L^2(\Omega)} \|\nabla w_s\|_{L^2(\Omega)}.$$

For the last term on the right-hand side of (4.1.20), we have

$$\left| \int_\Omega \vec{f}(x, t, \vec{Z}_N(x, t)) w_s(x) dx \right|$$
$$\leq \|w_s\|_{L^\infty(\Omega)} \int_\Omega |\vec{f}(x, t, \vec{Z}_N(x, t))| dx$$
$$\leq \|w_s\|_{L^\infty(\Omega)} \left(A \int_\Omega |\vec{Z}_N(x, t)|^l dx + B \right)$$
$$\leq C_{11} \|w_s\|_{L^\infty(\Omega)} \leq C_{12} \|w_s\|_{H^{2 + [\frac{m}{2}]}(\Omega)},$$

where we have used

$$\int_\Omega |\vec{Z}_N(x, t)|^l dx \leq C_{13} \|\vec{Z}_N(\cdot, t)\|_{L^2(\Omega)}^{(1-\alpha)l} \|\vec{Z}_N(\cdot, t)\|_{H^1(\Omega)}^{\alpha l},$$

$\alpha = \frac{m}{2} - \frac{m}{l}$, $2 \leq l \leq \frac{2m}{m-2}$.

In summary we have

$$\left| \int_\Omega \vec{Z}_{Nt} w_s \right| \le C_{14} \|w_s\|_{H^{2+[\frac{m}{2}]}(\Omega)},$$

where constant C_{14} depends on $\sup_{0 \le t \le T} \|\vec{Z}_N(\cdot, t)\|_{H^1(\Omega)}$ but is independent of ε, $0 \le t \le T$, and N.

Similarly, we also have

$$\left| \int_\Omega \vec{Z}_{Nt} \nu_N \right| \le C_{14} \|\nu_N\|_{H^{2+[\frac{m}{2}]}(\Omega)},$$

and hence

$$\left| \int_\Omega \vec{Z}_{Nt} \nu \right| \le C_{14} \|\nu\|_{H^{2+[\frac{m}{2}]}(\Omega)}, \quad \forall \ \nu \in H_0^{2+[\frac{m}{2}](\Omega)}.$$

The proof is complete.

4. *Existence of global weak solution*

Applying the fixed point theorem and the *a priori* estimates on the approximate solutions, we can prove the existence of global solution $\alpha_{sN}(t)$ of system (4.1.6) and (4.1.7) ($0 \le t \le T$, $s = 1, 2, \ldots, N$).

Lemma 4.1.4 *Let (1)–(3) hold. Then the ordinary differential system (4.1.6) and (4.1.7) admits at least one global continuously differentiable solution $\alpha_{sN}(t)$ ($0 \le t \le T$, $s = 1, 2, \ldots, N$).*

Lemma 4.1.5 *Let (1)–(3) hold. Then for the approximate solution,*

$$\|\vec{Z}_N(\cdot, t + \Delta t) - \vec{Z}_N(\cdot, t)\|_{L^2(\Omega)} \le K_4 \Delta t^{\frac{1}{3+[\frac{m}{2}]}} \tag{4.1.21}$$

holds, where K_4 is independent of ε, N, and $0 \le t$, $t + \Delta t \le T$.

Proof. It follows from the Sobolev inequality with negative index that

$$\|\vec{Z}_N(\cdot, t)\|_{L^2(\Omega)} \le C_{15} \|\vec{Z}_N(\cdot, t)\|_{H^{-(2+[\frac{m}{2}])}(\Omega)}^{\frac{1}{3+[\frac{m}{2}]}} \|\vec{Z}_N(\cdot, t)\|_{H^1(\Omega)}^{\frac{2+[\frac{m}{2}]}{3+[\frac{m}{2}]}}.$$

Applying this inequality to $\vec{Z}_N(x, t + \Delta t) - \vec{Z}_N(x, t)$, we get

$$\|\vec{Z}_N(\cdot, t + \Delta t) - \vec{Z}_N(\cdot, t)\|_{L^2(\Omega)}$$

$$\le C_{15} \|\vec{Z}_N(\cdot, t + \Delta t) - \vec{Z}_N(\cdot, t)\|_{H^{-(2+[\frac{m}{2}])}(\Omega)}^{\frac{1}{3+[\frac{m}{2}]}} \cdot \|\vec{Z}_N(\cdot, t + \Delta t) - \vec{Z}_N(\cdot, t)\|_{H^1(\Omega)}^{\frac{2+[\frac{m}{2}]}{3+[\frac{m}{2}]}},$$

where $0 \le t, t + \Delta t \le T$.

Thus,

$$\|\vec{Z}_N(\cdot, t + \Delta t) - \vec{Z}_N(\cdot, t)\|_{L^2(\Omega)}$$

$$\le C_{15} \Delta t^{\frac{1}{3+[\frac{m}{2}]}} \sup_{0 \le t \le T} \|\vec{Z}_N(\cdot, t)\|_{H^{-(2+[\frac{m}{2}])}(\Omega)}^{\frac{1}{3+[\frac{m}{2}]}} \cdot \sup_{0 \le t \le T} \|\vec{Z}_N(\cdot, t)\|_{H^1(\Omega)}^{\frac{2+[\frac{m}{2}]}{3+[\frac{m}{2}]}}$$

holds. This proves the lemma.

4.1.2 Weak Solution to Multidimensional System of Ferromagnetic Spin Chain

1. *Convergence of approximate solutions*

In order to get a weak solution to system (4.1.4) or (4.1.1) with homogeneous boundary condition (4.1.2) and initial condition (4.1.3), we allow $N \to \infty$.

Definition 4.1.1 *A vector-valued function $\vec{Z}(x,t) \in L^\infty(0,T;H^1(\Omega))$ is called a weak solution to system (4.1.4) ($\varepsilon > 0$) or (4.1.1) ($\varepsilon = 0$) with homogeneous boundary condition (4.1.2) and initial condition (4.1.3), if for any test function $\nu(x,t) \in C^1(Q_T)$: $\nu(x,t) = 0$ for $x \in \overline{\Omega}$ and $t = T$ and $x \in \partial\Omega$ and $0 \le t \le T$, then*

$$\int_0^T \int_\Omega [\nu_t(x,t)\vec{Z}(x,t) - \nabla\nu(x,t) * (\vec{Z}(x,t) \times \nabla\vec{Z}(x,t))$$
$$- \varepsilon\nabla\nu(x,t) * \nabla\vec{Z}(x,t) + \nu(x,t)\vec{f}(x,t,\vec{Z}(x,t))]dxdt$$
$$+ \int_\Omega \nu(x,0)\varphi(x)dx = 0, \quad \varepsilon \ge 0 \tag{4.1.22}$$

holds and $\vec{Z}(x,t)$ satisfies the homogeneous boundary condition on manifold $\partial\Omega \times [0,T]$.

The Galerkin approximate function of $\nu(x,t)$ is

$$\nu_N(x,t) = \sum_{n=1}^{N} \beta_{nN}(t)w_n(x),$$

which uniformly converges in $C^1(Q_T)$ to $\nu(x,t)$ as $N \to \infty$, where

$$\beta_{sN}(t) = \int_\Omega \nu(x,t)w_s(x)dx, \quad s = 1,2,\ldots,N.$$

It follows from (4.1.6) that the following integral equality holds:

$$\int_0^T \int_\Omega [\nu_{Nt}(x,t)\vec{Z}_N(x,t) - \nabla\nu_N(x,t) * (\vec{Z}_N(x,t) \times \nabla\vec{Z}_N(x,t))$$
$$- \varepsilon\nabla\nu_N(x,t) * \nabla\vec{Z}_N(x,t) + \nu_N(x,t)\vec{f}(x,t,\vec{Z}_N(x,t))]dxdt$$
$$+ \int_\Omega \nu_N(x,0)\varphi_N(x)dx = 0, \quad \varepsilon \ge 0. \tag{4.1.23}$$

2. *Estimates of uniform boundedness and convergence*

It follows from former estimates that $\vec{Z}_N(x,t)$ is uniformly bounded in the space:

$$G = L^\infty(0,T;H_0^1(\Omega)) \bigcap W_\infty^{(1)}(0,T;H^{-(2+[\frac{m}{2}])}(\Omega)).$$

Then we may choose a subsequence $\vec{Z}_{N_i}(x,t)$ from $\vec{Z}_N(x,t)$ such that $\vec{Z}_{N_i}(x,t)$ converges in $L^p(0,T;L^2(\Omega))$ to a vector $\vec{Z}(x,t) \in L^p(0,T;H^1(\Omega))$ $(1 < p < \infty)$, and $\nabla \vec{Z}_{N_i}(x,t)$ converges in $L^p(0,T;L^2(\Omega))$ to $\nabla \vec{Z}(x,t)$. On the other hand, since $\vec{Z}_{N_i t}(x,t)$ is uniformly bounded in $L^p(0,T;H^{-(2+[\frac{m}{2}])}(\Omega))$ $(1 < p < \infty)$, we may assume that $\vec{Z}_{N_i t}(x,t)$ weakly converges to a vector $w(x,t) \in L^p(0,T;H^{-(2+[\frac{m}{2}])}(\Omega))$ $(1 < p < \infty)$. Now we claim $w(x,t) = \vec{Z}_t(x,t)$. In fact, we have

$$\int_0^T \int_\Omega \vec{Z}_{N_i t}\nu(x,t)dxdt = -\int_0^T \int_\Omega \vec{Z}_{N_i}\nu_t(x,t)dxdt,$$

where $\nu(x,t)$ is any test function with compact support set in Q_T. Let $N_i \to \infty$, we get

$$\int_0^T \int_\Omega w(x,t)\nu(x,t)dxdt = -\int_0^T \int_\Omega \vec{Z}(x,t)\nu_t(x,t)dxdt.$$

The claim follows.

This means that

$$\vec{Z}(x,t) \in G_p = L^p(0,T;H^1(\Omega)) \bigcap W_p^{(1)}(0,T;H^{-(2+[\frac{m}{2}])}(\Omega)), \quad 1 < p < \infty,$$

and the norm of $\vec{Z}(x,t)$ in G_p is uniformly bounded in p. Then $\vec{Z}(x,t) \in G_\infty$. We have

Lemma 4.1.6 *Suppose that (1)–(3) hold. Then for the limit function $\vec{Z}_\varepsilon(x,t)$ of the approximate solution $\vec{Z}_N(x,t)$,*

$$\sup_{0 \le t \le T} \|\vec{Z}_\varepsilon(\cdot,t)\|_{H^1(\Omega)} \le K_5, \tag{4.1.24}$$

$$\sup_{0 \le t \le T} \|\vec{Z}_{\varepsilon t}(\cdot,t)\|_{H^{-(2+[\frac{m}{2}])}(\Omega)} \le K_6, \tag{4.1.25}$$

$$\|\vec{Z}_\varepsilon(\cdot,t+\Delta t) - \vec{Z}_\varepsilon(\cdot,t)\|_{L^2(\Omega)} \le K_7 \Delta t^{\frac{1}{3+[\frac{m}{2}]}} \tag{4.1.26}$$

hold, where K_5, K_6, K_7 are independent of ε and $0 \le t$, $t + \Delta t \le T$.

Choose a suitable subsequence from $\{\vec{Z}_{N_i}(x,t)\}$, still denoted by $\{\vec{Z}_{N_i}(x,t)\}$, such that for some vector-valued function $\vec{Z}(x,t)$ $\vec{Z}_{N_i}(x,t) \to \vec{Z}(x,t)$ holds strongly in $L^2(\Omega)$ for every $0 \le t \le T$, and $\|\vec{Z}_{N_i}(\cdot,t)\|_{L^2(\Omega)} \to \|\vec{Z}(\cdot,t)\|_{L^2(\Omega)}$ uniformly in $0 \le t \le T$. More precisely, $\vec{Z}_{N_i}(x,t)$ converges to $\vec{Z}(x,t)$ in $C^{(0,\frac{1}{3+[\frac{m}{2}]+\delta})}(0,T;L^2(\Omega))$, $\delta > 0$.

It is clear that $\vec{Z}(x,t)$ satisfies the homogeneous boundary condition almost everywhere on $\partial\Omega \times [0,T]$.

Now we consider the limit of the integral (4.1.23) as $N_i \to \infty$. Since as $N_i \to \infty$, $\{\nu_{N_i t}(x,t)\}$ and $\{\nabla\nu_{N_i}(x,t)\}$ uniformly converge in Q_T to $\nu_t(x,t)$ and $\nabla\nu(x,t)$, $\{\nu_{N_i}(x,0)\}$ uniformly converge in Ω to $\nu(x,0)$; $\{\varphi_{N_i}(x)\}$ converge in $L^2(\Omega)$ to $\varphi(x)$, for the second term of (4.1.23), we have

$$\left| \int_0^T \int_\Omega [\nabla\nu_{N_i} * (\vec{Z}_{N_i} \times \nabla\vec{Z}_{N_i}) - \nabla\nu * (\vec{Z} \times \nabla\vec{Z})] \right|$$

$$\le \left| \int_0^T \int_\Omega \nabla(\nu_{N_i} - \nu) * (\vec{Z}_{N_i} \times \nabla\vec{Z}_{N_i}) \right|$$

$$+\left|\int_0^T \int_\Omega \nabla\nu * ((\vec{Z}_{N_i} - \vec{Z}) \times \nabla\vec{Z}_{N_i})\right|$$

$$+\left|\int_0^T \int_\Omega \nabla\nu * (\vec{Z} \times \nabla(\vec{Z}_{N_i} - \vec{Z}))\right|$$

$$\leq \|\nabla(\nu_{N_i} - \nu)\|_{L^\infty(Q_T)}\|\vec{Z}_{N_i}\|_{L^2(Q_T)}\|\nabla\vec{Z}_{N_i}\|_{L^2(Q_T)}$$

$$+ \|\nabla\nu\|_{L^\infty(Q_T)}\|\vec{Z}_{N_i} - \vec{Z}\|_{L^2(Q_T)}\|\nabla\vec{Z}_{N_i}\|_{L^2(Q_T)}$$

$$+\left|\int_0^T \int_\Omega \nabla\nu * (\vec{Z} \times \nabla(\vec{Z}_{N_i} - \vec{Z}))\right|.$$

Every term of the above inequality tends to zero as $N_i \to \infty$.

Since $\{\vec{Z}_{N_i}(x,t)\}$ strongly converges to $\vec{Z}(x,t)$, $\{\vec{Z}_{N_i}(x,t)\}$ converges to $\vec{Z}(x,t)$ a.e. on Q_T, and $\{\vec{f}(x,t,\vec{Z}_{N_i}(x,t))\}$ converges to $\vec{f}(x,t,\vec{Z}(x,t))$ a.e. on Q_T, too. Therefore it follows from (4.1.10) that

$$\|\vec{f}(x,t,\vec{Z}_{N_i}(\cdot,t))\|_{L^q(\Omega)}^q = \int_\Omega |\vec{f}(x,t,\vec{Z}_{N_i}(x,t))|^q dx$$

$$\leq C_{17}\int_\Omega |\vec{Z}_{N_i}(x,t)|^{2+\frac{4}{m-2}} dx + C_{18},$$

where $q = (2 + \frac{4}{m-2})/l > 1$ since $l < 2 + \frac{4}{m-2}$. Hence $\{\vec{f}(x,t,\vec{Z}_{N_i}(x,t))\}$ weakly converges to $\vec{f}(x,t,\vec{Z}(x,t))$ in $L^\infty(0,T;L^q(\Omega))$ as $N_i \to \infty$.

Now we may send $N_i \to \infty$ so that (4.1.23) tends to (4.1.22). This means that the limit vector $\vec{Z}(x,t)$ is a weak solution of (4.1.4) ($\varepsilon > 0$) or (4.1.1) ($\varepsilon = 0$) and satisfies the homogeneous boundary condition (4.1.2) and initial condition (4.1.3).

3. *The global weak solution with homogeneous boundary condition*

Theorem 4.1.1 *Let* (1)–(3) *hold. Then Eqs.* (4.1.4) *($\varepsilon > 0$) and* (4.1.1) *($\varepsilon = 0$) with conditions* (4.1.2) *and* (4.1.3) *admit at least one global weak solution*

$$\vec{Z}(x,t) \in L^\infty(0,T;H_0^1(\Omega)) \bigcap C^{(0,\frac{1}{3+[\frac{m}{2}]})}(0,T;L^2(\Omega)).$$

Theorem 4.1.2 *Let* (1)–(3) *hold. Let $\vec{Z}(x,t)$ be any limit function of any convergent sequence of weak solutions $\{\vec{Z}_\varepsilon(x,t)\}$ to the spin equation* (4.1.4) *($\varepsilon > 0$) with conditions* (4.1.2) *and* (4.1.3) *in $C^{(0,\frac{1}{3+[\frac{m}{2}]})}(0,T;L^2(\Omega)) \cap L^\infty(0,T;H_0^1(\Omega))$ as $\varepsilon \to 0$. Then $\vec{Z}(x,t)$ is a weak solution of problem* (4.1.1)–(4.1.3).

Proof. For any $\varepsilon > 0$, there exists at least one $\vec{Z}_\varepsilon(x,t) \in G$ such that

$$\int_0^T \int_\Omega [\nu_t(x,t)\vec{Z}_\varepsilon(x,t) - \nabla\nu(x,t) * (\vec{Z}_\varepsilon(x,t) \times \nabla\vec{Z}_\varepsilon(x,t))$$

$$- \varepsilon\nabla\nu(x,t) * \nabla\vec{Z}_\varepsilon(x,t) + \nu(x,t)\vec{f}(x,t,\vec{Z}_\varepsilon(x,t))]dxdt$$

$$+ \int_\Omega \nu(x,0)\varphi(x)dx = 0, \quad \varepsilon \geq 0, \tag{4.1.27}$$

where $\nu(x,t)$ is any test function. Since

$$\varepsilon \int_0^T \int_\Omega \nabla \nu(x,t) * \vec{Z}_\varepsilon(x,t)dxdt \to 0, \quad \text{as } \varepsilon \to 0,$$

limit $\vec{Z}(x,t) \in G$ meets

$$\int_0^T \int_\Omega [\nu_t(x,t)\vec{Z}(x,t) - \nabla \nu(x,t) * (\vec{Z}(x,t) \times \nabla \vec{Z}(x,t)) + \nu(x,t)\vec{f}(x,t,\vec{Z}_\varepsilon(x,t))]dxdt$$
$$+ \int_\Omega \nu(x,0)\varphi(x)dx = 0.$$

That is,

$$\vec{Z}(x,t) \in L^\infty(0,T; H_0^1(\Omega)) \bigcap C^{(0,\frac{1}{3+[\frac{m}{2}]})}(0,T; L^2(\Omega))$$

is a weak solution of problem (4.1.1)–(4.1.3).

4. *Global weak solution in the cylinder with infinite length*

In former discussion, T may be arbitrary. We denote the cylinder with infinite length by

$$Q_\infty = \{x \in \Omega, 0 \le t \le \infty\}.$$

We see that the ordinary system (4.1.6) and (4.1.7) of coefficients $\alpha_{nN}(t)$ admits at least one continuously differentiable solution $\alpha_{nN}(t)$ which exists on $[0,\infty]$ so that the approximate solution $\vec{Z}_N(x,t)$ exists on Q_∞ too.

Let T_k be a number sequence such that $T_k \to \infty$ as $k \to \infty$. Let $\{\vec{Z}_{N_k,i}(x,t)\}$ $(k = 1,2,\ldots; i = 1,2,\ldots)$ be the kth subsequence of $\vec{Z}_N(x,t)$ such that

(1) $\{\vec{Z}_{N_{k+1},i}(x,t)\} \subset \{\vec{Z}_{N_k,i}(x,t)\}$;

(2) there is a vector-valued function $\vec{Z}(x,t)$ defined on Q_∞ such that the subsequence $\{\vec{Z}_{N_k,i}(x,t)\}$ weakly converges to $\vec{Z}(x,t)$ in

$$G(T_k) = L^\infty(0,T_k; H_0^1(\Omega)) \bigcap W_\infty^{(1)}(0,T_k; H^{-(2+[\frac{m}{2}])}(\Omega))$$
$$\bigcap C^{(0,\frac{1}{3+[\frac{m}{2}]})}(0,T_k; L^2(\Omega)), \quad k = 1,2,\ldots.$$

Hence we may take the diagonal method to choose a subsequence $\vec{Z}_{N_k,k}(x,t)$ that, on any cylinder Q_T, weakly converges in G to $\vec{Z}(x,t)$. Hence $\vec{Z}(x,t)$ is in the space

$$G_\infty = L_{\text{loc}}^\infty(0,\infty; H_0^1(\Omega)) \bigcap H_{\text{loc}}^1(0,\infty; H^{-(2+[\frac{m}{2}])}(\Omega))$$
$$\subset L_{\text{loc}}^\infty(0,\infty; H_0^1(\Omega)) \bigcap C^{(0,\frac{1}{3+[\frac{m}{2}]})}(0,\infty; L^2(\Omega)),$$

and meets the integral relation (4.1.22).

Theorem 4.1.3 *Suppose*

(i) *the 3×3 Jacobi matrix $\vec{f}_{\vec{Z}}(x,t,\vec{Z})$ is semi-bounded, i.e. for $(x,t) \in Q_\infty$ and $\vec{Z} \in R^3$, (4.1.9) holds.*

(ii) *for any $0 < T < \infty$ and $(x,t) \in Q_T$ and $\vec{Z} \in R^3$,*

$$\begin{cases} |\vec{f}(x,t,\vec{Z})| \le A(T)|\vec{Z}|^l + B(T), \\ |\nabla_x \vec{f}(x,t,\vec{Z})| \le A(T)|\vec{Z}|^{1+\frac{2}{m}} + B(T), \\ \vec{f}(x,t,0) \equiv 0 \end{cases} \qquad (4.1.28)$$

holds, where $A(T)$, $B(T)$ are positive constant depending on T, and

$$2 \le l \le 2 + \frac{4}{m-2}, \quad m \ge 2.$$

(iii) *Vector $\varphi(x) \in H_0^1(\Omega)$.*

Then both the spin equation (4.1.4) ($\varepsilon > 0$) and the ferromagnetic spin chain equation (4.1.1) ($\varepsilon = 0$) with conditions (4.1.2) and (4.1.3) have a global weak solution

$$\vec{Z}(x,t) \in L^\infty_{\text{loc}}(0,\infty; H_0^1(\Omega)) \bigcap C_{\text{loc}}^{(0, \frac{1}{3+[\frac{m}{2}]})}(0,\infty; L^2(\Omega)).$$

If $b < 0$, it follows from (4.1.14) that

$$\frac{d}{dt}\|\vec{Z}_N(\cdot,t)\|^2_{L^2(\Omega)} \le -2|b|\|\vec{Z}_N(\cdot,t)\|^2_{L^2(\Omega)},$$

and if (4.1.4) and (4.1.1) are both homogeneous, i.e. $\vec{f}(x,t,0) \equiv 0$, then

$$\|\vec{Z}_N(\cdot,t)\|_{L^2(\Omega)} \le \|\varphi\|_{L^2(\Omega)} e^{-|b|t}, \; 0 \le t < \infty.$$

We conclude that the similar inequality also holds for limit $\vec{Z}(x,t)$, that is, we have

Theorem 4.1.4 *Let* (i)–(iii) *hold. Then for the weak solution $\vec{Z}(x,t)$ to the spin equation (4.1.4) ($\varepsilon > 0$) or the ferromagnetic spin chain equation (4.1.1) ($\varepsilon = 0$) in Q_∞ with conditions (4.1.2) and (4.1.3),*

$$\lim_{t \to \infty} \|\vec{Z}(\cdot,t)\|_{L^2(\Omega)} = 0$$

holds.

5. *Uniqueness of smooth solution*

Theorem 4.1.5 *Assume that $\vec{f}(x,t,\vec{Z})$ is twice continuously differentiable with respect to x, and $\vec{Z}$, and the Jacobi matrix $\vec{f}_{\vec{Z}}(x,t,\vec{Z})$ is semi-bounded. Then the classical solution to problem (4.1.1)–(4.1.3) is unique.*

Proof. Let $\vec{u}(x,t), \vec{Z}(x,t)$ be two solutions. Set $w(x,t) = \vec{u}(x,t) - \vec{Z}(x,t)$. We have

$$\frac{d}{dt}\|w(\cdot,t)\|^2_{L^2(\Omega)} = 2\int_{\partial\Omega}(\nabla w \cdot (w \times \vec{u})) * \nu_m$$

$$-2\int_\Omega \nabla w * (w \times \nabla\vec{u}) + 2\int_\Omega w \cdot \frac{\partial \tilde{f}}{\partial \vec{Z}} \cdot w,$$

$$\frac{d}{dt}\|\nabla w(\cdot,t)\|_{L^2(\Omega)}^2 = 2\int_{\partial\Omega}(\nabla w\cdot(w\times\Delta\vec{Z}))*\nu_m - 2\int_\Omega \nabla w*(w\times\nabla\Delta\vec{Z})$$

$$+ 2\int_{\partial\Omega}\left(\nabla w\cdot\frac{\partial\tilde{f}}{\partial\vec{Z}}\cdot w\right)*\nu_m - 2\int_\Omega \nabla w*\nabla\frac{\partial\tilde{f}}{\partial\vec{Z}}\cdot\nabla w$$

$$- 2\int_\Omega \nabla w*\left(\frac{\partial^2\tilde{f}}{\partial\vec{Z}^2}\cdot w\right)\cdot\nabla\vec{u} - 2\int_\Omega \nabla w*\frac{\partial\vec{f}(x,t,\vec{Z})}{\partial\vec{Z}}\cdot\nabla w,$$

where

$$\frac{\partial\tilde{f}}{\partial\vec{Z}} = \int_0^1 \frac{\partial\vec{f}(x,t,\tau\vec{u}+(1-\tau)\vec{Z})}{\partial\vec{Z}}d\tau,$$

$$\frac{\partial^2\tilde{f}}{\partial\vec{Z}^2} = \int_0^1 \frac{\partial^2\vec{f}(x,t,\tau\vec{u}+(1-\tau)\vec{Z})}{\partial\vec{Z}^2}d\tau,$$

$$\nabla\frac{\partial\tilde{f}}{\partial\vec{Z}} = \int_0^1 \nabla\frac{\partial\vec{f}(x,t,\tau\vec{u}+(1-\tau)\vec{Z})}{\partial\vec{Z}}d\tau.$$

Hence $w(x,t)\in C^{(3,1)}(Q_T)$ satisfies the homogeneous equation and the homogeneous boundary and initial conditions. Thus

$$\frac{d}{dt}\|\nabla w(\cdot,t)\|_{H^1(\Omega)}^2 \leq C_{18}\|w(\cdot,t)\|_{H^1(\Omega)}^2.$$

This proves the theorem.

Finally we consider the "blow-up" problem for the weak solution $\vec{Z}(x,t)$ of (4.1.1).

Theorem 4.1.6 *If*

$$\vec{Z}\cdot\vec{f}(x,t.,\vec{Z}) \geq C_0|\vec{Z}|^{2+\delta}, \quad (x,t)\in Q_T,\ \vec{Z}\in R^3, \tag{4.1.29}$$

holds, where $C_0 > 0$, $\delta > 0$, and $\|\varphi(x)\|_{L^2(\Omega)} > 0$, then the weak solution $\vec{Z}(x,t)\in W_2^{(2,1)}(Q_T)$ to (4.1.1) blows up at finite time, i.e. for a finite $t_0 > 0$

$$\lim_{t\to t_0-0}\|\vec{Z}(\cdot,t)\|_{L^p(\Omega)} \to \infty, \quad 2\leq p < \infty.$$

Proof. we have by multiplying (4.1.1) by $|\vec{Z}|^{p-2}\vec{Z}$ and integrating over $[0,l]$ with respect to x that

$$\frac{1}{p}\frac{d}{dt}\|\vec{Z}(\cdot,t)\|_{L^p(\Omega)}^p = \int_\Omega(|\vec{Z}(x,t)|^{p-2}\vec{Z}(x,t)\cdot\vec{f}(x,t,\vec{Z}))dx$$

$$\geq C_0\int_\Omega |\vec{Z}(x,t)|^{p+\delta}dx \geq C_0(\text{mes }\Omega)^{-\frac{\delta}{p}}\|\vec{Z}(\cdot,t)\|_{L^p(\Omega)}^{p+\delta},$$

or

$$\frac{d}{dt}\|\vec{Z}(\cdot,t)\|_{L^p(\Omega)} \geq C_0(\text{mes }\Omega)^{-\frac{\delta}{p}}\|\vec{Z}(\cdot,t)\|_{L^p(\Omega)}^{1+\delta},$$

and then

$$\|\vec{Z}(\cdot,t)\|_{L^p(\Omega)} \geq (\|\varphi\|_{L^p(\Omega)}^{-\delta} - C_0 t\delta(\text{mes}\,\Omega)^{-\frac{\delta}{p}})^{-\frac{1}{\delta}}, \quad 2 \leq p < \infty.$$

This proves the theorem.

In the following we consider a more general blow-up result.

For example, we consider a general system

$$u_t = \nabla * A(x,t,u) \cdot \nabla u + f(x,t,u), \tag{4.1.30}$$

where $u(x,t) = (u_1(x,t),\ldots,u_n(x,t))$ is an n-dimensional vector-valued unknown defined on cylinder $Q_T = \{x \in \Omega \subset R^m, 0 \leq t \leq T\}$ and $\cdot$ and $*$ denote the inner product in R^n and R^m, respectively. $A(x,t,u)$ is a nonsingular and zero-definite matrix, and $f(x,t,u)$ is an n-dimensional vector-valued function satisfying

$$u \cdot f(x,t,u) \geq C_0|u|^{2+\delta} \tag{4.1.31}$$

for $(x,t) \in Q_T$ and $u \in R^n$, $C_0 > 0$, $\delta > 0$.

Multiplying (4.1.30) by u and integrating over Ω, one has

$$\frac{1}{2}\frac{d}{dt}\|u(\cdot,t)\|_{L^2(\Omega)}^2 = \int_{\partial\Omega}(u \cdot A \cdot \nabla u) * \nu_m$$
$$- \int_{\Omega} \nabla u * A \cdot \nabla u + \int_{\Omega} u \cdot f.$$

Consider the first initial boundary condition

$$u(x,t) = 0, \quad x \in \partial\Omega, \quad 0 \leq t \leq T, \tag{4.1.32}$$
$$u(x,0) = \varphi(x), \quad x \in \Omega; \tag{4.1.33}$$

or the second boundary condition

$$A(x,t,u(x,t)) \cdot \nabla u(x,t) * \nu_m = 0, \quad x \in \partial\Omega, \quad 0 \leq t \leq T, \tag{4.1.34}$$
$$u(x,0) = \varphi(x), \quad x \in \Omega. \tag{4.1.35}$$

We have

Theorem 4.1.7 *Under above condition on A, f, for solution $u(x,t)$ to problem (4.1.30), (4.1.32) and (4.1.33), or problem (4.1.30), (4.1.34) and (4.1.35), we have*

$$\|u(x,t)\|_{L^2(\Omega)} \to \infty, \quad t \to t_1 - 0,$$

where $t_1 > 0$ is finite and $\|\varphi\|_{L^2(\Omega)} > 0$.

4.2 Landau–Lifshitz Equations on Riemannian Manifold and Harmonic Maps

4.2.1 Landau–Lifshitz Equations and Harmonic Maps

1. *Landau–Lifshitz equations on Riemannian manifold*

Let M be an m-dimensional Riemannian manifold of metric γ without boundary, and S^2 be the surface of a unit sphere in R^3 centered at $x = 0$. In local coordinate $(x^1, x^2, \ldots, x^m)$ on M, the Laplace–Beltrami operator and norm $|Du(x)|$ are expressed by

$$\Delta_M := \frac{1}{\sqrt{\gamma}} \frac{\partial}{\partial x^\beta} \left(\gamma^{\alpha\beta} \sqrt{\gamma} \frac{\partial}{\partial x^\alpha} \right) = \gamma^{\alpha\beta} \left(\frac{\partial^2}{\partial x^\alpha \partial x^\beta} - \Gamma^k_{\alpha\beta} \frac{\partial}{\partial x^k} \right)$$

and

$$|Du(x)|^2 = \sum_{\alpha,\beta} \sum_i \gamma^{\alpha\beta} \frac{\partial u^i}{\partial x^\alpha} \frac{\partial u^i}{\partial x^\beta},$$

where $(\gamma^{\alpha\beta})$ is the inverse of $(\gamma_{\alpha\beta})$. The Landau–Lifshitz equations with the Gilbert damping term on M is given by

$$\vec{u}_t = -\alpha_1 \vec{u} \times \left(\vec{u} \times \frac{1}{\sqrt{\gamma}} \frac{\partial}{\partial x^\beta} \left(\gamma^{\alpha\beta} \sqrt{\gamma} \frac{\partial \vec{u}}{\partial x^\alpha} \right) \right)$$

$$+ \alpha_2 \vec{u} \times \frac{1}{\sqrt{\gamma}} \frac{\partial}{\partial x^\beta} \left(\gamma^{\alpha\beta} \sqrt{\gamma} \frac{\partial \vec{u}}{\partial x^\alpha} \right), \tag{4.2.1}$$

where $\alpha_1 > 0$ is the Gilbert constant and α_2 is a constant too, with initial value $\vec{u}_0(x)$ such that

$$|\vec{u}_0(x)|^2 = 1, \quad \forall \ x \in M. \tag{4.2.2}$$

2. *Harmonic maps*

Lemma 4.2.1 (Gronwall Inequality) *Let u, v, w be continuous functions for $t \geq 0$ and suppose that $w \geq 0$. Then the inequality*

$$u(t) \leq v(t) + \int_0^t w(s)u(s)ds$$

implies the inequality

$$u(t) \leq v(t) + \int_0^t w(s)v(s)\exp\left(\int_0^t w(r)dr \right) ds.$$

Now we can consider an equivalent form of (4.2.1) under condition (4.2.2)

$$\vec{u}_t = \alpha_1 \frac{1}{\sqrt{\gamma}} \frac{\partial}{\partial x^\beta} \left(\gamma^{\alpha\beta} \sqrt{\gamma} \frac{\partial \vec{u}}{\partial x^\alpha} \right)$$

$$+ \alpha_1 |\nabla \vec{u}|^2 \vec{u} + \alpha_2 \vec{u} \times \frac{1}{\sqrt{\gamma}} \frac{\partial}{\partial x^\beta} \left(\gamma^{\alpha\beta} \sqrt{\gamma} \frac{\partial \vec{u}}{\partial x^\alpha} \right), \tag{4.2.3}$$

with initial value $\vec{u}_0(x)$ such that

$$|\vec{u}_0(x)|^2 = 1, \quad \forall \ x \in M. \tag{4.2.4}$$

The equivalence follows from

Theorem 4.2.1 *In the classical sense, $\vec{u}$ is a solution of* (4.2.1) *and* (4.2.2) *if and only if $\vec{u}$ is a solution of* (4.2.3) *and* (4.2.4).

Proof. Suppose that $\vec{u}$ is a solution of (4.2.1) and (4.2.2). By the vector product formula

$$\vec{a} \times (\vec{b} \times \vec{c}) = (\vec{a} \cdot \vec{c})\vec{b} - (\vec{a} \cdot \vec{b})\vec{c},$$

we have

$$\vec{u} \times (\vec{u} \times \Delta_M \vec{u}) = (\vec{u} \cdot \Delta_M \vec{u}) - |\vec{u}|^2 \Delta_M \vec{u}.$$

Multiplying (4.2.1) by $\vec{u}$, we have

$$\vec{u} \cdot \vec{u}_t = 0, \quad \forall \ (x,t) \in M \times [0,\infty).$$

Then from (4.2.2) we get

$$|\vec{u}(x,t)|^2 = 1, \quad \forall \ (x,t) \in M \times [0,\infty).$$

This implies

$$\vec{u} \cdot D\vec{u} = 0, \quad \vec{u} \cdot \Delta_M u = -|D\vec{u}|^2.$$

Thus

$$\vec{u} \times (\vec{u} \times \Delta_M \vec{u}) = -\Delta_M \vec{u} - |D\vec{u}|^2 \vec{u}.$$

This proves that $\vec{u}$ solves (4.2.3) and (4.2.4).

Suppose that $\vec{u}$ is a solution of (4.2.3) and (4.2.4). Set $z(x,t) = |\vec{u}(x,t)|^2$ on $M_T := M \times [0,T]$, where T is a finite time $T < \infty$. We have

$$z_t = 2\vec{u} \cdot \vec{u}_t, \quad Dz = 2\vec{u} \cdot D\vec{u}.$$

By simple calculations we have

$$\Delta_M z = 2\vec{u} \cdot \Delta_M \vec{u} + 2|D\vec{u}|^2.$$

Thus we have

$$z_t = \alpha_1 \Delta_M z + 2\alpha_1 |D\vec{u}|^2(|\vec{u}|^2 - 1), \quad z(x,0) = 1, \ x \in M.$$

Setting $w = z - 1$, we have

$$w_t = \alpha_1 \Delta_M w + 2\alpha_1 |D\vec{u}|^2 w, \tag{4.2.5}$$

$$w(x,0) = 0. \tag{4.2.6}$$

Multiplying (4.2.5) by w and integrating it on M we get

$$\frac{1}{2}\frac{d}{dt}\int_M |w|^2 dx + \alpha_1 \int_M |Dw|^2 dx \leq 2\alpha_1 \max_{x,t} |D\vec{u}|^2 \int_M |w|^2 dx.$$

It follows from Gronwall's and Hölder's inequalities and (4.2.6) that $w(x,t) = 0$ for any $(x,t) \in M_T$, i.e. $|\vec{u}(x,t)|^2 = 1$. This proves that $\vec{u}$ solves (4.2.1) and (4.2.2).

Corollary 4.2.1 *Let $\alpha_2 = 0$ in (4.2.1). In the classical sense the Landau–Lifshitz equation is equivalent to the heat flow of harmonic maps*

$$\vec{u}_t = \Delta_M \vec{u} + |\Delta \vec{u}|^2 \vec{u}.$$

Now we establish a relation between harmonic maps and the solutions to the elliptic type Landau–Lifshitz equations.

Theorem 4.2.2 *In the classical sense, $\vec{u} : M \to S^2$ is a harmonic map if and only if $\vec{u}$ solves (4.2.1) and $\vec{u}_t(x,t) = 0$ for $t \geq 0$.*

Proof. First, suppose $\vec{u} : M \to S^2$ is a harmonic map, i.e.

$$-\Delta_M \vec{u} = |D\vec{u}|^2 \vec{u}.$$

By the property of vector cross products, we get

$$\vec{u}_t = -\alpha_1 \vec{u} \times (\vec{u} \times \Delta_M \vec{u}) + \alpha_2 \vec{u} \times \Delta_M \vec{u} = 0.$$

Second, suppose that $\vec{u}(x,t)$ solves (4.2.1) and $\vec{u}_t = 0$, i.e.

$$\alpha_1 \Delta_M \vec{u} + \alpha_1 |\Delta \vec{u}|^2 \vec{u} + \alpha_2 \vec{u} \times \Delta_M \vec{u} = 0. \tag{4.2.7}$$

By the cross product of $\vec{u}$ and (4.2.7), we have

$$\alpha_1 \vec{u} \times \Delta_M \vec{u} + \alpha_2 \vec{u} \times (\vec{u} \times \Delta_M \vec{u}) = 0.$$

By an argument similar to that of Theorem 4.2.1, we have

$$\alpha_1 \vec{u} \times \Delta_M \vec{u} - \alpha_2 \Delta_M \vec{u} - \alpha_2 |D\vec{u}|^2 \vec{u} = 0.$$

It implies that

$$\vec{u} \times \Delta_M \vec{u} = \frac{\alpha_2}{\alpha_1}(\Delta_M \vec{u} + |D\vec{u}|^2 \vec{u}). \tag{4.2.8}$$

Combining (4.2.8) with (4.2.7), we get

$$\alpha_1 \Delta_M \vec{u} + \alpha_1 |D\vec{u}|^2 \vec{u} + \frac{\alpha_2}{\alpha_1}(\Delta_M \vec{u} + |D\vec{u}|^2 \vec{u}) = 0, \tag{4.2.9}$$

which implies that $\vec{u}$ is a harmonic map. This proves Theorem 4.2.2.

Corollary 4.2.2. *Suppose that $\vec{u}_0 : M \to S^2$ is a given initial data. Then in the classical sense, solution $\vec{u}$ of the Landau–Lifshitz equation (4.2.1) ($\alpha_2 \neq 0$) is usually not equal to a solution of the equation of the harmonic map heat flow. It equals to one solution of the equation of the harmonic map heat flow if and only if $\vec{u}(x,t) \equiv \vec{u}_0(x)$ for all $(x,t) \in M \times R^+$ and $\vec{u}_0(x)$ is a harmonic map from M to S^2.*

Suppose that $\vec{u}(x,t) \equiv \vec{u}_0(x)$ for $(x,t) \in M \times R^+$ and $\vec{u}_0(x)$ is a harmonic map, i.e.

$$-\Delta_M \vec{u} = |D\vec{u}|^2 \vec{u}.$$

Then it is easy to see that $\vec{u}_0(x)$ is a solution to both (4.2.1) and the equation of harmonic map heat flow.

Suppose that $\vec{u}_0(x)$, a solution to the equation of harmonic map heat flow, is also a solution of (4.2.1) with $\alpha_2 \neq 0$. Then we have

$$(\alpha_1 - 1)\Delta_M \vec{u} + (\alpha_1 - 1)|D\vec{u}|^2 \vec{u} + \alpha_2 \times \Delta_M \vec{u} = 0.$$

Taking the cross product of $\vec{u}$ with the above equation, we can prove as that of Theorem 4.2.2 that

$$\Delta_M \vec{u} + |D\vec{u}|^2 \vec{u} = 0.$$

Then it follows from the equation of heat flow that

$$\vec{u}_t = 0.$$

Therefore, $\vec{u}(x,t) \equiv \vec{u}_0(x)$ is a harmonic map.

4.2.2   Local Smooth Solution of L–L Equation

1. *Strong parabolicity*

Now we consider a family of second order parabolic differential operators of the form

$$\vec{u}_t = \sum_{\alpha,\beta=1}^{m} D_\alpha(a_{\alpha\beta}(x,t,\vec{u})D_\beta\vec{u}) + f(x,t,\vec{u},D\vec{u}), \quad 0 \leq t \leq T < \infty, \qquad (4.2.10)$$

where $D_\alpha = \frac{\partial}{\partial x^\alpha}$, acting on N-vector valued real functions $\vec{u} : \Omega \to R^N$. We assume that the coefficient functions are smooth, that is,

$$a_{\alpha\beta} = a_{\beta\alpha}, \quad a_{\alpha\beta} \in C^\infty(\Omega \times [0,T] \times R^N, \mathcal{Z}(R^N)),$$

where $\mathcal{Z}(R^N)$ is the space of all endomorphisms of R^N. Like in [74], we say that Eq. (4.2.10) is a strongly parabolic system, i.e.

$$\sum_{i,j=1}^{N} \sum_{\alpha,\beta=1}^{m} a_{\alpha\beta}^{ij}(x,t,\eta)\xi^\alpha\xi^\beta\zeta_i\zeta_j > 0$$

for all $(x, t, \eta) \in \Omega \times [0, T] \times R^N$, for all $\xi = (\xi^1, \xi^2, \ldots, \xi^m) \in R^m \setminus \{0\}$, and for all $\zeta = (\zeta_1, \zeta_2, \ldots, \zeta_N) \in R^N \setminus \{0\}$, where $a^{ij}_{\alpha\beta}$ are the elements of matrices a_{jk}.

We denote matrix $g = (g_{ij})$ by

$$\begin{pmatrix} g_{11} & g_{12} & g_{13} \\ g_{21} & g_{22} & g_{23} \\ g_{31} & g_{32} & g_{33} \end{pmatrix} = \begin{pmatrix} \alpha_1 & -\alpha_2 u^3 & \alpha_2 u^2 \\ \alpha_2 u^3 & \alpha_1 & -\alpha_2 u^1 \\ -\alpha_2 u^2 & \alpha_2 u^1 & \alpha_1 \end{pmatrix}.$$

By Theorem 4.2.1, Eq. (4.2.1) is equivalent to (4.2.3) if the initial data $\vec{u}_0(x)$ satisfies condition (4.2.2). For the Landau–Lifshitz equation (4.2.3), the principle part is given by

$$\sum_{\alpha,\beta=1}^{m} D_\alpha(a_{\alpha\beta}(x, t, \vec{u}) D_\beta \vec{u}) = \sum_{\alpha,\beta=1}^{m} D_\alpha(\gamma^{\alpha\beta} D_\beta \vec{u} + \vec{u} \times \gamma^{\alpha\beta} D_\beta \vec{u}).$$

The corresponding coefficients $a^{ij}_{\alpha\beta}$ are given by

$$a^{ij}_{\alpha\beta} = \gamma^{\alpha\beta} g_{ij}.$$

Therefore the Landau–Lifshitz equation is strongly parabolic by the following relation:

$$\sum_{i,j=1}^{N} \sum_{\alpha,\beta=1}^{m} a^{ij}_{\alpha\beta}(x, t, \eta)\xi^\alpha \xi^\beta \zeta_i \zeta_j = \alpha_1 |\zeta|^2 \sum_{\alpha,\beta=1}^{m} \gamma^{\alpha\beta} \xi^\alpha \xi^\beta > 0$$

for all $(x, t, \eta) \in \Omega \times [0, T] \times R^N$, for all $\xi = (\xi^1, \xi^2, \xi^3) \in R^3 \setminus \{0\}$, and for all $\zeta = (\zeta_1, \zeta_2, \zeta_3) \in R^3 \setminus \{0\}$.

2. *Local smooth solution for L–L equation*

Now consider the Landau–Lifshitz equation from $R^2 \to S^2$.

First, let us consider the Cauchy problem of the Landau–Lifshitz equation from the flat torus. In order to formulate the problem, let Q be the flat torus R^2/Z^2. We suppose that $\vec{u}_0$ is a map from Q to S^2.

Consider the Cauchy problem Q into R^3

$$\vec{u}_t = -\alpha_1 \vec{u} \times (\vec{u} \times \Delta \vec{u}) + \alpha_2 \vec{u} \times \Delta \vec{u}, \tag{4.2.11}$$

with initial condition

$$\vec{u}|_{t=0} = \vec{u}(x, 0) = \vec{u}_0(x), \quad \forall\ x \in R^2. \tag{4.2.12}$$

It is easy to see from Theorem 4.2.1 that if $\vec{u}$ is a smooth solution of (4.2.11) and (4.2.12), then

$$|\vec{u}(x, t)|^2 = 1.$$

Lemma 4.2.2 *Suppose that $\vec{u}$ is a smooth solution of the Cauchy problem* (4.2.11) *and* (4.2.12). *Then*

$$\|\nabla \vec{u}(\cdot, t)\|_{L^2(Q)} \le \|\nabla \vec{u}_0\|_{L^2(Q)}, \quad \forall\ t \ge 0.$$

Proof. Multiplying (4.2.11) by $\Delta\vec{u}$, we have

$$\Delta\vec{u}\cdot\vec{u}_t = -\alpha_1\Delta\vec{u}\cdot(\vec{u}\times(\vec{u}\times\Delta\vec{u}))$$
$$= -\alpha_1(\vec{u}\times\Delta\vec{u})\cdot(\Delta\vec{u}\times\vec{u}) = \alpha_1|\vec{u}\times\Delta\vec{u}|^2.$$

Then

$$\frac{1}{2}\frac{d}{dt}\|\nabla\vec{u}\|_{L^2(Q)} + \alpha_1\|\vec{u}\times\Delta\vec{u}\|_{L^2(Q)} = 0.$$

This implies that

$$\frac{1}{2}\frac{d}{dt}\|\nabla\vec{u}\|_{L^2(Q)} \le 0.$$

The lemma follows.

Lemma 4.2.3 (Gagliardo–Nirenberg Inequality) *Let Ω be R^n or a bounded Lipschitz domain in R^n with boundary $\partial\Omega$, and let u be any function in $W^{m,r}(\Omega)\cap L^q(\Omega)$, $1\le r,\, q\le\infty$. For any integer j: $0\le j < m$, and for any number α in the interval $\frac{j}{m}\le\alpha\le 1$, set*

$$\frac{1}{p} = \frac{j}{n} + \alpha\left(\frac{1}{r} - \frac{m}{n}\right) + (1-\alpha)\frac{1}{q}.$$

If $m - j - \frac{n}{r}$ is not a nonnegative integer, then

$$\|\nabla^j u\|_{0,p} \le C(\|u\|_{m,r})^\alpha(\|u\|_{0,q})^{1-\alpha}. \tag{4.2.13}$$

If $m - j - \frac{n}{r}$ is a nonnegative integer, then (4.2.13) holds for $\alpha = \frac{j}{m}$, where $\|\cdot\|_{h,k} = \|\cdot\|_{W^{h,k}(\Omega)}$. Constant C depends only on r, q, m, j, α and the shape of Ω.

Lemma 4.2.4 *Suppose that $\nabla\vec{u}_0 \in H^{1,2}(Q)$. Let $T > 0$ be a constant. Let $\vec{u}$ be a smooth solution of (4.2.11) and (4.2.12) on $Q^T = Q\times[0,T]$. Then there is a constant $C > 0$ such that*

$$\|\nabla^2\vec{u}(\cdot,t)\|_{L^2(Q)} \le C, \quad \forall\; T\ge t\ge 0; \quad \|\nabla^3\vec{u}\|_{L^2(Q^T)} \le C,$$

provided that $\|\nabla\vec{u}_0\|_{L^2(Q)} \le \lambda$ for some small $\lambda > 0$.

Proof. Denote $D^\alpha = \partial_{x_1}^{\alpha_1}\partial_{x_2}^{\alpha_2}$, $|\alpha| = \alpha_1 + \alpha_2$, where α_1, α_2 are nonnegative integers. For simplicity, let D^k denote any kind of D^α, where $|\alpha| = k$.

Differentiating (4.2.11) with D^2, multiplying it by $D^2\vec{u}$, and integrating with respect to x on Q, we have

$$(\partial_t D^2\vec{u}, D^2\vec{u}) = \alpha_1(\Delta D^2\vec{u}, D^2\vec{u}) + \alpha_1(D^2(|\nabla\vec{u}|^2\vec{u}), D^2\vec{u})$$
$$+ \alpha_2(D^2(\vec{u}\times\Delta\vec{u}), D^2\vec{u}). \tag{4.2.14}$$

Since Q has no boundary we get

$$\begin{cases} (\partial_t D^2\vec{u}, D^2\vec{u}) = \dfrac{1}{2}\dfrac{d}{dt}\|D^2\vec{u}\|_{L^2(Q)}, \\[2mm] (\Delta D^2\vec{u}, D^2\vec{u}) = -\|\nabla D^2\vec{u}\|_{L^2(Q)}, \\[2mm] |(D^2(|\nabla\vec{u}|^2\vec{u}), D^2\vec{u})| = |(D(|\nabla\vec{u}|^2\vec{u}), D^3\vec{u})|. \end{cases} \tag{4.2.15}$$

Due to

$$D(|\nabla\vec{u}|^2\vec{u}) = |\nabla\vec{u}|^2 D\vec{u} + 2\vec{u}\nabla\vec{u}\cdot D\nabla\vec{u},$$

we have

$$\|D(|\nabla\vec{u}|^2\vec{u})\|_{L^2(Q)} \leq \|\nabla\vec{u}\|^2_{L^\infty(Q)}\|D\vec{u}\|_{L^2(Q)}$$
$$+ 2\|\vec{u}\|_{L^\infty(Q)}\|\nabla\vec{u}\|_{L^\infty(Q)}\|D\nabla\vec{u}\|_{L^2(Q)}.$$

From the Gagliardo–Nirenberg inequality, we have

$$\|\nabla\vec{u}\|_{L^4(Q)} \leq C_1\|\nabla\vec{u}\|^{1/4}_{H^{2,2}(Q)}\|\nabla\vec{u}\|^{3/4}_{L^2(Q)}, \tag{4.2.16}$$

$$\|\nabla^2\vec{u}\|_{L^4(Q)} \leq C_1\|\nabla\vec{u}\|^{3/4}_{H^{2,2}(Q)}\|\nabla\vec{u}\|^{1/4}_{L^2(Q)}, \tag{4.2.17}$$

$$\|\nabla\vec{u}\|_{L^6(Q)} \leq C_1\|\nabla\vec{u}\|^{1/3}_{H^{2,2}(Q)}\|\nabla\vec{u}\|^{2/3}_{L^2(Q)}. \tag{4.2.18}$$

Then from (4.2.16)–(4.2.18)

$$|(D^2(|\nabla\vec{u}|^2\vec{u}), D^2\vec{u})|$$
$$\leq \|\nabla^3\vec{u}\|_{L^2(Q)}(2\|\vec{u}\|_{L^\infty(Q)}\|\nabla\vec{u}\|_{L^4(Q)}\|D\nabla^2\vec{u}\|_{L^4(Q)} + \|\nabla\vec{u}\|^3_{L^6(Q)})$$
$$\leq C_2(C_1\|\nabla\vec{u}\|_{L^2(Q)} + C_1^3\|\nabla\vec{u}\|^2_{L^2(Q)})\|\nabla\vec{u}\|_{H^{2,2}(Q)}$$
$$\leq C_2(2C_1\|\nabla\vec{u}_0\|_{L^2(Q)} + C_1^3\|\nabla\vec{u}_0\|^2_{L^2(Q)})\|\nabla\vec{u}\|_{H^{2,2}(Q)}, \tag{4.2.19}$$

where $C_2 > 0$ is a constant.

On the other hand, from (4.2.16)–(4.2.18)

$$|(D^2(\vec{u}\times\Delta\vec{u}), D^2\vec{u})|$$
$$= |(D^2\nabla\cdot(\vec{u}\times\nabla\vec{u}), D^2\vec{u})|$$
$$= |(D^2(\vec{u}\times\nabla\vec{u}), \nabla D^2\vec{u})|$$
$$= |(D^2\vec{u}\times\nabla\vec{u} + 2D\vec{u}\times D\nabla\vec{u} + \vec{u}\times\nabla D^2\vec{u}, \nabla D^2\vec{u})|$$
$$\leq C_3\|\nabla\vec{u}\|_{L^4(Q)}\|\nabla^2\vec{u}\|_{L^4(Q)}\|\nabla^3\vec{u}\|_{L^4(Q)}$$
$$\leq C_1^2 C_3\|\nabla\vec{u}\|_{L^2(Q)}\|\nabla\vec{u}\|_{H^{2,2}(Q)}$$
$$\leq C_1^2 C_3\|\nabla\vec{u}_0\|_{L^2(Q)}\|\nabla\vec{u}\|_{H^{2,2}(Q)}, \tag{4.2.20}$$

where $C_3 > 0$ is an absolute constant and $\vec{u}\times\nabla\vec{u} := (\vec{u}\times\nabla_1\vec{u}, \vec{u}\times\nabla_2\vec{u})$.

Therefore from (4.2.14)–(4.2.20) we have

$$\frac{d}{dt}\|\nabla^2\vec{u}(\cdot,t)\|^2_{L^2(Q)} + \alpha_1\|\nabla^3\vec{u}(\cdot,t)\|^2_{L^2(Q)}$$
$$\leq \alpha_1\|\nabla^3\vec{u}(\cdot,t)\vec{u}\|^2_{L^2(Q)}(C_1^3 C_2\|\nabla\vec{u}_0\|^2_{L^2(Q)}$$
$$+ 2C_1 C_2\|\nabla\vec{u}_0\|_{L^2(Q)} + \frac{\alpha_2}{\alpha_1}C_1^2 C_3\|\nabla\vec{u}_0\|_{L^2(Q)})$$
$$+ C\|\nabla^2\vec{u}(\cdot,t)\|^2_{L^2(Q)} + C, \tag{4.2.21}$$

where $C > 0$ depends on $\|\nabla \vec{u}_0\|_{L^2(Q)}$. When λ is chosen small enough such that

$$\varepsilon := \left(C_1^3 C_2 \lambda^2 + 2C_1 C_2 \lambda + \frac{\alpha_2}{\alpha_1} C_1^2 C_3 \lambda \right) < 1,$$

we have

$$\frac{d}{dt}|\nabla^2 \vec{u}(\cdot, t)|_{L^2(Q)}^2 + \alpha_1(1 - \varepsilon)\|\nabla^3 \vec{u}(\cdot, t)\|_{L^2(Q)}^2 \leq C\|\nabla^2 \vec{u}(\cdot, t)\|_{L^2(Q)}^2 + C.$$

By the Gronwall inequality we have

$$\|\nabla^2 \vec{u}(\cdot, t)\|_{L^2(Q)}^2 \leq C, \quad \forall \; T \geq t > 0,$$

and

$$\|\nabla^3 \vec{u}\|_{L^2(Q^T)}^2 \leq C.$$

The lemma follows.

Lemma 4.2.5 *Let Ω be R^2 or bounded Lipschitz domain in R^2 with boundary Ω. Let u be any function in $H^{2,2}(\Omega)$. Then we have*

$$\|u\|_{L^\infty(\Omega)} \leq C\|u\|_{H^{2,2}(\Omega)}^{3/4}\|u\|_{L^2(\Omega)}^{1/4},$$

where $C > 0$ is a constant.

Proof. Without loss of generality, we assume that Ω be R^2, $u \in C_0^\infty(R^2)$. Then for any $x = (x_1, x_2) \in R^2$ we have

$$|u(x_1, x_2)| \leq \int \left| \frac{\partial u(x_1, x_2)}{\partial x_1} \right| dx_1.$$

For any fixed x_1, we define

$$v(x_2) = \frac{\partial u(x_1, x_2)}{\partial x_1}$$

which is a one-dimensional function. By using Theorem 2.2 of Chapter 2 of [39], we have

$$|v(x_2)| \leq C\|v_{x_2}\|_{L2(R)}^{1/2}\|v\|_{L^2(R)}^{1/2}.$$

Therefore using the above inequalities and the Gagliardo–Nirenberg inequality we have

$$\begin{aligned}
|u(x_1, x_2)| &\leq C \int \left\| \frac{\partial^2 u(\cdot, x_2)}{\partial x_1 \partial x_2} \right\|_{L^2}^{1/2} \left\| \frac{\partial u(\cdot, x_2)}{\partial x_1} \right\|_{L^2}^{1/2} dx_1 \\
&\leq C\|\nabla^2 u\|_{L^2(\Omega)}^{1/2}\|\nabla u\|_{L^2(\Omega)}^{1/2} \\
&\leq C\|\nabla^2 u\|_{L^2(\Omega)}^{3/4}\|u\|_{L^2(\Omega)}^{1/4}.
\end{aligned}$$

This proves the lemma.

Lemma 4.2.6 *Suppose that* $\nabla \vec{u}_0 \in H^{k,2}(Q; S^2)$, $k \geq 2$ *and the conditions of Lemma 4.2.4 hold. Then there exists a constant C such that*

$$\sup_{0 \leq t \leq T} \|\nabla \vec{u}(\cdot, t)\|^2_{H^{m,2}(Q)} \leq C \quad for \quad 2 \leq m \leq k.$$

Proof. We prove this lemma by induction.

First, we know from Lemma 4.2.4 that $\|\nabla \vec{u}(\cdot, t)\|_{H^{1,2}(Q)} \leq C$ for all $t \leq T$. Second, from (4.2.3) we have

$$\frac{d}{dt} D^3 \vec{u} = \alpha_1 \Delta D^3 \vec{u} + \alpha_1 D^3(|\nabla \vec{u}|^2 \vec{u}) + \alpha_2 D^3(\vec{u} \times \Delta \vec{u}).$$

Since Q has no boundary, we have

$$\begin{aligned}
(D^3 \vec{u}_t, D^3 \vec{u}) &= \frac{1}{2} \frac{d}{dt} \|D^3 \vec{u}(\cdot, t)\|^2_{L^2(Q)}, \\
(\Delta D^3 \vec{u}, D^3 \vec{u}) &= -\|\nabla D^3 \vec{u}(\cdot, t)\|^2_{L^2(Q)}.
\end{aligned} \tag{4.2.22}$$

Since

$$\begin{aligned}
D^2(|\nabla \vec{u}|^2 \vec{u}) &= |\nabla \vec{u}|^2 D^2 \vec{u} + 4 \nabla \vec{u} D \nabla \vec{u} \cdot D \vec{u} \\
&\quad + 2|D \nabla \vec{u}|^2 \vec{u} + 2 \nabla \vec{u} D^2 \nabla \vec{u} \cdot \vec{u},
\end{aligned}$$

we have from Lemmas 4.2.3 and 4.2.5

$$\begin{aligned}
|(D^3(|\nabla \vec{u}|^2 \vec{u}), D^3 \vec{u})| &= |(D^2(|\nabla \vec{u}|^2 \vec{u}), D^4 \vec{u})| \\
&\leq C(\|\nabla \vec{u}\|^2_{L^\infty(Q)} \|D \nabla \vec{u}\|_{L^2(Q)} + \|\nabla \vec{u}\|^2_{L^2(Q)} \\
&\quad + \|\nabla \vec{u}\|_{L^\infty(Q)} \|D^2 \nabla \vec{u}\|^2_{L^2(Q)}) \|D^4 \vec{u}\|_{L^2(Q)} \\
&\leq \frac{\varepsilon}{2} \|\nabla \vec{u}\|^2_{H^{3,2}(Q)} + C \|\nabla^3 \vec{u}\|^2_{L^2(Q)} + C
\end{aligned}$$

and

$$\begin{aligned}
&|\alpha_2(D^3(\vec{u} \times \Delta \vec{u}), D^3 \vec{u})| \\
&= |\alpha_2(\nabla \cdot D^3(\vec{u} \times \nabla \vec{u}), D^3 \vec{u})| \\
&= |\alpha_2(D^3(\vec{u} \times \nabla \vec{u}), \nabla D^3 \vec{u})| \\
&= |\alpha_2(D^3 \vec{u} \times \nabla \vec{u} + 5 D^2 \vec{u} \times D \nabla \vec{u} + 4 D \vec{u} \times D^2 \nabla \vec{u} + \vec{u} \times \nabla D^3 \vec{u}, \nabla D^3 \vec{u})| \\
&\leq C(\|\nabla \vec{u}\|^2_{L^\infty(Q)} \|D^3 \vec{u}\|_{L^2(Q)} + \|D^2 \vec{u}\|_{L^4(Q)} \|D \nabla \vec{u}\|_{L^4(Q)} + \|\nabla \vec{u}\|_{L^\infty(Q)}) \|D^3 \nabla \vec{u}\|_{L^2(Q)} \\
&\leq \frac{\alpha_1 \varepsilon}{2} \|\nabla \vec{u}\|^2_{H^{3,2}(Q)} + C \|\nabla^3 \vec{u}\|^2_{L^2(Q)} + C.
\end{aligned}$$

Choosing ε small enough, we have

$$\frac{d}{dt} \|\nabla^3 \vec{u}\|^2_{L^2(Q)} + \frac{\alpha_1}{2} \|\nabla^4 \vec{u}\|^2_{L^2(Q)} \leq C \|\nabla^3 \vec{u}\|^2_{L^2(Q)} + C.$$

By the Gronwall inequality, we have

$$\|\nabla\vec{u}(\cdot,t)\|^2_{H^{2,2}(Q)} \le C, \quad \forall \ 0 \le t \le T,$$

where C depends on $\|\nabla\vec{u}_0\|_{H^{2,2}(Q)}$. In fact, $\|\nabla\vec{u}\|_{L^\infty(Q)} \le C$.

Second, suppose that $\|\nabla\vec{u}\|_{H^{m,2}(Q)} \le C$, $m \ge 2$. From (4.2.3) we have

$$D^{m+1}\vec{u}_t = \alpha_1\Delta D^{m+1}\vec{u} + \alpha_1 D^{m+1}(|\nabla\vec{u}|^2\vec{u}) + \alpha_2 D^{m+1}(\vec{u}\times\Delta\vec{u}).$$

Since Q has no boundary, we have

$$(D^{m+1}\vec{u}_t, D^{m+1}\vec{u}) = \frac{1}{2}\frac{d}{dt}\|D^{m+1}\vec{u}(\cdot,t)\|^2_{L^2(Q)},$$

$$(\Delta D^{m+1}\vec{u}, D^{m+1}\vec{u}) = -\|\nabla D^{m+1}\vec{u}(\cdot,t)\|^2_{L^2(Q)}.$$

By using the well-known Kato's inequality (see Lemma 3.1 of [97]), we have

$$\|D^s(fg) - fD^sg\|_{L^p(Q)}$$
$$\le C(\|Df\|_{L^\infty(Q)}\|D^{s-1}g\|_{L^p(Q)} + \|D^sf\|_{L^p(Q)}\|g\|_{L^\infty(Q)})$$

for any two functions f, g.

Setting $f = \vec{u}$, $g = |\nabla\vec{u}|^2$, $s = m$, we have

$$\|D^m(|\nabla\vec{u}|^2\vec{u})\|_{L^2(Q)}$$
$$\le \|\vec{u}D^m(|\nabla\vec{u}|^2)\|_{L^2(Q)} + C(\|\nabla\vec{u}\|_{L^\infty(Q)}\|D^{m-1}|\nabla\vec{u}|^2\|_{L^2(Q)}$$
$$+ \|D^m\vec{u}\|_{L^2(Q)}\||\nabla\vec{u}|^2\|_{L^\infty(Q)})$$
$$\le \|D^m|\nabla\vec{u}|^2\|_{L^2(Q)} + C(\|D^{m-1}|\nabla\vec{u}|^2\|_{L^2(Q)} + \|D^m\vec{u}\|_{L^2(Q)})$$
$$\le C_1\|\nabla^{m+1}\vec{u}\|_{L^2(Q)}) + C_2.$$

Then

$$|(D^{m+1}(|\nabla\vec{u}|^2\vec{u}), D^{m+1}\vec{u})| = |(D^m(|\nabla\vec{u}|^2\vec{u}), D^{m+2}\vec{u})|$$
$$\le \frac{1}{4}\|\nabla D^{m+1}\vec{u}\|^2_{L^2(Q)} + \|\nabla^{m+1}\vec{u}\|^2_{L^2(Q)}.$$

On the other hand, we have

$$|(D^{m+1}(\vec{u}\times\Delta\vec{u}), D^{m+1}\vec{u})| = |(D^{m+1}(\vec{u}\times\nabla\vec{u}), \nabla D^{m+1}\vec{u})|.$$

Since

$$D^{m+1}(\vec{u}\times\nabla\vec{u}) = D^{m+1}\vec{u}\times\nabla\vec{u} + \vec{u}\times D^{m+1}\nabla\vec{u} + \sum_{h=1}^m c_h D^h\vec{u}\times D^{m+1-h}\nabla\vec{u},$$

$$\alpha_2|(D^{m+1}(\vec{u}\times\nabla\vec{u}), \nabla D^{m+1}\vec{u})|$$
$$\le \alpha_2|(D^{m+1}\vec{u}\times\nabla\vec{u}, \nabla D^{m+1}\vec{u})| + \alpha_2\sum_{h=1}^m c_h|(D^h\vec{u}\times D^{m+1-h}\nabla\vec{u}, \nabla D^{m+1}\vec{u})|$$
$$\le \|\nabla\vec{u}\|^2_{L^\infty(Q)}\|D^{m+1}\vec{u}\|_{L^2(Q)} + C\|D^m\vec{u}\|_{L^4(Q)}\|D^m\nabla\vec{u}\|_{L^4(Q)}\|\nabla D^{m+1}\vec{u}\|_{L^2(Q)}$$
$$\le \frac{\alpha_1}{4}\|\nabla D^{m+1}\vec{u}\|^2_{L^2(Q)} + C\|\nabla^{m+1}\vec{u}\|^2_{L^2(Q)}.$$

Therefore, we have

$$\frac{d}{dt}\|\nabla^{m+1}\vec{u}(\cdot,t)\|^2_{L^2(Q)} + \frac{\alpha_1}{2}\|\nabla^{m+2}\vec{u}(\cdot,t)\|^2_{L^2(Q)} \le C\|\nabla^{m+1}\vec{u}(\cdot,t)\|^2_{L^2(Q)}.$$

By the Gronwall inequality, we have

$$\|\nabla\vec{u}(\cdot,t)\|^2_{H^{m+1,2}(Q)} \le C, \quad m \ge 2, \quad \forall\ 0 \le t \le T.$$

This prove the lemma.

4.2.3 Global Smooth Solution to L–L Equation

Due to the strong parabolicity of the Landau-lifshitz equation with Gilbert damping (4.2.3), the local existence of smooth solution has been proved by Amann in [5]. Combining this local existence with Lemma 4.2.2–Lemma 4.2.6, we conclude that the local smooth solution can be extended to a large t to obtain the existence of global solution.

Theorem 4.2.3 *Let Q be the flat torus R^2/Z^2. Suppose that $\nabla\vec{u}_0(x)$ is a given initial value in $H^k(Q;S^2)$, where k is large enough. Then there exists a constant λ such that the periodic value problem (4.2.11) and (4.2.12) with the initial value $\vec{u}_0$ admits a smooth global solution $\vec{u}(x,t)$ provided that $\|\nabla\vec{u}_0\|_2 \le \lambda$.*

4.2.4 On the Landau–Lifshitz Equation on Riemannian Surface

1. *Local estimates*

Now we consider the Landau–Lifshitz equations on Riemannian manifold.

Let M be a closed Riemannian surface, $B_R^M(x) = \{y \in M : |y - x|_M < R\}$ for a domain Ω and $-\infty < s < t < \infty$, and let $\Omega_s^t = \Omega \times [s,t]$. The energy on $B_R^M(x)$ is given by

$$E_R(\vec{u};x) = \int_{B_R^M(x)} e(\vec{u})dM, \quad \text{where}\ \ e(\vec{u}) = \frac{1}{2}|\nabla\vec{u}|^2.$$

We denote

$$V(M_\tau^T;S^2) = \Big\{\vec{u} : M \times [\tau,T] \to S^2 \mid \vec{u}\ \text{is measurable and}$$

$$\text{ess sup}_{\tau \le t \le T} \int_M |\nabla\vec{u}(\cdot,t)|^2 dM + \int_\tau^T (|\nabla^2\vec{u}|^2 + |\vec{u}_t|^2)dMdt < \infty\Big\}.$$

Like in [133], we have

Lemma 4.2.7 *There exist constants C and R_0 such that for any $u \in V(M^T;S^2)$, and any $R \in (0, R_0)$ there holds the estimate*

$$\int_{M^T} |\nabla u|^4 dMdt \le C \underset{(x,t)\in M^T}{\text{ess sup}} \int_{B_R^M(x)} |\nabla u(x,t)|^2 dM$$

$$\times \left(\int_{M^T} |\nabla^2\vec{u}|^2 dMdt + R^{-2}\int_{M^T} |\nabla\vec{u}|^2 dMdt\right).$$

Lemma 4.2.8 *There exists a constant $C = C(S^2)$ such that for any solution $\vec{u} \in V(M^T; S^2)$ of (4.2.11) there holds the estimate*

$$\int_{M^T} |\vec{u}_t|^2 dMdt \le CE(\vec{u}_0).$$

Moreover, $E(\vec{u}(\cdot, t))$ is absolutely continuous on $[0, T]$ and non-increasing.

Proof. By Lemma 4.2.7, we may multiply (4.2.3) by $\vec{u}_t$ and integrate. Then for any $s, t \in [0, T]$

$$\int_{M_s^t} |\vec{u}_t|^2 dMdt + \alpha_1 \int_s^t \frac{d}{dt} E(\vec{u}(\cdot, t)) dt - \alpha_2 \int_{M_s^t} \vec{u}_t \cdot (\vec{u} \times \Delta_M \vec{u}) dMdt = 0. \quad (4.2.23)$$

In the distribution sense we have

$$\vec{u}_t = \alpha_1(\Delta_M \vec{u} + |D\vec{u}|^2 \vec{u}) + \alpha_2(\vec{u} \times \Delta_M \vec{u}).$$

By the cross product of the above equation with $\vec{u}$, we get

$$\vec{u} \times \vec{u}_t = \alpha_1(\vec{u} \times \Delta_M \vec{u}) + \alpha_2(\vec{u} \times (\vec{u} \times \Delta_M \vec{u}))$$
$$= \alpha_1(\vec{u} \times \Delta_M \vec{u}) - \alpha_2 \Delta_M \vec{u} - \alpha_2 |Du|^2 \vec{u}.$$

Since

$$-\Delta_M \vec{u} - |Du|^2 \vec{u} = -\frac{1}{\alpha_1} \vec{u}_t + \frac{\alpha_2}{\alpha_1}(\vec{u} \times \Delta_M \vec{u}),$$

we have

$$\vec{u} \times \vec{u}_t + \frac{\alpha_2}{\alpha_1} \vec{u}_t = \left(\alpha_1 + \frac{\alpha_2^2}{\alpha_1}\right)(\vec{u} \times \Delta_M \vec{u}). \quad (4.2.24)$$

Multiplying the above equation by $\vec{u}_t$, we have

$$\vec{u}_t \cdot (\vec{u} \times \Delta_M \vec{u}) = \alpha_2(\alpha_1^2 + \alpha_2^2)^{-1} |\vec{u}_t|^2.$$

Then we have

$$\alpha_2 \int_{M^T} \vec{u}_t(\vec{u} \times \Delta_M \vec{u}) dMdt = \frac{\alpha_2^2}{\alpha_1^2 + \alpha_2^2} \int_{M^T} |\vec{u}_t|^2 dMdt. \quad (4.2.25)$$

The conclusion of the lemma follows by (4.2.24) and (4.2.25).

Corollary 4.2.3 *Combining Lemma 4.2.7 and Lemma 4.2.8 we obtain the estimate*

$$\int_{M^T} |\nabla \vec{u}|^4 dMdt \le C \sup_{(x,t) \in M^T} E_R(\vec{u}(\cdot, t), x) \left(\int_{M^T} |\nabla^2 \vec{u}|^2 dMdt + \frac{T}{R^2} E(\vec{u}_0)\right)$$

for any solution $\vec{u} \in V(M^T; S^2)$ of (4.2.3) and any $R \in (0, R_0]$.

This corollary makes it important to control energy locally.

2. *Estimates for the solution to the equivalent equation*

Like in [133], we have

Lemma 4.2.9 *There exists a constant $C_1 = C_1(M; S^2)$ such that for any solution $\vec{u} \in V(M^T; S^2)$ of (4.2.3), any $R \in (0, R_0]$, and any $(x, t) \in M^T$, there holds the estimate*

$$E_R(\vec{u}(\cdot, t), x) \le E_{2R}(\vec{u}(\cdot, 0), x) + C_1 \frac{t}{R^2} E(\vec{u}_0).$$

Proof. Let $\phi \in C_0^\infty(B_{2R}^M(x_0))$ satisfy $0 \le \phi \le 1$, $\phi \equiv 1$ on $B_R^M(x_0)$, $|\nabla \phi| \le \frac{C}{R}$. By multiplying Eq. (4.2.3) by $\phi^2 \vec{u}_t$, and by the same arguments used in Lemma 4.2.8, we have

$$\alpha_2 \int_{M^T} \vec{u}_t \cdot (\vec{u} \times \Delta_M \vec{u}) \phi^2 dM dt = \frac{\alpha_2^2}{\alpha_1^2 + \alpha_2^2} \int_{M^T} |\vec{u}_t|^2 \phi^2 dM dt.$$

Hence by the same steps of Lemma 3.6 in [133], we obtain

$$E_R(\vec{u}(\cdot, t), x) \le \int_M e(\vec{u}(\cdot, t)) \phi^2 dM \le E_{2R}(\vec{u}_0, x) + C \frac{t}{R^2} E(\vec{u}_0)$$

as claimed.

For a solution $\vec{u} \in V(M^T; S^2)$ of (4.2.3) and $R \in (0, R_0]$, let

$$\varepsilon(R) := \varepsilon(R; \vec{u}, T) = \sup_{(x,t) \in M^T} E_R(\vec{u}(\cdot, t); x).$$

We give *a priori* estimates for the V-norm and the Hölder norm of $\vec{u}$ in terms of initial energy $E(\vec{u}_0)$, T, and the number

$$R = \sup\{R > 0 : \varepsilon(R; \vec{u}, T) \le \varepsilon_1\},$$

where $\varepsilon_1 > 0$ is a parameter depending on M which is to be determined.

Then we have

Lemma 4.2.10 *There is a constant $\varepsilon_1 > 0$ such that for any solution $\vec{u} \in V(M^T; S^2)$ of (4.2.3) and any number $R \in (0, R_0]$, there holds the estimate*

$$\int_{M^T} |\nabla^2 \vec{u}|^2 dM dt \le C E(\vec{u}_0)(1 + TR^{-2}),$$

provided $\varepsilon(R) \le \varepsilon_1$.

Proof. We only note that

$$\int_{M^T} \Delta_M \vec{u} \cdot (\alpha_1 \vec{u} \times \Delta_M \vec{u}) dM dt = 0.$$

The other steps follow as those of Lemma 3.7 in [133].

Corollary 4.2.4 *In the above proof we have the estimate*

$$\int_{M^T} |\Delta_M \vec{u}|^2 dM dt \ge \int_{M^T} |\nabla^2 \vec{u}|^2 dM dt - C \int_{M^T} |\nabla \vec{u}|^2 dM dt. \tag{4.2.26}$$

By the same proof as that of Lemma 3.8 in [133], we obtain

Lemma 4.2.11 *For any number ε, τ, $E_0 > 0$, $R_1 \in (0, R_0]$, there exists a number $\delta > 0$ such that for any solution $\vec{u} \in V(M^T; S^2)$ of (4.2.3) and any $I \subset [\tau, T]$ with measure $|I| < \delta$ there holds the estimate*

$$\int_I \int_M |\nabla \vec{u}|^4 dM dt < \varepsilon,$$

provided $\varepsilon(R_1) \leq \varepsilon_1$, $E(\vec{u}_0) \leq E_0$.

Lemma 4.2.12 *For numbers $E_0 > 0$, $R_0 > 0$, let $\vec{u}_m \in V(M^T; S^2)$ be solutions to Eq. (4.2.3) with initial value $\vec{u}_m(0)$ provided that $\varepsilon(R; \vec{u}_m, T) \leq \varepsilon_1$ for $R \in (0, R_0]$, and $\vec{u}_m(0) \to \vec{u}_0$ as $m \to \infty$. Then $\vec{u}_m$ converges to a solution of (4.2.3) in $V(M^T; S^2)$ as $m \to \infty$.*

Proof. The proof of Lemma 4.2.10 and [133, Lemma 3.8] then show that

$$\int \left(\int_M |\Delta_M \vec{u}_m|^4 dM \right) dt$$

is uniformly absolutely continuous on $[0, T]$. We may suppose that $\vec{u}_m \to \vec{u}$ a.e. and $\partial_t \vec{u}_m \rightharpoonup \partial_t \vec{u}$, $\Delta^2 \vec{u}_m \rightharpoonup \Delta^2 \vec{u}$ weakly in $L^2(M^T)$, and $\nabla \vec{u}_m \to \nabla \vec{u}$ strongly in $L^2(M^T)$. Let $v_m := \vec{u}_m - \vec{u}$; $|\nabla U_m| := |\nabla \vec{u}_m| + |\nabla \vec{u}|$. Then

$$(\partial_t v_m - \alpha_1 \Delta_M v_m + \alpha_2 \vec{u}_m \times \Delta_M v_m + \alpha_2 v_m \times \Delta_M \vec{u}_m)$$
$$\leq C_m(|v_m||\nabla U_m|^2 + |\nabla v_m||\nabla U_m|).$$

Multiplying it by $\Delta_M v_m$ and integrating it give

$$\frac{1}{2} \int_M \frac{d}{dt}(\nabla v_m, \nabla v_m)_M dM dt + \alpha_1 \int_{M^T} |\Delta_M v_m|^2 dM dt$$
$$\leq \frac{\alpha_1}{2} \int_{M^T} |\Delta_M v_m|^2 dM dt + \int_{M^T} |v_m|^2 |\Delta_M \vec{u}|^2 dM dt$$
$$+ C \int_{M^T} (|v_m|^2 |\nabla U_m|^4 + |\nabla v_m|^2 |\nabla U_m|^2) dM dt.$$

Then by (4.2.26) and Vitali's theorem, we have

$$\sup_{0 \leq t \leq T} \int_M |\nabla v_m(\cdot, t)|^2 dM + \int_{M^T} |\nabla^2 v_m|^2 dM dt \to 0$$

since $\nabla v_m(\cdot, 0) \to 0$ in $L^2(M)$.

Similarly,

$$\int_{M^T} |\partial_t v_m|^2 dM dt \leq C \int_{M^T} (|\Delta_M v_m|^2 + |v_m|^2 |\nabla U_m|^4$$
$$+ |\nabla v_m|^2 |\nabla U_m|^2 + |v_m|^2 |\Delta_M \vec{u}|^2) dM dt \to 0, \quad (m \to \infty),$$

i.e. $\vec{u}_m \to \vec{u}$ strongly in $V(M^T; S^2)$.

3. Regularity of solutions to L–L equation

Lemma 4.2.13 *Let $\vec{u} \in V(M^T; S^2) \cap_{\tau>0} C^2(M_\tau^T; S^2)$ be a regular solution to (4.2.3). Then for any $\tau > 0$ the Hölder norms of $\vec{u}$ and its derivatives may be estimated uniformly on M_τ^T by quantities involving only $E(\vec{u}_0)$, τ, T and R, provided $\varepsilon(R) \le \varepsilon_1$.*

Proof. Multiplying (4.2.3) by $\Delta_M \vec{u}$ and integrating it give

$$\int_M |\Delta_M \vec{u}(\cdot, t)|^2 dM \le \frac{1}{2} \int_M |\Delta_M \vec{u}(\cdot, t)|^2 dM$$
$$+ C \int_M |\partial_t \vec{u}(\cdot, t)|^2 dM + C \int_M |\nabla \vec{u}(\cdot, t)|^4 dM,$$

which implies

$$\int_M |\Delta_M \vec{u}(\cdot, t)|^2 dM \le C \int_M (|\nabla \vec{u}(\cdot, t)|^4 + |\partial_t \vec{u}(\cdot, t)|^2) dM, \quad \text{a.e.} \ \ t \in [0, T].$$

From

$$\vec{u} \times \Delta_M \vec{u} = \nabla \cdot (\vec{u} \times \nabla \vec{u}),$$

we have

$$\int_{M^T} (\vec{u} \times \Delta_M \vec{u})_t \cdot \vec{u}_t dM dt$$
$$= \int_{M^T} (\nabla \cdot [\partial_t \vec{u} \times \nabla \vec{u} + \vec{u} \times \nabla \partial_t \vec{u}]) \cdot \vec{u}_t dM dt$$
$$= - \int_{M^T} (\partial_t \vec{u} \times \nabla \vec{u}) \cdot \nabla \vec{u}_t dM dt$$
$$\le \frac{1}{4} \int_{M^T} |\nabla \partial_t \vec{u}|^2 dM dt + C \int_{M^T} |\partial_t \vec{u}|^2 |\nabla \vec{u}|^2 dM dt.$$

In order to estimate the right-hand side of the above inequality, we differentiate (4.2.3) with respect to t, and multiply with $\partial_t \vec{u}$, and then integrate over M_s^T, where $\tau \le s \le t \le T$, to give

$$\frac{1}{2} \int_{M_s^T} \partial_t |\vec{u}_t|^2 dM dt + \alpha_1 \int_{M_s^T} |\nabla \partial_t \vec{u}|^2 dM dt$$
$$\le C \int_{M_s^T} (|\partial_t \vec{u}|^2 |\nabla \vec{u}|^2 + |\partial_t \vec{u}||\nabla \vec{u}||\nabla \partial_t \vec{u}|) dM dt$$
$$+ \left| \alpha_2 \int_{M_s^T} (\vec{u} \times \Delta_M \vec{u})_t \cdot \vec{u}_t dM dt \right|$$
$$\le \frac{\alpha}{2} \int_{M_s^T} |\nabla \partial_t \vec{u}|^2 dM dt + C \int_{M_s^T} |\partial_t \vec{u}|^2 |\nabla \vec{u}|^2 dM dt.$$

Repeating the proof of Lemma 3.10 of [133], we have

$$\begin{cases} \sup_{\tau \le t \le T} \int_M |\partial_t \vec{u}(\cdot, t)|^2 dM \le C(1 + \tau^{-1}) E(\vec{u}_0); \\ \int_M |\nabla^2 \vec{u}(\cdot, t)|^2 dM \le C(1 + \tau^{-1} + R^{-2}) E(\vec{u}_0), \end{cases} \tag{4.2.27}$$

where $\varepsilon(R) \leq \varepsilon_1$. On the other hand, it follows from Lemma 4.2.5 and (4.2.27) that

$$\sup_{x \in M} \|\vec{u}(\cdot, t_1) - \vec{u}(\cdot, t_2)\|_{L^\infty(M)}$$

$$\leq C\|\vec{u}(\cdot, t_1) - \vec{u}(\cdot, t_2)\|_{H^{2,2}(M)}^{3/4} \|\vec{u}(\cdot, t_1) - \vec{u}(\cdot, t_2)\|_{L^2(M)}^{1/4}$$

$$\leq 2C \sup_t \|\vec{u}(\cdot, t)\|_{H^{2,2}(M)}^{3/4} \left\| \int_{t_1}^{t_2} \partial_t \vec{u} dt \right\|_{L^2(M)}^{1/4}$$

$$\leq C \sup_t \|\vec{u}(\cdot, t)\|_{H^{2,2}(M)}^{3/4} |t_1 - t_2|^{1/4} \left(\int_{M_\tau^T} |\vec{u}_t|^2 dM dt \right)^{1/4}$$

$$\leq C|t_1 - t_2|^{1/4}. \tag{4.2.28}$$

Combining (4.2.27) with (4.2.28), we know that the $C^{\frac{1}{2}, \frac{1}{4}}$-Hölder norm of $\vec{u}$ can be estimated uniformly on M_τ^T by quantities in $E(\vec{u}_0)$ and $\varepsilon(R) \leq \varepsilon_1$. Since (4.2.3) is parabolic, using [101, Chap. VII, Theorem 10.4] and by the standard bootstrap method, we may get the higher regularity about the solution.

4. *Uniqueness of solution to L–L equation*

Now we consider the uniqueness of solution to (4.2.3) in $V(M^T; S^2)$.

Theorem 4.2.4 *Suppose* $\vec{u}_1$, $\vec{u}_2 \in V(M^T; S^2)$ *are two solutions to* (4.2.3) *with* $\vec{u}_1(\cdot, 0) = \vec{u}_2(\cdot, 0) = \vec{u}_0(\cdot)$. *Then* $\vec{u}_1 = \vec{u}_2$ *in* M^T.

Proof. Let $v = \vec{u}_1 - \vec{u}_2$ and $|\nabla U| := |\nabla \vec{u}_1| + |\nabla \vec{u}_2|$. From (4.2.1) we obtain

$$|v_t - \alpha_1 \Delta_M v + \alpha_2 \vec{u}_1 \times \Delta_M \vec{u}_1 - \alpha_2 \vec{u}_2 \times \Delta_M \vec{u}_2| \leq C(|v||\nabla U|^2 + |\nabla v||\nabla U|).$$

If we multiply this by v and integrate over M^T, we have

$$\frac{1}{2} \int_M |v(\cdot, t)|^2 dM + \alpha_1 \int_{M^T} (\nabla v, \nabla v)_M dM dt$$

$$\leq C \int_{M^T} (|v|^2 |\nabla U|^2 + |v||\nabla v||\nabla U|) dM dt$$

$$+ C \left| \int_{M^T} (\vec{u}_1 \times \Delta_M \vec{u}_1 - \vec{u}_2 \times \Delta_M \vec{u}_2) v dM dt \right|.$$

Since

$$(\vec{u}_1 \times \Delta_M \vec{u}_1 - \vec{u}_2 \times \Delta_M \vec{u}_2)v = (v \times \Delta_M \vec{u}_1 + \vec{u}_2 \times \Delta_M v)v$$

$$= (\vec{u}_2 \times \Delta_M v) \cdot v = -(\vec{u}_2 \times v) \cdot \Delta_M v,$$

$$\left| \int_{M^T} (\vec{u}_1 \times \Delta_M \vec{u}_1 - \vec{u}_2 \times \Delta_M \vec{u}_2) v dM dt \right|$$

$$= \left| \int_{M^T} (\vec{u}_2 \times v) \Delta_M v dM dt \right|$$

$$= \left| \int_{M^T} (\nabla \vec{u}_2 \times v + \vec{u}_2 \times \nabla v) \cdot \nabla v dM dt \right|$$

$$= \left| \int_{M^T} (\nabla \vec{u}_2 \times v) \cdot \nabla v dM dt \right|$$

$$\leq C \int_{M^T} |v|^2 |\nabla U|^2 dM dt + \frac{1}{4} \int_{M^T} |\nabla v|^2 dM dt.$$

Therefore

$$\frac{1}{2}\int_M |v(\cdot,t)|^2 dM + \alpha_1 \int_{M^T} (\nabla v, \nabla v)_M dM dt$$
$$\leq \frac{\alpha_1}{2}\int_{M^T}(\nabla v, \nabla v)_M dM dt + C\int_{M^T}|v|^2|\nabla U|^2 dM dt.$$

By the same steps as that of the proof of Lemma 3.12 in [133], we complete the proof of the theorem.

In order to prove a local existence of solution to (4.2.3), we improve Hamilton's idea and derive the following

Lemma 4.2.14 *For every function $g \in L^p(M \times [\alpha,\omega])$ there exists a unique $f \in W^{2,p}(M \times [\alpha,\omega])$ such that*

$$\partial_t f = \alpha_1 \Delta_M f + a\nabla f + bf + \alpha_2 z \times \Delta_M f + g$$

on $M \times [\alpha,\omega]$, where a, b, z are given smooth functions.

Proof. Let $Hf = \partial_t f - \alpha_1 \Delta_M f - \alpha_2 z \times \Delta_M f$ and $Kf = a\nabla f + bf$. Then map $f \to Hf$ defines an isomorphism from $W^{2,p}(M \times [\alpha,\omega])$ onto $L^p(M \times [\alpha,\omega])$ since H is a strongly parabolic linear operator. Moreover, since K has weight 1, map $K : W^{2,p}(M \times [\alpha,\omega]) \to L^p$ is compact as it factors through the compact inclusion of $W^{2,p}$ into L^p. By the theory of Fredholm mappings map $W^{2,p} \to L^p$ given by $f \to (Hf - Kf)$ has finite dimensional kernel and cokernel. Moreover, its index is zero. Then to show that it is an isomorphism it suffices to show that its kernel is zero. Let $f \in W^{2,p}(M \times [\alpha,\omega])$ we know that $f|_{[M \times \alpha]} = 0$ since

$$\partial_t f - \alpha_1 \Delta_M f - \alpha_2 z \times \Delta_M f - a \cdot \nabla f - bf = 0.$$

Taking the scalar product of the above equation with f, we obtain

$$\int_M \partial_t |f|^2 dM + \alpha_1 |\nabla f|^2 dM - \alpha_2 \int_M f \cdot (z \times \Delta_M f) dM$$
$$= \int_M af \cdot \nabla f dM + bf^2 dM.$$

Since

$$\int_M f\nabla \cdot (z \times \nabla f) dM = -\int_M \nabla f \cdot (z \times \nabla f) dM = 0,$$
$$\int_M f \cdot (z \times \Delta_M f) dM = -\int_M f \cdot \nabla z \times \nabla f dM,$$

we have

$$\int_M \partial_t |f|^2 dM + \alpha_1 |\nabla f|^2 dM \leq \frac{\alpha_1}{2}\int_M |\nabla f|^2 dM + C\int_M bf^2 dM.$$

Using the Gronwall inequality, we have $f = 0$. This proves the lemma.

From Proposition 2.1 of [25], we have

Lemma 4.2.15 *Let M be a two-dimensional compact Riemannian manifold. Then for all $u \in W_p^{1,2}(M^T)$, $4 < p < \infty$, there exists constant $C > 0$ such that*

$$\sup_{t \in [0,T]} \|u(\cdot, t)\|_{C^{1+\alpha}(M)} \leq C \|u\|_{W_p^{1,2}(M^T)},$$

where $\alpha = 1 - \frac{4}{p}$.

We now prove the existence of solutions of (4.2.3) for short periods of time.

Theorem 4.2.5 *Let $\vec{u}_0 : M \to S^2$ be a smooth map. There exists a constant $\varepsilon > 0$ and a map $\vec{u} : M \times [0, \varepsilon] \to S^2$ with $\vec{u} \in L_2^p(M^\varepsilon)$ solving the equation*

$$\partial_t \vec{u} = \alpha_1 \Delta_M \vec{u} + \alpha_1 |\nabla \vec{u}|^2 \vec{u} + \alpha_2 \vec{u} \times \Delta_M \vec{u}, \quad on\ M \times [0, \varepsilon],$$
$$\vec{u} = \vec{u}_0, \quad on\ M \times \{0\}.$$

Moreover, $\vec{u}$ is unique and smooth.

Proof. First, we define a nonlinear map $L : W^{2,p}(M \times [0, \omega]) \to L^p(M \times [0, \omega])$ as follows:

$$L := \alpha_1 \Delta_M \vec{u} + \alpha_1 |\nabla \vec{u}|^2 \vec{u} + \alpha_2 \vec{u} \times \Delta_M \vec{u}.$$

For the smooth map $\vec{u}$, the derivative of L at $\vec{u}$ is given by

$$DL(\vec{u})k = \alpha_1 \Delta_M k + \alpha_2 \vec{u} \times \Delta_M k + a(\vec{u})k + b(\vec{u})k,$$

where a and b are smooth matrices of functions. Choosing $\vec{u}_b : M \times [0, \omega] \to R^3$ to be a smooth map with $\vec{u}_b = \vec{u}_0$ on $M \times \{0\}$, we denote $\vec{u}$ as sum $\vec{u}_b + \vec{u}_*$. Let $H(\vec{u}) = \partial_t \vec{u} - L(\vec{u})$. The derivative of $H(\vec{u})$ is given by

$$DH(\vec{u})k = \partial_t k - \alpha_1 \Delta_M k - \alpha_2 \vec{u} \times \Delta_M k - a\nabla k - bk.$$

Fix $\vec{u}_b$ and consider $\vec{u}_*$ as a variable function. Then $\vec{u}_* \to H(\vec{u}_b + \vec{u}_*)$ defines a continuously differentiable map of $W^{2,p}(M \times [0, \omega]) \to L^p(M \times [0, \omega])$. Its derivative at $\vec{u}_* = 0$, $DH(\vec{u}_b) : W^{2,p}(M \times [0, \omega]) \to L^p(M \times [0, \omega])$ is given by

$$DH(\vec{u}_b)k = \partial_t k - \alpha_1 \Delta_M k - \alpha_2 \vec{u} \times \Delta_M k - a\nabla k - bk,$$

which by Lemma 4.2.14 is an isomorphism. Therefore by the inverse function theorem, the set of all $H(\vec{u}_b + \vec{u}_*)$ for $\vec{u}_*$ is a neighborhood of $H(\vec{u}_b)$ in $L^p(M \times [0, \omega])$. If we choose $\varepsilon > 0$ small enough, the function equals to 0 for $0 \leq t \leq \varepsilon$, and equals to $H(\vec{u}_b)$ for $\varepsilon < t \leq \omega$ will be in this neighborhood. So $\vec{u}|_{M^\varepsilon}$ is in $W^{2,p}(M \times [0, \varepsilon])$, and it solves the Landau–Lifshitz equation (4.2.3).

Second, using [133, Chap. VII, Theorem 10.4] and the standard bootstrap method, we conclude that $\vec{u}$ is smooth.

5. *Uniqueness*

Using Lemma 4.2.15, we have

$$\sup_{0\le t\le s}\|\vec{u}(\cdot,t)\|_{C^{1+\alpha}(M)}\le C.$$

From the above estimate and Lemma 4.2.5 we know that the local solution is unique.
We thus have

Theorem 4.2.6 *For any initial value $\vec{u}_0\in H^{1,2}(M;S^2)$ there exists a unique solution $\vec{u}$ of (4.2.3) on $M\times[0,\infty)$ which is regular on $M\times[0,\infty)$ with exception of at most finitely many points (x^l,T^l), $1\le l\le L$, characterized by the condition that*

$$\lim_{T\to T^l}\sup_{T<T^l} E_R(\vec{u}(\cdot,T),x^l)>\varepsilon_1\quad\text{for all }\ R\in(0,R_0].$$

The proof of this theorem, which is based on Lemma 4.2.8, Lemma 4.2.9, Lemma 4.2.12, Theorem 4.2.4, and Theorem 4.2.5, is the same as the proof of Theorem 4.2 of [133].

4.2.5 The Landau–Lifshitz Equation in Higher Dimensions

1. *Global weak solution to L–L equation in higher dimensions*

Let M be a compact m-dimensional Riemannian manifold without boundary and $m\ge 3$.

Using a similar way as before, we can prove that Eq. (4.2.11) is equivalent to the following equation:

$$\frac{\alpha_1}{\alpha_1^2+\alpha_2^2}\partial_t\vec{u}-\frac{\alpha_2}{\alpha_1^2+\alpha_2^2}\vec{u}\times\partial_t\vec{u}=\Delta\vec{u}+|\nabla\vec{u}|^2\vec{u}. \tag{4.2.29}$$

So we define a global weak solution for (4.2.29) as follows:

Definition 4.2.1 *A vector function $\vec{u}$ is said to be a global weak solution to (4.2.29), if $\vec{u}$ is defined a.e. on $M\times R_+$ such that*
(1) $\vec{u}\in L^\infty((0,\infty);H^{1,2}(M))$ and $\partial_t\vec{u}\in L^2((0,\infty);L^2(M))$;
(2) $|\vec{u}(x,t)|^2=1$ a.e. on $M\times R_+$;
(3) (4.2.29) holds in the sense of distribution;
(4) $\vec{u}(x,0)=\vec{u}_0(x)$ in the trace sense.

We consider the following penalized equation:

$$\frac{\alpha_1}{\alpha_1^2+\alpha_2^2}\partial_t\vec{u}^k-\frac{\alpha_1}{\alpha_1^2+\alpha_2^2}\vec{u}^k\times\partial_t\vec{u}^k-\Delta\vec{u}^k+k^2(|\vec{u}^k|^2-1)\vec{u}^k=0, \tag{4.2.30}$$

where $t>0$, $x\in M$.

We construct approximate solutions of (4.2.30) by using the Galerkin method. Let $w_j(x)$ $(j=1,2,\ldots)$ be the normalized eigenfunction of equation $\Delta\vec{u}+\lambda\vec{u}=0$ on M corresponding to eigenvalue λ_j $(j=1,2,\ldots)$. $\{w_j(x)\}$ forms a normalized orthogonal basis in $H^{1,2}(M,R^3)$.

Denote the approximate solutions $\vec{u}_N^k(x,t)$ of Eq. (4.2.30) by the following form:

$$\vec{u}_N^k(x,t) = \sum_{j=1}^N \alpha_{kj}(t) w_j(x),$$

where $\alpha_{kj}(t)$ $(j = 1, 2, \ldots)$ are vector-valued functions of $t \in R_+$. By the standard procedure of the Galerkin approximation, there exists a global weak solution of (4.2.30) that satisfies

$$\int_M \left[\frac{\alpha_1}{\alpha_1^2 + \alpha_2^2} \partial_t \vec{u}_N^k \cdot w_s - \frac{\alpha_1}{\alpha_1^2 + \alpha_2^2} (\vec{u}_N^k \times \partial_t \vec{u}_N^k) \cdot w_s \right.$$
$$\left. + \nabla \vec{u}_N^k \cdot \nabla w_s + k^2 (|\vec{u}_N^k|^2 - 1) \vec{u}_N^k \cdot w_s \right] dM = 0, \qquad (4.2.31)$$

for all $s = 1, 2, \ldots, N$, with the initial value $\vec{u}_N(x,0)$ satisfying

$$\int_M \vec{u}_N(x,0) w_s(x) dM = \int_M \vec{u}_0(x) w_s(x) dM, \quad s = 1, 2, \ldots N, \qquad (4.2.32)$$

where $\vec{u}_0(x) \in H^{1,2}(M; S^2)$ is the initial value. Moreover $\vec{u}_N^k$ satisfies

$$\int_0^t \|\partial_t \vec{u}_N^k(\tau)\|_{L^2(M)}^2 d\tau + \|\nabla \vec{u}_N^k(\cdot, t)\|_{L^2(M)}^2$$
$$+ \frac{k^2}{4} \int_M (|\vec{u}_N^k(\cdot, t)|^2 - 1)^2 dM \leq C \qquad (4.2.33)$$

for all $t \in R_+$.

Since k is fixed, using (4.2.31) and (4.2.33) and letting $N \to \infty$, there exists a function $f^k(x) \in (H^{1,2}(M^T))^*$ such that

$$\lim_{N \to \infty} \int_{M^T} \vec{u}_N^k \times \partial_t \vec{u}_N^k \cdot \phi dx = \int_{M^T} f^k(x) \cdot \phi dx, \ \forall \ \phi \in H^{1,2}(M^T; R^3).$$

On the other hand, from (4.2.33) we have

$$\lim_{N \to \infty} \int_{M^T} \vec{u}_N^k \times \partial_t \vec{u}_N^k \cdot \phi dx = \int_{M^T} \vec{u}^k \times \partial_t \vec{u}^k \cdot \phi dx, \ \forall \ \phi \in H^{1,2}(M^T; R^3).$$

By a density argument, we get

$$f^k(x) = \vec{u}^k \times \partial_t \vec{u}^k.$$

Therefore using (4.2.31) and (4.2.33) there exists a global weak solution $\vec{u}^k$ of (4.2.30) such that

$$\int_0^t \|\partial_\tau \vec{u}^k(\tau)\|_{L^2(M)}^2 d\tau + \|\nabla \vec{u}^k(\cdot, t)\|_{L^2(M)}^2 + \frac{k^2}{4} \int_M (|\vec{u}^k(\cdot, t)|^2 - 1)^2 dM \leq C \quad (4.2.34)$$

for all $t \in R_+$.

From (4.2.31) we have

Lemma 4.2.16 *Let $\vec{u}^k$ be a global weak solution of* (4.2.30). *Then we have*

$$|\vec{u}^k(x,t)| \leq 1.$$

Proof. Let $\phi = \vec{u}^k - \frac{\vec{u}^k}{|\vec{u}^k|}\min\{1, |\vec{u}^k|\}$ be a test function for Eq. (4.2.30) like in [73]. Then by a simple calculation, we get

$$\frac{1}{2}\frac{\alpha_1}{\alpha_1^2 + \alpha_2^2}\partial_t \int_{|\vec{u}^k|\geq 1} |\vec{u}^k|^2\left(1 - \frac{1}{|\vec{u}^k|}\right)dM + \int_{|\vec{u}^k|\geq 1} |\nabla\vec{u}^k|^2\left(1 - \frac{1}{|\vec{u}^k|}\right)dM \leq 0. \quad (4.2.35)$$

Since $|\vec{u}^k(x,0)| = |\vec{u}_0| = 1$, we get that $|\vec{u}^k| \leq 1$ for all $t \geq 0$ and a.e. on M.

Using Lemma 4.2.16 and (4.2.34), we prove the following:

Theorem 4.2.7 *Let $\alpha_1 > 0$ in* (4.2.29). *For all $\vec{u}_0 \in H^{1,2}(M; S^2)$, there exists a global weak solution of Eq.* (4.2.29).

Proof. From (4.2.34) and Lemma 4.2.16 we can choose a subsequence of $\vec{u}^k$, still denoted by $\vec{u}^k$, such that as $k \to \infty$

$$
\begin{aligned}
\vec{u}^k &\rightharpoonup \vec{u}, &&\text{weakly-* in } L^\infty(0,\infty; H^{1,2}(M)); \\
\partial_t\vec{u}^k &\rightharpoonup \partial_t\vec{u}, &&\text{weakly in } L^2(0,\infty; L^2(M)); \\
\vec{u}^k \times \partial_t\vec{u}^k &\rightharpoonup \vec{u} \times \partial_t\vec{u}, &&\text{weakly in } L^2(0,\infty; L^2(M)); \\
\vec{u}^k &\to \vec{u}, &&\text{strongly in } L^2(0,\infty; L^2(M));
\end{aligned}
$$

and

$$|\vec{u}| = 1 \quad \text{a.e. on } R_+ \times M.$$

By taking the edge product of (4.2.30) with $\vec{u}^k$, we get

$$0 = \left(\frac{\alpha_1}{\alpha_1^2 + \alpha_2^2}\partial_t\vec{u}^k - \frac{\alpha_2}{\alpha_1^2 + \alpha_2^2}\vec{u}^k \times \partial_t\vec{u}^k - \Delta\vec{u}^k\right) \times \vec{u}^k.$$

Then the conclusion of theorem follows by allowing $k \to \infty$.

2. *Examples of solution to L–L equation in higher dimensions*

From the theory of harmonic maps we give the examples of solution to (4.2.29).

Example 4.2.1 Let $M = R^3$. Let $\vec{u} = \frac{x}{|x|} : R^3 \to S^2$. By the well-known result, we have

$$-\Delta\vec{u} = |\nabla\vec{u}|^2\vec{u}, \ \forall \ x \in R^3 \setminus \{0\}.$$

Then from the well-known result of De Giorgi, $\vec{u}$ is a solution of (4.2.3) from $R^3 \to S^2$, but it is not smooth at the origin in R^3.

Example 4.2.2 Let $M = R^3$. For any constant K, we set

$$\vec{u}(x,t) \equiv \vec{u}(x) = (\cos Kx^3, \sin Kx^3, 0),$$

where $x = (x^1, x^2, x^3)$. Then $\vec{u}$ is a smooth solution of (4.2.3) from $R^3 \times R_+$ into S^2.

4.3   Generalized L–L Systems and Harmonic Maps

4.3.1   Generalized Landau–Lifshitz Systems

1. *Generalized Landau–Lifshitz systems and harmonic maps*

Let (M, γ) be an m-dimensional Riemannian manifold of metric γ. Then generalized Landau–Lifshitz system from M to S^{n-1} is of the form

$$\vec{u}_t = \alpha_1(\Delta_M \vec{u} + |\nabla \vec{u}|^2 \vec{u})$$
$$+ \alpha_2 * [\vec{u} \wedge a_2(\vec{u}) \wedge \cdots \wedge a_{n-2}(\vec{u}) \wedge \Delta_M \vec{u}], \tag{4.3.1}$$

where "$\wedge$" denotes the outer product in R^n, "$*$" is the Hodge-$*$ operator in Euclid space R^n, $a_i(\vec{u}) : S^{n-1} \to R^n$ ($i = 1, 2, \ldots, n-2$) are smooth vector-valued functions which are linear independent. If $n = 3$ and M is a bounded domain, (4.3.1) is the usual Landau–Lifshitz system discussed in above section. Note that

$$(\Delta_M \vec{u} + |\nabla \vec{u}|^2 \vec{u}) \cdot (*[\vec{u} \wedge a_2(\vec{u}) \wedge \cdots \wedge a_{n-2}(\vec{u}) \wedge \Delta_M \vec{u}]) = 0,$$

that is, vector $\Delta_M \vec{u} + |\nabla \vec{u}|^2 \vec{u}$ is orthogonal to vector $*[\vec{u} \wedge a_2(\vec{u}) \wedge \cdots \wedge a_{n-2}(\vec{u}) \wedge \Delta_M \vec{u}]$. We may see that the harmonic map from M into S^{n-1}, i.e. the solution of

$$\Delta_M \vec{u} + |\nabla \vec{u}|^2 \vec{u} = 0$$

is also a solution from M to S^{n-1} of the following elliptic Landau–Lifshitz system:

$$\alpha_1 \Delta_M \vec{u} + |\nabla \vec{u}|^2 \vec{u} + \alpha_2 * [\vec{u} \wedge a_2(\vec{u}) \wedge \cdots \wedge a_{n-2}(\vec{u}) \wedge \Delta_M \vec{u}] = 0.$$

This establishes the relation between harmonic maps and the generalized Landau–Lifshitz system and makes that some approaches for harmonic maps can be applied to the generalized Landau–Lifshitz system.

First, we want to prove the existence of a global smooth solution to (4.3.1) from a closed Riemannian manifold into S^{n-1} with initial data of small energy. Then we discuss the global existence of a weak solution to (4.3.1) if the initial data $\vec{u}_0 \in H^1(M; S^{n-1})$ and $a_i(\vec{u}) : S^{n-1} \to R^n$ ($i = 1, 2, \ldots, n-2$) are linear independent. If $m = 2$, we have known that such weak solutions are in fact regular with exception of at most finitely many points.

In local coordinate $x = (x^1, x^2, \ldots, x^m)$ on M, (4.3.1) can be expressed as follows:

$$\vec{u}_t = \alpha_1 \frac{1}{\sqrt{\gamma}} \frac{\partial}{\partial x^\beta} \left(\gamma^{\alpha\beta} \sqrt{\gamma} \frac{\partial \vec{u}}{\partial x^\alpha} \right) + \alpha_1 |\nabla \vec{u}|^2 \vec{u}$$
$$+ \alpha_2 * \left[\vec{u} \wedge a_2(\vec{u}) \wedge \cdots \wedge a_{n-2}(\vec{u}) \wedge \frac{1}{\sqrt{\gamma}} \frac{\partial}{\partial x^\beta} \left(\gamma^{\alpha\beta} \sqrt{\gamma} \frac{\partial \vec{u}}{\partial x^\alpha} \right) \right], \tag{4.3.2}$$

where $*[\vec{u} \wedge a_2(\vec{u}) \wedge \cdots \wedge a_{n-2}(\vec{u}) \wedge \Delta_M \vec{u}]$ is

$$\begin{pmatrix} b_{11} & b_{12} & b_{13} & \cdots & b_{1n} \\ b_{21} & b_{22} & b_{23} & \cdots & b_{2n} \\ b_{31} & b_{32} & b_{33} & \cdots & b_{3n} \\ \vdots & \vdots & \vdots & \ddots & \vdots \\ b_{n1} & b_{n2} & b_{n3} & \cdots & b_{nn} \end{pmatrix} \begin{pmatrix} \Delta_M u^1 \\ \Delta_M u^2 \\ \Delta_M u^3 \\ \vdots \\ \Delta_M u^n \end{pmatrix}.$$

For simplicity, let $\alpha_1 = \alpha_2 = 1$. From the definitions of outer product and Hodge-$*$ operator, we have

$$*[\vec{u} \wedge a_2(\vec{u}) \wedge \cdots \wedge a_{n-2}(\vec{u}) \wedge \Delta_M \vec{u}]$$

$$= \begin{pmatrix} e_1 & u^1 & a_{21}(\vec{u}) & \cdots & a_{n-2,1}(\vec{u}) & \Delta_M u^1 \\ e_2 & u^2 & a_{22}(\vec{u}) & \cdots & a_{n-2,2}(\vec{u}) & \Delta_M u^2 \\ e_3 & u^3 & a_{23}(\vec{u}) & \cdots & a_{n-2,3}(\vec{u}) & \Delta_M u^3 \\ e_4 & u^4 & a_{24}(\vec{u}) & \cdots & a_{n-2,4}(\vec{u}) & \Delta_M u^4 \\ \vdots & \vdots & \vdots & \ddots & \vdots & \vdots \\ e_{n-1} & u^{n-1} & a_{2,n-1}(\vec{u}) & \cdots & a_{n-2,n-1}(\vec{u}) & \Delta_M u^{n-1} \\ e_n & u^n & a_{2n}(\vec{u}) & \cdots & a_{n-2,n}(\vec{u}) & \Delta_M u^{n-2} \end{pmatrix},$$

where $\{e_1, e_2, \ldots, e_n\}$ is a normal orthogonal basis of R^n, $\vec{u} = (u^1, u^2, \ldots, u^n)$. Direct computation shows that (b_{ij}) is antisymmetric. Let $S = (s_{ij})$ denote the following matrix:

$$S = (s_{ij}) = \begin{pmatrix} 1 & b_{12} & b_{13} & \cdots & b_{1n} \\ -b_{12} & 1 & b_{23} & \cdots & b_{2n} \\ -b_{13} & -b_{23} & 1 & \cdots & b_{3n} \\ \vdots & \vdots & \vdots & \ddots & \vdots \\ -b_{1n} & -b_{2n} & -b_{3n} & \cdots & 1 \end{pmatrix}.$$

It is easy to see that matrix (s_{ij}) is an $n \times n$ nondegenerate matrix. In fact,

$$\sum_{i,j=1}^{n} s_{ij} \eta_i \eta_j = \sum_{i=1}^{n} \eta_i^2, \quad \forall \ \eta \in R^n \setminus \{0\}.$$

Therefore in the local coordinates, the principle part of the generalized Landau–Lifshitz system is

$$\sum_{i,j=1}^{m} D_i(a_{ij}(\cdot, \vec{u}) D_j \vec{u}) = \sum_{i,j=1}^{m} D_i(\gamma^{ij} D_j \vec{u} + *[\vec{u} \wedge a_2(\vec{u}) \wedge \cdots \wedge a_{n-2}(\vec{u}) \wedge \gamma^{ij} D_j \vec{u}]).$$

The corresponding coefficients $a_{ij}^{\alpha\beta}$ are given by

$$a_{ij}^{\alpha\beta} = \gamma^{ij} s_{\alpha\beta}.$$

It is clear that $(a_{ij}^{\alpha\beta})$ satisfies the Legender–Hadamard condition:

$$\sum_{\alpha,\beta=1}^{n} \sum_{i,j=1}^{m} a_{ij}^{\alpha\beta}(x, p) \xi^i \xi^j \eta_\alpha \eta_\beta = |\eta|^2 \sum_{i,j=1}^{m} \gamma^{ij} \xi^i \xi^j > 0$$

for all $(x, p) \in M \times R^n$, $\eta \in R^n \setminus \{0\}$, and $\xi \in R^m \setminus \{0\}$.

    2. *Initial boundary value problem for the generalized Landau–Lifshitz equations*

Consider the following initial boundary value problem for the Landau–Lifshitz equation

$$\begin{cases} \vec{u}_t = \Delta_M \vec{u} + |\nabla \vec{u}|^2 \vec{u} + *[\vec{u} \wedge a_2(\vec{u}) \wedge \cdots \wedge a_{n-2}(\vec{u}) \wedge \Delta_M \vec{u}], \\ \vec{u} = \psi, \ x \in \partial M, \ t \geq 0; \quad \vec{u}(\cdot, 0) = \vec{u}_0, \ x \in M, \end{cases} \tag{4.3.3}$$

where M is a Riemannian manifold with boundary, and $\psi : \partial M \to S^n$ is a smooth map. $\vec{u}_0 : M \to S^n$ belongs to $C^2_\psi(M; S^n) \equiv \{\vec{u} : \vec{u} \in C^2(M; S^n), \vec{u}|_{\partial M} = \psi\}$.

In order to prove the local existence of solutions for (IBP) we shall use the results of Amann [5].

Let Ω be a bounded smooth domain in R^m and consider the second order parabolic systems of the form

$$\vec{u}_t = \sum_{i,j=1}^m D_i(a_{ij}(x,t,u)D_j u) + f(x,t,Du), 0 < t < T, x \in \Omega, \tag{4.3.4}$$

where $D_j = \frac{\partial}{\partial x_j}$ is acting on the N-dimensional vector $u : \Omega \to R^N$. Suppose that the coefficients of system (4.3.4) are smooth, that is,

$$a_{ij} = a_{ji}, \quad a_{ij} \in C^\infty(\Omega \times [t, T] \times R^N; l(R^N)),$$

where $l(R^N)$ is the self-isomorphism space. Furthermore, suppose that

$$\sum_{r,s=1}^n \sum_{j,k=1}^m a_{jk}^{rs} \xi^i \xi^j \eta_r \eta_s > 0, \ \forall \ (x,t,p) \in \overline{\Omega} \times [t,T] \times R^N, \ \eta \in R^n \setminus \{0\}, \ \xi \in R^m \setminus \{0\},$$

where a_{jk}^{rs} are the elements of matrix a_{jk}. Let

$$f \in C^\infty(\overline{\Omega} \times [t,T] \times R^m \times R^{mN}; R^N).$$

3. *Initial boundary value problem for the quasilinear parabolic systems*
Consider the following initial boundary value problem:

$$\begin{cases} \vec{u}_t = \sum_{j,k=1}^m D_j(a_{jk}(x,t,u)D_k u) + f(x,t,u,Du), \ s < t \leq T, x \in \Omega, \\ u = 0, \ \text{on} \ \partial\Omega \times (s,T]; \quad u(\cdot,s) = u_0, \end{cases} \tag{4.3.5}$$

where $0 \leq s < T$. The classical solution u of this problem is defined on interval $J \subset [s,T]$ such that

$$u \in C(\overline{\Omega} \times J; R^N) \cap C^1(\overline{\Omega} \times J; R^N) \cap C^{2,0}(\Omega \times J; R^N),$$

and it satisfies (4.3.5) pointwise.

Let $m < p < \infty$, and denote $W^r_p(\Omega, R^N)$ by W^r_p, the usual Sobolev space, where $r \in [0,\infty]$. H. Amann in [5] proved the following local existence result:

Theorem 4.3.1 *Let* $0 \leq s < T$, $\frac{m}{p} < \tau < \infty$, $u_0 \in W^\tau_p$. *Then the initial-boundary value problem* (4.3.5) *admits a classical solution* u *defined on an open interval* $J \subset [s,T]$. *If*

$$\sup_{t \in J} \|u((t)\|_{W^\tau_p} < \infty,$$

then $J = [s,T]$ *and* u *is a global solution.*

4. *Local smooth solution*

Applying Amann Theorem 4.3.1 to problem (4.3.3), we have

Theorem 4.3.2 *Let* $\psi(x,t) \in C^\infty(\partial M; S^n)$, $\vec{u}_0 \in C^2_\psi(M; S^n)$ *and* $a_i(\vec{u})$ ($i = 2, 3, \ldots, n-2$) *be smooth functions. Then the initial boundary value problem (4.3.3) admits a unique classical local solution* $\vec{u}(x,t)$ *defined on* $[0, \omega) \subset [0, T]$.

If M is a Riemannian manifold without boundary, we can consider the initial value problem (IP) for the generalized Landau–Lifshitz systems from M to S^{n-1} as follows:

$$\vec{u}_t = \Delta_M \vec{u} + |\nabla \vec{u}|^2 \vec{u} + *[\vec{u} \wedge a_2(\vec{u}) \wedge \cdots \wedge a_{n-2}(\vec{u}) \wedge \Delta_M \vec{u}], \ t > 0, \ x \in M,$$
$$\vec{u}(x,0) = \vec{u}_0, \ x \in M, \tag{4.3.6}$$

where $\vec{u}_0 \in C^2(M, S^{n-1})$ and $|\vec{u}_0| = 1$.

Theorem 4.3.3 *Let M be a closed smooth Riemannian manifold, $\vec{u}_0 \in C^2(M, S^{n-1})$, and $a_i(\vec{u})$ ($i = 2, 3, \ldots, n-2$) be smooth functions. Then the initial boundary value problem (4.3.6) admits a unique classical solution $\vec{u}(x,t)$ defined on $t \in [0, \omega) \subset [0, T]$.*

Proof. From Theorem 4.3.2, it suffices to prove that the smooth solution of (4.3.6) satisfies $|\vec{u}(x,t)| = 1$. In fact, setting $z(x,t) = |\vec{u}(x,t)|^2$ on $M_T := M \times [0, T]$, where T is a finite time $T < \infty$, we have

$$z_t = 2\vec{u} \cdot \vec{u}_t, \quad \nabla z = 2\vec{u} \cdot \nabla \vec{u}.$$

By simple calculations we have

$$\Delta_M z = 2\vec{u} \cdot \Delta_M \vec{u} + 2|D\vec{u}|^2.$$

Thus we have

$$z_t = \Delta_M z + 2|D\vec{u}|^2(|\vec{u}|^2 - 1), \quad z(x,0) = 1, \ x \in M.$$

Setting $w = z - 1$, we have

$$w_t = \Delta_M w + 2|D\vec{u}|^2 w, \tag{4.3.7}$$
$$w(x,0) = 0. \tag{4.3.8}$$

Multiplying (4.3.7) by w and integrating it on M, we get

$$\frac{1}{2}\frac{d}{dt}\int_M |w|^2 dx + \int_M |Dw|^2 dx \leq 2\max_{x,t}|D\vec{u}|^2 \int_M |w|^2 dx.$$

It follows from Gronwall's and (4.3.8) that $w(x,t) = 0$ for any $(x,t) \in M_T$, i.e. $|\vec{u}(x,t)|^2 = 1$. This proves the theorem.

Now consider the following Cauchy problem for the Landau–Lifshitz equation from the flat torus $T^2 = R^2/Z \oplus Z$ to S^{n-1}. Let $\vec{u}_0 : T^2 \to S^{n-1}$ be a mapping. Consider the following problem:

$$\vec{u}_t = \Delta_M \vec{u} + |\nabla \vec{u}|^2 \vec{u} + *[\vec{u} \wedge a_2(\vec{u}) \wedge \cdots \wedge a_{n-2}(\vec{u}) \wedge \Delta_M \vec{u}], \qquad (4.3.9)$$

$$\vec{u}(x,0) = \vec{u}_0(x), \ x \in T^2, \qquad (4.3.10)$$

where $a_i(\vec{u})$ $(i = 2, 3, \ldots, n-2)$ are smooth functions and $|a_i(\vec{u})| = 1$.

Lemma 4.3.1 *Let $\vec{u} \in C^\infty(T^2 \times R^+; S^{n-1})$ be a smooth solution to the Cauchy problem* (4.3.9) *and* (4.3.10). *Then we have*

$$E(\vec{u}(\cdot,t)) \leq E_0(\vec{u}_0) = E_0, \quad \forall \ t \geq 0, \qquad (4.3.11)$$

where $E(\vec{u}) = \int_{T^2} e(\vec{u})dx$, $e(\vec{u}) = \frac{1}{2}|\nabla \vec{u}|^2$.

Proof. Multiplying (4.3.9) by $\partial_t \vec{u}$ and integrating it over $T^2 \times [0,T]$, we have

$$\int_0^T \int_{T^2} |\vec{u}_t|^2 dx dt - \int_0^T \int_{T^2} \vec{u}_t \cdot (\Delta \vec{u} + |\nabla \vec{u}|^2 \vec{u}) dx dt$$

$$= \int_0^T \int_{T^2} \vec{u}_t \cdot \{*[\vec{u} \wedge a_2(\vec{u}) \wedge \cdots \wedge a_{n-2}(\vec{u}) \wedge \Delta \vec{u}]\} dx dt. \qquad (4.3.12)$$

Since $\Delta \vec{u} + |\nabla \vec{u}|^2 \vec{u}$ and $*[\vec{u} \wedge a_2(\vec{u}) \wedge \cdots \wedge a_{n-2}(\vec{u}) \wedge \Delta \vec{u}]$ are orthogonal with each other, we have

$$|\vec{u}_t|^2 = \left|\Delta \vec{u} + |\nabla \vec{u}|^2 \vec{u}\right|^2 |*[\vec{u} \wedge a_2(\vec{u}) \wedge \cdots \wedge a_{n-2}(\vec{u}) \wedge \Delta \vec{u}]|^2.$$

Then it follows from the definition of outer product that

$$\vec{u}_t \cdot \{*[\vec{u} \wedge a_2(\vec{u}) \wedge \cdots \wedge a_{n-2}(\vec{u}) \wedge \Delta \vec{u}]\}$$

$$= |*[\vec{u} \wedge a_2(\vec{u}) \wedge \cdots \wedge a_{n-2}(\vec{u}) \wedge \Delta \vec{u}]|^2$$

$$= \left|*[\vec{u} \wedge a_2(\vec{u}) \wedge \cdots \wedge a_{n-2}(\vec{u}) \wedge (\Delta \vec{u} + |\nabla \vec{u}|^2 \vec{u})]\right|^2$$

$$\leq |\Delta \vec{u} + |\nabla \vec{u}|^2 \vec{u}|^2,$$

where we have used $|a_i(\vec{u})| = 1$ $(i = 2, \ldots, n-2)$. Then we have

$$|\vec{u}_t \cdot *\{[\vec{u} \wedge a_2(\vec{u}) \wedge \cdots \wedge a_{n-2}(\vec{u}) \wedge \Delta \vec{u}]\}| \leq \frac{1}{2}|\vec{u}_t|^2, \qquad (4.3.13)$$

and therefore

$$\int_0^T \int_{T^2} \vec{u}_t \cdot (\Delta \vec{u} + |\nabla \vec{u}|^2 \vec{u}) dx dt = -\int_0^T \frac{d}{dt} E(\vec{u}(t)) dt. \qquad (4.3.14)$$

It then follows from (4.3.12)–(4.3.14) that

$$\int_0^T \int_{T^2} |\vec{u}_t|^2 dx dt + 2 \int_0^T \frac{d}{dt} E(\vec{u}(t)) dt \leq 0.$$

The lemma follows.

Lemma 4.3.2 *Suppose that $\vec{u}_0 \in H^2(T^2)$. Let $\vec{u}$ be a smooth solution of (4.3.9) and (4.3.10) on $T^2 \times [0, T]$. Then there is a constant $C > 0$ depending only on E_0 (if E_0 is small enough) such that*

$$\|\nabla^2 \vec{u}(\cdot, t)\|_{L^2(T^2)} \leq C, \quad \forall \ 0 < t \leq T; \quad \|\nabla^3 \vec{u}(x, t)\|_{L^2(T^2 \times [0,T])} \leq C.$$

Proof. Acting on (4.3.9) by Δ, multiplying it by $\Delta \vec{u}$, and integrating it with respect to x on T^2, we have

$$\int_{T^2} \Delta \vec{u}_t \cdot \Delta \vec{u} \, dx = \int_{T^2} \Delta^2 \vec{u} \cdot \Delta \vec{u} \, dx + \int_{T^2} \Delta(|\nabla \vec{u}|^2 \vec{u}) \cdot \Delta \vec{u} \, dx$$
$$+ \int_{T^2} \Delta\{*[\vec{u} \wedge a_2(\vec{u}) \wedge \cdots \wedge a_{n-2}(\vec{u}) \wedge \Delta \vec{u}]\} \Delta \vec{u} \, dx.$$

It is clear that

$$\int_{T^2} \Delta \vec{u}_t \cdot \Delta \vec{u} \, dx = \frac{1}{2} \frac{d}{dt} \|\Delta \vec{u}\|_{L^2(T^2)}^2;$$
$$\int_{T^2} \Delta^2 \vec{u} \cdot \Delta \vec{u} \, dx = -\|\nabla \Delta \vec{u}\|_{L^2(T^2)}^2.$$

Hence

$$\left| \int_{T^2} \Delta(|\nabla \vec{u}|^2 \vec{u}) \cdot \Delta \vec{u}) dx \right| = \left| \int_{T^2} \nabla(|\nabla \vec{u}|^2 \vec{u}) \cdot \nabla \Delta \vec{u}) dx \right|$$
$$\leq \|\nabla \Delta \vec{u}\|_{L^2(T^2)} (2\|\vec{u}\|_{L^\infty(T^2)} \|\nabla \vec{u}\|_{L^4(T^2)}$$
$$\cdot \|\Delta \vec{u}\|_{L^4(T^2)} + \|\nabla \vec{u}\|_{L^6(T^2)}^3). \qquad (4.3.15)$$

From the following Gagliardo–Nirenberg inequalities

$$\|\nabla \vec{u}\|_{L^4(T^2)} \leq C_1 \|\nabla \vec{u}\|_{L^2(T^2)}^{3/4} \|\nabla \Delta \vec{u}\|_{L^2(T^2)}^{1/4}, \qquad (4.3.16)$$
$$\|\nabla \vec{u}\|_{L^6(T^2)} \leq C_1 \|\nabla \Delta \vec{u}\|_{L^2(T^2)}^{1/3} \|\nabla \vec{u}\|_{L^2(T^2)}^{2/3}, \qquad (4.3.17)$$
$$\|\nabla^2 \vec{u}\|_{L^4(T^2)} \leq C_1 \|\nabla \vec{u}\|_{L^2(T^2)}^{1/4} \|\nabla \Delta \vec{u}\|_{L^2(T^2)}^{3/4}, \qquad (4.3.18)$$

we have

$$\left| \int_{T^2} \nabla(|\nabla \vec{u}|^2 \vec{u}) \cdot \nabla \Delta \vec{u} \, dx \right| \leq (2^{3/2} C_1^2 \sqrt{E_0} + 2C_1^3 E_0) \|\nabla \Delta \vec{u}\|_{L^2(T^2)}^2.$$

On the other hand, using (4.3.15) and (4.3.17), we have

$$\left| \int_{T^2} \Delta \vec{u} \cdot \Delta\{*[\vec{u} \wedge a_2(\vec{u}) \wedge \cdots \wedge a_{n-2}(\vec{u}) \wedge \Delta \vec{u}]\} dx \right|$$

$$\leq \left| \int_{T^2} \nabla \Delta \vec{u} \cdot \{*[\nabla \vec{u} \wedge a_2(\vec{u}) \wedge \cdots \wedge a_{n-2}(\vec{u}) \wedge \Delta \vec{u}]\} dx \right|$$

$$+ \sum_{i=2}^{n-2} \left| \int_{T^2} \nabla \Delta \vec{u} \cdot \{*[\vec{u} \wedge a_2(\vec{u}) \wedge \cdots \wedge a_i{}'(\vec{u}) \nabla \vec{u} \wedge \cdots \wedge a_{n-2}(\vec{u}) \wedge \Delta \vec{u}]\} dx \right|$$

$$+ \left| \int_{T^2} \nabla\Delta\vec{u} \cdot \{*[\vec{u} \wedge a_2(\vec{u}) \wedge \cdots \wedge a_{n-2}(\vec{u}) \wedge \nabla\Delta\vec{u}]\}dx \right|$$

$$\leq \|\nabla\Delta\vec{u}\|_{L^2(T^2)}(C_1^2\|\nabla\vec{u}\|_{L^2(T^2)}\|\nabla\Delta\vec{u}\|_{L^2(T^2)}$$

$$+ (n-2)C_1^2 \max_i \|a_i'(\vec{u})\|_{L^\infty(T^2\times(0,T])}\|\nabla\vec{u}\|_{L^2(T^2)}\|\nabla\Delta\vec{u}\|_{L^2(T^2)}).$$

We finally get

$$\frac{d}{dt}\|\Delta\vec{u}(\cdot,t)\|_{L^2(T^2)}^2 + \|\nabla\Delta\vec{u}(\cdot,t)\|_{L^2(T^2)}^2 \leq \varepsilon\|\nabla\Delta\vec{u}\|_{L^2(T^2)}^2,$$

where

$$\varepsilon := C_1^2\left(2^{\frac{3}{2}} + 2^{\frac{1}{2}} + 2^{\frac{1}{2}}(n-2)\max_i\|a_i'(\vec{u})\|_{L^\infty}\right)\sqrt{E_0} + 2C_1^3 E_0.$$

We have that when E_0 is small enough such that $\varepsilon < 1$

$$\frac{d}{dt}\|\Delta\vec{u}(\cdot,t)\|_{L^2(T^2)}^2 + (1-\varepsilon)\|\nabla\Delta\vec{u}\|_{L^2(T^2)}^2 \leq 0.$$

By the Gronwall inequality we have

$$\|\nabla^2\vec{u}(\cdot,t)\|_{L^2(T^2)}^2 \leq C, \quad \forall\ T \geq t > 0,$$

and

$$\|\nabla\Delta\vec{u}\|_{L^2(T^2\times(0,T))} \leq C.$$

The lemma follows.

Lemma 4.3.3 *Let $\vec{u}_0 \in H^3(T^2; S^{n-1})$ and the conditions of Lemma* 4.3.2 *hold. Then there exists a constant C such that*

$$\sup_{0\leq t\leq T}\|\nabla(\cdot,t)\|_{L^2(T^2)} \leq C. \tag{4.3.19}$$

Proof. Acting on (4.3.9) by $\nabla\Delta$, multiplying it by $\nabla\Delta\vec{u}$, and integrating it over T^2, we have

$$\frac{1}{2}\frac{d}{dt}\|\nabla\Delta\vec{u}\|_{L^2(T^2)}^2 + \|\Delta^2\vec{u}\|_{L^2(T^2)}^2 = \int_{T^2}\nabla\Delta(|\nabla\vec{u}|^2\vec{u})\cdot\nabla\Delta\vec{u}dx$$

$$+ \int_{T^2}\nabla\Delta\vec{u}\cdot\nabla\Delta\{*[\vec{u}\wedge a_2(\vec{u})\wedge\cdots\wedge a_{n-2}(\vec{u})\wedge\Delta\vec{u}]\}dx. \tag{4.3.20}$$

It follows from Lemma 4.3.4 and the Sobolev inequality that the first term on the right-hand side of (4.3.20) can be estimated as follows:

$$\left|\int_{T^2}\nabla\Delta(|\nabla\vec{u}|^2\vec{u})\cdot\nabla\Delta\vec{u}dx\right| = \left|\int_{T^2}\Delta(|\nabla\vec{u}|^2\vec{u})\cdot\Delta^2\vec{u}dx\right|$$

$$= \left|\int_{T^2}\{5|\nabla\vec{u}|^2\Delta\vec{u} + 2|\nabla\vec{u}|^2\vec{u} + 2(\Delta\vec{u}\cdot\nabla\Delta\vec{u})\}\Delta^2\vec{u}dx\right|$$

$$\leq C(\|\nabla\vec{u}\|_{L^\infty(T^2)}^2\|\Delta\vec{u}\|_{L^2(T^2)} + \|\Delta\vec{u}\|_{L^2(T^2)}$$

$$+ \|\nabla\vec{u}\|_{L^\infty(T^2)}\|\nabla\Delta\vec{u}\|_{L^2(T^2)})\|\Delta^2\vec{u}\|_{L^2(T^2)}$$

$$\leq \frac{1}{2}\|\Delta^2\vec{u}\|_{L^2(T^2)}^2 + C\|\nabla\Delta\vec{u}\|_{L^2(T^2)}^2 + C_1.$$

On the other hand, we estimate the second term on the right-hand side of (4.3.20) as follows

$$\left| \int_{T^2} \nabla\Delta\vec{u} \cdot \nabla\Delta\{*[\vec{u} \wedge a_2(\vec{u}) \wedge \cdots \wedge a_{n-2}(\vec{u}) \wedge \Delta\vec{u}]\}dx \right|$$

$$= \left| \int_{T^2} \Delta^2\vec{u} \cdot \Delta\{*[\nabla\vec{u} \wedge a_2(\vec{u}) \wedge \cdots \wedge a_{n-2}(\vec{u}) \wedge \Delta\vec{u}]\}dx \right|$$

$$\leq C(\|\nabla\vec{u}\|^2_{L^\infty(T^2)}\|\Delta\vec{u}\|^2_{L^2(T^2)} + \|\nabla\vec{u}\|_{L^\infty(T^2)}\|\nabla\Delta\vec{u}\|_{L^2(T^2)}$$

$$+ \|\Delta\vec{u}\|^2_{L^4(T^2)})\|\Delta^2\vec{u}\|_{L^2(T^2)}$$

$$\leq \frac{1}{2}\|\Delta^2\vec{u}\|^2_{L^2(T^2)} + C\|\nabla\Delta\vec{u}\|^2_{L^2(T^2)} + C_1.$$

Since E_0 is small enough, we finally get

$$\frac{d}{dt}\|\nabla\Delta\vec{u}\|^2_{L^2(T^2)} \leq C\|\nabla\Delta\vec{u}\|^2_{L^2(T^2)} + C_1.$$

By the Gronwall inequality, we have

$$\sup_{0 \leq t \leq T} \|\nabla\Delta\vec{u}(\cdot, t)\|^2_{L^2(T^2)} \leq C_2,$$

where C_2 depends on $\|\nabla\Delta\vec{u}_0\|_{L^2(T^2)}$ and E_0. The lemma follows.

5. *Global smooth solution*

Combining the local existence Theorem 4.3.2 and the *a priori* estimate Lemma 4.3.6, we know that the local solution can be extended to the global solution.

Theorem 4.3.4 *Suppose that* $\vec{u}_0 \in H^k(T^2; S^{n-1})$, $k \geq 4$, *and* E_0 *is small enough. Then the initial value problem* (4.3.9) *and* (4.3.10) *admits at least one global classical solution.*

4.3.2 The Global Weak Solution to the Generalized L–L Equations

Now we consider the global weak solution to the generalized Landau–Lifshitz equations from a two-dimensional Riemannian manifold to S^{n-1}. We first give some notions.

Let M be a compact Riemannian manifold of metric $\gamma = (\gamma_{ij})_{1 \leq i,j \leq 2}$, $\vec{u} : M \to S^{n-1}$ be a smooth map, and

$$e(\vec{u}) = \frac{1}{2}\gamma^{ij}(x)\frac{\partial u^\alpha}{\partial x_i}\frac{\partial u^\alpha}{\partial x_j}$$

be the density of energy on M where $(\gamma^{ij}) = (\gamma_{ij})^{-1}$. The energy on M is given by

$$E(\vec{u}) = \int_M e(\vec{u})dM.$$

The stationary point of $E(\vec{u})$ is called the harmonic map from M to S^{n-1} which satisfies the Euler–Lagrange equation

$$\Delta_M \vec{u} + |\nabla \vec{u}|^2 \vec{u} = 0.$$

Let L^p, H_p^m, and C^∞ be the Lebesgue, Sobolev, and Hölder spaces, respectively. $H_p^m(M, S^{n-1})$ is the space of function $\vec{u} : M \to S^{n-1}$, $\vec{u}|_\Omega \in H_p^m(\Omega, R^n)$, $\Omega \subset M$, and $|\vec{u}| = 1$, a.e. $x \in M$. $d_M(\cdot, \cdot)$ is the geodesic distance,

$$B_R^M(x) = \{y : y \in M, d_M(x, y) < R\},$$
$$E_R(\vec{u}; x) = \int_{B_R^M(x)} e(\vec{u}) dM.$$

1. *Generalized Landau–Lifshitz equations on Riemannian manifold*
Now consider the following problem:

$$\vec{u}_t = \Delta_M \vec{u} + |\nabla \vec{u}|^1 \vec{u} + \{*[\nabla \vec{u} \wedge a_2(\vec{u}) \wedge \cdots \wedge a_{n-2}(\vec{u}) \wedge \Delta \vec{u}]\}, \quad (4.3.21)$$
$$\vec{u}(x, 0) = \vec{u}_0(x), \quad x \in M, \quad (4.3.22)$$

where $a_i(\vec{u})$ $(i = 2, \ldots, n-2)$ are n-dimensional linear independent vectors, $\vec{u}_0 \in C^2(M)$, and M is a compact closed Riemannian manifold.

Lemma 4.3.4 ([133]) *For any $\vec{u} \in H_{\mathrm{loc}}^1(R^2)$ and any function $\varphi \in C_0^\infty(B_R)$, $0 \leq \varphi \leq 1$ and $|\nabla \varphi| \leq \frac{2}{R}$,*

$$\int_{R^2} |\vec{u}|^4 \varphi^2 dx \leq C \left(\int_{B_R} |\vec{u}|^2 dx \right) \cdot \left(\int_{B_R} |\Delta \vec{u}|^2 \varphi^2 dx + R^{-2} \int_{B_R} |\vec{u}|^2 dx \right)$$

holds, where C is independent of $\vec{u}$ and R.

Lemma 4.3.5 *Let $\vec{u} \in C^2(M \times [0, T]; S^{n-1})$ be a solution to problem (4.3.21) and (4.3.22). Then*

$$\frac{1}{2} \sup_{0 \leq t \leq T} \int_0^t \int_M |\vec{u}_t|^2 dx dt + \sup_{0 \leq t \leq T} E(\vec{u}(t)) \leq E(\vec{u}_0).$$

The proof of this lemma is just the same as that of Lemma 4.3.1; hence we omit it.

Lemma 4.3.6 *There exists an absolute constant $C_3 > 0$ such that for any $R < \frac{1}{2} i_M$*

$$E(\vec{u}(T); B_R^M) \leq E(\vec{u}_0; B_{2R}^M) + C_3 \frac{T}{R^2} E(\vec{u}_0).$$

for any solution of (4.3.21) and (4.3.22).

Proof. Multiplying (4.3.21) by $\varphi^2 \vec{u}_t$, where $\varphi \in C_0^\infty(B_{2R}^M)$, $0 \le \varphi \le 1$, $\varphi(x) \equiv 1$ in B_R^M and $|\nabla \varphi| \le \frac{2}{R}$, and integrating by parts by noting

$$\frac{1}{2}|\vec{u}_t|^2 \varphi^2 \ge \vec{u}_t \varphi^2 \cdot \{*[\vec{u} \wedge a_2(\vec{u}) \wedge \cdots \wedge a_{n-2}(\vec{u}) \wedge \Delta_M \vec{u}]\},$$

we have

$$\frac{1}{2} \int_0^T \int_{B_{2R}^M} \left\{ |\vec{u}_t|^2 \varphi^2 + \frac{d}{dt}(|\nabla \vec{u}|^2 \varphi^2) \right\} dM dt$$

$$\le -2 \int_0^T \int_{B_{2R}^M} \nabla \vec{u} \cdot \vec{u}_t \nabla \varphi \cdot \varphi dM dt$$

$$\le \frac{1}{2} \int_0^T \int_{B_{2R}^M} |\vec{u}_t|^2 \varphi^2 dM dt + 2 \int_0^T \int_{B_{2R}^M} |\nabla \vec{u}|^2 \varphi^2 dM dt.$$

Hence we have

$$E(\vec{u}(T); B_R^M) - E(\vec{u}_0; B_{2R}^M) \le \frac{1}{2} \int_{B_{2R}^M} |\nabla \vec{u}|^2 \varphi^2 dM|_0^T \le C_2 \frac{T}{R^2} E(\vec{u}_0).$$

This completes the proof.

Lemma 4.3.7 *There exists $\varepsilon_1 = \varepsilon_1(M; S^{n-1}) > 0$ such that if $\vec{u} \in C^2(B_{2R}^M \times (0, T); S^{n-1})$ be a solution of (4.3.21) and (4.3.22) on $B_{2R}^M \times [0, T)$ and if for some $0 < R < \frac{1}{2} i_M$*

$$\sup_{x \in B_{2R}^M; 0 \le t < T} E(\vec{u}(t); B_R^M(x)) < \varepsilon_1,$$

then

$$\int_0^T \int_{B_R^M} |\nabla^2 \vec{u}|^2 dM dt \le CE(\vec{u}_0)\left(1 + \frac{T}{R^2}\right),$$

where C depends only on M and S^{n-1}.

The proof is similar to that of Lemma 3.7 of [133] if we note that

$$\Delta_M \vec{u} \cdot \{*[\vec{u} \wedge a_2(\vec{u}) \wedge \cdots \wedge a_{n-2}(\vec{u}) \wedge \Delta \vec{u}]\}.$$

2. *Existence of global weak solution*

Let

$$V^0(M \times [\tau, T); S^{n-1})$$

$$= \Big\{ \vec{u} : M \times [\tau, T) \to S^{n-1}, \ \vec{u} \text{ is measurable}$$

$$\underset{0 \le t < T}{\text{ess sup}} \int_M |\nabla \vec{u}(\cdots, t)|^2 dM + \int_\tau^T \int_M (|\nabla^2 \vec{u}|^2 + |\vec{u}_t|^2) dM dt < \infty \Big\}.$$

Modifying the proof of Lemma 3.8 in [133], we can get

Lemma 4.3.8 *For any numbers ε, τ, $E_0^1 > 0$, $0 < R < \frac{1}{2}i_M$, there exists a positive number δ such that if $\vec{u} \in V^0(B_{2R}^M \times [0, T]; S^{n-1})$ is a solution of* (4.3.21) *and* (4.3.22), *then for any interval $I \subset [\tau, T)$ with $|I| < \delta$ there holds the estimate*

$$\int_I \int_{B_R^M} |\Delta \vec{u}|^4 dM dt < \varepsilon,$$

where $E_0 < E_0^1$ and

$$\varepsilon(2R; x) = \varepsilon(\vec{u}; 2R, x, T) = \sup_{0 \leq t < T} E(\vec{u}(\cdot, t); B_{2R}^M(x)) < \varepsilon_1.$$

Lemma 4.3.9 *Let $\vec{u} \in C^2(B_{2R}^M(x) \times [\tau, T]; S^{n-1}) \cap V^0(B_{2R}^M(x) \times [\tau, T]; S^{n-1})$ be a regular solution of* (4.3.21) *and* (4.3.22), *where $0 < R \leq \frac{1}{2}i_M$. Then for any $\tau > 0$ when $\varepsilon(4R, x) < \varepsilon_1$, the norms of $\vec{u}$ and its derivatives on $B_R^M \times [\tau, T]$ can be uniformly estimated by a quantity depending only on $E(\vec{u}_0)$, τ, T, and R.*

Proof. Multiplying (4.3.21) by $\varphi^2 \Delta_M \vec{u}$, where $\varphi \in C_0^\infty(B_{2R}^M)$, $0 \leq \varphi \leq 1$, $\varphi(x) \equiv 1$ in B_R^M, and $|\nabla \varphi| \leq \frac{2}{R}$, and integrating by parts, we have

$$\int_{B_{2R}^M} |\Delta_M \vec{u}|^2 \varphi^2 dM \leq C \int_{B_{2R}^M} (|\nabla \vec{u}(\cdot, t)|^4 \varphi^2 + |\vec{u}_t|^2 \varphi^2) dM. \qquad (4.3.23)$$

In order to estimate the bound for the right-hand side of above inequality, differentiating (4.3.21) with respect to t and then multiplying it by $\varphi^2 \vec{u}_t$ and integrating it over $B_{2R}^M \times [s, t]$, we have

$$\frac{1}{2} \int_s^t \int_{B_{2R}^M} \partial_t |\vec{u}_t|^2 \varphi^2 dM dt + \int_s^t \int_{B_{2R}^M} |\nabla \vec{u}_t|^2 \varphi^2 dM dt$$

$$\leq C \Bigg\{ \int_s^t \int_{B_{2R}^M} (|\nabla \vec{u}|^2 |\vec{u}_t|^2 \varphi^2 + |\vec{u}_t| |\nabla \vec{u}| |\nabla \vec{u}_t| \varphi^2) dM dt$$

$$+ \int_s^t \int_{B_{2R}^M} |\vec{u}_t| |\nabla \varphi| |\nabla \vec{u}_t| \varphi dM dt$$

$$+ \left| \int_s^t \int_{B_{2R}^M} \varphi^2 \vec{u}_t \cdot \partial_t \{ *[\vec{u} \wedge a_2(\vec{u}) \wedge \cdots \wedge a_{n-2}(\vec{u}) \wedge \Delta_M \vec{u}] \} dM dt \right| \Bigg\}. \qquad (4.3.24)$$

The last term in the above inequality can be estimated by

$$\left| \int_s^t \int_{B_{2R}^M} \varphi^2 \vec{u}_t \cdot \partial_t \{ *[\vec{u} \wedge a_2(\vec{u}) \wedge \cdots \wedge a_{n-2}(\vec{u}) \wedge \Delta_M \vec{u}] \} dM dt \right|$$

$$\leq C \left(\int_s^t \int_{B_{2R}^M} (2|\nabla \varphi| |\vec{u}_t| |\nabla \vec{u}_t| \varphi + |\vec{u}_t| |\nabla \vec{u}| \nabla \vec{u}_t| \varphi^2) dM dt \right)$$

$$+ C \int_s^t \int_{B_{2R}^M} (|\vec{u}_t|^2 + |\nabla \vec{u}|^2) dM dt$$

$$\leq \frac{1}{4} \int_s^t \int_{B_{2R}^M} |\nabla \vec{u}_t|^2 \varphi^2 + C \int_s^t \int_{B_{2R}^M} |\nabla \vec{u}|^2 |\vec{u}_t|^2 \varphi^2 dM dt + C. \qquad (4.3.25)$$

Hence

$$\frac{1}{2}\int_s^t \int_{B_{2R}^M} \partial_t |\vec{u}_t|^2 \varphi^2 dM dt + \frac{1}{2}\int_s^t \int_{B_{2R}^M} |\nabla \vec{u}_t|^2 \varphi^2 dM dt$$

$$\leq C \int_s^t \int_{B_{2R}^M} |\vec{u}_t|^2 |\nabla \vec{u}|^2 \varphi^2 dM dt + C.$$

Since

$$\int_s^t \int_{B_{2R}^M} |\vec{u}_t|^2 |\nabla \vec{u}|^2 \varphi^2 dM dt \leq \left(\int_s^t \int_{B_{2R}^M} |\nabla \vec{u}|^4 dM dt \int_s^t \int_{B_{2R}^M} |\vec{u}_t|^4 \varphi^2 dM dt \right)^{\frac{1}{2}}$$

$$\leq \left(\int_s^t \int_{B_{2R}^M} |\nabla \vec{u}|^4 dM dt \right)^{\frac{1}{2}} \cdot$$

$$\cdot \left(\int_s^t \left(\operatorname*{ess\,sup}_{s \leq \theta \leq t} \int_{B_{2R}^M} |\vec{u}_t(\cdot,\theta)|^2 \varphi^2 dM \int_{B_{2R}^M} |\nabla(\varphi|\vec{u}_t|)|^2 dM \right) dt \right)^{\frac{1}{2}}$$

$$\leq C \left(\int_s^t \int_{B_{2R}^M} |\nabla \vec{u}|^4 dM dt \right)^{\frac{1}{2}} \left(\int_s^t \operatorname*{ess\,sup}_{s \leq \theta \leq t} \int_{B_{2R}^M} |\vec{u}_t(\cdot,\theta)|^2 \varphi^2 dM \right)$$

$$+ \int_{B_{2R}^M} \varphi^2 |\nabla \vec{u}_t|^2 dM dt) + C.$$

It follows from Lemma 4.3.8 that if $t - s < \delta$ is small enough, the right-hand side of (4.3.25) can be controlled by the left-hand side. Therefore,

$$\int_{B_{2R}^M} |\vec{u}_t(\cdot,t)|^2 \varphi^2 dM \leq \inf_{t-\delta \leq s \leq t} \int_{B_{2R}^M} \varphi^2 |\vec{u}_t(\cdot,s)|^2 dM + C,$$

where C depends only on $E(\vec{u}_0)$, τ, T, and R, and then

$$\operatorname*{ess\,sup}_{\tau \leq t < T} \int_{B_{2R}^M} |\vec{u}_t(\cdot,t)|^2 \varphi^2 dM \leq C(1 + \tau^{-1}) \int_\tau^T \int_{B_{2R}^M} |\vec{u}_t|^2 dM dt + C,$$

that is

$$\operatorname*{ess\,sup}_{\tau \leq t < T} \int_{B_{2R}^M} |\vec{u}_t(\cdot,t)|^2 dM \leq C(1 + \tau^{-1}) E_0 + C(E_0, R, T), \qquad (4.3.26)$$

where $C(E_0, R, T)$ depends only on E_0, R, T. Applying Lemma 4.3.8 to $v = \nabla \vec{u}(\cdot,t)$, we have

$$\int_{B_R^M} |\nabla \vec{u}|^4 dM \leq \int_{B_{2R}^M} |\nabla \vec{u}|^4 \varphi^2 dM \leq C \operatorname*{ess\,sup}_{\tau \leq t < T} \int_{B_{2R}^M} |\nabla \vec{u}(\cdot,t)|^2 dM \cdot$$

$$\cdot \left(\int_{B_{2R}^M} \varphi^2 |\nabla^2 \vec{u}(\cdot,t)|^2 dM + R^{-2} \int_{B_{2R}^M} |\nabla \vec{u}(\cdot,t)|^2 dM \right)$$

$$\leq C \operatorname*{ess\,sup}_{\tau \leq t < T} E_R(\vec{u}(\cdot,t),x) \left(\int_{B_{2R}^M} \varphi^2 |\nabla^2 \vec{u}(\cdot,t)|^2 dM + R^{-2} E(\vec{u}_0) \right).$$

$$(4.3.27)$$

If $\varepsilon(4R; x) \leq \varepsilon_1$ is small enough, then from (4.3.23), (4.3.26), and (4.3.27) we have the uniform estimate in $t \in [\tau, T)$ as follows:

$$\int_{B_R^M} |\nabla^2 \vec{u}(\cdot, t)|^2 dM \leq CE(\vec{u}_0) \left(1 + \tau^{-1} + R^{-2}\right), \qquad (4.3.28)$$

where we have used in (4.3.28) the following inequality:

$$\int_{B_{2R}^M} |\Delta \vec{u}|^2 \varphi^2 dM \geq \frac{1}{2} \int_{B_{2R}^M} |\nabla^2 \vec{u}|^2 \varphi^2 dM - C \int_{B_{2R}^M} |\nabla \vec{u}|^2 |\nabla \varphi|^2 dM.$$

On the other hand, it follows from the Gagliardo–Nirenberg inequality and (4.3.28) that

$$\sup_{x' \in B_R^M} \|\vec{u}(x', t_2) - \vec{u}(x', t_1)\|_{L^\infty}$$

$$\leq C \|\vec{u}(\cdot, t_2) - \vec{u}(\cdot, t_1)\|_{H^2(B_R^M)}^{\frac{3}{4}} \cdot \|\vec{u}(\cdot, t_2) - \vec{u}(\cdot, t_1)\|_{L^2(B_R^M)}^{\frac{1}{4}}$$

$$\leq C \sup_{\tau \leq t < T} \|\vec{u}(\cdot, t)\|_{H^2(B_R^M)}^{\frac{3}{4}} \left(\int_\tau^T \int_{B_R^M} |\partial_t \vec{u}| dM dt\right)^{\frac{1}{4}} |t_2 - t_1|^{\frac{1}{4}}$$

$$\leq C |t_2 - t_1|^{\frac{1}{4}}; \quad (t_1, t_2) \subset [\tau, T). \qquad (4.3.29)$$

Then we deduce from (4.3.28) and (4.3.29) that the $C^{1-\delta, \frac{1}{4}}$-Hölder norm on $B_R^M \times [\tau, T)$ of $\vec{u}$ can be uniformly estimated by $E(\vec{u}_0)$, R, τ, T and ε_1. This proves the lemma.

Lemma 4.3.10 *Let* $\vec{u} \in V^0(B_{2R}^M \times [\tau, T); S^{n-1})$ *be a weak solution of* (4.3.21) *and* (4.3.22), *where* $\tau > 0$, $0 < R \leq \frac{1}{4} i_M$. *Then* $\vec{u} \in C^{1-\delta, \frac{1}{4}}(B_R^M \times [\tau, T); S^{n-1})$ *and its Hölder norm on* $B_R^M \times [\tau, T)$ *are uniformly bounded.*

Lemma 4.3.11 *Let* $Q = B_{2R}^M \times (0, T)$, *where* $0 < R \leq \frac{1}{4} i_M$. *If* $\vec{u} \in C^\infty(Q; S^{n-1}) \bigcap_{\tau > 0} V^0(B_{2R}^M \times [\tau, T); S^{n-1})$ *is a weak solution of* (4.3.21) *and* (4.3.22), *and if* $\varepsilon(4R; x) < \varepsilon_1$, *then* $\vec{u}$ *can be smoothly extended to* $B_R^M \times (0, T)$, *where* x *is the center of* B_{2R}^M.

Proof. From Lemma 4.3.9 and the imbedding of $H^2(B_{2R}^M)$ into $W_p^1(B_{2R}^M)$ for any $p < \infty$, we have

$$|\vec{u}_t - \Delta_M \vec{u} - \{*[\vec{u} \wedge a_2(\vec{u}) \wedge \cdots \wedge a_{n-2}(\vec{u}) \wedge \Delta \vec{u}]\}| \in L^p(B_{2R}^M \times [\tau, T)).$$

On the other hand, we know from Lemma 4.3.9 that $\vec{u} \in C^{1-\delta, \frac{1}{4}}(B_R^M \times [\tau, T); S^{n-1})$ and its Hölder norm on $B_R^M \times [\tau, T)$ is uniformly bounded. Hence it follows from the L^p theory of a strongly parabolic system [23, 24] that $\vec{u} \in W_p^{2,1}(B_R^M \times [\tau, T)$ for any $p < \infty$. The higher regularity can be proved by Eq. (4.3.21) and the bootstrap method.

Lemma 4.3.12 *Let*

$$V(M \times [0,T]; S^{n-1}) = \left\{ \vec{u} : M \times [0,T) \to S^{n-1}, \ \vec{u} \text{ is measurable} \right.$$

$$\left. \operatorname*{ess\,sup}_{\tau \leq t \leq T} \int_M |\nabla \vec{u}(\cdot, t)|^2 dM + \int_0^T \int_M (|\nabla^2 \vec{u}|^2 + |\vec{u}_t|^2) dM dt < \infty \right\}.$$

Let $\vec{u}_1, \vec{u}_2 \in V(M \times [0,T]; S^{n-1})$ *be two weak solutions of* (4.3.21) *with* $\vec{u}_1(\cdot, 0) = \vec{u}_2(\cdot, 0) = \vec{u}_0$. *Then* $\vec{u}_1 = \vec{u}_2 = \vec{u}_0$ *on* $M \times [0,T]$.

Proof. Set $v = \vec{u}_1 - \vec{u}_2$, $|\nabla U| = |\nabla \vec{u}_1| + |\nabla \vec{u}_2|$. It follows from (4.3.21) that

$$|v_t - \Delta_M v + \{*[\vec{u}_1 \wedge a_2(\vec{u}_1) \wedge \cdots \wedge a_{n-2}(\vec{u}_1) \wedge \Delta \vec{u}_1]\}$$
$$- \{*[\vec{u}_2 \wedge a_2(\vec{u}_2) \wedge \cdots \wedge a_{n-2}(\vec{u}_2) \wedge \Delta \vec{u}_2]\}|$$
$$\leq C(|v||\nabla U|^2 + |\nabla v||\nabla U|).$$

It follows from (4.3.21) that

$$\frac{1}{2} \int_M |v(\cdot, t)|^2 dM + \int_0^t \int_M (\nabla v, \nabla v)_M dM dt$$
$$\leq C \int_0^t \int_M (|v||\nabla v||\nabla U| + |v|^2 |\nabla U|^2) dM dt$$
$$+ C \left| \int_0^t \int_M v \cdot (\{*[\vec{u}_1 \wedge a_2(\vec{u}_1) \wedge \cdots \wedge a_{n-2}(\vec{u}_1) \wedge \Delta \vec{u}_1]\} \right.$$
$$\left. - \{*[\vec{u}_2 \wedge a_2(\vec{u}_2) \wedge \cdots \wedge a_{n-2}(\vec{u}_2) \wedge \Delta \vec{u}_2]\}) dM dt \right|.$$

Therefore, we have

$$\frac{1}{2} \int_M |v(\cdot, t)|^2 dM + \frac{1}{2} \int_0^t \int_M |\nabla v|^2 dM dt \leq C \int_0^t \int_M |v|^2 |\nabla U|^2 dM dt.$$

The remaining of the proof is similar to that of Lemma 3.12 of [133].

Lemma 4.3.13 *Let* $\vec{u}_m \in V(M \times [0,T]; S^{n-1})$ *be weak solutions of* (4.3.21) *with* $\vec{u}_m(\cdot, 0) = \vec{u}_m(0)$. *If* $\varepsilon(\vec{u}_m, R, T) = \sup_{(x,t) \in M \times [0,T]} E_R(\vec{u}_m) \leq \varepsilon_1$, $R \in [0, R_0]$, *where* R_0 *be a positive constant, and* $\vec{u}_m(0) \to \vec{u}_0$ *in* $H^1(M)$ $(m \to \infty)$, *then* $\vec{u}_m \to \vec{u}$ $(m \to \infty)$ *in* $V(M \times [0,T]; S^{n-1})$, *where* $\vec{u}$ *is a solution of* (4.3.21).

Theorem 4.3.5 *For any* $\vec{u}_0 \in H^1(M; S^{n-1})$ *there exists a unique global solution to* (4.3.21) *and* (4.3.22) *on* $M \times [0, \infty)$ *which is regular with exception of at least finitely many points* (x^l, T^l) $(1 \leq l \leq L)$ *characterized by*

$$\limsup_{T \to T^l, T < T^l} E_R(\vec{u}(\cdot, T), x^l) > \varepsilon, \ \forall \ R \in (0, R_0].$$

Theorem 4.3.6 *Let* $b = \inf\{E(\vec{u}) \,|\, \vec{u} : M \to S^{n-1} \text{ is nonconstant smooth harmonic map}\}$. *If* $\vec{u}_0 \in H^1(M; S^{n-1})$ *with* $E(\vec{u}_0) < b$, *then the solution to* (4.3.21) *and* (4.3.22) *is globally regular on* $M \times [0, \infty)$.

4.4 Regularity of Weak Solutions to the Two-Dimensional Landau–Lifshitz Equations

In the above sections, we have obtained the global existence of weak solutions to higher-dimensional Landau–Lifshitz equations. The relations with harmonic maps were proposed, and the partially regular solution to two-dimensional problem was constructed.

In this section, we will show that all weak solutions, with finite energy, to two-dimensional Landau–Lifshitz equations must be partially regular and then unique.

4.4.1 Cauchy Problem to Two-Dimensional L–L Equation

1. *The problem*

Let M be a two-dimensional compact Riemannian manifold without boundary. Consider the following Cauchy problem of the Landau–Lifshitz equation on $M \times R^+$:

$$\partial_t u = \alpha u \times \partial_t u + \beta \Delta_M u + \beta |du|^2 u, \qquad x \in M,\ t > 0, \tag{4.4.1}$$

$$u(x, 0) = u_0, \qquad x \in M, \tag{4.4.2}$$

where $u_0 \in H^1(M, S^2)$, and $\dim(M) = 2$.

We want to prove that any weak solution with finite energy, i.e.,

$$E(u(t)) = \int_M |\nabla u(\cdot, t)|^2 dM \leq E_0, \tag{4.4.3}$$

is "almost smooth."

The weak solution of this problem is defined as follows:

Definition 4.4.1 *A map $u : M \times [0, T] \to S^2$ is said to be a weak solution of* (4.4.1) *and* (4.4.2) *if $u \in W_T$, where*

$$W_T =: \{u :\ \partial_t u \in L^2(0, T; L^2(M)),\ du \in L^\infty((0, T), L^2(M)),$$
$$|u| = 1,\ a.e.\ on\ M \times [0, T]\},$$

and if u satisfies (4.4.1) *and* (4.4.2) *in the sense of distribution.*

The existence, uniqueness, and regularity for the strong solution have been obtained in Sec. 4.2 (Theorems 4.2.4 and 4.2.6); now we prove the following theorem:

Theorem 4.4.1 *Let M be a compact two-dimensional Riemannian manifold without boundary. For any $T > 0$, if $u \in W_T$ is a weak solution of* (4.4.1) *and* (4.4.2) *with finite energy $E(u(t)) = \int_M |du|^2(\cdot, t) \leq E_0$ for some constant E_0 and for a.e. $t \in [0, T]$, then u is unique and is smooth on $M \times [0, T]$ with the exception of at most finitely many points.*

From now on, we call such solution as the **"almost smooth"** solution.

Our proof is completed by first showing that any weak solution of (4.4.1) and (4.4.2) with finite energy is in space $L^2(0, T; W^{1,4}(M))$, and second by proving that such a solution in fact is in space $L^4(0, T; W^{1,4}(M))$. Then, we can prove the uniqueness for the weak solutions of (4.4.1) and (4.4.2). At last, the regularity result is the consequence of the uniqueness result and the existence result.

The Sobolev space $W^{k,p}(M, S^2)$ is defined by

$$W^{k,p}(M, S^2) = \{u \in W^{k,P}(M, R^3) | u \in S^2 \text{ a.e. } x \in M\}.$$

Norm $\|u\|_{W^{k,p}(M,S^2)}$ is simply denoted by $\|u\|_{W^{k,p}}$; here domain M and target S^2 are usually omitted. We use $W^{k,p}(M, \wedge^\ell)$ to denote the space of differential ℓ-forms on M with coefficients in the Sobolev space $W^{k,p}(M)$. The exterior derivative operator is denoted by d and the conjugate operator of the exterior differential operator d in the metric of M is denoted by d^*. It is well known that $dd = 0$, $d^*d^* = 0$, and the Laplace operator for differential forms is given by $\Delta_M = dd^* + d^*d$.

Lemma 4.4.1 (Hodge decomposition theorem) [92] *Let M be an m-dimensional compact Riemannian manifold without boundary and W be an ℓ-form on M in $L^p(M, \wedge^\ell)$. Then there is an $(\ell - 1)$-form A and $(\ell + 1)$-form B such that*

$$W = dA + d^*B,$$

and $A \in W^{1,p}(M, \wedge^{\ell-1}), B \in W^{1,p}(M, \wedge^{\ell+1})$ satisfy

$$\|A\|_{W^{1,p}} + \|B\|_{W^{1,p}} \le C\|W\|_{L^p}, \tag{4.4.4}$$

where $C > 0$ is a constant depending on M and p. Moreover A and B are unique and satisfy

$$d^*A = dB = 0.$$

This result is due to Iwaniec and Martin [92], and the original proof is for the case that $M = R^m$, but it is not difficult to see that the lemma is also true in this case.

Lemma 4.4.2 (Wente's theorem) [144] *Let M be a two-dimensional compact Riemannian manifold without boundary. If $g \in H^1(M, \wedge^2)$ and $h \in H^1(M, \wedge^0)$, then $\langle d^*g, dh \rangle \in H^{-1}(M, \wedge^0)$ and*

$$\|\langle d^*g, dh \rangle\|_{H^{-1}} \le C\|d^*g\|_{L^2}\|dh\|_{L^2}, \tag{4.4.5}$$

where $C > 0$ is a constant depending only on M.

This result is due to Wente; the original proof is for the case that $M = R^2$, but the argument can also be applied to this case.

2. *Regularity and uniqueness to the Cauchy problem*

The next lemma is on the uniqueness and regularity for weak solutions to the Cauchy problem of the following system:

$$\partial_t u^i - \beta \Delta_M u^i = \beta \sum_{j=1}^{3} d^* g^{ij} \cdot du^i + f^i, \qquad x \in M, t > 0, \qquad (4.4.6)$$

$$u^i(x,0) = u_0^i(x), \qquad x \in M, \qquad i = 1,2,3, \qquad (4.4.7)$$

where $\beta > 0$ is a absolute constant, $g^{ij}(i,j = 1,2,3)$ are differential two-forms on M for each $t \in (0,T)$, and f^i $(i = 1,2,3)$ are functions of x and t.

Let

$$X_T = \{u : M \to R^3 \mid u \in L^2([0,T], H^{1,2}(M)), \ \partial_t u \in L^2([0,T], H^{-1}(M))\},$$

and for $u \in X_T$, define the norm of u in X_T by

$$\|u\|_{X_T} = \|\partial_t u\|_{L^2([0,T],H^{-1})} + \|u\|_{L^2([0,T],H^{1,2})},$$

and

$$Y_T = \{u : M \to R^3 \mid u \in L^2([0,T], W^{1,4}(M)), \ \partial_t u \in L^2([0,T], L^{\frac{4}{3}}(M))\}, \qquad (4.4.8)$$

and for $u \in Y_T$, define the norm of u in Y_T by

$$\|u\|_{Y_T} = \|\partial_t u\|_{L^2([0,T],L^{\frac{4}{3}})} + \|u\|_{L^2([0,T],W^{1,4})}.$$

We have the following lemma:

Lemma 4.4.3 *Let M be a two-dimensional compact Riemannian manifold without boundary, and in (4.4.6) and (4.4.7) f^i $(i = 1,2,3) \in L^2([0,T], L^{\frac{4}{3}}(M, \wedge^0))$, g^{ij} $(1 \le i,j \le 3) \in L^\infty([0,T], H^1(M, \wedge^2))$, and $u_0 \in H^1(M, S^2)$. Then there exists an $\varepsilon_0 > 0$, depending only on M, T, and β, such that if*

$$\|g\|_{L^\infty([0,T],H^1)} = \max_{1 \le i,j \le 3} \|g^{ij}\|_{L^\infty([0,T],H^1)} < \varepsilon_0, \qquad (4.4.9)$$

then any solution u of (4.4.6) and (4.4.7) in space X_T must belong to space Y_T.

Proof. It is sufficient to prove

Claim 1: Problem (4.4.6) and (4.4.7) has a unique solution u in X_T.

Claim 2: Problem (4.4.6) and (4.4.7) has a unique solution w in Y_T.

In fact, from the second claim, the problem (4.4.6) and (4.4.7) admits a unique solution w in Y_T. Since $Y_T \subset X_T$, so $w \in X_T$. Then, from the first claim, if u is a solution of (4.4.6) and (4.4.7) in X_T, u must coincide with w. Therefore, $u \in Y_T$.

Now let us prove these claims by the contraction mapping principle. Endowed with metric $\rho_1(u_1, u_2) = \|u_1 - u_2\|_{X_T}$ on X_T and with metric $\rho_2(u_1, u_2) = \|u_1 - u_2\|_{Y_T}$

on Y_T, X_T and Y_T are nonempty complete metric spaces. For any $v \in X_T$, by solving the Cauchy problem of the following parabolic system

$$\partial_t u^i - \beta \Delta_M u^i = \beta \sum_{j=1}^{3} d^* q^{ij} \cdot dv^j + f^i, \text{ in } M \times (0, T], \qquad (4.4.10)$$

$$u^i(x, 0) = u_0^i(x) \text{ on } M, \qquad i = 1, 2, 3, \qquad (4.4.11)$$

we define a map $L : v \to u = Lv$.

By the theory for linear parabolic equations [101], problem (4.4.10) and (4.4.11) admits a unique solution $u \in X_T$ and

$$\|u\|_{X_T} = \|\partial_t u\|_{L^2([0,T],H^{-1})} + \|du\|_{L^2([0,T],L^2)}$$

$$\leq C \left(\sum_{j=1}^{3} \|d^* g^{ij} \cdot dv^j\|_{L^2([0,T],H^{-1})} + \|f\|_{L^2([0,T],H^{-1})} + \|u_0\|_{L^2} \right).$$

Applying Lemma 4.4.3 and the Sobolev embedding theorem to the right of this inequality, we obtain that

$$\|u\|_{X_T} \leq C(\|d^* g\|_{L^\infty([0,T],L^2)}\|dv\|_{L^2([0,T],L^2)}$$

$$+ \|f\|_{L^2([0,T],L^{\frac{4}{3}})} + \|u_0\|_{H^1}), \qquad (4.4.12)$$

where $C > 0$ is a constant depending only on M and β. Therefore, the mapping L maps X_T into itself. Moreover, we would like to show that L is a contraction mapping with respect to the metric of X_T. For any $\bar{v}, \bar{\bar{v}} \in X_T$, let $\bar{u} = L\bar{v}$, $\bar{\bar{u}} = L\bar{\bar{v}}$, and $u = \bar{u} - \bar{\bar{u}}$, $v = \bar{v} - \bar{\bar{v}}$. Since L maps X_T into itself, $\bar{u}, \bar{\bar{u}} \in X_T$, and u satisfies

$$\partial_t u^i - \beta \Delta_M u^i = \beta \sum_{j=1}^{3} d^* g^{ij} \cdot dv^j, \text{ in } M \times (0, T], \qquad (4.4.13)$$

$$u^i(x, 0) = 0 \qquad \text{on } M, \qquad i = 1, 2, 3. \qquad (4.4.14)$$

Applying estimate (4.4.12) to problem (4.4.13) and (4.4.14) and noticing that

$$\|dv\|_{L^2([0,T],L^2)} \leq T\|v\|_{X_T},$$

we find that under assumption (4.4.9),

$$\|u\|_{X_T} \leq C\varepsilon_0 \|v\|_{X_T},$$

where $C > 0$ is a constant depending only on M, T, and β. Therefore, if ε_0 is chosen to be equal to $\frac{1}{2C}$, L is a contraction mapping on X_T. By the contraction mapping principle, there exists a unique solution $u \in X_T$ to problem (4.4.6) and (4.4.7). The proof of Claim 1 is complete.

Claim 2 can be proved by the same method. In fact, by the regularity theory for linear parabolic equations [101], problem (4.4.10) and (4.4.11) admits a unique solution $u \in Y_T$ and

$$\begin{aligned}
\|u\|_{Y_T} &\leq \|\partial_t u\|_{L^2([0,T],L^{4/3})} + \|d^2 u\|_{L^2([0,T],L^{4/3})} \\
&\leq C(\|d^* g\|_{L^\infty([0,T],L^2)} \|dv\|_{L^2([0,T],L^4)} \\
&\quad + \|f\|_{L^2([0,T],L^{\frac{4}{3}})} + \|u_0\|_{H^1}).
\end{aligned} \tag{4.4.15}$$

By estimate (4.4.15), it easy to see that if ϵ_0 in (4.4.9) is sufficiently small, then map L is a contraction mapping on Y_T. Therefore, there exists a unique solution $u \in Y_T$ to problem (4.4.6) and (4.4.7).

The last lemma in this subsection is regarding the following Cauchy problem:

$$\partial_t u^i - G_1(x,t)\Delta u = G_2(x,t)\Delta u + g(x,t) \qquad x \in M, t > 0, \tag{4.4.16}$$

$$u(x,0) = u_0(x), \qquad x \in M, \tag{4.4.17}$$

where $G_i(x,t)$ $(i = 1,2)$ are matrices and $g(x,t)$ is a vector.

Lemma 4.4.4 *Let M be a two-dimensional compact Riemannian manifold without boundary. Suppose that in (4.4.16) and (4.4.17)*

(1) $G_1(x,t)\Delta u$ is strongly elliptic.

(2) $G_1 \in C^\infty(M \times (0,T))$, $G_2 \in L^\infty(M \times (0,T))$, $g \in L^4(0,T;L^{4/3}(M))$ and $u_0 \in H^1(M)$.

Then there exists a constant $\epsilon_1 > 0$ depending only on M and T such that if

$$\|G_2\|_{L^\infty(M \times (0,T))} \leq \epsilon_1, \tag{4.4.18}$$

then problem (4.4.16) and (4.4.17) has a unique solution in $L^s(0,T;W^{2,4/3}(M))$ for any $s \in [2,4]$.

Proof. Again we will use the contraction mapping principle to prove this lemma. Fix an $s \in [2,4]$. For any $v \in L^s(0,T;W^{2,4/3}(M))$ by solving the following linear strongly parabolic system

$$\partial_t u - G_1(x,t)\Delta u = G_2(x,t)\Delta v + g(x,t), \qquad x \in M, t > 0, \tag{4.4.19}$$

with the initial condition (4.4.17), we define a map $L : v \to u = Lv$.

By the theory for linear parabolic equations (Theorem 9.3, Remark 9.14, and Remark 9.15 of [39]), problem (4.4.19) and (4.4.17) admits a unique solution $u \in L^s(0,T;W^{2,4/3}(M))$ and

$$\begin{aligned}
\|u\|_{L^s(0,T;W^{2,4/3}(M))} &= C\{\|G_2\|_{L^\infty(M \times [0,T])}\|v\|_{L^s(0,T;W^{2,4/3}(M))} \\
&\quad + \|g\|_{L^4(0,T;L^{4/3}(M))} + \|u_0\|_{H^1(M)}\} \\
&\leq C\{\epsilon_1\|v\|_{L^s(0,T;W^{2,4/3}(M))} + \|g\|_{L^4(0,T;L^{4/3}(M))} + \|u_0\|_{H^1(M)}\},
\end{aligned} \tag{4.4.20}$$

where $C > 0$ depends only on M and T, and in the last inequality we have used (4.4.18). Therefore, mapping L maps $L^s(0, T; W^{2,4/3}(M))$ into itself. Next we show that L is a contraction mapping with respect to the metric of $L^s(0, T; W^{2,4/3}(M))$. For any $\bar{v}, \bar{\bar{v}} \in L^s(0, T; W^{2,4/3}(M))$, let $\bar{u} = L\bar{v}$, $\bar{\bar{u}} = L\bar{\bar{v}}$, and $u = \bar{u} - \bar{\bar{u}}$, $v = \bar{v} - \bar{\bar{v}}$. Since L maps $L^s(0, T; W^{2,4/3}(M))$ into itself, $\bar{u}, \bar{\bar{u}} \in L^s(0, T; W^{2,4/3}(M))$, and u satisfies (4.4.19) and (4.4.17) with $g = 0$ and $u_0 = 0$. It is easy to see from (4.4.20) that

$$\|u\|_{L^s(0,T;W^{2,4/3}(M))} \le \frac{1}{2}\|v\|_{L^s(0,T;W^{2,4/3}(M))}$$

if ϵ_1 is sufficiently small. Therefore, L is a contraction mapping. By the contraction mapping principle, there exists a unique solution $u \in L^s(0, T; W^{2,4/3}(M))$ to problem (4.4.16) and (4.4.17) for any $s \in [2, 4]$.

4.4.2   Proof of Regularity of Solution to L–L Equation

1. *Weak solution with finite energy*

In this section we shall prove Theorem 4.4.1. In the first step, we will show that if $u \in W_T$ is a weak solution of (4.4.1) and (4.4.2) with finite energy, then $u \in Y_T \cap W_T$.

Let

$$W^{ij} = u^i du^j - u^j du^i, \qquad i, j = 1, 2, 3. \tag{4.4.21}$$

Since $E(u(t)) \le E_0$ for a.e. $t \in [0, T]$, we have

$$W^{ij} \in L^\infty([0, T], L^2(M, \wedge^1))$$

and

$$\|W^{ij}\|_{L^\infty([0,T],L^2)} \le 2\sqrt{E_0}.$$

By using the fact that $|u| = 1$ and Hélein's trick [90], Eq. (4.4.1) can be written as

$$\partial_t u^i - \beta \Delta_M u^i = \alpha(u \times \partial_t u)^i + \beta \sum_{j=1}^3 W^{ij} \cdot du^j. \tag{4.4.22}$$

Applying the Hodge decomposition theorem (Lemma 4.4.1) to W^{ij} at each time slice $t \in (0, T)$, one may find that $A^{ij} \in L^\infty([0, T], H^1(M, \wedge^0))$ and $B^{ij} \in L^\infty([0, T], H^1(M, \wedge^2))$ such that

$$W^{ij} = dA^{ij} + d^*B^{ij} \qquad \text{for a.e. } t \in [0, T], \tag{4.4.23}$$

and

$$\|A^{ij}\|_{L^\infty([0,T],H^1)} + \|B^{ij}\|_{L^\infty([0,T],H^1)} \le C\|W^{ij}\|_{L^\infty([0,T],L^2)} \le C\sqrt{E_0}. \tag{4.4.24}$$

From (4.4.23), (4.4.21) and (4.4.1), we get

$$\Delta A^{ij} = d^*W^{ij} = u^i \Delta_M u^j - u^j \Delta_M u^i$$
$$+ \beta^{-1}(u^i \partial_t u^j - u^j \partial_t u^i) - \frac{\alpha}{\beta}(u^i(u \times \partial_t u)^j - u^j(u \times \partial_t u)^i)$$
$$\in L^2([0, T], L^2(M, \wedge^0)).$$

By the Calderon–Zygmund inequality, $dA^{ij} \in L^2([0,T], H^1(M, \wedge^0))$ and

$$\|dA^{ij}\|_{L^2([0,T],H^1)} \le C\|\partial_t u\|_{L^2(M\times[0,T])}, \tag{4.4.25}$$

where $C > 0$ is a constant depending only on α, β and M.

On the other hand, since $B^{ij} \in L^\infty([0,T], H^1(M, \wedge^2))$, for any $\varepsilon > 0$, we always can find $B_1^{ij} \in L^\infty([0,T], C^\infty(M, \wedge^2))$ and $B_2^{ij} \in L^\infty([0,T], H^1(M, \wedge^2))$ such that

$$B^{ij} = B_1^{ij} + B_2^{ij}, \tag{4.4.26}$$

and

$$\|B_2^{ij}\|_{L^\infty([0,T],H^1)} < \varepsilon. \tag{4.4.27}$$

Now, by using (4.4.23) and (4.4.26), we may rewrite Eq. (4.4.22) in the following form:

$$\partial_t u^i - \beta \Delta_M u^i = \beta \sum_{j=1}^3 d^* B_2^{ij} \cdot du^j + f^i, \tag{4.4.28}$$

where

$$f^i = \alpha(u \times \partial_t u)^i + \beta \sum_{j=1}^3 dA^{ij} \cdot du^j + \beta \sum_{j=1}^3 d^* B_1^{ij} \cdot du^j. \tag{4.4.29}$$

By the Sobolev embedding theorem and (4.4.25),

$$\left\| \sum_{j=1}^3 dA^{ij} \cdot du^j \right\|_{L^2([0,T],L^{\frac{4}{3}})}$$

$$\le \sum_{j=1}^3 \|dA^{ij}\|_{L^2([0,T],L^4)} \|du^j\|_{L^\infty([0,T],L^2)}$$

$$\le \sum_{j=1}^3 \|dA^{ij}\|_{L^2([0,T],H^1)} \|du^j\|_{L^\infty([0,T],L^2)} \le C_1 \sqrt{E_0}, \tag{4.4.30}$$

where $C_1 > 0$ is a constant depending only on M and $\|\partial_t u\|_{L^2(M\times[0,T])}$.

On the other hand, noticing that $B_1^{ij} \in L^\infty([0,T], C^\infty(M, \wedge^2))$, we have

$$\|d^* B_1^{ij}\|_{L^2([0,T],L^4)} \le C(T,M)\|d^* B_1^{ij}\|_{L^\infty(M\times[0,T])} \le C(T,M).$$

This leads to the estimate

$$\left\| \sum_{j=1}^3 d^* B_1^{ij} \cdot du^j \right\|_{L^2([0,T],L^{\frac{4}{3}})}$$

$$\le \sum_{j=1}^3 \|d^* B_1^{ij}\|_{L^2([0,T],L^4)} \|du^j\|_{L^\infty([0,T],L^2)} \le C_2 \sqrt{E_0}, \tag{4.4.31}$$

where $C_2 > 0$ is a constant depending only on M, T.

It is obvious that

$$u \times \partial_t u \in L^2([0,T], L^2) \subset L^2([0,T], L^{\frac{4}{3}}). \tag{4.4.32}$$

From (4.4.29)–(4.4.32), we get that

$$f^i \in L^2([0,T], L^{\frac{4}{3}}), \qquad i = 1,2,3, \tag{4.4.33}$$

and $\|f^i\|_{L^2([0,T],L^{\frac{4}{3}})}$ is bounded by a constant depending only on M, T, ε, E_0 and $\|\partial_t u\|_{L^2(M \times [0,T])}$.

Applying Lemma 4.4.4 to problem (4.4.28) and (4.4.2), from (4.4.27) and (4.4.33), we can conclude that if $\varepsilon \leq \varepsilon_0$, where ε_0 is determined in Lemma 4.4.4, then any solution of (4.4.28) and (4.4.2) in X_T is in Y_T. On the other hand, if u is a weak solution of (4.4.28) and (4.4.2), then u is also a solution of (4.4.28) and (4.4.2) in X_T. Therefore, $u \in Y_T$.

2. *Uniqueness of weak solution*

In this step we will show that if $u \in Y_T \cap W_T$ is a weak solution of (4.4.1) and (4.4.2), then $u \in L^4(0,T;W^{2,4/3}) \cap W_T$.

For $u \in Y_T \cap W_T$, u is defined a.e. on $M \times [0,T]$. From (4.4.1), one can have

$$u \times \partial_t u = \beta u \times \Delta u - \alpha \partial_t u.$$

Inserting this into (4.4.1), we have

$$(1 + \alpha^2)\partial_t u = \beta \Delta u + \beta |du|^2 u + \alpha \beta u \times \Delta u,$$

i.e.

$$\partial_t u - G(u)\Delta u = \tilde{\beta}|du|^2 u, \tag{4.4.34}$$

where $\tilde{\beta} = \beta/(1+\alpha^2)$ and

$$G(u) = \frac{1}{1+\alpha^2}\begin{pmatrix} \beta & -\alpha\beta u^3 & \alpha\beta u^2 \\ \alpha\beta u^3 & \beta & -\alpha\beta u^1 \\ -\alpha\beta u^2 & \alpha\beta u^1 & \beta \end{pmatrix}. \tag{4.4.35}$$

Since $|u| = 1$, for any $\epsilon > 0$ we can decompose u into the form

$$u = m + n, \tag{4.4.36}$$

where $m \in C^\infty(M \times (0,T))$, $n \in L^\infty(M \times (0,T))$ with

$$\|m\|_{L^\infty(M \times (0,T))} \leq C\|u\|_{L^\infty(M \times (0,T))}, \tag{4.4.37}$$

$$\|n\|_{L^\infty(M \times (0,T))} \leq \epsilon. \tag{4.4.38}$$

Inserting (4.4.36) into (4.4.34), we get that

$$\partial_t u - G(m)\Delta u = G(n)\Delta u + \tilde{\beta}|du|^2 u. \tag{4.4.39}$$

Letting $G_1(x,t) = G(m(x,t))$, $G_2(x,t) = G(n(x,t))$, and $g(x,t) = \tilde{\beta}|du|^2 u$, (4.4.39) takes the same form as (4.4.16). Moreover,

$$\|G_2\|_{L^\infty(M \times (0,T))} \leq C\|n\|_{L^\infty(M \times (0,T))},$$

$$\|g\|_{L^4(0,T;L^{4/3})} \leq C\|du\|_{L^\infty([0,T],L^2)}\|du\|_{L^2([0,T],L^4)}.$$

We can see from (4.4.35), (4.4.37) and (4.4.38), that if ϵ is sufficiently small, then the assumptions in Lemma 4.4.5 and (4.4.18) are satisfied. Therefore, by Lemma 4.4.4, problem (4.4.39) with the initial condition (4.4.2) admits an unique solution $v \in L^4(0, T; W^{2,4/3}) \subset L^2(0, T; W^{2,4/3})$ and the solution in space $L^2(0, T; W^{2,4/3})$ is also unique. On the other hand, it is obvious that if $u \in Y_T \cap W_T$ is a weak solution of (4.4.1) and (4.4.2), then u is a solution of (4.4.39) and (4.4.2) in $L^2(0, T; W^{2,4/3})$. Therefore, $u = v \in L^4(0, T; W^{2,4/3})$.

    3. *Almost everywhere smoothness*

In this step we will complete our proof for Theorem 4.4.1 by showing the following lemma:

Lemma 4.4.5 *Let* $u \in L^4(0, T; W^{2,4/3}) \cap W_T$ *and* $v \in L^4(0, T; W^{2,4/3}) \cap W_T$ *be two solutions of* (4.4.1) *and* (4.4.2). *Then,* $u = v$ *a.e. on* $M \times [0, T]$.

Proof. For simplicity we assume $\alpha = 1$, $\beta = 1$. Now $w = u - v$ solves the equation

$$w_t - \Delta_M w = u \times \Delta_M w + w \times \Delta_M u + u(|du|^2 - |dv|^2) + |dv|^2 w, \qquad (4.4.40)$$

with the condition $w(x, 0) = 0$. Let $|dU|^2 = |du|^2 + |dv|^2$. Testing (4.4.40) by w and integrating it by parts, we obtain

$$\frac{1}{2}\frac{d}{dt}\|w\|^2_{L^2(M)} + \frac{1}{2}\int_M |dw|^2 \leq C \int_M |w|^2 |dU|^2. \qquad (4.4.41)$$

Using the Sobolev inequality

$$\|w\|^4_{L^4} \leq 4\|w\|^2_{L^2}\|dw\|^2_{L^2},$$

we have

$$\int_M |w|^2 |dU|^2 \leq C\left(\int_M |dU|^4\right)^{1/2}\left(\int_M |w|^2\right)^{1/2}\left(\int_M |dw|^2\right)^{1/2}$$

$$\leq \frac{1}{4}\int_M |dw|^2 + C\int_M |dU|^4 \int_M |w|^2.$$

This combined with (4.4.41) implies

$$f'(t) + \frac{1}{2}g(t) \leq Cp(t)f(t)$$

where $f(t) = \int_M |w|^2$, $g(t) = \int_M |dw|^2$ and $p(t) = \int_M |dU|^4$. Since $u, v \in L^4(0, T; W^{2,4/3})$, $p(t) \in L^1(0, T)$. We have from the Gronwall inequality that $f(t) = 0$ for a.e. $t \in [0, T]$, and hence $w = 0$ a.e. on $M \times [0, T]$. The proof of the lemma is complete.

    The proof of Theorem 4.4.1 is complete.

4.5 Ginzburg–Landau Approximation to Landau–Lifshitz Equations

In this section, we will discuss the Ginzburg–Landau approximation to Landau–Lifshitz equations to obtain the existence and partial regularity for the global weak solutions to the initial and boundary value problem. We will consider the weighted Landau–Lifshitz flow in two dimensions and follow the ideas of Paul Harpes' [89] for $a(x) = 1$.

$$\frac{1}{2}\partial_t u - \frac{1}{2}u \times \partial_t u - \nabla \cdot (a(x)\nabla u) = a(x)|\nabla u|^2 u \quad \text{in } \Omega \times R_+, \qquad (4.5.1)$$

$$u = u_0 \qquad \text{on } \Omega \times \{0\} \bigcup \partial\Omega \times R_+ \qquad (4.5.2)$$

where domain $\Omega \subset R^2$ is open, bounded and smooth. The initial and boundary data u_0 is assumed to be a smooth map into the standard sphere $S^2 \subset R^3$.

In the classical sense, Eq. (4.5.1) is equivalent to

$$u_t = u \times \nabla \cdot (a(x)\nabla u) - u \times (u \times \nabla \cdot (a(x)\nabla u)). \qquad (4.5.3)$$

The Ginzburg–Landau appoximations $u_\epsilon : \bar{\Omega} \times R_+ \to R^3$ to Landau–Lifshitz flow (4.5.1) are solutions of

$$\frac{1}{2}\partial_t u_\epsilon - \frac{1}{2}u_\epsilon \times \partial_t u_\epsilon - \nabla \cdot (a(x)\nabla u_\epsilon) = \frac{1}{\epsilon^2}a(x)(1 - |u_\epsilon|^2)u_\epsilon \quad \text{in } \Omega \times R_+, \quad (4.5.4)$$

$$u_\epsilon = u_0 \quad \text{on } (\Omega \times \{0\}) \bigcup (\partial\Omega \times R_+), \qquad (4.5.5)$$

where $a(x)$ is a positive smooth function satisfying $0 < m \leq a(x) \leq M$. For small $\epsilon > 0$, we can see that maps $\{u_\epsilon\}_\epsilon$ approximate the weighted Landau–Lifshitz flow in $\Omega \times R_+$. For fixed $\epsilon > 0$, a smooth solution to (4.5.4) and (4.5.5) on $\Omega \times R_+$ exists and if $u_0 \in H^{1,2}(\Omega; S^2) \cap H^{\frac{3}{2}}(\partial\Omega; S^2)$, it is unique in $H^{1,2}_{\text{loc}} \cap L^\infty(H^{1,2}) := H^{1,2}_{\text{loc}}(\Omega \times R_+; R^3) \cap L^\infty(H^{1,2})$. Existence is obtained by Galerkin's method. C^∞ regularity follows from a standard bootstrap argument. The total energy of the approximate flow at time $t \geq 0$ is defined by

$$G_\epsilon(u_\epsilon(t)) := \int_\Omega g_\epsilon(u_\epsilon(x, t))dx, \qquad (4.5.6)$$

where

$$g_\epsilon(u_\epsilon(x, t)) = a(x)\left[\frac{1}{2}|\nabla u_\epsilon|^2 + \frac{1}{4\epsilon^2}(1 - |u_\epsilon|^2)^2\right]. \qquad (4.5.7)$$

We will see that the total energy of the ϵ approximation always decreases.

We define the local energy by

$$G_\epsilon(u_\epsilon(t), B_R^\Omega(x_0)) := \int_{B_R(x_0)\bigcap\Omega} g_\epsilon(u_\epsilon(x, t))dx \qquad (4.5.8)$$

which may concentrate at space–time points (x_0, t_0) as $\epsilon \searrow 0$, either for fixed $t = t_0$ or for variable $t \nearrow t_0$ or $t \searrow t_0$. It characterizes the local "asymptotic behavior" of the weighted flow. Here asymptotic refers to limit $\epsilon \searrow 0$. We will show that all the derivatives of the family of maps $\{u_\epsilon\}_{\epsilon>0}$ are locally uniformly bounded on the regular set $\text{Reg}\{u_\epsilon\}_{\epsilon>0}$ consisting of all points $z_0 = (x_0, t_0) \in \overline{\Omega} \times (0, \infty)$ for which there is $R_0 = R_0(z_0)$ such that

$$\limsup_{\epsilon \searrow 0} \sup_{t_0 - R_0^2 < t < t_0} G_\epsilon(u_\epsilon(t), B_{R_0}^\Omega(x_0)) < \epsilon_0 \qquad (4.5.9)$$

for a constant $\epsilon_0 > 0$ that will be determined later in Lemmas 4.5.9 and 4.5.11. Complement $S(\{u_\epsilon\}_{\epsilon>0}) := \overline{\Omega} \times R_+ \backslash \text{Reg}\{u_\epsilon\}_{\epsilon>0}$ is referred to as the energy-concentration set. We will show that the approximation solutions converge to the global weak solution of (4.5.1) with the Dirichlet condition. This convergence is smooth in $\text{Reg}\{u_\epsilon\}_{\epsilon>0}$, while the energy-concentration set is closed, with the locally finite parabolic Hausdorff measure. Delicate energy inequality shows that, in fact, the singular set consists of finitely many points as observed in [29, 89]. Such Ginzburg–Landau penalty method was first used to study the harmonic map heat flow in higher dimensions by Chen and Struwe in [34]. So we call the solution obtained the Chen–Struwe solution.

4.5.1   Estimates for Strong Parabolic System

In this section, we will show that under the uniform smallness condition (4.5.9) on the local energy, all higher derivatives of u_ϵ are locally and uniformly bounded.

We first recall some facts about L^p estimates for a strongly parabolic system and C^α estimates for the parabolic system in the divergence form. Then, we derive the L^∞ and L^p bounds for the right-hand side of (4.5.4) which are necessary to get the uniform bounds of $\frac{1}{\epsilon^2}a(x)(1 - |u_\epsilon|^2)$ for L^p estimates. Finally, we prove that all the derivatives of u_ϵ and $\frac{1}{\epsilon^2}a(x)(1 - |u_\epsilon|^2)$ are locally uniformly bounded if the energy density satisfies the condition $\limsup_{\epsilon \searrow 0} \sup_{P_R(z_0)} g_\epsilon(u_\epsilon(x,t)) < C_0$, which may be verified under the uniformly smallness condition (4.5.9). For $z_0 = (x_0, t_0)$, denote $P_R(z_0) = B_R(x_0) \times (t_0 - R^2, t_0)$ and $P_{\delta R}^\Omega := (B_{\delta R}(x_0) \cap \Omega) \cap (t_0 - \delta^2 R^2, t_0)$.

We rewrite Eq. (4.5.4) in the form

$$\partial_t u_\epsilon - M(u_\epsilon)a(x)\triangle u_\epsilon - M(u_\epsilon)\nabla a(x) \cdot \nabla u_\epsilon = M(u_\epsilon)\frac{1}{\epsilon^2}a(x)(1 - |u_\epsilon|^2)u_\epsilon := f_\epsilon(u_\epsilon),$$

where $M(u_\epsilon)$ satisfies

$$m|\xi|^2 \leq \xi^T a(x) M(u_\epsilon)\xi = \frac{2a(x)}{(1 + |u_\epsilon|^2)}\{|\xi|^2 + (u_\epsilon \cdot \xi)^2\} \leq 2M|\xi|^2, \quad \forall \; \xi \in R^3. \qquad (4.5.10)$$

We may therefore write (4.5.4) as

$$L_\epsilon(u_\epsilon) := \partial_t u_\epsilon - M(u_\epsilon)a(x)\triangle u_\epsilon - M(u_\epsilon)\nabla a(x) \cdot \nabla u_\epsilon = f_\epsilon(u_\epsilon),$$

where the coefficient-matrix $M(u_\epsilon)$ is smooth with respect to u_ϵ. Note that L_ϵ defines a strongly parabolic system in the Petrovskii sense [101]. So L^p global and local estimates hold for such system. We list two *a priori* L^p estimates concerning the strongly parabolic system in the Petrovskii sense.

Lemma 4.5.1 (Global L^p estimates) *Let $f_\epsilon \in L^p(\Omega \times [0,T]; R^3)$ and $u_0 \in H^{2,p}(\Omega; R^3)$. A solution of $L_\epsilon(v) = f_\epsilon$ in $(\Omega \times (0,T); R^3)$ with $v = u_0$ on $(\Omega \times \{0\}) \cup (\partial\Omega \times (0,T))$ satisfies*

$$\|v\|_{H_p^{2,1}}(\Omega \times [0,T]) \leq C_p(\Omega, T, \omega_{u_\epsilon})(\|f_\epsilon\|_{L^p(\Omega \times [0,T])} + \|u_0\|_{H^{2,p}(\Omega)}). \qquad (4.5.11)$$

Lemma 4.5.2 (Local L^p estimates) *Let $f_\epsilon \in L^p(\Omega \times [0,T]; R^3)$ and $u_0 \in H^{2,p}(\Omega; R^3)$. A solution of $L_\epsilon(v) = f_\epsilon$ in $(\Omega \times (0,T); R^3)$ with $v = u_0$ on $(\Omega \times \{0\}) \cup (\partial\Omega \times (0,T))$ satisfies*

$$\|v\|_{H_p^{2,1}(P_{\delta R}^\Omega(z_0))} \leq \tilde{C}_p(R, \Omega, T, \omega_{u_\epsilon})\Big(\|f_\epsilon\|_{L^p(P_R^\Omega(z_0))}$$
$$+ \|v\|_{L^q(P_{\delta R}^\Omega(z_0))} + \delta_{B_R \bigcap \partial\Omega}\|u_0\|_{H^{2-\frac{1}{p},p}(B_R^\Omega \bigcap \partial\Omega)}\Big) \qquad (4.5.12)$$

for all $1 \leq q \leq p$. Here, $\delta_{B_R \bigcap \partial\Omega} = 1$ if $B_R \bigcap \partial\Omega \neq \emptyset$ and 0 otherwise. The trace theorem of course implies $\|u_0\|_{H^{2-\frac{1}{p},p}(\partial\Omega)} \leq \|u_0\|_{H^{2,p}(\Omega)}$. Constants C_p and $\tilde{C}_p$ depend on the indicated quantities and additionally on the uniform lower and upper bounds for the eigenvalues of $a(x)M(u_\epsilon)$, that is m and $2M$, which are chosen to be independent of $\epsilon > 0$. Note that C_p and $\tilde{C}_p$ also depend on the modulus of continuity of the coefficients of the leading term-, i.e. the modulus of continuity ω_{u_ϵ} of u_ϵ.

The equation can also be written in the divergence form

$$L_\epsilon(v) := \partial_t v - a(x)\nabla \cdot (M(u_\epsilon)\nabla v) + a(x)(\partial_k M(u_\epsilon)\partial_k u_\epsilon)\partial_k v - M(u_\epsilon)\nabla a(x) \cdot \nabla v = f_\epsilon(u_\epsilon).$$

Lemma 4.5.3 *If we assume*

$$\limsup_{\epsilon \searrow 0} \sup_{P_R^\Omega} |\nabla u_\epsilon| < \infty, \qquad (4.5.13)$$

then $v \in C^{\gamma, \frac{\gamma}{2}}(P_{\delta R}^\Omega; R^3)$ for some $\gamma \in (0,1)$ and any $\delta \in (0,1)$. If the right-hand side $f_\epsilon \in L^p(P_R^\Omega; R^3)$ with $p > 2$, we have the following estimate for the mixed Hölder-norm of v on $P_{\delta R}^\Omega$

$$\|v\|_{C^{\gamma, \frac{\gamma}{2}}(P_{\delta R}^\Omega; R^3)} \leq C(f_\epsilon), \qquad (4.5.14)$$

where bound $C(f_\epsilon)$ depends on the parabolicity constants, δ, $\sup_{P_R^\Omega} |u_\epsilon|$, $\|f_\epsilon\|_{L^p(P_R^\Omega)}$, and also depends on $\|u_0\|_{C^\gamma(B_R \bigcap \partial\Omega)}$ if $B_R \bigcap \partial\Omega \neq \emptyset$.

If (4.5.13) holds and $\|f_\epsilon\|_{L^p(P_R^\Omega)}$ or $\sup_{P_R^\Omega} |\nabla u_\epsilon|$ is uniformly bounded with respect to $\epsilon > 0$, then estimate (4.5.14) holds for u_ϵ and is uniform in $\epsilon > 0$. Assumption $\sup_{P_R^\Omega} |\nabla u_\epsilon| \leq C$, however, does not include the time derivatives. (4.5.14) enables us to obtain bounds on the modulus of continuity with respect to time variable. Thus the modulus of continuity of u_ϵ on $P_{\delta R}^\Omega$ is bounded from above independent of $\epsilon > 0$. Therefore estimates (4.5.11) and (4.5.12) are uniform in $\epsilon > 0$.

4.5.2 L^∞ and L^p Bounds for $\frac{1}{\epsilon^2}a(x)(1 - |u_\epsilon|^2)$

Multiplying (4.5.4) by $-u_\epsilon$, we obtain

$$\frac{1}{4}\partial_t\rho_\epsilon - \frac{1}{2}a(x)\triangle\rho_\epsilon - \frac{1}{2}\nabla a(x)\nabla\rho_\epsilon + \frac{1}{\epsilon^2}a(x)\rho_\epsilon = a(x)|\nabla u_\epsilon|^2 + a(x)\frac{1}{\epsilon^2}\rho_\epsilon^2, \quad (4.5.15)$$

where $\rho_\epsilon = 1 - |u_\epsilon|^2$. On the parabolic boundary, $\rho_\epsilon = 1 - |u_0|^2 = 0$. We get $\rho_\epsilon \geq 0$ in $\overline{\Omega} \times R_+$ by using the maximum principle, i.e. $|u_\epsilon| \leq 1$.

Consider the following auxiliary problem:

$$\partial_t f - a(x)\triangle f - \nabla a(x) \cdot \nabla f + \frac{1}{\epsilon^2}a(x)f = a(x)g \qquad \text{in } P_R, \qquad (4.5.16)$$

$$|f| \leq a \qquad \text{on } \partial P_R, \qquad (4.5.17)$$

where $\partial P_R = B_R(0) \times \{-R^2\} \cup \partial B_R(0) \times [-R^2, 0]$.

Lemma 4.5.4 *Let* $a > 0$, $\epsilon \in (0,1)$, $g \in C^0(\overline{P_R})$, *with* $\epsilon^2 \sup_{P_R} |g| \leq a$. *Let* $f \in C^0(\overline{P_R}) \cap C^2(P_R)$ *be a solution of* (4.5.16) *and* (4.5.17). *Then there exists* R_0 *depending on* m, M, *and* $M_1 = \max_{x\in\bar\Omega} |\nabla a(x)|$ *such that for any* $\delta \in (0,1)$, $R \in (0, R_0)$ *we have*

$$\frac{1}{\epsilon^2}|f| \leq \sup_{P_R^\Omega} |g| + \frac{2a}{\epsilon^2} \exp\left(-\frac{1}{\epsilon}(1 - \delta^2)^2 R^4\right) \quad \text{on } P_{\delta R}. \qquad (4.5.18)$$

Proof. Taking $w(x,t) = 2a\exp[-\frac{1}{\epsilon}(R^2 - x^2)(R^2 + t)]$, we have

$$\epsilon^2[\partial_t w - a(x)\triangle w - \nabla a(x) \cdot \nabla w] + a(x)w$$
$$= w[a(x) - \epsilon(R^2 - x^2) - a(x)4|x|^2(R^2 + t)^2$$
$$- a(x)\epsilon \cdot 4(R^2 + t) - \epsilon\nabla a(x) \cdot 2x(R^2 + t)]$$
$$\geq w(m - \epsilon R^2 - 4MR^4 - 4\epsilon MR^2 - 2R\epsilon M_1 R^2).$$

Therefore, there exists R_0 depending on m, M, M_1 such that if $R \in (0, R_0)$, $\epsilon \in (0,1)$

$$\epsilon^2[\partial_t w - a(x)\triangle w - \nabla a(x) \cdot \nabla w] + a(x)w > 0 \qquad \text{in } P_R,$$
$$w = 2a \qquad \text{on } \partial P_R$$

holds.

For $f_1 = f - \epsilon^2 \sup_{P_R} |g|$ and $f_2 = f + \epsilon^2 \sup_{P_R} |g|$, we have $|f_1| \leq 2a$ and $|f_2| \leq 2a$ on ∂P_R and

$$\epsilon^2[\partial_t f_1 - a(x)\triangle f_1 - \nabla a(x) \cdot \nabla f_1] + a(x)f_1$$
$$= \epsilon^2 a(x)g - a(x)\epsilon^2 \sup_{P_R} |g| \leq 0$$
$$\leq \epsilon^2[\partial_t w - a(x)\triangle w - \nabla a(x) \cdot \nabla w] + a(x)w.$$

Therefore, we obtain by the comparison principle that

$$f_1 - w \leq 0 \qquad \text{in } P_R. \tag{4.5.19}$$

Similarly, we have

$$f_2 + w \geq 0 \qquad \text{in } P_R. \tag{4.5.20}$$

Combining (4.5.19) and (4.5.20), one gets

$$-w - \epsilon^2 \sup_{P_R^\Omega} |g| \leq f \leq w + \epsilon^2 \sup_{P_R^\Omega} |g|,$$

which yields the desired conclusion

$$\frac{1}{\epsilon^2}|f| \leq \sup_{P_R^\Omega} |g| + \frac{1}{\epsilon^2} w \leq \sup_{P_R^\Omega} |g| + \frac{2a}{\epsilon^2} \exp\left(-\frac{1}{\epsilon}(1 - \delta^2)^2 R^4 \right).$$

If $B_R \cap \Omega \neq \emptyset$ and $f \equiv 0$ on $B_R \cap \partial\Omega$, we have the local estimate near the boundary.

Lemma 4.5.5 *Consider a smooth domain $\Omega \subset R^2$, $a > 0$, $\epsilon \in (0,1)$, $g \in C^0(\overline{P_R})$, with $\epsilon^2 \sup_{P_R} |g| \leq a$. Let $f \in C^0(\overline{P_R}) \cap C^2(P_R)$ be a solution of (4.5.16) and (4.5.17) and $f = 0$ on $\partial\Omega \cap P_R$. Then there exist R_0, depending on m, M and M_1 such that for any $\delta \in (0,1)$, $R \in (0, R_0)$, we have*

$$\frac{1}{\epsilon^2}|f| \leq \sup_{P_R^\Omega} |g| + \frac{2a}{\epsilon^2} \exp\left(-\frac{1}{\epsilon}(1 - \delta^2)^2 R^4 \right) \quad \text{on } P_{\delta R}^\Omega. \tag{4.5.21}$$

In the sequel, we will derive *a priori* L^p estimates for Eqs. (4.5.16) and (4.5.17). First we give L^1 estimates.

Lemma 4.5.6 *Let $\Omega \subset R^2$ be as bounded smooth domain, $a > 0$, $\epsilon \in (0,1)$, $g \in C^0(\overline{P_R})$. For any nonnegative function $f \in C^1(\overline{\Omega} \times (0,T)) \cap C^2(\Omega \times (0,T))$ satisfying*

$$\partial_t f - a(x)\triangle f - \nabla a(x) \cdot \nabla f + \frac{1}{\epsilon^2}a(x)f \leq a(x)g \qquad \text{in } \Omega \times (0,T),$$
$$f = 0 \quad \text{on } (\Omega \times \{0\}) \cup (\partial\Omega \times (0,T)),$$

(1) there exists some constant $c > 0$ only depending on M and m such that

$$\left\| \frac{1}{\epsilon^2} f \right\|_{L^1(\Omega \times (0,T))} \leq c \|g\|_{L^1(\Omega \times (0,T))}; \tag{4.5.22}$$

(2) for any $R, \rho > 0$ and $z_0 = (x_0, t_0) \in \Omega \times (0,T)$ with $R^2 + \rho^2 < t_0$, we have

$$\int_{P_R^\Omega(z_0)} \frac{1}{\epsilon^2}|f| \, dz \leq c \int_{P_{R+\rho}^\Omega(z_0)} \left(|g| + \frac{c}{\rho^2}|f| \right) dz. \tag{4.5.23}$$

Proof of (1). Multiplying Eq. (4.5.16) by $\frac{f}{\sqrt{f^2+\delta^2}}$, we attain

$$\partial_t f \cdot \frac{f}{\sqrt{f^2+\delta^2}} - a(x)\triangle f \cdot \frac{f}{\sqrt{f^2+\delta^2}} - \nabla a(x) \cdot \nabla f \frac{f}{\sqrt{f^2+\delta^2}}$$

$$+ \frac{1}{\epsilon^2} a(x) f \frac{f}{\sqrt{f^2+\delta^2}} = a(x) g \cdot \frac{f}{\sqrt{f^2+\delta^2}}. \qquad (4.5.24)$$

(4.5.24) can be written as

$$\partial_t |f| \cdot \frac{|f|}{\sqrt{f^2+\delta^2}} + a(x) \cdot \frac{|\nabla f|^2}{\sqrt{f^2+\delta^2}} \left(1 - \frac{f^2}{f^2+\delta^2}\right) + \frac{1}{\epsilon^2} a(x) \frac{f^2}{\sqrt{f^2+\delta^2}}$$

$$= a(x) g \cdot \frac{f}{\sqrt{f^2+\delta^2}} + \operatorname{div}\left(a(x)\nabla f \frac{f}{\sqrt{f^2+\delta^2}}\right). \qquad (4.5.25)$$

Integrating (4.5.25) over $\Omega \times (0,t)$, letting $\delta \to 0$, and using the monotone convergence theorem, we have

$$\sup_{0 \le t \le T} \int_\Omega |f(x,t)|dx + \int_0^T \int_\Omega a(x) \frac{1}{\epsilon^2} |f(x,t)|dxdt \le \int_0^T \int_\Omega a(x)|g(x,t)|dxdt,$$

where we have used the fact that $\int_0^T \int_\Omega \operatorname{div}(a(x)\nabla f \frac{f}{\sqrt{f^2+\delta^2}})dxdt = 0$ by the divergence theorem. From the above inequality, we obtain

$$\int_0^T \int_\Omega \frac{1}{\epsilon^2}|f(x,t)|dxdt \le \frac{M}{m} \int_0^T \int_\Omega |g(x,t)|dxdt.$$

Taking $C = \frac{M}{m}$, we have

$$\left\| \frac{1}{\epsilon^2} f \right\|_{L^1(\Omega \times (0,T))} \le c\|g\|_{L^1(\Omega \times (0,T))}.$$

Proof of (2). Multiplying Eq. (4.5.16) by $\frac{f}{\sqrt{f^2+\delta^2}}(x,t)\phi^2(x)\eta(t)$ where the cutoff function $\phi(x)$ satisfies $0 \le \phi(x) \in C^\infty(\Omega)$ with $\operatorname{spt}\phi \subset B_{R+\rho}(x_0)$ and $\phi \equiv 1$ on $B_R(x_0)$, $\eta(t)$ satisfies $\eta(t) \in C^\infty(R_+)$ with $0 \le \eta(t) \le 1$, $\eta(t_0 - R^2 - \rho^2) = 1$ and $\eta(t) \equiv 1$, $|\nabla \phi| \le \frac{c}{\rho}$, $|\nabla^2 \phi| \le \frac{c}{\rho^2}$, and $|\partial_t \eta| \le \frac{c}{\rho^2}$, we get

$$\partial_t f \cdot \frac{f}{\sqrt{f^2+\delta^2}}\phi^2\eta - a(x)\triangle f \cdot \frac{f}{\sqrt{f^2+\delta^2}}\phi^2\eta - \nabla a(x) \cdot \nabla f \frac{f}{\sqrt{f^2+\delta^2}}\phi^2\eta$$

$$+ \frac{1}{\epsilon^2} a(x) f \frac{f}{\sqrt{f^2+\delta^2}}\phi^2\eta = a(x) g \cdot \frac{f}{\sqrt{f^2+\delta^2}}\phi^2\eta. \qquad (4.5.26)$$

(4.5.26) can be written as

$$\partial_t(|f|\phi^2\eta)\frac{|f|}{\sqrt{f^2+\delta^2}} + a(x) \cdot \frac{|\nabla f|^2\phi^2\eta}{\sqrt{f^2+\delta^2}}\left(1 - \frac{f^2}{f^2+\delta^2}\right) + \frac{1}{\epsilon^2}a(x)\frac{f^2\phi^2\eta}{\sqrt{f^2+\delta^2}}$$

$$= a(x)g \cdot \frac{f\phi^2\eta}{\sqrt{f^2+\delta^2}} + \operatorname{div}\left(a(x)\nabla f \frac{f\phi^2\eta}{\sqrt{f^2+\delta^2}}\right)$$

$$- a(x)\nabla f \frac{f}{\sqrt{f^2+\delta^2}}2\phi\nabla\phi\eta + |f|\phi^2\partial_t\eta. \qquad (4.5.27)$$

Integrating (4.5.27) over $\Omega \times (0, t)$ and letting $\delta \to 0$, we get

$$\sup_{t_0 - (R^2 + \rho^2) \leq t \leq t_0} \int_{B_\Omega^R} |f(x, t)| dx + \int_{P_R^\Omega} a(x) \frac{1}{\epsilon^2} |f(x, t)|$$

$$\leq \int_{P_{R+\rho}^\Omega} \left(a(x)|g(x, t)| + \frac{c}{\rho^2} |f(x, t)| \right).$$

Recalling the assumption $m = \min_{x \in \overline{\Omega}} |a(x)| \leq a(x) \leq \max_{x \in \overline{\Omega}} |a(x)| = M$, we arrive at the conclusion

$$\int_{P_R^\Omega(z_0)} \frac{1}{\epsilon^2} |f| dz \leq c \int_{P_{R+\rho}^\Omega(z_0)} \left(|g| + \frac{c}{\rho^2} |f| \right) dz.$$

Lemma 4.5.7 *Let $\Omega \subset R^2$ be as above, $g \in L^1 \cap L^p(\overline{\Omega} \times (0, T))$ for $p \geq 2$ and $\epsilon > 0$. For any nonnegative function $f \in C^1(\overline{\Omega} \times (0, T)) \cap C^2(\Omega \times (0, T))$ satisfying*

$$\partial_t f - a(x) \triangle f - \nabla a(x) \cdot \nabla f + \frac{1}{\epsilon^2} a(x) f \leq a(x) g \qquad \text{in } \Omega \times (0, T),$$

$$f = 0 \qquad \text{on } (\Omega \times \{0\}) \bigcup (\partial \Omega \times (0, T))$$

$\forall \ \delta \in (0, 1)$, $z_0 = (x_0, t_0) \in \Omega \times (0, T]$ with $0 < R^2 < t_0$, we have

$$\left\| \frac{1}{\epsilon^2} f \right\|_{L^p(P_{\delta R}^\Omega(z_0))} \leq c_1 \|g\|_{L^p(P_{\delta R}^\Omega(z_0))} + c_2 \epsilon^{2/p}, \qquad (4.5.28)$$

where $c_1 = c_1(p, m, M)$ and $c_2 = c_2(\|g\|_{L^p(P_R^\Omega)}, \|f\|_{L^{2p-1}(P_R(z_0))}, p, \delta, R, m, M)$.

Proof. Multiplying the equation by $f|f|^{2s-2} \phi^2 \eta$, where $s \geq 1$, and taking cutoff functions ϕ and η as in the proof of Lemma 4.5.6, we have

$$\partial_t f f |f|^{2s-2} \phi^2 \eta - a(x) \triangle f f |f|^{2s-2} \phi^2 \eta - \nabla a(x) \cdot \nabla f f |f|^{2s-2} \phi^2 \eta$$

$$+ \frac{1}{\epsilon^2} a(x) f f |f|^{2s-2} \phi^2 \eta = a(x) g f |f|^{2s-2} \phi^2 \eta. \qquad (4.5.29)$$

(4.5.29) can be written as

$$\frac{1}{2s} \partial_t (|f|^{2s} \phi^2 \eta) + a(x) \frac{2s-1}{s^2} |\nabla|f|^s|^2 \phi^2 \eta + \frac{1}{\epsilon^2} a(x) |f|^{2s} \phi^2 \eta$$

$$= \operatorname{div}(a(x) \nabla f f |f|^{2s-2} \phi^2 \eta) + \frac{1}{2s} |f|^{2s} \phi^2 \partial_t \eta$$

$$+ a(x) g f |f|^{2s-2} \phi^2 \eta - a(x) \nabla |f| |f|^{2s-1} 2 \nabla \phi \phi \eta. \qquad (4.5.30)$$

We now estimate the last two terms of (4.5.30). By the Young inequality we have

$$||g f |f|^{2s-2}| \leq \frac{1}{2\epsilon^2} \frac{2s-1}{2s} |f|^{2s} + (2\epsilon^2)^{2s-1} \frac{1}{2s} |g|^{2s}$$

and

$$|\nabla|f||f|^{2s-1}2\nabla\phi\phi\eta| = 2\left|\left(\frac{1}{s}\nabla|f|^s\phi\sqrt{\eta}\right)\cdot(|f|^s\nabla\phi\sqrt{\eta}\right|$$

$$\leq \frac{2s-1}{2s^2}|\nabla|f|^s|^2\phi^2\eta + \frac{2}{2s-1}|f|^{2s}|\nabla\phi|^2\eta$$

which lead to

$$\frac{1}{2s}\partial_t(|f|^{2s}\phi^2\eta) + a(x)\frac{2s-1}{2s^2}|\nabla|f|^s|^2\phi^2\eta + \frac{1}{2\epsilon^2}\frac{1}{2s}a(x)|f|^{2s}\phi^2\eta$$

$$\leq \operatorname{div}(a(x)\nabla ff|f|^{2s-2}\phi^2\eta) - a(x)(2\epsilon^2)^{2s-1}\frac{1}{2s}|g|^{2s}\phi^2\eta$$

$$+ \frac{2}{2s-1}|f|^{2s}[a(x)|\nabla\phi|^2\eta + \phi^2|\partial_t\eta|]. \tag{4.5.31}$$

Setting $p = 2s$, multiplying (4.5.31) by $(2s)\cdot(\frac{1}{\epsilon^2})^{p-1}$, and integrating it over $P_{R+\rho}^\Omega$, we get, for $p \geq 2$,

$$\sup_{t\geq -R^2-\rho^2}\int_{B_{R+\rho}^\Omega}\left(\frac{1}{\epsilon^2}\right)^{p-1}|f|^p\phi^2\eta dx$$

$$+ \int_{P_{R+\rho}^\Omega}\left(\frac{1}{\epsilon^2}\right)^{p-1}|\nabla|f|^{\frac{p}{2}}|^2\phi^2\eta dz + \int_{P_{R+\rho}^\Omega}\left(\frac{1}{\epsilon^2}\right)^p|f|^p\phi^2\eta dz$$

$$\leq C(p,m,M)\left\{\int_{P_{R+\rho}^\Omega}|g|^p\phi^2\eta + \int_{P_{R+\rho}^\Omega}\left(\frac{1}{\epsilon^2}\right)^{p-1}|f|^p[|\nabla\phi|^2\eta + \phi^2|\partial_t\eta|]\right\}dz. \tag{4.5.32}$$

Hence we obtain

$$\int_{P_R^\Omega}\left(\frac{1}{\epsilon^2}\right)^p|f|^pdz \leq C(p,m,M)\left\{\int_{P_{R+\rho}^\Omega}|g|^p + \epsilon^2\frac{c}{\rho^2}\int_{P_{R+\rho}^\Omega}\left(\frac{1}{\epsilon^2}\right)^p|f|^p\right\}dz. \tag{4.5.33}$$

Let $p = \frac{k}{2} + 1$ for $k \in N$. Then Hölder's inequality for $q_1 = \frac{2p-1}{2p-2}$ and $q_2 = 2p-1$ implies

$$\int_{P_{R+\rho}^\Omega}\left(\frac{1}{\epsilon^2}|f|\right)^{p-1}|f|dz \leq \left\|\frac{1}{\epsilon^2}f\right\|_{L^{p-\frac{1}{2}}}^{p-1}\|f\|_{L^{2p-1}}.$$

Now (4.5.33) yields

$$\left\|\frac{1}{\epsilon^2}f\right\|_{L^p(P_R^\Omega)}^p \leq C(p,m,M)\left[\|g\|_{L^p(P_{R+\rho}^\Omega)}^p + \frac{c}{\rho^2}\left\|\frac{1}{\epsilon^2}f\right\|_{L^{p-\frac{1}{2}}(P_{R+\rho}^\Omega)}^{(p-\frac{1}{2})\cdot\frac{2p-2}{2p-1}}\|f\|_{L^{2p-1}(P_{R+\rho}^\Omega)}\right]. \tag{4.5.34}$$

Noticing that $\frac{2p-2}{2p-1} < 1$, we can iterate (4.5.34) for k steps and use Lemma 4.5.6 to derive that

$$\left\|\frac{1}{\epsilon^2}f\right\|_{L^p(P_R^\Omega)}^p \leq C\left(\|g\|_{L^p(P_{R+(k+1)\rho}^\Omega)}^p, \|f\|_{L^{2p-1}(P_{R+(k+1)\rho}^\Omega)}, p, \rho, R\right).$$

If we use new cutoff function $\bar{\phi}(x)$ and $\bar{\eta}(t)$ satisfying $0 \leq \bar{\phi}(x) \in C^\infty(\Omega)$ with $\mathrm{spt}\,\bar{\phi}(x) \subset B_{R+2\rho}(x_0)$ and $\bar{\phi} \equiv 1$ on $B_{R+\rho}(x_0)$ and $\bar{\eta}(t)$ satisfying $\bar{\eta}(t) \in C^\infty(R_+)$ with $0 \leq \bar{\eta}(t) \leq 1$, $\bar{\eta}(t_0 - (R+\rho)^2 - \rho^2) = 1$, $\bar{\eta}(t) \equiv 1$ if $t \geq t_0 - (R+\rho)^2$, $|\nabla\bar{\phi}| \leq \frac{c}{\rho}$, $|\nabla^2\bar{\phi}| \leq \frac{c}{\rho^2}$, and $|\partial_t\bar{\eta}| \leq \frac{c}{\rho^2}$, we get, by the same iteration, that

$$\left\| \frac{1}{\epsilon^2} f \right\|_{L^p(P_{R+\rho}^\Omega)}^p \leq C\left(\|g\|_{L^p(P_{R+\rho+(k+1)\rho}^\Omega)}^p, \|f\|_{L^{2p-1}(P_{R+\rho+(k+1)\rho}^\Omega)}, p, \rho, R \right). \tag{4.5.35}$$

Taking $\rho := \frac{\delta R}{k+2}$ and then substituting (4.5.35) into (4.5.34), one deduces

$$\left\| \frac{1}{\epsilon^2} f \right\|_{L^p(P_R^\Omega)}^p \leq C(p, m, M)\left[\|g\|_{L^p(P_{(1+\delta)R}^\Omega)}^p + \epsilon^2 C_2 \right], \tag{4.5.36}$$

where $c_2 = c_2(\|g\|_{L^p(P_{(1+\delta)R}^\Omega)}, \|f\|_{L^{2p-1}(P_{(1+\delta)R}^\Omega)}, p, \delta, R)$. The lemma follows if one sets $R_{\mathrm{new}} = (1+\delta)R$ and $\delta_{\mathrm{new}} = \frac{1}{1+\delta}$, i.e. $R_{\mathrm{new}}\delta_{\mathrm{new}} = R$.

4.5.3 Higher Order Interior Estimates

We intend to claim that the higher derivatives of u_ϵ and $\frac{1}{\epsilon^2}a(x)(1-|u_\epsilon|^2)$ are locally and uniformly bounded in the interior point under the uniform smallness condition (4.5.9), where u_ϵ are the solution to approximating equation.

Lemma 4.5.8 *Let u_ϵ be a solution of (4.5.4) and assume*

$$\limsup_{\epsilon \searrow 0} \sup_{P_R^\Omega} g_\epsilon(u_\epsilon) \leq C_0, \tag{4.5.37}$$

where $B_R(x_0) \subset \Omega$, $0 < R^2 < t_0$. Then for any $\delta \in (0,1)$, we have

$$\limsup_{\epsilon \searrow 0} \|u_\epsilon\|_{C^k(P_{\delta R}(z_0))} \leq C_k \quad and \quad \limsup_{\epsilon \searrow 0} \left\| \frac{1}{\epsilon^2}(1-|u_\epsilon|^2) \right\|_{C^k(P_{\delta R}(z_0))} \leq \tilde{C}_k \tag{4.5.38}$$

for all integers $k \geq 0$. Constants C_k and $\tilde{C}_k$ depend on C_0, k, R, $\delta > 0$, $\|a(x)\|_{C^k(\Omega)}$.

Proof. We prove this lemma by induction. If $k = 0$, we have proved $\|u_\epsilon\|_{L^\infty} \leq 1$. Using assumption (4.5.37), we obtain $\sup_{P_R} \sqrt{a(x)}|\nabla u_\epsilon| \leq C_0$, and $\frac{1}{\epsilon^2}a(x)(1-|u_\epsilon|^2)^2 \leq C_0$. Therefore, $g = |\nabla u_\epsilon|^2 + \frac{1}{\epsilon^2}\rho_\epsilon^2$ can be controlled by a multiple of $C_0(m)$. Lemma 4.5.4 implies $\|\frac{1}{\epsilon^2}(1-|u_\epsilon|^2)\|_{L^\infty} \leq \tilde{C}_0$. For $k = 1$, since the conclusions for $k = 0$ hold, i.e., $\limsup_{\epsilon \searrow 0} \|u_\epsilon\|_{L^p(P_{\delta R}(z_0))} \leq C_k$ and $\limsup_{\epsilon \searrow 0} \|\frac{1}{\epsilon^2}(1-|u_\epsilon|^2)\|_{L^p(P_{\delta R}(z_0))} \leq \tilde{C}_k$, using the $C^{\alpha,\frac{\alpha}{2}}$ estimate of strongly parabolic systems in the divergence form, we know there exists $\gamma \in (0,1)$ such that $\|u_\epsilon\|_{C^{\gamma,\frac{\gamma}{2}}} \leq C(\|\frac{1}{\epsilon^2}(1-|u_\epsilon|^2)\|_{L^p})$. From Lemma 4.5.4, we know that $\|\frac{1}{\epsilon^2}1-|u_\epsilon|^2\|_{L^\infty} \leq C_0$. Using the $W_p^{2,1}$ estimate, we get

$$\|u_\epsilon\|_{W_p^{2,1}(P_{\delta R})} \leq C_P(\Omega, T, w_{u_\epsilon})\left(\left\| \frac{1}{\epsilon^2}a(x)(1-|u_\epsilon|^2) \right\|_{L^p(P_R)} + \|u_\epsilon\|_{L^q(P_R)} \right)$$

for any $1 \leq q \leq p$. From the Sobolev inequality, we have when $p > 2 + 2 = 4$,

$$\|\nabla u_\epsilon\|_{C^\alpha(P_{\delta R})} \leq C(m, p, \alpha, \delta)\|u_\epsilon\|_{W_p^{2,1}(P_{\delta R})}.$$

We can take derivatives in (4.5.16) with respect to x to obtain, $\nabla(a(x)[|\nabla u_\epsilon|^2 + \frac{1}{\epsilon^2}(1 - |u_\epsilon|^2)^2]) \in L^p$. Using (4.5.16) and Lemma 4.5.5, we get $\nabla(\frac{1}{\epsilon^2}(1 - |u_\epsilon|^2)^2) \in L^p$. Taking derivatives with respect to x in (4.5.4), we can get $u_\epsilon \in W_p^{3,1}$. By the Sobolev embedding theorem, we know that $u_\epsilon \in C^{1,1}$. Using (4.5.16) again, we get $\frac{1}{\epsilon^2}(1 - |u_\epsilon|^2) \in C^{1,1}$. By the standard bootstrap argument, we complete the proof.

The following Lemma proves that the boundedness of (4.5.37) can be verified by the smallness of the energy, i.e. the energy density is uniformly bounded in the regularity points of u_ϵ.

Lemma 4.5.9 *There are constants $C_1 > 0$, $\epsilon_0 > 0$, $R_0 \in (0, \min\{1, \sqrt{t_0}\})$ such that for solution u_ϵ of (4.5.4) satisfying*

$$\sup_{t_0 - R_0^2 < t < t_0} \int_{B_{R_0}(x_0)} g_\epsilon(u_\epsilon(x, t))dx < \epsilon_0,$$

there holds, $\forall\ \delta \in (0, 1)$,

$$\sup_{P_{\delta R_0}(z_0)} g_\epsilon(u_\epsilon) \leq \frac{C}{(1 - \delta)^2 R_0^2}$$

where $x_0 \in \Omega$ and $B_{R_0} \subset \Omega$.

Proof. Without loss of generality, let $(x_0, t_0) = 0$. We set $P_R := P_R(0)$. For fixed $\epsilon > 0$, since the solution u_ϵ of (4.5.4) is smooth, there exist $\sigma_\epsilon \in [0, R_0)$ such that

$$(R_0 - \sigma_\epsilon)^2 \sup_{P_{\sigma_\epsilon}} g_\epsilon = \max_{0 \leq \sigma \leq R_0} (R_0 - \sigma)^2 \sup_{P_\sigma} g_\epsilon,$$

and there is some $z_\epsilon = (x_\epsilon, t_\epsilon) \in P_R(z_0) \in \overline{P_{\sigma_\epsilon}}$ such that $e_\epsilon := g_\epsilon(u_\epsilon(z_\epsilon)) = \sup_{P_{\sigma_\epsilon}} g_\epsilon$. Setting $\rho_\epsilon := \frac{1}{2}(R_0 - \sigma_\epsilon)$ such that $P_{\rho_\epsilon}(z_\epsilon) \subset P_{\sigma_\epsilon + \rho_\epsilon} \subset P_{R_0}$, we have

$$\sup_{P_{\rho_\epsilon}(z_\epsilon)} g_\epsilon \leq \frac{1}{[R_0 - (\sigma_\epsilon + \rho_\epsilon)^2]}[R_0 - (\sigma_\epsilon + \rho_\epsilon)^2] \sup_{P_{\rho_\epsilon + \sigma_\epsilon}(z_\epsilon)} g_\epsilon \leq \frac{4}{(R_0 - \sigma_\epsilon)^2}(R_0 - \sigma_\epsilon)^2 e_\epsilon \leq 4e_\epsilon.$$

Setting $r_\epsilon = \sqrt{e_\epsilon}\rho_\epsilon$, we can consider a rescaled map $v_\epsilon = v(y, s) = u(x_\epsilon + e_\epsilon^{-\frac{1}{2}}y, t_\epsilon + e_\epsilon^{-1}s)$, $(y, s) \in P_{r_\epsilon}$. Thus v_ϵ satisfies Eq. (4.5.4) with $\tilde{\epsilon} := \sqrt{e_\epsilon}\epsilon$. By computation, $g_{\sqrt{e_\epsilon}\epsilon}(v_\epsilon)(0, 0) = 1$ and $\sup_{P_{r_\epsilon}} g_{\sqrt{e_\epsilon}\epsilon}(v_\epsilon) = 4$. We now claim that $r_\epsilon \leq 2$. If it holds, we can use the definition of r_ϵ and set $\sigma = \delta R_0$ to finish the proof:

$$r_\epsilon^2 = \left[\frac{1}{2}(R_0 - \sigma_\epsilon)\sqrt{e_\epsilon}\right]^2 \leq 4$$

and

$$(R_0 - \delta R_0)^2 \sup_{P_{\delta R_0}} g_\epsilon(u_\epsilon) = (R_0 - \sigma_\epsilon)^2 \sup_{P_{\sigma_\epsilon}} g_\epsilon(u_\epsilon) \leq 16,$$

which implies

$$\sup_{P_{\delta R_0}} g_\epsilon(u_\epsilon) \le \frac{C}{(1-\delta)^2 R_0^2}.$$

We prove the claim by the contradiction argument. Suppose $r_\epsilon > 2$. Since $B_{R_0}(x_0) \subset \Omega$, all the derivatives of v_ϵ are then bounded on P_1 independently of $\epsilon > 0$. Indeed, if $\liminf_{\epsilon \searrow 0} \sqrt{e_\epsilon}\epsilon > 0$, from the equation

$$\frac{1}{2}\partial_t v_\epsilon - \frac{1}{2}v_\epsilon \times \partial_t v_\epsilon - \nabla \cdot (a(x)\nabla v_\epsilon) = \frac{1}{\bar\epsilon^2}a(x)(1-|v_\epsilon|^2)v_\epsilon$$

the claim holds by using the L^p estimates and the fact that $|v_\epsilon| \le 1$. If $\liminf_{\epsilon \searrow 0} \sqrt{e_\epsilon}\epsilon = 0$, then the claim follows from the fact that $\sup_{P_{r_\epsilon}} g_{\sqrt{e_\epsilon}\epsilon} \le 4$ and Lemma 4.5.8. In particular, all the derivatives of v_ϵ are uniformly bounded. Thus

$$\sqrt{\partial_t g_{\bar\epsilon}(v_\epsilon)}, \quad |\nabla g_{\bar\epsilon}(v_\epsilon)| \le C < \infty \quad \text{in } P_1.$$

Therefore, if we choose $r_0 = \min\{\frac{1}{4C}, 1\}$, we have

$$|g_{\bar\epsilon}(v_\epsilon)(x,t) - g_{\bar\epsilon}(v_\epsilon)(0,0)| = |\partial_t g_{\bar\epsilon}(v_\epsilon)(x',t')||t| + |\nabla g_{\bar\epsilon}(v_\epsilon)(x',t')||x| < \frac{1}{2}.$$

Using the differential mean-value theorem, we get

$$g_{\bar\epsilon}(v_\epsilon)(x,t) > g_{\bar\epsilon}(v_\epsilon)(0,0) - \frac{1}{2} > \frac{1}{2},$$

which implies

$$1 = g_{\sqrt{e_\epsilon}\epsilon}(v_\epsilon)(0,0) \le \frac{2}{\pi r_0^2} \sup_{-r_0^2 < s < 0} \int_{B_{r_0}} g_{\sqrt{e_\epsilon}\epsilon}(v_\epsilon)dy$$

$$\le C_* \sup_{t_\epsilon - r_0^2 e_\epsilon^{-1} < t < t_\epsilon} \int_{B_{\sqrt{e_\epsilon}r_0}(x_\epsilon)} g_\epsilon(u_\epsilon)dx$$

$$\le C_* \sup_{-(\frac{r_0^2}{e_\epsilon}+\sigma_\epsilon^2) < t < 0} \int_{B_{\frac{r_0}{\sqrt{e_\epsilon}}+\sigma_\epsilon(x_0)}(x_\epsilon)} g_\epsilon(u_\epsilon)dx. \qquad (4.5.39)$$

Setting $\epsilon_1 = \min\{\frac{1}{2}, \frac{1}{2C_*}\}$, since $r_\epsilon = \sqrt{e_\epsilon}\rho_\epsilon > 2 > r_0$, we get $\frac{r_0}{\sqrt{e_\epsilon}} + \sigma_\epsilon \le \rho_\epsilon + \sigma_\epsilon \le R_0$ and $(\frac{r_0}{\sqrt{e_\epsilon}})^2 + \sigma_\epsilon^2 \le (\rho_\epsilon + \sigma_\epsilon)^2 \le R_0^2$. Hence, the right-hand side of (4.5.39) $\le \epsilon_1 \le \frac{1}{2}$. This leads to a contradiction. Therefore, $r_\epsilon \le 2$.

4.5.4 Boundary Estimates

In this section, we will derive local boundary sup-estimates for the energy density, thereby the $W_p^{2,1}$-estimates for u_ϵ and L^p-estimates for $\frac{1}{\epsilon^2}a(x)(1-|u_\epsilon|^2)$ near the boundary.

Lemma 4.5.10 *Let u_ϵ be a solution of (4.5.4) and (4.5.5) with $u_0 \in H^{1,2}(\Omega; S^2) \cap H^{2,P}(\partial\Omega; S^2)$. Assume*

$$\sup_{P_R^\Omega} g_\epsilon(u_\epsilon) \leq C_0 \tag{4.5.40}$$

and $B_R(x_0) \cap \partial\Omega \neq \emptyset$, $0 < R^2 < t_0$. Then for any $\delta \in (0,1)$, we have

$$\|u_\epsilon\|_{W_p^{2,1}(P_{\delta R}^\Omega(z_0))}$$
$$\leq C_1 \left(\left\| \frac{1}{\epsilon^2}(1 - |u_\epsilon|^2) \right\|_{L^p(P_R^\Omega(z_0))} + \|u_\epsilon\|_{L^2(P_R^\Omega(z_0))} + \|u_0\|_{H^{2 - \frac{1}{p}, p}(B_R^\Omega(z_0) \cap \partial\Omega)} \right),$$

where constant C_1 depends on C_0, δ, R, p, Ω and $\|a(x)\|_{L^\infty(\Omega)}$. Furthermore, for any $\delta \in (0,1)$ we have

$$\left\| \frac{1}{\epsilon^2}(1 - |u_\epsilon|^2) \right\|_{L^p(P_{\delta R}^\Omega(z_0))}$$
$$\leq C(p, \|a(x)\|_{L^\infty}) \|g_\epsilon\|_{L^p(P_R^\Omega)} + \epsilon^{\frac{2}{p}} C(\|g_\epsilon\|_{L^p(P_R^\Omega)}, p, \delta, R, \|a(x)\|_{L^\infty(\Omega)})$$

and

$$\left\| \frac{1}{\epsilon^2}(1 - |u_\epsilon|^2) \right\|_{L^\infty(P_{\delta R}^\Omega(z_0))} \leq 4C_0 + O_\delta(\epsilon),$$

where $\epsilon \mapsto O_\delta(\epsilon)$ is a function that depends on $\delta \in (0,1)$ and $\lim_{\epsilon \searrow 0} \epsilon^{-k} O_\delta(\epsilon) = 0$ for all $k \in N$. All the constants also depend on the parabolic constants.

Proof. Assumption $\sup_{P_R^\Omega} g_\epsilon(u_\epsilon) \leq C_0$ implies that $\limsup_{\epsilon \searrow 0} \sup_{P_{\delta R}^\Omega} \|\nabla u_\epsilon\| \leq C_0(m) < \infty$. Therefore from the C^α estimate, there exists $\gamma \in (0,1)$ such that $\|u_\epsilon\|_{C^{\gamma, \frac{\gamma}{2}}(P_{\delta R}^\Omega)} \leq C(f_\epsilon)$. Furthermore, $\sup_{P_R^\Omega} \frac{1}{\epsilon^2} a(x)(1 - |u_\epsilon|^2)^2 \leq C$. Therefore $g = \frac{1}{2}a(x)(|\nabla u_\epsilon| + \frac{1}{\epsilon^2}(1 - |u_\epsilon|^2)^2) \in L^p(P_R^\Omega)$. From the $W_p^{2,1}$ estimate, we have

$$\|u_\epsilon\|_{W_p^{2,1}(P_{\delta R}^\Omega(z_0))} \leq C_1 \left(\left\| \frac{1}{\epsilon^2} a(x)(1 - |u_\epsilon|^2) \right\|_{L^p(P_R^\Omega(z_0))} \right.$$
$$\left. + \|u_\epsilon\|_{L^2(P_R^\Omega(z_0))} + \|u_0\|_{H^{2 - \frac{1}{p}, p}(B_R^\Omega(z_0) \cap \partial\Omega)} \right).$$

From Lemma 4.5.7, we have

$$\left\| \frac{1}{\epsilon^2}(1 - |u_\epsilon|^2) \right\|_{L^p(P_{\delta R}^\Omega(z_0))}$$
$$\leq C(p, \|a(x)\|_{L^\infty}) \|g_\epsilon\|_{L^p(P_R^\Omega)} + \epsilon^{\frac{2}{p}} C(\|g_\epsilon\|_{L^p(P_R^\Omega)}, p, \delta, R, \|a(x)\|_{L^\infty(\Omega)}).$$

From Lemma 4.5.5, we have

$$\left\| \frac{1}{\epsilon^2}(1 - |u_\epsilon|^2) \right\|_{L^\infty(P^\Omega_{\delta R}(z_0))} \leq \sup_{P^\Omega_R} |g(u_\epsilon)| + \frac{2a}{\epsilon^2} \exp\left(-\frac{1}{\epsilon}(1 - \delta^2)^2 R^4 \right).$$

Noting that

$$g(u_\epsilon) = \left[|\nabla u_\epsilon|^2 + \frac{1}{\epsilon^2}(1 - |u_\epsilon|^2)^2 \right]$$
$$\leq 4\left[\frac{1}{2}|\nabla u_\epsilon|^2 + \frac{1}{4\epsilon^2}(1 - |u_\epsilon|^2)^2 \right] + O_\delta(\epsilon) \leq C_0(m) + O_\delta(\epsilon),$$

where $\lim_{\epsilon \to 0} \epsilon^{-k} \cdot \frac{2a}{\epsilon^2} \exp(-\frac{1}{\epsilon}(1 - \delta^2)^2 R^4 = 0, \ \forall \ k \in N$. The lemma follows.

The following lemma proves that the boundedness of (4.5.40) may be verified by the smallness of the energy, i.e. the energy density is uniformly bounded in the regularity points of u_ϵ.

Lemma 4.5.11 *Let u_ϵ be a solution to (4.5.4) and (4.5.5) with $u_0 \in H^{1,2}(\Omega; S^2) \cap C^2(\partial\Omega; S^2)$. There are constants $C_0 = C_0(\|u_0\|_{C^2(\partial\Omega)}, E_0, \Omega)$ and $\epsilon_0 = \epsilon_0(\|u_0\|_{C^2(\partial\Omega)}, E_0, \Omega) > 0$, such that if for some $z_0 = (x_0, t_0)$ and $R_0 \in (0, \min\{1, \sqrt{t_0}\})$,*

$$\limsup_{\epsilon \searrow 0} \sup_{t_0 - R_0^2 < t < t_0} \int_{B_{R_0}(x_0) \bigcap \Omega} g_\epsilon(u_\epsilon(x, t))dx < \epsilon_0,$$

then for any $\delta \in (0, 1)$, we have

$$\limsup_{\epsilon \searrow 0} \sup_{P^\Omega_{\delta R_0}(z_0)} g_\epsilon(u_\epsilon) \leq \frac{C}{(1 - \delta)^2 R_0^2}.$$

The proof is similar to the interior case, though the computation is much more complicate. For more detail, we refer to [89].

4.5.5 Energy Estimates

In this section, we will prove that the total energy of the smooth weighted flow of (4.5.4) and (4.5.5) is decreasing. Recall that in the first section we have defined the total energy $G_\epsilon(u_\epsilon(t)) := \int_\Omega g_\epsilon(u_\epsilon(x, t))dx$ and the local energy $G_\epsilon(u_\epsilon(t), B^\Omega_R(x_0)) := \int_{B_R(x_0) \bigcap \Omega} g_\epsilon(u_\epsilon(x, t))dx$, respectively.

Lemma 4.5.12 *Let u_ϵ be a solution of (4.5.4) and (4.5.5). Then, we have*

$$G_\epsilon(u_\epsilon(T)) + \frac{1}{2}\int_0^T \int_\Omega |\partial_t u_\epsilon|^2 dx dt = G_\epsilon(u_\epsilon(0) = E(u_0) = E_0. \tag{4.5.41}$$

Proof. We multiply Eq. (4.5.4) by $\partial_t u_\epsilon$ and integrate it over Ω to get

$$\int_\Omega \frac{1}{2}|\partial_t u_\epsilon|^2 dx - \int_\Omega \nabla \cdot (a(x)\nabla u_\epsilon) \cdot \partial_t u_\epsilon dx = \int_\Omega \frac{1}{\epsilon^2} a(x)(1-|u_\epsilon|^2) u_\epsilon \cdot \partial_t u_\epsilon dx$$

and then

$$\int_\Omega \frac{1}{2}|\partial_t u_\epsilon|^2 dx + \frac{\partial}{\partial t}\int_\Omega \frac{1}{2}a(x)|\nabla u_\epsilon|^2 = -\frac{\partial}{\partial t}\int_\Omega \frac{1}{4\epsilon^2}a(x)(1-|u_\epsilon|^2)^2.$$

Integrating the above equality over $[0, T]$ leads to

$$\int_0^T \int_\Omega \frac{1}{2}|\partial_t u_\epsilon|^2 dx dt + \int_\Omega a(x)\Big[\frac{1}{2}|\nabla u_\epsilon|^2 + \frac{1}{4\epsilon^2}(1-|u_\epsilon|^2)^2\Big](T)$$

$$= \int_\Omega a(x)\Big[\frac{1}{2}|\nabla u_0|^2 + \frac{1}{4\epsilon^2}(1-|u_0|^2)^2\Big] = G_\epsilon(u_\epsilon(0)) := E_0.$$

(4.5.41) follows.

The following lemma is about the estimate for the local energy.

Lemma 4.5.13 *Let u_ϵ be a solution of (4.5.4) and (4.5.5). Then, for $0 \le T_1 < T_2$ we have*

$$G_\epsilon(u_\epsilon(T_2), B_R^\Omega(x_0)) \le G_\epsilon(u_\epsilon(T_1), B_{2R}^\Omega(x_0)) + \frac{C}{R^2}\int_{T_1}^{T_2} G_\epsilon(u_\epsilon(t), B_{2R}^\Omega(x_0))dt. \quad (4.5.42)$$

Proof. We multiply Eq. (4.5.4) by $\partial_t u_\epsilon \phi^2$, where ϕ is a cutoff function satisfying $\phi(x) \in C_c^\infty(\Omega)$, $0 \le \phi(x) \le 1$, $\phi \equiv 1$ on $B_R(x_0)\bigcap\Omega$, $\phi \equiv 1$ on $B_R(x_0)\bigcap\Omega$; $\phi \equiv 0$ on $(B_{2R}(x_0)\bigcap\Omega)^C$, $|\nabla\phi| \le \frac{C}{R^2}$, and then integrate over $B_{2R}^\Omega = B_{2R}(x_0)\bigcap\Omega$ to derive

$$\int_{B_{2R}^\Omega} \frac{1}{2}|\partial_t u_\epsilon|^2\phi^2 dx - \int_{B_{2R}^\Omega} \nabla \cdot (a(x)\nabla u_\epsilon) \cdot \partial_t u_\epsilon \phi^2 dx$$

$$= \int_{B_{2R}^\Omega} \frac{1}{\epsilon^2}a(x)(1-|u_\epsilon|^2)u_\epsilon \cdot \partial_t u_\epsilon \phi^2 dx.$$

Integrating by parts, we have

$$\int_{B_{2R}^\Omega} \frac{1}{2}|\partial_t u_\epsilon|^2\phi^2 dx + \frac{\partial}{\partial t}\int_{B_{2R}^\Omega} \frac{1}{2}a(x)|\nabla u_\epsilon|^2\phi^2 dx + \frac{\partial}{\partial t}\int_{B_{2R}^\Omega} \frac{1}{4\epsilon^2}a(x)(1-|u_\epsilon|^2)^2\phi^2 dx$$

$$= -2\int_{B_{2R}^\Omega} a(x)\nabla u_\epsilon \partial_t u_\epsilon \nabla\phi\phi dx$$

$$\le \frac{1}{2}\int_{B_{2R}^\Omega} |\partial_t u_\epsilon|^2\phi^2 dx + 8\int_{B_{2R}^\Omega} a(x)^2|\nabla u_\epsilon|^2|\nabla\phi|^2 dx.$$

Integrating over $[T_1, T_2]$, using the assumption that $a(x) \le \max_{x\in\bar\Omega}|a(x)| = M$ and the property of the cut-off function ϕ, we get

$$G_\epsilon(u_\epsilon(T_2), B_R^\Omega(x_0)) \le G_\epsilon(u_\epsilon(T_1), B_{2R}^\Omega(x_0)) + \frac{C}{R^2}\int_{T_1}^{T_2} G_\epsilon(u_\epsilon(t), B_{2R}^\Omega(x_0))dt.$$

The lemma follows.

Lemma 4.5.14 $\forall\ \eta > 0$, $\exists\ T_0 > 0$, $R_0 > 0$, such that $\forall\ x_0 \in \Omega$ and $\forall\ \epsilon > 0$,

$$\sup_{0 \le t \le T_0} G_\epsilon(u_\epsilon(t), B_{R_0}^\Omega(x_0)) \le \eta \tag{4.5.43}$$

holds.

Proof. For each fixed $\eta > 0$, using the absolute continuity of integration, we can choose R_0 small enough to guarantee

$$G_\epsilon(u_\epsilon(0), B_{2R_0}^\Omega(x_0)) = \int_{B_{2R_0}(x_0)\bigcap\Omega} \frac{1}{2} a(x)|\nabla u_0|^2 dx \le \frac{\eta}{2}.$$

Setting $T_1 = 0$, $T_2 = T_0 = \frac{R_0^2 \eta}{2CE_0}$ in (4.5.42) and choosing T_0 small enough we attain

$$G_\epsilon(u_\epsilon(t), B_{R_0}^\Omega(x_0)) \le G_\epsilon(u_\epsilon(0), B_{2R_0}^\Omega(x_0))$$
$$+ \frac{C}{R_0^2} \int_0^t G_\epsilon(u_\epsilon(t), B_{2R_0}^\Omega(x_0)) dt \le \frac{\eta}{2} + \frac{C}{R_0^2} T_0 E_0 \le \eta.$$

Taking supremum for t over $[0, T_0]$, one has $\sup_{0 \le t \le T_0} G_\epsilon(u_\epsilon(t), B_{R_0}^\Omega(x_0)) \le \eta$. (4.5.43) is proved.

4.5.6 Hausdorff Measure Estimate for Singularity

First of all, it follows from energy estimates (4.5.41) and (4.5.42) that for any $0 \le s < t$

$$G_\epsilon(u_\epsilon(t), B_R^\Omega(x_0)) \le G_\epsilon(u_\epsilon(s), B_{2R}^\Omega(x_0)) + \frac{C(t-s)E_0}{R^2}. \tag{4.5.44}$$

Define $\delta_0 := \frac{\epsilon_0}{2CE_0}$, where ϵ_0 is the constant from (4.5.9) and (4.5.11). We may assume $0 < \delta_0 < 1$, otherwise we can choose a larger C.

Lemma 4.5.15 *Let u_ϵ be a solution of (4.5.4) and (4.5.5) with $u_0 \in H^{1,2}(\Omega; S^2) \bigcap C^2(\partial\Omega; S^2)$. Then the following assertions are equivalent:*
(1) $z_0 = (x_0, t_0) \in \mathrm{Reg}(\{u_\epsilon\}_{\epsilon>0})$,
(2) $\exists\ \delta,\ R > 0$ such that $\limsup_{\epsilon\searrow 0} \sup_{t_0 - \delta < t < t_0} G_\epsilon(u_\epsilon(t), B_R^\Omega(x_0)) < \epsilon_0$,
(3) $\exists\ \delta > 0$ such that $\lim_{R\searrow 0} \limsup_{\epsilon\searrow 0} \sup_{t_0 - \delta < t < t_0} G_\epsilon(u_\epsilon(t), B_R^\Omega(x_0)) = 0$,
(4) $\exists\ R > 0$ such that $\limsup_{\epsilon\searrow 0} \frac{1}{R^2} \int_{t_0 - R^2}^{t_0} \int_{B_R^\Omega(x_0)} g_\epsilon(u_\epsilon) dx dt < \frac{1}{4}\delta_0\epsilon_0$,
(5) $\exists\ \delta,\ R > 0$ such that $\limsup_{\epsilon\searrow 0} \sup_{t_0 - \delta < t < t_0 + \delta} G_\epsilon(u_\epsilon(t), B_R^\Omega(x_0)) = 0$.

Sketch of the proof. We can easily prove them by using the energy estimates Lemma 4.5.9 and 4.5.11, which characterize the supnorm of energy density under the smallness energy condition. For the details, we refer to [89].

Corollary 4.5.1 *Let u_ϵ be a solution of (4.5.4) and (4.5.5) with $u_0 \in H^{1,2}(\Omega; S^2) \bigcap C^2(\partial\Omega; S^2)$. Let $\{\epsilon_i\}_i$ be a sequence with $\epsilon_i \searrow 0$ as $i \to \infty$. Then the following hold:*
(1) $\mathrm{Reg}(\{u_\epsilon\}_\epsilon)$ and $\mathrm{Reg}(\{u_{\epsilon_i}\}_i)$ are open in $\overline{\Omega} \times R_+$.
(2) *There exists some $T_0 > 0$ such that $\overline{\Omega} \times (0, T_0) \subset \mathrm{Reg}(\{u_\epsilon\}_\epsilon)$.*

Proof of Corollary 4.5.1.

(1) follows from Lemma 4.5.15(5) which is the characterization of a regularity point.

(2) follows from Lemma 4.5.12 and Lemma 4.5.14. We can set $\eta = \epsilon_0$, where ϵ_0 is determined in Lemma 4.5.9 and Lemma 4.5.11 to obtain a corresponding T_0. Then Lemma 4.1(2) implies that T_0 satisfies the conclusion. This completes the proof.

Set $Q_R(z) := B_R(x) \times (t - R^2, t + R^2)$ for $z = (x, t)$. Let $\mathcal{H}^2$ denotes the two-dimensional parabolic Hausdorff measure.

We give in the following theorem the Hausdorff measure estimate for a singularity set.

Theorem 4.5.1 *Let u_ϵ be a solution of (4.5.4) and (4.5.5) with $u_0 \in H^{1,2}(\Omega; S^2) \cap C^2(\partial\Omega; S^2)$. Let $\{\epsilon_i\}_i$ be a sequence with $\epsilon_i \searrow 0$ as $i \to \infty$. Then the following hold:*

(1) $S(\{u_\epsilon\}_\epsilon)$ has locally finite two-dimensional parabolic Hausdorff measure. More precisely there is a constant $K_1 = K_1(E_0, \epsilon_0) > 0$, such that for any compact interval $I \subset R_+$, $\mathcal{H}^2(S(\{u_\epsilon\}_\epsilon) \cap (\overline{\Omega} \times I)) \leq K_1 |I|$.

(2) There is a constant $K_2 = K_2(E_0, \epsilon_0) > 0$ such that for any $t > 0$, set $S^t(\{u_\epsilon\}_\epsilon) := S(\{u_\epsilon\}_\epsilon) \cap (\overline{\Omega} \times \{t\})$ has at most K_2 points.

Proof. (1) By (4) of Lemma 4.5.15, we have for any $z_0 = (x_0, t_0) \in S(\{u_\epsilon\}_\epsilon)$, any $R > 0$, and sufficiently small $0 < \epsilon \leq \epsilon(z_0)$

$$\frac{1}{R^2} \int_{t_0 - R^2}^{t_0} \int_{B_R^\Omega(x_0)} g_\epsilon(u_\epsilon) > \frac{1}{4} \delta_0 \epsilon_0. \tag{4.5.45}$$

Fix a compact interval $I \subset R_+$ and $\delta > 0$. By compactness and Vitali's Covering theorem [7], any covering of $S(\{u_\epsilon\}_\epsilon) \cap (\overline{\Omega} \times I)$ by parabolic cylinders Q_R^Ω with $0 < R^2 < \delta$ and $z \in S(\{u_\epsilon\}_\epsilon) \cap (\overline{\Omega} \times I)$ contains a finite covering $\bigcup_j Q_{5R_j}^\Omega(z_j) \supset S(\{u_\epsilon\}_\epsilon) \cap (\overline{\Omega} \times I)$ such that cylinders $Q_{R_j}^\Omega(z_j)$ are pairwise disjoint. Since $S(\{u_\epsilon\}_\epsilon) \cap (\overline{\Omega} \times I)$ is a closed subset of the compact set $\overline{\Omega} \times I$, it is compact. And therefore the countable subcovering stated in the Vitali theorem is indeed finite. By (4.5.45) and the energy estimate, we obtain for $0 < \epsilon \leq \min_j \{\epsilon(z_j)\}$,

$$\sum_j \omega_2(5R_j^2) \leq 25 \frac{4\omega_2}{\delta_0 \epsilon_0} \Sigma_j \int_{t_j - R_j^2}^{t_j} \int_{B_R^\Omega(x_j)} g_\epsilon(u_\epsilon(x, t)) dx dt < \frac{100\omega_2}{\delta_0 \epsilon_0} (|I| + \delta) E_0.$$

By letting $\delta \searrow 0$, and by the definition of the Hausdorff measure we find that

$$\mathcal{H}^2(S(\{u_\epsilon\}_\epsilon) \cap (\overline{\Omega} \times I)) \leq \frac{100\omega_2 E_0}{\delta_0 \epsilon_0} |I|.$$

(2) Pick any $(x_1, T), \ldots, (x_k, T) \in S(\{u_\epsilon\})$. By assumption we conclude that $\forall\, R, \delta, \gamma > 0$, $\exists\, \epsilon \in (0, \gamma)$ such that

$$\sup_{T - \delta < t < T} \int_{B_R^\Omega(x_l)} g_\epsilon(u_\epsilon(x, t)) dx > \frac{\epsilon_0}{2} \quad \text{for } 1 \leq l \leq k,$$

where we have used Lemma 4.5.15. We may choose $R > 0$ such that $B_{2R}^{\Omega}(x_l)$ $(1 \leq l \leq k)$ are pairwise disjoint. Choose $\delta \in (0, \frac{R^2 \epsilon_0}{4CE_0})$, where C is the constant from Lemma 4.5.13 and $\epsilon \in (0, \gamma)$ as above. Since $t \mapsto \int_{B_R^{\Omega}(x_l)} g_\epsilon(u_\epsilon(x, t)) dx$ is continuous, we may find $t_\delta^l \in (T - \delta, T)$ such that $\int_{B_R^{\Omega}(x_l)} g_\epsilon(u_\epsilon(x, t_\delta^l)) dx \geq \frac{\epsilon_0}{2}$ for $1 \leq l \leq k$. The energy estimates and the local energy inequality now imply

$$E_0 \geq \sum_{l=1}^{k} \int_{B_{2R}^{\Omega}(x_l)} g_\epsilon(u_\epsilon(x, T - \delta)) dx$$

$$\geq \sum_{l=1}^{k} \int_{B_R^{\Omega}(x_l)} g_\epsilon(u_\epsilon(x, t_\delta^l)) dx - \frac{C}{R^2} \int_{T-\delta}^{T} \int_{B_{2R}^{\Omega}(x_l)} g_\epsilon(u_\epsilon, B_{2R}^{\Omega}(x_l)) dt.$$

Thus $E_0 \geq k(\frac{\epsilon_0}{2} - \frac{CE_0}{R^2}\delta)$. Since $\delta < \frac{R^2 \epsilon_0}{4CE_0}$, we finally get $k \leq \frac{4E_0}{\epsilon_0} := K_2$. Theorem 4.5.1 is proved.

4.5.7 Passing to the Limits

In this section, we prove

Theorem 4.5.2 *Let u_ϵ be a solution of (4.5.4) and (4.5.5) with u_0 in $H^{1,2}(\Omega; S^2) \cap H^{\frac{3}{2},2}(\partial\Omega; S^2)$. Then there is at least one sequence $\{u_{\epsilon_i}\}_i$ and $u_* \in H^{1,2}_{\text{loc}}(\overline{\Omega} \times R_+; S^2) \cap L^\infty(R^+; H^{1,2}(\Omega; S^2))$ such that $u_{\epsilon_i} \rightharpoonup u_*$ weakly in $H^{1,2}_{\text{loc}}(\overline{\Omega} \times R_+; R^3)$ and weak* in $L^\infty(R_+; H^{1,2}(\Omega; R^3))$.*

Moreover the following hold:

(1) For any such sequence $\{u_{\epsilon_i}\}_i$, we have $\lim_{i\to\infty} u_{\epsilon_i} = u_*$ and $\frac{1}{\epsilon^2}(1 - |u_\epsilon|^2) \to |\nabla u_*|^2$ in $C^\infty(\text{Reg}(\{u_{\epsilon_i}\}_i) \cap (\Omega \times R_+))$.

(2) u_* is a smooth solution of (4.5.1) in $\text{Reg}(\{u_{\epsilon_i}\}_i) \cap (\Omega \times R_+)$ and a global distributional solution in $H^{1,2}_{\text{loc}}(\overline{\Omega} \times R_+) \cap L^\infty(R_+; H^{1,2}(\Omega; R^3))$. Furthermore, u_* satisfies the initial and boundary condition in the sense $\lim_{t\searrow 0} u_*(\cdot, t) = u_0$ in $H^{2,2}(\Omega; R^3)$ and $u_*|_{\partial\Omega} = u_0|_{\partial\Omega}$ as an $H^{2,2}(\Omega; R^3)$-trace for a.e. $t > 0$, respectively.

(3) If u_* is regular at $z_0 = (x_0, t_0)$ in the sense $\lim_{R\searrow 0} \sup_{t_0 - R^2 \leq t \leq t_0} \int_{B_R(x_0)} a(x)|\nabla u_*|^2 dx = 0$, then z_0 is parabolic isolated for $\{u_{\epsilon_i}\}_i$, i.e. $B_R(x_0) \times (t_0 - R_0^2, t_0) - \{z_0\} \subset \text{Reg}(\{u_{\epsilon_i}\})$ for some $R_0 > 0$.

Proof. From the local energy estimates in Lemma 4.5.13, we see that $\{u_{\epsilon_i}\}$ is uniformly bounded in $H^{1,2}_{\text{loc}}(\overline{\Omega} \times R_+) \cap L^\infty(R_+; H^{1,2}(\Omega; R^3))$. From the weak compactness, and using the diagonal method, we can see that, there is $u_* \in H^{1,2}_{\text{loc}}(\overline{\Omega} \times R_+; R^3) \cap L^\infty(R_+; H^{1,2}(\Omega; R^3))$, and a subsequence $\{\epsilon_i\}_i$ such that $u_{\epsilon_i} \rightharpoonup u_*$ weakly in $H^{1,2}_{\text{loc}}(\overline{\Omega} \times R_+; R^3)$ and weakly* in $L^\infty(R_+; H^{1,2}(\Omega; R^3))$.

(1) From Lemma 4.5.9, we can see that $\forall$ $z_0 \in \text{Reg}(\{u_{\epsilon_i}\}_i) \cap (\Omega \times R_+)$, $\sup_{P_{\delta R}(z_0)} g_\epsilon(u_\epsilon) \leq \frac{C_1}{(1-\delta)^2 R_0^2}$. By Lemma 4.5.8, we obtain $\lim_{i\to\infty} u_{\epsilon_i} = u_*$ in $P_{\delta R(z_0)}$. Since point z_0 is arbitrary, we get $\lim_{i\to\infty} u_{\epsilon_i} = u_*$ in $C^\infty(\text{Reg}(\{u_{\epsilon_i}\}_i) \cap (\Omega \times R_+); R^3)$.

(2) We multiply Eq. (4.5.4) by u_{ϵ_i} to get

$$u_{\epsilon_i} \times \frac{1}{2}\partial_t u_{\epsilon_i} - u_{\epsilon_i} \times \left(\frac{1}{2}u_{\epsilon_i} \times \partial_t u_{\epsilon_i}\right) - u_{\epsilon_i} \times \nabla \cdot (a(x)\nabla u_{\epsilon_i}) = 0.$$

Using Lemma 4.5.8 that $\frac{1}{\epsilon_i^2}(1 - |u_{\epsilon_i}|^2)$ is uniformly bounded in $\mathrm{Reg}(\{u_{\epsilon_i}\}_i) \cap (\Omega \times R_+)$, we get $(1 - |u_{\epsilon_i}|^2) \to 0$ smoothly, i.e. $|u_{\epsilon_*}(x,t)| = 1$ in $\mathrm{Reg}(\{u_{\epsilon_i}\}_i) \cap (\Omega \times R_+)$. Now from

$$\vec{a} \times (\vec{b} \times \vec{c}) = (\vec{a} \cdot \vec{c})\vec{b} - (\vec{a} \cdot \vec{b})\vec{c}$$

and the fact that $|u_*(x,t)| = 1$, we can let ϵ_i to approach 0 to obtain

$$\frac{1}{2}u_* \times \partial_t u_* - \frac{1}{2}u_* \times (u_* \times \partial_t u_*) - u_* \times \nabla \cdot (a(x)\nabla u_*) = 0$$

in the sense of distribution. It is easy to get

$$\frac{1}{2}\partial_t u_* - \frac{1}{2}u_* \times \partial_t u_* - \nabla \cdot (a(x)\nabla u_*) = a(x)|\nabla u_*|^2 u_*. \qquad (4.5.46)$$

From Eq. (4.5.4)

$$\frac{1}{2}\partial_t u_{\epsilon_i} - \frac{1}{2}u_{\epsilon_i} \times \partial_t u_{\epsilon_i} - \nabla \cdot (a(x)\nabla u_{\epsilon_i}) = \frac{1}{\epsilon_i^2}a(x)(1 - |u_{\epsilon_i}|^2)u_{\epsilon_i}.$$

We can see that the left side of the above equation converges to that of (4.5.46) and correspondingly we obtain

$$\frac{1}{\epsilon_i^2}(1 - |u_{\epsilon_i}|^2) \to |\nabla u_*|^2 \text{ in } C^\infty(\mathrm{Reg}(\{u_{\epsilon_i}\}_i) \cap (\Omega \times R_+)).$$

In the sequel, we will rigorously prove that u_* is the distributional $H^{1,2}_{\mathrm{loc}} \cap L^\infty(H^{1,2})$ solution of (4.5.1)

$$\frac{1}{2}\partial_t u_* - \frac{1}{2}u_* \times \partial_t u_* - \nabla \cdot (a(x)\nabla u_*) = a(x)|\nabla u_*|^2 u_* \quad \text{in } \Omega \times R_+.$$

Note that sequence $\{u_{\epsilon_i}\}_i$ converges weakly in $H^{1,2}(\Omega \times R_+; S^2)$ and smoothly on $\mathrm{Reg}(\{u_{\epsilon_i}\}_i) \cap (\Omega \times R_+)$. Furthermore, since $S^t(\{u_{\epsilon_i}\}_i) := S(\{u_{\epsilon_i}\}) \cap (\overline{\Omega} \times \{t\})$ is finite for all $t \geq 0$, we have both $u_{\epsilon_i} \to u_*$ pointwise a.e. in $\Omega \times R_+$ and $u_{\epsilon_i}(\cdot, t) \to u_*(\cdot, t)$ pointwise a.e. in Ω for all $t \in R^+$.

From the energy estimates in Lemma 4.5.12, we have $\int_0^\infty \int_\Omega |\partial_t u_{\epsilon_i}|^2 dx\,dt \leq E_0$. By Fatou's lemma, the complement of $F := \{t \geq 0 | \liminf_{\epsilon_i \searrow 0} \int_\Omega |\partial_t u_{\epsilon_i}|^2(x,t)dx < \infty\}$ has measure zero. For $t_0 \in F$, there is a subsequence, still denoted by u_{ϵ_i}, such that $\partial_t u_{\epsilon_i}(\cdot, t_0) \rightharpoonup \partial_t u_*(\cdot, t_0)$ weakly in $L^2(\Omega; R^3)$. By the local energy estimate, we may assume that, for the same subsequence, we also have $\partial_t u_{\epsilon_i}(\cdot, t_0) \rightharpoonup \partial_t u_*(\cdot, t_0)$ weakly in $H^{1,2}(\Omega; S^2)$. By the uniqueness of the limit, the whole sequence converges. Hence, $u_*(\cdot, t_0) \in H^{1,2}(\Omega; S^2)$ and $\partial_t u_*(\cdot, t_0) \in L^2(\Omega; R^3)$ for all $t_0 \in F$.

Moreover, since $S^{t_0}(\{u_{\epsilon_i}\}_i)$ consists of finitely many points, it has zero 2-capacity in R^2, i.e.

$$\mathrm{Cap}_2(S^{t_0}(\{u_{\epsilon_i}\}_i)) = 0.$$

Therefore, from the definition of capacity, there exists a sequence $\{\eta_k\}_k = \{\eta_{k,q}\}_k \subset C_c^\infty(R^2)$ such that $\eta_k(x) = 1$, $\forall\, x \in S^{t_0}(\{u_{\epsilon_i}\}_i)$ and $\|\eta_k\|_{H^{1,2}(R^2)} \to 0$ as $k \to \infty$.

For $\phi \in C_c^\infty(\Omega)$, we multiply Eq. (4.5.4) by $(1-\eta_k(t))\phi(x)$, with its support contained in $\text{Reg}(\{u_{\epsilon_i}\}_i)$. After passing to limit $k \to \infty$ and using the property of the test function η, we see that for any $t \in F$

$$\int_\Omega \frac{1}{2}\partial_t u_*(x,t)\phi(x)dx - \frac{1}{2}u_*(x,t) \times \partial_t u_*(x,t)\phi(x)$$
$$+ a(x)\nabla u_*(x,t)\nabla\phi(x)dx = \int_\Omega a(x)|\nabla u_*|^2 u_*(x,t)\phi(x)dx.$$

The above equation holds for a.e. $t \geq 0$. On the other hand, we have $u_* \in H^{1,2}(\Omega \times [0,t]; S^2)$ for any $t \geq 0$. Therefore both sides of the above equation are locally integrable on R_+. Multiplying the above equation by $\psi \in C_c^\infty[0,\infty)$ and integrating it over R_+, and noticing that linear combinations of $\Sigma_k a_k \phi_k(x)\psi_k(t)$ with $\phi_k(x) \in C_c^\infty(\Omega)$ and $\psi_k(t) \in C_c^\infty([0,\infty))$ being dense in $C_c^\infty(\Omega \times [0,\infty))$), we therefore get

$$\int_0^\infty \int_\Omega \frac{1}{2}\partial_t u_*(x,t)\phi(x,t)dx - \frac{1}{2}u_*(x,t) \times \partial_t u_*(x,t)\phi(x,t)$$
$$+ a(x)\nabla u_*(x,t)\nabla\phi(x,t)dxdt = \int_0^\infty \int_\Omega a(x)|\nabla u_*|^2 u_*(x,t)\phi(x,t)dxdt$$

for any $\phi(x,t) \in C_c^\infty(\Omega \times [0,\infty))$. Thus we have proved that u_* is a distributional solution to (4.5.1). We still need to verify that u_* satisfies the initial and boundary condition. Now the equation can be written as

$$-a(x)\triangle u_*(\cdot,t_0) = a(x)|\nabla u_*|^2 u_*(\cdot,t_0) + f,$$

where

$$f = -\frac{1}{2}\partial_t u_*(\cdot,t_0) + \frac{1}{2}u_* \times \partial_t u_*(\cdot,t_0) + \nabla a(x)\nabla u_*(\cdot,t_0) \in L^2(\Omega; R^3).$$

By a regularity result due to Rivière (see [126]), we have $u_*(\cdot,t_0) \in H^{2,2}(\Omega; S^2)$ if $u_0 \in H^{\frac{3}{2},2}(\partial\Omega; S^2) \bigcap H^{2,2}(\Omega; S^2)$. This implies $u_*(\cdot,t)|_{\partial\Omega} = u_0|_{\partial\Omega}$ as an $H^{2,2}$-trace for any $t \in F$. As for the initial condition, we have

$$\lim_{t \searrow 0} u_*(\cdot,t) = 0 \quad \text{in } H^{1,2}(\Omega; S^2).$$

As a matter of fact, it follows from the following commutative diagram:

$$u_\epsilon(x,t) \longrightarrow u_*(x,t) \quad \text{as } \epsilon \searrow 0 \text{ in } C^\infty(\text{Reg}(\{u_\epsilon\}) \bigcap(\Omega \times R_+))$$
$$\downarrow t \to 0 \qquad \downarrow t \to 0$$
$$u_\epsilon(x,0) \longrightarrow u_0(x) \quad \text{as } \epsilon \searrow 0 \text{ in } C^\infty,$$

where $u_0 \in H^{1,2} \bigcap H^{\frac{3}{2},2}(\partial\Omega)$ is the boundary value of $u_* \in H^{2,2}$ in the trace sense. Thus we have proved statement (2) of the theorem.

(3) From the regularity assumption, $\exists\, R > 0$ such that

$$\sup_{t_0-R^2 \leq t \leq t_0} \int_{B_R(x_0)} \frac{1}{2}a(x)|\nabla u_*(x,t)|^2 dx < \frac{\epsilon_0}{4}.$$

Set $\delta := \min\{R^2, \frac{R^2 \epsilon_0}{2CE_0}\}$ and $s_0 = t_0 - \frac{1}{2}\delta$. We may also assume that for same $R > 0$, $P_{2R}^{\Omega}(z_0) \backslash \{z_0\} \subset \text{Reg}(\{u_{\epsilon_i}\})$. Using Lemmas 4.5.8 and 4.5.11 and the above steps, we have $|u_{\epsilon_i}| \to |u_*| = 1$ in $\text{Reg}(\{u_{\epsilon_i}\}) \cap \Omega$ and

$$\lim_{i \to \infty} \int_{B_R(x_0) \cap \Omega} g_{\epsilon_i}(u_{\epsilon_i(x,s_0)}) dx = \int_{B_R(x_0) \cap \Omega} \frac{1}{2} a(x) |\nabla u_*(x, s_0)|^2 dx < \frac{\epsilon_0}{4}.$$

From Lemma 4.5.13 we have

$$\sup_{s_0 \leq t \leq s_0 + \delta} \int_{B_{\frac{1}{2}R}(x_0) \cap \Omega} g_{\epsilon_i}(u_{\epsilon_i}(x, t)) dx$$

$$\leq \int_{B_R(x_0) \cap \Omega} g_{\epsilon_i}(u_{\epsilon_i}(x, s_0)) dx + \frac{\delta C E_0}{R^2} < \frac{\epsilon_0}{2} + \frac{\epsilon_0}{2} = \epsilon_0$$

for sufficiently large i. By construction, $t_0 \in (s_0, s_0 + \delta)$; hence the claim follows.

Corollary 4.5.2 *There is $T_0 > 0$, such that $\lim_{\epsilon \searrow 0} u_\epsilon = u_*$ in $C^\infty(\Omega \times (0, T_0); S^2)$, where u_* is the smooth solution of (4.5.1) with initial and boundary data u_0.*

Proof. From above discussions we know that there exists $T_0 > 0$ such that $\overline{\Omega} \times [0, T_0) \subset \text{Reg}(\{u_\epsilon\})$. Using Theorem 4.5.2, we obtain

$$u_\epsilon(x, t) \to u_*(x, t), \quad \text{in } C^\infty(\text{Reg}(\{u_{\epsilon_i}\}_i)).$$

4.5.8   Chen–Struwe Solution

The main purpose of this section is to show that the solution to (4.5.1) is indeed a Struwe solution. In Theorem 4.5.1 of Sec. 4.5.6, we have proved that for each $t > 0$, there are at most finitely many singularities; in particular, so is the maximal existence time T_0. By iterating the short time existence result, a short time smooth solution can be extended to a global weak solution which is smooth except for finitely many point singularities and has decreasing energies $t \to E(u_*(t))$. This extension solution is referred to as a Struwe solution. First, we give a lemma describing the energy of sublimits u_*.

Lemma 4.5.16 *Let u_ϵ be a solution of (4.5.4) and (4.5.5) for each fixed $\epsilon > 0$, and assume u_* to be weak $H^{1,2}$ limits of u_{ϵ_i} for a sequence $0 < \epsilon_i \searrow 0$. If $s < t$ and $S^s(\{u_{\epsilon_i}\}) := (\overline{\Omega} \times \{s\}) \cap S(\{u_{\epsilon_i}\}) = \emptyset$ and $S^t(\{u_{\epsilon_i}\}) \neq \emptyset$, then*

$$\int_\Omega \frac{1}{2} a(x) |\nabla u_*|^2(x, s) dx \geq \int_\Omega \frac{1}{2} a(x) |\nabla u_*|^2(x, \tau) dx \qquad \forall \ \tau > s$$

and

$$\int_\Omega \frac{1}{2} a(x) |\nabla u_*|^2(x, s) dx \geq \int_\Omega \frac{1}{2} a(x) |\nabla u_*|^2(x, t) dx + \epsilon_0 \qquad \forall \ t > s \qquad (4.5.47)$$

where ϵ_0 is the constant from Lemma 4.5.11.

Proof. Set $\overline{x} := (x_1, x_2, \ldots, x_k)$ if $S^t(\{u_{\epsilon_i}\}) = (x_1, x_2, \ldots, x_k)$ and $B_R(\overline{x}) := \bigcup_{j=1}^{k}(B_R(x_j))$. Using the fact that $u_*(x,t) = 1$, on $\mathrm{Reg}(\{u_\epsilon\})$, we have

$$E(u_*(s), \Omega) := \int_\Omega \frac{1}{2}a(x)|\nabla u_*|^2(x,s)dx = \lim_{i \to \infty} \int_\Omega g_{\epsilon_i}(u_{\epsilon_i})(x,s)dx$$

$$\geq \limsup_{i \to \infty} \int_\Omega g_{\epsilon_i}(u_{\epsilon_i})(x,\tau)dx \quad (\forall \ \tau > s) \quad \text{(by Lemma 4.5.12)}$$

$$\geq \int_\Omega \frac{1}{2}a(x)|\nabla u_*|^2(x,\tau)dx \quad \text{(by weak lower semi-continuity).}$$

Also

$$E(u_*(s), \Omega) \geq \limsup_{i \to \infty} \left(\int_{\Omega - B_R(\overline{x})} g_{\epsilon_i(u_{\epsilon_i})}(x,\tau)dx + \int_{B_R(\overline{x})} g_{\epsilon_i(u_{\epsilon_i})}(x,\tau)dx \right)$$

for all $\tau \in (s,t)$. Now from Lemma 4.5.15, for all $\delta \in (0,1)$ and $R > 0$, there are sequences $s < t_i \nearrow t$ and $0 < \delta_i \searrow 0$ such that

$$\int_{B_R(\overline{x})} g_{\epsilon_i(u_{\epsilon_i})}(x,t_i)dx = \sup_{t-\delta_i < \tau < t} \int_{B_R(\overline{x})} g_{\epsilon_i(u_{\epsilon_i})}(x,\tau)dx \geq \delta\epsilon_0$$

and so

$$E(u_*(s)) = \limsup_{i \to \infty} \int_{\overline{\Omega} - B_R(\overline{x})} g_{\epsilon_i(u_{\epsilon_i})}(x,t_i)dx + \delta\epsilon_0 \geq \int_{\overline{\Omega} - B_R(\overline{x})} g_{\epsilon_i(u_{\epsilon_i})}(x,\tau)dx + \delta R_0$$

$\forall \ R > 0$, and $\delta \in (0,1)$. Since the last inequality holds for any $R > 0$ and $\delta \in (0,1)$, the claim follows.

Theorem 4.5.3 *Let $u_0 \in H^{1,2}(\Omega; S^2)$. Then there is a global distributional solution $u \in H^{1,2}_{\mathrm{loc}}(\overline{\Omega} \times (0,\infty); S^2) \cap L^\infty((0,\infty); H^{1,2}(\Omega; S^2))$ with $\partial_t u \in L^2(\Omega \times (0,\infty); R^3)$ of (4.5.1) with initial and boundary data u_0. u is smooth on $\Omega \times (0,\infty)$ except at finitely many points and has decreasing and right continuous energy. If in addition $u_0 \in H^{\frac{3}{2},2}(\partial\Omega; S^2)$, then u is unique among the solutions of (4.5.1) with initial and boundary data u_0. Also u is smooth except for isolated singular points, with $\lim_{t \searrow s} E(u(t)) < E(u(s)) + \epsilon_0$ for all $s \geq 0$.*

Proof. By Theorem 4.5.2, the ϵ-approximation scheme provides a smooth short time solution $u \in C^\infty(\Omega \times (0, T_0); S^2)$ to (4.5.1) with boundary data u_0 and $\lim_{t \nearrow 0} u(\cdot, t) = u_0$ in $H^{1,2}(\Omega; R^3)$. Also, there are $\{x_1, \ldots, x_k\} \subset \Omega$ such that $\lim_{t \nearrow T_0} u(\cdot, t) = u(\cdot, T_0) \in C^\infty(\Omega \backslash \{x_1, \ldots, x_k\}, R^3)$ and

$$\|\sqrt{a(x)}\nabla u(\cdot, T_0)\|^2_{L^2(\Omega)} \leq \liminf_{t \nearrow T_0} \|\sqrt{a(x)}\nabla u(\cdot, t)\|^2_{L^2(\Omega)} \leq 2E_0.$$

We have used the maximum principle to prove that $u(\cdot, t) \leq 1$; therefore, $u(\cdot, T_0) \in H^{1,2}(\Omega)$. If we now set $\tilde{u}_0 := u(\cdot, T_0)$ and repeat the same procedure with $\tilde{u}_0$ instead of u_0, we obtain step by step a global solution with point singularities. To see that

$\partial_t u \in L^2(\Omega \times R_+; R^3)$ we first use Lemma 4.5.12 to interval (t_k, t_{k+1}): for each $t \in (t_k, t_{k+1})$,

$$G_\epsilon(u_\epsilon(t)) + \frac{1}{2} \int_{t_k}^t \int_\Omega |\partial_t u_\epsilon| dx dt = G_\epsilon(u_\epsilon(t_k))$$

Setting $t \nearrow t_{k+1}$, we have

$$\limsup_{t \nearrow t_{k+1}} G_\epsilon(u_\epsilon(t)) + \frac{1}{2} \int_{t_k}^{t_{k+1}} \int_\Omega |\partial_t u_\epsilon| dx dt = G_\epsilon(u_\epsilon(t_k)).$$

By Lemma 4.5.16 we get

$$\limsup_{t \nearrow t_{k+1}} E(u(t)) \geq E(u(t_{k+1})) + \epsilon_0.$$

This implies that $\frac{1}{2} \int_0^\infty \int_\Omega |\partial_t u|^2 dx dt \leq E_0 - \Sigma_k \epsilon_0$ and also shows there can only be finitely many singular time t_k. Now assume we have two solutions u_1 and u_2 of (4.5.1) with initial and boundary data u_0 and both with finitely many singularities and $\lim_{t \searrow s} E(u(t)) < E(u(s)) + \epsilon_0$ for all $s \geq 0$. We have $u_1 = u_2$ on $\Omega \times (0, T_1)$, where T_1 is the maximal common smooth existence time, i.e., either u_1 or u_2 has point singularities at T_1. However, if u_1 admits a smooth extension up to T_1, so does u_2 and vice versa. Moreover, since the criterion for the existence of a smooth extension is local, both the solutions have the same singularities $x_1, \ldots, x_k$ at time T_1 and $u_1(\cdot, T_1) = u_2(\cdot, T_1)$ on $\Omega \backslash \{x_1, \ldots, x_k\}$. By Theorem 6 in [34], and the assumption on the energy, the extension of u_1 and u_2 after T_1 is again unique for a short time, and an iteration of the previous argument leads to the claimed uniqueness.

4.6   Smooth Solution and Decay Estimates to the L–L System with Small Initial Data in Higher Dimensions

4.6.1   Initial Value Problem to the L–L System in Higher Dimensions

1. *Existence and uniqueness of the global smooth solution to the Cauchy problem*

Consider the existence and uniqueness of a smooth solution of the Cauchy problem to the Landau–Lifshitz system with the Gilbert term in higher dimensions:

$$\vec{Z}_t = \lambda_1 \vec{Z} \times \Delta \vec{Z} - \lambda_2 \vec{Z} \times (\vec{Z} \times \Delta \vec{Z}), \ x \in R^n, \ t > 0, \qquad (4.6.1)$$

where $n \geq 2$, and $\vec{Z}(x, t) = (Z_1(x, t), Z_2(x, t), Z_3(x, t))$. For simplicity, we let $\lambda_1 = \lambda_2 = 1$. Under condition $|\vec{Z}(x, t)| = 1$, system (4.6.1) is equivalent to the following system:

$$\vec{Z}_t = \Delta \vec{Z} + |\nabla \vec{Z}|^2 \vec{Z} + \vec{Z} \times \Delta \vec{Z}, \ x \in R^n, \ t > 0. \qquad (4.6.2)$$

Taking the mirror mapping from surface $B = \{\vec{Z} : |\vec{Z}(x,t)| = 1\}$ onto the complex plane by

$$w(x,t) = \frac{Z_1(x,t) + iZ_2(x,t)}{1 + Z_3(x,t)}, \qquad (4.6.3)$$

we may rewrite (4.6.2) into a nonlinear Schrödinger equation as follows:

$$w_t - (1+i)\Delta w + \frac{2(1+i)w^*}{1+|w|^2}|\nabla w|^2 = 0, \qquad (4.6.4)$$

where w^* is the conjugate of w. Then our problem is changed to prove the existence and uniqueness of a smooth solution of the Cauchy problem with small initial data of nonlinear Schrödinger equation (4.6.4) with the following initial condition:

$$w(x,0) = w_0(x), \quad x \in R^n, \quad n \geq 2. \qquad (4.6.5)$$

If this is proved, then making inverse transformation

$$Z_1 = \frac{2\Re w(x,t)}{1+|w|^2}, \quad Z_2 = \frac{2\Im w(x,t)}{1+|w|^2}, \quad Z_3 = \frac{1-|w|^2}{1+|w|^2}, \qquad (4.6.6)$$

we may get the existence and uniqueness of a global smooth solution to the Cauchy problem

$$\vec{Z}(x,0) = \vec{Z}_0(x), \ x \in R^n, \ n \geq 2 \qquad (4.6.7)$$

of Eq. (4.6.2), where $\Re w$ and $\Im w$ are the real and the image part of a complex number w.

To do this we need to introduce some preliminary knowledge which can be found in [121] and [133].

2. *Heat conduct equation with complex coefficient*

For the Cauchy problem of complex heat conduct equation

$$u_t = a\Delta u, \quad \Re a > 0, \ x \in R^n, \ t > 0,$$
$$u|_{t=0} = \varphi(x), \quad x \in R^n, \ n \geq 2, \qquad (4.6.8)$$

we also have the following expression for the solution:

$$u(x,t) = S(t)\varphi = \frac{1}{(\sqrt{2a\sqrt{\pi t}})^n} \int_{R^n} \exp\left\{-\frac{|x-\xi|^2}{4at}\right\}\varphi(\xi)d\xi, \qquad (4.6.9)$$

since

$$\|u(\cdot,t)\|_2 \leq \|\varphi\|_2 \qquad (4.6.10)$$

and

$$\|u(\cdot,t)\|_\infty \leq Ct^{-\frac{n}{2}}\|\varphi\|_{L^1(R^n)} \qquad (4.6.11)$$

and the Riesz–Therin interpolation inequality

$$\|u(\cdot,t)\|_{W^{N,q}(R^n)} \leq Ct^{-\frac{n}{2}(\frac{n}{p}-\frac{1}{2})}\|\varphi\|_{W^{N,p}(R^n)}, \qquad (4.6.12)$$

where

$$1 < p, q < \infty, \quad \frac{1}{p} + \frac{1}{q} = 1.$$

If $q = \infty$, we have

$$\|u(\cdot, t)\|_{W^{N,\infty}(R^n)} \leq C t^{-\frac{n}{2}} \|\varphi\|_{W^{N,l}(R^n)}, \ \forall \ l > 0, \tag{4.6.13}$$

$$\|u(\cdot, t)\|_{W^{N,\infty}(R^n)} \leq C(1+t)^{-\frac{n}{2}} \|\varphi\|_{W^{N+n+1,1}(R^n)}, \ \forall \ l > 0. \tag{4.6.14}$$

3. *Smoothness of vector-valued functions*

Lemma 4.6.1 *Suppose that $F = F(w)$ is smooth enough and $F(0) = 0$, where $w = (w_1, w_2, \ldots, w_n)$. For any integer $s \geq 0$ if*

$$w = w(x) \in W^{s,p}(R^n), \quad 1 \leq p \leq \infty,$$

and

$$\|w\|_{L^\infty(R^n)} \leq M,$$

then $F(w) \in W^{s,p}(R^n)$ and

$$\|F(w)\|_{W^{s,p}(R^n)} \leq C(M) \|w\|_{W^{s,p}(R^n)}. \tag{4.6.15}$$

Lemma 4.6.2 *Suppose that $F = F(w)$ is smooth enough, where $w = (w_1, w_2, \ldots, w_n)$. If for any $|w| \leq \nu_0$:*

$$F(w) = O(|w|^{1+\alpha}), \quad \alpha \geq 1 \ is \ integer,$$

and if for any integer $s \geq 0$:

$$\|w\|_{L^\infty(R^n)} \leq \nu_0,$$

and the integer appearing in the following formula makes sense, then

$$\|F(w)\|_{W^{s,r}(R^n)} \leq C \|w\|_{W^{s,q}(R^n)} \|w\|_{L^p(R^n)} \|w\|_{L^\infty(R^n)}^{\alpha-1}, \tag{4.6.16}$$

where

$$1 \leq p, q, r \leq \infty, \quad \frac{1}{p} + \frac{1}{q} = \frac{1}{r}.$$

Taking

$$r = 2, \ q = 2, \ p = \infty;$$

or

$$r = 1, \ p = 2, \ q = 2;$$

or

$$r = 1, \ q = 1, \ p = \infty;$$

respectively, we have

Corollary 4.6.1 *Under the condition of Lemma 4.6.2,*

$$\|F(w)\|_{H^s(R^n)} \le C\|w\|_{H^s(R^n)}\|w\|_{L^\infty(R^n)}^\alpha, \tag{4.6.17}$$

$$\|F(w)\|_{W^{s,1}(R^n)} \le C\|w\|_{W^{s,2}(R^n)}\|w\|_{L^2(R^n)}^\alpha\|w\|_{L^\infty(R^n)}^{\alpha-1}, \tag{4.6.18}$$

$$\|F(w)\|_{W^{s,1}(R^n)} \le C\|w\|_{W^{s,1}(R^n)}\|w\|_{L^\infty(R^n)}^\alpha \tag{4.6.19}$$

hold.

Lemma 4.6.3 *Let* $1 \le p,\, q,\, r \le \infty$, $\frac{1}{p} + \frac{1}{q} = \frac{1}{r}$, *for any given integer* $s \ge 0$,

$$f \in W^{s,p}(R^n), \quad g \in W^{s,q}(R^n).$$

Then

$$\|D^s(fg)\|_{L^r(R^n)} \le C(\|f\|_{L^p(R^n)}\|D^s g\|_{L^q(R^n)} + \|D^s f\|_{L^p(R^n)}\|g\|_{L^q(R^n)}). \tag{4.6.20}$$

If $s \ge 1$, *then*

$$\|D^s(fg) - f(D^s g)\|_{L^r(R^n)}$$
$$\le C(\|Df\|_{L^p(R^n)}\|D^{s-1} g\|_{L^q(R^n)} + \|D^s f\|_{L^p(R^n)}\|Dg\|_{L^s(R^n)}). \tag{4.6.21}$$

If $s \ge 0$, *then*

$$\|D^s(fg)\|_{L^r(R^n)} \le C(\|f\|_{L^p(R^n)}\|g\|_{W^{s,q}(R^n)} + \|f\|_{W^{s,p}(R^n)}\|g\|_{L^p(R^n)}). \tag{4.6.22}$$

Let

$$F(v) = \frac{2(1+i)v^*}{1+|v|^2}|\nabla v|^2. \tag{4.6.23}$$

Take space $X_{s,E}$ as

$$X_{s,E} = \{u(x,t) : D_s(u) \le E\}, \tag{4.6.24}$$

where

$$D_s(u) = \sup_{t \ge 0}(1+t)^{\frac{n}{2}}\|u(\cdot,t)\|_{W^{s-n-3,\infty}(R^n)} + \sup_{t \ge 0}\|u(\cdot,t)\|_{W^{s,l}(R^n)}$$
$$+ \left(\int_0^\infty \sum_{|k|\le 2}\|D_x^k u(\cdot,t)\|_{H^s(R^n)}^2\right)^{\frac{1}{2}}, \tag{4.6.25}$$

where $s \ge n + 4$.

Construct a mapping $\hat{T}$ defined on $X_{s,E}$, $\hat{T} : \vec{v} \in X_{s,E} \to \hat{T}v = w$ by

$$\begin{cases} w_t - (1+i)\Delta w + F(v) = 0, \\ w|_{t=0} = \varphi(x). \end{cases} \tag{4.6.26}$$

Lemma 4.6.4 *If* $n \ge 2$ *and* E *is small, then mapping* $\hat{T}$ *maps* $X_{s,E}$ *into itself.*

Proof. From Poisson formula (4.6.9), we have

$$w(x,t) = S(t)\varphi + \int_0^t S(t-\tau)F(v(\tau))d\tau.$$

It follows from (4.6.14) that

$$\|w\|_{S^{s-n-3,\infty}(R^n)} \le C(1+t)^{-\frac{n}{2}}\|\varphi\|_{W^{s-2,1}(R^n)} + \int_0^t (1+t-\tau)^{-\frac{n}{2}}\|F(v(\tau))\|_{W^{s-2,1}(R^n)}d\tau.$$

(i) It follows from Lemma 4.6.1 and Lemma 4.6.2 that

$$
\begin{aligned}
\|F(v)\|_{W^{s,r}(R^n)} &\le C\Big[\Big\|\frac{v^*}{1+|v|^2}\Big\|_{L^r}\|(\nabla v)^2\|_{W^{s,q}} + \Big\|\frac{v^*}{1+|v|^2}\Big\|_{W^{s,q}}\|(\nabla v)^2\|_{L^p}\Big]\\
&\le C\big[\|v\|_{L^p}\|\nabla v\|_{W^{s,q}}\|\nabla v\|_{L^\infty} + \|v\|_{W^{s,q}}\|\nabla v\|_{L^\infty}\|\nabla v\|_{L^p}\big]\\
&\le \overline{C}\big[\|v\|_{L^p}\|v\|_{W^{s+1,q}} + \|v\|_{W^{s,q}}\|v\|_{W^{1,p}}\|\nabla v\|_{L^\infty}\big]\|\nabla v\|_{L^\infty}. \quad (4.6.27)
\end{aligned}
$$

Taking $r=1$, $p=\infty$, we have

$$
\begin{cases}
\|F(v)\|_{W^{s,1}(R^n)} \le \overline{C}[\|v\|_{L^\infty}\|\nabla v\|_{L^\infty}\|v\|_{W^{s+1,1}} + \|v\|_{W^{s,1}}\|\nabla v\|^2_{L^\infty}],\\
\|F(v)\|_{W^{s,2}(R^n)} \le \overline{C}\|v\|_{W^{s-1,1}(R^n)} \cdot \|v\|^2_{W^{1,\infty}(R^n)}.
\end{cases}
\quad (4.6.28)
$$

When $s \ge n+4$, it follows from the definition of $X_{s,E}$ that

$$\|v\|_{W^{1,\infty}} \le \|v(\cdot,\tau)\|_{W^{s-n-3,\infty}} \le E(1+\tau)^{-\frac{n}{2}};$$

$$\|v\|_{W^{s-1,1}(R^n)} \le E.$$

Hence, if

$$\|\varphi\|_{W^{s,1}} + \|\varphi\|_{H^{s+1}}, \le \delta E, \quad (4.6.29)$$

then from (4.6.26) and (4.6.28)

$$
\begin{aligned}
\|w(\cdot,t)&\|_{W^{s-n-3,\infty}(R^n)}\\
&\le C\|\varphi\|_{W^{s-2,1}}(1+t)^{-\frac{n}{2}} + CE^3\int_0^t (1+t-\tau)^{-\frac{n}{2}}(1+\tau)^{-n}d\tau\\
&\le C\delta E(1+t)^{-\frac{n}{2}} + CE^3(1+t)^{-\frac{n}{2}}.
\end{aligned}
$$

That is

$$\sup_{t\ge 0}(1+t)^{\frac{n}{2}}\|w(\cdot,t)\|_{W^{s-n-3,\infty}(R^n)} \le C_1(\delta E + E^3). \quad (4.6.30)$$

(ii) In (4.6.27), taking $r=1$, $p=q=2$, we have

$$\|F(v)\|_{W^{s,1}(R^n)} \le C\|v\|^2_{H^{s+1}(R^n)}\|v\|_{W^{s-n-3,\infty}(R^n)}.$$

Therefore it follows from (4.6.26) that

$$
\begin{aligned}
\|w(\cdot,t)\|_{W^{s,1}(R^n)} &\le \|\varphi\|_{W^{s,1}(R^n)} + \int_0^t \|F(v)\|_{W^{s,1}(R^n)}dt\\
&\le \|\varphi\|_{W^{s,1}(R^n)} + C\int_0^t \|v\|^2_{H^{s+1}(R^n)}\|v\|_{W^{1,\infty}(R^n)}d\tau. \quad (4.6.31)
\end{aligned}
$$

Noting that

$$\|v\|^2_{H^{s+1}(R^n)} = \|v\|^2_{H^1} + \sum_{|k|=2} \|D^k_x v(\cdot,\tau)\|^2_{H^{s-1}(R^n)}$$

$$\leq C\|v(\cdot,\tau)\|_{W^{1,\infty}}\|v(\cdot,\tau)\|_{W^{1,1}} + \sum_{|k|=2} \|D^k_x v(\cdot,\tau)\|^2_{H^{s-1}(R^n)}$$

$$\leq CE^2(1+\tau)^{-\frac{n}{2}} + \sum_{|k|=2} \|D^k_x v(\cdot,\tau)\|^2_{H^{s-1}(R^n)},$$

and substituting this inequality into (4.6.31), we have

$$\|w(\cdot,t)\|_{W^{s,1}(R^n)} \leq \delta E + +CE^3 \int_0^t (1+\tau)^{-n}d\tau + CE^2 \sup_{t\geq 0}\|v\|_{W^{1,\infty}(R^n)}.$$

When $s \geq n+4$, we have

$$\sup_{t\geq 0}\|v\|_{W^{1,\infty}(R^n)} \leq C \sup_{t\geq 0}\|v\|_{W^{s,1}(R^n)} \leq CE.$$

Therefore

$$\|w(\cdot,t)\|_{W^{s,1}(R^n)} \leq \delta E + C_2 E^3. \tag{4.6.32}$$

(iii) For the complex heat conduct equation

$$u_t - a\Delta u = F(x,t) \tag{4.6.33}$$

with initial condition

$$u|_{t=0} = \varphi(x), \tag{4.6.34}$$

where $\Re a > 0$, we have the following theorem:

Theorem 4.6.1 *Let $\varphi(x) \in H^{s+1}(R^n)$, $F \in L^2(0,T;H^s(R^n))$, where $s \geq 0$ is an integer. Then, the Cauchy problem (4.6.33) and (4.6.34) admits a unique solution*

$$u \in L^2(0,T;H^{s+2}(R^n)), \quad u_t \in L^2(0,T;H^s(R^n)),$$

and there holds the estimate

$$\int_0^t \sum_{|k|=2} \|D^k_x u(\cdot,\tau)\|^2_{H^s(R^n)}d\tau \leq C_0\left(\|\varphi\|^2_{H^{s+1}(R^n)} + \int_0^t \|F(\cdot,\tau)\|^2_{H^s(R^n)}d\tau\right),$$

where C_0 is independent of T, and k is a multi-index: $k = (k_1, k_2, \ldots, k_n)$, $|k| = k_1 + k_2 + \cdots + k_n$, $D^k_x = \frac{\partial^k}{\partial x_1^{k_1}\cdots\partial x_n^{k_n}}$.

Now we finish the proof of Lemma 4.6.4.

It follows from (4.6.25) that

$$\int_0^t \sum_{|k|=2} \|D^k_x u(\cdot,\tau)\|^2_{H^s(R^n)}d\tau \leq C\left(\|\varphi\|^2_{H^{s+1}(R^n)} + \int_0^t \|F(v)\|^2_{H^s(R^n)}d\tau\right). \tag{4.6.35}$$

Let $r = q = 2$, $p = \infty$ in (4.6.27). We have

$$\|F(v)\|_{H^s(R^n)} \leq C\|v(\cdot, t)\|_{H^{s+2}}\|v(\cdot, t)\|_{W^{1,\infty}}^2.$$

Note that

$$\|v(\cdot, t)\|_{H^{s+2}}^2 = \|v\|_{H^1}^2 + \sum_{|k|=2} \|D_x^k v(\cdot, t)\|_{H^s}^2$$

$$\leq CE^2(1 + \tau)^{-\frac{n}{2}} + \sum_{|k|=2} \|D_x^k v(\cdot, t)\|_{H^s}^2.$$

Substituting this into (4.6.35), we have

$$\left(\int_0^t \sum_{|k|=2} \|D_x^k u(\cdot, \tau)\|_{H^s(R^n)}^2 d\tau \right)^{\frac{1}{2}}$$

$$\leq C\left[\delta E + \left(\int_0^t \|v(\cdot, t)\|_{H^{s+2}}^2 \|v(\cdot, t)\|_{W^{1,\infty}}^4 d\tau \right)^{\frac{1}{2}} \right]$$

$$\leq C_3(\delta E + E^3). \tag{4.6.36}$$

It follows from (4.6.30), (4.6.32) and (4.6.36) that

$$D_s(w) \leq \max\{1, C_1, C_2, C_3\}(\delta E + E^3) = C(\delta E + E^3). \tag{4.6.37}$$

Taking $\delta = \delta_0 = \frac{1}{\sqrt{2C}}$, when $E \leq E_0 = \frac{1}{2C}$, we have

$$D_s(w) \leq E(C\delta_0 + CE^2) \leq E. \tag{4.6.38}$$

This implies that $\hat{T} : v = \hat{T}v$ maps $X_{s,E}$ into itself. The lemma is proved.

    5. *Contracting mapping*

In the following we want to prove that map $\hat{T}$ is contracting.

Lemma 4.6.5 *If E is small enough, then $\hat{T}$ is contracting in the metric of $X_{s,E}$.*

Proof. For any v_1, $v_2 \in X_{s,E}$, from Theorem 4.6.1, when E is small enough, we have $w_i = \hat{T}v_i \in X_{s,E}$, $i = 1, 2$. Set $v = v_1 - v_2$, $w = w_1 - w_2$. It suffices to prove that if E is small, then there exists a constant $0 < \eta < 1$ such that

$$\rho(w_1, w_2) \leq \eta \rho(v_1, v_2). \tag{4.6.39}$$

(i) From the definition of $\hat{T}$ we know

$$\begin{cases} w_t - (1 + i)\Delta w = F(v_1) - F(v_2), \\ w|_{t=0} = 0. \end{cases} \tag{4.6.40}$$

It follows from (4.6.26) that

$$\|w(\cdot, t)\|_{W^{s-n-3,\infty}(R^n)}$$

$$\leq C \int_0^t (1 + t - \tau)^{-\frac{n}{2}} \|F(v_1) - F(v_2)\|_{W^{s-2,1}(R^n)} d\tau. \tag{4.6.41}$$

Since

$$F(v_1) - F(v_2) = 2(1+i)\left[\frac{v_1^*}{1+|v_1|^2}|\nabla v_1|^2 - \frac{v_2^*}{1+|v_2|^2}|\nabla v_2|^2\right]$$

$$= 2(1+i)\frac{v_1^*}{1+|v_1|^2}(|\nabla v_1|^2 - |\nabla v_2|^2)$$

$$+ 2(1+i)|\nabla v_2|^2\left[\frac{v_1^*}{1+|v_1|^2} - \frac{v_2^*}{1+|v_2|^2}\right]$$

$$= 2(1+i)\frac{v_1^*}{1+|v_1|^2}(\nabla v_1 + \nabla v_2)\nabla v$$

$$+ 2(1+i)|\nabla v_2|^2\frac{(1+|v_1|^2)v^* + (|v_1| + |v_2|)^2 v_1^*}{(1+|v_1|^2)(1+|v_2|^2)},$$

we have

$$\|F(v_1) - F(v_2)\|_{W^{s,r}(R^n)}$$

$$\leq \|\frac{v_1^*}{1+|v_1|^2}(\nabla v_1 + \nabla v_2)\|_{L^p(R^n)}\|\nabla v\|_{W^{s,q}(R^n)}$$

$$+ \left\|\frac{v_1^*}{1+|v_1|^2}(\nabla v_1 + \nabla v_2)\right\|_{W^{s,q}(R^n)}\|\nabla v\|_{L^p(R^n)}$$

$$\left\|\frac{|\nabla v_2|^2(1+|v_2|^2 + (|v_1| + |v_2|)v_1)}{(1+|v_1|^2)(1+|v_2|^2)}\right\|_{L^p(R^n)}\|v\|_{W^{s,q}(R^n)}$$

$$+ \left\|\frac{|\nabla v_2|^2(1+|v_2|^2 + (|v_1| + |v_2|)v_1)}{(1+|v_1|^2)(1+|v_2|^2)}\right\|_{W^{s,q}(R^n)}\|v\|_{L^p(R^n)}, \qquad (4.6.42)$$

where

$$\left\|\frac{v_1^*}{1+|v_1|^2}(\nabla v_1 + \nabla v_2)\right\|_{L^p(R^n)}$$

$$\leq (\|\nabla v_1\|_{L^\infty(R^n)} + \|\nabla v_2\|_{L^\infty(R^n)})\|v_1\|_{L^p(R^n)};$$

$$\left\|\frac{v_1^*}{1+|v_1|^2}(\nabla v_1 + \nabla v_2)\right\|_{W^{s,q}(R^n)}$$

$$\leq C[\|\nabla v_1\|_{L^{p'}}(\|\nabla v_1\|_{W^{s,q'}} + \|\nabla v_2\|_{W^{s,q'}})$$

$$+ \|\nabla v_1\|_{W^{s,q'}}(\|\nabla v_1\|_{L^{p'}} + \|\nabla v_2\|_{L^{p'}})];$$

$$\left\|\frac{|\nabla v_2|^2(1+|v_2|^2 + (|v_1| + |v_2|)v_1)}{(1+|v_1|^2)(1+|v_2|^2)}\right\|_{L^p(R^n)}$$

$$\leq C\||\nabla v_2|^2\|_{L^p(R^n)} \leq C\|\nabla v_2\|_{L^\infty}\|\nabla v_2\|_{L^p};$$

and

$$\left\|\frac{|\nabla v_2|^2(1+|v_2|^2+(|v_1|+|v_2|)|v_1|)}{(1+|v_1|^2)(1+|v_2|^2)}\right\|_{W^{s,q}}$$
$$\leq C\|\nabla v_2\|_{L^{p'}}\|\nabla v_2\|_{W^{s,q'}},$$

where $\frac{1}{q}=\frac{1}{p'}+\frac{1}{q'}$.

Therefore, we have

$$\|F(v_1)-F(v_2)\|_{W^{s,r}(R^n)}$$
$$\leq C[(\|\nabla v_1\|_{L^\infty}+\|\nabla v_2\|_{L^\infty})\|v_1\|_{L^p}\|v\|_{W^{s+1,q}}$$
$$+\{\|\nabla v_1\|_{L^{p'}}(\|v_1\|_{W^{s+1,q'}}+\|v_2\|_{W^{s+1,q'}})$$
$$+\|v_1\|_{W^{s,q'}}(\|\nabla v_1\|_{L^{p'}}+\|\nabla v_2\|_{L^{p'}})\}\|\nabla v\|_{L^p}$$
$$+\|\nabla v_2\|_{L^\infty}^2\|v_2\|_{L^{p'}}\|v\|_{W^{s,q}}$$
$$+\|\nabla v_2\|_{L^{p'}}\|v_2\|_{W^{s+1,q'}}\|v\|_{L^p}]. \tag{4.6.43}$$

In the above inequality, we take s as $s-2$, $r=1$, $p=\infty$, $q=1$ to give

$$\|F(v_1)-F(v_2)\|_{W^{s-2,1}(R^n)}$$
$$\leq C[(\|\nabla v_1\|_{L^\infty}+\|\nabla v_2\|_{L^\infty})\|v_1\|_{L^\infty}\|v\|_{W^{s-1,1}}$$
$$+\{\|v_1\|_{L^\infty}(\|v_1\|_{W^{s-1,1}}+\|v_2\|_{W^{s-1,1}})$$
$$+\|v_1\|_{W^{s,1}}(\|\nabla v_1\|_{L^\infty}+\|\nabla v_2\|_{L^\infty})\}\|\nabla v\|_{L^\infty}$$
$$+\|v_2\|_{L^\infty}^2\|v_1\|_{L^\infty}\|v\|_{W^{s-2,1}}$$
$$+\|\nabla v_1\|_{L^\infty}\|v_2\|_{W^{s-1,1}}\|v\|_{L^\infty}]$$
$$\leq C\|v\|_{W^{s,1}}(\|v_1\|_{W^{1,\infty}}+\|v_2\|_{W^{1,\infty}})\|v_1\|_{W^{1,\infty}}$$
$$+C\|v\|_{W^{s-n-3,\infty}}(\|v_1\|_{W^{s,1}}+\|v_2\|_{W^{s,1}})\|v_1\|_{L^\infty}$$
$$+C\|v\|_{W^{s-n-3,\infty}}(\|\nabla v_1\|_{L^\infty}+\|\nabla v_2\|_{L^\infty})\}\|v_1\|_{W^{s,1}}$$
$$+C\|v\|_{W^{s,1}}\|v_2\|_{W^{s-n-3,\infty}}^2\|v_1\|_{W^{s-n-3,\infty}}$$
$$+C\|v\|_{W^{s-n-3,\infty}}\|v_1\|_{W^{s-n-3,\infty}}\|v_2\|_{W^{s,1}}.$$

Since for $i=1,2$

$$\|v_i\|_{W^{1,\infty}}\leq C\|v_i\|_{W^{s-n-3,\infty}}\leq E(1+\tau)^{-\frac{n}{2}},$$
$$\|\nabla v_i\|_{L^\infty}\leq\|v_i\|_{W^{1,\infty}},$$

we have

$$\|F(v_1)-F(v_2)\|_{W^{s-2,1}}\leq CE^2(1+\tau)^{-n}D_s(v).$$

On the other hand, since for $n>1$

$$\int_0^t(1+t-\tau)^{-\frac{n}{2}}(1+\tau)^{-n}d\tau=C(1+t)^{-\frac{n}{2}},$$

we have from (4.6.41) that

$$\|w(\cdot,t)\|_{W^{s-n-3,\infty}} \le CE^3(1+t)^{-\frac{n}{2}}D_s(v).$$

That is

$$\sup_{t>0}(1+t)^{-\frac{n}{2}}\|w(\cdot,t)\|_{W^{s-n-3,\infty}} \le CE^3 D_s(v).$$

(ii) Similarly to (4.6.31) we have

$$\|w(\cdot,t)\|_{W^{s,1}} \le \int_0^t \|F(v_1) - F(v_2)\|_{W^{s,1}}d\tau. \tag{4.6.44}$$

Taking $r = 1$, $p = q = 2$ in (4.6.43), we have

$$\begin{aligned}
\|F(v_1) - F(v_2)\|_{W^{s,1}(R^n)} \\
\le C\|v\|_{H^{s+1}}(\|\nabla v_1\|_{L^\infty} + \|\nabla v_2\|_{L^\infty})\|v_1\|_{L^2} \\
+ C\|\nabla v\|_{L^\infty}[\|v_1\|_{L^\infty}(\|v_1\|_{H^{s+1}} + \|v_2\|_{H^{s+1}}) \\
+ \|v_1\|_{H^s}(\|\nabla v_1\|_{L^\infty} + \|\nabla v_2\|_{L^\infty})] \\
+ C\|v\|_{H^s}\|\nabla v_1\|_{L^\infty}^2\|v_1\|_{L^2} \\
+ C\|v\|_{L^2}\|\nabla v_1\|_{L^\infty}\|v_2\|_{H^{s+1}}.
\end{aligned} \tag{4.6.45}$$

However,

$$\begin{aligned}
\|v\|_{H^{s+2}}^2 &\le C\|v(\cdot,\tau)\|_{W^{s-n-3,\infty}}\|v(\cdot,\tau)\|_{W^{s,1}} + \sum_{|k|=2}\|D_x^k v(\cdot,\tau)\|_{H^s}^2 \\
&\le C(1+\tau)^{-\frac{n}{2}}D_s^2(v) + \sum_{|k|=2}\|D_x^k v(\cdot,\tau)\|_{H^s}^2
\end{aligned} \tag{4.6.46}$$

and

$$\begin{aligned}
\|v\|_{H^1}^2 &\le C\|v(\cdot,\tau)\|_{W^{1,\infty}}\|v(\cdot,\tau)\|_{W^{1,1}} \\
&\le C\|v(\cdot,\tau)\|_{W^{s-n-3,\infty}}\|v\|_{W^{s,1}} \\
&\le C(1+\tau)^{-\frac{n}{2}}D_s^2(v),
\end{aligned} \tag{4.6.47}$$

we have

$$\begin{cases}
\|v\|_{H^{s+2}(R^n)} \le C(1+\tau)^{-\frac{n}{4}}D_s(v) + \left(\sum_{|k|=2}^{\infty}\|D_x^k v(\cdot,\tau)\|_{H^s(R^n)}^2\right)^{\frac{1}{2}}, \\
\|v\|_{H^1(R^n)} \le C(1+\tau)^{-\frac{n}{4}}D_s(v).
\end{cases} \tag{4.6.48}$$

Similarly,

$$\|v_1\|_{H^{s+2}} + \|v_2\|_{H^{s+2}} \le CE(1+\tau)^{-\frac{n}{4}} + \left(\sum_{|k|=2,i=1,2}\|D_x^k v_i(\cdot,\tau)\|_{H^s}^2\right)^{\frac{1}{2}}, \tag{4.6.49}$$

$$\sum_{i=1,2}\|v_i(\cdot,\tau)\|_{H^1} \le CE(1+\tau)^{-\frac{n}{4}} \tag{4.6.50}$$

and

$$\sum_{i=1,2} \|v_i(\cdot,\tau)\|_{W^{1,\infty}} \le CE(1+\tau)^{-\frac{n}{2}}. \tag{4.6.51}$$

Hence

$$\|F(v_1) - F(v_2)\|_{W^{s,1}(R^n)}$$
$$\le C\left[(1+\tau)^{-\frac{n}{4}} D_s(v) + \left(\sum_{|k|=2} \|D_x^k v(\cdot,\tau)\|_{H^s}^2\right)^{\frac{1}{2}}\right] E^2(1+\tau)^{-(\frac{n}{2}+\frac{n}{4})}$$
$$+ C(1+\tau)^{-n} D_s(v) E^2$$
$$+ C\left[(1+\tau)^{-\frac{n}{4}} D_s(v) + \left(\sum_{|k|=2} \|D_x^k v(\cdot,\tau)\|_{H^s}^2\right)^{\frac{1}{2}}\right] E^3(1+\tau)^{-\frac{5n}{4}}$$
$$\times C D_s(v)[E^2(1+\tau)^{-n} + E^3(1+\tau)^{-\frac{3n}{2}} + E^2(1+\tau)^{-\frac{3n}{4}}]$$
$$+ C\left(\sum_{|k|=2} \|D_x^k v(\cdot,\tau)\|_{H^s}^2\right)^{\frac{1}{2}} (E^3(1+\tau)^{-\frac{5n}{4}} + E^2(1+\tau)^{-\frac{3n}{4}}).$$

Applying this inequality to (4.6.44) and using the Hölder inequality, we have

$$\|w(\cdot,t)\|_{W^{s,1}} \le C D_s(v)(E^2 + E^3) + C(E^2 + E^3),$$
$$\left(\int_0^\infty \sum_{|k|=2} \|D_x^k v(\cdot,\tau)\|_{H^s}^2 d\tau\right)^{\frac{1}{2}} \le 2C D_s(v)(E^2 + E^3). \tag{4.6.52}$$

(iii) It follows from (4.6.35) that

$$\int_0^t \sum_{|k|=2} \|D_x^k v(\cdot,\tau)\|_{H^s}^2 d\tau \le C \int_0^t \|F(v_1) - F(v_2)\|_{H^s} d\tau. \tag{4.6.53}$$

Taking $r = q = 2$, $p = \infty$ in (4.6.43), we have

$$\|F(v_1) - F(v_2)\|_{H^s(R^n)}$$
$$\le C\|v\|_{H^{s+1}}\|v_1\|_{L^\infty}(\|\nabla v_1\|_{L^\infty} + \|\nabla v_2\|_{L^\infty})$$
$$+ C\|\nabla v\|_{L^\infty}[\|v_1\|_{L^\infty}(\|v_1\|_{H^{s+1}} + \|v_2\|_{H^{s+1}})$$
$$+ \|v_1\|_{H^s}(\|\nabla v_1\|_{L^\infty} + \|\nabla v_2\|_{L^\infty})]$$
$$+ C\|v\|_{H^s}\|\nabla v_1\|_{L^\infty}^2\|v_2\|_{L^\infty}$$
$$+ C\|v\|_{L^\infty}\|\nabla v_1\|_{L^\infty}\|v_2\|_{H^{s+1}}$$
$$\le C\|v\|_{H^{s+2}}\|v_1\|_{W^{1,\infty}}(\|v_1\|_{W^{1,\infty}} + \|\nabla v_2\|_{W^{1,\infty}})$$
$$+ C\|v\|_{W^{1,\infty}}\|v_1\|_{H^{s+2}}(\|v_1\|_{W^{1,\infty}} + \|v_2\|_{W^{1,\infty}})$$
$$+ C\|v\|_{H^{s+2}}\|v_1\|_{H^{s+2}}(\|v_1\|_{W^{1,\infty}} + \|v_2\|_{W^{1,\infty}}$$
$$\le C E^2 (1+\tau)^{-\frac{n}{2}}\left[(1+\tau)^{-\frac{n}{4}} D_s(v) + \left(\sum_{|k|=2} \|D_x^k v(\cdot,\tau)\|_{H^s}^2\right)^{\frac{1}{2}}\right]$$

$$+ C\left[E(1+\tau)^{-\frac{n}{4}} + \left(\sum_{|k|=2}\|D_x^k v_1(\cdot,\tau)\|_{H^s}^2\right)^{\frac{1}{2}}\right]E(1+\tau)^{-n}$$

$$+ CE^3(1+\tau)^{-\frac{3}{2}n}\left[(1+\tau)^{-\frac{n}{4}}D_s(v) + \left(\sum_{|k|=2}\|D_x^k v(\cdot,\tau)\|_{H^s}^2\right)^{\frac{1}{2}}\right]$$

$$+ CED_s(v)(1+\tau)^{-n}\left[E(1+\tau)^{-\frac{n}{4}} + \left(\sum_{|k|=2}\|D_x^k v(\cdot,\tau)\|_{H^s}^2\right)^{\frac{1}{2}}\right].$$

Substituting this into (4.6.53) and applying the Hölder inequality, we have

$$\int_0^\infty \sum_{|k|=2}\|D_x^k v(\cdot,\tau)\|_{H^s}^2 d\tau \le CD_s^2(E^2 + E^3)^2. \tag{4.6.54}$$

In summary, we get

$$D_s(w) \le C_0 D_s(E^2 + E^3),$$

where $C_0 = C_0(s,n)$ is a positive constant. Choose E_1 such that

$$C_0(E_1^2 + E_1^3) = \eta < 1. \tag{4.6.55}$$

Letting

$$E \le \min\{E_0, E_1\}, \tag{4.6.56}$$

we get

$$D_s(w) \le \eta D_s(v),$$

where E_0 is given in Lemma 4.6.4. The lemma is proved.

4.6.2 Decay of Solution to the Higher-Dimensional L–L Equations

1. *Decay of the Cauchy problem of nonlinear Schrödinger equations*
   From Lemmas 4.6.4 and 4.6.5, we have

Theorem 4.6.2 *Consider the Cauchy problem of the following nonlinear Schrödinger system:*

$$\begin{cases} w_t - (1+i)\Delta w + \frac{2(1+i)w^*}{1+|w|^2}|\nabla w|^2 = 0, \\ w(x,0) = w_0(x), \ x \in R^n, \ n \ge 2. \end{cases} \tag{4.6.57}$$

If for $s \ge n+4$ there exists small constants δ_0 and E_0 such that for $\delta \le \delta_0$, $E \le E_0$ the initial data $w_0(x)$ satisfies

$$w_0(x) \in W^{s,1}(R^n) \cap H^{s+1}(R^n), \tag{4.6.58}$$

and

$$\|w_0(x)\|_{W^{s,1}(R^n)} + \|w_0(x)\|_{H^{s+1}(R^n)} \le \delta E, \tag{4.6.59}$$

then problem (4.6.57) admits a unique smooth solution $w(x,t) \in X_{s,E}$, that is

$$w(x,t) \in L^2(0,T;H^{s+2}(R^n)) \cap L^\infty(0,T;H^{s+1}(R^n)),$$
$$w_t(x,t) \in L^2(0,T;H^s(R^n)) \cap L^\infty(0,T;H^{s-2}(R^n)),$$

and the decay estimate holds, i.e.

$$\|w(\cdot,t)\|_{W^{s-n-3,\infty}(R^n)} \le C(1+t)^{-\frac{n}{2}}, \ \forall \ t \ge 0,$$

where C is independent of t.

Now we return to the Cauchy problem (4.6.2). Note that

$$w_0(x) = \frac{Z_1^0(x) + iZ_2^0(x)}{1 + Z_3^0(x)}.$$

If there exists a constant $\varepsilon_0 > 0$ such that

$$\varepsilon_0 \le 1 + Z_3^0(x) < 2, \quad \forall \ x \in R^n,$$

since

$$\begin{aligned}
|w_0(x)|^2 &= \frac{(Z_1^0(x))^2 + (Z_2^0(x))^2}{(1 + Z_3^0(x))^2} \\
&= \frac{1 - (Z_3^0(x))^2}{(1 + Z_3^0(x))^2} = \frac{1 - Z_3^0(x)}{1 + Z_3^0(x)},
\end{aligned} \tag{4.6.60}$$

and

$$\vec{Z}_0(x) \cdot \frac{\partial \vec{Z}_0(x)}{\partial x_j} = 0, \ j = 1, 2, \ldots, n,$$

we have

$$\begin{aligned}
\left| \frac{\partial w_0(x)}{\partial x_j} \right|^2 &= \sum_{i=1}^{3} \left(\frac{\partial}{\partial x_j} Z_i^0(1 + Z_3^0) - \frac{\partial}{\partial x_j} Z_i^0 Z_3^0 \right)^2 / (1 + Z_3^0)^4 \\
&= \left(\frac{\partial}{\partial x_j}(Z_1^0)^2 + \frac{\partial}{\partial x_j}(Z_2^0)^2 \right) / (1 + Z_3^0)^2 \\
&\quad + \frac{\partial}{\partial x_j}(Z_3^0)^2((Z_1^0)^2 + (Z_2^0)^2)/(1 + Z_3^0)^4 \\
&\quad - 2\frac{\partial}{\partial x_j} Z_3^0 \left(\frac{\partial}{\partial x_j} Z_1^0 \cdot Z_1^0 + \frac{\partial}{\partial x_j} Z_2^0 Z_2^0 \right) / (1 + Z_3^0)^4 \\
&= \sum_{i=1}^{3} \left(\frac{\partial}{\partial x_j} Z_i^0 \right)^2 / (1 + Z_3^0)^2.
\end{aligned}$$

Then

$$|\nabla w_0|^2 = \frac{|\nabla \vec{Z}_0|^2}{(1 + Z_3^0)^2}, \quad |\nabla w_0| \le \frac{1}{\varepsilon_0}|\nabla \vec{Z}_0|. \tag{4.6.61}$$

For $k > 1$ it is easily verified that

$$|\nabla^k w_0| \leq C_k(\varepsilon_0)\left(|\nabla^k \vec{Z}_0| + \sum_{i=1}^{k-1} |\nabla^i \vec{Z}_0||\nabla^{k-i}\vec{Z}_0|\right).$$

This implies

$$\|\nabla^k w_0\|_{L^1(R^n)} \leq C_k(\varepsilon_0)\left(\|\nabla^k \vec{Z}_0\|_{L^1(R^n)} + \sum_{i=1}^{k-1} \|\nabla^i \vec{Z}_0\|_{L^2(R^n)}\|\nabla^{k-i}\vec{Z}_0\|_{L^2(R^n)}\right)$$

$$\leq C_k(\varepsilon_0)(\|\nabla^k \vec{Z}_0\|_{L^1(R^n)} + \|\nabla \vec{Z}_0\|_{H^{k-1}(R^n)}^2). \tag{4.6.62}$$

It follows from (4.6.60) and (4.6.61) that

$$\|\nabla w_0\|_{W^{s-1,1}(R^n)} \leq C(\varepsilon_0)(\|\nabla \vec{Z}_0\|_{W^{s-1,1}(R^n)} + \|\nabla \vec{Z}_0\|_{H^{s-1}(R^n)}^2).$$

It is easy to verify that

$$\|\nabla w_0\|_{H^s} \leq C(\varepsilon_0, n, s)(\|\nabla \vec{Z}_0\|_{H^s} + \|\nabla \vec{Z}_0\|_{H^s}^2).$$

In summary, we have for $s \geq n + 4$,

$$\|w_0\|_{W^{s,1}} + \|w_0\|_{H^{s+1}}$$

$$\leq \int_{R^n} \left[\left(\frac{1 - Z_3^0}{1 + Z_3^0}\right)^{\frac{1}{2}} + \frac{1 - Z_3^0}{1 + Z_3^0}\right] dx$$

$$+ C(\varepsilon_0, n, s)(\|\nabla \vec{Z}_0\|_{W^{s-1,1}} + \|\nabla \vec{Z}_0\|_{H^s} + \|\nabla \vec{Z}_0\|_{H^s}^2). \tag{4.6.63}$$

Hence we only need to choose small $\vec{Z}_0 = (Z_1^0, Z_2^0, Z_3^0)$ such that

$$\int_{R^n} \left[\left(\frac{1 - Z_3^0}{1 + Z_3^0}\right)^{\frac{1}{2}} + \frac{1 - Z_3^0}{1 + Z_3^0}\right] dx$$

$$+ C(\varepsilon_0, n, s)(\|\nabla \vec{Z}_0\|_{W^{s-1,1}} + \|\nabla \vec{Z}_0\|_{H^s} + \|\nabla \vec{Z}_0\|_{H^s}^2) \leq \delta E. \tag{4.6.64}$$

2. *Global smooth solution to the Cauchy problem*
We have from Theorem 4.6.2 that

Theorem 4.6.3 *Let $\vec{Z}_0(x) = (Z_1^0(x), Z_2^0(x), Z_3^0(x))$ such that*
(i) $|\vec{Z}_0(x)| = 1$, $\inf_{x \in R^n} Z_3^0(x) > -1$, $Z_3^0(x) < 1$;
(ii) $\nabla \vec{Z}_0(x) \in W^{s-1,1}(R^n) \cap H^s(R^n)$;
(iii)

$$\int_{R^n} \left[\left(\frac{1 - Z_3^0}{1 + Z_3^0}\right)^{\frac{1}{2}} + \frac{1 - Z_3^0}{1 + Z_3^0}\right] dx$$

$$+ C(\varepsilon_0, n, s)(\|\nabla \vec{Z}_0\|_{W^{s-1,1}} + \|\nabla \vec{Z}_0\|_{H^s} + \|\nabla \vec{Z}_0\|_{H^s}^2) \leq d,$$

where $d = d(n, s)$ is a small positive number, $s \geq n + 4$, $n \geq 2$.

Then the Cauchy problem for the Landau–Lifshitz equations

$$\vec{Z}_t = \Delta\vec{Z} + |\nabla\vec{Z}|^2\vec{Z} + \vec{Z} \times \Delta\vec{Z}, \ x \in R^n, \ t > 0, \tag{4.6.65}$$

$$\vec{Z}|_{t=0} = \vec{Z}_0(x), \ x \in R^n, \ n \geq 2, \tag{4.6.66}$$

has a unique global smooth solution $\vec{Z}(x,t)$ such that
(i) $|\vec{Z}(x,t)| = 1, \ \forall \ (x,t) \in R^n \times R^+$;
(ii) $\nabla\vec{Z} \in L^2(0,T;H^s(R^n)) \cap L^\infty(0,T;H^{s-1}(R^n))$;
(iii) $\vec{Z}_t \in L^2(0,T;H^s(R^n)) \cap L^\infty(0,T;H^{s-2}(R^n))$;
(iv) $\|\nabla\vec{Z}(\cdot,t)\|_{W^{s-n-4,\infty}(R^n)} \leq C(1+t)^{-\frac{1}{2}}, \ \forall \ t \geq 0.$

Proof. It follows from (4.6.63), (4.6.64) and Theorem 4.6.2 that the Cauchy problem (4.6.57) admits a unique solution $w(x,t) \in X_{s,E}$. Define $\vec{Z}(x,t) = (Z_1(x,t), Z_2(x,t), Z_3(x,t))$ by

$$Z_1 = \frac{2\Re w}{1+|w|^2}, \quad Z_1 = \frac{2\Im w}{1+|w|^2}, \quad Z_1 = \frac{1-|w|^2}{1+|w|^2}.$$

It is easy to verify that $|\vec{Z}(x,t)| = 1$ for all $(x,t) \in R^n \times R^+$, and

$$\nabla Z_1 = \frac{2\Re\nabla w}{1+|w|^2} - \frac{4w\nabla w\Re w}{(1+|w|^2)^2},$$

$$\nabla Z_2 = \frac{2\Im\nabla w}{1+|w|^2} - \frac{4w\nabla w\Im w}{(1+|w|^2)^2},$$

$$\nabla Z_3 = \frac{-4w\nabla w}{(1+|w|^2)^2}.$$

Since

$$\|w(\cdot,t)\|_{L^\infty} \leq C\|D^s w\|_{L^1}^{\frac{n}{s}}\|w\|_{L^1}^{1-\frac{n}{s}} \leq C\|w\|_{W^{s,1}},$$

we have

$$\sup_{t\geq 0}\|w(\cdot,t)\|_{L^\infty} \leq C\sup_{t\geq 0}\|w(\cdot,t)\|_{W^{s,1}} \leq CE.$$

Applying Lemmas 4.6.1 and 4.6.2, we have

$$\left\|\frac{w}{1+|w|^2}\right\|_{W^{k,p}(R^n)} \leq C\|w\|_{W^{k,p}(R^n)},$$

$$\left\|\frac{1}{1+|w|^2}\right\|_{W^{k,p}(R^n)} \leq C\|w\|_{W^{k,p}(R^n)}\|w\|_\infty \leq C\|w\|_{W^{k,p}}.$$

Hence

$$\|\nabla\vec{Z}\|_{W^{k,p}(R^n)} \leq C[(\|\Re\nabla w\|_{W^{k,p}} + \|\Im\nabla w\|_{W^{k,p}})\left\|\frac{1}{1+|w|^2}\right\|_\infty$$

$$+ (\|\Re\nabla w\|_\infty + \|\Im\nabla w\|_\infty)\left\|\frac{1}{1+|w|^2}\right\|_{W^{k,p}}$$

$$
\begin{aligned}
&+ \left(\left\| \frac{w}{(1+|w|^2)^2} \right\|_{W^{k,p}} + \left\| \frac{w\Im w}{(1+|w|^2)^2} \right\|_{W^{k,p}} \right. \\
&\left. + \left\| \frac{w\Re w}{(1+|w|^2)^2} \right\|_{W^{k,p}} \right) \|\nabla w\|_\infty \\
&+ \left(\left\| \frac{w}{(1+|w|^2)^2} \right\|_\infty + \left\| \frac{w\Im w}{(1+|w|^2)^2} \right\|_\infty \right. \\
&\left. + \left\| \frac{w\Re w}{(1+|w|^2)^2} \right\|_\infty \right) \|\nabla w\|_{W^{k,p}} \bigg] \\
&\le C_1 \big[\|\Re\nabla w\|_{W^{k,p}} + \|\Im\nabla w\|_{W^{k,p}} \\
&\quad + (\|\Re\nabla w\|_\infty + \|\Im\nabla w\|_\infty) \cdot \|w\|_{W^{k,p}} \big] \\
&\quad + C_2 \|w\|_{W^{k,p}} \|\nabla w\|_\infty + C_3 \|\nabla w\|_{W^{k,p}} \\
&\le d_1 \|\nabla w\|_{W^{k,p}} + d_2 \|\nabla w\|_\infty \|w\|_{W^{k,p}}.
\end{aligned}
\tag{4.6.67}
$$

Taking $k = s - 1$, $p = 1$ in (4.6.67), we have

$$
\begin{aligned}
\sup_{t\ge0} \|\nabla\vec{Z}(\cdot,t)\|_{W^{s-1,1}(R^n)} &\le d_1 \sup_{t\ge0} \|\nabla w\|_{W^{s-1,1}} \\
&+ d_2 \sup_{t\ge0} \|w(\cdot,t)\|_{W^{s-n-3,\infty}} \sup_{t\ge0} \|w(\cdot,t)\|_{W^{s-1,1}} < \infty.
\end{aligned}
$$

Taking $k = s + 1$, $p = 2$ in (4.6.67), we have

$$
\begin{aligned}
\|\nabla\vec{Z}(\cdot,t)\|_{H^{s+1}} &\le d_1 \|\nabla w\|_{H^{s+1}} \\
&+ d_2 E(1+t)^{-\frac{n}{2}} \|w\|_{H^{s+1}} \le d\|w\|_{W^{s+2}},
\end{aligned}
\tag{4.6.68}
$$

and

$$
\int_0^T \|\nabla\vec{Z}(\cdot,t)\|_{H^{s+1}}^2 dt \le d^2 \int_0^T \|w(\cdot,t)\|_{W^{s+2}}^2 dt \le d^2 E^2 < \infty.
\tag{4.6.69}
$$

Finally, we take $k = s - n - 4$, $p = \infty$ in (4.6.67) to give

$$
\begin{aligned}
\|\nabla\vec{Z}(\cdot,t)\|_{W^{s-n-4,\infty}} &\le d_1 \|w(\cdot,t)\|_{W^{s-n-3,\infty}} \\
&+ d_2 \|w(\cdot,t)\|_{W^{s-n-3,\infty}} \|w(\cdot,t)\|_{W^{s-n-4,\infty}} \\
&\le d_1 E(1+t)^{-\frac{n}{2}} + d_2 E^2(1+t)^{-n} \\
&< (d_1 E + d_2 E^2)(1+t)^{-\frac{n}{2}}, \quad \forall\ t > 0.
\end{aligned}
\tag{4.6.70}
$$

Direct computations show that $\vec{Z}(x,t)$ satisfies the equation and the initial condition. The theorem is proved.

4.7 Radial Solution

The result of this section shows that the radial symmetric Landau–Lifshitz equations always admits smooth solution outside a cylinder centered at $r = 0$. In other words, the only possible blowing-up must occur at the axis $r = 0$ for the radial symmetric problem.

4.7.1   Two-Dimensional Radial Symmetric Landau–Lifshitz Equation

This section is devoted to the radial solutions of the Cauchy problem

$$u_t = u \times u_{rr} + \frac{1}{r} u \times u_r, \tag{4.7.1}$$

$$u(r, t = 0) = \phi(r), \tag{4.7.2}$$

where $u : \Omega \times R^+ \longrightarrow S^2$, $r = |x|$, $x \in \Omega = \{x \in R^2 \mid |x| > R_0\}$ and u satisfies the following boundary condition:

$$\partial_r u|_{r=R_0} = 0, \tag{4.7.3}$$

where constant $R_0 > 0$. The following results will be established.

Theorem 4.7.1 *Let initial data $\phi(r) \in S^2$ and $\phi_r \in H^m(\Omega)$ ($\forall \, m \geq 1$). Then for all time $T > 0$, there exists a solution of problem (4.7.1)–(4.7.3) such that $u \in S^2$ and $u_{r^{k_1} t^{k_2}}(x, t) \in L^\infty(0, T; \, L^2(\Omega))$, where $1 \leq 2k_2 + k_1 \leq m+1$. Moreover, for all integer $m \geq 3$, the solution is unique.*

4.7.2   *A Priori* Estimates

To prove Theorem 4.7.1, we will construct the local (in time t) solutions of problem (4.7.1)–(4.7.3), and then extend it to large time t by using *a priori* estimates.

Proposition 4.7.1 *There exists a positive constant C such that the following inequality holds for all $u \in H_0^1(R_0, \, \infty)$ and $r^\alpha u$, $r^\alpha u_r \in L^2(R_0, \, \infty)$*

$$\|r^\beta u\|_{L_r^p} \leq C \|r^\alpha u_r\|_{L_r^2}^b \|r^\alpha u\|_{L_r^2}^{1-b} \tag{4.7.4}$$

if and only if the following relations hold:

$$\frac{1}{p} + \beta = \frac{1}{2} + \alpha - b,$$

$$\alpha - \sigma \geq 0 \quad \text{if } b > 0,$$

$$\alpha - \sigma \leq 1 \quad \text{if } b > 0 \text{ and } \alpha - \frac{1}{2} = \frac{1}{p} + \beta,$$

where $p > 0$, $0 \leq b \leq 1$, $\alpha > -\frac{1}{2}$, $\beta > -\frac{1}{p}$ and $\beta = b\sigma + (1 - b)\alpha$.

Lemma 4.7.1 *For all solutions of problem (4.7.1)–(4.7.3), we have the following estimations for all time $T > 0$:*

$$\|r^{1/2} u_r\|_{L^\infty(0,T; \, L_r^2)} = \|r^{1/2} \phi_r\|_{L_r^2}, \tag{4.7.5}$$

$$\|r^{1/2} u_t\|_{L^\infty(0,T;L_r^2)} + \|r^{1/2} u_{rr}\|_{L^\infty(0,T;L_r^2)} \leq C(T, \, R_0, \, \|r^{1/2}\phi_r\|_{L_r^2}, \, \|r^{1/2}\phi_{rr}\|_{L_r^2}), \tag{4.7.6}$$

$$\|u_r\|_{L^\infty(0,T;L^\infty(R_0,\infty))} \leq C\left(T, \, R_0, \, \|r^{1/2}\phi_r\|_{L_r^2}, \, \|r^{1/2}\phi_{rr}\|_{L_r^2}\right), \tag{4.7.7}$$

where $\| \cdot \|_{L_r^2} = \| \cdot \|_{L^2(R_0,\infty)}$.

Proof. Taking the scalar product of ru_{rr} and Eq. (4.7.1), and then integrating the result over $(R_0, \infty) \times [0, T]$, we have (4.7.5).

To prove (4.7.6), we first differentiate (4.7.1) with respect to t, and get

$$u_{tt} = u_t \times u_{rr} + u \times u_{rrt} + \frac{1}{r}u_t \times u_r + \frac{1}{r}u \times u_{rt} \tag{4.7.8}$$

Taking the scalar product of ru_t and Eq. (4.7.8), and then integrating the result over (R_0, ∞), we get

$$\int_{R_0}^{\infty} ru_t \cdot u_{tt}dr = -\int_{R_0}^{\infty} ru_t \cdot (u_r \times u_{rt})dr. \tag{4.7.9}$$

We differentiate (4.7.1) with respect to r, and get

$$u_{rt} = u_r \times u_{rr} + u \times u_{rrr} + \frac{1}{r}u \times u_{rr} - \frac{1}{r^2}u \times u_r. \tag{4.7.10}$$

Substituting (4.7.10) into (4.7.9), we have

$$\int_{R_0}^{\infty} ru_t \cdot u_{tt}dr = \int_{R_0}^{\infty} \left(r(u_r \cdot u_r)u_t \cdot u_{rr} - r(u_r \cdot u_{rr})u_r \cdot u_t \right) dr$$

$$= \frac{1}{8}\frac{d}{dt} \int_{R_0}^{\infty} r(u_r \cdot u_r)^2 dr + \int_{R_0}^{\infty} u_r \cdot u_r(u \times u_r) \cdot u_{rr}dr. \tag{4.7.11}$$

By using (4.7.1), one has

$$\int_{R_0}^{\infty} ru_t \cdot u_t dr = \int_{R_0}^{\infty} \left(ru_{rr} \cdot u_{rr} - r(u_r \cdot u_r)^2 + \frac{1}{r}u_r \cdot u_r \right) dr. \tag{4.7.12}$$

Substituting (4.7.12) into (4.7.11), we get

$$\frac{d}{dt} \int_{R_0}^{\infty} \left(ru_{rr} \cdot u_{rr} + \frac{1}{r}u_r \cdot u_r - \frac{5}{4}r(u_r \cdot u_r)^2 \right) dr = 2 \int_{R_0}^{\infty} u_r \cdot u_r(u \times u_r) \cdot u_{rr}dr. \tag{4.7.13}$$

The right-hand side of (4.7.13) is bounded from above by

$$2 \left(\int_{R_0}^{\infty} r |u_{rr}|^2 \, dr \right)^{1/2} \left(\int_{R_0}^{\infty} \frac{1}{r} |u_r|^6 \, dr \right)^{1/2}$$

$$\leq C\|r^{1/2}u_{rr}\|_{L^2}\|r^{1/2}u_r\|_{L^6}^3 \leq C\|r^{1/2}u_{rr}\|_{L^2}^2\|r^{1/2}u_r\|_{L^2}^2, \tag{4.7.14}$$

where we have used Proposition 4.7.1. The third term on the left-hand side of (4.7.13) is bounded by

$$\frac{5}{4}R_0^{-1}\|r^{1/2}u_r\|_{L^4}^4 \leq C\|r^{1/2}u_{rr}\|_{L^2}\|r^{1/2}u_r\|_{L^2}^3 \leq \frac{1}{2}\|r^{1/2}u_{rr}\|_{L^2}^2 + C\|r^{1/2}u_r\|_{L^2}^6, \tag{4.7.15}$$

where we have used Proposition 4.7.1. Putting (4.7.13)–(4.7.15) together and using Gronwall's inequality, we get

$$\|r^{1/2}u_{rr}\|_{L^\infty(0,T;L_r^2)} \leq C\left(T, R_0, \|r^{1/2}\phi_r\|_{L_r^2}, \|r^{1/2}\phi_{rr}\|_{L_r^2}\right). \tag{4.7.16}$$

Putting (4.7.12), (4.7.15) and (4.7.16) together, we have (4.7.6).

By using Proposition 4.7.1, we have

$$\|u_r\|_{L^\infty(0,T;L^\infty(R_0,\infty))} \le R_0^{-1/4}\|r^{1/4}u_r\|_{L^\infty(0,T;L^\infty(R_0,\infty))}$$
$$\le CR_0^{-1/4}\|r^{1/2}u_{rr}\|_{L^2}^{3/4}\|r^{1/2}u_r\|_{L^2}^{1/4}. \tag{4.7.17}$$

Putting (4.7.5), (4.7.6) and (4.7.17) together, we get (4.7.7).

Lemma 4.7.2 *For all solutions of problem* (4.7.1)–(4.7.3), *we have the following estimations for all time* $T > 0$:

$$\|r^{1/2}u_{tt}\|_{L^\infty(0,T;L_r^2)} + \|r^{1/2}u_{rrt}\|_{L^\infty(0,T;L_r^2)} + \|r^{1/2}u_{r^4}\|_{L^\infty(0,T;L_r^2)}$$
$$\le C\left(T,\ R_0,\ \|r^{1/2}\phi_r\|_{L_r^2},\ldots,\|r^{1/2}\phi_{r^4}\|_{L_r^2}\right), \tag{4.7.18}$$
$$\|u_t\|_{L^\infty(\Omega_T)} + \|u_{rt}\|_{L^\infty(\Omega_T)} + \|u_{rr}\|_{L^\infty(\Omega_T)} + \|u_{r^3}\|_{L^\infty(\Omega_T)} \le C, \tag{4.7.19}$$

where $\|\cdot\|_{L_r^2} = \|\cdot\|_{L^2(R_0,\infty)}$ *and* $\Omega_T = [0,T] \times \Omega$.

Proof. To prove (4.7.18), we first differentiate (4.7.8) with respect to t, and get

$$u_{ttt} = u_{tt} \times \Delta u + 2u_t \times \Delta u_t + u \times \Delta u_{tt}, \tag{4.7.20}$$

where $\Delta w = w_{rr} + \frac{w_r}{r}$. Taking the scalar product of ru_{tt} and Eq. (4.7.20), and then integrating the result over (R_0,∞), we get

$$\int_{R_0}^\infty ru_{ttt} \cdot u_{tt}dr = \int_{R_0}^\infty \{ru_{rtt} \cdot (u_r \times u_{tt}) - 2ru_{rtt} \cdot (u_t \times u_{rt})\}\, dr. \tag{4.7.21}$$

After substituting (4.7.8) and (4.7.10) in (4.7.21), the right-hand side of (4.7.21) becomes

$$\int_{R_0}^\infty ru_r \cdot u_t\, \Delta u \cdot u_{rtt}dr + \int_{R_0}^\infty ru_r \cdot \Delta u\, u_t \cdot u_{rtt}dr$$
$$+ \int_{R_0}^\infty r\left(u_r \cdot \Delta u_t - 2u_t \cdot (\Delta u)_r\right) u \cdot u_{rtt}dr. \tag{4.7.22}$$

Using integration by parts and estimates (4.7.5)–(4.7.7), the first term of (4.7.22) is equal to

$$-\int_{R_0}^\infty r\left\{u_r \cdot u_{rt}\, \Delta u \cdot u_{tt} + u_r \cdot u_t\, (\Delta u)_r \cdot u_{tt}\right\} dr$$

and is bounded from above by

$$C\left(\|r^{1/3}u_{rt}\|_{L^3}\|r^{1/6}\Delta u\|_{L^6} + \|r^{1/6}u_t\|_{L^6}\|r^{1/3}(\Delta u)_r\|_{L^3}\right)\|r^{1/2}u_{tt}\|_{L^2}; \tag{4.7.23}$$

the second term of (4.7.22) is equal to

$$-\int_{R_0}^\infty r\left\{u_r \cdot \Delta u\, u_{rt} \cdot u_{tt} + u_r \cdot (\Delta u)_r\, u_t \cdot u_{tt}\right\} dr,$$

and is also bounded from above by (4.7.23). Using the fact $u \cdot u = 1$, we have

$$u \cdot u_{rtt} = -u_{tt} \cdot u_r - 2u_t \cdot u_{rt}. \tag{4.7.24}$$

Substituting (4.7.24) in (4.7.22), the third term of (4.7.22) is bounded from above by

$$C\left(\|r^{1/2}u_{tt}\|_{L^2} + \|r^{1/3}u_{rt}\|_{L^3}\|r^{1/6}u_t\|_{L^6}\right)\left(\|r^{1/2}u_{rrt}\|_{L^2} + 1\right)$$
$$+ C\|r^{1/6}u_t\|_{L^6}\|r^{1/3}(\Delta u)_r\|_{L^3}\|r^{1/2}u_{tt}\|_{L^2} + C\|r^{1/6}u_t\|_{L^6}^2\|r^{1/3}(\Delta u)_r\|_{L^3}\|r^{1/3}u_{rt}\|_{L^3}, \tag{4.7.25}$$

where we have used the following inequality:

$$\|r^{1/2}u_{rt}\|_{L^2} \le C\left(\|r^{1/2}u_{rrt}\|_{L^2}^{1/2} + 1\right). \tag{4.7.26}$$

In fact, we have

$$\|r^{1/2}u_{rt}\|_{L^2}^2 = \int_{R_0}^{\infty} r u_{rt} \cdot u_{rt} dr = -\int_{R_0}^{\infty}\left(u_t \cdot u_{rt} + r u_t \cdot u_{rrt}\right) dr$$
$$\le \frac{1}{2}\|r^{1/2}u_{rt}\|_{L^2}^2 + C\|r^{1/2}u_t\|_{L^2}^2 + \|r^{1/2}u_{rrt}\|_{L^2}\|r^{1/2}u_t\|_{L^2}.$$

Using Proposition 4.7.1 and (4.7.25), one has

$$\|r^{1/3}u_{rt}\|_{L^3} \le C\|r^{1/2}u_{rrt}\|_{L^2}^{1/3}\|r^{1/2}u_{rt}\|_{L^2}^{2/3} \le C\left(\|r^{1/2}u_{rrt}\|_{L^2}^{2/3} + 1\right). \tag{4.7.27}$$

By using (4.7.1), we get

$$u_t \cdot u_t = \Delta u \cdot \Delta u - (u_r \cdot u_r)^2. \tag{4.7.28}$$

Using (4.7.5)–(4.7.7), (4.7.28) and Sobolev's imbedding theorems, we have

$$\|r^{1/6}u_t\|_{L^6} \le C\left(\|r^{1/6}\Delta u\|_{L^6} + 1\right) \le C\left(\|r^{1/6}u_{rr}\|_{L^6} + 1\right). \tag{4.7.29}$$

Similarly, using (4.7.10), (4.7.5)–(4.7.7) and Sobolev's imbedding theorems, one has

$$(\Delta u)_r \cdot (\Delta u)_r = u_{rt} \cdot u_{rt} + (u_r \cdot \Delta u)^2 + \left(u_r \cdot \Delta u - \frac{2}{r}u_r \cdot u_r\right)^2 - u_r \cdot u_r\, \Delta u \cdot \Delta u,$$

$$\|r^{1/3}(\Delta u)_r\|_{L^3} \le C\left(\|r^{1/3}u_{rt}\|_{L^3} + \|r^{1/3}\Delta u\|_{L^3} + 1\right)$$
$$\le C\left(\|r^{1/2}u_{rrt}\|_{L^2}^{2/3} + \|r^{1/6}u_{rr}\|_{L^6}^{1/2} + 1\right). \tag{4.7.30}$$

Using integration by parts, (4.7.27) and estimations (4.7.5)–(4.7.7), we get

$$\|r^{1/6}u_{rr}\|_{L^6}^6 = \int_{R_0}^{\infty} r\left(u_{rr} \cdot u_{rr}\right)^3 dr$$
$$= -\int_{R_0}^{\infty}\left\{u_r \cdot u_{rr}(u_{rr} \cdot u_{rr})^2 + r u_r \cdot u_{rrr}(u_{rr} \cdot u_{rr})^2\right.$$
$$\left. + 2r u_r \cdot u_{rr} u_{rr} \cdot u_{rr} u_{rr} \cdot u_{rrr}\right\} dr$$
$$\le \frac{1}{2}\|r^{1/6}u_{rr}\|_{L^6}^6 + C + C\|r^{1/6}u_{rr}\|_{L^6}^4\|r^{1/3}u_{rrr}\|_{L^3},$$

$$\|r^{1/6}u_{rr}\|_{L^6} \le C(\|r^{1/3}u_{rrr}\|_{L^3}^{1/2} + 1) \le C\left(\|r^{1/3}u_{rt}\|_{L^3}^{1/2} + \|r^{1/3}u_{rr}\|_{L^3}^{1/2} + 1\right)$$

$$\le \frac{1}{2}\|r^{1/6}u_{rr}\|_{L^6} + C\left(\|r^{1/2}u_{rrt}\|_{L^2}^{1/3} + 1\right). \tag{4.7.31}$$

Substituting estimates (4.7.22) and (4.7.23), (4.7.25)–(4.7.27) and (4.7.29)–(4.7.31) into (4.7.21), we get

$$2\frac{d}{dt}\int_{R_0}^{\infty} ru_{tt} \cdot u_{tt}\, dr \le C\left(\|r^{1/2}u_{tt}\|_{L^2}^2 + \|r^{1/2}u_{rrt}\|_{L^2}^2 + 1\right). \tag{4.7.32}$$

Using Eq. (4.7.8), one has

$$u_{tt} \cdot u_{tt} = \Delta u_t \cdot \Delta u_t + 2u_r \cdot u_r\, u_t \cdot \Delta u_t + u_t \cdot u_t\, \Delta u \cdot \Delta u - 4\left(u_r \cdot u_{rt}\right)^2, \tag{4.7.33}$$

where we have used the facts $\Delta u \cdot u_t = 0$, $u \cdot u = 1$, $u \cdot \Delta u = -u_r \cdot u_r$ and $u \cdot \Delta u_t = -2u_r \cdot u_{rt}$. Substituting (4.7.33) into (4.7.32), the left-hand side of (4.7.32) reads

$$\frac{d}{dt}\int_{R_0}^{\infty} r\left\{|u_{tt}|^2 + |\Delta u_t|^2 + 2\,|u_r|^2\, u_t \cdot \Delta u_t + |u_t|^2\,|\Delta u|^2 - 4\left(u_r \cdot u_{rt}\right)^2\right\} dr. \tag{4.7.34}$$

Using (4.7.5)–(4.7.7) and (4.7.26), the second term of (4.7.34) is rewritten as

$$\int_{R_0}^{\infty} r\,|\Delta u_t|^2\, dr = \int_{R_0}^{\infty}\left(r\,|u_{rrt}|^2 + \frac{1}{r}\,|u_{rt}|^2\right) dr \ge \|r^{1/2}u_{rrt}\|_{L^2}^2, \tag{4.7.35}$$

and the third and fifth terms of (4.7.34) is bounded from above by

$$\frac{1}{5}\|r^{1/2}u_{rrt}\|_{L^2}^2 + C. \tag{4.7.36}$$

The fourth term of (4.7.34) is bounded from above by

$$C\left(\|r^{1/4}u_{rr}\|_{L^4}^4 + 1\right) \le C\left(\|r^{1/6}u_{rr}\|_{L^6}^3 + 1\right)$$

$$\le C\left(\|r^{1/2}u_{rrt}\|_{L^2} + 1\right) \le \frac{1}{5}\|r^{1/2}u_{rrt}\|_{L^2}^2 + C, \tag{4.7.37}$$

where we have used (4.7.5)–(4.7.7), (4.7.28) and (4.7.31). Putting (4.7.32) and (4.7.34)–(4.7.37) together and using Gronwall's inequality, we get

$$\|r^{1/2}u_{tt}\|_{L^\infty(0,T;L^2_r)} + \|r^{1/2}u_{rrt}\|_{L^\infty(0,T;L^2_r)} \le C. \tag{4.7.38}$$

Using (4.7.10), (4.7.26) and (4.7.38), one has

$$\|r^{1/2}u_{rt}\|_{L^\infty(0,T;L^2_r)} + \|r^{1/2}(\Delta u)_r\|_{L^\infty(0,T;L^2_r)} \le C. \tag{4.7.39}$$

Differentiating (4.7.10) with respect to r, we get

$$u_{rrt} = \frac{1}{r}u_{rr} \times u_r + 2u_r \times (\Delta u)_r + u \times (\Delta u)_{rr} = I_1 + u \times (\Delta u)_{rr},$$

$$(\Delta u)_{rr} \cdot (\Delta u)_{rr} \le 2u_{rrt} \cdot u_{rrt} + 2I_1 \cdot I_1$$
$$+ \{2u_r \cdot (u_{rr} + \Delta u)_r + u_{rr} \cdot (2u_{rr} + \Delta u)\}^2, \qquad (4.7.40)$$

where we have used the fact

$$u \cdot (\Delta u)_{rr} = -2u_r \cdot (u_{rr} + \Delta u)_r - u_{rr} \cdot (2u_{rr} + \Delta u).$$

Putting together (4.7.38)–(4.7.40), we have (4.7.18). By using (4.7.18) and Proposition 4.7.1, we have

$$\|u_{rt}\|_{L^\infty(0,T;L^\infty(R_0,\infty))} + \|u_{rrr}\|_{L^\infty(0,T;L^\infty(R_0,\infty))} \le C. \qquad (4.7.41)$$

Using (4.7.18) and Sobolev's imbedding theorems, one has

$$\|u_t\|_{L^\infty(0,T;L^\infty(R_0,\infty))} + \|u_{rr}\|_{L^\infty(0,T;L^\infty(R_0,\infty))} \le C. \qquad (4.7.42)$$

Putting (4.7.41) and (4.7.42) together, we get (4.7.19).

Lemma 4.7.3 *For all solutions of problem* (4.7.1)–(4.7.3), *we have the following estimations for all time $T > 0$ and all integer $m \ge 2$:*

$$\|r^{1/2}u_{t^m}\|_{L^\infty(0,T;L_r^2)} + \|r^{1/2}u_{rrt^{m-1}}\|_{L^\infty(0,T;L_r^2)} + \|r^{1/2}u_{r^4t^{m-2}}\|_{L^\infty(0,T;L_r^2)}$$
$$\le C\left(T, R_0, \|r^{1/2}\phi_r\|_{L_r^2}, \ldots, \|r^{1/2}\phi_{r^{2m}}\|_{L_r^2}\right), \qquad (4.7.43)$$

$$\|u_{t^{m-1}}\|_{L^\infty(\Omega_T)} + \|u_{rt^{m-1}}\|_{L^\infty(\Omega_T)} + \|u_{rrt^{m-2}}\|_{L^\infty(\Omega_T)} + \|u_{r^3t^{m-2}}\|_{L^\infty(\Omega_T)} \le C, \quad (4.7.44)$$

where $\|\cdot\|_{L_r^2} = \|\cdot\|_{L^2(R_0,\infty)}$ and Ω_T is defined in Lemma 4.7.2.

Proof. This lemma is proved by mathematical induction as follows. For $m = 2$, estimates (4.7.43) and (4.7.43) have been proved in Lemma 4.7.2. Suppose that (4.7.43) and (4.7.43) are valid for all $2 \le m \le M$.

Differentiating (4.7.1), we have

$$u_{t^{M+2}} = \sum_{k=0}^{M+1} \binom{M+1}{k} u_{t^{M+1-k}} \times \Delta u_{t^k}, \qquad (4.7.45)$$

where $\Delta w = w_{rr} + \frac{w_r}{r}$. Taking the scalar product of $ru_{t^{M+1}}$ and Eq. (4.7.45), and then integrating the result over (R_0, ∞), we get

$$\int_{R_0}^\infty ru_{t^{M+2}} \cdot u_{t^{M+1}} dr = (M+1)\int_{R_0}^\infty ru_{t^{M+1}} \cdot (u_{t^M} \times \Delta u_t)\, dr$$
$$+ \int_{R_0}^\infty \sum_{k=2}^M \binom{M+1}{k} ru_{t^{M+1}} \cdot (u_{t^{M+1-k}} \times \Delta u_{t^k})\, dr$$
$$+ \int_{R_0}^\infty ru_{t^{M+1}} \cdot (u \times \Delta u_{t^{M+1}})\, dr = I_1 + I_2 + I_3. \quad (4.7.46)$$

By the induction assumption and Sobolev's imbedding theorems, we obtain

$$I_1 \leq C\|u_{t^M}\|_{L^\infty(\Omega)}\|r^{1/2}\Delta u_t\|_{L^2}\|r^{1/2}u_{t^{M+1}}\|_{L^2}$$
$$\leq C\left(1 + \|r^{1/2}\Delta u_{t^M}\|_{L^2}\right)\|r^{1/2}u_{t^{M+1}}\|_{L^2}, \tag{4.7.47}$$

$$I_2 \leq C\sum_{k=2}^{M}\|u_{t^{M+1-k}}\|_{L^\infty(\Omega)}\|r^{1/2}\Delta u_{t^k}\|_{L^2}\|r^{1/2}u_{t^{M+1}}\|_{L^2} \leq C\|r^{1/2}u_{t^{M+1}}\|_{L^2}. \tag{4.7.48}$$

Integrating by parts, one has

$$I_3 = \int_{R_0}^{\infty} r\left(u_r \times u_{t^{M+1}}\right) \cdot u_{rt^{M+1}} \, dr. \tag{4.7.49}$$

Differentiating (4.7.1), we have

$$u_{t^{M+1}} = \sum_{k=0}^{M}\binom{M}{k} u_{t^{M-k}} \times \Delta u_{t^k}, \tag{4.7.50}$$

$$u_r \times u_{t^{M+1}} = \sum_{k=0}^{M}\binom{M}{k}\left(u_r \cdot \Delta u_{t^k}\, u_{t^{M-k}} - u_r \cdot u_{t^{M-k}}\, \Delta u_{t^k}\right). \tag{4.7.51}$$

Using the fact that $u \cdot u = 1$, one has

$$u \cdot u_{rt^{M+1}} = -\sum_{k=0}^{M}\binom{M+1}{k} u_{t^{M+1-k}} \cdot u_{rt^k}. \tag{4.7.52}$$

Substituting (4.7.51) and (4.7.52) into (4.7.49) and integrating by parts, $-I_3$ becomes

$$\int_{R_0}^{\infty}\sum_{k=0}^{M-1}\binom{M}{k}\left(u_r \cdot \Delta u_{t^k}\, u_{t^{M-k}} + r u_{rr} \cdot \Delta u_{t^k}\, u_{t^{M-k}} + r u_r \cdot \Delta u_{t^k}\, u_{rt^{M-k}}\right.$$
$$\left. - u_r \cdot u_{t^{M-k}}\, \Delta u_{t^k} - r u_{rr} \cdot u_{t^{M-k}}\, \Delta u_{t^k} - r u_r \cdot u_{rt^{M-k}}\, \Delta u_{t^k}\right) \cdot u_{t^{M+1}} \, dr$$
$$+ \int_{R_0}^{\infty}\sum_{k=0}^{M-2}\binom{M}{k}\left(r u_r \cdot (\Delta u)_{rt^k}\, u_{t^{M-k}} - r u_r \cdot u_{t^{M-k}}\, (\Delta u)_{rt^k}\right) \cdot u_{t^{M+1}} \, dr$$
$$+ \int_{R_0}^{\infty}\left(r u_r \cdot (\Delta u)_{rt^{M-1}}\, u_t - r u_r \cdot u_t\, (\Delta u)_{rt^{M-1}}\right) \cdot u_{t^{M+1}} \, dr$$
$$+ \int_{R_0}^{\infty}\sum_{k=0}^{M}\binom{M+1}{k} r u_r \cdot \Delta u_{t^M}\, u_{t^{M+1-k}} \cdot u_{rt^k} \, dr. \tag{4.7.53}$$

By the induction assumption, the integrations corresponding to the first integral operator and the second integral operator of (4.7.53) is bounded from above by

$$C\|r^{1/2}u_{t^{M+1}}\|_{L^2}; \tag{4.7.54}$$

and the integration corresponding to the fourth integral operator of (4.7.53) is bounded from above by

$$C\|r^{1/2}u_{rrt^M}\|_{L^2}\left(\|r^{1/2}u_{t^{M+1}}\|_{L^2} + \|r^{1/2}u_{rt^M}\|_{L^2} + 1\right). \tag{4.7.55}$$

Differentiating (4.7.1), we have

$$u_{rt^M} = u \times (\Delta u)_{rt^{M-1}} + u_r \times \Delta u_{t^{M-1}} + \sum_{k=0}^{M-2} \binom{M-1}{k} \left(u_{t^{M-1-k}} \times \Delta u_{t^k}\right)_r, \quad (4.7.56)$$

$$\left(u \times (\Delta u)_{rt^{M-1}}\right) \cdot \left(u \times (\Delta u)_{rt^{M-1}}\right) = |(\Delta u)_{rt^{M-1}}|^2 - \left(u \cdot (\Delta u)_{rt^{M-1}}\right)^2. \quad (4.7.57)$$

By the fact $u \cdot u = 1$, we have

$$u \cdot (\Delta u)_{rt^{M-1}} = -\left(2u_r \cdot u_{rr} + u_r \cdot \Delta u\right)_{t^{M-1}} - \sum_{k=0}^{M-2} \binom{M-1}{k} u_{t^{M-1-k}} \times (\Delta u)_{rt^k}. \quad (4.7.58)$$

Putting (4.7.56)–(4.7.58) together and by the induction assumption, one has

$$\|r^{1/2}(\Delta u)_{rt^{M-1}}\|_{L^2} \leq C \left(1 + \|r^{1/2}u_{rt^M}\|_{L^2}\right). \quad (4.7.59)$$

By the same argument as in the proof of (4.7.26), we obtain

$$\|r^{1/2}u_{rt^M}\|_{L^2} \leq C \left(1 + \|r^{1/2}u_{rrt^M}\|_{L^2}^{1/2}\right). \quad (4.7.60)$$

Using (4.7.59) and (4.7.60), the integration corresponding to the third integral operator of (4.7.53) is bounded from above by

$$C\|r^{1/2}u_{t^{M+1}}\|_{L^2}\|r^{1/2}(\Delta u)_{rt^{M-1}}\|_{L^2} \leq C\|r^{1/2}u_{t^{M+1}}\|_{L^2}\left(1 + \|r^{1/2}u_{rrt^M}\|_{L^2}^{1/2}\right). \quad (4.7.61)$$

Substituting all estimates (4.7.47)–(4.7.49), (4.7.53)–(4.7.55), (4.7.60) and (4.7.61) in (4.7.46), we get

$$2\frac{d}{dt}\int_{R_0}^{\infty} r u_{t^{M+1}} \cdot u_{t^{M+1}} \, dr \leq C \left(\|r^{1/2}u_{t^{M+1}}\|_{L^2}^2 + \|r^{1/2}u_{rrt^M}\|_{L^2}^2 + 1\right). \quad (4.7.62)$$

By the fact $u \cdot u = 1$, we obtain

$$\left(u \times \Delta u_{t^M}\right) \cdot \left(u \times \Delta u_{t^M}\right) = \Delta u_{t^M} \cdot \Delta u_{t^M} - \left(u \cdot \Delta u_{t^M}\right)^2, \quad (4.7.63)$$

$$u \cdot \Delta u_{t^M} = -(u_r \cdot u_r)_{t^M} - \sum_{k=0}^{M-1} \binom{M}{k} \left(u_{t^{M-k}} \cdot \Delta u_{t^k}\right). \quad (4.7.64)$$

Using Eqs. (4.7.50), (4.7.63) and (4.7.64), and the induction assumption, one has

$$\|r^{1/2}u_{rrt^M}\|_{L^2}^2 \leq C \left(\|r^{1/2}u_{t^{M+1}}\|_{L^2}^2 + \|r^{1/2}u_{rt^M}\|_{L^2}^2 + 1\right). \quad (4.7.65)$$

By substituting (4.7.60) and (4.7.65) into (4.7.62), (4.7.62) can be rewritten as

$$\frac{d}{dt}\int_{R_0}^{\infty} r \left(|u_{t^{M+1}}|^2 + |u_{rrt^M}|^2\right) dr \leq C \left(\|r^{1/2}u_{t^{M+1}}\|_{L^2}^2 + \|r^{1/2}u_{rrt^M}\|_{L^2}^2 + 1\right). \quad (4.7.66)$$

Using Gronwall's inequality, we get

$$\|r^{1/2}u_{t^{M+1}}\|_{L^2}^2 + \|r^{1/2}u_{rrt^M}\|_{L^2}^2 \leq C \left(T, \, R_0, \, \|r^{1/2}\phi_r\|_{L_r^2}, \ldots, \|r^{1/2}\phi_{r^{2M+2}}\|_{L_r^2}\right). \quad (4.7.67)$$

Using (4.7.59) and (4.7.60), one has

$$\|r^{1/2}u_{rt^M}\|_{L^\infty(0,T;L^2_r)} + \|r^{1/2}(\Delta u)_{rt^{M-1}}\|_{L^\infty(0,T;L^2_r)} \leq C. \qquad (4.7.68)$$

Differentiating (4.7.56) with respect to r, we get

$$u_{rrt^M} = u \times (\Delta u)_{rrt^{M-1}} + 2u_r \times (\Delta u)_{rt^{M-1}} + u_{rr} \times \Delta u_{t^{M-1}}$$
$$+ \sum_{k=0}^{M-2} \binom{M-1}{k} (u_{t^{M-1-k}} \times \Delta u_{t^k})_{rr}, \qquad (4.7.69)$$

$$(u \times (\Delta u)_{rrt^{M-1}}) \cdot (u \times (\Delta u)_{rrt^{M-1}}) = |(\Delta u)_{rrt^{M-1}}|^2 - (u \cdot (\Delta u)_{rrt^{M-1}})^2. \qquad (4.7.70)$$

By the fact $u \cdot u = 1$, we have

$$u \cdot (\Delta u)_{rrt^{M-1}} = -\left(2u_r \cdot u_{rrr} + 2u_{rr} \cdot u_{rr} + 2u_r \cdot (\Delta u)_r + u_{rr} \cdot \Delta u\right)_{t^{M-1}}$$
$$- \sum_{k=0}^{M-2} \binom{M-1}{k} u_{t^{M-1-k}} \times (\Delta u)_{rrt^k}. \qquad (4.7.71)$$

Putting (4.7.68)–(4.7.71) together and by the induction assumption, one has

$$\|r^{1/2}(\Delta u)_{rrt^{M-1}}\|_{L^2} \leq C\left(1 + \|r^{1/2}u_{rrt^M}\|_{L^2}\right) \leq C. \qquad (4.7.72)$$

By using (4.7.67), (4.7.68), (4.7.72) and the same arguments in the proof of (4.7.41) and (4.7.42), we have

$$\|u_{t^M}\|_{L^\infty(\Omega_T)} + \|u_{rt^M}\|_{L^\infty(\Omega_T)} + \|u_{rrt^{M-1}}\|_{L^\infty(\Omega_T)} + \|u_{r^3t^{M-1}}\|_{L^\infty(\Omega_T)} \leq C. \qquad (4.7.73)$$

Thus putting (4.7.67), (4.7.72) and (4.7.73) together and using the induction procedure, one finishes the proof.

Corollary 4.7.1 *For all solutions of problem* (4.7.1)–(4.7.3), *we have the following estimations for all time $T > 0$ and all integer $m \geq 0$:*

$$\|u_{r^{k_1}t^{k_2}}\|_{L^\infty(0,T;\,L^2(\Omega))} \leq C\left(R_0,\ T,\ \|\phi_r\|_{H^m(\Omega)}\right), \qquad (4.7.74)$$

where $1 \leq 2k_2 + k_1 \leq m + 1$.

By using the results of Lemmas 4.7.1–4.7.3, repeating the same procedure used in the proof of 4.7.73, and employing the method of induction, Corollary 4.7.1 can be established. We shall not reproduce the procedure here.

4.7.3   Existence of Local Solutions

In the above section we have done the *a priori* estimate. To obtain the global smooth solution, we only need to construct the local (in time t) solution u of problem (4.7.1)–(4.7.3). This local solution u is the limit of sequence $\{u_h\}$ when h tends to zero, where sequence $\{u_h\}$ satisfies the following ordinary differential-difference system

$$\frac{du_j}{dt} = u_j \times \frac{D_+D_-u_j}{h^2} + \frac{1}{r_j}u_j \times \frac{D_+u_j}{h}, \quad j = 1, \ldots, \qquad (4.7.75)$$

with initial data

$$u_j(0) = \phi_j = \phi(r_j), \quad j = 0, 1, \ldots, \tag{4.7.76}$$

and boundary conditions

$$D_+ u_0 - 0, \tag{4.7.77}$$

where h is step size, $0 < h < 1$, $r_j = R_0 + jh$, $u_j = u(r_j, t)$ $(j = 0, 1, \ldots)$, and D_+ and D_- denote the forward and backward difference operators, respectively. It is well-known that there exists a local smooth solution of problem (4.7.75)–(4.7.77), $u_h = \{u_j = u(r_j, t) | j = 0, 1, \ldots\}$. In order to verify the local existence of problem (4.7.1)–(4.7.3), it suffices to get the uniform $a\ priori$ estimates of u_h with respect to h.

Let us define $\delta u_h = \{\frac{D_+ u_j}{h} | j = 0, 1, \ldots\} = \{\frac{D_- u_j}{h} | j = 1, \ldots\}$. Similarly, the discrete functions $\delta^k u_h$ $(k \geq 2)$ can be defined. The norms of discrete functions $\delta^k u_h$ $(k \geq 0)$ are defined as follows:

$$\|r^\beta \delta^k u_h\|_p = \left(\sum_{j=0}^{\infty} \left| r_j^\beta \frac{D_+^k u_j}{h^k} \right|^p h \right)^{1/p} = \left(\sum_{j=l}^{\infty} \left| r_j^\beta \frac{D_+^{k-l} D_-^l u_j}{h^k} \right|^p h \right)^{1/p},$$

$$\|r^\beta \delta^k u_h\|_\infty = \sup_{j=0,1,\ldots} \left| r_j^\beta \frac{D_+^k u_j}{h^k} \right| = \sup_{j \geq l} \left| r_j^\beta \frac{D_+^{k-l} D_-^l u_j}{h^k} \right|,$$

where $1 \leq p < \infty$ and $\beta \geq 0$.

First we are going to derive the following discrete versions of Sobolev and Gagliardo–Nirenberg type inequalities.

Lemma 4.7.4 *For any discrete function $v_h = \{v_0, v_1, \ldots\}$ with $v_0 = 0$, the following inequality holds:*

$$\|r^\beta v_h\|_p \leq C \|r^\beta v_h\|_2^{\frac{1}{2} + \frac{1}{p}} \left(\|r^\beta v_h\|_2 + \|r^\beta \delta v_h\|_2 \right)^{\frac{1}{2} - \frac{1}{p}}, \tag{4.7.78}$$

where $2 \leq p \leq \infty$, $\beta \geq 0$, constant C is independent of the discrete functions v_h and the step size h.

Proof. When $\beta = 0$, (4.7.78) is well known, which has been obtained in [6, 15, 16]. When $\beta > 0$, for any $m \in \{0, 1, \ldots\}$, we have

$$r_m^{2\beta} v_m \cdot v_m = \sum_{j=0}^{m-1} \left(r_{j+1}^\beta v_{j+1} + r_j^\beta v_j \right) D_+ \left(r_j^\beta v_j \right) \leq C \|r^\beta v_h\|_2 \{ \|r^\beta \delta v_h\|_2 + \|r^\beta v_h\|_2 \}.$$

This completes the proof of (4.7.78) for $p = \infty$. For $2 \leq p < \infty$, it is clear that

$$\|r^\beta v_h\|_p \leq \|r^\beta v_h\|_\infty^{1 - \frac{2}{p}} \|r^\beta v_h\|_2^{\frac{2}{p}} \leq C \|r^\beta v_h\|_2^{\frac{1}{2} + \frac{1}{p}} \left(\|r^\beta v_h\|_2 + \|r^\beta \delta v_h\|_2 \right)^{\frac{1}{2} - \frac{1}{p}}.$$

The proof of this lemma is complete.

Lemma 4.7.5 *For any discrete function* $v_h = \{v_0, v_1, \ldots\}$ *with* $\|r^\beta v_h\|_2 < \infty$, *the following inequality holds:*

$$\|r^\beta v_h\|_p \leq C\|r^\beta v_h\|_2^{\frac{1}{2}+\frac{1}{p}} \left(\|r^\beta v_h\|_2 + \|r^\beta \delta v_h\|_2\right)^{\frac{1}{2}-\frac{1}{p}}, \qquad (4.7.79)$$

where $2 \leq p \leq \infty$, $\beta \geq 0$, *constant* C *is independent of the discrete functions* v_h *and the step size* h.

Proof. Since $\|r^\beta v_h\|_2 < \infty$, for any $m \in \{0, 1, \ldots\}$, we have

$$r_m^{2\beta} v_m \cdot v_m = -\sum_{j=m}^{\infty} \left(r_{j+1}^\beta v_{j+1} + r_j^\beta v_j\right) D_+ \left(r_j^\beta v_j\right) \leq C\|r^\beta v_h\|_2\{\|r^\beta \delta v_h\|_2 + \|r^\beta v_h\|_2\}.$$

Thus, repeating the same procedure in the proof of Lemma 4.7.4, this lemma is proved.

Lemma 4.7.6 *Let* $\phi(x) = \phi(|x|) \in S^2$, $x \in \Omega$, $\nabla\phi(x) \in L^2(\Omega)$, *and suppose that* u_h *is the solution of problem* (4.7.75)–(4.7.77). *Then* $u_j(t) \in S^2$ *for all* $t \geq 0$ *and*

$$\|r^{1/2}\delta u_h\|_2 \leq C, \qquad \forall\ t \geq 0, \qquad (4.7.80)$$

where constant C *is independent of* h.

Proof. Taking the scalar product of (4.7.75) and u_j, we obtain that $u_j \cdot u_{jt} = 0$ $(j = 1, \ldots)$. Using that $\phi \in S^2$, we have $u_j \in S^2$ $(j = 0, 1, \ldots)$.

In addition, taking the scalar product of (4.7.75) and $r_j \frac{D_+ D_- u_j}{h^2}$, and then summing the result over j from 1 to ∞, we have

$$\sum_{j=1}^{\infty} r_j \frac{D_+ D_- u_j}{h^2} \cdot u_{jt} h = \sum_{j=1}^{\infty} u_j \times \frac{D_+ u_j}{h} \cdot \frac{D_+ D_- u_j}{h^2} h. \qquad (4.7.81)$$

By direct calculation, the left-hand side of (4.7.81) is equal to

$$-\frac{1}{2}\frac{d}{dt}\sum_{j=1}^{\infty} r_{j+1}\frac{D_+ u_j}{h} \cdot \frac{D_+ u_j}{h} h - \sum_{j=1}^{\infty} u_j \times \frac{D_+ D_- u_j}{h^2} \cdot \frac{D_+ u_j}{h} h. \qquad (4.7.82)$$

Putting (4.7.81) and (4.7.82) together, we obtain (4.7.80).

Lemma 4.7.7 *Suppose that the conditions of Lemma* 4.7.7 *are satisfied and* $\nabla\phi(x) \in H^2(\Omega)$. *Then there exists a constant* $T_0 = T_0(\|\nabla\phi\|_{H^2}) > 0$ *such that*

$$\sup_{0 \leq t \leq T_0} \left(\|r^{1/2}u_{ht}\|_2 + \|r^{1/2}\delta^2 u_h\|_2\right) \leq C, \qquad (4.7.83)$$

$$\sup_{0 \leq t \leq T_0} \left(\|r^{1/2}\delta u_{ht}\|_2 + \|r^{1/2}\delta^3 u_h\|_2\right) \leq C, \qquad (4.7.84)$$

$$\sup_{0 \leq t \leq T_0} \left(\|r^{1/2}\delta u_h\|_\infty + \|r^{1/2}u_{ht}\|_\infty + \|r^{1/2}\delta^2 u_h\|_\infty\right) \leq C, \qquad (4.7.85)$$

where constants C *and* T_0 *are independent of* h.

Proof. We first differentiate (4.7.75) with respect to t

$$u_{jtt} = u_j \times \left(\frac{D_+ D_- u_{jt}}{h^2} + \frac{D_+ u_{jt}}{r_j h} \right) + u_{jt} \times \left(\frac{D_+ D_- u_j}{h^2} + \frac{D_+ u_j}{r_j h} \right), \quad j = 1, 2, \dots. \quad (4.7.86)$$

Taking the scalar product of (4.7.86) and $r_j u_{jt}$, and then summing the result over j from 1 to ∞, we have

$$\sum_{j=1}^{\infty} r_j u_{jt} \cdot u_{jtt} h = \sum_{j=1}^{\infty} r_j u_j \times \frac{D_+ D_- u_{jt}}{h^2} \cdot u_{jt} h + \sum_{j=1}^{\infty} u_j \times \frac{D_+ u_{jt}}{h} \cdot u_{jt} h$$

$$= -\sum_{j=1}^{\infty} r_{j+1} \frac{D_+ u_j}{h} \times \frac{D_+ u_{jt}}{h} \cdot u_{jt} h \le C \|\delta u_h\|_{\infty} \|r^{1/2} u_{ht}\|_2 \|r^{1/2} \delta u_{ht}\|_2. \quad (4.7.87)$$

Taking the scalar product of (4.7.86) and $r_j \frac{D_+ D_- u_{jt}}{h^2}$, and then summing the result over j from 1 to ∞, we have

$$\sum_{j=1}^{\infty} r_j u_{jtt} \cdot \frac{D_+ D_- u_{jt}}{h^2} h$$

$$= \sum_{j=1}^{\infty} r_j u_{jt} \times \frac{D_+ D_- u_j}{h^2} \cdot \frac{D_+ D_- u_{jt}}{h^2} h + \sum_{j=1}^{\infty} u_{jt} \times \frac{D_+ u_j}{h} \cdot \frac{D_+ D_- u_{jt}}{h^2} h$$

$$+ \sum_{j=1}^{\infty} u_j \times \frac{D_+ u_{jt}}{h} \cdot \frac{D_+ D_- u_{jt}}{h^2} h. \quad (4.7.88)$$

Using (4.7.86), the left-hand side of (4.7.88) is equal to

$$-\sum_{j=1}^{\infty} r_{j+1} \frac{D_+ u_{jt}}{h} \cdot \frac{D_+ u_{jtt}}{h} h - \sum_{j=1}^{\infty} \frac{D_+ u_{jt}}{h} \cdot u_{jtt} h$$

$$= -\frac{1}{2} \frac{d}{dt} \sum_{j=1}^{\infty} r_{j+1} \frac{D_+ u_{jt}}{h} \cdot \frac{D_+ u_{jt}}{h} h - \sum_{j=1}^{\infty} u_{jt} \times \frac{D_+ D_- u_j}{h^2} \cdot \frac{D_+ u_{jt}}{h} h$$

$$+ \sum_{j=1}^{\infty} u_j \times \frac{D_+ u_{jt}}{h} \cdot \frac{D_+ D_- u_{jt}}{h^2} h - \sum_{j=1}^{\infty} \frac{1}{r_j} u_{jt} \times \frac{D_+ u_j}{h} \cdot \frac{D_+ u_{jt}}{h} h; \quad (4.7.89)$$

the first term on the right-hand side of (4.7.88) is equal to

$$-\sum_{j=1}^{\infty} r_{j+1} u_{j+1\,t} \times \frac{D_+^2 D_- u_j}{h^3} \cdot \frac{D_+ u_{jt}}{h} h - \sum_{j=1}^{\infty} u_{jt} \times \frac{D_+ D_- u_j}{h^2} \cdot \frac{D_+ u_{jt}}{h} h; \quad (4.7.90)$$

the second term on the right-hand side of (4.7.88) is equal to

$$-\sum_{j=1}^{\infty} u_{jt} \times \frac{D_+^2 u_j}{h^2} \cdot \frac{D_+ u_{jt}}{h} h. \quad (4.7.91)$$

Substituting (4.7.89)–(4.7.91) into (4.7.88), we have

$$
\frac{1}{2}\frac{d}{dt}\sum_{j=1}^{\infty} r_{j+1}\frac{D_+ u_{jt}}{h}\cdot\frac{D_+ u_{jt}}{h}h
$$
$$
= \sum_{j=1}^{\infty} r_{j+1}u_{j+1\,t}\times\frac{D_+^2 D_- u_j}{h^3}\cdot\frac{D_+ u_{jt}}{h}h + \sum_{j=1}^{\infty} u_{jt}\times\frac{D_+^2 u_j}{h^2}\cdot\frac{D_+ u_{jt}}{h}h
$$
$$
- \sum_{j=1}^{\infty}\frac{1}{r_j}u_{jt}\times\frac{D_+ u_j}{h}\cdot\frac{D_+ u_{jt}}{h}h. \tag{4.7.92}
$$

The third term on the right-hand side of (4.7.92) is bounded from above by

$$
C\|\delta u_h\|_\infty\|r^{1/2}u_{ht}\|_2\|r^{1/2}\delta u_{ht}\|_2. \tag{4.7.93}
$$

Substituting (4.7.75) into (4.7.92), the second term on the right-hand side of (4.7.92) is bounded from above by

$$
C\left(\|\delta u_h\|_\infty\|r^{1/2}\delta^2 u_h\|_2 + \|r^{1/2}\delta^2 u_h\|_4^2\right)\|r^{1/2}\delta u_{ht}\|_2. \tag{4.7.94}
$$

Using (4.7.75), one has

$$
\frac{D_+ u_{jt}}{h} = u_{j+1}\times\frac{D_+^2 D_- u_j}{h^3} + \frac{D_+ u_j}{h}\times\frac{D_+ D_- u_j}{h^2}
$$
$$
+ \frac{u_{j+1}}{r_{j+1}}\times\frac{D_+ D_- u_j}{h^2} - \frac{u_j}{r_j r_{j+1}}\times\frac{D_+ u_j}{h}
$$
$$
= u_{j+1}\times\frac{D_+^2 D_- u_j}{h^3} + I_1^j, \tag{4.7.95}
$$
$$
\frac{D_+^2 D_- u_j}{h^3}\times\frac{D_+ u_{jt}}{h} = \frac{D_+^2 D_- u_j}{h^3}\cdot\frac{D_+^2 D_- u_j}{h^3}u_{j+1}
$$
$$
- u_{j+1}\cdot\frac{D_+^2 D_- u_j}{h^3}\frac{D_+^2 D_- u_j}{h^3} + \frac{D_+^2 D_- u_j}{h^3}\times I_1^j. \tag{4.7.96}
$$

Using the fact $u_j\in S^2$, we get

$$
u_{j+1}\cdot\frac{D_+^2 D_- u_j}{h^3} = -\frac{D_+ u_{j+1}+D_- u_{j+1}}{2h}\cdot\frac{D_+ D_- u_{j+1}}{h^2} - \frac{3D_+ u_j + D_- u_j}{2h}\cdot\frac{D_+ D_- u_j}{h^2}. \tag{4.7.97}
$$

Substituting (4.7.75), (4.7.96), and (4.7.97) in (4.7.92), the first term on the right-hand side of (4.7.92) is bounded from above by

$$
C\left(\|\delta u_h\|_\infty + 1\right)\|r^{1/2}\delta^2 u_h\|_4^2\|r^{1/2}\delta^3 u_h\|_2
$$
$$
+ C\left(\|\delta u_h\|_\infty^2 + \|\delta u_h\|_\infty\right)\|r^{1/2}\delta^2 u_h\|_2\|r^{1/2}\delta^3 u_h\|_2
$$
$$
+ C\|\delta u_h\|_\infty\|r^{1/2}u_{ht}\|_2\|r^{1/2}\delta^3 u_h\|_2. \tag{4.7.98}
$$

Using the fact $u_j \in S^2$, we get

$$u_j \cdot \frac{D_+ D_- u_j}{h^2} = -\frac{1}{2}\left(\frac{D_+ u_j}{h} \cdot \frac{D_+ u_j}{h} + \frac{D_- u_j}{h} \cdot \frac{D_- u_j}{h}\right). \tag{4.7.99}$$

Using (4.7.75) and (4.7.99), we infer that

$$u_{jt} \cdot u_{jt} \geq \frac{1}{2}\frac{D_+ D_- u_j}{h^2} \cdot \frac{D_+ D_- u_j}{h^2} - C\left|\frac{D_+ u_j}{h}\right|^4 - C\left|\frac{D_- u_j}{h}\right|^4 - \frac{C}{r_j^2}\left|\frac{D_+ u_j}{h}\right|^2. \tag{4.7.100}$$

Similarly, using (4.7.95) and (4.7.97), we have

$$\frac{D_+ u_{jt}}{h} \cdot \frac{D_+ u_{jt}}{h} \geq \frac{1}{2}\frac{D_+^2 D_- u_j}{h^3} \cdot \frac{D_+^2 D_- u_j}{h^3} - C\sum_{k=j}^{j+1}\left(\left|\frac{D_+ u_k}{h}\right|^2 + \left|\frac{D_- u_k}{h}\right|^2\right)\left|\frac{D_+ D_- u_k}{h^2}\right|^2$$

$$- \frac{C}{r_{j+1}^2}\left|\frac{D_+ D_- u_j}{h^2}\right|^2 - \frac{C}{r_j^2 r_{j+1}^2}\left|\frac{D_+ u_j}{h}\right|^2. \tag{4.7.101}$$

Using (4.7.77) and Lemma 4.7.5, we get

$$\|\delta u_h\|_\infty \leq C\|\delta u_h\|_2^{1/2}\|\delta^2 u_h\|_2^{1/2} \leq C\|r^{1/2}\delta u_h\|_2^{1/2}\|r^{1/2}\delta^2 u_h\|_2^{1/2}. \tag{4.7.102}$$

By direct calculation, we have

$$\|r^{1/2}\delta^2 u_h\|_2^2 = -\sum_{j=1}^\infty D_+\left(r_j\frac{D_+ D_- u_j}{h^2}\right) \cdot \frac{D_+ u_j}{h}h$$

$$\leq C\left(\|r^{1/2}\delta^2 u_h\|_2 + \|r^{1/2}\delta^3 u_h\|_2\right), \tag{4.7.103}$$

$$\|r^{1/2}\delta^2 u_h\|_4^4 = -\sum_{j=1}^\infty D_+\left(r_j^2\frac{D_+ D_- u_j}{h^2} \cdot \frac{D_+ D_- u_j}{h^2}\right)\frac{D_+ D_- u_{j+1}}{h^2} \cdot \frac{D_+ u_j}{h}h$$

$$- \sum_{j=1}^\infty r_j^2\frac{D_+ D_- u_j}{h^2} \cdot \frac{D_+ D_- u_j}{h^2}\frac{D_+^2 D_- u_j}{h^3} \cdot \frac{D_+ u_j}{h}h$$

$$\leq C\|r^{1/2}\delta^2 u_h\|_4^2\left(\|r^{1/2}\delta^2 u_h\|_2 + \|r^{1/2}\delta^3 u_h\|_2\right)^{5/4}, \tag{4.7.104}$$

$$\sum_{j=1}^\infty r_j\left(\left|\frac{D_+ u_j}{h}\right|^2 + \left|\frac{D_- u_j}{h}\right|^2\right)\left|\frac{D_+ D_- u_j}{h^2}\right|^2$$

$$\leq C\|\delta u_h\|_\infty^2\|r^{1/2}\delta^2 u_h\|_2^2 \leq C\left(\|r^{1/2}\delta^2 u_h\|_2 + \|r^{1/2}\delta^3 u_h\|_2\right)^{3/2}, \tag{4.7.105}$$

where we have used (4.7.77), (4.7.80), (4.7.102) and (4.7.103). Putting estimates (4.7.87), (4.7.92)–(4.7.94), (4.7.98) and (4.7.100)–(4.7.105) together, for all $t \geq 0$, we finally obtain

$$\|r^{1/2}u_{ht}\|_2^2 + \|r^{1/2}\delta^2 u_h\|_2^2 + \|r^{1/2}\delta u_{ht}\|_2^2 + \|r^{1/2}\delta^3 u_h\|_2^2$$

$$\leq C + C\int_0^t\left(\|r^{1/2}u_{ht}\|_2 + \|r^{1/2}\delta^2 u_h\|_2 + \|r^{1/2}\delta u_{ht}\|_2 + \|r^{1/2}\delta^3 u_h\|_2\right)^3 d\tau.$$

Thus there exists a constant $T_0 = T_0(\|\nabla\phi\|_{H^2}) > 0$ such that

$$\sup_{0 \le t \le T_0} \left(\|r^{1/2}u_{ht}\|_2 + \|r^{1/2}\delta^2 u_h\|_2 + \|r^{1/2}\delta u_{ht}\|_2 + \|r^{1/2}\delta^3 u_h\|_2 \right) \le C.$$

Moreover, estimates (4.7.83) and (4.7.84) are proved. Employing (4.7.80), (4.7.83), (4.7.84), and Lemma 4.7.5, (4.7.85) is proved. This completes the proof of the lemma.

Lemma 4.7.8 *Suppose that the conditions of Lemma 4.7.6 are satisfied and $\nabla\phi(x) \in H^3(\Omega)$. Then*

$$\sup_{0 \le t \le T_0} \left(\|r^{1/2}u_{htt}\|_2 + \|r^{1/2}\delta^2 u_{ht}\|_2 + \|r^{1/2}\delta^4 u_h\|_2 \right) \le C, \tag{4.7.106}$$

$$\sup_{0 \le t \le T_0} \left(\|r^{1/2}\delta u_{ht}\|_\infty + \|r^{1/2}\delta^3 u_h\|_\infty \right) \le C, \tag{4.7.107}$$

where T_0 is defined in Lemma 4.7.7, and constant C is independent of h.

Proof. In what follows, let $0 \le t \le T_0$. We first differentiate (4.7.86) with respect to t

$$u_{jttt} = u_j \times \frac{D_+ D_- u_{jtt}}{h^2} + 2u_{jt} \times \frac{D_+ D_- u_{jt}}{h^2} + u_{jtt} \times \frac{D_+ D_- u_j}{h^2}$$

$$+ \frac{u_j}{r_j} \times \frac{D_+ u_{jtt}}{h} + 2\frac{u_{jt}}{r_j} \times \frac{D_+ u_{jt}}{h} + \frac{u_{jtt}}{r_j} \times \frac{D_+ u_j}{h}, \quad j = 1, 2, \ldots . \tag{4.7.108}$$

Taking the scalar product of (4.7.108) and $r_j u_{jtt}$, and then summing the result over j from 1 to ∞, we have

$$\sum_{j=1}^\infty r_j u_{jtt} \cdot u_{jttt} h = 2\sum_{j=1}^\infty r_j u_{jt} \times \frac{D_+ D_- u_{jt}}{h^2} \cdot u_{jtt} h + 2\sum_{j=1}^\infty u_{jt} \times \frac{D_+ u_{jt}}{h} \cdot u_{jtt} h$$

$$- \sum_{j=1}^\infty r_{j+1} \frac{D_+ u_j}{h} \times \frac{D_+ u_{jtt}}{h} \cdot u_{jtt} h. \tag{4.7.109}$$

Substituting (4.7.75) into (4.7.109) and using (4.7.80), (4.7.83)–(4.7.85), the first term on the right-hand side of (4.7.109) is bounded from above by

$$\|u_{ht}\|_\infty \|r^{1/2}u_{htt}\|_2 \|r^{1/2}\delta^2 u_{ht}\|_2 \le C\|r^{1/2}u_{htt}\|_2 \|r^{1/2}\delta^2 u_{ht}\|_2, \tag{4.7.110}$$

and the second term on the right-hand side of (4.7.109) is bounded from above by

$$\|u_{ht}\|_\infty \|r^{1/2}\delta u_{ht}\|_2 \|r^{1/2}u_{htt}\|_2 \le C\|r^{1/2}u_{htt}\|_2. \tag{4.7.111}$$

Using (4.7.75), we have

$$\frac{D_+ u_{jtt}}{h} = \frac{u_{j+1} + u_j}{2} \times \frac{D_+^2 D_- u_{jt}}{h^3} + \frac{u_{j+1\,t} + u_{jt}}{2} \times \frac{D_+^2 D_- u_j}{h^3} + I_{2t}^j; \tag{4.7.112}$$

$$I_2^j = \frac{D_+ u_j}{h} \times \left(\frac{D_+ D_- (u_{j+1} + u_j)}{2h^2} + \frac{u_j}{r_j r_{j+1}} + \frac{D_+ u_{j+1}}{2r_{j+1} h} \right) + \frac{u_{j+1} + u_j}{2r_{j+1}} \times \frac{D_+^2 u_j}{h^2};$$

$$(4.7.113)$$

$$\frac{D_+ u_j}{h} \times \frac{D_+ u_{jtt}}{h} = \frac{D_+ u_j}{h} \cdot \frac{D_+^2 D_- u_{jt}}{h^3} \frac{u_{j+1} + u_j}{2} + \frac{D_+ u_j}{h} \cdot \frac{D_+^2 D_- u_j}{h^3} \frac{u_{j+1\,t} + u_{jt}}{2}$$
$$- \frac{D_+ u_j}{h} \cdot \frac{u_{j+1\,t} + u_{jt}}{2} \frac{D_+^2 D_- u_j}{h^3} + \frac{D_+ u_j}{h} \times I_{2t}^j. \qquad (4.7.114)$$

Substituting (4.7.113) and (4.7.114) in (4.7.109) and using (4.7.80) and (4.7.83)–(4.7.85), the third term on the right-hand side of (4.7.109) is bounded from above by

$$-\sum_{j=1}^\infty r_{j+1} \frac{D_+ u_j}{h} \cdot \frac{D_+^2 D_- u_{jt}}{h^3} \frac{u_{j+1} + u_j}{2} \cdot u_{jtt} h + C \|r^{1/2} u_{htt}\|_2 \left(\|r^{1/2} \delta^2 u_{ht}\|_2 + 1 \right).$$

$$(4.7.115)$$

By direct calculation, the first term of (4.7.115) is equal to

$$\sum_{j=1}^\infty r_{j+1} \frac{D_+ u_j}{h} \cdot \frac{D_+^2 u_{jt}}{h^2} \frac{u_{j+1} + u_j}{2} \cdot \frac{D_+ u_{jtt}}{h} h$$
$$+ \sum_{j=1}^\infty \frac{D_+ u_{j+1}}{h} \cdot \frac{D_+^2 u_{jt}}{h^2} \frac{u_{j+1} + u_{j+2}}{2} \cdot u_{j+1\,tt} h$$
$$+ \sum_{j=1}^\infty r_{j+1} \frac{D_+ u_{j+1}}{h} \cdot \frac{D_+^2 u_{jt}}{h^2} \frac{D_+ (u_{j+1} + u_j)}{2h} \cdot u_{j+1\,tt} h$$
$$+ \sum_{j=1}^\infty r_{j+1} \frac{D_+^2 u_j}{h^2} \cdot \frac{D_+^2 u_{jt}}{h^2} \frac{u_j + u_{j+1}}{2} \cdot u_{j+1\,tt} h. \qquad (4.7.116)$$

Substituting (4.7.112) and (4.7.113) into (4.7.116) and using (4.7.80), (4.7.83)–(4.7.85), the first term of (4.7.115) is bounded from above by

$$C\|r^{1/2} \delta^2 u_{ht}\|_2 \left(\|r^{1/2} u_{htt}\|_2 + \|r^{1/2} \delta^2 u_{ht}\|_2 + 1 \right). \qquad (4.7.117)$$

Using the fact $u_j \in S^2$, we get

$$u_j \cdot \frac{D_+ D_- u_{jt}}{h^2} = -\left(\frac{u_j}{r_j} \times \frac{D_+ u_j}{h} \right) \cdot \frac{D_+ D_- u_j}{h^2} - \left(\frac{D_+ u_j}{h} \cdot \frac{D_+ u_{jt}}{h} + \frac{D_- u_j}{h} \cdot \frac{D_- u_{jt}}{h} \right).$$

$$(4.7.118)$$

Using (4.7.86) and (4.7.118), we infer that

$$u_{jtt} \cdot u_{jtt} \geq \frac{1}{2} \frac{D_+ D_- u_{jt}}{h^2} \cdot \frac{D_+ D_- u_{jt}}{h^2} - C \left| \frac{D_+ D_- u_j}{h^2} \right|^4 - C \left| \frac{D_+ u_j}{h} \right|^2 \left| \frac{D_+ u_{jt}}{h} \right|^2$$
$$- C \left| \frac{D_- u_j}{h} \right|^2 \left| \frac{D_- u_{jt}}{h} \right|^2 - \frac{C}{r_j^2} \left| \frac{D_- u_{jt}}{h} \right|^2 - \frac{C}{r_j^2} \left| \frac{D_- u_j}{h} \right|^4 - \frac{C}{r_j^4} \left| \frac{D_+ u_j}{h} \right|^4. \qquad (4.7.119)$$

Putting (4.7.110), (4.7.111), (4.7.115), (4.7.117) and (4.7.119) together, and using (4.7.80), (4.7.83)–(4.7.85), we finally obtain

$$\|r^{1/2}u_{htt}\|_2^2 + \|r^{1/2}\delta^2 u_{ht}\|_2^2 \leq C + C\int_0^t \left(\|r^{1/2}u_{htt}\|_2^2 + \|r^{1/2}\delta^2 u_{ht}\|_2^2\right) d\tau.$$

Thus by Gronwall's inequality, one has

$$\sup_{0\leq t\leq T_0} \left(\|r^{1/2}u_{htt}\|_2 + \|r^{1/2}\delta^2 u_{ht}\|_2\right) \leq C. \tag{4.7.120}$$

Using the fact $u_j \in S^2$, we get

$$u_j \cdot \frac{D_+^2 D_-^2 u_j}{h^4} = -\frac{D_+ u_j}{h} \cdot \frac{D_+^2 D_- u_j}{h^3} - \frac{D_- I_3^j}{h}, \tag{4.7.121}$$

$$I_3^j = \frac{D_+(u_{j+1}+u_j)}{2h} \cdot \frac{D_+ D_- u_{j+1}}{h^2} + \frac{D_-(3u_{j+1}+u_j)}{2h} \cdot \frac{D_+ D_- u_j}{h^2}. \tag{4.7.122}$$

Using (4.7.95), one has

$$\frac{D_+ D_- u_{jt}}{h^2} = u_j \times \frac{D_+^2 D_-^2 u_j}{h^4} + \frac{D_+ u_j}{h} \times \frac{D_+^2 D_- u_j}{h^3} + \frac{D_- I_1^j}{h}. \tag{4.7.123}$$

By using (4.7.121)–(4.7.123), we have

$$\frac{D_+^2 D_-^2 u_j}{h^4} \cdot \frac{D_+^2 D_-^2 u_j}{h^4}$$
$$\leq C\left(\frac{D_+ D_- u_{jt}}{h^2} \cdot \frac{D_+ D_- u_{jt}}{h^2} + \left|\frac{D_+ u_j}{h}\right|^2 \left|\frac{D_+^2 D_- u_j}{h^3}\right|^2 + \left|\frac{D_- I_3^j}{h}\right|^2 + \left|\frac{D_- I_1^j}{h}\right|^2\right). \tag{4.7.124}$$

Using (4.7.80), (4.7.83)–(4.7.85), (4.7.120) and (4.7.124), we infer that

$$\sup_{0\leq t\leq T_0} \|r^{1/2}\delta^4 u_h\|_2 \leq C. \tag{4.7.125}$$

Putting (4.7.84), (4.7.120) and (4.7.125) together and using Lemma 4.7.5, this lemma is proved.

Lemma 4.7.9 *Suppose that the conditions of Lemma 4.7.6 are satisfied and $\nabla\phi(x) \in H^{2m-1}(\Omega)$. Then for all integer $m \geq 2$*

$$\sup_{0\leq t\leq T_0} \left(\|r^{1/2}u_{ht^m}\|_2 + \|r^{1/2}\delta^2 u_{ht^{m-1}}\|_2 + \|r^{1/2}\delta^4 u_{ht^{m-2}}\|_2\right) \leq C, \tag{4.7.126}$$

$$\sup_{0\leq t\leq T_0} \left(\|r^{1/2}u_{ht^{m-1}}\|_\infty + \|r^{1/2}\delta u_{ht^{m-1}}\|_\infty \right.$$
$$\left. + \|r^{1/2}\delta^2 u_{ht^{m-2}}\|_\infty + \|r^{1/2}\delta^3 u_{ht^{m-2}}\|_\infty\right) \leq C, \tag{4.7.127}$$

where T_0 is defined in Lemma 4.7.7, and constant C is independent of h.

Proof. In what follows, let $0 \leq t \leq T_0$. This lemma is proved by mathematical induction as follows. For $m = 2$, estimates (4.7.126) and (4.7.127) have been proved in Lemma 4.7.7 and Lemma 4.7.8. Suppose that (4.7.126) and (4.7.127) are valid for all $2 \leq m \leq M - 1$.

Differentiating (4.7.75), we have

$$u_{jtM+1} = u_j \times \frac{D_+ D_- u_{jtM}}{h^2} + \frac{u_j}{r_j} \times \frac{D_+ u_{jtM}}{h}$$
$$+ \sum_{k=0}^{M-1} \binom{M}{k} u_{jtM-k} \times \left(\frac{D_+ D_- u_{jtk}}{h^2} + \frac{D_+ u_{jtk}}{hr_j} \right), \quad j = 1, 2, \ldots . \quad (4.7.128)$$

Taking the scalar product of (4.7.128) and $r_j u_{jtM}$, and then summing the result over j from 1 to ∞, we have

$$\sum_{j=1}^{\infty} r_j u_{jtM+1} \cdot u_{jtM} h$$

$$= \sum_{j=1}^{\infty} r_j u_j \times \frac{D_+ D_- u_{jtM}}{h^2} \cdot u_{jtM} h + \sum_{j=1}^{\infty} u_j \times \frac{D_+ u_{jtM}}{h} \cdot u_{jtM} h$$

$$+ \sum_{j=1}^{\infty} r_j \sum_{k=1}^{M-1} \binom{M}{k} u_{jtM-k} \times \left(\frac{D_+ D_- u_{jtk}}{h} + \frac{D_+ u_{jtk}}{hr_j} \right) \cdot u_{jtM} h$$

$$\leq - \sum_{j=1}^{\infty} r_{j+1} \frac{D_+ u_j}{h} \times \frac{D_+ u_{jtM}}{h} \cdot u_{jtM} h$$

$$+ C \sum_{k=1}^{M-1} \| u_{htM-k} \|_\infty \left(\| r^{1/2} \delta^2 u_{htk} \|_2 + \| r^{1/2} \delta u_{htk} \|_2 \right) \| r^{1/2} u_{htM} \|_2$$

$$\leq - \sum_{j=1}^{\infty} r_{j+1} \frac{D_+ u_j}{h} \times \frac{D_+ u_{jtM}}{h} \cdot u_{jtM} h$$

$$+ C \left(\| u_{htM-1} \|_\infty + \| r^{1/2} \delta u_{htM-1} \|_2 + \| r^{1/2} \delta^2 u_{htM-1} \|_2 + 1 \right) \| r^{1/2} u_{htM} \|_2. \quad (4.7.129)$$

By direct calculation, we get

$$\| r^{1/2} \delta u_{htM-1} \|_2^2 = - \sum_{j=1}^{\infty} D_- \left(r_j \frac{D_+ u_{jtM-1}}{h} \right) \cdot u_{jtM-1}$$

$$\leq \frac{1}{2} \| r^{1/2} \delta u_{htM-1} \|_2^2 + C \left(\| r^{1/2} \delta^2 u_{htM-1} \|_2 + 1 \right). \quad (4.7.130)$$

Employing Lemma 4.7.5, one has

$$\| u_{htM-1} \|_\infty \leq C \left(\| r^{1/2} \delta u_{htM-1} \|_2 + 1 \right)^{1/2} \leq C \left(\| r^{1/2} \delta^2 u_{htM-1} \|_2 + 1 \right)^{1/4}, \quad (4.7.131)$$

where we have used the induction assumption in (4.7.129)–(4.7.131). Using (4.7.75), we have

$$
\frac{D_+ u_{jt^M}}{h} = \frac{u_{j+1} + u_j}{2} \times \frac{D_+^2 D_- u_{jt^{M-1}}}{h^3}
$$
$$
+ \sum_{k=0}^{M-2} \binom{M-1}{k} \frac{u_{j+1\,t^{M-1-k}} + u_{jt^{M-1-k}}}{2} \times \frac{D_+^2 D_- u_{jt^k}}{h^3} + I_{2t^{M-1}}^j,
$$

$$(4.7.132)$$

$$
\frac{D_+ u_j}{h} \times \frac{D_+ u_{jt^M}}{h} = \frac{D_+ u_j}{h} \cdot \frac{D_+^2 D_- u_{jt^{M-1}}}{h^3} \frac{u_{j+1} + u_j}{2} + \frac{D_+ u_j}{h} \times I_{2t^{M-1}}^j
$$
$$
+ \sum_{k=0}^{M-2} \binom{M-1}{k} \frac{D_+ u_j}{h} \times \left(\frac{u_{j+1\,t^{M-1-k}} + u_{jt^{M-1-k}}}{2} \times \frac{D_+^2 D_- u_{jt^k}}{h^3} \right).
$$

$$(4.7.133)$$

where I_2^j is defined in (4.7.113). Substituting (4.7.132) and (4.7.132) into (4.7.129) and using the induction assumption, the first term on the right-hand side of (4.7.129) is bounded from above by

$$
-\sum_{j=1}^{\infty} r_{j+1} \frac{D_+ u_j}{h} \cdot \frac{D_+^2 D_- u_{jt^{M-1}}}{h^3} \frac{u_{j+1} + u_j}{2} \cdot u_{jt^M} h
$$
$$
+ C \| r^{1/2} u_{ht^M} \|_2 \left(\| r^{1/2} \delta^2 u_{ht^{M-1}} \|_2 + \| r^{1/2} \delta u_{ht^{M-1}} \|_2 + \| r^{1/2} \delta^3 u_{ht^{M-2}} \|_2 + 1 \right).
$$

$$(4.7.134)$$

By direct calculation, the first term of (4.7.134) is equal to

$$
\sum_{j=1}^{\infty} r_{j+1} \frac{D_+ u_j}{h} \cdot \frac{D_+^2 u_{jt^{M-1}}}{h^2} \frac{u_{j+1} + u_j}{2} \cdot \frac{D_+ u_{jt^M}}{h} h
$$
$$
+ \sum_{j=1}^{\infty} \frac{D_+ u_{j+1}}{h} \cdot \frac{D_+^2 u_{jt^{M-1}}}{h^2} \frac{u_{j+1} + u_{j+2}}{2} \cdot u_{j+1\,t^M} h
$$
$$
+ \sum_{j=1}^{\infty} r_{j+1} \frac{D_+ u_{j+1}}{h} \cdot \frac{D_+^2 u_{jt^{M-1}}}{h^2} \frac{D_+(u_{j+1} + u_j)}{2h} \cdot u_{j+1\,t^M} h
$$
$$
+ \sum_{j=1}^{\infty} r_{j+1} \frac{D_+^2 u_j}{h^2} \cdot \frac{D_+^2 u_{jt^{M-1}}}{h^2} \frac{u_j + u_{j+1}}{2} \cdot u_{j+1\,t^M} h. \qquad (4.7.135)
$$

Substituting (4.7.132) into (4.7.135) and using the induction assumption, the first term of (4.7.134) is bounded from above by

$$
C \| r^{1/2} \delta^2 u_{ht} \|_2 \left(\| r^{1/2} u_{ht^M} \|_2 + \| r^{1/2} \delta^2 u_{ht^{M-1}} \|_2 \right.
$$
$$
\left. + \| r^{1/2} \delta u_{ht^{M-1}} \|_2 + \| r^{1/2} \delta^3 u_{ht^{M-2}} \|_2 + 1 \right). \qquad (4.7.136)
$$

Using the fact $u_j \in S^2$, we get

$$u_j \cdot \frac{D_+ D_- u_{jt^{M-1}}}{h^2} = -\sum_{k=0}^{M-2} \binom{M-1}{k} u_{jt^{M-1-k}} \cdot \frac{D_+ D_- u_{jt^k}}{h^2}$$
$$-\frac{1}{2}\left(\frac{D_+ u_j}{h} \cdot \frac{D_+ u_j}{h} + \frac{D_- u_j}{h} \cdot \frac{D_- u_j}{h}\right)_{t^{M-1}}, \quad (4.7.137)$$

$$u_{j+1} \cdot \frac{D_+^2 D_- u_{jt^{M-2}}}{h^3}$$
$$= -\sum_{k=0}^{M-3} \binom{M-2}{k} u_{j+1\,t^{M-2-k}} \cdot \frac{D_+^2 D_- u_{jt^k}}{h^3}$$
$$-\left(\frac{D_+ u_{j+1} + D_- u_{j+1}}{2h} \cdot \frac{D_+ D_- u_{j+1}}{h^2} + \frac{3D_+ u_j + D_- u_j}{2h} \cdot \frac{D_+ D_- u_j}{h^2}\right)_{t^{M-2}}.$$
$$(4.7.138)$$

Using (4.7.137) and (4.7.138), the induction assumption, and the following equations,

$$u_{jt^M} = u_j \times \frac{D_+ D_- u_{jt^{M-1}}}{h^2} + \frac{u_j}{r_j} \times \frac{D_+ u_{jt^{M-1}}}{h}$$
$$+ \sum_{k=0}^{M-2} \binom{M-1}{k} u_{jt^{M-1-k}} \times \left(\frac{D_+ D_- u_{jt^k}}{h^2} + \frac{D_+ u_{jt^k}}{hr_j}\right), \quad j = 1, 2, \ldots,$$
$$(4.7.139)$$

$$\frac{D_+ u_{jt^{M-1}}}{h} = u_{j+1} \times \frac{D_+^2 D_- u_{jt^{M-2}}}{h^3}$$
$$+ \sum_{k=0}^{M-3} \binom{M-2}{k} u_{jt^{M-2-k}} \times \frac{D_+^2 D_- u_{jt^k}}{h^3} + I_{1t^{M-2}}^j, \quad j = 1, 2, \ldots,$$
$$(4.7.140)$$

we infer that

$$\|r^{1/2}\delta^2 u_{ht^{M-1}}\|_2^2 \leq C\left(\|r^{1/2}u_{ht^M}\|_2^2 + \|r^{1/2}\delta u_{ht^{M-1}}\|_2^2 + 1\right)$$
$$\leq C\left(\|r^{1/2}u_{ht^M}\|_2^2 + \|r^{1/2}\delta^2 u_{ht^{M-1}}\|_2 + 1\right), \quad (4.7.141)$$
$$\|r^{1/2}\delta^3 u_{ht^{M-2}}\|_2^2 \leq C\left(\|r^{1/2}\delta u_{ht^{M-1}}\|_2^2 + 1\right) \leq C\left(\|r^{1/2}\delta^2 u_{ht^{M-1}}\|_2 + 1\right), \quad (4.7.142)$$

where I_1^j is defined in (4.7.95) and (4.7.130) has been used. Putting (4.7.129)–(4.7.131), (4.7.134), (4.7.136), (4.7.141) and (4.7.142) together and using the induction assumption, we finally obtain

$$\|r^{1/2}u_{ht^M}\|_2^2 + \|r^{1/2}\delta^2 u_{ht^{M-1}}\|_2^2 \leq C + C\int_0^t \left(\|r^{1/2}u_{ht^M}\|_2^2 + \|r^{1/2}\delta^2 u_{ht^{M-1}}\|_2^2\right) d\tau.$$

Thus by Gronwall's inequality, one has

$$\sup_{0 \leq t \leq T_0} \left(\|r^{1/2}u_{ht^M}\|_2 + \|r^{1/2}\delta^2 u_{ht^{M-1}}\|_2\right) \leq C. \quad (4.7.143)$$

Putting (4.7.130), (4.7.142) and (4.7.143) together, one has

$$\sup_{0 \le t \le T_0} \left(\|r^{1/2} \delta u_{htM-1}\|_2 + \|r^{1/2} \delta^3 u_{htM-2}\|_2 \right) \le C. \tag{4.7.144}$$

Using the fact $u_j \in S^2$, we get

$$u_j \cdot \frac{D_+^2 D_-^2 u_{jtM-2}}{h^4} = -\sum_{k=0}^{M-3} \binom{M-2}{k} u_{jtM-2-k} \cdot \frac{D_+^2 D_-^2 u_{jtk}}{h^4} \\ - \left(\frac{D_+ u_j}{h} \cdot \frac{D_+^2 D_- u_j}{h^3} \right)_{tM-2} - \frac{D_- I_{3tM-2}^j}{h}, \tag{4.7.145}$$

where I_3^j is defined in (4.7.122). Employing (4.7.123), one has

$$\frac{D_+ D_- u_{jtM-1}}{h^2} = u_j \times \frac{D_+^2 D_-^2 u_{jtM-2}}{h^4} + \sum_{k=0}^{M-3} \binom{M-2}{k} u_{jtM-2-k} \times \frac{D_+^2 D_-^2 u_{jtk}}{h^4} \\ + \left(\frac{D_+ u_j}{h} \times \frac{D_+^2 D_- u_j}{h^3} \right)_{tM-2} + \frac{D_- I_{1tM-2}^j}{h}. \tag{4.7.146}$$

Using (4.7.143)–(4.7.146) and the induction assumption, we infer that

$$\sup_{0 \le t \le T_0} \|r^{1/2} \delta^4 u_{htM-2}\|_2 \le C \left(\|r^{1/2} \delta^2 u_{htM-1}\|_2 + 1 \right) \le C. \tag{4.7.147}$$

Putting (4.7.143), (4.7.144) and (4.7.147) together, and from the induction assumption and Lemma 4.7.5, we obtain

$$\sup_{0 \le t \le T_0} \left(\|r^{1/2} u_{htM-1}\|_\infty + \|r^{1/2} \delta u_{htM-1}\|_\infty + \|r^{1/2} \delta^2 u_{htM-2}\|_\infty + \|r^{1/2} \delta^3 u_{htM-2}\|_\infty \right) \le C.$$

Thus by the induction procedure, this lemma is proved.

Using the results of Lemmas 4.7.5–4.7.9, one can achieve the local existence of solution for problem (4.7.1)–(4.7.3). Thus, we propose

Proposition 4.7.2 *For problem* (4.7.1)–(4.7.3), *suppose that the initial function* $\phi(x) \in S^2$ *and* $\nabla \phi(x) \in H^{2m-1}(\Omega)$, $m \ge 1$. *Then there exists a constant* $T_0 = T_0 \left(\|\nabla \phi\|_{H^{2m-1}} \right) > 0$ *and a solution such that*

$$\partial_t^m u(x,t), \ \Delta \partial_t^{m-1} u(x,t), \ \Delta^2 \partial_t^{m-2} u(x,t) \in L^\infty(0, T_0; \ L^2(\Omega)).$$

Using Lemmas 4.7.1–4.7.3 and Corollary 4.7.1, the local smooth solution $u(x,t)$ constructed in Proposition 4.7.2 can be extended to all time $t > 0$. Thus theorem 4.7.1 is proved.

4.8 Bibliography Comments

This chapter is mainly regarding the global solution theory for the higher-dimensional Landau–Lifshitz equation on the Riemannian manifold.

The studies on higher-dimensional Landau–Lifshitz equations began in the early 1980s by Yulin Zhou, Boling Guo and Hesheng Sun (see [150–157]). They were mainly concerned with the global existence of weak solutions by the Galerkin method and the difference method. At the same time, the same problem was also discussed by Sulem and Sulem [132]. F. Alouger and A. Soyeur in 1992 [4] obtained a similar result to that by Zhou and Guo *et al.* [150–156].

After Zhou, Guo and Tan obtained smooth solution for the one-dimensional problem in 1991 [158], Guo and Hong, in 1993 [77], first discussed the two-dimensional problem. They discovered the relationships between the Landau–Lifshitz equations and the harmonic maps and established the partial regularity theory by using the method dealing with harmonic maps, that is, they proved that two-dimensional Landau–Lifshitz equation admits a global almost smooth solution which is now called the Chen–Struwe solution in the field of harmonic map heat flow.

In [87], Guo and Wang proved the above results for the generalized Landau–Lifshitz systems which are from $R^n \to S^2$, $n \geq 2$.

Later, Guo and Ding applied the penalty method to the initial boundary value problems of saturated or unsaturated or the case with applied field Landau–Lifshitz systems in two dimensions. The existence theory of global solution to the L–L systems in higher dimensions can be found in [59].

In [29, 31], Chen, Guo and Ding proved that any weak solution with finite energy, decreasing in time, must be almost smooth for the two-dimensional problem.

In 2003, Xiangao Liu [109] proved a partial regularity theorem for the stationary weak solution using the similar method applied to the harmonic map heat flow by Feldman [58]. And, Roger Moser [115] obtained a similar result for the lower-dimensional problem ($n \leq 4$). In 2004, Ding and Guo [49] proved a similar result for the Landau–Lifshitz–Maxwell equations, see the next chapter.

Recently, Melcher [113] proved that there exists a partially regular solution for the three-dimensional Cauchy problem of Landau–Lifshitz equations. This result extends the regularity theorem from a stationary weak solution to a weak solution. However, Melcher's method does not apply to higher-dimensional problems. In 2006, Changyou Wang [143] extended Melcher's result to three and four dimensions by the Pohozaev method.

In 2007, Ding and Wang [52] proved that for some initial data, the finite time blow-up for the solutions to the Dirichlet problem and the Neumann problem does occur. These results answer the long unsolved questions: Do Landau–Lifshitz equations really blow up at finite time?

For the other problems, such as vortices and domain walls, as well as steady state problems, we refer to [18–24, 44–46, 116, 117, 120].

Chapter 5

Landau–Lifshitz–Maxwell Equations

5.1 Global Weak Solution in Three Dimensions

5.1.1 The Periodic Initial Value Problem

In this chapter, we first study the following coupled Landau–Lifshitz–Maxwell equations:

$$\vec{Z}_t = \alpha_1 \vec{Z} \times (\Delta \vec{Z} + \vec{H}) - \alpha_2 \vec{Z} \times (\vec{Z} \times (\Delta \vec{Z} + \vec{H})), \tag{5.1.1}$$

$$\nabla \times \vec{H} = \frac{\partial \vec{E}}{\partial t} + \sigma \vec{E}, \tag{5.1.2}$$

$$\nabla \times \vec{E} = -\frac{\partial \vec{H}}{\partial t} - \beta \frac{\partial \vec{Z}}{\partial t}, \tag{5.1.3}$$

$$\nabla \cdot \vec{H} + \beta \nabla \cdot \vec{Z} = 0, \qquad \nabla \cdot \vec{E} = 0, \tag{5.1.4}$$

where α_1, α_2, σ and β are constants, $\alpha_2 \geq 0$, $\sigma \geq 0$, $\vec{H} = (H_1(x,t), H_2(x,t), H_3(x,t))$ is the magnetic field, $\vec{E}(x,t) = (E_1(x,t), E_2(x,t), E_3(x,t))$ is the electric field and $\vec{H}^e = \Delta \vec{Z} + \vec{H}$ is the effective magnetic field.

We study the existence of the global generalized solutions for the three-dimensional Landau–Lifshitz–Maxwell system (5.1.1)–(5.1.4) with the periodic initial value condition:

$$\vec{Z}(x + 2D\vec{e}_i, t) = \vec{Z}(x, t), \quad \vec{H}(x + 2D\vec{e}_i, t) = \vec{H}(x, t),$$

$$\vec{E}(x + 2D\vec{e}_i, t) = \vec{E}(x, t), \quad (x \in \Omega \subset R^3, t \geq 0) \tag{5.1.5}$$

$$\vec{Z}(x, 0) = \vec{Z}_0(x), \quad \vec{H}(x, 0) = \vec{H}_0(x), \quad \vec{E}(x, 0) = \vec{E}_0(x), \quad (x \in \Omega \subset R^3) \tag{5.1.6}$$

or with the initial value condition:

$$\vec{Z}(x, 0) = \vec{Z}_0(x), \quad \vec{H}(x, 0) = \vec{H}_0(x), \quad \vec{E}(x, 0) = \vec{E}_0(x), \quad x \in R^3, \tag{5.1.7}$$

where $x + 2D\vec{e}_i = (x_1, \ldots, x_{i-1}, x_i + 2D, x_{i+1}, \ldots, x_n)$, $(i = 1, 2, 3)$, $D > 0$, $\Omega \subset R^3$ represents the three-dimensional cube with width $2D$ along each direction, i.e. $\overline{\Omega} = \{x = (x_1, x_2, x_3), |x_i| \leq D, i = 1, 2, 3\}$, $Q_T = \{(x, t), x \in \overline{\Omega}, 0 \leq t \leq T\}$.

5.1.2   Approximate Solutions and the *A Priori* Estimates for the Periodic Initial Value Problem

We use Galerkin method to establish the *a priori* estimate for the approximate solutions of (5.1.1)–(5.1.6).

Let $w_n(x)$ $(n = 1, 2, \ldots)$ be the unit eigenfunctions satisfying the equation $\Delta w_n + \lambda_n w_n = 0$, with periodicity $w_n(x - D\vec{e}_i) = w_n(x + D\vec{e}_i)$ $(i = 1, 2, 3)$, and λ_n $(n = 1, 2, \ldots)$ the corresponding eigenvalues different from each other. $\{w_n(x)\}$ consists of the orthonormal base in L^2.

Denote the approximate solution of the problem (5.1.1)–(5.1.6) by $\vec{Z}_N(x, t)$, $\vec{H}_N(x, t)$ and $\vec{E}_N(x, t)$ in the following form

$$\begin{cases} \vec{Z}_N(x, t) = \displaystyle\sum_{s=1}^{N} \vec{\alpha}_{sN}(t) w_s(x), \\[2mm] \vec{H}_N(x, t) = \displaystyle\sum_{s=1}^{N} \vec{\beta}_{sN}(t) w_s(x), \\[2mm] \vec{E}_N(x, t) = \displaystyle\sum_{s=1}^{N} \vec{\gamma}_{sN}(t) w_s(x), \end{cases} \tag{5.1.8}$$

where $\vec{\alpha}_{sN}(t)$, $\vec{\beta}_{sN}(t)$, $\vec{\gamma}_{sN}(t)$, $(t \in R^+)$, $(s = 1, 2, \ldots, N; N = 1, 2, \ldots)$ are the three-dimensional vector-valued functions satisfying the following system of ordinary differential equations of first order:

$$\int_\Omega \vec{Z}_{Nt} w_s(x) dx = \alpha_1 \int_\Omega \vec{Z}_N \times (\Delta \vec{Z}_N + \vec{H}_N) \cdot w_s(x) dx$$

$$- \alpha_2 \int_\Omega \vec{Z}_N \times (\vec{Z}_N \times (\Delta \vec{Z}_N + \vec{H}_N)) w_s(x) dx, \tag{5.1.9}$$

$$\int_\Omega (\vec{H}_{Nt} + \beta \vec{Z}_{Nt}) w_s(x) dx = -\int_\Omega \nabla \times \vec{E}_N \cdot w_s(x) dx, \tag{5.1.10}$$

$$\int_\Omega \vec{E}_{Nt} w_s(x) dx + \sigma \int_\Omega \vec{E}_N \cdot w_s(x) dx = \int_\Omega \nabla \times \vec{H}_N \cdot w_s(x) dx, \tag{5.1.11}$$

and the initial condition

$$\int_\Omega \vec{Z}_N(x, 0) w_s(x) dx = \int_\Omega \vec{Z}_0(x) w_s(x) dx,$$

$$\int_\Omega \vec{H}_N(x, 0) w_s(x) dx = \int_\Omega \vec{H}_0(x) w_s(x) dx, \tag{5.1.12}$$

$$\int_\Omega \vec{E}_N(x, 0) w_s(x) dx = \int_\Omega \vec{E}_0(x) w_s(x) dx.$$

Obviously there holds

$$\int_\Omega \vec{Z}_{Nt} w_s(x) dx = \alpha'_{sN}(t), \quad \int_\Omega \vec{H}_{Nt} w_s(x) dx = \beta'_{sN}(t), \quad \int_\Omega \vec{E}_{Nt} w_s(x) dx = \gamma'_{sN}(t).$$

$$\tag{5.1.13}$$

For the sake of simplicity, denote $\| \cdot \|_{L^p(\Omega)} = \| \cdot \|_p, p \geq 2$.

Lemma 5.1.1 *Assume* $(\vec{Z}_0(x), \vec{H}_0(x), \vec{E}_0(x)) \in (H^1(\Omega), L^2(\Omega), L^2(\Omega))$, $\alpha_2 \geq 0$. *Then for the solution of the initial value problem (5.1.9)–(5.1.12) we have the following estimates*:

$$\sup_{0\leq t\leq T} \left[\|\vec{Z}_N(\cdot,t)\|^2_{H^1(\Omega)} + \|\vec{E}_N(\cdot,t)\|^2_{L^2(\Omega)} + \|\vec{H}_N(\cdot,t)\|^2_{L^2(\Omega)} \right] \leq K_0,$$

$$\sup_{0\leq t\leq T} \|\vec{Z}_N(\cdot,t)\|_6 \leq K_0, \qquad \sup_{0\leq t\leq T} \|\vec{Z}_N(\cdot,t) \times \nabla \vec{Z}_N(\cdot,t)\|_{\frac{3}{2}} \leq K_0, \tag{5.1.14}$$

where the constant K_0 is independent of N, α_2 and D. When $\alpha_0 > 0$, there is

$$\|\vec{Z}_N \times (\Delta\vec{Z}_N + \vec{H}_N)\|_{L^2(0,T;L^2(\Omega))} \leq K_0', \tag{5.1.15}$$

where the constant K_0' is independent of N and D.

Proof. (1) Multiplying (5.1.9) by $\vec{\alpha}_{sN}(t)$, and summing up the products for $s = 1, 2, \ldots, N$, we get

$$\frac{d}{dt} \int_\Omega |\vec{Z}_N(x,t)|^2 dx = 0.$$

Then

$$\|\vec{Z}_N(\cdot,t)\|^2_2 \leq \|\vec{Z}_N(\cdot,0)\|^2_2 \leq \|\vec{Z}_0(x)\|^2_2. \tag{5.1.16}$$

(2) Making the scalar product of $(-\lambda_s \vec{\alpha}_{sN}(t) + \vec{\beta}_{sN}(t))$ with (5.1.9), and summing up the resulting product for $s = 1, 2, \ldots, N$, and noticing that

$$\Delta\vec{Z}_N + \vec{H}_N = \sum_{s=1}^{N}(-\lambda_s\vec{\alpha}_{sN} + \vec{\beta}_{sN})w_s(x),$$

we have

$$\int_\Omega \vec{Z}_{Nt}(\Delta\vec{Z}_N + \vec{H}_N)dx = -\alpha_2 \int_\Omega \vec{Z}_N \times (\vec{Z}_N \times (\Delta\vec{Z}_N + \vec{H}_N)) \cdot (\Delta\vec{Z}_N + \vec{H}_N)dx$$

$$= \alpha_2 \int_\Omega |\vec{Z}_N \times (\Delta\vec{Z}_N + \vec{H}_N)|^2 dx,$$

i.e.

$$\frac{1}{2}\frac{d}{dt} \int_\Omega |\nabla\vec{Z}_N(x,t)|^2 dx + \alpha_2\|\vec{Z}_N \times (\Delta\vec{Z}_N + \vec{H}_N)\|^2_2 - \int_\Omega \vec{Z}_{Nt} \cdot \vec{H}_N dx = 0. \tag{5.1.17}$$

Making the scalar product of $\vec{\beta}_{sN}(t)$ with (5.1.10), and $\vec{\gamma}_{sN}(t)$ with (5.1.11), respectively, and adding the two equalities obtained, and then summing up the resulting products for $s = 1, 2, \ldots, N$, we get

$$\frac{1}{2}\frac{d}{dt} \int_\Omega (|\vec{E}_N(x,t)|^2 + |\vec{H}_N(x,t)|^2)dx + \sigma\|\vec{E}_N(\cdot,t)\|^2_2 + \beta \int_\Omega \vec{Z}_{Nt} \cdot \vec{H}_N dx = 0. \tag{5.1.18}$$

From (5.1.17) and (5.1.18) (multiply (5.1.18) by δ_0) it follows that

$$\frac{d}{dt} \int_\Omega \left[\frac{1}{2}|\nabla\vec{Z}_N|^2 + \frac{\delta_0}{2}(|\vec{E}_N|^2 + |\vec{H}_N|^2) \right] dx + \sigma\delta_0\|\vec{E}_N\|^2_2$$

$$+ \alpha_2\|\vec{Z}_N \times (\Delta\vec{Z}_N + \vec{H}_N)\|^2_2 + (\beta\delta_0 - 1) \int_\Omega \vec{Z}_{Nt} \cdot \vec{H}_N dx = 0. \tag{5.1.19}$$

Multiplying (5.1.10) by $(\beta\delta_0 - 1)\vec{\alpha}_{sN}(t)$, and summing up the products for $s = 1, 2, \ldots, N$, we obtain

$$(\beta\delta_0 - 1)\int_\Omega \vec{H}_{Nt} \cdot \vec{Z}_N dx + \frac{1}{2}\beta(\beta\delta_0 - 1)\frac{d}{dt}\int_\Omega |\vec{Z}_N|^2 dx$$
$$+ (\beta\delta_0 - 1)\int_\Omega \nabla \times \vec{E}_N \cdot \vec{Z}_N dx = 0. \tag{5.1.20}$$

Adding (5.1.19) and (5.1.20) gives

$$\frac{1}{2}\frac{d}{dt}\int_\Omega \left[|\nabla\vec{Z}_N|^2 + \delta_0(|\vec{E}_N|^2 + |\vec{H}_N|^2)\right] dx + \frac{1}{2}\beta(\beta\delta_0 - 1)\frac{d}{dt}\int_\Omega |\vec{Z}_N|^2 dx$$
$$+ (\beta\delta_0 - 1)\frac{d}{dt}\int_\Omega \vec{Z}_N \cdot \vec{H}_N dx + \alpha_2\|\vec{Z}_N \times (\Delta\vec{Z}_N + \vec{H}_N)\|_2^2$$
$$= -\sigma\delta_0\|\vec{E}_N(\cdot, t)\|_2^2 + (1 - \beta\delta_0)\int_\Omega \nabla \times \vec{E}_N \cdot \vec{Z}_N dx$$
$$= -\sigma\delta_0\|\vec{E}_N(\cdot, t)\|_2^2 + (1 - \beta\delta_0)\int_\Omega \nabla \times \vec{Z}_N \cdot \vec{E}_N dx$$
$$\leq -\sigma\delta_0\|\vec{E}_N(\cdot, t)\|_2^2 + \varepsilon_0\|\vec{E}_N(\cdot, t)\|_2^2 + \frac{(1 - \beta\delta_0)^2}{4\varepsilon_0}\|\nabla\vec{Z}_N(\cdot, t)\|_2^2, \quad (\varepsilon_0 > 0).$$

Integrating the above inequality with respect to t we get

$$\frac{1}{2}\|\nabla\vec{Z}_N(\cdot, t)\|_2^2 + \frac{1}{2}\delta_0(\|\vec{E}_N(\cdot, t)\|_2^2 + \|\vec{H}_N(\cdot, t)\|_2^2)$$
$$+ \frac{1}{2}\beta(\beta\delta_0 - 1)\|\vec{Z}_N(\cdot, t)\|_2^2 + (\beta\delta_0 - 1)\int_\Omega \vec{H}_N(x, t) \cdot \vec{Z}_N(x, t)dx$$
$$+ \alpha_2\int_0^t \|\vec{Z}_N \times (\Delta\vec{Z}_N + \vec{H}_N)\|_2^2 dt$$
$$\leq \frac{1}{2}\|\nabla\vec{Z}_0\|_2^2 + \frac{1}{2}\delta_0(\|\vec{E}_0\|_2^2 + \|\vec{H}_0\|_2^2) + \frac{1}{2}|\beta(\beta\delta_0 - 1)|\|\vec{Z}_0\|_2^2$$
$$+ |\beta\delta_0 - 1|\|\vec{H}_0\|_2\|\vec{Z}_0\|_2 + (\varepsilon_0 - \sigma\delta_0)\int_0^t \|\vec{E}_N(\cdot, t)\|_2^2 dt$$
$$+ \frac{(1 - \beta\delta_0)^2}{4\varepsilon_0}\int_0^t \|\nabla\vec{Z}_N(\cdot, t)\|_2^2 dt. \tag{5.1.21}$$

Denote

$$c_0 = \frac{1}{2}\|\nabla\vec{Z}_0\|_2^2 + \frac{1}{2}\delta_0(\|\vec{E}_0\|_2^2 + \|\vec{H}_0\|_2^2)$$
$$+ \frac{1}{2}|\beta(\beta\delta_0 - 1)|\|\vec{Z}_0\|_2^2 + |\beta\delta_0 - 1|\|\vec{H}_0\|_2\|\vec{Z}_0\|_2.$$

Then, by (5.1.21) and (5.1.16)

$$\frac{1}{2}\|\nabla \vec{Z}_N(\cdot,t)\|_2^2 + \frac{1}{2}\delta_0(\|\vec{E}_N(\cdot,t)\|_2^2 + \|\vec{H}_N(\cdot,t)\|_2^2) + \alpha_2 \int_0^t \|\vec{Z}_N \times (\Delta \vec{Z}_N + \vec{H}_N)\|_2^2 dx$$

$$\leq c_0 + \frac{|\beta(\beta\delta_0 - 1)|}{2}\|\vec{Z}_N(\cdot,t)\|_2^2 + |\beta\delta_0 - 1|\|\vec{H}_N(\cdot,t)\|_2\|\vec{Z}_N(\cdot,t)\|_2$$

$$+ (\varepsilon_0 - \sigma\delta_0)\int_0^t \|\vec{E}_N(\cdot,t)\|_2^2 dt + \frac{(1 - \beta\delta_0)^2}{4\varepsilon_0}\int_0^t \|\nabla \vec{Z}_N(\cdot,t)\|_2^2 dt$$

$$\leq c_0 + \frac{\delta_0}{4}\|\vec{H}_N(\cdot,t)\|_2^2 + \left(\frac{|\beta(\beta\delta_0 - 1)|}{2} + \frac{(\beta\delta_0 - 1)^2}{\delta_0}\right)\|\vec{Z}_0\|_2^2$$

$$+ (\varepsilon_0 - \sigma\delta_0)\int_0^t \|\vec{E}_N\|_2^2 dt + \frac{(1 - \beta\delta_0)^2}{4\varepsilon_0}\int_0^t \|\nabla \vec{Z}_N\|_2^2 dt.$$

Therefore,

$$\frac{1}{2}\|\nabla \vec{Z}_N(\cdot,t)\|_2^2 + \frac{\delta_0}{4}\|\vec{H}_N(\cdot,t)\|_2^2 + \frac{\delta_0}{2}\|\vec{E}_N(\cdot,t)\|_2^2$$

$$+ \alpha_2 \int_0^t \|\vec{Z}_N \times (\Delta \vec{Z}_N + \vec{H}_N)\|_2^2 dt$$

$$\leq d_0 + (\varepsilon_0 - \sigma\delta_0)\int_0^t \|\vec{E}_N\|_2^2 + \frac{(1 - \beta\delta_0)^2}{4\varepsilon_0}\int_0^t \|\nabla \vec{Z}_N\|_2^2 dt, \qquad (5.1.22)$$

where

$$d_0 = c_0 + \left(\frac{|\beta(\beta\delta_0 - 1)|}{2} + \frac{(\beta\delta_0 - 1)^2}{\delta_0}\right)\|\vec{Z}_0\|_2^2.$$

When $\beta > 0, \sigma \geq 0$, we let $\delta_0 = \frac{1}{\beta}$, $\varepsilon_0 = \sigma\delta_0 + \frac{1}{4}\delta_0$. Then, from (5.1.22) it follows that

$$\frac{1}{2}\|\nabla \vec{Z}_N(\cdot,t)\|_2^2 + \frac{1}{4\beta}(\|\vec{E}_N(\cdot,t)\|_2^2 + \|\vec{H}_N(\cdot,t)\|_2^2)$$

$$+ \alpha_2 \int_0^t \|\vec{Z}_N \times (\Delta \vec{Z}_N + \vec{H}_N)\|_2^2 dt \leq d_0, \qquad \forall\, t \geq 0. \qquad (5.1.23)$$

When σ and β are arbitrary constants, we take $\varepsilon_0 = \delta_0 = 1$. By (5.1.22) there is

$$\frac{1}{2}\|\nabla \vec{Z}_N(\cdot,t)\|_2^2 + \frac{1}{4}\|\vec{H}_N(\cdot,t)\|_2^2 + \frac{1}{2}\|\vec{E}_N(\cdot,t)\|_2^2 + \alpha_2 \int_0^t \|\vec{Z}_N \times (\Delta \vec{Z}_N + \vec{H}_N)\|_2^2 dt$$

$$\leq d_0 + |1 - \sigma|\int_0^t \|\vec{E}_N(\cdot,t)\|_2^2 dt + \frac{(1 - \beta)^2}{4}\int_0^t \|\nabla \vec{Z}_N(\cdot,t)\|_2^2 dt$$

$$\leq d_0 + e_0 \int_0^t (\|\vec{E}_N\|_2^2 + \|\nabla \vec{Z}_N\|_2^2) dt, \qquad (5.1.24)$$

where

$$e_0 = \max\left\{|1 - \sigma|, \frac{(1 - \beta)^2}{4}\right\}.$$

Applying Gronwall inequality for (5.1.24) yields

$$\sup_{t\in[0,T]} (\|\nabla \vec{Z}_N(\cdot,t)\|_2^2 + \|\vec{H}_N(\cdot,t)\|_2^2 + \|\vec{E}_N(\cdot,t)\|_2^2) \leq K_0, \tag{5.1.25}$$

where K_0 is independent of N, α_2, D. By Sobolev imbedding theorem and Hölder inequality, we have

$$\sup_{0\leq t\leq T} \|\vec{Z}_N(\cdot,t)\|_6^2 \leq K_0, \quad \sup_{0\leq t\leq T} \|\vec{Z}_N(\cdot,t) \times \nabla \vec{Z}_N(\cdot,t)\|_{\frac{3}{2}} \leq K_0. \tag{5.1.26}$$

When $\alpha_2 > 0$, using (5.1.23), (5.1.24) and (5.1.25), we get

$$\|\vec{Z}_N \times (\Delta \vec{Z}_N + \vec{H}_N)\|_{L^2(0,T;L^2(\Omega))} \leq K_0', \tag{5.1.27}$$

where K_0' is independent of N. The proof of the lemma is complete.

Lemma 5.1.2 *Under the conditions of Lemma 5.1.1, for the solution $(\vec{Z}_N(x,t),$ $\vec{H}_N(x,t), \vec{E}_N(x,t))$ of the initial value problem (5.1.9)–(5.1.12) there are*
(1) when $\alpha_2 = 0$,

$$\sup_{t\in[0,T]} \left[\|\vec{Z}_{Nt}\|_{H^{-2}(\Omega)} + \|\vec{E}_{Nt}\|_{H^{-2}(\Omega)} + \|\vec{H}_{Nt}\|_{H^{-2}(\Omega)} \right] \leq K_1; \tag{5.1.28}$$

(2) when $\alpha_2 > 0$,

$$\|\vec{Z}_{Nt}\|_{L^{\frac{3}{2}}(Q_T)} + \|\vec{E}_{Nt}\|_{L^2(0,T;H^{-1}(\Omega))} + \|\vec{H}_{Nt}\|_{L^2(0,T;H^{-1}(\Omega))} \leq K_2, \tag{5.1.29}$$

where the constants K_1 and K_2 are independent of N and D.

Proof. (1) When $\alpha_2 = 0$, for any $\varphi \in H_0^2$, φ can be represented by

$$\varphi = \varphi_N + \bar{\varphi}_N,$$

where

$$\varphi_N = \sum_{s=1}^{N} \beta_s w_s(x), \qquad \bar{\varphi}_N = \sum_{s=N+1}^{\infty} \beta_s w_s(x).$$

For $s \geq N+1$,

$$\int_\Omega \vec{Z}_{Nt} \cdot w_s(x) dx = 0.$$

Then, by Lemma 5.1.1 there are

$$\int_\Omega \vec{Z}_{Nt} \varphi dx = \int_\Omega \vec{Z}_{Nt} \varphi_N(x) dx = \alpha_1 \int_\Omega \vec{Z}_N \times (\Delta \vec{Z}_N + \vec{H}_N)\varphi_N(x) dx$$

$$= \alpha_1 \int_\Omega \nabla \vec{Z}_N \times \vec{Z}_N \cdot \nabla \varphi_N dx + \alpha_1 \int_\Omega \vec{Z}_N \times \vec{H}_N \cdot \varphi_N dx$$

$$\leq |\alpha_1|(\|\nabla \vec{Z}_N\|_2\|\vec{Z}_N\|_2 + \|\vec{Z}_N\|_2\|\vec{H}_N\|_2)(\|\nabla \varphi_N\|_2 + \|\varphi_N\|_\infty)$$

$$\leq C\|\varphi_N\|_{H^2(\Omega)} \leq C\|\varphi\|_{H^2(\Omega)},$$

$$\int_\Omega \vec{E}_{Nt}\varphi dx \le C_1(\|\vec{E}_N\|_2 + \|\vec{H}_N\|_2)(\|\nabla\varphi\|_2 + \|\varphi\|_2) \le C\|\varphi\|_{H^2(\Omega)},$$

$$\int_\Omega \vec{H}_{Nt}\varphi dx \le |\beta|\left|\int_\Omega \vec{Z}_{Nt}\varphi dx\right| + \int_\Omega |\nabla\varphi||\vec{E}_N|dx$$

$$\le |\beta|\left|\int_\Omega \vec{Z}_{Nt}\varphi dx\right| + \|\nabla\varphi\|_2\|\vec{E}_N\|_2$$

$$\le C\|\varphi\|_{H^2(\Omega)} + C_2\|\nabla\varphi\|_2 \le C_3\|\varphi\|_{H^2}.$$

Then there holds

$$\|\vec{Z}_{Nt}\|_{H^{-2}(\Omega)} + \|\vec{E}_{Nt}\|_{H^{-2}(\Omega)} + \|\vec{H}_{Nt}\|_{H^{-2}(\Omega)} \le K_1.$$

(2) For $\alpha_2 > 0$,

$$\left|\int_0^T \int_\Omega (\vec{Z}_{Nt}\cdot\varphi)dxdt\right| \le |\alpha_1|\left|\iint_{Q_T} \vec{Z}_N \times (\Delta\vec{Z}_N + \vec{H}_N)\varphi dxdt\right|$$

$$+ \alpha_2\left|\iint_{Q_T} \vec{Z}_N \times (\vec{Z}_N \times (\Delta\vec{Z}_N + \vec{H}_N))\varphi dxdt\right|$$

$$\le |\alpha_1|\|\vec{Z}_N \times (\Delta\vec{Z}_N + \vec{H}_N)\|_{L^2(Q_T)}\|\varphi\|_{L^2(Q_T)}$$

$$+ \alpha_2\|\vec{Z}_N\|_{L^6(Q_T)}\|\vec{Z}_N \times (\Delta\vec{Z}_N + \vec{H}_N)\|_{L^2(Q_T)}\|\varphi\|_{L^3(Q_T)}$$

$$\le C_4\|\varphi\|_{L^3(Q_T)},$$

$$\left|\int_\Omega (\vec{Z}_{Nt}\cdot\varphi)dx\right| \le C_5\|\varphi\|_{H^1(\Omega)},$$

$$\left|\iint_{Q_T} \vec{E}_{Nt}\varphi dxdt\right| \le |\sigma|\int_0^T \|\vec{E}_N\|_2\|\varphi\|_2 dt + \int_0^T \|\vec{H}_N\|_2\|\nabla\varphi_N\|_2 dt$$

$$\le C_6 T^{\frac{1}{2}}\|\varphi\|_{L^2(0,T;H^1(\Omega))},$$

$$\left|\iint_{Q_T} \vec{H}_{Nt}\varphi dxdt\right| = \left|-\beta\iint_{Q_T} \vec{Z}_{Nt}\varphi_N(x,t)dxdt - \iint_{Q_T} \nabla \times \vec{E}_N\varphi_N(x,t)dxdt\right|$$

$$\le C_7\|\varphi_N\|_{L^2(0,T;H^1(\Omega))} + C_8\int_0^T \|\vec{E}_N\|_2\|\nabla\varphi_N\|dt$$

$$\le C_7\|\varphi\|_{L^2(0,T;H^1(\Omega))} + C_8 T^{\frac{1}{2}}\|\varphi_N\|_{L^2(0,T;H^1(\Omega))}$$

$$\le C_9\|\varphi\|_{L^2(0,T;H^1(\Omega))}.$$

The lemma is proved.

Lemma 5.1.3 *Under the conditions of Lemma 5.1.1, for the solution* $(\vec{Z}_N(x,t),$ $\vec{H}_N(x,t), \vec{E}_N(x,t))$ *of the initial value problem* (5.1.9)–(5.1.12) *there are*
   (1) *when* $\alpha_2 = 0$,

$$\|\vec{Z}_N(\cdot,t_1) - \vec{Z}_N(\cdot,t_2)\|_2 + \|\vec{H}_N(\cdot,t_1) - \vec{H}_N(\cdot,t_2)\|_2 + \|\vec{E}_N(\cdot,t_1) - \vec{E}_N(\cdot,t_2)\|_2$$

$$\le K_3|t_1 - t_2|^{\frac{1}{3}}, \tag{5.1.30}$$

where the constant K_3 is independent of N, D.

    (2) *when $\alpha_2 > 0$,*

$$\|\vec{Z}_N(\cdot, t_1) - \vec{Z}_N(\cdot, t_2)\|_3 \le K_4 |t_1 - t_2|^{\frac{2}{3}}, \tag{5.1.31}$$

$$\|\vec{H}_N(\cdot, t_1) - \vec{H}_N(\cdot, t_2)\|_{H^{-1}(\Omega)} + \|\vec{E}_N(\cdot, t_1) - \vec{E}_N(\cdot, t_2)\|_{H^{-1}(\Omega)} \le K_5 |t_1 - t_2|^{\frac{1}{2}}, \tag{5.1.32}$$

where the constant K_4 is independent of N and D.

Proof. (1) When $\alpha_2 = 0$, by the Sobolev interpolation inequality of negative order, there is

$$\|\vec{Z}_N(\cdot, t_1) - \vec{Z}_N(\cdot, t_2)\|_2 \le C\|\vec{Z}_N(\cdot, t_1) - \vec{Z}_N(\cdot, t_2)\|_{H^{-2}(\Omega)}^{\frac{1}{3}} \|\vec{Z}_N(\cdot, t_1) - \vec{Z}_N(\cdot, t_2)\|_{H^1(\Omega)}^{\frac{2}{3}}$$

$$\le C' \left\| \int_{t_1}^{t_2} \frac{\partial \vec{Z}_N}{\partial t} dt \right\|_{H^{-2}(\Omega)}^{\frac{1}{3}} \le C'' |t_2 - t_1|^{\frac{1}{3}}.$$

Similarly, above inequality holds for $\|\vec{H}_N(\cdot, t_1) - \vec{H}_N(\cdot, t_2)\|_2$, and $\|\vec{E}_N(\cdot, t_1) - \vec{E}_N(\cdot, t_2)\|_2$.

    (2) When $\alpha_2 > 0$,

$$\|\vec{Z}_N(\cdot, t_1) - \vec{Z}_N(\cdot, t_2)\|_3 = \left\| \int_{t_1}^{t_2} \frac{\partial \vec{Z}_N}{\partial t} dt \right\|_3$$

$$\le |t_2 - t_1|^{\frac{2}{3}} \left(\iint_{Q_T} \left| \frac{\partial \vec{Z}_N}{\partial t} \right|^3 dx dt \right)^{\frac{1}{3}}$$

$$\le K_4 |t_2 - t_1|^{\frac{2}{3}},$$

$$\|\vec{H}_N(\cdot, t_1) - \vec{H}_N(\cdot, t_2)\|_{H^{-1}(\Omega)} = \left\| \int_{t_1}^{t_2} \frac{\partial \vec{H}_N}{\partial t} dt \right\|_{H^{-1}(\Omega)}$$

$$\le |t_2 - t_1|^{\frac{1}{2}} \left(\int_0^T \left\| \frac{\partial \vec{H}_N}{\partial t} \right\|_{H^{-1}(\Omega)}^2 dt \right)^{\frac{1}{2}}$$

$$\le K_5 |t_2 - t_1|^{\frac{1}{2}}.$$

For $\vec{E}_N(x, t_1) - \vec{E}_N(x, t_2)$ similar inequality holds. The lemma is proved.

    By using the above integral estimates of the approximate solution, we have

Lemma 5.1.4 *Under the conditions of Lemma 5.1.1, the initial value problem for the system of the ordinary differential equation (5.1.9)–(5.1.12) has at least one continuously differentiable and global solution $\vec{\alpha}_{sN}(t)$, $\vec{\beta}_{sN}(t)$, $\vec{\gamma}_{sN}(t)$ ($s = 1, 2, \ldots, N$, $t \in [0, T]$).*

5.1.3 Existence of Generalized Solution

Definition 5.1.5 *A triple of three-dimensional vector functions $\vec{Z}(x,t) \in L^{\infty}(0,T; H^1(\Omega))$, $\vec{H}(x,t) \in L^{\infty}(0,T; L^2(\Omega))$, $\vec{E}(x,t) \in L^{\infty}(0,T; L^2(\Omega))$ is called the generalized solution for the periodic initial value problem (5.1.5) and (5.1.6) of the system (5.1.1)–(5.1.4), if for any vector-valued test function $\vec{\psi}(x,t)$, and test function $\zeta(x,t) \in C^1(Q_T)$ with $\vec{\psi}(x,T) = 0$, $\vec{\psi}(x + De_i, t) = \vec{\psi}(x - De_i, t)$, $i = 1, 2, 3$, the following equalities hold*

$$\iint_{Q_T} \vec{Z}_t \cdot \vec{\psi}\,dxdt - \alpha_1 \iint_{Q_T} \vec{Z} \times (\Delta\vec{Z} + \vec{H}) \cdot \vec{\psi}\,dxdt$$

$$- \alpha_2 \iint_{Q_T} \vec{Z} \times (\Delta\vec{Z} + \vec{H}) \cdot \vec{Z} \times \vec{\psi}\,dxdt = 0, \quad (\alpha_2 > 0) \qquad (5.1.33)$$

or

$$\iint_{Q_T} \vec{Z}_t \cdot \vec{\psi}\,dxdt + \alpha_1 \iint_{Q_T} \vec{Z} \times \nabla\vec{Z} \cdot \nabla\vec{\psi}\,dxdt$$

$$-\alpha_1 \iint_{Q_T} (\vec{Z} \times \vec{H}) \cdot \vec{\psi}\,dxdt = 0, \quad (\alpha_2 = 0),$$

$$\iint_{Q_T} \vec{E} \cdot \vec{\psi}_t e^{\sigma t}\,dxdt + \iint_{Q_T} e^{\sigma t}\nabla \times \vec{\psi} \cdot \vec{H}(x,t)\,dxdt$$

$$+ \int_{\Omega} \vec{E}_0(x) \cdot \vec{\psi}(x,0)\,dx = 0, \qquad (5.1.34)$$

$$\iint_{Q_T} (\vec{H}(x,t) + \beta\vec{Z}(x,t))\vec{\psi}_t\,dxdt - \iint_{Q_T} \nabla \times \vec{\psi} \cdot \vec{E}(x,t)\,dxdt$$

$$+ \int_{\Omega} (\vec{H}_0(x) + \beta\vec{Z}_0(x)) \cdot \vec{\psi}(x,0)\,dx = 0, \qquad (5.1.35)$$

$$\iint_{Q_T} \nabla\zeta \cdot (\vec{H} + \beta\vec{Z})\,dxdt = 0, \qquad (5.1.36)$$

$$\iint_{Q_T} \nabla\zeta \cdot \vec{E}\,dxdt = 0, \qquad (5.1.37)$$

$$\vec{Z}(x,0) = \vec{Z}_0(x), \quad a.e. \ x \in \Omega. \qquad (5.1.38)$$

Lemma 5.1.5 *Let the initial value vector functions $\vec{E}_0(x), \vec{H}_0(x)$ satisfy the condition*

$$\int_{\Omega} \nabla\xi \cdot \vec{E}_0\,dx = 0, \qquad \int_{\Omega} \nabla\xi \cdot (\vec{H}_0(x) + \beta\vec{Z}_0(x))\,dx = 0, \qquad (5.1.39)$$

for all $\xi(x) \in C^1(\Omega)$. Then for any $\zeta(x,t) \in C^1(Q_T)$ with $\zeta(x,T) = 0$, $\zeta_0 = \zeta(x,0)$, we have, from (5.1.36), (5.1.37) that

$$\iint_{Q_T} \nabla\zeta \cdot \vec{E}(x,t)\,dxdt = 0, \qquad \iint_{Q_T} \nabla\zeta \cdot (\vec{H}(x,t) + \beta\vec{Z}(x,t))\,dxdt = 0.$$

Proof. Take

$$\vec{\psi}_t(x,t) = \int_0^t e^{-\sigma_\tau}\nabla\zeta(x,\tau)d\tau - \int_0^T e^{-\sigma_\tau}\nabla\zeta(x,\tau)d\tau, \qquad \zeta \in C^1(Q_T).$$

By (5.1.34) we get

$$\iint_{Q_T} \vec{E}\cdot\nabla\zeta dxdt + \int_0^T e^{-\sigma_\tau}d\tau \int_\Omega \nabla\zeta(x,\tau)\cdot\vec{E}_0(x)dx = 0.$$

Since

$$\int_\Omega \nabla\zeta(x,\tau)\cdot\vec{E}_0(x)dx = 0,$$

then

$$\iint_{Q_T} \vec{E}\cdot\nabla\zeta dxdt = 0.$$

By letting $\vec{\psi} = \int_0^t \nabla\zeta dt - \int_0^T \nabla\zeta dt$, we have, from (5.1.35)

$$\iint_{Q_T} (\vec{H}+\beta\vec{Z})\nabla\zeta dxdt - \int_0^T \int_\Omega (\vec{H}_0 + \beta\vec{Z}_0(x))\cdot\nabla\zeta dxd\tau = 0.$$

From

$$\int_\Omega (\vec{H}_0 + \beta\vec{Z}_0(x))\nabla\zeta dx = 0,$$

it follows that

$$\iint_{Q_T} (\vec{H}+\beta\vec{Z})\cdot\nabla\zeta dxdt = 0.$$

Theorem 5.1.1 *Assume $\vec{Z}_0(x) \in H^1(\Omega)$, $\vec{H}_0(x) \in L^2(\Omega)$, $\vec{E}_0(x) \in L^2(\Omega)$, and they are periodic functions with periodicity D, and satisfy (5.1.36). Then the periodic initial value problem (5.1.5) and (5.1.6) for the system of Landau–Lifshitz–Maxwell (5.1.1)–(5.1.4) has at least one generalized solution $\vec{Z}(x,t)$, $\vec{H}(x,t)$, $\vec{E}(x,t)$. When $\alpha_2 > 0$,*

$$\vec{Z}(x,t) \in L^\infty(0,T;H^1(\Omega)) \bigcap C^{(0,\frac{2}{3})}(0,T;L^3(\Omega)),$$
$$\vec{E}(x,t),\vec{H}(x,t) \in L^\infty(0,T;L^2(\Omega)) \bigcap C^{(0,\frac{1}{2})}(0,T;H^{-1}(\Omega)). \tag{5.1.40}$$

When $\alpha_2 = 0$,

$$\vec{Z}(x,t) \in L^\infty(0,T;H^1(\Omega)) \bigcap C^{(0,\frac{1}{2})}(0,T;L^2(\Omega)),$$
$$\vec{E}(x,t),\vec{H}(x,t) \in L^\infty(0,T;L^2(\Omega)) \bigcap C^{(0,\frac{1}{3})}(0,T;L^2(\Omega)). \tag{5.1.41}$$

Proof. For any vector-valued test function $\vec{\psi}(x,t) \in C^1(Q_T)$, $\vec{\psi}(x,T) = 0$, we define an approximate sequence

$$\vec{\psi}_N(x,t) = \sum_{n=1}^N \vec{\beta}_n(t)w_n(x),$$

where $\vec{\beta}_n(t) = \int_\Omega \vec{\psi}(x,t)w_n(x)dx$. We know that $\vec{\psi}_N$ is uniformly convergent to $\vec{\psi}(x,t)$ in $C^1(Q_T)$.

From the uniform estimate for the approximate solution $\left\{\vec{Z}_N(x,t), \vec{H}_N(x,t), \vec{E}_N(x,t)\right\}$ in Sec. 5.1.2, it follows that

$$
\begin{aligned}
\vec{Z}_N(x,t) &\rightarrow \vec{Z}(x,t) &&\text{weakly in } L^6(Q_T), \\
\vec{Z}_N(x,t) &\rightarrow \vec{Z}(x,t) &&\text{strongly in } L^{6-\eta}(Q_T), \quad (\eta > 0), \\
\vec{Z}_N(x,t) &\rightarrow \vec{Z}(x,t) &&\text{strongly in } L^\infty(0,T;H^1(\Omega)), \\
\vec{H}_N(x,t) &\rightarrow \vec{H}(x,t) &&\text{weakly in } L^\infty(0,T;L^2(\Omega)), \\
\vec{E}_N(x,t) &\rightarrow \vec{E}(x,t) &&\text{weakly in } L^\infty(0,T;L^2(\Omega)), \\
\vec{Z}_{Nt}(x,t) &\rightarrow \vec{Z}_t(x,t) &&\text{weakly in } L^{\frac{3}{2}}(Q_T), \quad (\alpha_2 > 0), \\
\vec{Z}_{Nt}(x,t) &\rightarrow \vec{Z}_t(x,t) &&\text{weakly in } L^\infty(0,T;H^{-2}(\Omega)), \quad (\alpha_2 = 0).
\end{aligned}
\tag{5.1.42}
$$

Making the scalar products of $\vec{\beta}_s(t)$ with (5.1.9), and $e^{\sigma t}\vec{\beta}_s$ with (5.1.10), and summing up the products for $s = 1, 2, \ldots$, we get

$$
\iint_{Q_T} \vec{Z}_{Nt} \cdot \vec{\psi}_N dxdt - \alpha_1 \iint_{Q_T} (\vec{Z}_N \times (\Delta\vec{Z}_N + \vec{H}_N)) \cdot \vec{\psi}_N dxdt
$$
$$
- \alpha_2 \iint_{Q_T} \vec{Z}_N \times (\Delta\vec{Z}_N + \vec{H}_N) \cdot (\vec{Z}_N \times \vec{\psi}_N)dxdt = 0, \tag{5.1.43}
$$

$$
\iint_{Q_T} (\vec{H}_{Nt} + \beta\vec{Z}_{Nt}) \cdot \vec{\psi}_N(x,t)dxdt = -\iint_{Q_T} \nabla \times \vec{E}_N \cdot \vec{\psi}_N(x)dxdt, \tag{5.1.44}
$$

$$
\iint_{Q_T} \frac{d}{dt}(e^{\sigma t}\vec{E}_N) \cdot \vec{\psi}_N(x,t)dxdt = \iint_{Q_T} e^{\sigma t}\nabla \times \vec{H}_N \cdot \vec{\psi}_N(x)dxdt. \tag{5.1.45}
$$

Rewriting (5.1.44), we get

$$
\iint_{Q_T} (\vec{H}_N + \beta\vec{Z}_N(x,t))\vec{\psi}_{Nt}dxdt + \int_\Omega (\vec{H}_N(x,0) + \beta\vec{Z}_N(x,0)) \cdot \vec{\psi}_N(x,0)dxdt
$$
$$
- \iint_{Q_T} \nabla \times \vec{\psi}_N \cdot \vec{E}_N(x,t)dxdt = 0. \tag{5.1.46}
$$

Rewriting (5.1.45), we get

$$
\iint_{Q_T} \vec{E}_N \cdot (\vec{\psi}_{Nt}e^{\sigma t})dxdt + \iint_{Q_T} e^{\sigma t}\nabla \times \vec{\psi}_N \cdot \vec{H}_N(x,t)dxdt
$$
$$
+ \int_\Omega \vec{E}_N(\cdot,0) \cdot \vec{\psi}_N(\cdot,0)dx = 0. \tag{5.1.47}
$$

(1) By Lemmas 5.1.1 and 5.1.2, we get

$$
\iint_{Q_T} \vec{Z}_{Nt} \cdot \vec{\psi}_N dxdt = \iint_{Q_T} \vec{Z}_{Nt} \cdot \vec{\psi}dxdt \rightarrow \iint_{Q_T} \vec{Z}_t\vec{\psi}dxdt,
$$

$$\iint_{Q_T}(\vec{H}_{Nt}+\beta\vec{Z}_{Nt})\vec{\psi}_N dxdt \to \iint_{Q_T}(\vec{H}_t+\beta\vec{Z}_t)\cdot\vec{\psi}dxdt,$$

$$\int_\Omega(\vec{H}_N(x,0)+\beta\vec{Z}_N(x,0))\vec{\psi}_N(x,0)dx \to \int_\Omega(\vec{H}_0(x)+\beta\vec{Z}_0(x))\vec{\psi}(x,0)dx.$$

(2) (i)

$$\iint_{Q_T}\nabla\times\vec{\psi}_N\cdot\vec{E}_N(x,t)dxdt \to \iint_{Q_T}\nabla\times\vec{\psi}\cdot\vec{E}dxdt,$$

(ii)

$$\iint_{Q_T}e^{\sigma t}\nabla\times\vec{\psi}_N\cdot\vec{H}_N dxdt \to \iint_{Q_T}e^{\sigma t}\nabla\times\vec{\psi}\cdot\vec{H}dxdt.$$

In fact

$$\iint_{Q_T}\nabla\times\vec{\psi}_N\cdot\vec{E}_N dxdt$$

$$=\iint_{Q_T}\nabla\times(\vec{\psi}_N-\vec{\psi})\cdot\vec{E}_N dxdt + \iint_{Q_T}(\nabla\times\vec{\psi})\cdot\vec{E}_N dxdt$$

$$=\iint_{Q_T}\nabla\times(\vec{\psi}_N-\vec{\psi})\cdot\vec{E}_N dxdt + \iint_{Q_T}(\nabla\times\vec{\psi})\cdot\vec{E}dxdt$$

$$+\iint_{Q_T}(\nabla\times\vec{\psi})(\vec{E}_N-\vec{E})dxdt.$$

As

$$\left|\iint_{Q_T}\nabla\times(\vec{\psi}_N-\vec{\psi})\cdot\vec{E}_N dxdt\right| \leq \left(\iint_{Q_T}|\nabla(\vec{\psi}_N-\vec{\psi})|^2 dxdt\right)^{\frac{1}{2}}\|\vec{E}_N\|_{L^2(Q_T)} \to 0,$$

$$\left|\iint_{Q_T}(\nabla\times\vec{\psi})\cdot(\vec{E}_N-\vec{E})dxdt\right| \to 0,$$

(i) is proved. Similarly (ii) can be proved.

(3) There exists a subsequence $\{\vec{Z}_N(x,t)\}$ such that, as $N\to+\infty$

(i)

$$\vec{Z}_N\times\frac{\partial\vec{Z}_N}{\partial x_i} \to \vec{Z}\times\frac{\partial\vec{Z}}{\partial x_i}\quad \text{weakly-}* \text{ in}\quad L^\infty(0,T;L^{\frac{3}{2}}(\Omega)),\quad (i=1,2,3). \qquad (5.1.48)$$

(ii)

$$\left(\vec{Z}_N\times\frac{\partial\vec{Z}_N}{\partial x_i}\right)_{x_i} \to \left(\vec{Z}\times\frac{\partial\vec{Z}}{\partial x_i}\right)_{x_i}\quad \text{weakly-}* \text{ in}\quad L^2(Q_T),\quad (i=1,2,3). \qquad (5.1.49)$$

In fact, for any test function $\vec{\psi}(x,t)\in C^1(Q_T)$, there is

$$\iint_{Q_T}(\vec{Z}_N\times\frac{\partial\vec{Z}_N}{\partial x_i}-\vec{Z}\times\frac{\partial\vec{Z}}{\partial x_i})\cdot\vec{\psi}dxdt$$

$$=\iint_{Q_T}\left[(\vec{Z}_N-\vec{Z})\times\frac{\partial\vec{Z}_N}{\partial x_i}\right]\cdot\vec{\psi}dxdt$$

$$+\iint_{Q_T}\left[\vec{Z}\times\left(\frac{\partial\vec{Z}_N}{\partial x_i}-\frac{\partial\vec{Z}}{\partial x_i}\right)\right]dxdt.$$

Since $\vec{Z}_N(x,t)$ is strongly convergent to $\vec{Z}(x,t)$ in $L^2(Q_T)$ and $\frac{\partial \vec{Z}_N}{\partial x_i}$ is weakly convergent to $\frac{\partial \vec{Z}}{\partial x_i}$ in $L^2(Q_T)$, and $\vec{Z}_N$ and $\frac{\partial \vec{Z}_N}{\partial x_i}$ are uniformly bounded in $L^2(Q_T)$ with respect to N, it is easy to prove that the right-hand side of the above equality is convergent to zero as $N \to +\infty$.

In the following we prove (ii). By Lemma 5.1.1, when $\alpha_2 > 0$, $\vec{Z}_N \times (\Delta \vec{Z}_N + \vec{H}_N)$ is uniformly bounded in $L^2(Q_T)$ with respect to N. Then there exists a vector function $\vec{V}(x,t) \in L^2(Q_T)$ such that for any test function $\vec{\psi}(x,t) \in C^1(Q_T)$, there holds

$$\left(\iint_{Q_T} (\vec{Z}_N \times (\Delta \vec{Z}_N + \vec{H}_N) \cdot \vec{\psi})dxdt \right.$$

$$\left. - \iint_{Q_T} (\vec{V}(x,t) + \vec{Z}_N \times \vec{H}_N) \cdot \vec{\psi}(x,t)dxdt \right) \to 0.$$

On the other hand, as $N \to +\infty$,

$$\iint_{Q_T} (\vec{Z}_N \times (\Delta \vec{Z}_N + \vec{H}_N)) \cdot \vec{\psi}dxdt$$

$$= - \iint_{Q_T} (\vec{Z}_N \times \nabla \vec{Z}_N \cdot \nabla \vec{\psi} + \vec{Z}_N \times \vec{H}_N \cdot \vec{\psi})dxdt$$

$$\to - \iint_{Q_T} (\vec{Z} \times \nabla \vec{Z}) \cdot \nabla \vec{\psi}dxdt + \iint_{Q_T} (\vec{Z} \times \vec{H}) \cdot \vec{\psi}dxdt,$$

where we have used (i) and the fact that

$$\left| \iint_{Q_T} (\vec{Z}_N \times \vec{H}_N - \vec{Z} \times \vec{H}) \cdot \vec{\psi}dxdt \right|$$

$$\leq \left| \iint_{Q_T} (\vec{Z}_N - \vec{Z}) \times \vec{H}_N \cdot \vec{\psi}dxdt \right| + \left| \iint_{Q_T} \vec{Z} \times (\vec{H}_N - \vec{H}) \cdot \vec{\psi}dxdt \right|$$

$$\leq \int_0^T \|\vec{Z}_N - \vec{Z}\|_5 \|\vec{H}_N\|_2 \|\vec{\psi}\|_{\frac{10}{3}} dt + \left| \iint_{Q_T} \vec{Z} \times (\vec{H}_N - \vec{H}) \cdot \vec{\psi}dxdt \right|$$

$$\to 0, \quad (N \to +\infty). \tag{5.1.50}$$

Then

$$\iint_{Q_T} \vec{V} \cdot \vec{\psi}dxdt = - \iint_{Q_T} (\vec{Z} \times \nabla \vec{Z}) \cdot \nabla \vec{\psi}dxdt,$$

$$\vec{Z} \times \Delta \vec{Z} = \vec{V} \in L^2(Q_T).$$

To prove the existence of the generalized solution it remains to prove

$$\iint_{Q_T} \vec{Z}_N \times (\Delta \vec{Z}_N + \vec{H}_N) \cdot (\vec{Z}_N \times \vec{\psi}_N)dxdt$$

$$\to \iint_{Q_T} \vec{Z} \times (\Delta \vec{Z} + \vec{H}) \cdot (\vec{Z} \times \vec{\psi})dxdt.$$

In fact

$$\iint_{Q_T} \vec{Z}_N \times (\Delta\vec{Z}_N + \vec{H}_N) \cdot (\vec{Z}_N \times \vec{\psi}_N) dxdt$$

$$- \iint_{Q_T} \vec{Z} \times (\Delta\vec{Z} + \vec{H}) \cdot (\vec{Z} \times \vec{\psi}) dxdt$$

$$= \iint_{Q_T} \left[(\vec{Z}_N \times \Delta\vec{Z}_N) - (\vec{Z} \times \Delta\vec{Z}) \right] \cdot \vec{Z} \times \vec{\psi} dxdt$$

$$+ \iint_{Q_T} \left[(\vec{Z}_N \times \Delta\vec{Z}_N) \cdot (\vec{Z}_N \times \vec{\psi}_N - \vec{Z} \times \vec{\psi}) \right] dxdt$$

$$+ \iint_{Q_T} (\vec{Z}_N \times \vec{H}_N - \vec{Z} \times \vec{H}) \cdot \vec{Z} \times \vec{\psi} dxdt$$

$$+ \iint_{Q_T} \vec{Z}_N \times \vec{H}_N \cdot (\vec{Z}_N \times \vec{\psi}_N - \vec{Z} \times \vec{\psi}) dxdt$$

$$= I_1^N + I_2^N + I_3^N + I_4^N.$$

By (5.1.49), $I_1^N \to 0$ as $N \to \infty$. Moreover,

$$|I_2^N| \le \|\vec{Z}_N \times \Delta\vec{Z}_N\|_{L^2(Q_T)} \left(\iint_{Q_T} |\vec{Z}_N \times \vec{\psi} - \vec{Z} \times \vec{\psi}|^2 dxdt \right)^{\frac{1}{2}}$$

$$\le C \iint_{Q_T} |\vec{Z}_N \times (\vec{\psi}_N - \vec{\psi}) + (\vec{Z}_N - \vec{Z}) \times \vec{\psi}|^2 dxdt \to 0.$$

$I_3^N \to 0$ $(N \to \infty)$, which is similar to (5.1.50).

$$|I_4^N| \le \|\vec{H}_N\|_{L^2(Q_T)} \|\vec{Z}_N\|_{L^4(Q_T)} \|\vec{Z}_N \times \vec{\psi}_N - \vec{Z} \times \vec{\psi}\|_{L^4(Q_T)} \to 0.$$

Therefore, by taking $N \to \infty$ in (5.1.43), (5.1.44) and (5.1.45), we obtain the limit function $\vec{Z}(x,t)$, $\vec{H}(x,t)$, $\vec{E}(x,t)$ satisfying the integral equality (5.1.33)–(5.1.35). Obviously (5.1.36)–(5.1.38) hold. The generalized solution of the periodic initial value problem (5.1.1)–(5.1.6) is obtained.

5.1.4   Existence of Solution for the Initial Problem

Note that the above *a priori* estimates are independent of D. By using the diagonal method and letting $D \to \infty$, we can get

Theorem 5.1.2 *Assume that the conditions of Theorem 5.1.1 hold in R^3. Then for the initial value problem (5.1.1)–(5.1.4) there exists one generalized solution satisfying*

$$\vec{Z}(x,t) \in L^\infty(0,T; H^1(R^3)) \bigcap C_{\text{loc}}^{(0,\frac{2}{3})}(0,T; L^3(R^3)), \quad (\alpha_2 > 0),$$

$$\vec{H}(x,t), \vec{E}(x,t) \in L^\infty(0,T; L^2(R^3)) \bigcap C_{\text{loc}}^{(0,\frac{1}{3})}(0,T; H^{-1}(R^3)),$$

$$\vec{Z}(x,t) \in L^\infty(0,T; H^1(R^3)) \bigcap C_{\text{loc}}^{(0,\frac{1}{2})}(0,T; H^{-1}(R^3)),$$

$$\vec{H}(x,t), \vec{E}(x,t) \in L^\infty(0,T; L^2(R^3)) \bigcap C_{\text{loc}}^{(0,\frac{1}{3})}(0,T; L^2(R^3)).$$

5.2 Global Smooth Solution in One or Two Dimensions with Small Initial Data

5.2.1 The Problem

In the above subsection, we have obtained the global weak solution to the periodic initial boundary value problem (5.1.1)–(5.1.6), or Cauchy problem (5.1.1)–(5.1.5) and (5.1.7) for Landau–Lifshitz–Maxwell equations. When $n \leq 2$, in this subsection, we prove the global existence of smooth solution to the periodic problem (5.1.1)–(5.1.6), or to the initial (5.1.1)–(5.1.5) and (5.1.7) with small initial data. For this aim we make the *a priori* estimates.

5.2.2 *A Priori* Estimates

Lemma 5.2.1 *Let $|\vec{Z}_0(x)| = 1$. Then for the smooth solution of the periodic initial value problem of the system* (5.1.1)–(5.1.6)

$$|\vec{Z}(x,t)| = 1, \qquad x \in \Omega, \qquad t \geq 0. \tag{5.2.1}$$

Proof. Making the scalar product of $\vec{Z}$ with (5.1.1), we get

$$\vec{Z} \cdot \vec{Z}_t = 0, \qquad |\vec{Z}(x,t)|_t^2 = 0.$$

Then

$$|\vec{Z}(x,t)| = |\vec{Z}_0(x)| = 1.$$

Lemma 5.2.2 *Assume $\beta \geq 0$, $\alpha_2 > 0$, $\sigma \geq 0$, $\nabla \vec{Z}_0(x) \in L_2(\Omega)$, $\vec{E}_0(x) \in L_2(\Omega)$, $\vec{H}_0(x) \in L_2(\Omega)$. Then for the smooth solution of the periodic initial value problem of the system* (5.1.1)–(5.1.6)

$$\sup_{0 \leq t \leq T} \left[\|\nabla \vec{Z}(\cdot,t)\|_{L_2(\Omega)}^2 + \|\vec{E}(\cdot,t)\|_{L_2(\Omega)}^2 + \|\vec{H}(\cdot,t)\|_{L_2(\Omega)}^2 \right] \leq K_1,$$

$$\int_0^T \|\vec{Z} \times (\Delta \vec{Z} + \vec{H})\|_{L_2}^2 dt \leq K_2, \tag{5.2.2}$$

where the constants K_1 and K_2 depend only on $\|\nabla \vec{Z}_0(x)\|_{L_2}$, $\|\vec{E}_0(x)\|_{L_2}$ and $\|\vec{H}_0(x)\|_{L_2}$.

Proof. First assume $\beta > 0$. Making the scalar product of $\vec{E}$ with (5.1.2), and making the scalar product of $-\vec{H}$ with (5.1.3), and then adding these two equalities obtained, we obtain

$$(\nabla \times \vec{H}) \cdot \vec{E} - (\nabla \times \vec{E}) \cdot \vec{H} = \frac{\partial \vec{E}}{\partial t} \cdot \vec{E} + \frac{\partial \vec{H}}{\partial t} \cdot \vec{H} + \beta \frac{\partial \vec{Z}}{\partial t} \cdot \vec{H} + \sigma |\vec{E}|^2. \tag{5.2.3}$$

By using the formula

$$(\nabla \times \vec{H}) \cdot \vec{E} - (\nabla \times \vec{E}) \cdot \vec{H} = \nabla \cdot (\vec{H} \times \vec{E}), \tag{5.2.4}$$

and integrating (5.2.3) over $x \in \Omega$, we get

$$\frac{1}{2}\frac{d}{dt}\left(\|\vec{E}\|^2_{L_2} + \|\vec{H}\|^2_{L_2}\right) + \sigma\|\vec{E}\|^2_{L_2} + \beta\int_\Omega \frac{\partial\vec{Z}}{\partial t}\cdot\vec{H}dx = 0. \qquad (5.2.5)$$

Making the scalar product of $(\Delta\vec{Z} + \vec{H})$ with (5.1.1), we get

$$(\Delta\vec{Z} + \vec{H})\cdot\frac{\partial\vec{Z}}{\partial t} = -\alpha_2(\Delta\vec{Z} + \vec{H})\cdot[\vec{Z}\times(\vec{Z}\times(\Delta\vec{Z} + \vec{H}))], \qquad (5.2.6)$$

where

$$-(\Delta\vec{Z} + \vec{H})\cdot[\vec{Z}\times(\vec{Z}\times(\Delta\vec{Z} + \vec{H}))] = |\vec{Z}\times(\Delta\vec{Z} + \vec{H})|^2. \qquad (5.2.7)$$

Integrating (5.2.6) over $x \in \Omega$, one gets

$$\int_\Omega(\Delta\vec{Z} + \vec{H})\cdot\frac{\partial\vec{Z}}{\partial t}dx = \alpha_2\int_\Omega |\vec{Z}\times(\Delta\vec{Z} + \vec{H})|^2dx,$$

and then

$$\int_\Omega \frac{\partial\vec{Z}}{\partial t}\cdot\vec{H}dx = -\int_\Omega \Delta\vec{Z}\cdot\frac{\partial\vec{Z}}{\partial t}dx + \alpha_2\int_\Omega |\vec{Z}\times(\Delta\vec{Z} + \vec{H})|^2dx$$

$$= \frac{1}{2}\frac{d}{dt}\|\nabla\vec{Z}\|^2_{L_2} + \alpha_2\int_\Omega |\vec{Z}\times(\Delta\vec{Z} + \vec{H})|^2dx. \qquad (5.2.8)$$

Adding (5.2.5) and (5.2.8), we obtain

$$\frac{1}{2}\frac{d}{dt}(\|\vec{E}\|^2_{L_2} + \|\vec{H}\|^2_{L_2}) + \sigma\|\vec{E}\|^2_{L_2}$$

$$+ \frac{\beta}{2}\frac{d}{dt}\|\nabla\vec{Z}\|^2_{L_2} + \beta\alpha_2\|\vec{Z}\times(\Delta\vec{Z} + \vec{H})\|^2_{L_2} = 0. \qquad (5.2.9)$$

Integrating the above equality with respect to $t \in [0, T]$, we have

$$\mathcal{E}(t) \equiv \frac{1}{2}(\|\vec{E}(\cdot, t)\|^2_{L_2} + \|\vec{H}(\cdot, t)\|^2_{L_2}) + \sigma\int_0^t \|\vec{E}(\cdot, t)\|^2_{L_2}dt$$

$$+ \frac{\beta}{2}\|\nabla\vec{Z}(\cdot, t)\|^2_{L_2} + \beta\alpha_2\int_0^t \|\vec{Z}\times(\Delta\vec{Z} + \vec{H})\|^2_{L_2}dt$$

$$= \mathcal{E}(0) = \frac{1}{2}(\|\vec{E}_0(x)\|^2_{L_2} + \|\vec{H}_0(x)\|^2_{L_2}) + \frac{\beta}{2}\|\nabla\vec{Z}_0(x)\|^2_{L_2}. \qquad (5.2.10)$$

It follows that the estimate (5.2.2) holds.

Now assume $\beta = 0$, then we get

$$\frac{1}{2}\frac{d}{dt}(\|\vec{E}\|^2_{L_2} + \|\vec{H}\|^2_{L_2}) + \sigma\|\vec{E}\|^2_{L_2} \doteq 0.$$

Therefore,

$$\|\vec{E}(\cdot, t)\|^2_{L_2} + \|\vec{H}(\cdot, t)\|^2_{L_2} \leq \|\vec{E}_0(x)\|^2_{L_2} + \|\vec{H}_0(x)\|^2_{L_2}. \qquad (5.2.11)$$

Note that

$$\int_\Omega \vec{Z}\cdot\vec{H}_t dx = -\int_\Omega \nabla\times\vec{E}\cdot\vec{Z}dx = -\int_\Omega \nabla\times\vec{Z}\cdot\vec{E}dx. \tag{5.2.12}$$

Adding (5.2.8) to (5.2.12) yields

$$\frac{d}{dt}\int_\Omega \vec{Z}\cdot\vec{H}dx = \frac{1}{2}\frac{d}{dt}\|\nabla\vec{Z}\|_{L_2}^2 + \alpha_2\int_\Omega |\vec{Z}\times(\Delta\vec{Z}+\vec{H})|^2 dx - \int_\Omega (\nabla\times\vec{Z},\vec{F})dx.$$

So,

$$\|\nabla\vec{Z}(\cdot,t)\|_{L_2}^2 + 2\alpha_2\int_0^t \|\vec{Z}\times(\Delta\vec{Z}+\vec{H})\|_{L_2}^2 dt$$

$$\leq 2\int_\Omega \vec{Z}(x,t)\cdot\vec{H}(x,t)dx - 2\int_\Omega \vec{Z}_0(x)\cdot\vec{H}_0(x)dx$$

$$+ \int_0^t \|\nabla\vec{Z}(\cdot,t)\|_{L_2}^2 dt + \int_0^t \|\vec{E}(\cdot,t)\|_{L_2}^2 dt$$

$$\leq \frac{1}{2}\|\nabla\vec{Z}(\cdot,t)\|_2^2 + C\int_0^t \|\vec{H}\|_2^2 dt + C_1$$

$$\leq C\int_0^t \|\nabla\vec{Z}\|_2^2 dt + C_2.$$

From the Gronwall inequality (5.2.2) is proved.

Lemma 5.2.3 (Gagliardo–Nirenberg Inequality) *Assume that $u\in L_q(\Omega)$, $D^m u\in L_r(\Omega)$, $\Omega\subset R^n$, $1\leq q$, $r\leq\infty$, $0\leq j\leq m$. Then*

$$\|D^j u\|_{L_p(\Omega)} \leq C\|u\|_{W_r^m(\Omega)}^a \|u\|_{L_q(\Omega)}^{1-a}, \tag{5.2.13}$$

where C is a positive constant, and

$$\frac{1}{p} = \frac{j}{n} + a\left(\frac{1}{r}-\frac{m}{n}\right) + (1-a)\frac{1}{q}, \qquad \frac{j}{m}\leq a\leq 1.$$

Lemma 5.2.4 *Assume that the conditions of Lemmas 5.2.1 and 5.2.2 hold, and $\vec{Z}_0(x)\in H^2(\Omega)$, $\vec{E}_0(x)$, $\vec{H}_0(x)\in H^1(\Omega)$, $\Omega\subset R^n$, $n=1,2$. When $n=2$, $\beta>0$, there is a small $\delta>0$ such that*

$$\|\nabla\vec{Z}_0\|_2^2 + \|\vec{E}_0(x)\|_2^2 + \|\vec{H}_0(x)\|_2^2 \leq \delta, \tag{5.2.14}$$

then

$$\sup_{0\leq t\leq T}\left[\|\Delta\vec{Z}(\cdot,t)\|_2^2 + \|\nabla\vec{H}(\cdot,t)\|_2^2 + \|\nabla\vec{E}(\cdot,t)\|_2^2\right]$$

$$+ \int_0^T \|\nabla\Delta\vec{Z}\|_2^2 dt \leq K_3 \tag{5.2.15}$$

holds, where K_3 depends only on $\|\vec{Z}_0(x)\|_{H^2(\Omega)}$, $\|\vec{H}_0(x)\|_{H^1(\Omega)}$ and $\|\vec{E}_0(x)\|_{H^1(\Omega)}$.

When $\beta=0$, $\vec{E}_0(x)\in H^2(\Omega)$, $\vec{H}_0(x)\in H^2(\Omega)$, there are

$$\sup_{0\leq t\leq T}\left[\|\Delta\vec{E}(\cdot,t)\|_2^2 + \|\Delta\vec{H}(\cdot,t)\|_2^2\right] \leq K_3', \tag{5.2.16}$$

where K_3' also depends on $\|\Delta\vec{E}_0\|_2$ and $\|\Delta\vec{H}_0\|_2$.

Proof. We know that under the condition $|\vec{Z}_0(x)| = 1$, Eq. (5.1.1) in the classic sense is equivalent to the equation

$$\vec{Z}_t = \alpha_2 \Delta\vec{Z} + \alpha_2|\nabla\vec{Z}|^2\vec{Z} + \alpha_1\vec{Z} \times (\Delta\vec{Z} + \vec{H}) - \alpha_2\vec{Z} \times (\vec{Z} \times \vec{H}). \tag{5.2.17}$$

Making the scalar product of $\Delta(\Delta\vec{Z} + \vec{H})$ with (5.2.17), and integrating the resulting equality with respect to $x \in \Omega$, we have

$$\frac{1}{2}\frac{\partial}{\partial t}\int_\Omega |\Delta\vec{Z}|^2 dx + \alpha_2\|\nabla\Delta\vec{Z}(\cdot, t)\|_2^2 + \int_\Omega \vec{Z}_t \cdot \Delta\vec{H}dx$$

$$= -\alpha_1 \int_\Omega \nabla\vec{Z} \times (\Delta\vec{Z} + \vec{H}) \cdot (\nabla\Delta\vec{Z} + \nabla\vec{H})dx$$

$$- \alpha_2 \int_\Omega (|\nabla\vec{Z}|^2\nabla\vec{Z} + 2(\nabla\vec{Z} \cdot \nabla^2\vec{Z})\vec{Z}) \cdot (\nabla\Delta\vec{Z} + \nabla\vec{H})dx$$

$$+ \alpha_2 \int_\Omega \nabla(\vec{Z} \times (\vec{Z} \times \vec{H})) \cdot (\nabla\Delta\vec{Z} + \nabla\vec{H})dx + \alpha_2 \int_\Omega \Delta\vec{Z} \cdot \Delta\vec{H}dx$$

$$= -\alpha_1 \int_\Omega ((\nabla\vec{Z} \times \Delta\vec{Z}) \cdot \nabla\Delta\vec{Z} + (\nabla\vec{Z} \times \vec{H}) \cdot \nabla\Delta\vec{Z} + (\nabla\vec{Z} \times \vec{H}) \cdot \nabla\vec{H})dx$$

$$+ \alpha_1 \int_\Omega (\nabla\vec{Z} \times \nabla\Delta\vec{Z} \cdot \vec{H})dx$$

$$- \alpha_2 \int_\Omega (|\nabla\vec{Z}|^2\nabla\vec{Z} \cdot \nabla\Delta\vec{Z} + 2(\nabla\vec{Z} \cdot \nabla^2\vec{Z})\vec{Z} \cdot \nabla\Delta\vec{Z}$$

$$+ |\nabla\vec{Z}|^2\nabla\vec{Z} \cdot \nabla\vec{H} + 2(\nabla\vec{Z} \cdot \Delta\vec{Z})\vec{Z} \cdot \nabla\vec{H})dx$$

$$+ \alpha_2 \int_\Omega \nabla(\vec{Z} \times (\vec{Z} \times \vec{H}))(\nabla\Delta\vec{Z} + \nabla\vec{H})dx - \alpha_2 \int_\Omega \nabla\Delta\vec{Z} \cdot \nabla\vec{H}dx$$

$$\leq (2\alpha_2 + |\alpha_1|)\|\nabla\vec{Z}\|_\infty\|\Delta\vec{Z}\|_2\|\nabla\Delta\vec{Z}\|_2 + 2|\alpha_1|\|\nabla\vec{Z}\|_\infty\|\vec{H}\|_2\|\nabla\Delta\vec{Z}\|_2$$

$$+ \alpha_1\|\vec{H}\|_2\|\nabla\vec{H}\|_2\|\nabla\vec{Z}\|_\infty + \alpha_2\|\nabla\vec{Z}\|_{L_6}^3\|\nabla\Delta\vec{Z}\|_2 + \alpha_2\|\nabla\vec{Z}\|_{L_6}^3\|\nabla\vec{H}\|_2$$

$$+ 2\alpha_2\|\nabla\vec{Z}\|_\infty\|\Delta\vec{Z}\|_2\|\nabla\vec{H}\|_2 + 2\alpha_2\|\vec{H}\|_2\|\nabla\vec{Z}\|_\infty\|\nabla\Delta\vec{Z}\|_2$$

$$+ 2\alpha_2\|\vec{H}\|_2\|\nabla\vec{Z}\|_\infty\|\nabla\vec{H}\|_2 + \alpha_2\|\nabla\vec{H}\|_2^2 + \alpha_2\|\nabla\vec{H}\|_2\|\nabla\Delta\vec{Z}\|_2. \tag{5.2.18}$$

By using Lemma 5.2.3 and interpolation formulas and Poincaré inequality

$$\|\nabla\vec{Z}\|_{L_6}^6 \leq C\|\nabla\Delta\vec{Z}\|_2^n\|\Delta\vec{Z}\|_2^{6-n},$$

$$\|\Delta\vec{Z}\|_2^2 \leq C\|\nabla\Delta\vec{Z}\|_2\|\nabla\vec{Z}\|_2, \tag{5.2.19}$$

$$\|\nabla\vec{Z}\|_\infty \leq C\|\nabla\Delta\vec{Z}\|_2^{\frac{n}{4}}\|\nabla\vec{Z}\|_2^{1-\frac{n}{4}},$$

where

$$\int_\Omega \nabla\vec{Z}dx = 0,$$

and substituting (5.2.19) into (5.2.18), we get

$$\frac{1}{2}\frac{\partial}{\partial t}\int_\Omega |\Delta\vec{Z}|^2 dx + \alpha_2\|\nabla\Delta\vec{Z}\|_2^2 + \int_\Omega \vec{Z}_t\cdot\Delta\vec{H}dx$$

$$\leq (2\alpha_2 + |\alpha_1|)\|\nabla\vec{Z}\|_2^{\frac{1}{2}-\frac{n}{4}}\|\nabla\Delta\vec{Z}\|_2^{\frac{3}{2}+\frac{n}{4}} + C\alpha_2\|\nabla\vec{Z}\|_2^{3-\frac{n}{2}}\|\nabla\Delta\vec{Z}\|_2^{1+\frac{n}{2}}$$

$$+ C\alpha_1\|\nabla\vec{Z}\|_2^{1-\frac{n}{4}}\|\nabla\Delta\vec{Z}\|_2^{1+\frac{n}{4}} + C\alpha_2\|\nabla\vec{Z}\|_2^{3-\frac{n}{2}}\|\nabla\Delta\vec{Z}\|_2^{\frac{n}{2}}\|\nabla\vec{H}\|_2$$

$$+ C|\alpha_1|\|\nabla\vec{H}\|_2\|\nabla\Delta\vec{Z}\|_2^{\frac{n}{4}}\|\nabla\vec{Z}\|_2^{1-\frac{n}{4}}$$

$$+ C\alpha_2\|\nabla\vec{Z}\|_2^{1-\frac{n}{4}}\|\nabla\Delta\vec{Z}\|_2^{1+\frac{n}{4}} + C\alpha_2\|\nabla\vec{H}\|_2\|\nabla\Delta\vec{Z}\|_2^{\frac{n}{4}}$$

$$+ \alpha_2\|\nabla\vec{H}\|_2^2 + \alpha_2\|\nabla\vec{H}\|_2\|\nabla\Delta\vec{Z}\|_2$$

$$\leq \begin{cases} \frac{7\alpha_2}{M}\|\nabla\Delta\vec{Z}\|_2^2 + C_1\|\nabla\vec{H}\|_2^2 + d_1, & n = 1 \\ \left[(2\alpha_2 + |\alpha_1|)\|\nabla\vec{Z}\|_2 + C\alpha_2\|\nabla\vec{Z}\|_2^2\right]\|\nabla\Delta\vec{Z}\|_2^2 \\ \quad + \frac{5\alpha_2}{M}\|\nabla\Delta\vec{Z}\|_2^2 + C_2\|\nabla\vec{H}\|_2^2 + d_2, & n = 2, \end{cases} \tag{5.2.20}$$

where C_i and d_i $(i = 1, 2)$ are the constants independent of t, D and M is a constant to be chosen later.

Here the inequality (5.2.20) will be discussed for $n = 1$ and 2, respectively.

(1) For $n = 1$, letting $M = 14$, then (5.2.20) yields

$$\frac{1}{2}\frac{d}{dt}\int_\Omega |\Delta\vec{Z}|^2 dx + \frac{\alpha_2}{2}\int_\Omega |\nabla\Delta\vec{Z}|^2 dx + \int_\Omega \vec{Z}_t\cdot\vec{H}dx \leq C\|\nabla\vec{H}\|_2^2 + d. \tag{5.2.21}$$

(2) For $n = 2$, take $M = 10$. If $\|\nabla\vec{Z}_0\|_2 + \|\vec{E}_0(x)\|_2 + \|\vec{H}_0(x)\|_2$ is small enough, then by Lemma 5.2.2 there holds

$$(2\alpha_2 + |\alpha_1|)\|\nabla\vec{Z}\|_2 + C\alpha_2\|\nabla\vec{Z}\|_2^2 \leq \frac{\alpha_2}{2}. \tag{5.2.22}$$

So, (5.2.20) is still true.

Next making the products of $\Delta\vec{E}$ with (5.1.2) and $\Delta\vec{H}$ with (5.1.3), respectively, and summing the two equalities, and then integrating the resulting equality with respect to $x \in \Omega$, we obtain $(\beta > 0)$

$$\frac{1}{2}\frac{d}{dt}\int_\Omega (|\nabla\vec{E}|^2 + |\nabla\vec{H}|^2)dx + \frac{\sigma}{\beta}\int_\Omega |\nabla\vec{E}|^2 dx - \int_\Omega \vec{Z}_t\cdot\Delta\vec{H}dx = 0, \tag{5.2.23}$$

where

$$\int_\Omega (\nabla\times\vec{H}\cdot\Delta\vec{E} - \nabla\times\vec{E}\cdot\Delta\vec{H})dx$$

$$= -\int_\Omega (\nabla\times\nabla\vec{H}\cdot\nabla\vec{E} - \nabla\times\nabla\vec{E}\cdot\nabla\vec{H})dx = 0. \tag{5.2.24}$$

Combining (5.2.21) with (5.2.23) we have

$$\frac{d}{dt}\int_\Omega |\Delta\vec{Z}|^2 dx + \frac{1}{\beta}\frac{d}{dt}\int_\Omega (|\nabla\vec{Z}|^2 + |\nabla\vec{H}|^2)dx$$
$$+ \alpha_2 \int_\Omega |\nabla\Delta\vec{Z}|^2 dx \le C\|\nabla\vec{H}\|_2^2 + d. \tag{5.2.25}$$

From the Gronwall inequality it follows that (5.2.15) is proved.

When $\beta = 0$, by making the products of $\Delta^2\vec{E} - \Delta\vec{E}$ with (5.1.2) and $\Delta^2\vec{H} - \Delta\vec{H}$ with (5.1.3), respectively, and then integrating the resulting equalities with respect to $x \in \Omega$, we obtain

$$\int_\Omega \vec{E}_t(\Delta^2\vec{E} - \Delta\vec{E})dx + \sigma\int_\Omega \vec{E}(\Delta^2\vec{E} - \Delta\vec{E})dx = \int_\Omega \nabla\times\vec{H}(\Delta^2\vec{E} - \Delta\vec{E})dx,$$
$$\int_\Omega \vec{H}_t(\Delta^2\vec{H} - \Delta\vec{H})dx = -\int_\Omega \nabla\times\vec{E}(\Delta^2\vec{H} - \Delta\vec{H})dx.$$

It is easy to see that

$$\frac{1}{2}\frac{d}{dt}\int_\Omega (|\Delta\vec{E}|^2 + |\nabla\vec{E}|^2 + |\Delta\vec{H}|^2 + |\nabla\vec{H}|^2)dx \le 0,$$

i.e. there are

$$\sup_{t\in[0,T]} \left[\|\Delta\vec{E}(\cdot,t)\|_2^2 + \|\nabla\vec{E}(\cdot,t)\|_2^2 + \|\Delta\vec{H}(\cdot,t)\|_2^2\|\nabla\vec{H}(\cdot,t)\|_2^2\right]$$
$$\le \|\Delta\vec{E}_0\|_2^2 + \|\nabla\vec{E}_0\|_2^2 + \|\Delta\vec{H}_0\|_2^2 + \|\nabla\vec{H}_0\|_2^2 \le K'. \tag{5.2.26}$$

Note that

$$|\vec{Z}_t|^2 = \alpha_1^2|\vec{Z}\times(\Delta\vec{Z}+\vec{H})|^2 + \alpha_2^2|\vec{Z}\times(\vec{Z}\times(\Delta\vec{Z}+\vec{H}))|^2$$
$$\le (\alpha_1^2 + \alpha_2^2)|\Delta\vec{Z}+\vec{H}|^2 \le 2(\alpha_1 + \alpha_2^2)(|\Delta\vec{Z}|^2 + |\vec{H}|^2),$$

$$\left|\int_\Omega \vec{Z}_t\cdot\Delta\vec{H}dx\right| \le (\alpha_1^2 + \alpha_2^2)(\|\Delta\vec{Z}\|_2^2 + \|\vec{H}\|_2^2) + 2\|\Delta\vec{H}\|_2^2.$$

From (5.2.20) it follows that

$$\frac{1}{2}\frac{d}{dt}\int_\Omega |\Delta\vec{Z}|^2 dx + \frac{\alpha_2}{2}\|\nabla\Delta\vec{Z}\|_2^2 \le C_3\|\nabla\Delta\vec{Z}\|_2^2 + d_3. \tag{5.2.27}$$

By the Gronwall inequality and (5.2.25), we get (5.2.16).

Lemma 5.2.5 *Assume that $(\vec{Z}(x,t), \vec{H}(x,t), \vec{E}(x,t))$ is the smooth solution of the problem (5.1.1)–(5.1.6). If the conditions of Lemma 5.2.4 hold and $(\vec{Z}_0(x),\ \vec{H}_0(x),\ \vec{F}_0(x)) \in (H^m(\Omega), H^{m-1}(\Omega), H^{m-1}(\Omega))\ (m \geq 2)$, then for any $T > 0$,*

$$\sup_{t\in[0,T]} \left[\|\vec{Z}(\cdot,t)\|^2_{H^m(\Omega)} + \|\vec{H}(\cdot,t)\|^2_{H^{m-1}(\Omega)} + \|\vec{E}(\cdot,t)\|^2_{H^{m-1}(\Omega)} \right] \leq K_m, \qquad (5.2.28)$$

where the constant K_m is independent of D, $\Omega \subset R^n$, $n \leq 2$.

Proof. The lemma will be proved by the induction for m. Obviously, when $m = 2$, (5.2.28) is true. Now suppose that (5.2.28) hold for $m = K(\geq 2)$, i.e.

$$\sup_{t\in[0,T]} \left(\sum_{l\leq K-2} \|\nabla^l \vec{Z}\|^2_\infty + \sum_{l\leq K} \|\nabla^l \vec{Z}\|^2_2 + \sum_{l\leq K-1} (\|\nabla^l \vec{E}\|^2_2 + \|\nabla^l \vec{H}\|^2_2) \right) \leq E_l. \qquad (5.2.29)$$

We shall prove (5.2.28) holds for $m = K + 1$.

(1) Making the scalar product of $\Delta^K(\Delta\vec{Z} + \vec{H})$ with (5.2.17), and integrating the resulting equality with respect to $x \in \Omega$, we have

$$\begin{aligned}
(\vec{Z}_t, \Delta^K \vec{H}) &+ \frac{1}{2}(-1)^{K+1}\frac{d}{dt}\int_\Omega |\nabla^{K+1}\vec{Z}|^2 dx \\
&= (-1)^K \alpha_2 \int_\Omega |\nabla^{K+2}\vec{Z}|^2 dx \\
&\quad + (-1)^K \alpha_1 \int_\Omega \nabla^K(\vec{Z} \times (\Delta\vec{Z} + \vec{H}))(\nabla^{K+2}\vec{Z} + \nabla^K \vec{H})dx \\
&\quad + \alpha_2(-1)^K \int_\Omega \nabla^K(|\nabla\vec{Z}|^2\vec{Z}) \cdot (\nabla^{K+2}\vec{Z} + \nabla^K \vec{H})dx \\
&\quad - \alpha_2(-1)^K \int_\Omega \nabla^K(\vec{Z} \times (\vec{Z} \times \vec{H})) \cdot (\nabla^{K+2}\vec{Z} + \nabla^K \vec{H})dx.
\end{aligned}$$

Then

$$\begin{aligned}
\frac{1}{2}\frac{d}{dt}&\int_\Omega |\nabla^{K+1}\vec{Z}|^2 dx + \alpha_2\|\nabla^{K+2}\vec{Z}\|^2_2 - (-1)^K \int_\Omega \vec{Z}_t \cdot \nabla^K \vec{H} dx \\
&\leq \frac{\alpha_2}{4}\|\nabla^{K+2}\vec{Z}\|^2_2 + C\|\nabla^K \vec{H}\|^2_2 \\
&\quad + \alpha_1^2 \int_\Omega |\nabla^K(\vec{Z} \times (\Delta\vec{Z} + \vec{H})) - \vec{Z} \times \nabla^K(\Delta\vec{Z} + \vec{H})|^2 dx \\
&\quad + \alpha_2 \int_\Omega |\nabla^K(|\nabla\vec{Z}|^2\vec{Z})|^2 dx + \alpha_2 \int_\Omega |\nabla^K(\vec{Z} \times (\vec{Z} \times \vec{H}))|^2 dx, \qquad (5.2.30)
\end{aligned}$$

where for the third term on the right-hand side of the inequality (5.2.30)

$$\begin{aligned}
\nabla^K(\vec{Z} \times (\Delta\vec{Z} + \vec{H})) &- \vec{Z} \times \nabla^K(\Delta\vec{Z} + \vec{H}) \\
&= \sum_{i=0}^{K-1} C_K^i \nabla^{K-i}\vec{Z} \times (\nabla^{i+2}\vec{Z} + \nabla^i \vec{H})
\end{aligned}$$

$$= K\nabla\vec{Z} \times (\nabla^{K+1}\vec{Z} + \nabla^{K-1}\vec{H}) + \nabla^K\vec{Z} \times (\Delta\vec{Z} + \vec{H})$$

$$+ \sum_{i=1}^{K-2} C_K^i \nabla^{K-i}\vec{Z} \times (\nabla^{i+2}\vec{Z} + \nabla^i\vec{H})$$

$$\leq K|\nabla\vec{Z}||\nabla^{K+1}\vec{Z}| + K|\nabla\vec{Z}||\nabla^{K-1}\vec{H}| + |\nabla^K\vec{Z}||\Delta\vec{Z}|$$

$$+ |\nabla^K\vec{Z}||\vec{H}| + C\sum_{i=1}^{K-2} |\nabla^{K-i}\vec{Z}||\nabla^{i+2}\vec{Z}| + C\sum_{i=1}^{K-2} |\nabla^{K-i}\vec{Z}||\nabla^i\vec{H}|,$$

$$\|\nabla^K((\vec{Z} \times (\Delta\vec{Z} + \vec{H})) - \vec{Z} \times \nabla^K(\Delta\vec{Z} + \vec{H})\|_2^2$$

$$\leq C_1\|\nabla\vec{Z}\|_\infty^2\|\nabla^{K+1}\vec{Z}\|_2^2 + C_2\|\nabla\vec{Z}\|_\infty^2 + C_3\|\nabla^K\vec{Z}\|_\infty^2$$

$$+ C_3\sum_{i=1}^{K-2} \|\nabla^{K-i}\vec{Z}\|_\infty^2\|\nabla^{i+2}\vec{Z}\|_2^2 + C_4\sum_{i=1}^{K-2} \|\nabla^{K-i}\vec{Z}\|_\infty^2\|\nabla^i\vec{H}\|_2^2$$

$$\leq C_5\left[\|\nabla^{K+2}\vec{Z}\|_2^{1+\frac{n}{K+1}} + \|\nabla^{K+2}\vec{Z}\|_2^{\frac{n}{K+1}} + \|\nabla^{K+2}\vec{Z}\|_2^{\frac{n}{2}} + \|\nabla^{K+2}\vec{Z}\|_2^{1+\frac{n}{3}}\right] + C_6$$

$$\leq \frac{\alpha_2}{M}\|\nabla^{K+2}\vec{Z}\|_2^2 + C_7, \quad (M \geq 3), \tag{5.2.31}$$

where the Sobolev interpolation inequality in Lemma 5.2.3 have been used

$$\|\nabla^l\vec{Z}\|_\infty^2 \leq C\|\nabla^{K+2}\vec{Z}\|_2^{\frac{n}{2(K+2-l)}}\|\nabla^l\vec{Z}\|_2^{1-\frac{n}{2(K+2-l)}}, \quad (l \leq K),$$

$$\|\nabla^{K+1}\vec{Z}\|_2^2 \leq C\|\nabla^{K+2}\vec{Z}\|_2\|\nabla^K\vec{Z}\|_2.$$

For the fourth term on the right-hand side of the inequality (5.2.30)

$$\nabla^K(|\nabla\vec{Z}|^2\vec{Z}) = \sum_{i=0}^{K} C_K^i \nabla^i\vec{Z}\nabla^{K-i}(\nabla\vec{Z} \cdot \nabla\vec{Z})$$

$$= \sum_{i=0}^{K}\sum_{j=0}^{K-i} C_K^i C_{K-i}^j \nabla^i\vec{Z}(\nabla^{K+1-i-j}\vec{Z} \cdot \nabla^{j+1}\vec{Z})$$

$$= \sum_{j=0}^{K} C_K^j \vec{Z}(\nabla^{K+1-j}\vec{Z}, \nabla^{j+1}\vec{Z})$$

$$+ \sum_{i=1}^{K}\sum_{j=0}^{K-i} C_K^i C_{K-i}^j \nabla^i\vec{Z}(\nabla^{K+1-i-j} \cdot \nabla^{j+1}\vec{Z})$$

$$= 2\vec{Z}(\nabla^{K+1}\vec{Z} \cdot \nabla\vec{Z}) + \sum_{j=1}^{K-1} C_K^j (\nabla^{K+1-j}\vec{Z} \cdot \nabla^{j+1}\vec{Z})\vec{Z}$$

$$+ \sum_{i=1}^{K}\sum_{j=0}^{K-i} C_K^i C_{K-i}^j \nabla^i\vec{Z}(\nabla^{K+1-i-j}\vec{Z} \cdot \nabla^{j+1}\vec{Z}).$$

Then

$$\|\nabla^K(|\nabla\vec{Z}|^2\vec{Z})\|_2^2 \leq 2\|\nabla\vec{Z}\|_\infty^2\|\nabla^{K+1}\vec{Z}\|_\infty^2 + C\sum_{j=1}^{K-1}\|\nabla^{K+1-j}\vec{Z}\|_\infty^2\|\nabla^{j+1}\vec{Z}\|_2^2$$

$$+ C\sum_{i=1}^{K}\sum_{j=0}^{K-i}\|\nabla^i\vec{Z}\|_\infty^2\|\nabla^{j+1}\vec{Z}\|_\infty^2\|\nabla^{K+1-i-j}\vec{Z}\|_2^2$$

$$\leq C_1\|\nabla^{K+2}\vec{Z}\|_2^{1+\frac{n}{K+1}} + C_2(\|\nabla^K\vec{Z}\|_\infty^2 + \|\nabla^{K-1}\vec{Z}\|_\infty^2)$$

$$+ C_3\sum_{i=1}^{K}\sum_{j=0}^{K-i}\|\nabla^{K+2}\vec{Z}\|_2^{\frac{n}{K+2-i}+\frac{n}{K+1-j}}. \qquad (5.2.32)$$

Notice that

(i) when $K \geq 2$, and $n \leq 2$, there is $1 + \frac{n}{K+1} < 2$;

(ii) when $1 \leq i \leq K$ and $0 \leq j \leq K - i$, there is

$$\frac{1}{K+2-i} + \frac{1}{K+1-j} < \frac{1}{2} + \frac{1}{i+1} < 2;$$

(iii) $\|\nabla^K\vec{Z}\|_\infty^2 \leq C\|\nabla^K\vec{Z}\|_2^{2-\frac{n}{2}}\|\nabla^{K+2}\vec{Z}\|_2^{\frac{n}{2}} \leq C\|\nabla^{K+1}\vec{Z}\|_2^{\frac{n}{2}}$;

(iv) $\|\nabla^{K-1}\vec{Z}\|_\infty^2 \leq C\|\nabla^{K+2}\vec{Z}\|_2^{\frac{n}{3}}$.

From (5.2.32) we obtain

$$\|\nabla^K(|\nabla\vec{Z}|^2\vec{Z})\|_2^2 \leq \frac{\alpha_2}{M}\|\nabla^{K+2}\vec{Z}\|_2^2 + C. \qquad (5.2.33)$$

By the same way we can prove for the last term on the right-hand side of the inequality (5.2.30)

$$\|\nabla^K(\vec{Z}\times(\vec{Z}\times\vec{H}))\|_2^2 \leq \frac{\alpha_2}{M}\|\nabla^{K+2}\vec{Z}\|_2^2 + C\|\nabla^K\vec{H}\|_2^2 + C'. \qquad (5.2.34)$$

Combining (5.2.30), (5.2.31), (5.2.33), and (5.2.34), and taking $M \geq 3$, we get

$$\frac{1}{2}\frac{d}{dt}\int_\Omega |\nabla^{K+1}\vec{Z}|^2 dx - (-1)^K\int_\Omega \vec{Z}_t\cdot\Delta^K\vec{H}dx \leq C\|\nabla^K\vec{H}\|_2^2 + C''. \qquad (5.2.35)$$

(2) Making the scalar products of $\Delta^K\vec{E}$ with (5.1.2) and $\Delta^K\vec{H}$ with (5.1.3), respectively, and integrating the resulting equalities with respect to $x \in \Omega$, we have

$$\frac{1}{2}(-1)^K\frac{d}{dt}\int_\Omega|\nabla^K\vec{E}|^2 dx + (-1)^K\sigma\|\nabla^K\vec{E}\|_2^2 = \int_\Omega\nabla\times\vec{H}\cdot\Delta^K\vec{E}dx,$$

$$\frac{1}{2}(-1)^K\frac{d}{dt}\int_\Omega|\nabla^K\vec{H}|^2 dx + \beta\int_\Omega\vec{Z}_t\cdot\Delta^K\vec{H}dx = -\int_\Omega\nabla\times\vec{E}\cdot\Delta^K\vec{H}dx.$$

Summing up the two equalities above gives

$$\frac{1}{2}\frac{d}{dt}\int_\Omega(|\nabla^K\vec{Z}|^2 + |\nabla^K\vec{H}|^2)dx + \sigma\|\nabla^K\vec{E}\|_2^2 + (-1)^K\beta\int_\Omega\vec{Z}_t\cdot\Delta^K\vec{H}dx$$

$$= \int_\Omega(\nabla\times\nabla^K\vec{H})\cdot\nabla^K\vec{E} - \nabla\times(\nabla^K\vec{E})\cdot\nabla^K\vec{H})dx$$

$$= \int_\Omega\nabla\cdot(\nabla^K\vec{H}\times\nabla^K\vec{E})dx = 0. \qquad (5.2.36)$$

From (5.2.35) and (5.2.36) it follows that

$$\frac{1}{2}\frac{d}{dt}\left[\int_{\Omega}|\nabla^{K+1}\vec{Z}|^2 dx + \frac{1}{\beta}\int_{\Omega}(|\nabla^K\vec{Z}|^2 + |\nabla^K\vec{H}|^2)dx\right]$$
$$\leq C\|\nabla^K\vec{H}\|_2^2 + C'.$$

Then (5.2.28) is true for $m = K + 1$. The proof of the lemma is complete.

5.2.3 Existence of Global Smooth Solution

By using the following result about the local existence of the smooth solution for (5.1.1)–(5.1.6) and the *a priori* estimates of the smooth solution, we can obtain the global existence of the smooth solution for the problem (5.1.1)–(5.1.6) from the extension principle.

By the Galerkin method, we can prove

Theorem 5.2.1 (Local existence of the smooth solution) *Assume the constants* $\alpha_2 > 0$, $\beta \geq 0$, $\sigma \geq 0$, $(\vec{Z}_0(x), \vec{H}_0(x), \vec{E}_0(x)) \in (H^k(\Omega), H^{k-1}(\Omega), H^{k-1}(\Omega))$ $k \geq 3 + \left[\frac{n}{3}\right]$, $n \leq 2$, $\nabla \cdot (\vec{H}_0 + \beta\vec{Z}_0) = 0$, $\nabla \cdot \vec{E}_0 = 0$. *Then the periodic initial value problem* (5.1.1)–(5.1.6) *has a local smooth solution*

$$\vec{Z}(x,t) \in \cap_{s=0}^{\left[\frac{k}{2}\right]}W_{\infty}^s(0,T_0;H^{k-2s}(\Omega)),$$

$$\vec{H}(x,t), \vec{E}(x,t) \in \cap_{s=0}^{k-1}W_{\infty}^s(0,T_0;H^{k-1-s}(\Omega)).$$

Remark. From the relations

$$\nabla \cdot \vec{H}_0 + \beta\nabla \cdot \vec{Z}_0 = 0, \qquad \nabla \cdot \vec{E}_0 = 0,$$

we can deduce that

$$\nabla \cdot \vec{H} + \beta\nabla \cdot \vec{Z} = 0, \qquad \nabla \cdot \vec{E} = 0.$$

In fact, by applying the divergence operator $\nabla\cdot$ for (5.1.2) and (5.1.3), we get

$$\frac{\partial}{\partial t}\nabla \cdot \vec{E} + \sigma\nabla \cdot \vec{E} = 0, \qquad \frac{\partial}{\partial t}\nabla \cdot (\vec{H} + \beta\vec{Z}) = 0,$$

and so the assertions above hold.

From the Theorem 5.2.1 and *a priori* estimates in Sec. 5.2.2 it follows that

Theorem 5.2.2 *Assume that the constants* $\alpha_2 > 0$, $\beta \geq 0$, $\sigma \geq 0$, *the initial value functions* $(\vec{Z}_0(x), \vec{H}_0(x), \vec{E}_0(x)) \in (H^k(\Omega), H^{k-1}(\Omega), H^{k-1}(\Omega))$, $k \geq 3 + \left[\frac{n}{2}\right]$, $\Omega \subset R^n$ $(n \leq 2)$ *is a bounded domain, and* $|\vec{Z}_0(x)| = 1$, $\nabla \cdot \vec{E}_0 = 0$, $\nabla \cdot (\vec{H}_0 + \beta\vec{Z}_0) = 0$. *Moreover, when* $n = 2$, *there is*

$$\|\nabla\vec{Z}_0\|_2 + \|\vec{E}_0\|_2 + \|\vec{H}_0\|_2 \leq \delta,$$

where δ is a suitable small constant. Then the periodic problem (5.1.1)–(5.1.6) has one global smooth solution such that $|\vec{Z}(z,t)| = 1$, $x \in \Omega$, $t \in R^+$, and

$$\vec{Z}(x,t) \in \cap_{s=0}^{\left[\frac{k}{2}\right]} W_\infty^s(0,T; H^{k-2s}(\Omega)),$$

$$\vec{H}(x,t) \in \cap_{s=0}^{k-1} W_\infty^s(0,T; H^{k-1-s}(\Omega)),$$

$$\vec{E}(x,t) \in \cap_{s=0}^{k-1} W_\infty^s(0,T; H^{k-1-s}(\Omega)).$$

Noticing that all the constants K of the estimates are independent of the periodic D, and letting $D \to \infty$, we can obtain the solution of the Cauchy problem.

Theorem 5.2.3 *Under the conditions of Theorem 5.2.2, the Cauchy problem (5.1.1)–(5.1.5) and (5.1.7) has one global smooth solution $\vec{Z}(x,t)$, $\vec{H}(x,t)$, $\vec{E}(x,t)$ such that*

$$|\vec{Z}(x,t)| = 1, \qquad \forall\ (x,t) \in R^n \times R^+, \quad (n \le 2),$$

$$\nabla \vec{Z}(x,t) \in \cap_{s=0}^{\left[\frac{k}{2}\right]} W_\infty^s(0,T; H^{k-1-2s}(\Omega)),$$

$$\vec{H}(x,t) \in \cap_{s=0}^{k-1} W_\infty^s(0,T; H^{k-1-s}(\Omega)),$$

$$\vec{E}(x,t) \in \cap_{s=0}^{k-1} W_\infty^s(0,T; H^{k-1-s}(\Omega)),$$

where $\Omega = R^2$.

Theorem 5.2.4 *Assume that the conditions of Theorem 5.2.2 are satisfied. Then the global smooth solution for the periodic initial value problem (the Cauchy problem) for Landau–Lifshitz–Maxwell system is unique.*

Proof. Let $(\vec{Z}_i, \vec{H}_i, \vec{E}_i)$ $(i = 1,2)$ be the smooth solutions for the problem (5.1.1)–(5.2.6), or (5.1.1)–(5.1.5) and (5.1.7). Denote

$$\begin{aligned}
\vec{Z}(x,t) &= \vec{Z}_1(x,t) - \vec{Z}_2(x,t), \\
\vec{H}(x,t) &= \vec{H}_1(x,t) - \vec{H}_2(x,t), \\
\vec{E}(x,t) &= \vec{E}_1(x,t) - \vec{E}_2(x,t).
\end{aligned} \tag{5.2.37}$$

Then $\vec{Z}(x,t)$, $\vec{H}(x,t)$ and $\vec{E}(x,t)$ satisfy the following system:

$$\begin{aligned}
\vec{Z}_t = {}& \alpha_1 \vec{Z} \times \Delta \vec{Z}_2 + \alpha_1 \vec{Z}_1 \times \Delta \vec{Z} + \alpha_1 \vec{Z} \times \vec{H}_2 + \alpha_1 \vec{Z}_1 \times \vec{H} + \alpha_2 \Delta \vec{Z} \\
& + \alpha_2 |\nabla \vec{Z}_2|^2 \vec{Z} + \alpha_2 (\nabla \vec{Z} \cdot \nabla(\vec{Z}_1 + \vec{Z}_2)) \vec{Z}_1 + \alpha_2 \vec{H} \\
& - \alpha_2 (\vec{Z}_2 \cdot \vec{H}_2) \vec{Z} - \alpha_2 (\vec{Z}_2 \cdot \vec{H} + \vec{H}_1 \cdot \vec{Z}) \vec{Z}_1,
\end{aligned} \tag{5.2.38}$$

$$\vec{E}_t + \sigma \vec{E} = \nabla \times \vec{H}, \tag{5.2.39}$$

$$\vec{H}_t + \beta \vec{Z}_t = -\nabla \times \vec{E}, \tag{5.2.40}$$

$$\vec{Z}(x,0) = \vec{H}(x,0) = \vec{E}(x,0) = 0. \tag{5.2.41}$$

(1) Making the scalar product of $\vec{Z}$ with (5.2.38) and integrating the equality obtained with respect to $x \in \Omega$, we get

$$\frac{1}{2}\frac{d}{dt}\int_{\Omega}|\vec{Z}|^2 dx + \alpha_2\|\nabla\vec{Z}\|_2^2$$

$$\leq C(|\alpha_1| + \alpha_2)(\|\nabla\vec{Z}\|_2^2 + \|\vec{Z}\|_2^2 + \|\vec{H}\|_2^2). \tag{5.2.42}$$

(2) Making the scalar product of $\Delta\vec{Z}$ with (5.2.38) and integrating the equality obtained with respect to $x \in \Omega$, we get

$$\frac{1}{2}\frac{d}{dt}\int_{\Omega}|\nabla\vec{Z}|^2 dx + \alpha_2\|\Delta\vec{Z}\|_2^2$$

$$= -\alpha_1\int_{\Omega}\vec{Z}\times\Delta\vec{Z}_2\cdot\Delta\vec{Z}dx - \alpha_1\int_{\Omega}\vec{Z}\times\vec{H}_2\cdot\Delta\vec{Z}dx$$

$$- \alpha_1\int_{\Omega}\vec{Z}_1\times\vec{H}\cdot\Delta\vec{Z}dx - \alpha_2\int_{\Omega}|\nabla\vec{Z}_2|^2\vec{Z}\cdot\Delta\vec{Z}dx$$

$$- \alpha_2\int_{\Omega}(\nabla\vec{Z}\cdot(\nabla(\vec{Z}_1 + \vec{Z}_2))\vec{Z}_1\cdot\Delta\vec{Z}dx + \alpha_2\int_{\Omega}\vec{H}\cdot\Delta\vec{Z}dx$$

$$+ \alpha_2\int_{\Omega}(\vec{Z}_2\cdot\vec{H}_2)\vec{Z}\cdot\Delta\vec{Z}dx + \alpha_2\int_{\Omega}(\vec{Z}_2\cdot\vec{H} + \vec{H}_1\cdot\vec{Z})\vec{Z}_1\cdot\Delta\vec{Z}dx. \tag{5.2.43}$$

Now, we estimate each term on the right-hand side of (5.2.43).

$$\left|-\alpha_1\int_{\Omega}\vec{Z}\times\Delta\vec{Z}_2\cdot\Delta\vec{Z}dx\right| = \left|\alpha_1\int_{\Omega}\vec{Z}\times\nabla\Delta\vec{Z}_2\cdot\nabla\vec{Z}dx\right|$$

$$\leq |\alpha_1|\|\nabla\Delta\vec{Z}\|_\infty\|\vec{Z}\|_2\|\nabla\vec{Z}\|_2 \leq C|\alpha_1|(\|\vec{Z}\|_2^2 + \|\nabla\vec{Z}\|_2^2),$$

$$\left|-\alpha_1\int_{\Omega}\vec{Z}\times\vec{H}_2\cdot\Delta\vec{Z}dx\right| = \left|\alpha_1\int_{\Omega}\vec{Z}\times\nabla\vec{H}_2\cdot\nabla\vec{Z}dx\right|$$

$$\leq |\alpha_1|\|\nabla\vec{H}_2\|_\infty\|\vec{Z}\|_2\|\nabla\vec{Z}\|_2 \leq C|\alpha_1|(\|\vec{Z}\|_2^2 + \|\nabla\vec{Z}\|_2^2),$$

$$\left|-\alpha_2\int_{\Omega}|\nabla\vec{Z}_2|^2\vec{Z}\cdot\Delta\vec{Z}dx\right| \leq \alpha_2\|\nabla\vec{Z}_2\|_\infty^2\|\vec{Z}\|_2\|\Delta\vec{Z}\|_2$$

$$\leq \frac{\alpha_2}{K}\|\nabla\vec{Z}\|_2^2 + C(K)\|\vec{Z}\|_2^2, \quad (K \geq 6),$$

$$\left|-\alpha_2\int_{\Omega}\nabla\vec{Z}\cdot(\nabla(\vec{Z}_1 + \vec{Z}_2))\vec{Z}_1\cdot\Delta\vec{Z}dx\right| \leq \alpha_2\|\nabla(\vec{Z}_1 + \vec{Z}_2)\|_\infty\|\nabla\vec{Z}\|_2\|\Delta\vec{Z}\|_2$$

$$\leq \frac{\alpha_2}{K}\|\Delta\vec{Z}\|_2^2 + C(K)\alpha_2\|\nabla\vec{Z}\|_2^2,$$

$$\left|\alpha_2\int_{\Omega}\vec{H}\cdot\Delta\vec{Z}dx\right| \leq \alpha_2\|\vec{H}\|_2\|\Delta\vec{Z}\|_2$$

$$\leq \frac{\alpha_2}{K}\|\Delta\vec{Z}\|_2^2 + C(K)\alpha_2\|\vec{H}\|_2^2,$$

$$\left|\alpha_2\int_{\Omega}(\vec{Z}_2\cdot\vec{H}_2)\vec{Z}\cdot\Delta\vec{Z}dx\right| \leq \alpha_2\|\vec{H}_2\|_\infty\|\vec{Z}\|_2\|\Delta\vec{Z}\|_2$$

$$\leq \frac{\alpha_2}{K}\|\Delta\vec{Z}\|_2^2 + C(K)\alpha_2\|\vec{Z}\|_2^2,$$

$$\left|\alpha_2 \int_\Omega (\vec{Z}_2 \cdot \vec{H} + \vec{H}_1 \cdot \vec{Z}) \cdot \Delta\vec{Z} dx\right| \leq \alpha_2(\|\vec{H}\|_2\|\Delta\vec{Z}\|_2 + \|\vec{H}_1\|_\infty\|\vec{Z}\|_2\|\Delta\vec{Z}\|_2)$$

$$\leq \frac{\alpha_2}{K}\|\Delta\vec{Z}\|_2^2 + C(K)\alpha_2(\|\vec{H}\|_2^2 + \|\vec{Z}\|_2^2).$$

Then (5.2.43) can be rewritten as

$$\frac{1}{2}\frac{d}{dt}\int_\Omega |\nabla\vec{Z}| dx + \alpha_2\|\Delta\vec{Z}\|_2^2 + \alpha_1 \int_\Omega \vec{Z}_1 \times \vec{H} \cdot \Delta\vec{Z} dx$$

$$\leq \frac{5}{K}\|\Delta\vec{Z}\|_2^2 + C(K)\alpha_2(\|\nabla\vec{Z}\|_2^2 + \|\vec{Z}\|_2^2 + \|\vec{H}\|_2^2)$$

$$+ C|\alpha_1|(\|\vec{Z}\|_2^2 + \|\nabla\vec{Z}\|_2^2). \tag{5.2.44}$$

(3) Making the scalar products of $\vec{E}$ with (5.2.39) and $\vec{H}$ with (5.2.40), respectively, and integrating the equalities obtained with respect to $x \in \Omega$, and then summing up the two equations, we get

$$\frac{1}{2\beta}\frac{d}{dt}\int_\Omega (|\vec{E}|^2 + |\vec{H}|^2) dx + \frac{\sigma}{\beta}\|\vec{E}\|_2^2 = -\int_\Omega \vec{Z}_t \cdot \vec{H} dx. \tag{5.2.45}$$

From (5.2.42), (5.2.44) and (5.2.45) it follows that

$$\frac{1}{2}\frac{d}{dt}\int_\Omega \left[|\vec{E}|^2 + |\vec{H}|^2 + \frac{1}{\beta}(|\vec{E}|^2 + |\vec{H}|^2)\right] dx + \alpha_2\|\Delta\vec{Z}\|_2^2$$

$$\leq -\int_\Omega (\vec{Z}_t \cdot \vec{H} + \alpha_1\vec{Z} \times \vec{H} \cdot \Delta\vec{Z}) dx + \frac{5}{K}\alpha_2\|\Delta\vec{Z}\|_2^2$$

$$+ C|\alpha_1|(\|\nabla\vec{Z}\|_2^2 + \|\vec{Z}\|_2^2 + \|\vec{H}\|_2^2)$$

$$+ C(K)\alpha_2(\|\nabla\vec{Z}\|_2^2 + \|\vec{Z}\|_2^2 + \|\vec{H}\|_2^2). \tag{5.2.46}$$

(4) Making the scalar product of $\vec{H}$ with (5.2.38), and integrating the equality obtained with respect to $x \in \Omega$, we get

$$\left|\int_\Omega (\vec{Z}_t \cdot \vec{H} + \alpha_1\vec{Z}_1 \times \vec{H} \cdot \Delta\vec{Z}) dx\right|$$

$$\leq |\alpha_1|\left|\int_\Omega (\vec{Z} \times \Delta\vec{Z}_2 + \vec{Z} \times \vec{H}_2) \cdot \vec{H} dx\right| + \alpha_2\left|\int_\Omega \Delta\vec{Z} \cdot \vec{H} dx\right|$$

$$+ \alpha_2\left|\int_\Omega |\nabla\vec{Z}_2|^2\vec{Z} \cdot \vec{H} dx\right| + \alpha_2\left|\int_\Omega (\nabla\vec{Z} \cdot \nabla(\vec{Z}_1 + \vec{Z}_2)\vec{Z}_1 \cdot \vec{H} dx\right|$$

$$+ \alpha_2\|\vec{H}\|_2^2 + \alpha_2\left|\int_\Omega (\vec{Z}_2 \cdot \vec{H}_2)\vec{Z} \cdot \vec{H} dx\right|$$

$$+ \alpha_2\left|\int_\Omega (\vec{Z}_2 \cdot \vec{H})(\vec{Z}_1 \cdot \vec{H}) dx\right| + \alpha_2\left|\int_\Omega (\vec{H}_1 \cdot \vec{Z})\vec{Z}_1 \cdot \vec{H} dx\right|$$

$$\leq \frac{\alpha_2}{K}\|\Delta\vec{Z}\|_2^2 + (C(K)\alpha_2 + C|\alpha_1|)(\|\nabla\vec{Z}\|_2^2 + \|\vec{Z}\|_2^2 + \|\vec{H}\|_2^2). \tag{5.2.47}$$

Substituting (5.2.47) into (5.2.46), and taking $K \geq 6$, we get

$$\frac{d}{dt} \int_\Omega \left[|\nabla \vec{Z}|^2 + |\vec{Z}|^2 + \frac{1}{\beta}(|\vec{E}|^2 + |\vec{H}|^2) \right] dx$$

$$\leq C(\alpha_2 + |\alpha_1|)(\|\nabla \vec{Z}\|_2^2 + \|\vec{Z}\|_2^2 + \|\vec{H}\|_2^2).$$

From the Gronwall inequality it follows that $\vec{Z}(x,t) = \vec{H}(x,t) = \vec{E}(x,t) \equiv 0$, $\forall \ (x,t) \in \Omega \times (0,T)$.

When $\beta = 0$, by (5.2.45) we obtain

$$\frac{d}{dt}(\|\vec{E}\|_2^2 + \|\vec{H}\|_2^2) + \sigma\|\vec{E}\|_2^2 = 0.$$

Then $\vec{E}(x,t) \equiv \vec{H}(x,t) \equiv 0$. From (5.2.42) and (5.2.44) it follows that $\vec{Z}(x,t) \equiv 0$. The proof of the theorem is complete. $\qquad\qquad\square$

5.3   Global Smooth Solution to One-Dimensional L–L–M with Large Data

In Sec. 5.1, we have obtained the global weak solutions to three-dimensional Landau–Lifshitz–Maxwell equations and, in Sec. 5.2, we proved the existence and uniqueness of global smooth solution to this system with small initial data. Now we are going to derive the existence and uniqueness of global smooth solution to this system with large initial data in one dimension. Especially, we deal with the equations without Gilbert damping, that is, $\alpha_2 = 0$ in (5.1.1). So it states as follows:

$$\vec{Z}_t = \alpha_1 \vec{Z} \times (\Delta \vec{Z} + \vec{H}), \tag{5.3.1}$$

$$\nabla \times \vec{H} = \frac{\partial \vec{E}}{\partial t} + \sigma \vec{E}, \tag{5.3.2}$$

$$\nabla \times \vec{E} = -\frac{\partial \vec{H}}{\partial t} - \beta \frac{\partial \vec{Z}}{\partial t}, \tag{5.3.3}$$

$$\nabla \cdot \vec{H} + \beta \nabla \cdot \vec{Z} = 0, \quad \nabla \cdot \vec{E} = 0, \tag{5.3.4}$$

with the periodic initial value condition:

$$\vec{Z}(x + 2D, t) = \vec{Z}(x,t), \quad \vec{H}(x + 2D, t) = \vec{H}(x,t),$$

$$\vec{E}(x + 2D, t) = \vec{E}(x,t), \quad (x \in R, t \geq 0) \tag{5.3.5}$$

$$\vec{Z}(x,0) = \vec{Z}_0(x), \quad \vec{H}(x,0) = \vec{H}_0(x), \quad \vec{E}(x,0) = \vec{E}_0(x), \quad (x \in R), \tag{5.3.6}$$

where $D > 0$; or with the initial value condition:

$$\vec{Z}(x,0) = \vec{Z}_0(x), \quad \vec{H}(x,0) = \vec{H}_0(x), \quad \vec{E}(x,0) = \vec{E}_0(x), \quad x \in R. \tag{5.3.7}$$

5.3.1 Viscosity Vanishing Method

In one space dimension we shall prove the existence of the global smooth solution of the periodic problem (5.3.1)–(5.3.6) or the initial problem (5.3.1)–(5.3.4) and (5.3.7) by the viscosity vanishing method. It means that we approximate (5.3.1) by an equation with viscosity term of small parameter, i.e. the following periodic initial value problem is considered:

$$\vec{Z}_t = \alpha_1 \vec{Z} \times (\Delta\vec{Z} + \vec{H}) - \varepsilon\vec{Z} \times (\vec{Z} \times (\Delta\vec{Z} + \vec{H})), \tag{5.3.8}$$

$$\nabla \times \vec{H} = \frac{\partial \vec{E}}{\partial t} + \sigma\vec{E}, \tag{5.3.9}$$

$$\nabla \times \vec{E} = -\frac{\partial \vec{H}}{\partial t} - \beta\frac{\partial \vec{Z}}{\partial t}, \tag{5.3.10}$$

$$\nabla \cdot \vec{H} + \beta\nabla \cdot \vec{Z} = 0, \quad \nabla \cdot \vec{E} = 0, \tag{5.3.11}$$

$$\vec{Z}(x + 2D, t) = \vec{Z}(x, t), \quad \vec{H}(x + 2D, t) = \vec{H}(x, t),$$

$$\vec{E}(x + 2D, t) = \vec{E}(x, t), \quad (x \in R, t \geq 0) \tag{5.3.12}$$

$$\vec{Z}(x, 0) = \vec{Z}_0(x), \quad \vec{H}(x, 0) = \vec{H}_0(x), \quad \vec{E}(x, 0) = \vec{E}_0(x), \quad (x \in R) \tag{5.3.13}$$

and the *a priori* estimate independent of ε is proved, then the solution of the problem (5.3.1)–(5.3.6) is obtained as $\varepsilon \to 0$.

Lemma 5.3.1 *Let* $|\vec{Z}_0(x)| = 1$. *Then for the smooth solution of the periodic initial value problem* (5.3.8)–(5.3.13) *there holds*

$$|\vec{Z}(x, t)| = 1, \qquad x \in \Omega \equiv [0, 2D], \quad t \geq 0. \tag{5.3.14}$$

Proof. Making the scalar product of $\vec{Z}$ with (5.3.8), we get

$$\frac{\partial}{\partial t}|\vec{Z}(x, t)|^2 = 0.$$

Then the conclusion of the lemma is proved.

Lemma 5.3.2 *Assume* $\beta \geq 0$, $\sigma \geq 0$, $\nabla\vec{Z}_0(x) \in L_2(\Omega)$, $\vec{E}_0(x) \in L_2(\Omega)$, $\vec{H}_0(x) \in L_2(\Omega)$. *Then for the smooth solution of the periodic initial value problem* (5.3.8)–(5.3.13)

$$\sup_{0 \leq t \leq T} \left[\|\nabla\vec{Z}(\cdot, t)\|_2^2 + \|\vec{E}(\cdot, t)\|_2^2 + \|\vec{H}(\cdot, t)\|_2^2\right] \leq K_1,$$

$$\varepsilon\int_0^T \|\vec{Z} \times (\Delta\vec{Z} + \vec{H})\|_2^2 dt \leq K_2, \tag{5.3.15}$$

where the constants K_1 *and* K_2 *depend only on* $\|\nabla\vec{Z}_0(x)\|_2$, $\|\vec{E}_0(x)\|_2$ *and* $\|\vec{H}_0(x)\|_2$, *but not on* ε *and* D.

Proof. First assume $\beta > 0$. Making the scalar product of $\vec{E}$ with (5.3.9), and making the scalar product of $-\vec{H}$ with (5.3.10), and then adding these two equalities obtained, we have

$$(\nabla \times \vec{H}) \cdot \vec{E} - (\nabla \times \vec{E}) \cdot \vec{H} = \frac{\partial \vec{E}}{\partial t} \cdot \vec{E} + \sigma |\vec{E}|^2 + \frac{\partial \vec{H}}{\partial t} \cdot \vec{H} + \beta \frac{\partial \vec{Z}}{\partial t} \cdot \vec{H}. \qquad (5.3.16)$$

By using the formula

$$(\nabla \times \vec{H}) \cdot \vec{E} - (\nabla \times \vec{E}) \cdot \vec{H} = \nabla \cdot (\vec{H} \times \vec{E}), \qquad (5.3.17)$$

and integrating (5.3.16) over $x \in \Omega$, we get

$$\frac{1}{2} \frac{d}{dt} \left(\|\vec{E}\|_2^2 + \|\vec{H}\|_2^2 \right) + \sigma \|\vec{E}\|_2^2 + \beta \int_\Omega \frac{\partial \vec{Z}}{\partial t} \cdot \vec{H} dx = 0. \qquad (5.3.18)$$

Making the scalar product of $(\Delta \vec{Z} + \vec{H})$ with (5.3.8), we get

$$(\Delta \vec{Z} + \vec{H}) \cdot \frac{\partial \vec{Z}}{\partial t} = -\varepsilon(\Delta \vec{Z} + \vec{H}) \cdot [\vec{Z} \times (\vec{Z} \times (\Delta \vec{Z} + \vec{H}))], \qquad (5.3.19)$$

where

$$-(\Delta \vec{Z} + \vec{H}) \cdot [\vec{Z} \times (\vec{Z} \times (\Delta \vec{Z} + \vec{H}))] = |\vec{Z} \times (\Delta \vec{Z} + \vec{H})|^2. \qquad (5.3.20)$$

From integrating (5.3.19) over $x \in \Omega$ it follows that

$$\int_\Omega (\Delta \vec{Z} + \vec{H}) \cdot \frac{\partial \vec{Z}}{\partial t} dx = \varepsilon \int_\Omega |\vec{Z} \times (\Delta \vec{Z} + \vec{H})|^2 dx,$$

and then

$$\int_\Omega \frac{\partial \vec{Z}}{\partial t} \cdot \vec{H} dx = -\int_\Omega \Delta \vec{Z} \cdot \frac{\partial \vec{Z}}{\partial t} dx + \varepsilon \int_\Omega |\vec{Z} \times (\Delta \vec{Z} + \vec{H})|^2 dx$$

$$= \frac{1}{2} \frac{d}{dt} \|\nabla \vec{Z}\|_2^2 + \varepsilon \int_\Omega |\vec{Z} \times (\Delta \vec{Z} + \vec{H})|^2 dx. \qquad (5.3.21)$$

By adding (5.3.18) and (5.3.21) we obtain

$$\frac{1}{2} \frac{d}{dt} (\|\vec{E}\|_2^2 + \|\vec{H}\|_2^2) + \sigma \|\vec{E}\|_2^2$$

$$+ \frac{\beta}{2} \frac{d}{dt} \|\nabla \vec{Z}\|_2^2 + \beta \varepsilon \|\vec{Z} \times (\Delta \vec{Z} + \vec{H})\|_2^2 = 0. \qquad (5.3.22)$$

Integrating the above equality with respect to $t \in [0, T]$, we have

$$\mathcal{E}(t) \equiv \frac{1}{2} (\|\vec{E}(\cdot, t)\|_2^2 + \|\vec{H}(\cdot, t)\|_2^2) + \sigma \|\vec{E}(\cdot, t)\|_2^2$$

$$+ \frac{\beta}{2} \|\nabla \vec{Z}(\cdot, t)\|_2^2 + \beta \varepsilon \int_0^t \|\vec{Z} \times (\Delta \vec{Z} + \vec{H})\|_2^2 dt$$

$$= \mathcal{E}(0) = \frac{1}{2} (\|\vec{E}_0(x)\|_2^2 + \|\vec{H}_0(x)\|_2^2)$$

$$+ \sigma \|\vec{E}_0(x)\|_2^2 + \frac{\beta}{2} \|\nabla \vec{Z}_0(x)\|_2^2. \qquad (5.3.23)$$

It follows that the estimate (5.3.15) holds. Now assume $\beta = 0$. Then Eq. (5.3.10) becomes

$$\nabla \times \vec{E} = -\frac{\partial \vec{H}}{\partial t}.$$

Combining the equality above with the equation (5.3.9), we get

$$\frac{1}{2}\frac{d}{dt}(\|\vec{E}\|_2^2 + \|\vec{H}\|_2^2) + \sigma\|\vec{E}\|_2^2 = 0.$$

Therefore,

$$\|\vec{E}(\cdot,t)\|_2^2 + \|\vec{H}(\cdot,t)\|_2^2 \leq \|\vec{E}_0(x)\|_2^2 + \|\vec{H}_0(x)\|_2^2. \tag{5.3.24}$$

Note that

$$\int_\Omega \vec{Z}\cdot\vec{H}_t dx = -\int_\Omega \nabla\times\vec{E}\cdot\vec{Z}dx = -\int_\Omega \nabla\times\vec{Z}\cdot\vec{E}dx. \tag{5.3.25}$$

Adding (5.3.21) and (5.3.25) yields

$$\frac{d}{dt}\int_\Omega \vec{Z}\cdot\vec{H}dx = \frac{1}{2}\frac{d}{dt}\|\nabla\vec{Z}\|_2^2 + \varepsilon\|\vec{Z}\times(\Delta\vec{Z}+\vec{H})\|_2^2 - \int_\Omega (\nabla\times\vec{Z})\cdot\vec{E}dx.$$

So,

$$\|\nabla\vec{Z}(\cdot,t)\|_2^2 + 2\varepsilon\int_0^t \|\vec{Z}\times(\Delta\vec{Z}+\vec{H})\|_2^2 dt$$

$$\leq 2\int_\Omega \vec{Z}(x,t)\cdot\vec{H}(x,t)dx - 2\int_\Omega \vec{Z}_0(x)\cdot\vec{H}_0(x)dx$$

$$+ \int_0^t \|\nabla\vec{Z}(\cdot,t)\|_2^2 dt + \int_0^t \|\vec{E}(\cdot,t)\|_2^2 dt$$

$$\leq \frac{1}{2}\|\nabla\vec{Z}(\cdot,t)\|_2^2 + C\int_0^t \|\nabla\vec{Z}(\cdot,t)\|_2^2 dt + C_1$$

$$\leq C\int_0^t \|\nabla\vec{Z}(\cdot,t)\|_2^2 dt + C_2.$$

From the Gronwall inequality (5.3.15) is proved.

Lemma 5.3.3 *Assume that the conditions of Lemmas 5.3.1 and 5.3.2 hold, and $\vec{Z}_0(x) \in H^2(\Omega)$, $\vec{E}_0(x)$, $\vec{H}_0(x) \in H^1(\Omega)$. Then there holds*

$$\sup_{0\leq t\leq T} \left[\|\Delta\vec{Z}(\cdot,t)\|_2^2 + \|\nabla\vec{H}(\cdot,t)\|_2^2 + \|\nabla\vec{E}(\cdot,t)\|_2^2\right] \leq K_3, \tag{5.3.26}$$

$$\sup_{0\leq t\leq T} \left[\|\vec{Z}_t(\cdot,t)\|_2^2 + \|\vec{H}_t(\cdot,t)\|_2^2 + \|\vec{E}_t(\cdot,t)\|_2^2\right] \leq K_4, \tag{5.3.27}$$

where K_3 and K_4 depend only on $\|\vec{Z}_0(x)\|_{H^2(\Omega)}$, $\|\vec{H}_0(x)\|_{H^1(\Omega)}$ and $\|\vec{E}_0(x)\|_{H^1(\Omega)}$, but not on ε and D.

Proof. The proof of this lemma is longer than others. In the following the constants C_i, d_j, C, d are the constants independent of ε and D. First, by the argument used in [20], we can prove that under the condition $|\vec{Z}_0(x)| = 1$, Eq. (5.3.8) in the classic sense is equivalent to the equation

$$\vec{Z}_t = \alpha_1 \vec{Z} \times (\Delta \vec{Z} + \vec{H}) + \varepsilon \Delta \vec{Z} + \varepsilon |\nabla \vec{Z}|^2 \vec{Z} - \varepsilon(\vec{Z} \cdot \vec{H})\vec{Z} + \varepsilon \vec{H}. \tag{5.3.28}$$

Differentiating (5.3.28) with respect to t we get

$$\vec{Z}_{tt} = \alpha_1 \vec{Z}_t \times (\Delta + \vec{H}) + \alpha_1 \vec{Z} \times (\Delta \vec{Z}_t + \vec{H}_t) + \varepsilon \Delta \vec{Z}_t + \varepsilon |\nabla \vec{Z}|_t^2 \vec{Z}$$
$$+ \varepsilon |\nabla \vec{Z}|^2 \vec{Z}_t - \varepsilon(\vec{Z} \cdot \vec{H})_t \vec{Z} - \varepsilon(\vec{Z} \cdot \vec{H})\vec{Z}_t + \varepsilon \vec{H}_t. \tag{5.3.29}$$

(I) Making the scalar product of $\vec{Z}_t$ with the above equality, and integrating the resulting equality with respect to $x \in \Omega$, we have

$$\frac{1}{2}\frac{d}{dt}\int_\Omega |\vec{Z}_t|^2 dx + \varepsilon \|\nabla \vec{Z}_t\|_2^2$$
$$= \alpha_1 \int_\Omega \vec{Z} \times \Delta \vec{Z}_t \cdot \vec{Z}_t dx + \alpha_1 \int_\Omega \vec{Z} \times \vec{H}_t \cdot \vec{Z}_t dx + \varepsilon \int_\Omega \vec{H}_t \cdot \vec{Z}_t dx$$
$$+ \varepsilon \int_\Omega |\nabla \vec{Z}|^2 |\vec{Z}_t|^2 dx - \varepsilon \int_\Omega (\vec{Z} \cdot \vec{H})|\vec{Z}_t|^2 dx. \tag{5.3.30}$$

By using the Sobolev inequality in Lemma 5.3.3, we estimate each term at the right-hand side of (5.3.30) except the first term.

(i)

$$\left| \alpha_1 \int_\Omega \vec{Z} \times \vec{H}_t \cdot \vec{Z}_t dx \right| = \left| -\alpha_1 \int_\Omega \vec{Z} \times (\nabla \times \vec{E}) \cdot \vec{Z}_t dx \right|$$
$$\leq C \left(\|\nabla \vec{E}\|_2^2 + \|\vec{Z}_t\|_2^2 \right),$$

(ii)

$$\left| \varepsilon \int_\Omega \vec{H}_t \cdot \vec{Z}_t dx \right| = \left| -\varepsilon \beta \int_\Omega |\vec{Z}_t|^2 dx - \varepsilon \int_\Omega \nabla \times \vec{E} \cdot \vec{Z}_t dx \right|$$
$$\leq C \left(\|\nabla \vec{E}\|_2^2 + \|\vec{Z}_t\|_2^2 \right),$$

(iii)

$$\varepsilon \int_\Omega |\nabla \vec{Z}|^2 |\vec{Z}_t|^2 dx \leq \varepsilon \|\nabla \vec{Z}\|_2^2 \|\vec{Z}_t\|_\infty^2 \leq C\varepsilon \|\nabla \vec{Z}_t\|_2 \|\vec{Z}_t\|_2$$
$$\leq \frac{\varepsilon}{K}\|\nabla \vec{Z}_t\|_2^2 + C\|\vec{Z}_t\|_2^2, \quad K \text{ is a constant to be chosen later,}$$

(iv)

$$\left| -\varepsilon \int_\Omega (\vec{Z} \cdot \vec{H})|\vec{Z}_t|^2 dx \right| \leq \varepsilon \|\vec{H}\|_2 \|\vec{Z}_t\|_4^2 \leq C\varepsilon \|\vec{H}\|_2 \|\vec{Z}_t\|_2^{\frac{3}{2}} \|\nabla \vec{Z}_t\|_2^{\frac{1}{2}}$$
$$\leq \frac{\varepsilon}{K}\|\nabla \vec{Z}_t\|_2^2 + C\|\vec{Z}_t\|_2^2. \tag{5.3.31}$$

(II) To estimate the first term on the right-hand side of (5.3.30), we apply the operator Δ to (5.3.28) and obtain

$$\Delta \vec{Z}_t = \alpha_1(\vec{Z} \times \Delta^2 \vec{Z} + 2\nabla\vec{Z} \times \nabla\Delta\vec{Z} + \Delta\vec{Z} \times \vec{H} + 2\nabla\vec{Z} \times \nabla\vec{H} + \vec{Z} \times \Delta\vec{H})$$
$$+ \varepsilon \left(\Delta^2 \vec{Z} + |\nabla\vec{Z}|^2 \Delta\vec{Z} + 2|\Delta\vec{Z}|^2 \vec{Z} \right.$$
$$+ 2(\nabla\vec{Z} \cdot \nabla\Delta\vec{Z})\vec{Z} + 4(\nabla\vec{Z} \cdot \Delta\vec{Z})\nabla\vec{Z} + \Delta\vec{H}$$
$$\left. - (\vec{Z} \cdot \vec{H})\Delta\vec{Z} - 2(\nabla\vec{Z} \cdot \vec{H})\nabla\vec{Z} - 2(\vec{Z} \cdot \nabla\vec{H})\nabla\vec{Z} \right). \qquad (5.3.32)$$

Making the cross product of $\vec{Z}$ with the equality above, and making the scalar products of $\vec{Z}_t$ with the resulting equality, and then using $\vec{Z} \cdot \vec{Z}_t = 0$, $\nabla\vec{Z} \cdot \vec{Z} = 0$, and $\vec{Z} \cdot \Delta\vec{Z} = -|\nabla\vec{Z}|^2$, we obtain

$$\alpha_1 \vec{Z} \times \Delta\vec{Z}_t \cdot \vec{Z}_t = \alpha_1^2(-\Delta^2\vec{Z} \cdot \vec{Z}_t + 2(\vec{Z} \cdot \nabla^3\vec{Z})(\nabla\vec{Z} \cdot \vec{Z}_t)$$
$$+ \vec{Z} \times (\Delta\vec{Z} + \vec{H}) \cdot \vec{Z}_t - \Delta\vec{H} \cdot \vec{Z}_t + 2(\vec{Z} \cdot \nabla\vec{H})(\nabla\vec{Z} \cdot \vec{Z}_t)$$
$$+ \alpha_1\varepsilon \left[\vec{Z} \times \Delta^2\vec{Z} \cdot \vec{Z}_t + |\nabla\vec{Z}|^2\vec{Z} \times \Delta\vec{Z} \cdot \vec{Z}_t \right.$$
$$+ 4(\nabla\vec{Z} \cdot \nabla^2\vec{Z})\vec{Z} \times \Delta\vec{Z} \cdot \vec{Z}_t + \vec{Z} \times \Delta\vec{H} \cdot \vec{Z}_t - (\vec{Z} \cdot \vec{H})\vec{Z} \times \Delta\vec{Z} \cdot \vec{Z}_t$$
$$\left. - 2(\nabla\vec{Z} \cdot \vec{H})\vec{Z} \times \nabla\vec{Z} \cdot \vec{Z}_t - 2(\vec{Z} \cdot \nabla\vec{H})\vec{Z} \times \nabla\vec{Z} \cdot \vec{Z}_t \right]. \qquad (5.3.33)$$

We estimate each term on the right-hand side of (5.3.32) except the fourth term, $-\Delta\vec{H} \cdot \vec{Z}_t$.

(1)
$$-\alpha_1^2 \int_\Omega \Delta^2\vec{Z} \cdot \vec{Z}_t dx = -\frac{\alpha_1^2}{2}\frac{d}{dt}\int_\Omega |\Delta\vec{Z}|^2 dx.$$

(2) Since $\vec{Z} \cdot \nabla^3\vec{Z} = -\frac{3}{2}\nabla(|\nabla\vec{Z}|^2)$, we have

$$2\alpha_1^2 \int_\Omega (\vec{Z} \cdot \nabla^3\vec{Z})(\nabla\vec{Z} \cdot \vec{Z}_t)dx$$
$$= -3\alpha_1^2 \int_\Omega \nabla(|\nabla\vec{Z}|^2)(\nabla\vec{Z} \cdot \vec{Z}_t)dx$$
$$= 3\alpha_1^2 \int_\Omega |\nabla\vec{Z}|^2(\nabla\vec{Z} \cdot \nabla\vec{Z}_t)dx + 3\alpha_1^2 \int_\Omega |\nabla\vec{Z}|^2\Delta\vec{Z} \cdot \vec{Z}_t dx$$
$$= \frac{3}{4}\alpha_1^2 \left(\int_\Omega |\nabla\vec{Z}|^4 dx \right)_t + 3\alpha_1^2 \int_\Omega |\nabla\vec{Z}|^2\Delta\vec{Z} \cdot [\varepsilon|\nabla\vec{Z}|^2\vec{Z}$$
$$+ \varepsilon\Delta\vec{Z} + \alpha_1\vec{Z} \times \vec{H} - \varepsilon\vec{Z} \times (\vec{Z} \times \vec{H})]dx$$
$$\leq \frac{3}{4}\alpha_1^2 \left(\int_\Omega |\nabla\vec{Z}|^4 dx \right)_t - 3\alpha_1^2\varepsilon \int_\Omega |\nabla\vec{Z}|^6 dx$$
$$+ 3C\alpha_1^2\varepsilon\|\nabla\vec{Z}\|_2^{2+\frac{1}{2}}\|\nabla\Delta\vec{Z}\|_2^{1+\frac{1}{2}} + C\|\nabla\vec{Z}\|_2\|\vec{H}\|_2\|\Delta\vec{Z}\|_2^2$$
$$\leq \frac{3}{4}\alpha_1^2 \left(\int_\Omega |\nabla\vec{Z}|^4 dx \right)_t - 3\alpha_1^2\varepsilon \int_\Omega |\nabla\vec{Z}|^6 dx$$
$$+ \frac{\varepsilon}{K}\|\nabla\Delta\vec{Z}\|_2^2 + C_3\|\Delta\vec{Z}\|_2^2 + d.$$

(3)

$$\alpha_1^2 \int_\Omega \vec{Z} \times (\Delta\vec{Z} + \vec{H}) \cdot \vec{Z}_t dx$$

$$= \alpha_1^2 \int_\Omega (\vec{Z} \cdot \vec{H})(\Delta\vec{Z} \cdot \vec{Z}_t) dx + \alpha_1^2 \int_\Omega |\nabla\vec{Z}|^2 \vec{Z}_t \cdot \vec{H} dx$$

$$= \alpha_1^2 \int_\Omega (\vec{Z} \cdot \vec{H})\Delta\vec{Z} \cdot (\alpha_1\vec{Z} \times \vec{H} + \varepsilon\Delta\vec{Z} + \varepsilon|\nabla\vec{Z}|^2\vec{Z} - \varepsilon\vec{Z} \times (\vec{Z} \times \vec{H})) dx$$

$$+ \alpha_1^2 \int_\Omega |\nabla\vec{Z}|^2 \vec{Z}_t \cdot \vec{H} dx$$

$$\leq \alpha_1^3 \|\Delta\vec{Z}\|_2 \|\vec{H}\|_4^2 + \varepsilon\alpha_1^2\|\vec{H}\|_2(\|\Delta\vec{Z}\|_4^2 + \|\nabla\vec{Z}\|_8^4)$$

$$+ \alpha_1^2\varepsilon\|\Delta\vec{Z}\|_2\|\vec{H}\|_4^2 + \alpha_1^2\|\vec{H}\|_2\|\vec{Z}_t\|_2\|\nabla\vec{Z}\|_\infty^2$$

$$\leq C\|\Delta\vec{Z}\|_2\|\nabla\vec{H}\|_2^{\frac{1}{2}} + C\varepsilon(\|\nabla\Delta\vec{Z}\|_2^{\frac{10}{8}} + \|\nabla\Delta\vec{Z}\|_2^{\frac{3}{4}}) + C\|\vec{Z}_t\|_2\|\Delta\vec{Z}\|_2$$

$$\leq \frac{\varepsilon}{K}\|\nabla\Delta\vec{Z}\|_2^2 + C_1\|\Delta\vec{Z}\|_2^2 + C_2\|\vec{Z}_t\|_2^2 + C_3\|\nabla\vec{H}\|_2^2 + d_1$$

$$\leq \frac{\varepsilon}{K}\|\nabla\Delta\vec{Z}\|_2^2 + C_4\|\Delta\vec{Z}\|_2^2 + C_3\|\nabla\vec{H}\|_2^2 + d_2.$$

(4)

$$2\alpha_1^2 \int_\Omega (\vec{Z} \cdot \nabla\vec{H})(\Delta\vec{Z} \cdot \vec{Z}_t) dx$$

$$= 2\alpha_1^2 \int_\Omega (\nabla \cdot (\vec{Z} \cdot \vec{H}) - \nabla\vec{Z} \cdot \vec{H})(\Delta\vec{Z} \cdot \vec{Z}_t) dx$$

$$= -2\alpha_1^2 \int_\Omega (\vec{Z} \cdot \vec{H})\nabla\vec{Z} \cdot \Delta\vec{Z}_t dx - 2\alpha_1^2 \int_\Omega (\vec{Z} \cdot \vec{H})\Delta\vec{Z} \cdot \vec{Z}_t dx$$

$$- 2\alpha_1^2 \int_\Omega (\nabla\vec{Z} \cdot \vec{H})\nabla\vec{Z} \cdot \vec{Z}_t dx$$

$$= \left(-\alpha_1^2 \int_\Omega (\vec{Z} \cdot \vec{H})|\nabla\vec{Z}|^2 dx\right)_t + \alpha_1^2 \int_\Omega (\vec{Z}_t \cdot \vec{H} + \vec{Z} \cdot \vec{H}_t)|\nabla\vec{Z}|^2 dx$$

$$- 2\alpha_1^2 \int_\Omega \left[(\vec{Z} \cdot \vec{H})\Delta\vec{Z} \cdot \vec{Z}_t + (\nabla\vec{Z} \cdot \vec{H})(\nabla\vec{Z} \cdot \vec{Z}_t)\right] dx.$$

From (3) and

$$\vec{Z} \cdot \vec{H}_t = -\beta\vec{Z} \cdot \vec{Z}_t - \vec{Z} \cdot \nabla \times \vec{E} = -\vec{Z} \cdot \nabla \times \vec{E},$$

it follows that

$$2\alpha_1^2 \int_\Omega (\vec{Z} \cdot \nabla\vec{H})(\Delta\vec{Z} \cdot \vec{Z}_t) dx$$

$$\leq \left(-\alpha_1^2 \int_\Omega (\vec{Z} \cdot \vec{H})|\nabla\vec{Z}|^2 dx\right)_t + \alpha_1^2(\|\nabla\vec{Z}\|_\infty^2\|\vec{H}\|_2\|\vec{Z}_t\|_2 + \|\nabla\vec{Z}\|_2\|\nabla\vec{Z}\|_4^2)$$

$$+ \frac{\varepsilon}{K}\|\nabla\Delta\vec{Z}\|_2^2 + C_3\|\nabla\vec{Z}\|_2^2 + C_4\|\nabla\vec{H}\|_2^2 + d_2$$

$$\leq \left(-\alpha_1^2 \int_\Omega (\vec{Z} \cdot \vec{H})|\nabla\vec{Z}|^2 dx\right)_t$$

$$+ \frac{\varepsilon}{K}\|\nabla\Delta\vec{Z}\|_2^2 + C(\|\Delta\vec{Z}\|_2^2 + \|\nabla\vec{H}\|_2^2 + \|\nabla\vec{E}\|_2^2) + d,$$

where the constants C and d are independent of ε and D.

(5)

$$\alpha_1\varepsilon \int_\Omega \vec{Z} \times \Delta^2 \vec{Z} \cdot \vec{Z}_t dx$$

$$= \alpha_1\varepsilon \int_\Omega \nabla^3 \vec{Z} \cdot \vec{Z} \times \nabla \vec{Z}_t dx + \alpha_1\varepsilon \int_\Omega \nabla^3 \vec{Z} \cdot \nabla \vec{Z} \times \vec{Z}_t dx.$$

Note that

$$\nabla \vec{Z}_t = \varepsilon \nabla \Delta \vec{Z} + \varepsilon |\nabla \vec{Z}|^2 \nabla \vec{Z} + 2\varepsilon(\nabla \vec{Z} \cdot \Delta \vec{Z})\vec{Z} + \alpha_1 \nabla \vec{Z} \times \Delta \vec{Z}$$

$$+ \alpha_1 \vec{Z} \times \nabla \Delta \vec{Z} + \alpha_1 \nabla(\vec{Z} \times \vec{H}) - \varepsilon \nabla(\vec{Z} \times (\vec{Z} \times \vec{H})). \qquad (5.3.34)$$

Then

$$\alpha_1\varepsilon \int_\Omega \vec{Z} \times \Delta^2 \vec{Z} \cdot \vec{Z}_t dx$$

$$= \alpha_1\varepsilon \int_\Omega \nabla^3 \vec{Z} \times \vec{Z} \cdot \left[\varepsilon \nabla^3 \vec{Z} + \varepsilon |\nabla \vec{Z}|^2 \nabla \vec{Z} - \varepsilon \nabla(\vec{Z} \times (\vec{Z} \times \vec{H}))\right.$$

$$\left. + \alpha_1 \nabla(\vec{Z} \times \vec{H}) + \alpha_1 \vec{Z} \times \nabla^3 \vec{Z} + \alpha_1 \nabla \vec{Z} \times \Delta \vec{Z}\right] dx + \alpha_1\varepsilon \int_\Omega \nabla^3 \vec{Z} \cdot \nabla \vec{Z} \times \vec{Z}_t dx$$

$$\leq -\alpha_1^2\varepsilon \|\nabla^3 \vec{Z} \times \vec{Z}\|_2^2 + C_1\varepsilon \|\nabla \Delta \vec{Z}\|_2(\|\nabla \vec{Z}\|_6^3 + \|\nabla \vec{Z}\|_\infty + \|\nabla \vec{H}\|_2)$$

$$+ C_1\varepsilon \|\nabla \Delta \vec{Z}\|_2 \|\nabla \vec{Z}\|_\infty(\|\Delta \vec{Z}\|_2 + \|\vec{Z}_t\|_2)$$

$$\leq -\alpha_1^2\varepsilon \|\nabla^3 \vec{Z} \times \vec{Z}\|_2^2 + C_2\varepsilon(\|\nabla \Delta \vec{Z}\|_2^{1+\frac{1}{2}} + \|\nabla \Delta \vec{Z}\|_2^{\frac{5}{4}}$$

$$+ \|\nabla \Delta \vec{Z}\|_2^{\frac{7}{4}} + \|\nabla \Delta \vec{Z}\|_2 \|\nabla \vec{H}\|_2)$$

$$\leq -\alpha_1^2\varepsilon \|\nabla^3 \vec{Z} \times \vec{Z}\|_2^2 + \frac{\varepsilon}{K} \|\nabla \Delta \vec{Z}\|_2^2 + C\|\nabla \vec{H}\|_2^2 + d.$$

(6)

$$\left|\alpha_1\varepsilon \int_\Omega |\nabla \vec{Z}|^2 \vec{Z} \times \Delta \vec{Z} \cdot \vec{Z}_t dx\right|$$

$$\leq |\alpha_1|\varepsilon \|\nabla \vec{Z}\|_\infty^2 \|\Delta \vec{Z}\|_2 \|\vec{Z}_t\|_2$$

$$\leq C_1\varepsilon \|\nabla \vec{Z}\|_\infty^2 \|\Delta \vec{Z}\|_2^2 + C_1\varepsilon \|\nabla \vec{Z}\|_\infty^2 \|\Delta \vec{Z}\|_2 \|\vec{H}\|_2$$

$$\leq C_2\varepsilon(\|\nabla \Delta \vec{Z}\|_2^{\frac{3}{2}} + \|\nabla \Delta \vec{Z}\|_2)$$

$$\leq \frac{\varepsilon}{K} \|\nabla \Delta \vec{Z}\|_2^2 + d.$$

(7) By (6) we have

$$\left|4\alpha_1\varepsilon \int_\Omega (\nabla \vec{Z} \cdot \nabla^2 \vec{Z})\vec{Z} \times \Delta \vec{Z} \cdot \vec{Z}_t dx\right|$$

$$\leq 4|\alpha_1|\varepsilon \|\nabla \vec{Z}\|_\infty^2 \|\Delta \vec{Z}\|_2 \|\vec{Z}_t\|_2$$

$$\leq \frac{\varepsilon}{K} \|\nabla \Delta \vec{Z}\|_2^2 + d.$$

(8)

$$\left| \alpha_1 \varepsilon \int_\Omega \vec{Z} \times \Delta \vec{H} \cdot \vec{Z}_t dx \right|$$

$$= \left| \alpha_1 \varepsilon \int_\Omega \nabla \vec{H} \cdot (\nabla \vec{Z} \times \vec{Z}_t + \vec{Z} \times \nabla \vec{Z}_t) dx \right|$$

$$\leq |\alpha_1| \varepsilon (\|\nabla \vec{H}\|_2 \|\vec{Z}_t\|_2 \|\nabla \vec{Z}\|_\infty + \|\nabla \vec{H}\|_2 \|\nabla \vec{Z}_t\|_2)$$

$$\leq C_1 \varepsilon \left[\|\nabla \vec{H}\|_2 \|\Delta \vec{Z}\|_2 \|\nabla \vec{Z}\|_\infty + \|\nabla \vec{H}\|_2 (\|\nabla \vec{Z}\|_\infty + \|\nabla \vec{Z}_t\|_2) \right]$$

$$\leq C_2 \varepsilon \|\nabla \vec{H}\|_2 (\|\nabla \Delta \vec{Z}\|_2^{\frac{3}{4}} + \|\nabla \Delta \vec{Z}\|_2^{\frac{1}{4}} + \|\nabla \vec{Z}_t\|_2)$$

$$\leq \frac{\varepsilon}{K} \|\nabla \Delta \vec{Z}\|_2^2 + \frac{\varepsilon}{K} \|\nabla \vec{Z}_t\|_2^2 + C\|\nabla \vec{H}\|_2^2 + d.$$

(9)

$$\left| -\alpha_1 \varepsilon \int_\Omega (\vec{Z} \cdot \vec{H}) \vec{Z} \times \Delta \vec{Z} \cdot \vec{Z}_t dx \right|$$

$$\leq |\alpha_1| \varepsilon \|\vec{H}\|_\infty \|\Delta \vec{Z}\|_2 \|\vec{Z}_t\|_2$$

$$\leq C_1 \varepsilon \|\nabla \vec{H}\|_2^{\frac{1}{2}} (\|\Delta \vec{Z}\|_2^2 + \|\Delta \vec{Z}\|_2)$$

$$\leq C_2 \varepsilon \|\nabla \vec{H}\|_2^{\frac{1}{2}} (\|\nabla \Delta \vec{Z}\|_2 + \|\nabla \Delta \vec{Z}\|_2^{\frac{1}{2}})$$

$$\leq \frac{\varepsilon}{K} \|\nabla \Delta \vec{Z}\|_2^2 + C\|\nabla \vec{H}\|_2^2 + d.$$

(10)

$$\left| -2\alpha_1 \varepsilon \int_\Omega (\nabla \vec{Z} \cdot \vec{H}) \vec{Z} \times \nabla \vec{Z} \cdot \vec{Z}_t dx \right|$$

$$\leq 2|\alpha_1| \varepsilon \|\nabla \vec{Z}\|_\infty^2 \|\vec{H}\|_2 \|\vec{Z}_t\|_2$$

$$\leq C_1 \varepsilon \|\Delta \vec{Z}\|_2 \|\vec{Z}_t\|_2$$

$$\leq C_2 \varepsilon (\|\Delta \vec{Z}\|_2^2 + \|\Delta \vec{Z}\|_2) \leq C\|\Delta \vec{Z}\|_2^2 + d.$$

(11)

$$\left| -2\alpha_1 \varepsilon \int_\Omega (\vec{Z} \cdot \nabla \vec{H}) \vec{Z} \times \nabla \vec{Z} \cdot \vec{Z}_t dx \right|$$

$$\leq 2|\alpha_1| \varepsilon \|\nabla \vec{Z}\|_\infty \|\nabla \vec{H}\|_2 \|\vec{Z}_t\|_2$$

$$\leq C_1 \varepsilon \|\nabla \vec{H}\|_2 \|\Delta \vec{Z}\|_2^{\frac{1}{2}} (\|\Delta \vec{Z}\|_2 + \|\vec{H}\|_2)$$

$$\leq C_2 \varepsilon \|\nabla \vec{H}\|_2 \|\Delta \vec{Z}\|_2^{\frac{1}{2}} + C_3 \varepsilon \|\nabla \vec{H}\|_2 \|\nabla \Delta \vec{Z}\|_2^{\frac{3}{4}}$$

$$\leq \frac{\varepsilon}{K} \|\nabla \Delta \vec{Z}\|_2^2 + C(\|\nabla \vec{H}\|_2^2 + \|\Delta \vec{Z}\|_2^2) + d.$$

Combining the results of (1)–(11) and using (5.3.31) we have

$$\alpha_1 \int_\Omega \vec{Z} \times \Delta \vec{Z}_t \cdot \vec{Z}_t dx$$

$$\leq -\alpha_1^2 \int_\Omega \vec{Z}_t \cdot \Delta \vec{H} dx - \frac{\alpha_1^2}{2} \left(\int_\Omega |\Delta \vec{Z}|^2 dx \right)_t$$

$$+ \left(\frac{3}{4} \alpha_1^2 \int_\Omega |\nabla \vec{Z}|^4 dx - \alpha_1^2 \int_\Omega (\vec{Z} \cdot \vec{H}) |\nabla \vec{Z}|^2 dx \right)_t$$

$$- 3\alpha_1^2 \varepsilon \|\nabla \vec{Z}\|_6^6 - \alpha_1^2 \varepsilon \|\nabla^3 \vec{Z} \times \vec{Z}\|_2^2$$

$$+ \frac{9\varepsilon}{K} \|\nabla \Delta \vec{Z}\|_2^2 + \frac{\varepsilon}{K} \|\nabla \vec{Z}_t\|_2^2$$

$$+ C(\|\Delta \vec{Z}\|_2^2 + \|\nabla \vec{H}\|_2^2 + \|\nabla \vec{E}\|_2^2) + d. \qquad (5.3.35)$$

Substituting (5.3.35) into (5.3.30) we get

$$\frac{1}{2} \left(\int_\Omega (|\vec{Z}_t|^2 + \alpha_1^2 |\Delta \vec{Z}|^2) dx \right)_t + \left(\int_\Omega (2\alpha_1^2 (\vec{Z} \cdot \vec{H}) |\nabla \vec{Z}|^2 - \frac{3}{2} \alpha_1^2 |\nabla \vec{Z}|^4) dx \right)_t$$

$$+ \varepsilon \|\nabla \vec{Z}_t\|_2^2 + \alpha_1^2 \int_\Omega \vec{Z}_t \cdot \vec{H} dx + \alpha_1^2 \varepsilon \int_\Omega (3|\nabla \vec{Z}|^6 + |\nabla^3 \vec{Z} \times \vec{Z}|^2) dx$$

$$\leq \frac{9\varepsilon}{K} \|\nabla \Delta \vec{Z}\|_2^2 + \frac{3\varepsilon}{K} \|\nabla \vec{Z}_t\|_2^2 + C_0(\|\Delta \vec{Z}\|_2^2 + \|\nabla \vec{H}\|_2^2 + \|\nabla \vec{E}\|_2^2) + d_0. \qquad (5.3.36)$$

Moreover, from (5.3.28)

$$\vec{Z} \times \nabla \vec{Z}_t = \varepsilon \vec{Z} \times \nabla \Delta \vec{Z} + \varepsilon |\nabla \vec{Z}|^2 \vec{Z} \times \nabla \vec{Z} + \alpha_1 \vec{Z} \times (\nabla \vec{Z} \times \Delta \vec{Z}) - \alpha_1 \nabla \Delta \vec{Z}$$

$$+ \alpha_1 (\vec{Z} \cdot \nabla^3 \vec{Z}) \vec{Z} + \alpha_1 \vec{Z} \times \nabla (\vec{Z} \times \vec{H}) - \varepsilon \vec{Z} \times \nabla (\vec{Z} \times (\vec{Z} \times \vec{H})).$$

Then, by $\vec{Z} \cdot \nabla^3 \vec{Z} = -3\nabla \vec{Z} \cdot \Delta \vec{Z}$ and the equality above, we have

$$|\alpha_1| |\nabla \Delta \vec{Z}| \leq |\nabla \vec{Z}_t| + \varepsilon |\nabla \Delta \vec{Z}| + \varepsilon |\nabla \vec{Z}|^3 + 4|\alpha_1| |\nabla \vec{Z}| |\Delta \vec{Z}|$$

$$+ (|\alpha_1| + \varepsilon)(|\nabla \vec{Z}| |\vec{H}| + |\nabla \vec{H}|).$$

Take $\varepsilon < \frac{|\alpha_1|}{2}$. The above inequality yields

$$|\alpha_1| |\nabla \Delta \vec{Z}| \leq C(|\nabla \vec{Z}_t| + |\nabla \vec{Z}| |\Delta \vec{Z}| + |\nabla \vec{Z}| |\vec{H}| + |\nabla \vec{H}|).$$

Therefore,

$$|\alpha_1| \|\nabla \Delta \vec{Z}\|_2 \leq C \|\nabla \vec{Z}_t\|_2 + C \|\nabla \vec{Z}\|_\infty (\|\Delta \vec{Z}\|_2 + \|\vec{H}\|_2) + C \|\nabla \vec{H}\|_2$$

$$\leq C \|\nabla \vec{Z}_t\|_2 + C_1(\|\nabla \Delta \vec{Z}\|_2^{\frac{3}{4}} + \|\nabla \Delta \vec{Z}\|_2^{\frac{1}{4}}) + C \|\nabla \vec{H}\|_2$$

$$\leq \frac{|\alpha_1|}{2} \|\nabla \Delta \vec{Z}\|_2 + C(\|\nabla \vec{Z}_t\|_2 + \|\nabla \vec{H}\|_2) + d,$$

and it follows that

$$\|\nabla \Delta \vec{Z}\|_2 \leq C_0(\|\nabla \vec{Z}_t\|_2 + \|\nabla \vec{H}\|_2) + d_0. \qquad (5.3.37)$$

Substituting (5.3.37) into (5.3.36) and choosing $K \geq 9C_0^2 + 3$, we obtain

$$\frac{d}{dt} \int_\Omega (|\vec{Z}_t(\cdot, t)|^2 + \alpha_1^2 |\Delta \vec{Z}(\cdot, t)|^2) dx$$

$$+ \frac{d}{dt} \int_\Omega \left[2\alpha_1^2 (\vec{Z} \cdot \vec{H}) |\nabla \vec{Z}|^2 - \frac{3}{2} \alpha_1^2 |\nabla \vec{Z}|^4 \right] dx + 2\alpha_1^2 \int_\Omega \vec{Z}_t \cdot \vec{H} dx$$

$$\leq C(\|\Delta \vec{Z}\|_2^2 + \|\nabla \vec{H}\|_2^2 + \|\nabla \vec{E}\|_2^2) + d. \tag{5.3.38}$$

(III) In the case of $\beta > 0$, we make the scalar products of $\Delta \vec{E}$ with (5.3.9) and $\Delta \vec{H}$ with (5.3.10), respectively, and then get

$$\frac{1}{2\beta} \frac{d}{dt} \int_\Omega (|\nabla \vec{E}|^2 + |\nabla \vec{H}|^2) dx + \frac{\sigma}{\beta} \|\nabla \vec{E}\|_2^2 = \int_\Omega \vec{Z}_t \cdot \vec{H} dx. \tag{5.3.39}$$

In the case of $\beta = 0$, from (5.3.9) and (5.3.10) it follows that

$$\|\vec{H}(\cdot, t)\|_{H^1(\Omega)}^2 + \|\vec{E}(\cdot, t)\|_{H^1(\Omega)}^2 \leq \|\vec{H}_0(x)\|_{H^1(\Omega)}^2 + \|\vec{E}_0(x)\|_{H^1(\Omega)}^2. \tag{5.3.40}$$

(1) Substituting (5.3.39) into (5.3.38) we get

$$\frac{d}{dt} \int_\Omega \left[|\vec{Z}_t|^2 + \alpha_1^2 |\Delta \vec{Z}|^2 + \frac{\alpha_1^2}{\beta} (|\nabla \vec{E}|^2 + |\nabla \vec{H}|^2) \right] dx$$

$$+ \frac{d}{dt} \int_\Omega \left[2\alpha_1^2 (\vec{Z} \cdot \vec{H}) |\nabla \vec{Z}|^2 - \frac{3}{2} \alpha_1^2 |\nabla \vec{Z}|^4 \right] dx$$

$$\leq C(\|\Delta \vec{Z}\|_2^2 + \|\nabla \vec{H}\|_2^2 + \|\nabla \vec{E}\|_2^2) + d. \tag{5.3.41}$$

Denote

$$A(t) = \|\vec{Z}_t(\cdot, t)\|_2^2 + \|\Delta \vec{Z}(\cdot, t)\|_2^2 + \frac{\alpha_1^2}{\beta} (\|\nabla \vec{E}(\cdot, t)\|_2^2 + \|\nabla \vec{H}(\cdot, t)\|_2^2)$$

$$= \alpha_1^2 \|\vec{Z} \times (\Delta \vec{Z} + \vec{H})\|_2^2 + \varepsilon^2 \|\vec{Z} \times (\vec{Z} \times (\Delta \vec{Z} + \vec{H}))\|_2^2$$

$$+ \|\Delta \vec{Z}\|_2^2 + \frac{\alpha_1^2}{\beta} (\|\nabla \vec{E}\|_2^2 + \|\nabla \vec{H}\|_2^2),$$

$$B(t) = \int_\Omega \left[2\alpha_1^2 (\vec{Z} \cdot \vec{H}) |\nabla \vec{Z}|^2 - \frac{3}{2} \alpha_1^2 |\nabla \vec{Z}|^4 \right] dx.$$

Integrating (5.3.41) over $(0, t)$ $(0 < t \leq T)$, we get

$$A(t) + B(t) \leq A(0) + B(0) + C \int_0^t (\|\Delta \vec{Z}(\cdot, t)\|_2^2 + \|\nabla \vec{H}(\cdot, t)\|_2^2 + \|\nabla \vec{E}(\cdot, t)\|_2^2) dt$$

and

$$A(t) \leq A(0) + B(0) + |B(t)| + \int_0^t (\|\Delta \vec{Z}\|_2^2 + \|\nabla \vec{H}\|_2^2 + \|\nabla \vec{E}\|_2^2) dt. \tag{5.3.42}$$

Note that

$$|B(t)| \leq 2\alpha_1^2 \int_\Omega |\vec{H}||\nabla\vec{Z}|^2 dx + \frac{3}{2}\alpha_1^2 \int_\Omega |\nabla\vec{Z}|^4 dx$$

$$\leq C_1\|\vec{H}\|_\infty + C_2\|\nabla\vec{Z}\|_4^4$$

$$\leq \frac{|\alpha_1|^2}{2}\|\Delta\vec{Z}\|_2^2 + \frac{\alpha_1^2}{2\beta}\|\nabla\vec{H}\|_2^2 + d_1. \tag{5.3.43}$$

Substituting (5.3.43) into (5.3.42) we obtain

$$\frac{\alpha_1^2}{2}\left[\|\Delta\vec{Z}(\cdot,t)\|_2^2 + \frac{1}{\beta}(\|\nabla\vec{H}(\cdot,t)\|_2^2 + \|\nabla\vec{E}(\cdot,t)\|_2^2\right]$$

$$\leq A_0 + B_0 + d_1 + dT + \int_0^t (\|\Delta\vec{Z}\|_2^2 + \|\nabla\vec{H}\|_2^2 + \|\nabla\vec{E}\|_2^2)dt.$$

Applying the Gronwall inequality for the above inequality gives

$$\sup_{0\leq t\leq T} \left(\|\Delta\vec{Z}(\cdot,t)\|_2^2 + \|\nabla\vec{H}(\cdot,t)\|_2^2 + \|\nabla\vec{E}(\cdot,t)\|_2^2\right) \leq C, \tag{5.3.44}$$

where the constant C is independent of D and ε.

(2) When $\beta = 0$, it follows that

$$\frac{d}{dt}\int_\Omega (|\vec{Z}_t|^2 + \alpha_1^2|\Delta\vec{Z}|^2)dx + \frac{d}{dt}B(t) \leq C\|\Delta\vec{Z}(\cdot,t)\|_2^2 + d.$$

Similar argument as (1) gives

$$\sup_{t\in[0,T]} \|\Delta\vec{Z}(\cdot,t)\|_2^2 \leq C,$$

where the constant C is independent of D and ε. By (5.3.39) and the above inequality we conclude that (5.3.26) and (5.3.27) are true still for $\beta = 0$. The proof of the lemma is complete.

By using the induction and the analogous derivation as that of Lemma 5.3.3, after complicated calculation we can get the following result.

Lemma 5.3.4 *Assume that $(\vec{Z}(x,t), \vec{H}(x,t), \vec{E}(x,t))$ is the smooth solution of the periodic value problem (5.3.8)–(5.3.13), and $|\alpha_1| > 0$, $\sigma \geq 0$, $\beta \geq 0$. If $(\vec{Z}_0(x), \vec{H}_0(x), \vec{E}_0(x)) \in (H^k(\Omega), H^{k-1}(\Omega), H^{k-1}(\Omega))$ $(k \geq 2)$, then there are estimates*

$$\sup_{t\in[0,T]} \left[\sum_{s=0}^{[\frac{k}{2}]}\|D_t^s D_x^{k-2s}\vec{Z}(\cdot,t)\|_2 + \sum_{s=0}^{k-1}\left(\|D_t^s D_x^{k-1-s}\vec{H}(\cdot,t)\|_2 + \|D_t^s D_x^{k-1-s}\vec{E}(\cdot,t)\|_2\right)\right] \leq C,$$

$$\tag{5.3.45}$$

where the constant C is independent of D and ε, and k and s are nonnegative integers.

5.3.2   Global Existence of Smooth Solution

From the *a priori* estimates independent of ε for the smooth solution of the periodic value problem (5.3.8)–(5.3.13), we prove the global existence of the smooth solution.

Theorem 5.3.1 [42] *Assume the constants $\varepsilon > 0$, $\beta \geq 0$, $\sigma \geq 0$, the periodic value functions $(\vec{Z}_0(x), \vec{H}_0(x), \vec{E}_0(x)) \in (H^k(\Omega), H^{k-1}(\Omega), H^{k-1}(\Omega))$ $k \geq 3 + \left[\frac{n}{3}\right]$, $\Omega \subset R^n$ $(n \leq 2)$ is bounded domain, and $|\vec{Z}_0(x)| = 1$, $\nabla \cdot (\vec{H}_0 + \beta \vec{Z}_0) = 0$, $\nabla \cdot \vec{E}_0 = 0$. Moreover, when $n = 2$, there is*

$$\|\nabla \vec{Z}_0\|_2 + \|\vec{E}_0\|_2 + \|\vec{H}_0\|_2 \leq \delta,$$

where δ is a suitable small constant. Then the periodic initial value problem (5.3.8)–(5.3.13) of the Landau–Lifshitz–Maxwell system has a unique global smooth solution such that $|\vec{Z}(z,t)| = 1$, $x \in \Omega$, $t \in R^+$, and

$$\vec{Z}(x,t) \in \cap_{s=0}^{\left[\frac{k}{2}\right]} W_\infty^s(0, T; H^{k-2s}(\Omega)),$$
$$\vec{H}(x,t) \in \cap_{s=0}^{k-1} W_\infty^s(0, T; H^{k-1-s}(\Omega)),$$
$$\vec{E}(x,t) \in \cap_{s=0}^{k-1} W_\infty^s(0, T; H^{k-1-s}(\Omega)).$$

By the standard principle of compactness and letting $\varepsilon \to 0$, we obtain

Theorem 5.3.2 *Under the conditions of Theorem 5.3.1, the one-dimensional periodic initial value problem (5.3.1)–(5.3.6) of the strong degenerate Landau–Lifshitz–Maxwell system admits a unique global smooth solution.*

Remark. *The following assumptions are imposed:*

$$\nabla \cdot \vec{H}_0 + \beta \nabla \cdot \vec{Z}_0 = 0, \quad \nabla \cdot \vec{E}_0 = 0,$$

to assure that

$$\nabla \cdot \vec{H} + \beta \nabla \cdot \vec{Z} = 0, \quad \nabla \cdot \vec{E} = 0.$$

Since all the *a priori* estimates are independent of the periodicity D, we let $D \to \infty$ and then obtain the following result.

Theorem 5.3.3 *Under the conditions of Theorem 5.3.1, the one-dimensional initial value problem (5.3.1)–(5.3.4) and (5.3.7) for strong degenerate Landau–Lifshitz–Maxwell system has a unique global smooth solution $(\vec{Z}(x,t), \vec{H}(x,t), \vec{E}(x,t))$ satisfying*

$$|\vec{Z}(x,t)| = 1, \qquad \forall \ (x,t) \in R^n \times R^+,$$

$$\nabla \vec{Z}(x,t) \in \cap_{s=0}^{\left[\frac{k}{2}\right]} W_\infty^s(0, T; H^{k-1-2s}(R)),$$
$$\vec{H}(x,t) \in \cap_{s=0}^{k-1} W_\infty^s(0, T; H^{k-1-s}(R)),$$
$$\vec{E}(x,t) \in \cap_{s=0}^{k-1} W_\infty^s(0, T; H^{k-1-s}(R)).$$

5.4 Global Weak Solution to L–L–M System on Riemannian Manifold

5.4.1 L–L–M System on Riemannian Manifold

Assume that (M, g) is an n-dimensional compact Riemannian manifold without boundary with metric g. In local coordinates $x = (x^1, \ldots, x^n)$, the Laplace–Beltrami operator and $|\nabla u(x)|$ read as

$$\Delta_M = \frac{1}{\sqrt{g}} \frac{\partial}{\partial x^\beta} \left(g^{\alpha\beta} \sqrt{g} \frac{\partial}{\partial x^\alpha} \right) = g^{\alpha\beta} \frac{\partial^2}{\partial x^\alpha \partial x^\beta} - \Gamma^k_{\alpha\beta} \frac{\partial}{\partial x^k},$$

$$|\nabla u(x)|^2 = \sum_{\alpha\beta} \sum_j g^{\alpha\beta} \frac{\partial u^j}{\partial x^\alpha} \frac{\partial u^j}{\partial x^\beta},$$

where $(g^{\alpha\beta})$ is the inverse of $(g_{\alpha\beta})$. The L–L–M system on Riemannian manifold M reads as

$$\vec{Z}_t = \alpha_1 \vec{Z} \times (\Delta_M \vec{Z} + \vec{H}) - \alpha_2 \vec{Z} \times (\vec{Z} \times (\Delta_M \vec{Z} + \vec{H})), \tag{5.4.1}$$

$$\nabla_j F_l^j = 0, \quad l = 0,$$

$$\nabla_j F_l^j = \sigma F_{0l}, \quad l \neq 0, \tag{5.4.2}$$

$$\nabla_l (F_{jk} + \beta Z_{jk}) + \nabla_j (F_{kl} + \beta Z_{kl}) + \nabla_k (F_{lj} + \beta Z_{lj}) = 0, \quad ljk \neq 0,$$

$$\nabla_l (F_{jk} + \beta Z_{jk}) + \nabla_j (F_{kl} + \beta Z_{kl}) + \nabla_k (F_{lj} + \beta Z_{lj})$$

$$= -\beta \frac{\partial Z_{lj}}{\partial t}, \quad \text{one of } l, j, k = 0, \tag{5.4.3}$$

where

$$F_l^j = \overline{g}^{jk} F_{kl}, \qquad \vec{Z} = (Z_1, Z_2, Z_3) = (Z_{23}, Z_{31}, Z_{12}).$$

F_{kl} is second order asymmetric covariant tensor energy:

$$(F_{kl}) = \begin{bmatrix} 0 & F_{01} & F_{02} & F_{03} \\ F_{10} & 0 & F_{12} & F_{13} \\ F_{20} & F_{21} & 0 & F_{23} \\ F_{30} & F_{31} & F_{32} & 0 \end{bmatrix}$$

$$= \begin{bmatrix} 0 & E_1 & E_2 & E_3 \\ -E_1 & 0 & H_3 & -H_2 \\ -E_2 & -H_3 & 0 & H_1 \\ -E_3 & H_2 & -H_1 & 0 \end{bmatrix}, \tag{5.4.4}$$

where $\vec{E} = (E_1, E_2, E_3)$, $\vec{H} = (H_1, H_2, H_3)$, $(\overline{g}^{jk})$ is the metric of the manifold $M(x^0, x^1, \ldots, x^n)$. In fact we have

$$(\overline{g}^{jk}) = \begin{pmatrix} 1 & 0 \\ 0 & -g^{jk} \end{pmatrix}, \tag{5.4.5}$$

where (g^{jk}) is the metric of the manifold $M(x^1, \ldots, x^n)$, D_j is the covariant derivative. Since

$$\nabla_j \bar{g}^{ij} = 0, \tag{5.4.6}$$

(5.4.2) is equivalent to

$$\bar{g}^{ik} \nabla_j F_{kl} = \begin{cases} 0, & l = 0, \\ -\sigma F_{0l}, & l \neq 0. \end{cases} \tag{5.4.7}$$

In particular, taking $\bar{g}^{00} = 1$, $\bar{g}^{11} = \bar{g}^{22} = \bar{g}^{33} = -1$ and $x^0 = -t$, we may get the L–L–M system on Euclid space R^n.

Now we calculate the electromagnetic field energy T_{ij} of (5.4.1)–(5.4.3). It follows from the definition that

$$T_{ij} = \frac{1}{4} \bar{g}_{il} F_{kl} F^{kl} - \bar{g}_{kl} F_i^l F_{lj}, \tag{5.4.8}$$

where

$$F^{kl} = \bar{g}^{ki} \bar{g}^{ls} F_{is}. \tag{5.4.9}$$

From

$$\nabla_l \bar{g}_{0k} = 0, \quad \det(\bar{g}_{jk}) \neq 0,$$

we have

$$\nabla_j \bar{g}^{ij} = 0,$$

$$\nabla_j T^{ij} = \frac{1}{4} \nabla_j (\bar{g}^{ij} F_{kl} F^{kl}) - \nabla_j (F^{il} F_l^j)$$

$$= \frac{1}{4} \bar{g}^{ij} \nabla F_{kl} \cdot F^{kl} + \frac{1}{4} \bar{g}^{ij} F_{kl} \nabla_j F^{kl} - (\nabla_j F^{il}) F_l^j - F_{jl} \nabla_j F_l^j$$

$$= \frac{1}{2} \bar{g}^{ij} F^{kl} \nabla_j F_{kl} - \bar{g}^{ij} (\nabla_j F_{kl}) F^{jl} - \sigma F^{il} F_{0l};$$

$$\bar{g}^{ik} (\nabla_j F_{kl}) F^{jl} = \frac{1}{2} \bar{g}^{ik} \nabla_j F_{kl} (F^{jl} - F^{lj})$$

$$= -\frac{1}{2} \bar{g}^{ik} F^{lj} (\nabla_j F_{kl} - \nabla_l F_{kj}),$$

$$\frac{1}{2} \bar{g}^{ij} F^{kl} \nabla_j F_{kl} = \frac{1}{2} \bar{g}^{ik} F^{lj} \nabla_k F_{lj},$$

therefore it follows from (5.5.8) that

$$\nabla_j T^{ij} = \frac{1}{2} \bar{g}^{ik} F^{lj} \nabla_j F_{kl} - \bar{g}^{ik} \nabla_j F_{kl} F^{jl} + \sigma F^{il} F_{0l}$$

$$= \frac{1}{2} \bar{g}^{ik} F_{lj} (\nabla_k F_{lj} + \nabla_j F_{kl} + \nabla_l F_{jk}) + \sigma F^{il} F_{0l},$$

$$\nabla_j T^{0j} = -\frac{\beta}{2}\bar{g}^{00}F^{lj}\frac{\partial Z_{lj}}{\partial t} - \sigma F^{0l}F_{0l}, \tag{5.4.10}$$

$$F^{lj}\frac{\partial Z_{lj}}{\partial t} = H\frac{\partial Z}{\partial t}.$$

Let n be the unit vector on $M \times \{t\}$. When $i > 0$, $n_i = 0$. Integrating (5.4.10) over $M \times [0, t]$, we have

$$\int_{M\times\{t\}} T^{00} - \int_{M\times\{0\}} T^{00} = -\sigma \int_{M\times[0,t]} F^{0l}F_{0l} + \frac{\beta}{2}\int_0^t \int_M \bar{g}^{00}\vec{H}\frac{\partial \vec{Z}}{\partial t}dMdt. \tag{5.4.11}$$

It follows from (5.4.1) that

$$(\Delta\vec{Z} + \vec{H}) \cdot \vec{Z}_t = -\alpha_2(\Delta_M\vec{Z} + \vec{H})[\vec{Z} \times (\vec{Z} \times (\Delta_M\vec{Z} + \vec{H}))]$$
$$= \alpha_2|\vec{Z} \times (\Delta\vec{Z} + \vec{H})|^2,$$

and hence

$$\int_M \vec{H} \cdot \vec{Z}_t dM = -\int_M \Delta_M\vec{Z} \cdot \vec{Z}_t dM$$
$$+ \alpha_2\int_M |\vec{Z} \times (\Delta\vec{Z} + \vec{H})|^2 dM - \int_M \Delta_M\vec{Z} \cdot \vec{Z}_t dM$$
$$= -\int_M \frac{1}{\sqrt{g}}\frac{\partial}{\partial x^\beta}\left(g^{\alpha\beta}\sqrt{g}\frac{\partial \vec{Z}}{\partial x^\alpha}\right) \cdot \frac{\partial \vec{Z}}{\partial t}\sqrt{g}dM$$
$$= \int_M g^{\alpha\beta}\sqrt{g}\frac{\partial \vec{Z}}{\partial x^\alpha} \cdot \frac{\partial^2 \vec{Z}}{\partial x^\beta \partial t}dM$$
$$= \frac{1}{2}\left(\int_M |\nabla\vec{Z}|^2 dM\right)_t,$$

$$T^{00} = \frac{1}{4}\bar{g}^{00}F_{kl}F^{kl} - F_{0l}F^{0l}$$
$$= \frac{1}{4}\bar{g}^{00}\sum_{i>0,j>0} F_{ij}F^{ij} + \frac{1}{2}\bar{g}^{00}F_{0l}F^{0l}$$
$$= \frac{1}{2}\bar{g}^{00}(|\vec{H}|^2 + |\vec{E}|^2),$$

where

$$|\vec{H}|^2 = \frac{1}{2}\sum_{i,j>0} F_{ij}F^{ij},$$
$$|\vec{E}|^2 = -g^{00}g^{lj}F_{0l}F_{0j} \geq 0.$$

It follows from (5.4.11) that

$$\frac{1}{2}\int_{M\times\{t\}}\bar{g}^{00}(|\vec{E}|^2+|\vec{H}|^2)dM+\sigma\int_{M\times[0,t]}F^{0l}F_{0l}dMdt$$

$$=\frac{1}{2}\int_{M\times\{t\}}|\nabla\vec{Z}|^2dM+\int_{M\times[0,t]}\alpha_2|\vec{Z}\times(\Delta\vec{Z}+\vec{H})|^2dM$$

$$+\frac{1}{2}\int_{M\times\{0\}}\bar{g}^{00}(|\vec{E}|^2+|\vec{H}|^2)dM+\frac{1}{2}\int_{M\times\{0\}}|\nabla\vec{Z}|^2dM. \tag{5.4.12}$$

In order to prove the existence of weak solution to the initial value problem of L–L–M equations, we derive the L–L–M equations on Riemannian manifold by much more simple method and get its energy conservation law.

Now we give the definitions of gradient operator and spin operator on Riemannian manifold.

For a scalar field, its covariant derivative is defined by

$$\nabla_j\varphi=\frac{\partial\varphi}{\partial x^j}$$

which is the first order covariant energy.

$$g_{ij}\nabla_j\varphi$$

or

$$\nabla^j\varphi$$

is the second order inverse variant energy. The gradient is defined by

$$\mathrm{grad}\varphi=g^{ij}\nabla_j\varphi\vec{e}_i, \tag{5.4.13}$$

where $\vec{e}_i$ is the covariant local coordinate.

Suppose that there is a vector $\vec{A}=a_i\vec{e}^i=a^i\vec{e}_i$. Its covariant derivatives $\nabla_i a_j$ form a covariant tensor

$$\nabla_i a_j-\nabla_j a_i, \tag{5.4.14}$$

which is the second order antisymmetry covariant energy. In three-dimensional space, there are only three independent elements of the second order antisymmetric tensor. We may construct a vector whose inverse variant elements are

$$\xi^i=\frac{1}{2}\epsilon^{ijk}(D_j a_k-D_k a_j)=\epsilon^{ijk}D_j a_k, \tag{5.4.15}$$

where

$$\epsilon^{ijk}=e^i\cdot(e^j\times e^k). \tag{5.4.16}$$

Vector $\vec{\xi}$ is the rotation of vector $\vec{A}$,

$$\begin{cases} \xi^1 = \frac{1}{\sqrt{g}}\left(\frac{\partial a_3}{\partial x^2} - \frac{\partial a_2}{\partial x^3}\right), \\ \xi^2 = \frac{1}{\sqrt{g}}\left(\frac{\partial a_1}{\partial x^3} - \frac{\partial a_3}{\partial x^1}\right), \\ \xi^3 = \frac{1}{\sqrt{g}}\left(\frac{\partial a_2}{\partial x^1} - \frac{\partial a_1}{\partial x^2}\right), \end{cases} \tag{5.4.17}$$

$$\operatorname{rot}\vec{A} = \xi^i \vec{e}^i. \tag{5.4.18}$$

Therefore, we may write L–L–M equations on manifold as

$$\begin{cases} \frac{\partial E^1}{\partial t} - \frac{1}{\sqrt{g}}\left(\frac{\partial H^3}{\partial x^2} - \frac{\partial H^2}{\partial x^3}\right) + \sigma E^1 = 0, \\ \frac{\partial E^2}{\partial t} - \frac{1}{\sqrt{g}}\left(\frac{\partial H^1}{\partial x^3} - \frac{\partial H^3}{\partial x^1}\right) + \sigma E^2 = 0, \\ \frac{H^3}{\partial t} - \frac{1}{\sqrt{g}}\left(\frac{\partial H^2}{\partial x^1} - \frac{\partial H^1}{\partial x^2}\right) + \sigma E^3 = 0, \end{cases} \tag{5.4.19}$$

$$\vec{Z}_t = \alpha_1 \vec{Z} \times (\Delta_M \vec{Z} + \vec{H}) - \alpha_2 \vec{Z} \times (\vec{Z} \times (\Delta \vec{Z} + \vec{H})), \tag{5.4.20}$$

$$\begin{cases} \frac{\partial H^1}{\partial t} + \beta\frac{\partial Z^1}{\partial t} + \frac{1}{\sqrt{g}}\left(\frac{\partial E^3}{\partial x^2} - \frac{\partial E^2}{\partial x^3}\right) = 0, \\ \frac{\partial H^2}{\partial t} + \beta\frac{\partial Z^2}{\partial t} + \frac{1}{\sqrt{g}}\left(\frac{\partial E^1}{\partial x^3} - \frac{\partial E^3}{\partial x^1}\right) = 0, \\ \frac{\partial H^3}{\partial t} + \beta\frac{\partial Z^3}{\partial t} + \frac{1}{\sqrt{g}}\left(\frac{\partial E^2}{\partial x^1} - \frac{\partial E^1}{\partial x^2}\right) = 0. \end{cases} \tag{5.4.21}$$

As for the divergence equation we have

$$\nabla \cdot (\vec{H} + \beta\vec{Z}) = 0, \qquad \nabla \cdot \vec{E} = 0, \tag{5.4.22}$$

where
$$\vec{Z} = (Z^1, Z^2, Z^3), \quad \vec{H} = (H^1, H^2, H^3), \quad \vec{E} = (E^1, E^2, E^3).$$

We first give the definition of the divergence. The divergence of a vector means the invariant quantity after contraction for the index of covariant derivative of its inverse variant components:

$$\operatorname{div}\vec{A} = \nabla_i a^i,$$

where $\vec{A} = a^i \vec{e}_i$. From

$$\nabla_i a^i = \frac{\partial a^i}{\partial x^i} + \Gamma^k_{ij} a^k, \qquad \Gamma^k_{ij} = \frac{\partial \ln\sqrt{g}}{\partial x^k},$$

one has

$$\begin{aligned} \nabla_i a^i &= \frac{\partial a^i}{\partial x^i} + \frac{\partial \ln\sqrt{g}}{\partial x^i} a^i \\ &= \frac{1}{\sqrt{g}}\left(\sqrt{g}\frac{\partial a^i}{\partial x^i} + \frac{\partial\sqrt{g}}{\partial x^i} a^i\right) \\ &= \frac{1}{\sqrt{g}}\frac{\partial\sqrt{g}a^i}{\partial x^i}. \end{aligned}$$

Therefore, the divergence of a vector is

$$\operatorname{div}\vec{A} = \frac{1}{\sqrt{g}}\frac{\partial(\sqrt{g}a^i)}{\partial x^i}. \tag{5.4.23}$$

On Riemannian manifold M, Eq. (5.4.22) read as

$$\frac{1}{\sqrt{g}}\frac{\partial(\sqrt{g}H^i)}{\partial x^i} + \frac{\beta}{\sqrt{g}}\frac{\partial(\sqrt{g}Z^i)}{\partial x^i} = 0, \tag{5.4.24}$$

$$\frac{1}{\sqrt{g}}\frac{\partial(\sqrt{g}E^i)}{\partial x^i} = 0. \tag{5.4.25}$$

Now we prove the existence of Eqs. (5.4.19)–(5.4.21), (5.4.24) and (5.4.25) with the initial conditions

$$E^i|_{t=0} = E_0^i, \quad H^i|_{t=0} = H_0^i, \quad Z^i|_{t=0} = Z_0^i, \tag{5.4.26}$$

where $i = 1, 2, 3$ and

$$\operatorname{div}\vec{E}_0 = \operatorname{div}(\vec{H}_0 + \beta\vec{Z}_0) = 0. \tag{5.4.27}$$

We will use Galerkin method. Let $\{\lambda_j\}$ be the eigenvalues of the operator $-\Delta_M$, that is

$$-\Delta_M\varphi_j = \lambda_j\varphi_j, \quad j = 1, 2, \ldots, \tag{5.4.28}$$

and the eigenfunctions $\{\varphi_j\}$ form the normal orthogonal basis of $H^1(M; R^3)$. Suppose that the approximate solutions of problem (5.4.19)–(5.4.21), (5.4.28) and the initial conditions (5.4.26) and (5.4.27) are

$$\begin{cases} E_m = \displaystyle\sum_{j=1}^{m}\alpha_{jm}(t)\varphi_j(x), \\[2mm] H_m = \displaystyle\sum_{j=1}^{m}\beta_{jm}(t)\varphi_j(x), \\[2mm] Z_m = \displaystyle\sum_{j=1}^{m}\gamma_{jm}(t)\varphi_j(x), \end{cases} \tag{5.4.29}$$

where

$$E_m = (E_m^1, E_m^2, E_m^3),$$
$$H_m = (H_m^1, H_m^2, H_m^3),$$
$$Z_m = (Z_m^1, Z_m^2, Z_m^3),$$

and $\alpha_m(t)$, $\beta_m(t)$, $\gamma_m(t)$ are three-dimensional vectors in $t \in R^+$ which solve the following ordinary differential systems:

$$(Z_{mt} - \alpha_1 Z_m \times (\Delta_M Z_m + H_m)$$
$$+ \alpha_2 Z_m \times (Z_m \times (\Delta_M Z_m + H_m)), \varphi_j) = 0, \tag{5.4.30}$$

$$\left(\frac{\partial E_m^1}{\partial t} - \frac{1}{\sqrt{g}} \left(\frac{\partial H_m^3}{\partial x^2} - \frac{\partial H_m^2}{\partial x^3} \right) + \sigma E_m^1, \varphi_j \right) = 0, \tag{5.4.31}$$

$$\left(\frac{\partial E_m^2}{\partial t} - \frac{1}{\sqrt{g}} \left(\frac{\partial H_m^1}{\partial x^3} - \frac{\partial H_m^3}{\partial x^1} \right) + \sigma E_m^2, \varphi_j \right) = 0, \tag{5.4.32}$$

$$\left(\frac{H_m^3}{\partial t} - \frac{1}{\sqrt{g}} \left(\frac{\partial H_m^2}{\partial x^1} - \frac{\partial H_m^1}{\partial x^2} \right) + \sigma E_m^3, \varphi_j \right) = 0, \tag{5.4.33}$$

$$\left(\frac{\partial H_m^1}{\partial t} + \beta \frac{\partial Z_m^1}{\partial t} + \frac{1}{\sqrt{g}} \left(\frac{\partial E_m^3}{\partial x^2} - \frac{\partial E_m^2}{\partial x^3} \right), \varphi_j \right) = 0, \tag{5.4.34}$$

$$\left(\frac{\partial H_m^2}{\partial t} + \beta \frac{\partial Z_m^2}{\partial t} + \frac{1}{\sqrt{g}} \left(\frac{\partial E_m^1}{\partial x^3} - \frac{\partial E_m^3}{\partial x^1} \right), \varphi_j \right) = 0, \tag{5.4.35}$$

$$\left(\frac{\partial H_m^3}{\partial t} + \beta \frac{\partial Z_m^3}{\partial t} + \frac{1}{\sqrt{g}} \left(\frac{\partial E_m^2}{\partial x^1} - \frac{\partial E_m^1}{\partial x^2} \right), \varphi_j \right) = 0, \tag{5.4.36}$$

with initial conditions

$$\begin{cases} \int_M Z_m(x,0)\varphi_j(x)dM = \int_M Z_0(x)\varphi_j(x)dM, \\[2mm] \int_M H_m(x,0)\varphi_j(x)dM = \int_M H_0(x)\varphi_j(x)dM, \\[2mm] \int_M E_m(x,0)\varphi_j(x)dM = \int_M E_0(x)\varphi_j(x)dM, \end{cases} \tag{5.4.37}$$

$j = 1, 2, 3, \ldots$.

From the following *a priori* estimates, we know that this problem admits a global solution on $[0, T]$.

In the following we give the *a priori* estimate for the approximate solutions. Multiplying (5.4.30)–(5.4.36) by γ_m, α_m^1, α_m^2, α_m^3, β_m^1, β_m^2 and β_m^3, respectively, and summing from $j = 1$ to m, we have

$$(Z_{mt} - \alpha_1 Z_m \times (\Delta_M Z_m + H_m)$$
$$+ \alpha_2 Z_m \times (Z_m \times (\Delta_M Z_m + H_m)), Z_m) = 0, \tag{5.4.38}$$

$$\left(\frac{\partial E_m^1}{\partial t} - \frac{1}{\sqrt{g}} \left(\frac{\partial H_m^3}{\partial x^2} - \frac{\partial H_m^2}{\partial x^3} \right) + \sigma E_m^1, E_m^1 \right) = 0, \tag{5.4.39}$$

$$\left(\frac{\partial E_m^2}{\partial t} - \frac{1}{\sqrt{g}}\left(\frac{\partial H_m^1}{\partial x^3} - \frac{\partial H_m^3}{\partial x^1}\right) + \sigma E_m^2, E_m^2\right) = 0, \tag{5.4.40}$$

$$\left(\frac{H_m^3}{\partial t} - \frac{1}{\sqrt{g}}\left(\frac{\partial H_m^2}{\partial x^1} - \frac{\partial H_m^1}{\partial x^2}\right) + \sigma E_m^3, E_m^3\right) = 0, \tag{5.4.41}$$

$$\left(\frac{\partial H_m^1}{\partial t} + \beta\frac{\partial Z_m^1}{\partial t} + \frac{1}{\sqrt{g}}\left(\frac{\partial E_m^3}{\partial x^2} - \frac{\partial E_m^2}{\partial x^3}\right), H_m^1\right) = 0, \tag{5.4.42}$$

$$\left(\frac{\partial H_m^2}{\partial t} + \beta\frac{\partial Z_m^2}{\partial t} + \frac{1}{\sqrt{g}}\left(\frac{\partial E_m^1}{\partial x^3} - \frac{\partial E_m^3}{\partial x^1}\right), H_m^2\right) = 0, \tag{5.4.43}$$

$$\left(\frac{\partial H_m^3}{\partial t} + \beta\frac{\partial Z_m^3}{\partial t} + \frac{1}{\sqrt{g}}\left(\frac{\partial E_m^2}{\partial x^1} - \frac{\partial E_m^1}{\partial x^2}\right), H_m^3\right) = 0. \tag{5.4.44}$$

Adding (5.4.39)–(5.4.44), we have

$$\frac{1}{2}\frac{\partial}{\partial t}(\|E_m\|_2^2 + \|H_m\|_2^2)$$
$$+ \sigma\|E_m\|_2^2 + \beta\int_M \frac{\partial Z_m}{\partial t}\cdot H_m dM = 0. \tag{5.4.45}$$

It follows from (5.4.38) that $\frac{d}{dt}\|Z_m\|_2^2 = 0$, then

$$\|Z_m(\cdot, t)\|_2 = \|Z_m(\cdot, 0)\|_2 = \|Z_0\|_2. \tag{5.4.46}$$

On the other hand, we have from (5.4.30) that

$$(Z_{mt} - \alpha_1 Z_m \times (\Delta_M Z_m + H_m)$$
$$+ \alpha_2 Z_m \times (Z_m \times (\Delta_M Z_m + H_m)), \Delta_m Z_m + H_m) = 0,$$

that is

$$\int_M Z_{mt}(\Delta_M Z_m + H_m)dM$$
$$- \alpha_2 \int_M |Z_m \times (\Delta_M Z_m + H_m)|^2 dM = 0. \tag{5.4.47}$$

Since

$$-\int_M Z_{mt}\Delta Z_m dM = -\int_M Z_{mt}\frac{\partial}{\partial x^\beta}\left(g^{\alpha\beta}\sqrt{g}\frac{\partial Z_m}{\partial x^\alpha}\right)dM$$
$$= \int_M g^{\alpha\beta}\sqrt{g}\frac{\partial Z_{mt}}{\partial x^\beta}\frac{\partial Z_m}{\partial x^\alpha}dM$$
$$= \int_M g^{\alpha\beta}\frac{\partial Z_{mt}}{\partial x^\beta}\frac{\partial Z_m}{\partial x^\alpha}dM$$
$$= \frac{1}{2}\left(\int_M |\nabla Z_m|^2 dM\right)_t,$$

we have from (5.4.47) that

$$\frac{1}{2}\frac{d}{dt}\|\nabla Z_m\|_2^2 + \alpha_2\|Z_m \times (\Delta_M Z_m + H_m)\|_2^2 = \int_M Z_{mt} \cdot H_m dM. \qquad (5.4.48)$$

From (5.4.45) and (5.4.48) and the method in Sec. 5.1, we have

Lemma 5.4.1 *Let $\alpha_2 \geq 0$ and $(Z_0(x), E_0(x), H_0(x)) \in (H^1(M), L^2(M), L^2(M))$. Then for the approximate solutions of problem (5.4.30)–(5.4.37) there holds*

$$(i) \quad \sup_{0\leq t\leq T} (\|Z_m(\cdot,t)\|_{H^1}^2 + \|E_m(\cdot,t)\|_2^2 + \|H_m(\cdot,t)\|_2^2) \leq C_1, \qquad (5.4.49)$$

where C_1 is independent of α_2 and m.
   When $\alpha_2 > 0$, there holds

$$(ii) \quad \|Z_m \times (\Delta_M Z_m + H_m)\|_{L^2(0,T;L^2(M))} \leq C_2,$$

where C_2 is independent of m.

5.4.2 Existence of Generalized Solution

It follows from Lemma 5.4.1 that we may choose a subsequence such that

$$Z_m(x,t) \to Z(x,t) \text{ in } L^\infty(0,T;H^1(M)) \quad \text{weakly-}*, \text{ as } m \to \infty;$$
$$E_m(x,t) \to E(x,t) \text{ in } L^\infty(0,T;L^2(M)) \quad \text{weakly-}*, \text{ as } m \to \infty;$$
$$H_m(x,t) \to H(x,t) \text{ in } L^\infty(0,T;L^2(M)) \quad \text{weakly-}*, \text{ as } m \to \infty.$$

If $\alpha_2 > 0$, $Z_m \times (\Delta Z_m + H_m)$ weakly converge in $L^2(0,T;L^2(M))$.

Definition 5.4.2 *Three-dimensional vectors $Z(x,t) \in L^\infty(0,T;H^1(M))$, $E(x,t) \in L^\infty(0,T;L^2(M))$, $H(x,t) \in L^\infty(0,T;L^2(M))$ are called global weak solutions to problem (5.4.19)–(5.4.21), and (5.4.24) and (5.4.25) if for any test function $v(x,t) \in C^1(Q_T)$ with $v(x,T) = 0$ there holds*

$$\int_0^T \int_M Z v_t dM dt + \alpha_1 \int_0^T \int_M \nabla Z \times Z \cdot \nabla v dM dt$$

$$+ \alpha_1 \int_0^T \int_M Z \times H \cdot v dM dt + \alpha_2 \int_0^T \int_M (Z \times H) \times Z \cdot \nabla v dM dt$$

$$+ \alpha_2 \int_0^T \int_M \nabla Z \times (Z \times \nabla Z) \cdot v dM dt$$

$$+ \alpha_2 \int_0^T \int_M (\nabla Z \times Z) \times Z \cdot \nabla v dM dt + \int_M Z_0 v_0 dM = 0, \qquad (5.4.50)$$

$$\int_0^T \int_M E v_t dM dt + \int_0^T \int_M \nabla \times v \cdot H dM dt$$

$$- \sigma \int_0^T \int_M E \cdot v dM dt + \int_M E_0 v_0 dM = 0, \qquad (5.4.51)$$

$$\int_0^T \int_M (H + \beta Z) v_t \, dM \, dt - \int_0^T \int_M \nabla \times v \cdot E_t \, dM \, dt$$

$$+ \int_0^T \int_M (H_0 + \beta Z_0) \, dM = 0, \tag{5.4.52}$$

and

$$\int_0^T \int_M \nabla \xi \cdot E \, dM \, dt = 0, \tag{5.4.53}$$

$$\int_0^T \int_M \nabla \eta \cdot (H + \beta Z) \, dM \, dt = 0, \tag{5.4.54}$$

for any test function ξ, $\eta \in L^2(0, T; C_0^\infty(M))$.

Theorem 5.4.1 *Let $\alpha_2 \geq 0$ and the initial data $(Z_0, E_0, H_0) \in (H^1(M), L^2(M), L^2(M))$, such that*

$$\nabla \cdot E_0, \quad \nabla \cdot (H_0 + \beta Z_0) = 0 \tag{5.4.55}$$

in the sense of distribution. Then problem (5.4.19)–(5.4.21), and (5.4.24) and (5.4.25) admits at least one global weak solution $(Z(x,t), H(x,t), E(x,t))$.

Proof. Similar to Sec. 5.1 we may verify that (Z, H, E) satisfy integral equalities (5.4.50)–(5.4.52). So we only need to prove that conditions (5.4.53) and (5.4.54) hold.

(i) For any $\xi \in L^2(0, T; C_0^\infty(M))$, take test function $v(x, t)$ as follows:

$$v(x, t) = \int_0^t \nabla \xi(x, \tau) d\tau - \int_0^T \nabla \xi(x, \tau) d\tau.$$

It is easy to verify that $v(x, t) \in C^1(Q_T)$, $v(x, T) = 0$, and

$$v_t(x, t) = \nabla \xi(x, t), \quad v(x, 0) = -\int_0^T \nabla \xi(x, \tau) d\tau.$$

Substituting them into (5.4.52) we have

$$\int_0^T \int_M (H + \beta Z) \nabla \xi \, dM \, dt = \int_0^T \int_M (H_0 + \beta Z_0) \nabla \xi \, dM \, dt = 0. \tag{5.4.56}$$

(ii) For any $\xi \in L^2(0, T; C_0^\infty(M))$, denote $\xi_1 = \xi e^{-\sigma t}$. Take test function $v(x, t)$ as follows

$$v(x, t) = \left(\int_0^t \nabla \xi_1(x, \tau) d\tau - \int_0^T \nabla \xi_1(x, \tau) d\tau \right) e^{\sigma t}.$$

It is easy to verify that $v(x, t) \in C^1(Q_T)$, $v(x, T) = 0$, and

$$v_t - \sigma v = \nabla \xi_1(x, t) e^{\sigma t} = \nabla \xi(x, t), \quad v(x, 0) = -\int_0^T \nabla \xi_1(x, \tau) d\tau.$$

Substituting them into (5.4.51) we have

$$\int_0^T \int_M E \cdot \nabla \xi \, dM \, dt = \int_0^T \int_M E_0 \cdot \nabla \xi_1 \, dM \, dt = 0. \tag{5.4.57}$$

conditions (5.4.53) and (5.4.54) follow from (5.4.56) and (5.4.57). The theorem is proved.

5.5 Partial Regularity for Stationary Solutions to L–L–M Equations

In this section, we discuss the equations as follows:

$$\frac{1}{2}u_t - \frac{1}{2}(u \times u_t) = \Delta u + u|\nabla u|^2 + H - \langle H, u \rangle u, \quad \text{in } B^3 \times R_+, \qquad (5.5.1)$$

where $H(u)$ is the nonlocal term that satisfies the following quasi-steady state Maxwell equations

$$\mathbf{curl}\, H = 0, \quad \text{in } \mathcal{D}'(R^3), \qquad (5.5.2)$$

$$\text{div}(H + \overline{u}) = 0, \quad \text{in } \mathcal{D}'(R^3). \qquad (5.5.3)$$

We impose to the equations (5.5.1)–(5.5.3) the initial condition

$$u(x, 0) = u_0(x) \qquad (5.5.4)$$

and boundary condition

$$\left. \frac{\partial u}{\partial \nu} \right|_{\partial B^3} = 0 \qquad (5.5.5)$$

in (5.5.4), $|u_0(x)| \equiv 1$, in (5.5.5), ν is the unit outer normal to the boundary of B^3, $\overline{u}$ is the zero extension of u from B^3 to R^3, $u = (u_1, u_2, u_3)$ is the spin vector, "$\times$" denotes the vector cross product in R^3. We should notice that this extension guarantees $\overline{u} \in L^\infty(R^3 \times R_+) \cap L^\infty(0, \infty; W^{-1,\infty}(R^3))$.

5.5.1 Quasi-Static Maxwell Equations

Lemma 5.5.1 *Let $u \in H^1(B^3, S^2)$. Let $H = \nabla\Phi \in L^2(R^3, R^3)$ be the solution of*

$$\mathbf{curl}\, H = 0, \quad \text{div}(H + \tilde{u}) = 0 \qquad (5.5.6)$$

in $\mathcal{D}'(R^3)$ where $\tilde{u}$ is equal to u in B^3 ($|u| = 1$ a.e. in B^3) and zero outside B^3. Then

$$H \in \cap_{1 \leq p < \infty} L^p(R^3) \qquad (5.5.7)$$

and for all $p \in (1, \infty)$ there exists a constant $K_p > 0$ such that

$$\|H\|_{L^p(R^3)} \leq K_p \|u\|_{L^p(B^3)}. \qquad (5.5.8)$$

Proof. This is because that $|\tilde{u}| = |u| = 1$ a.e in B^3 and $\tilde{u} = 0$ outside of B^3, then $\tilde{u} \in L^\infty(R^3)$ and $\text{div}\, u \in W^{-1,\infty}$. Consider $\Phi \in H^1(R^3)$ such that $H = \nabla\Phi$ and $\Delta\Phi = -\text{div}\, u$ in R^3. We have that $\Phi \in \cap_{1 \leq p < \infty} W^{1,p}$ and $\forall\ 1 < p < \infty$,

$$\|\nabla\Phi\|_{L^p(R^3)} = \|H\|_{L^p(R^3)} \leq K_p \|u\|_{L^p(B^3)}. \qquad (5.5.9)$$

The lemma is proved.

Remark 5.5.1 For $n = 3$, taking $p = 3^* = 6$, we get from (5.5.9)

$$\|\nabla\Phi\|_{L^6(R^3)} = \|H\|_{L^6(R^3)} \leq K\|u\|_{L^6(B^3)} \leq C\|u\|_{H^1} = C\|\nabla u\|_{L^2}. \tag{5.5.10}$$

Therefore, $\Phi \in W^{1,6}(R^3)$ with embedding $W^{1,6}(R^3) \subset C^{0,\frac{1}{2}}(R^3)$ (note that $n = 3 < 6$)

$$\|\Phi\|_{L^\infty} \leq C\|\nabla u\|_{L^2} \tag{5.5.11}$$

and

$$\|\Phi\|_{L^p} \leq C\|\nabla u\|_{L^2}, \quad \forall\ 1 \leq p < \infty. \tag{5.5.12}$$

Remark 5.5.2 We have

$$-\int_{R^3} u \cdot \nabla\Phi = \int_{R^3} \Phi\,\mathrm{div}\,u = -\int_{R^3} \Phi\Delta\Phi = \int_{R^3} |\nabla\Phi|^2 = \int_{R^3} |H|^2. \tag{5.5.13}$$

Remark 5.5.3 Next we recall some results concerning the singular integral transforms, Riesz transform.

Let $G(r) = \frac{C}{r}$ be the Laplace kernel in R^3, u be as above. Then the solution of the Poisson equation with parameter t

$$\Delta\Phi(x, t) = -\mathrm{div}\,u, \quad \text{in } \mathcal{D}'(R^3)$$

can be expressed by

$$\Phi(x, t) = -\int_{B^3} G(|x - y|)\mathrm{div}\,u(y, t)dy + \int_{\partial B^3} G(|x - y|)\langle u(y, t), \mathbf{n}(\mathbf{y})\rangle d\sigma(y).$$

On the other hand, since $\mathrm{div}\,u \in L^\infty(0, T; L^2(B^3))$, from the L^p theory of Riesz transforms we have

$$\frac{\partial^2\Phi}{\partial x_i \partial x_j} = -\mathcal{R}_i\mathcal{R}_j(-\mathrm{div}\,u), \quad i, j = 1, 2, 3,$$

where $\mathcal{R}_i$ are the Riesz transforms

$$\mathcal{R}_i(-\mathrm{div}\,u) = \frac{\Gamma((3+1)/2)}{\pi^{(3+2)/2}} \int_{R^2} \frac{(x_i - y_i)(-\mathrm{div}\,u(y, t))}{|x - y|^{(3+1)}}dy$$

and $\varphi \in W^{2,2}(R^3)$ with

$$\|\Phi\|_{W^{2,2}(R^3)}(t) \leq C\|\mathrm{div}\,u\|_{L^2(R^3)}(t).$$

This result can also be deduced from Coifman–Fefferman [36] for general singular integral operators. We should also note that (5.5.9)–(5.5.12) hold for $H = \nabla\Phi$ and Φ.

Remark 5.5.4 If

$$L_\varepsilon(x, t) = \int_{y \in B^3 \backslash B(x,\varepsilon)} \frac{x - y}{|x - y|} G'(|x - y|)\mathrm{div}\,u(y, t)dy \tag{5.5.14}$$

and $L(x,t) = \lim_{\varepsilon \to 0} L_\varepsilon(x,t) = -H + v$, then we have from the theory of Riesz transform that $H(x,t) \in L^\infty(0,\infty; L^p(B^3))$, $|v(x,t)| \leq K \int_{\partial B^3} \frac{d\sigma(y)}{|x-y|^2} \leq K \ln(1 - |x|) \in L^\infty(0,\infty; L^p(B^3))$ for any $p > 1$ and

$$\|L(x,t)\|_{L^2} \leq C\|\nabla u\|_{L^2}. \tag{5.5.15}$$

Remark 5.5.5 (Definition of weak solution) A function $u \in L^\infty(0,T; H^1(B^3))$ with $u_t \in L^2(0,T; L^2(B^3))$ is said a weak solution if (5.5.1)–(5.5.3) hold in the sense of distribution.

5.5.2 Definition of Stationary Solution

Definition 5.5.1 *A weak solution u of (5.5.1) is said a stationary solution if for any $\xi(x,t) \in C_0^1(B^3 \times R_+, R^3)$, $\theta(x,t) \in C_0^1(B^3 \times R_+, R)$ with $\xi(x,t)$, $\theta(x,t)$, $\nabla_{(x,t)}\xi$, $\nabla_{(x,t)}\theta$ bounded on $B^3 \times R_+$ and $\xi, \theta \equiv 0$ for $t = 0$ and $t \geq t^* > 0$ such that $x + \tau\xi|_{\partial B^3} = Id$, $t + \tau\theta|_{\partial B^3} = Id$, there holds*

$$\int_0^{+\infty} \int_{B^3} \left(\frac{1}{2}u_t - \frac{1}{2}u \times u_t\right)\left(\frac{\partial u^\tau}{\partial \tau}\right)_{\tau=0} + \partial_\tau^+ \int_0^{+\infty} \int_{B^3} e(u^\tau) + |H(u^\tau)|^2 dxdt \leq 0, \tag{5.5.16}$$

where $u^\tau(x,t) = u(x + \tau\xi(x,t), t + \tau\theta(x,t))$, $e(u) = \frac{1}{2}|\nabla u(x,t)|^2$.

We want to make the definition applicable in the following. To this aim we first compute the right derivative in the definition. For simplicity, we simply compute the derivative at $\tau = 0$ and without loss of generality, we simply assume that u is smooth.

It follows from (5.5.2) and (5.5.3) that $H_\tau = H(u^\tau)$ satisfies

$$\mathbf{curl}\, H_\tau = 0, \quad \text{in } \mathcal{D}'(R^3), \tag{5.5.17}$$

$$\text{div}(H_\tau + \overline{u}^\tau) = 0, \quad \text{in } \mathcal{D}'(R^3). \tag{5.5.18}$$

Let the potential $\alpha_\tau(x,t)$ be such that $H_\tau = \nabla\alpha_\tau$, then (5.5.17) and (5.5.18) become

$$\Delta\alpha_\tau = -\text{div}\,\overline{u}^\tau, \quad \text{in } \mathcal{D}'(R^3). \tag{5.5.19}$$

Therefore,

$$\alpha_\tau = -\int_{B^3} G(|x-y|)\text{div}\,u^\tau dy + \int_{\partial B^3} G(|x-y|)\langle u^\tau, \vec{n}(y)\rangle d\sigma(y), \tag{5.5.20}$$

where G is the fundamental solution in three dimensions. Denote $(x_\tau, t_\tau) = (x + \tau\xi(x,t), t + \tau\theta(x,t))$, $A_\tau = \frac{1}{2}\int_0^{+\infty} \int_{B^3} |\nabla u^\tau(x,t)|^2 dxdt$, $B_\tau = -\int_0^{+\infty} \int_{B^3} |H_\tau|^2 dxdt$. It is simple to compute

$$\text{div}\,u^\tau(x,t) = \text{div}\,u(x_\tau, t_\tau) + \tau u^i_{x_j}(x_\tau, t_\tau)\xi^j_{x_i} + \tau u^i_t(x_\tau, t_\tau)\theta_{x_i}. \tag{5.5.21}$$

We have

$$\left.\frac{dA_\tau}{d\tau}\right|_{\tau=0} = \int_0^{+\infty} \int_{B^3} (u^i_{x_j} u^i_{x_k}\xi^k_{x_j} + u^i_t u^i_{x_j}\theta_{x_j}) - \frac{1}{2}\int_0^{+\infty} \int_{B^3} |\nabla u|^2(\text{div}\,\xi + \theta_t). \tag{5.5.22}$$

On the other hand

$$B_\tau = \int_0^{+\infty} \int_{B^3} |H_\tau|^2 dx dt$$

$$= -\int_0^{+\infty} \int_{B^3} H_\tau \cdot u_\tau dx dt$$

$$= -\int_0^{+\infty} \int_{B^3} \nabla \alpha_\tau \cdot u_\tau dx dt$$

$$= \int_0^{+\infty} \int_{B^3} (\operatorname{div} u_\tau) \cdot \alpha_\tau dx dt - \int_0^{+\infty} \int_{\partial B^3} (u_\tau \cdot \vec{n}) \alpha_\tau d\sigma. \tag{5.5.23}$$

It follows from (5.5.20), (5.5.21) and (5.5.23) that

$$\frac{dB_\tau}{d\tau} = \frac{dB_1}{d\tau} + \frac{dB_2}{d\tau} + \frac{dB_3}{d\tau} + \frac{dB_4}{d\tau}, \tag{5.5.24}$$

where

$$-B_1 = \int_0^{+\infty} dt \int_{B^3} \int_{B^3} [\operatorname{div} u(x_\tau, t_\tau) + \tau u^i_{x_j}(x_\tau, t_\tau)\xi^j_{x_i} + \tau u^i_t(x_\tau, t_\tau)\theta_{x_i}]$$

$$\times \left[\operatorname{div} u(y_\tau, t_\tau) + \tau u^i_{y_j}(y_\tau, t_\tau)\xi^j_{y_i} + \tau u^i_t(y_\tau, t_\tau)\theta_{y_i} \right] G(|x - y|). \tag{5.5.25}$$

Hence we have

$$-\frac{dB_1}{d\tau}\bigg|_{\tau=0} = \int_0^{+\infty} dt \int_{B^3} \int_{B^3} [u^i_{x_i x_k}\xi^k + u^i_{x_i t}\theta + u^i_{x_j}\xi^j_{x_i} + u^i_t\theta_{x_i}][\operatorname{div} u(y,t)]G(|x - y|)dx dy$$

$$+ \int_0^{+\infty} dt \int_{B^3} \int_{B^3} [u^i_{y_i y_k}\xi^k + u^i_{y_i t}\theta + u^i_{y_j}\xi^j_{y_i} + u^i_t\theta_{y_i}][\operatorname{div} u(x,t)]G(|x - y|). \tag{5.5.26}$$

It follows from Carbou's result [21] that

$$\int_{B^3} \int_{B^3} [u^i_{x_i x_k}\xi^k + u^i_{x_j}\xi^j_{x_i}][\operatorname{div} u(y,t)]G(|x - y|)dx dy$$

$$+ \int_{B^3} \int_{B^3} [u^i_{y_i y_k}\xi^k + u^i_{y_j}\xi^j_{y_i}][\operatorname{div} u(x,t)]G(|x - y|)dx dy$$

$$= -2\int_{B^3} \int_{B^3} [\operatorname{div} u \operatorname{div} \xi - u^i_{x_j}\xi^j_{x_i}][\operatorname{div} u(y,t)]G(|x - y|)dx dy$$

$$+ \int_{B^3} \int_{B^3} G'(|x - y|)\left\langle \frac{x - y}{|x - y|}, \xi(x,t) - \xi(y,t) \right\rangle \operatorname{div} u(x,t)\operatorname{div} u(y,t). \tag{5.5.27}$$

Therefore,

$$-\frac{dB_1}{d\tau}\bigg|_{\tau=0} = -2\int_0^{+\infty} dt \int_{B^3} \int_{B^3} [\operatorname{div} u \operatorname{div} \xi - u^i_{x_j}\xi^j_{x_i}][\operatorname{div} u(y,t)]G(|x - y|)dx dy$$

$$+ \int_0^{+\infty} dt \int_{B^3} \int_{B^3} G'(|x - y|)$$

$$\times \left\langle \frac{x-y}{|x-y|}, \xi(x,t) - \xi(y,t) \right\rangle \operatorname{div} u(x,t) \operatorname{div} u(y,t) dx dy$$

$$+ \int_0^{+\infty} dt \int_{B^3} \int_{B^3} (u_{x_i t}^i \theta + u_t^i \theta_{x_i}) \operatorname{div} u(y,t) G(|x-y|) dx dy$$

$$+ \int_0^{+\infty} dt \int_{B^3} \int_{B^3} (u_{y_i t}^i \theta + u_t^i \theta_{y_i}) \operatorname{div} u(x,t) G(|x-y|) dx dy. \qquad (5.5.28)$$

The last two terms can be estimated as follows:

$$\int_{B^3} (u_{x_i t}^i \theta + u_t^i \theta_{x_i}) \operatorname{div} u(y,t) G(|x-y|) dx dy$$

$$= - \int_{B^3} u_t^i(x,t) \theta(x,t) \operatorname{div} u(y,t) \frac{x^i - y^i}{|x-y|} G'(|x-y|) dx dy, \qquad (5.5.29)$$

$$\int_{B^3} (u_{y_i t}^i \theta + u_t^i \theta_{y_i}) \operatorname{div} u(x,t) G(|x-y|) dx dy$$

$$= - \int_{B^3} u_t^i(y,t) \theta(y,t) \operatorname{div} u(x,t) \frac{x^i - y^i}{|x-y|} G'(|x-y|) dx dy. \qquad (5.5.30)$$

So, we have from (5.5.28)–(5.5.30) that

$$\frac{dB_1}{d\tau}\bigg|_{\tau=0} = 2 \int_0^{+\infty} dt \int_{B^3} \int_{B^3} [\operatorname{div} u \operatorname{div} \xi - u_{x_j}^i \xi_{x_i}^j][\operatorname{div} u(y,t)] G(|x-y|) dx dy$$

$$- \int_0^{+\infty} dt \int_{B^3} \int_{B^3} G'(|x-y|)$$

$$\times \left\langle \frac{x-y}{|x-y|}, \xi(x,t) - \xi(y,t) \right\rangle \operatorname{div} u(x,t) \operatorname{div} u(y,t) dx dy$$

$$+ 2 \int_0^{+\infty} dt \int_{B^3} u_t^i(x,t) \theta(x,t) \operatorname{div} u(y,t) \frac{x^i - y^i}{|x-y|} G'(|x-y|) dx dy. \qquad (5.5.31)$$

Moreover, since B_2 and B_3 in (5.5.24) are

$$B_2 = \int_0^{+\infty} dt \int_{x \in B^3} \int_{y \in \partial B^3} [\operatorname{div} u(x_\tau, t_\tau)$$

$$+ \tau u_{x_j}^i \xi_{x_i}^j + \tau u_t^i \theta_{x_i}](u(y,t), \vec{n}(y)) G(|x-y|) dx d\sigma(y), \qquad (5.5.32)$$

$$B_3 = \int_0^{+\infty} dt \int_{x \in \partial B^3} \int_{y \in B^3} [\operatorname{div} u(y_\tau, t_\tau)$$

$$+ \tau u_{y_j}^i \xi_{y_i}^j + \tau u_t^i \theta_{y_i}](u(x,t), \vec{n}(x)) G(|x-y|) dy d\sigma(x), \qquad (5.5.33)$$

we have

$$\frac{dB_2}{d\tau}\bigg|_{\tau=0} = \int_0^{+\infty} dt \int_{x \in B^3} \int_{y \in \partial B^3} [u_{x_i x_k}^i \xi^k$$

$$+ u_{x_i t}^i \theta + u_{x_j}^i \xi_{x_i}^j + u_t^i \theta_{x_i}](u(y,t), \vec{n}(y)) G(|x-y|) dx d\sigma(y), \qquad (5.5.34)$$

$$\frac{dB_3}{d\tau}\bigg|_{\tau=0} = \int_0^{+\infty} dt \int_{x \in \partial B^3} \int_{y \in B^3} [u_{y_i y_k}^i \xi^k$$

$$+ u_{y_i t}^i \theta + u_{y_j}^i \xi_{y_i}^j + u_t^i \theta_{y_i}](u(x,t), \vec{n}(x)) G(|x-y|) dy d\sigma(x), \qquad (5.5.35)$$

and similar to the above we have

$$\frac{dB_2}{d\tau}\bigg|_{\tau=0} + \frac{dB_3}{d\tau}\bigg|_{\tau=0}$$

$$= 2\int_0^{+\infty} dt \int_{x\in B^3}\int_{y\in\partial B^3} [-\operatorname{div} u(x,t)\operatorname{div}\xi(x,t)$$

$$+ u^i_{x_j}\xi^j_{x_i} + u^i x_i t + u^i_t\theta_{x_i}](u(y,t),\vec{n}(y))G(|x-y|)dxd\sigma(y). \quad (5.5.36)$$

We note that the term B_4 is determined only by the boundary data which is independent of τ, then $\frac{dB_4}{d\tau}\big|_{\tau=0} = 0$. We finally get

$$\frac{dB_\tau}{d\tau}\bigg|_{\tau=0} = -2\int_0^{+\infty} dt \int_{B^3} \operatorname{div} u(x,t)\operatorname{div}\xi(x,t)\Phi(x,t)dxdt$$

$$+ 2\int_0^{+\infty} dt \int_{B^3} \Phi(x,t)u^i_{x_j}(x,t)\xi^j_{x_i}(x,t)$$

$$+ \int_0^{+\infty} dt \int_{B^3}\int_{B^3}$$

$$\times \left\langle \frac{x-y}{|x-y|}, \xi(x,t)-\xi(y,t)\right\rangle G'(|x-y|)\operatorname{div} u(x,t)\operatorname{div} u(y,t)dxdy$$

$$- 2\int_0^{+\infty} dt \int_{B^3}\int_{B^3}$$

$$\times \left\langle \frac{x-y}{|x-y|}, u_t(x,t)\right\rangle \theta(x,t)\operatorname{div} u(y,t)G'(|x-y|)dxdy, \quad (5.5.37)$$

where

$$\Phi(x,t) = -\int_{B^3} G(|x-y|)\operatorname{div} u(y,t)dy + \int_{B^3} G(|x-y|)\langle u(y,t),\mathbf{n}(y)\rangle d\sigma(y)$$

which satisfies (5.5.12).

Combining (5.5.22) and (5.5.37), we may now simplify the definition of stationary weak solution for our problem (5.5.1)–(5.5.5) as follows.

Remark 5.5.6 A stationary solution u of (5.5.1) satisfies for any $\xi(x,t) \in C_0^1(B^3 \times R_+, R^3)$ and $\theta(x,t) \in C_0^1(B^3 \times R_+, R)$ as above, the following inequality

$$\int_0^{+\infty}\int_{B^3}\left(\frac{1}{2}u_t - \frac{1}{2}u\times u_t\right)(\nabla u\cdot\xi + u_t\theta) + \int_0^{+\infty} dt \int_{B^3}(u^i_{x_j}u^i_{x_k}\xi^k_{x_j} + u^i_t u^i_{x_j}\theta_{x_j})dx$$

$$- \frac{1}{2}\int_0^{+\infty} dt \int_{B^3}|\nabla u|^2(\operatorname{div}\xi + \theta_t)dx$$

$$- 2\int_0^{+\infty} dt \int_{B^3}\operatorname{div} u(x,t)\operatorname{div}\xi(x,t)\Phi(x,t)dxdt$$

$$+ 2\int_0^{+\infty} dt \int_{B^3}\Phi(x,t)u^i_{x_j}(x,t)\xi^j_{x_i}(x,t)$$

$$+ \int_0^{+\infty} dt \int_{B^3} \int_{B^3} \left\langle \frac{x-y}{|x-y|}, \xi(x,t) - \xi(y,t) \right\rangle G'(|x-y|) \operatorname{div} u(x,t) \operatorname{div} u(y,t) dx dy$$

$$- 2 \int_0^{+\infty} dt \int_{B^3} \int_{B^3} \left\langle \frac{x-y}{|x-y|}, u_t(x,t) \right\rangle \theta(x,t) \operatorname{div} u(y,t) G'(|x-y|) dx dy \le 0. \tag{5.5.38}$$

In what follows we want to derive some more applicable inequalities and equalities from (5.5.38) to be used in the future.

Now let $M_\varepsilon = B^3 \times B^3 \setminus \{|x-y| < \varepsilon\}$ for $\varepsilon > 0$ be small enough. Since $\xi(\cdot, t)$ is smooth, $\frac{\xi(x,t) - \xi(y,t)}{|x-y|}$ in (5.5.37) is uniformly bounded.

$$\int_0^{+\infty} dt \int_{B^3} \int_{B^3} \left\langle \frac{x-y}{|x-y|}, \xi(x,t) - \xi(y,t) \right\rangle G'(|x-y|) \operatorname{div} u(x,t) \operatorname{div} u(y,t) dx dy$$

$$= \lim_{\varepsilon \to 0} \int_0^{+\infty} dt \int_{M_\varepsilon} \left\langle \frac{x-y}{|x-y|}, \xi(x,t) - \xi(y,t) \right\rangle G'(|x-y|) \operatorname{div} u(x,t) \operatorname{div} u(y,t) dx dy$$

$$= \lim_{\varepsilon \to 0} 2 \int_0^{+\infty} dt \int_{M_\varepsilon} \xi(x,t) \frac{x-y}{|x-y|} G'(|x-y|) \operatorname{div} u(x,t) \operatorname{div} u(y,t) dx dy$$

$$= \lim_{\varepsilon \to 0} 2 \int_0^{+\infty} dt \int_{x \in B^3} \xi(x,t) \operatorname{div} u(x,t) \int_{y \in B^3 \setminus B(x,\varepsilon)} \frac{x-y}{|x-y|} G'(|x-y|) \operatorname{div} u(y,t). \tag{5.5.39}$$

Let

$$L_\varepsilon(x,t) = 2 \int_{y \in B^3 \setminus B(x,\varepsilon)} \frac{x-y}{|x-y|} G'(|x-y|) \operatorname{div} u(y,t) dy.$$

We have

$$\lim_{\varepsilon \to 0} L_\varepsilon(x,t) = L(x,t) = -H(x,t) + v(x,t) \tag{5.5.40}$$

with $H(x,t) \in L^\infty(0,\infty; L^p(B^3))$, $|v(x,t)| \le K \int_{\partial B^3} \frac{d\sigma(y)}{|x-y|^2} \le K \ln(1-|x|) \in L^\infty(0,\infty; L^p(B^3))$ for any $p > 1$ and there holds

$$\|L(x,t)\|_{L^2} \le C\|\nabla u\|_{L^2}.$$

Then (5.5.39) can be rewritten as

$$\int_0^{+\infty} dt \int_{B^3} \int_{B^3} \left\langle \frac{x-y}{|x-y|}, \xi(x,t) - \xi(y,t) \right\rangle G'(|x-y|) \operatorname{div} u(x,t) \operatorname{div} u(y,t) dx dy$$

$$= \int_0^\infty \int_{B^3} \xi(x,t) \cdot L(x,t) \operatorname{div} u(x,t) dx dt. \tag{5.5.41}$$

Similarly we deduce that

$$- \int_0^{+\infty} dt \int_{B^3} \int_{B^3} \left\langle \frac{x-y}{|x-y|}, u_t(x,t) \right\rangle \theta(x,t) \operatorname{div} u(y,t) G'(|x-y|) dx dy$$

$$= - \int_0^\infty \int_{B^3} u_t \cdot L(x,t) \theta(x,t) dx dt. \tag{5.5.42}$$

Finally, inequality (5.5.38) can be rewritten as

Assumption (S)

$$\int_0^{+\infty} \int_{B^3} \left(\frac{1}{2} u_t - \frac{1}{2} u \times u_t \right) (\nabla u \cdot \xi + u_t \theta)$$

$$+ \int_0^{+\infty} dt \int_{B^3} (u^i_{x_j} u^i_{x_k} \xi^k_{x_j} + u^i_t u^i_{x_j} \theta_{x_j}) dx$$

$$- \frac{1}{2} \int_0^{+\infty} dt \int_{B^3} |\nabla u|^2 (\operatorname{div} \xi + \theta_t) dx$$

$$- 2 \int_0^{+\infty} dt \int_{B^3} \operatorname{div} u(x,t) \operatorname{div} \xi(x,t) \Phi(x,t) dx dt$$

$$+ 2 \int_0^{+\infty} dt \int_{B^3} \Phi(x,t) u^i_{x_j}(x,t) \xi^j_{x_i}(x,t)$$

$$+ \int_0^{\infty} \int_{B^3} L(x,t)[\xi \operatorname{div} u(x,t) - u_t \theta] dx dt \le 0. \tag{5.5.43}$$

Formula (5.5.43), Assumption (S), is just the starting point of all the following discussions.

From Assumption (S), one can easily derive the following lemma as in [58] which will be used to get the generalized monotonicity inequality in the following section.

Lemma 5.5.2 *Let u be a stationary weak solution of (5.5.1)–(5.5.5) and ξ, θ as before. Then we have*

$$\int_{B^3 \times \{t\}} (u_t - u \times u_t) \nabla u \cdot \xi - |\nabla u|^2 \operatorname{div} \xi + 2 u_{x_j} u_{x_k} \xi^k_{x_j}$$

$$- 2 \operatorname{div} u(x,t) \operatorname{div} \xi(x,t) \Phi(x,t) + 2\Phi(x,t) u^i_{x_j} \xi^j_{x_i} + 2 \operatorname{div} u(x,t)(\xi \cdot L(x,t)) dx dt = 0 \tag{5.5.44}$$

and

$$\left(\int_{B^3 \times \{t_2\}} - \int_{B^3 \times \{t_1\}} \right) |\nabla u|^2 \theta dx$$

$$\le \int_{t_1}^{t_2} \int_{B^3} |\nabla u|^2 \theta_t - |u_t|^2 \theta - 2(\nabla u \cdot u_t) \nabla \theta + 2(u_t \cdot L(x,t)) \theta dx dt. \tag{5.5.45}$$

5.5.3   Estimates for Local Energy

In this section, we use inequalities (5.5.44) and (5.5.45) to derive the generalized monotonicity inequality which will be used to deduce the energy decay in the next section.

Denote $B_\rho = B_\rho(0)$, $P_\rho(z) = B_\rho(x) \times (t - \rho^2, t + \rho^2)$ for $z = (x,t)$. In the following, we always denote by u the stationary solution of (5.5.1)–(5.5.5).

Lemma 5.5.3 *For any $B_s(x_0) \subset B_{r/2}(0)$ we have for the stationary solution u that*

$$\frac{1}{s}\int_{B_s(x_0)} |\nabla u|^2 \le \frac{1}{r}\int_{B_r} |\nabla u|^2 + Kr\int_{B_r} |u_t|^2. \tag{5.5.46}$$

Proof. Taking $\xi(x) = \phi(|x|)x$ in inequality (5.5.44) where for any fixed $0 < r$, $r + h < 1$, $h > 0$

$$\phi(s) = \begin{cases} 1, & \text{if } s \le r \\ 1 + \frac{r-s}{h}, & \text{if } r \le s \le r + h \\ 0, & \text{if } r + h \le s < 1 \end{cases}$$

and noting that

$$\xi^k_{x_j}(x) = \delta_{jk}\phi(|x|) + \phi'(|x|)\frac{x_j x_k}{|x|}, \qquad \operatorname{div}\xi = 3\phi(|x|) + \phi'(|x|)|x| \tag{5.5.47}$$

where $\phi'(s) = 0$ if $s \le r$, $\phi'(s) = -\frac{1}{h}$ if $r < s < r + h$ and $\phi'(s) = 0$ if $r + h < s < 1$, we obtain from (5.5.44) that

$$\int_{B_{r+h}} (u_t - u \times u_t)(\nabla u \cdot x)\phi(|x|) - |\nabla u|^2(3\phi(|x|) + \phi'(|x|)|x|)$$

$$+ 2u_{x_j}u_{x_k}\left(\delta_{jk}\phi(|x|) + \phi'(|x|)\frac{x_j x_k}{|x|}\right)$$

$$- 2\Phi(x,t)\operatorname{div} u(x,t)(3\phi(|x|) + \phi'(|x|)|x|)$$

$$+ 2\Phi(x,t)u^j_{x_k}\left(\delta_{jk}\phi(|x|) + \phi'(|x|)\frac{x_j x_k}{|x|}\right)$$

$$+ 2\operatorname{div} u(x,t)\phi(|x|)(x \cdot L(x,t))dx = 0. \tag{5.5.48}$$

Sending $h \to 0$ we may estimate every term in (5.5.48) as follows.

$$\lim_{h\to 0}\int_{B_{r+h}} (u_t - u \times u_t)(\nabla u \cdot x)\phi(|x|) = \int_{B_r} (u_t - u \times u_t)(\nabla u \cdot x), \tag{5.5.49}$$

$$-\lim_{h\to 0}\int_{B_{r+h}} |\nabla u|^2(3\phi(|x|) + \phi'(|x|)|x|) = -3\int_{B_r} |\nabla u|^2 + r\int_{\partial B_r} |\nabla u|^2 d\sigma, \tag{5.5.50}$$

$$\lim_{h\to 0}\int_{B_{r+h}} 2u_{x_j}u_{x_k}\left(\delta_{jk}\phi(|x|) + \phi'(|x|)\frac{x_j x_k}{|x|}\right)$$

$$= 2\int_{B_r} |\nabla u|^2 - \frac{2}{r}\int_{\partial B_r} |x \cdot \nabla u|^2 d\sigma, \tag{5.5.51}$$

$$-\lim_{h\to 0}\int_{B_{r+h}} 2\Phi(x,t)\operatorname{div} u(x,t)(3\phi(|x|) + \phi'(|x|)|x|)$$

$$= -6\int_{B_r} \Phi(x,t)\operatorname{div} u(x,t) + 2r\int_{\partial B_r} \Phi \operatorname{div} u(x,t)d\sigma, \tag{5.5.52}$$

$$- \lim_{h \to 0} \int_{B_{r+h}} 2\Phi(x,t) u_{x_k}^j \left(\delta_{jk}\phi(|x|) + \phi'(|x|)\frac{x_j x_k}{|x|} \right)$$

$$= -2 \int_{B_r} \Phi(x,t)\mathrm{div}\, u(x,t) - \frac{2}{r} \int_{\partial B_r} \Phi(x \cdot \nabla u) \cdot x d\sigma, \qquad (5.5.53)$$

$$\lim_{h \to 0} \int_{B_{r+h}} 2\,\mathrm{div}\, u(x,t)\phi(|x|)(x \cdot L(x,t))$$

$$= 2 \int_{B_r} (x \cdot L(x,t))\mathrm{div}\, u(x,t). \qquad (5.5.54)$$

Substituting (5.5.49)–(5.5.54) into (5.5.48), we get

$$\int_{B_r} (u_t - u \times u_t)(\nabla u \cdot x) - \int_{B_r} |\nabla u|^2 + r \int_{\partial B_r} |\nabla u|^2 d\sigma$$

$$- \frac{2}{r} \int_{\partial B_r} |x \cdot \nabla u|^2 d\sigma - 8 \int_{B_r} \Phi(x,t)\mathrm{div}\, u(x,t) + 2r \int_{\partial B_r} \Phi\,\mathrm{div}\, u(x,t)d\sigma$$

$$- \frac{2}{r} \int_{\partial B_r} \Phi(x \cdot \nabla u) \cdot x d\sigma + 2 \int_{B_r} (x \cdot L(x,t))\mathrm{div}\, u(x,t) = 0. \qquad (5.5.55)$$

On the other hand, since

$$\frac{d}{dr}\left\{ \frac{1}{r} \int_{B_r} \left[|\nabla u|^2 - (u_t - u \times u_t)(\nabla u \cdot x) \right.\right.$$

$$\left.\left. + 8\Phi(x,t)\mathrm{div}\, u(x,t) - 2(x \cdot L(x,t))\mathrm{div}\, u(x,t) \right] \right\}$$

$$= -\frac{1}{r^2} \int_{B_r} \left[|\nabla u|^2 - (u_t - u \times u_t)(\nabla u \cdot x) \right.$$

$$\left. + 8\Phi(x,t)\mathrm{div}\, u(x,t) - 2(x \cdot L(x,t))\mathrm{div}\, u(x,t) \right]$$

$$+ \frac{1}{r} \int_{\partial B_r} \left[|\nabla u|^2 - (u_t - u \times u_t)(\nabla u \cdot x) \right.$$

$$\left. + 8\Phi(x,t)\mathrm{div}\, u - 2(x \cdot L(x,t))\mathrm{div}\, u \right] \qquad (5.5.56)$$

we get from (5.5.55) and (5.5.56) that

$$\frac{d}{dr}\left\{ \frac{1}{r} \int_{B_r} \left[|\nabla u|^2 - (u_t - u \times u_t)(\nabla u \cdot x) \right.\right.$$

$$\left.\left. + 8\Phi(x,t)\mathrm{div}\, u(x,t) - 2(x \cdot L(x,t))\mathrm{div}\, u(x,t) \right] \right\}$$

$$= \frac{1}{r} \int_{\partial B_r} [-(u_t - u \times u_t)(\nabla u \cdot x)$$

$$+ 8\Phi(x,t)\mathrm{div}\, u(x,t) - 2(x \cdot L(x,t))\mathrm{div}\, u(x,t)]$$

$$+ \int_{\partial B^3} \left[\frac{2}{r^3}|x \cdot \nabla u|^2 - \frac{2}{r}\Phi\,\mathrm{div}\, u(x,t) + \frac{2}{r^3}\Phi(x \cdot \nabla u) \cdot x \right]$$

$$= 2 \int_{\partial B^3} \left[\frac{|x \cdot \nabla u|^2}{|x|^3} - \frac{(x \cdot \nabla u)(u_t - u \times u_t)}{2|x|} \right.$$
$$\left. + \frac{3\Phi \operatorname{div} u}{|x|} - \frac{(x \cdot L)\operatorname{div} u}{|x|} + \frac{\Phi(x \cdot \nabla u) \cdot x}{|x|^3} \right]. \tag{5.5.57}$$

Denote

$$\psi(\rho, t) = \frac{1}{\rho} \int_{B_\rho} \left[|\nabla u|^2 - (u_t - u \times u_t)(\nabla u \cdot x) + 8\Phi(x, t)\operatorname{div} u - 2(x \cdot L)\operatorname{div} u \right]. \tag{5.5.58}$$

Then by integrating (5.5.57) from s to ρ ($s < \rho$), we obtain

$$\psi(\rho, t) - \psi(s, t) = 2 \int_{B_\rho \backslash B_s} \left[\frac{|x \cdot \nabla u|^2}{|x|^3} - \frac{(x \cdot \nabla u)(u_t - u \times u_t)}{2|x|} \right.$$
$$\left. + \frac{3\Phi \operatorname{div} u}{|x|} - \frac{(x \cdot L)\operatorname{div} u}{|x|} + \frac{\Phi(x \cdot \nabla u) \cdot x}{|x|^3} \right] \tag{5.5.59}$$

or

$$\psi(s, t) = \psi(\rho, t) - 2 \int_{B_\rho \backslash B_s} \left[\frac{|x \cdot \nabla u|^2}{|x|^3} - \frac{(x \cdot \nabla u)(u_t - u \times u_t)}{2|x|} \right.$$
$$\left. + \frac{3\Phi \operatorname{div} u}{|x|} - \frac{(x \cdot L)\operatorname{div} u}{|x|} + \frac{\Phi(x \cdot \nabla u) \cdot x}{|x|^3} \right], \tag{5.5.60}$$

that is

$$\frac{1}{s} \int_{B_s} |\nabla u|^2 = \frac{1}{s} \int_{B_s} [(u_t - u \times u_t)(\nabla u \cdot x)$$
$$- 8\Phi(x, t)\operatorname{div} u(x, t) + 2(x \cdot L(x, t))\operatorname{div} u(x, t)]$$
$$+ \psi(\rho, t) - 2 \int_{B_\rho \backslash B_s} \left[\frac{|x \cdot \nabla u|^2}{|x|^3} - \frac{(x \cdot \nabla u)(u_t - u \times u_t)}{2|x|} \right.$$
$$\left. + \frac{3\Phi \operatorname{div} u}{|x|} - \frac{(x \cdot L)\operatorname{div} u}{|x|} + \frac{\Phi(x \cdot \nabla u) \cdot x}{|x|^3} \right]. \tag{5.5.61}$$

By hole-filling method and note that $|x| \leq s$ if $x \in B_s$, we have

$$\frac{1}{s} \int_{B_s} |\nabla u|^2 \leq \psi(\rho, t) + 2 \int_{B_\rho} \left[\frac{|(x \cdot \nabla u)(u_t - u \times u_t)|}{2|x|} + \frac{4|\Phi \operatorname{div} u|}{|x|} \right.$$
$$\left. + \frac{|(x \cdot L)\operatorname{div} u|}{|x|} + \frac{|\Phi||x \cdot \nabla u|}{|x|^2} \right]. \tag{5.5.62}$$

Now we can estimate all the terms on the right-hand side of (5.5.62) and get the generalized monotonicity inequality. It follows from (5.5.62) and Hölder inequality that

$$\frac{1}{s} \int_{B_s} |\nabla u|^2 \leq \frac{1}{\rho} \int_{B_\rho} |\nabla u|^2 + K\rho \int_{B_\rho} [|u_t|^2 + |\Phi|^2 + |\Phi|^8 + |L|^2]. \tag{5.5.63}$$

In this inequality, we should note that $\Phi(x,t) \in L^\infty(0,T;W^{1,p}(\Omega))$, $L(x,t) \in L^\infty(0,T;L^p(\Omega))$ for any $p > 1$. Then it follows from Lemma 5.5.1, (5.5.12) and (5.5.15) that

$$\frac{1}{s}\int_{B_s} |\nabla u|^2 \le K \frac{1}{\rho} \int_{B_\rho} |\nabla u|^2 + K\rho \int_{B_\rho} |u_t|^2 \tag{5.5.64}$$

with K independent of t.

It is not difficult to get, for any $B_s(x_0) \subset B_{r/2}(0)$

$$\frac{1}{s}\int_{B_s(x_0)} |\nabla u|^2 \le \frac{1}{r}\int_{B_r} |\nabla u|^2 + Kr \int_{B_r} |u_t|^2. \tag{5.5.65}$$

Lemma 5.5.3 is proved.

In the following we want to prove

Lemma 5.5.4 (Generalized monotonicity inequality) *There exists a constant $C > 0$ such that for any $\theta \in (0, 1/16]$ there is $\varepsilon_0 > 0$, if $u \in H^1(P_r(z_0), S^2)$ is a stationary solution, we have*

$$r^{-3}\int_{P_{r/8}(0)} |\nabla u|^2 dz \le \delta\varepsilon^2 + C_1 r^{-5}\int_{P_r(0)} |u - (u)_{P_r(0)}|^2 dz$$

under the condition

$$r^{-3}\int_{P_r(0)} |\nabla u|^2 dz \le \varepsilon^2 \le \varepsilon_0^2.$$

Proof. It follows from (5.5.45) that

$$\left(\int_{B^3 \times \{t_2\}} - \int_{B^3 \times \{t_1\}}\right) |\nabla u|^2 \theta dx$$

$$\le \int_{t_1}^{t_2} \int_{B^3} |\nabla u|^2 \theta_t - |u_t|^2 \theta - 2(\nabla u \cdot u_t)\nabla\theta + 2(u_t \cdot L(x,t))\theta dx dt. \tag{5.5.66}$$

Therefore, we have

$$\left(\int_{B^3 \times \{t_2\}} - \int_{B^3 \times \{t_1\}}\right) |\nabla u|^2 \theta dx \le C \int_{t_1}^{t_2} \int_{B^3} |\nabla u|^2 (\theta_t + |\nabla\theta|^2) + |L(x,t)|^2 \theta.$$

Hence, if taking $\theta(x,t)$ such that $|\nabla\theta|^2$, $|\theta_t| \le \frac{C}{r^2}$, $\theta(x,t_1) = 0$ and $\theta \equiv 1$ for $x \in B_{r/2}, t \in ((-\frac{r}{2})^2, (\frac{r}{2})^2)$, we obtain from Lemma 5.5.1 for almost every $t = t_2 \in (-(\frac{r}{2})^2, (\frac{r}{2})^2)$

$$\frac{1}{r}\int_{B_{r/2}(0)\times\{t\}} |\nabla u|^2 \le K_1 \left(\frac{1}{r^3}\int_{P_r(0)} |\nabla u|^2 dx dt\right). \tag{5.5.67}$$

Claim. There exists a constant $C > 0$ such that for any given $0 < \lambda < 1$, there exists a set $\Lambda \subset (-r^2/2, r^2/2)$ with $|\Lambda| \le \lambda$ satisfying

$$\frac{1}{r}\int_{B_{\frac{r}{2}}(0)\times\{t\}} |u_t|^2 dx \le \frac{C\varepsilon^2}{\lambda} \tag{5.5.68}$$

for almost every $t \notin \Lambda$ where $\frac{1}{r^3}\int_{P_r} |\nabla u|^2 dz \le \varepsilon^2 \le \varepsilon_0^2$.

Proof. Taking smooth test function $\theta(x,t)$ such that $\theta \equiv 1$ in $P_{r/2}(0)$ but $\theta \equiv 0$ outside $P_r(0)$, $t_1 = -r^2$, $t_2 = r^2$, and noticing that $L(x,t) \in L^\infty(0,\infty;L^p(B^3))$ for any $p \geq 1$, we get from Lemma 5.5.1 that

$$\int_{P_{r/2}(0)} |u_t|^2 dz \leq C_0 \int_{P_r(0)} |\nabla u|^2(\theta_t + |\nabla\theta|^2) + \theta|L(x,t)|^2 dz$$

$$\leq C_0 r \left(\frac{1}{r^3}\int_{P_r(0)} |\nabla u|^2\right) \leq C_0 r \varepsilon^2. \tag{5.5.69}$$

If the Claim is false, then for any $C > 0$, there exists $0 < \lambda < 1$ such that for some set $\Lambda \subset (-r^2/2, r^2/2)$ with $|\Lambda| > \lambda$ there holds: if $t \in \Lambda$, then

$$\frac{1}{r}\int_{P_{r/2}(0)} |u_t|^2 dz \geq \frac{1}{r}\int_\Lambda \int_{B_{r/2}(0)} |u_t|^2 dz \geq C\varepsilon^2.$$

This contradicts (5.5.69) by taking $C > C_0$. The Claim follows.

We can even prove by (5.5.65) (Lemma 5.5.3) that if $t \notin \Lambda$ then

$$\sup_{B_s(x_0) \subset B_{r/4}(x_0)} \left(\frac{1}{s}\int_{B_s(x_0)\times\{t\}} |\nabla u|^2 dx\right) \leq \frac{C\varepsilon^2}{\lambda}. \tag{5.5.70}$$

In the sequence, we mean $t \notin \Lambda$.

In order to prove the lemma, we take $\xi \in C_0^\infty(B_{r/4}(0))$ such that $0 \leq \xi \leq 1$, $\xi \equiv 1$ in $B_{r/8}(0)$, $|\nabla\xi| \leq \frac{16}{r}$, to compute

$$\int_{B_r(0)\times\{t\}} \xi(x)|\nabla u|^2 dx = \int_{B_r(0)\times\{t\}} \xi(x)\nabla u \cdot \nabla(u - (u)_{P_r(0)})dx$$

$$= -\int_{B_r(0)\times\{t\}} (u - (u)_{P_r(0)})\nabla\xi \cdot \nabla u$$

$$- \int_{B_r(0)\times\{t\}} \xi(x)(u - (u)_{P_r(0)})\Delta u.$$

Using the equation, we have

$$\int_{B_r(0)\times\{t\}} \xi(x)|\nabla u|^2 dx = -\int_{B_r(0)\times\{t\}} (u - (u)_{P_r(0)})\nabla\xi \cdot \nabla u$$

$$- \int_{B_r(0)\times\{t\}} \xi(x)(u - (u)_{P_r(0)})\left(\frac{1}{2}u_t - \frac{1}{2}u \times u_t\right)$$

$$+ \int_{B_r(0)\times\{t\}} \xi(x)(u - (u)_{P_r(0)})u|\nabla u|^2$$

$$- \int_{B_r(0)\times\{t\}} \xi(x)(u - (u)_{P_r(0)})uH(u). \tag{5.5.71}$$

In the following we estimate every term on the right-hand side of (5.5.71).

Estimate of the first term:

$$\int_{B_r(0)\times\{t\}}(u-(u)_{P_r(0)})\nabla\xi\cdot\nabla u \le \frac{16}{r}\|\nabla u\|_{L^2(B_r)}\|u-(u)_{P_r(0)}\|_{L^2(B_r)}$$

$$\le \frac{16}{\sqrt{r}}\left(\frac{1}{r}\int_{B_r(0)}|\nabla u|^2 dx\right)^{1/2}\|u-(u)_{P_r(0)}\|_{L^2(B_r)}$$

$$\le C_0\varepsilon\frac{1}{\sqrt{r}}\|u-(u)_{P_r(0)}\|_{L^2(B_r)} \quad \text{(by (5.5.67))}$$

$$\le C\varepsilon^2\frac{\delta\lambda r}{2}+\frac{C}{\delta\lambda}\frac{1}{r^2}\|u-(u)_{P_r(0)}\|^2_{L^2(B_r)}. \tag{5.5.72}$$

Estimate of the second term: Since $t\notin\Lambda$, we get

$$\int_{B_r(0)\times\{t\}}\xi(x)(u-(u)_{P_r(0)})\left(\frac{1}{2}u_t-\frac{1}{2}u\times u_t\right)$$

$$\le C\|u_t\|_{L^2(B_r)}\|u-(u)_{P_r(0)}\|_{L^2(B_r)}$$

$$\le C\sqrt{r}\left(\frac{1}{r}\int_{B_r(0)}|u_t|^2 dx\right)^{1/2}\|u-(u)_{P_r(0)}\|_{L^2(B_r)}$$

$$\le \frac{C\varepsilon\sqrt{r}}{\sqrt{\lambda}}\|u-(u)_{P_r(0)}\|_{L^2(B_r)} \quad \text{(by (5.5.68))}$$

$$\le C\varepsilon^2\frac{\delta r}{2}+\frac{C}{\delta\lambda}\|u-(u)_{P_r(0)}\|^2_{L^2(B_r)}. \tag{5.5.73}$$

Estimate of the third term: To estimate this term, we use Hélein method to decompose it.

Since $|u|=1$ a.e. we have

$$u^i|\nabla u|^2 = \sum_{j=1}^{3}\nabla u^j(u^i\nabla u^j - u^j\nabla u^i)$$

and then

$$\int_{B_r(0)\times\{t\}}\xi(x)(u-(u)_{P_r(0)})u|\nabla u|^2$$

$$= \sum_{i=1}^{3}\sum_{j=1}^{3}\int_{B_r(0)\times\{t\}}(u^i-(u^i)_{P_r(0)})\nabla u^j(\xi(x)(u^i\nabla u^j - u^j\nabla u^i)).$$

On the other hand,

$$\text{div}(u^i\nabla u^j - u^j\nabla u^i) = u^i\Delta u^j - u^j\Delta u^i = u^iw^j - u^jw^i,$$

where $w = \frac{1}{2}u_t + \frac{1}{2}u \times u_t$, we obtain

$$\|\operatorname{div}(\xi(x)(u^i\nabla u^j - u^j\nabla u^i))\|_{L^2(B_r)}$$
$$\leq \|\nabla\xi(x)\cdot(u^i\nabla u^j - u^j\nabla u^i)\|_{L^2(B_r)} + \|\xi(r)(u^i w^j - u^j w^i)\|_{L^2(B_r)}$$
$$\leq \frac{32}{r}\|\nabla u\|_{L^2(B_{r/4}(0))} + 2\|u_t\|_{L^2(B_{r/4}(0))}$$
$$\leq \frac{C\varepsilon}{\sqrt{\lambda r}} \quad \text{(by (5.5.67) and (5.5.68))}. \tag{5.5.74}$$

To continue the proof, we recall a lemma by Feldman [58].

Lemma 5.5.5 ([58]) *Let f, $h \in H^1(R^n)$ and $g \in L^2(R^n, R^n)$ with $\operatorname{div} g \in L^2(R^n)$ in the distribution sense, and*

$$\sup_{x_0\in R^n,\ r>0}\left(r^{2-n}\int_{B_r(x_0)}|\nabla h|^2 dx\right) = A^2 < \infty.$$

Then

$$\left|\int_{R^n} fg\cdot\nabla h\right| \leq CA(\|\nabla f\|_{L^2}\|g\|_{L^2} + \|f\|_{L^2}\|\operatorname{div} g\|_{L^2})$$

for some universal constant C.

Now we apply Lemma 5.5.5 to $f = u^i - (u^i)_{P_r(0)}$, $h = u^j$ and $g = \xi(x)(u^i\nabla u^j - u^j\nabla u^i)$ by extending them properly to the whole space R^3. By (5.5.67) and (5.5.74)

$$\int_{B_r(0)\times\{t\}} \xi(x)(u - (u)_{P_r(0)})u|\nabla u|^2$$
$$= \int_{B_r(0)\times\{t\}} (u^i - (u^i)_{P_r(0)})\nabla u^j(\xi(x)(u^i\nabla u^j - u^j\nabla u^i))$$
$$\leq \frac{C\varepsilon}{\sqrt{\lambda}}\left[\|\nabla f\|_{L^2}\|g\|_{L^2} + \|f\|_{L^2}\|\operatorname{div} g\|_{L^2}\right]$$
$$\leq \frac{C\varepsilon}{\sqrt{\lambda}}\left[C\varepsilon^2 r + \frac{C\varepsilon}{\sqrt{\lambda r}}\|u - (u)_{P_r(0)}\|_{L^2}\right]$$
$$\leq \frac{C\varepsilon^3 r}{\lambda} + \frac{C\varepsilon^2}{\lambda r^2}\|u - (u)_{P_r(0)}\|_{L^2}^2. \tag{5.5.75}$$

Finally, we estimate the last term on the right-hand side of (5.5.71) as follows. Since $H(u) = \nabla\Phi$ and $\Delta\Phi = -\operatorname{div} u$, $\operatorname{div} u \in L^\infty(0,\infty;H^{-1})$, we know $\|H(u)\|_{L^2(B_r)}^2 \leq C_0\|\nabla u\|_{L^2(B_r)}^2 \leq Cr\varepsilon^2$. Hence we derive

$$\int_{B_r(0)\times\{t\}} \xi(x)(u - (u)_{P_r(0)})uH(u)$$
$$\leq C\sqrt{r}\varepsilon\|u - (u)_{P_r(0)}\|_{L^2(B_r(x_0))}$$
$$\leq C\varepsilon^2\frac{\delta\lambda r}{2} + \frac{C}{\delta\lambda}\|u - (u)_{P_r(0)}\|_{L^2(B_r(x_0))}^2. \tag{5.5.76}$$

Combining (5.5.71)–(5.5.76), we have obtained for $t \notin \Lambda$ $(|\Lambda| < \lambda)$

$$\int_{B_{r/8}(0)\times\{t\}} |\nabla u|^2 dx \leq \left(\frac{C\varepsilon}{\lambda} + \frac{\delta}{2}\right)\varepsilon^2 r + \frac{C}{\delta\lambda r^2}\|u - (u)_{P_r(0)}\|^2_{L^2(B_r(0))}. \quad (5.5.77)$$

Integrating over $(-r^2/2, r^2/2)$ and using (5.5.67), we get

$$\int_{-r^2/2}^{r^2/2}\int_{B_{r/8}(0)\times\{t\}} |\nabla u|^2 dx \leq \left(\frac{C\varepsilon}{\lambda} + C\lambda + \frac{\delta}{2}\right)\varepsilon^2 r^3 + \frac{C}{\delta\lambda r^2}\|u - (u)_{P_r(0)}\|^2_{L^2(P_r(0))}. \tag{5.5.78}$$

Hence, choosing C, λ and ε_0 properly we obtain

$$r^{-3}\int_{P_{r/8}(0)} |\nabla u|^2 dz \leq \delta\varepsilon^2 + C_1 r^{-5}\int_{P_r(0)} |u - (u)_{P_r(0)}|^2 dz \tag{5.5.79}$$

under the condition

$$r^{-3}\int_{P_r(0)} |\nabla u|^2 dz \leq \varepsilon^2 \leq \varepsilon_0^2.$$

Lemma 5.5.4 is proved.

5.5.4   Energy Decay and Partial Regularity

In this section we derive the energy decay which will be used to prove the regularity.

Lemma 5.5.6 *There exists a constant $C > 0$ such that for any $\theta \in (0, \frac{1}{2}]$, there is a number $\varepsilon_0 > 0$ such that for any stationary solution $u \in H^1(P_r(z_0); S^2)$ of (5.5.1)–(5.5.5) satisfying*

$$r^{-3}\int_{P_r(z_0)} |\nabla u|^2 dz \leq \varepsilon^2 \leq \varepsilon_0^2,$$

we have

$$(\theta r)^{-5}\int_{P_{\theta r}(z_0)} |u - (u)_{P_{\theta r}(z_0)}|^2 dz \leq C\theta^2\varepsilon^2. \tag{5.5.80}$$

Proof. Since the integrals are invariant under the transformation $(x, t) \to (rx + x_0, r^2 t + t_0)$, we may assume $P_r(z_0) = P_1(0)$.

We argue by contradiction: If the conclusion is untrue, then for any $C > 0$ we may find $\theta \in (0, 1/2]$ and stationary weak solution $u_k \in H^1(P_1(0); S^2)$ of the considered problem such that

$$\int_{P_1(0)} |\nabla u_k|^2 dz = \varepsilon_k^2 \to 0, \quad k \to \infty, \tag{5.5.81}$$

but

$$\int_{P_\theta(0)} |u_k - (u_k)_{P_\theta(0)}|^2 > C\theta^{3+4}\varepsilon_k^2. \tag{5.5.82}$$

It follows from these assumptions and the lemmas in the above sections that the sequence $\{v_k\} = \{\frac{1}{\varepsilon_k}(u_k - (u_k)_{P_\theta(0)})\}$ is bounded in $H^1(P_{1/2}(0))$ which allows us to assume that there is a map $v \in H^1(P_{1/2}(0); R^3)$ such that

$$v_k \to v, \text{ weakly in } H^1(P_{1/2}(0); R^3); \quad v_k \to v, \text{ strongly in } L^2(P_{1/2}(0); R^3)$$

with

$$\int_{P_\theta(0)} v\,dz = 0; \qquad \int_{P_{1/2}(0)} |\nabla v|^2 dz \le 1.$$

It is obvious that we can assume that $u_k \to p$ strongly in $L^2(P_{1/2}(0))$ for some constant map $p \in S^2$. Note that u_k solves the following equation:

$$\frac{1}{2}u_{kt} - \frac{1}{2}(u_k \times u_{kt}) = \Delta u_k + u_k|\nabla u_k|^2 + H(u_k) - H(u_k)u_k, \text{ in } B^3 \times R_+, \quad (5.5.83)$$

where

$$\mathbf{curl}\, H(u_k) = 0, \quad \text{in } \mathcal{D}'(R^3), \qquad\qquad (5.5.84)$$

$$\mathrm{div}(H(u_k) + \overline{u}_k) = 0, \quad \text{in } \mathcal{D}'(R^3). \qquad\qquad (5.5.85)$$

Let $H(u_k) = \nabla\Phi_k$. Then $\Delta\Phi_k = -\mathrm{div}\,u_k$. Note that $u_k = \varepsilon_k v_k + (u_k)_{P_\theta(0)}$.
Then for any $\phi \in C_0^\infty(P_{1/2}(0), R^3)$ we have by multiplying (5.5.83) by ϕ

$$\int_{P_{1/2}(0)} \left(\frac{1}{2}u_{kt} - \frac{1}{2}(u_k \times u_{kt})\right)\phi$$

$$= \int_{P_{1/2}(0)} (\Delta u_k + u_k|\nabla u_k|^2 + H(u_k) - H(u_k)u_k)\phi$$

$$= \int_{P_{1/2}(0)} (-\nabla u_k \cdot \nabla\phi + u_k|\nabla u_k|^2\phi + \phi\nabla\Phi_k - u_k\phi\nabla\Phi_k). \quad (5.5.86)$$

Note that $u_k = \varepsilon_k v_k + (u_k)_{P_\theta(0)}$. We have from (5.5.86) that

$$\varepsilon \int_{P_{1/2}(0)} \left(\frac{1}{2}v_{kt} - \frac{1}{2}(u_k \times v_{kt})\right)\phi$$

$$= \int_{P_{1/2}(0)} (-\varepsilon_k\nabla v_k \cdot \nabla\phi + u_k|\nabla u_k|^2\phi + \phi\nabla\Phi_k - u_k\phi\nabla\Phi_k) \quad (5.5.87)$$

with $\Delta\Phi_k = -\varepsilon_k \mathrm{div}\,v_k$. Divide both sides of (5.5.87) by ε_k and send $k \to \infty$ to give

$$\int_{P_{1/2}(0)} \left(\frac{1}{2}v_t - \frac{1}{2}(p \times v_t)\right)\phi$$

$$= \int_{P_{1/2}(0)} (-\nabla v \cdot \nabla\phi + \phi\nabla\Phi_\infty - p\phi\nabla\Phi_\infty) \quad (5.5.88)$$

since $\lim_{k\to\infty} \frac{1}{\varepsilon_k}\int_{P_{1/2}(0)} |\nabla u_k|^2 u_k\phi = 0$ from (5.5.81), where $\Delta\Phi_\infty = -\mathrm{div}\,v$. Since $v \in H^1(P_{1/2}(0))$ we know $\mathrm{div}\,v \in L^2(P_{1/2}(0))$, then $\Phi_\infty \in W_2^{2,2}$. Denote $H_\infty = \nabla\Phi_\infty$.

It follows from (5.5.88) that v satisfies

$$\frac{1}{2}v_t - \frac{1}{2}(p \times v_t) = \Delta v + H_\infty - pH_\infty. \tag{5.5.89}$$

By standard estimates and the boot-strapping method, we know that v is smooth and there holds

$$\int_{P_\theta(0)} |v|^2 dz \leq \int_{P_\theta(0)} |v - v_{P_\theta(0)}|^2 dz \leq C_0 \theta^{3+4} \tag{5.5.90}$$

by Poincaré inequality. Inequality (5.5.90) contradicts (5.5.82) from the strong L^2-convergence of v_k to v if one chooses $C > C_0$.

Combining these lemmas we get by iteration method

Proposition 5.5.1 *There is a constant $C > 0$ such that for any $\theta \in (0, 1/16]$, there exists a number $\varepsilon_0 > 0$ such that if $u \in H^1(P_r(z_0), S^2)$ is a stationary solution of (5.5.1)–(5.5.5) satisfying the small energy condition*

$$r^{-3} \int_{P_r(z_0)} |\nabla u|^2 dz \leq \varepsilon^2 \leq \varepsilon_0^2, \tag{5.5.91}$$

then

$$(\theta r)^{-3} \int_{P_{\theta r}(z_0)} |\nabla u|^2 dz \leq C\theta^2 \varepsilon^2. \tag{5.5.92}$$

Proof. Given $0 < \theta < 1/16$, taking k such that $\frac{1}{8^{k+2}} \leq \theta \leq \frac{1}{8^{k+1}}$. Denote $\Phi(r) = r^{-3} \int_{P_r(z_0)} |\nabla u|^2 dz$, $\Psi(r) = r^{-5} \int_{P_r(z_0)} |u - (u)_{P_r(z_0)}|^2$. It follows from (5.5.91) and Lemma 5.5.4 that for any $\delta > 0$

$$\Phi\left(\frac{r}{8}\right) \leq \delta\varepsilon^2 + C_1\Psi(r). \tag{5.5.93}$$

Similarly we have

$$\Phi\left(\frac{r}{8^2}\right) \leq \delta\Phi\left(\frac{r}{8}\right) + C_1\Psi\left(\frac{r}{8}\right) \leq \delta(\delta\varepsilon^2 + C_1\Psi(r)) + C_1\Psi\left(\frac{r}{8}\right). \tag{5.5.94}$$

Iterating, applying Lemma 5.5.6 and choosing δ properly, we get

$$\Phi\left(\frac{r}{8^{k+1}}\right) \leq \delta^{k+1}\varepsilon^2 + C_1\sum_{j=1}^{k}\delta^j\Psi\left(\frac{r}{8^j}\right) + \Psi\left(\frac{r}{8^k}\right) \leq C\delta^{k+1}\varepsilon^2 + C\left(\frac{1}{8^k}\right)^2\varepsilon^2 \leq C\theta^2\varepsilon^2, \tag{5.5.95}$$

where we have used Lemma 5.5.6 again which shows $\Psi(\frac{r}{8^j}) \leq C(\frac{1}{8^j})^2\varepsilon^2$. Then we have

$$\Phi(\theta r) = (\theta r)^{-3}\int_{P_{\theta r}} |\nabla u|^2 dz \leq \theta^{-3}\left(\frac{r}{8^{k+1}}\right)^{-3}(8^{k+1})^{-3}\int_{P_{\frac{r}{8^{k+1}}}} |\nabla u|^2 dz$$

$$\leq (8^{k+1}\theta)^{-3}\left(\frac{r}{8^{k+1}}\right)^{-3}\int_{P_{\frac{r}{8^{k+1}}}} |\nabla u|^2 dz \leq 8^3\Phi\left(\frac{r}{8^{k+1}}\right) \leq C\theta^2\varepsilon^2 \quad \text{(by (5.5.95))}.$$

Remark 5.5.1 By virtue of (5.5.70), Proposition 5.5.1 also holds if replacing $P_{\theta r}(z_0)$ by $P_s(z_1)$ for any $z_1 \in P_{15r/16}(z_0)$ and $s \in (0, r/16)$.

Theorem 5.5.1 *There exist constants $\varepsilon_0 > 0$ and $C_{kl} > 0$ such that any stationary solution $u \in H^1(P_r(z_0))$ of (5.5.1)–(5.5.4) satisfying the small energy condition (5.5.89) is smooth in $P_{r/2}(z_0)$ and*

$$\|\partial_t^l \nabla^k u\|_{L^\infty(P_{r/2}(z_0))} \leq C_{kl} r^{-k-2l} \varepsilon, \quad k, l = 0, 1, 2, \ldots. \tag{5.5.96}$$

Proof. For any given $P_r(z_0) \subset B^3 \times (0, T)$, Proposition 5.5.1 shows that for any $\lambda \in (0, 1)$, if ε_0 is small enough, we have

$$\int_{P_s(z_1)} (|\nabla u|^2 + s^2 |u_t|^2) dz \leq C_1 s^{n+2\lambda} \tag{5.5.97}$$

for any $z_1 \in P_{15r/16}(z_0)$ and $s \in (0, r/16)$ with C_1 only depending on λ. In fact, let $\theta r = s$, then $\theta = \frac{s}{r}$. Substituting this θ into (5.5.92), we have

$$s^{-3} \int_{P_s(z_1)} |\nabla u|^2 dz \leq \left(\frac{s}{r}\right)^2 \varepsilon_0^2. \tag{5.5.98}$$

On the other hand, we may control $\int_{P_s(z_1)} s^2 |u_t|^2 dz$ as in Feldman [58]. (5.5.97) is proved.

By standard method we deduce that u is smooth in $P_{15r/16}(z_0)$ by Morrey's Lemma which was done by Feldman [58]. The estimates (5.5.96) can be obtained by a scaling argument.

By the standard method as in Giaquinta's book [64], it is easy to conclude.

Theorem 5.5.2 *Let $u \in H^1(\Omega \times (0, T); S^2)$ be a stationary solution of (5.5.1)–(5.5.4). There is an open set $Q \subset \Omega \times (0, T)$ such that u is smooth in Q and*

$$\mathcal{H}^3(\Omega \times (0, T) \setminus Q) = 0 \tag{5.5.99}$$

where

$$\Omega \times (0, T) \setminus Q = \{z = (x, t) \mid \liminf_{r \to 0} r^{-3} \int_{P_r(z)} |\nabla u|^2 dz \geq \varepsilon_0\}. \tag{5.5.100}$$

5.6 Weak Solutions to Landau–Lifshitz–Maxwell Equations with Polarization

5.6.1 The Problem and Physics Background

In this section, we study the three-dimensional Landau–Lifshitz–Maxwell equations coupling with polarization as follows

$$\vec{Z}_t = \alpha_1 \vec{Z} \times \left(\triangle \vec{Z} + \vec{H}\right) - \alpha_2 \vec{Z} \times \left(\vec{Z} \times \left(\triangle \vec{Z} + \vec{H}\right)\right), \tag{5.6.1}$$

$$\nabla \times \vec{H} = \frac{\partial(\vec{E} + \vec{P})}{\partial t} + \sigma\vec{E}, \tag{5.6.2}$$

$$\nabla \times \vec{E} = -\frac{\partial\vec{H}}{\partial t} - \beta\frac{\partial\vec{Z}}{\partial t}, \tag{5.6.3}$$

$$\frac{\partial^2\vec{P}}{\partial t^2} + \lambda^2 \operatorname{curl}^2 \vec{P} + \mu\frac{\partial\vec{P}}{\partial t} = \nu\left(\vec{E} - 2\vec{P}\Phi'(|\vec{P}|^2)\right), \tag{5.6.4}$$

where $\vec{P}(x,t) = (P_1(x,t), P_2(x,t), P_3(x,t))$ denotes the electric polarization, $\widehat{E}(\vec{P}) = 2\vec{P}\Phi'(|\vec{P}|^2)$ the equilibrium electric field, $\operatorname{curl}^2 \vec{P} = \operatorname{curl}(\operatorname{curl} \vec{P}) = \nabla \times (\nabla \times \vec{P})$, $\alpha_2 \geq 0$ is the Gilbert damping coefficient, $\lambda > 0$ denotes the speed of light for the internal fields, $\sigma \geq 0$ denotes the constant conductivity, constant β can be viewed as the magnetic permeability of free space. The physical meanings of parameters μ, ν can be found in [69].

We assume that $\Phi : R^+ \rightarrow R$ is a C^2 convex function such that

$$|\Phi'(r)| \leq C_0, \qquad r\Phi''(r) \leq C_1 \tag{5.6.5}$$

for all $r \geq 0$. We also assume that function $\Phi(r^2)$ has unique minimum at some point r_0^2. These assumptions guarantee that $r\Phi'(r^2) \leq C_2$ for all $r \geq 0$, where $C_2 = C_0 + 2C_1$. Therefore, we have

$$\left|\vec{X}\Phi'(|\vec{X}|^2) - \vec{Y}\Phi'(|\vec{Y}|^2)\right| \leq C_2\left|\vec{X} - \vec{Y}\right| \quad \text{for all } \vec{X}, \vec{Y} \in R^3. \tag{5.6.6}$$

Much more about the equilibrium relation of Φ may be found in Landau and Lifshitz [104], pp. 84–91.

System (5.6.1)-(5.6.4) models the dynamics of magnetization, magnetic field, electric field. and electric polarization for the ferromagnetic–ferroelectric materials, which includes a new equation for polarization $\vec{P}$. As we know that, some ferromagnetic substances, such as ferrites, are not only ferromagnetic materials, but also ferroelectric ones (such as $LiFePO_4$), we call them the ferromagnetic–ferroelectrics [148].

If an electric field is applied to a medium (such as a dielectric one) made up of a large number of atoms or molecules, the charges bound in each molecule will respond to the applied field and will execute perturbed motions: the molecular charge density will be distorted. The multipole moments of each molecule will be different from what they were in the absence of the field. In simple substances, when there is no applied field the multipole moments are all zero, at least when they averaged over many molecules. The dominant molecular multipole with the applied fields is the dipole. There is thus produced in the medium an electric polarization $\vec{P}$ (the dipole moment per unit volume). A dielectric in which $\vec{P}$ differs from zero is said to be polarized. The vector $\vec{P}$ determines not only the volume charge density but also the density of the charge on the surface of the polarized dielectric [93]. One can learn more about polarization in [17, 43, 62, 104].

The coupling of this classical Landau–Lifshitz–Maxwell system with $\vec{P}$ and Eq. (5.6.4) for $\vec{P}$ can be derived from the full Maxwell system as follows:

$$\frac{\partial \vec{B}}{\partial t} = -\operatorname{curl} \vec{E} \quad \text{and} \quad \frac{\partial \vec{D}}{\partial t} + \sigma \vec{E} = \operatorname{curl} \vec{H}, \tag{5.6.7}$$

where $\vec{E}$ and $\vec{H}$ are the electric and magnetic fields, $\sigma \geq 0$ is the conductivity, $\vec{D}$ and $\vec{B}$ the electric and magnetic displacements defined by

$$\vec{D} = \epsilon_0 \vec{E} + \vec{P}, \quad \vec{B} = \mu_0(\vec{H} + \vec{Z}),$$

where ϵ_0 is the permittivity of free space, μ_0 is the magnetic permeability of free space, $\vec{Z}$ is the magnetization, and $\vec{P}$ is the electric polarization. Substituting these definitions into (5.6.7), one may couple $\vec{Z}$, $\vec{E}$, $\vec{H}$ and $\vec{P}$ by systems like (5.6.1)–(5.6.3). For the derivation of (5.6.4), we refer to [69].

For the system (5.6.1)–(5.6.4), we impose the following periodic initial conditions

$$(\vec{Z}(x,0), \vec{H}(x,0), \vec{E}(x,0), \vec{P}(x,0), \vec{P}_t(x,0))$$
$$= (\vec{Z}_0(x), \vec{H}_0(x), \vec{E}_0(x), \vec{P}_0(x), \vec{P}_1(x)), \tag{5.6.8}$$

where we call a function $f(x)$ is $2D$-periodic if $f(x + 2De_i) = f(x)$, $(i = 1,2,3)$, where (e_1, e_2, e_3) forms the unit orthogonal basis of R^3, $D > 0$ is a constant. And, we always assume that $Z_0(x)$, $H_0(x)$, $E_0(x)$, $P_0(x)$ and $P_1(x)$ are $2D$-periodic. We denote by $\Omega \subset R^3$ the three-dimensional cube with width $2D$ along each direction, i.e. $\Omega = \{x = (x_1, x_2, x_3) | \ |x_i| < D; (i = 1,2,3)\}$ and $Q_T = \{(x,t) | \ x \in \Omega, 0 < t \leq T\}$.

Definition 5.6.1 *A $2D$-periodic vector $(\vec{Z}(x,t), \vec{E}(x,t), \vec{H}(x,t), \vec{P}(x,t)) \in (L^\infty(0,T; H^1(\Omega))$, $L^\infty(0,T; L^2(\Omega))$, $L^\infty(0,T; L^2(\Omega))$, $W^{1,\infty}(0,T; L^2(\Omega)) \cap L^\infty(0,T; H^1(\Omega))$ is called a weak solution to problem (5.6.1)–(5.6.4) and (5.6.8), if for any $2D$-periodic vector-valued test function $\vec{\Psi}(x,t) \in C^1(Q_T)$ such that $\vec{\Psi}(x,T) = 0$, the following equalities hold*

$$\iint_{Q_T} \vec{Z} \cdot \vec{\Psi}_t + \alpha_1 \iint_{Q_T} (\vec{Z} \times \nabla \vec{Z}) \cdot \nabla \vec{\Psi} - \alpha_1 \iint_{Q_T} (\vec{Z} \times \vec{H}) \cdot \vec{\Psi}$$

$$+ \alpha_2 \iint_{Q_T} \left(\vec{Z} \times \nabla \vec{Z}\right) \cdot \nabla \left(\vec{Z} \times \vec{\Psi}\right)$$

$$- \alpha_2 \iint_{Q_T} \left(\vec{Z} \times \vec{H}\right) \cdot \nabla \left(\vec{Z} \times \vec{\Psi}\right) + \int_\Omega \vec{Z}_0 \cdot \vec{\Psi}(x,0) = 0, \tag{5.6.9}$$

$$\iint_{Q_T} \left(\vec{E} + \vec{P}\right) \cdot \vec{\Psi}_t e^{\sigma t} + \sigma \iint_{Q_T} e^{\sigma t} \vec{P} \cdot \vec{\Psi}$$

$$+ \iint_{Q_T} e^{\sigma t} \nabla \times \vec{\Psi} \cdot \vec{H} + \int_\Omega \left(\vec{E}_0 + \vec{P}_0\right) \cdot \Psi(x,0) = 0, \tag{5.6.10}$$

$$\iint_{Q_T} \left(\vec{H} + \beta \vec{Z}\right) \cdot \vec{\Psi}_t - \iint_{Q_T} \left(\nabla \times \vec{\Psi}\right) \cdot \vec{E} + \int_\Omega \left(\vec{H}_0 + \beta \vec{Z}_0\right) \cdot \vec{\Psi}(x,0) = 0, \tag{5.6.11}$$

$$\iint_{Q_T} \vec{P}_t \cdot \vec{\Psi}_t - \lambda^2 \iint_{Q_T} \operatorname{curl} \vec{P} \cdot \operatorname{curl} \vec{\Psi} - \mu \iint_{Q_T} \vec{P}_t \cdot \vec{\Psi} + \nu \iint_{Q_T} \vec{E} \cdot \vec{\Psi}$$

$$- 2\nu \iint_{Q_T} \Phi'(|\vec{P}|^2)\vec{P} \cdot \vec{\Psi} + \int_\Omega \vec{P}_1 \cdot \vec{\Psi}(x,0) = 0. \tag{5.6.12}$$

5.6.2   Viscosity Approximation

Since Eq. (5.6.1) is strongly coupled, it is not easy to obtain weak solutions by use of the theory of semigroups as Habib Ammari in [6]. We are going to use Galerkin method here.

However, it is easy to see that Eq. (5.6.4) lacks the compactness which we need to get the estimate of H^1-norm for $\vec{P}$. In fact, from this equation, we will only be able to obtain the $L^\infty(0, T; L^2(\Omega))$ estimates for curl $\vec{P}$, but do not have the $L^\infty(0, T; L^2(\Omega))$ estimates for div $\vec{P}$. To overcome this difficulty, we firstly apply the viscosity vanishing argument to get the weak solution for the viscosity problem and then, we will set the viscosity constant to zero. But the lack of compactness reappears in the limit procedure. Therefore, we secondly consider more regular (than energy ones) class weak solution and then obtain the additional *a priori* estimates for the div-component of Maxwell fields (cf. Lemma 5.6.9). Finally we obtain the desired weak solution to the original problem.

Replacing (5.6.4) by the following viscosity approximation

$$\frac{\partial^2 \vec{P}}{\partial t^2} + \lambda^2 \operatorname{curl}^2 \vec{P} + \mu \frac{\partial \vec{P}}{\partial t} - \epsilon \triangle \vec{P} = \nu(\vec{E} - 2\vec{P}\Phi'(|\vec{P}|^2)), \qquad (5.6.13)$$

we get a viscosity system (5.6.1)–(5.6.3) and (5.6.13) with the 2D-periodic initial conditions (5.6.8).

5.6.3   Solutions to the Viscosity Problem

Definition 5.6.2 *The space $H_P(\operatorname{curl}, \Omega)$ is defined by*

$$H_P(\operatorname{curl}, \Omega) = \{\vec{V} \in L^2(\Omega); \ \vec{V} \ is \ 2D\text{-}periodic \ and \ \operatorname{curl} \vec{V} \in L^2(\Omega)\},$$

and is provided with the norm

$$\|\vec{V}\|_{H_P(\operatorname{curl}, \Omega)} = \{\|\vec{V}\|_{L^2(\Omega)}^2 + \|\operatorname{curl} \vec{V}\|_{L^2(\Omega)}^2\}^{1/2}.$$

The space $H_P(\operatorname{div}, \Omega)$ is defined by

$$H_P(\operatorname{div}, \Omega) = \{\vec{V} \in L^2(\Omega); \ \vec{V} \ is \ 2D\text{-}periodic \ and \ \operatorname{div} \vec{V} \in L^2(\Omega)\},$$

and is provided with the norm

$$\|\vec{V}\|_{H_P(\operatorname{div}, \Omega)} = \{\|\vec{V}\|_{L^2(\Omega)}^2 + \|\operatorname{div} \vec{V}\|_{L^2(\Omega)}^2\}^{1/2}.$$

Finally, we set

$$X_P(\Omega) = H_P(\operatorname{curl}, \Omega) \cap H_P(\operatorname{div}, \Omega)$$

with the norm

$$\|\vec{V}\|_{X_P(\Omega)} = \{\|\vec{V}\|_{L^2(\Omega)}^2 + \|\operatorname{curl} \vec{V}\|_{L^2(\Omega)}^2 + \|\operatorname{div} \vec{V}\|_{L^2(\Omega)}^2\}^{1/2}.$$

Lemma 5.6.1 *Assume $X \subset E \subset Y$ are Banach spaces and $X \hookrightarrow\hookrightarrow E$. Then the following imbedding are compact, if $1 \leq q \leq \infty$, or $1 < r \leq \infty$*

$$\text{(i)} \quad L^q(0,T;X) \cap \left\{ \varphi : \frac{\partial \varphi}{\partial t} \in L^1(0,T;Y) \right\} \hookrightarrow\hookrightarrow L^q(0,T;E), \qquad (5.6.14)$$

$$\text{(ii)} \quad L^\infty(0,T;X) \cap \left\{ \varphi : \frac{\partial \varphi}{\partial t} \in L^r(0,T;Y) \right\} \hookrightarrow\hookrightarrow C([0,T];E). \qquad (5.6.15)$$

Let $\omega_n(x), (n = 1,2,3,\ldots)$ be the unit eigenfunctions satisfying the equation $-\triangle\omega_n = \lambda_n\omega_n$, with periodicity $\omega_n(x - De_i) = \omega_n(x + De_i)$ and $\lambda_n, (n = 1,2,3,\ldots)$ the corresponding eigenvalues different from each other. Denote the approximate solutions of the problem (5.6.1)–(5.6.3), (5.6.13) and (5.6.8) by $\vec{Z}_N^\epsilon(x,t)$, $\vec{H}_N^\epsilon(x,t)$, $\vec{E}_N^\epsilon(x,t)$, $\vec{P}_N^\epsilon(x,t)$ in the following form:

$$\vec{Z}_N^\epsilon(x,t) = \sum_{s=1}^{N} \vec{\alpha}_{sN}^\epsilon(t)\omega_s(x), \quad \vec{H}_N^\epsilon(x,t) = \sum_{s=1}^{N} \vec{\beta}_{sN}^\epsilon(t)\omega_s(x),$$

$$\vec{E}_N^\epsilon(x,t) = \sum_{s=1}^{N} \vec{\gamma}_{sN}^\epsilon(t)\omega_s(x), \quad \vec{P}_N^\epsilon(x,t) = \sum_{s=1}^{N} \vec{\delta}_{sN}^\epsilon(t)\omega_s(x),$$

where $\vec{\alpha}_{sN}^\epsilon(t)$, $\vec{\beta}_{sN}^\epsilon(t)$, $\vec{\gamma}_{sN}^\epsilon(t)$, $\vec{\delta}_{sN}^\epsilon(t)$, $(t \in R^+)$, $(s = 1,2,\ldots,N; N = 1,2,\ldots)$ are three-dimensional vector-valued functions satisfying the following system of ordinary differential equations:

$$\int_\Omega \vec{Z}_{Nt}^\epsilon\omega_s = \alpha_1 \int_\Omega \vec{Z}_N^\epsilon \times \left(\triangle\vec{Z}_N^\epsilon + \vec{H}_N^\epsilon\right)\omega_s$$
$$- \alpha_2 \int_\Omega \vec{Z}_N^\epsilon \times \left(\vec{Z}_N^\epsilon \times \left(\triangle\vec{Z}_N^\epsilon + \vec{H}_N^\epsilon\right)\right)\omega_s, \qquad (5.6.16)$$

$$\int_\Omega \left(\vec{H}_{Nt}^\epsilon + \beta\vec{Z}_{Nt}^\epsilon\right)\omega_s = -\int_\Omega \left(\nabla \times \vec{E}_N^\epsilon\right)\omega_s(x), \qquad (5.6.17)$$

$$\int_\Omega \left(\vec{E}_{Nt}^\epsilon + \vec{P}_{Nt}^\epsilon\right)\omega_s(x) + \sigma \int_\Omega \left(\vec{E}_N^\epsilon + \vec{P}_N^\epsilon\right)\omega_s(x)$$
$$= \int_\Omega \left(\nabla \times \vec{H}_N^\epsilon\right)\omega_s(x) + \sigma \int_\Omega \vec{P}_N^\epsilon\omega_s(x), \qquad (5.6.18)$$

$$\int_\Omega \vec{P}_{Ntt}^\epsilon\omega_s + \lambda^2 \int_\Omega \text{curl}^2\vec{P}_N^\epsilon\omega_s(x) + \mu \int_\Omega \vec{P}_{Nt}^\epsilon\omega_s(x) - \epsilon \int_\Omega \triangle\vec{P}_N^\epsilon\omega_s(x)$$
$$= \nu \int_\Omega \vec{E}_N^\epsilon\omega_s(x) - 2\nu \int_\Omega \vec{P}_N^\epsilon\Phi'(|\vec{P}_N^\epsilon|^2)\omega_s(x) \qquad (5.6.19)$$

with initial conditions

$$\int_\Omega \vec{Z}_N^\epsilon(x,0)\omega_s(x) = \int_\Omega \vec{Z}_0(x)\omega_s(x), \qquad (5.6.20)$$

$$\int_\Omega \vec{H}_N^\epsilon(x,0)\omega_s(x) = \int_\Omega \vec{H}_0(x)\omega_s(x), \qquad (5.6.21)$$

$$\int_\Omega \vec{E}_N^\epsilon(x,0)\omega_s(x) = \int_\Omega \vec{E}_0(x)\omega_s(x), \qquad (5.6.22)$$

$$\int_\Omega \vec{P}_N^\epsilon(x,0)\omega_s(x)dx = \int_\Omega \vec{P}_0(x)\omega_s(x)dx, \qquad (5.6.23)$$

$$\int_\Omega \vec{P}_{Nt}^\epsilon(x,0)\omega_s(x)dx = \int_\Omega \vec{P}_1(x)\omega_s(x)dx. \qquad (5.6.24)$$

It follows from the standard theory on nonlinear ordinary differential equations that the problem (5.6.16)–(5.6.24) admits unique local solution. The following *a priori* estimates enable us to take the limit $N \to \infty$ in (5.6.16)–(5.6.24) to obtain the global solution to the viscosity problem.

For the sake of simplicity, we denote $\| \cdot \|_{L^p(\Omega)} = \| \cdot \|_p$, $p \geq 2$.

Lemma 5.6.2 *Assume* $(\vec{Z}_0(x), \vec{H}_0(x), \vec{E}_0(x), \vec{P}_0(x), \vec{P}_1(x)) \in (H^1(\Omega), L^2(\Omega), L^2(\Omega),$ $H^1(\Omega), L^2(\Omega))$. *Then for the solution of the initial value problem* (5.6.16)–(5.6.24), *we have the following estimates:*

$$\sup_{0 \leq t \leq T} [\|\vec{Z}_N^\epsilon(\cdot,t)\|_{H^1(\Omega)}^2 + \|\vec{E}_N^\epsilon(\cdot,t)\|_2^2 + \|\vec{H}_N^\epsilon(\cdot,t)\|_2^2 + \|\vec{P}_N^\epsilon(\cdot,t)\|_2^2$$

$$+ \|\mathrm{curl}\, \vec{P}_N^\epsilon(\cdot,t)\|_2^2 + \epsilon\|\nabla \vec{P}_N^\epsilon(\cdot,t)\|_2^2 + \|\vec{P}_{Nt}^\epsilon(\cdot,t)\|_2^2] \leq C_3, \qquad (5.6.25)$$

$$\sup_{0 \leq t \leq T} \|\vec{Z}_N^\epsilon(\cdot,t)\|_6^2 \leq C_4, \qquad \sup_{0 \leq t \leq T} \|\vec{Z}_N^\epsilon(\cdot,t) \times \nabla \vec{Z}_N^\epsilon(\cdot,t)\|_{L^{\frac{3}{2}}(\Omega)} \leq C_5, \qquad (5.6.26)$$

where the constants C_3, C_4 *and* C_5 *are independent of* N, α_2 *and* D. *When* $\alpha_2 > 0$, *there is*

$$\|\vec{Z}_N^\epsilon \times (\triangle \vec{Z}_N^\epsilon + \vec{H}_N^\epsilon)\|_{L^2(0,T;L^2(\Omega))} \leq C_6, \qquad (5.6.27)$$

where the constant C_6 *is independent of* N *and* D.

Proof. 1. Multiplying (5.6.16) by $\vec{\alpha}_{sN}^\epsilon(t)$, summing up the products for $s = 1, 2, \ldots, N$, we get

$$\frac{d}{dt} \int_\Omega |\vec{Z}_N^\epsilon(\cdot,t)|^2 dx = 0.$$

Then we have

$$\|\vec{Z}_N^\epsilon(\cdot,t)\|_2^2 = \|\vec{Z}_N^\epsilon(\cdot,0)\|_2^2 \leq \|\vec{Z}_0(x)\|_2^2, \quad \forall\, t \geq 0. \qquad (5.6.28)$$

2. Making the scalar product of $\left(-\lambda_s\vec{\alpha}_{sN}^\epsilon(t) + \vec{\beta}_{sN}^\epsilon(t)\right)$ with (5.6.16), summing up the resulting product for $s = 1, 2, \ldots, N$ and then integrating by parts, we have

$$\frac{1}{2}\frac{d}{dt} \int_\Omega \left|\nabla \vec{Z}_N^\epsilon(\cdot,t)\right|^2 dx + \alpha_2\left\|\vec{Z}_N^\epsilon \times \left(\triangle \vec{Z}_N^\epsilon + \vec{H}_N^\epsilon\right)\right\|_2^2 - \int_\Omega \vec{Z}_{Nt}^\epsilon \cdot \vec{H}_N^\epsilon dx = 0. \qquad (5.6.29)$$

Multiplying (5.6.17) by $\vec{\beta}^{\epsilon}_{sN}(t)$ and (5.6.18) by $\vec{\gamma}^{\epsilon}_{sN}(t)$, summing up, and integrating by parts, we obtain

$$\frac{1}{2}\frac{d}{dt}\int_{\Omega}\left(\left|\vec{H}^{\epsilon}_N(\cdot,t)\right|^2+\left|\vec{E}^{\epsilon}_N(\cdot,t)\right|^2\right)$$
$$+\sigma\int_{\Omega}\left|\vec{E}^{\epsilon}_N\right|^2+\int_{\Omega}\vec{P}^{\epsilon}_{Nt}\cdot\vec{E}^{\epsilon}_N+\beta\int_{\Omega}\vec{Z}^{\epsilon}_{Nt}\cdot\vec{H}^{\epsilon}_N=0. \tag{5.6.30}$$

Multiplying (5.6.19) by $(\vec{\gamma}^{\epsilon}_{sN}(t)+\vec{\delta}^{\epsilon}_{sN}(t))$, summing up, and integrating by parts, we get

$$\frac{1}{2}\frac{d}{dt}\int_{\Omega}\left|\vec{E}^{\epsilon}_N(\cdot,t)+\vec{P}^{\epsilon}_N(\cdot,t)\right|^2 dx+\sigma\int_{\Omega}\left|\vec{E}^{\epsilon}_N+\vec{P}^{\epsilon}_N\right|^2 dx$$
$$=\sigma\int_{\Omega}\vec{P}^{\epsilon}_N\cdot\left(\vec{E}^{\epsilon}_N+\vec{P}^{\epsilon}_N\right) dx+\int_{\Omega}\left(\nabla\times\vec{H}^{\epsilon}_N\right)\cdot\left(\vec{E}^{\epsilon}_N+\vec{P}^{\epsilon}_N\right) dx. \tag{5.6.31}$$

Putting these equalities together, we have

$$\frac{1}{2}\frac{d}{dt}\int_{\Omega}\left[\left|\vec{E}^{\epsilon}_N(\cdot,t)+\vec{P}^{\epsilon}_N(\cdot,t)\right|^2+2\left|\vec{H}^{\epsilon}_N(\cdot,t)\right|+\left|\vec{E}^{\epsilon}_N(\cdot,t)\right|\right] dx$$
$$+\sigma\int_{\Omega}\left|\vec{E}^{\epsilon}_N+\vec{P}^{\epsilon}_N\right|^2 dx+\sigma\int_{\Omega}\left|\vec{E}^{\epsilon}_N\right|^2 dx$$
$$+2\beta\int_{\Omega}\vec{Z}^{\epsilon}_{Nt}\cdot\vec{H}^{\epsilon}_N dx+\int_{\Omega}\vec{P}^{\epsilon}_{Nt}\cdot\vec{E}^{\epsilon}_N dx$$
$$=\sigma\int_{\Omega}\vec{P}^{\epsilon}_N\cdot\left(\vec{E}^{\epsilon}_N+\vec{P}^{\epsilon}_N\right) dx+\int_{\Omega}\left(\nabla\times\vec{H}^{\epsilon}_N\right)\cdot\vec{P}^{\epsilon}_N dx. \tag{5.6.32}$$

Multiplying (5.6.19) by $\vec{\delta}^{\epsilon}_{sN}{}'(t)$, summing up the product for $s=1,2,\ldots,N$ and integrating by parts, one has, by noticing that $\vec{P}^{\epsilon}_N$ is periodic,

$$\frac{1}{2}\frac{d}{dt}\int_{\Omega}\left|\vec{P}^{\epsilon}_{Nt}(\cdot,t)\right|^2 dx+\frac{\lambda^2}{2}\frac{d}{dt}\int_{\Omega}\left|\nabla\times\vec{P}^{\epsilon}_N(\cdot,t)\right|^2 dx$$
$$+\frac{\epsilon}{2}\frac{d}{dt}\int_{\Omega}\left|\nabla\vec{P}^{\epsilon}_N(\cdot,t)\right|^2 dx+\mu\int_{\Omega}\left|\vec{P}^{\epsilon}_{Nt}\right|^2 dx$$
$$=\nu\int_{\Omega}\vec{E}^{\epsilon}_N\cdot\vec{P}^{\epsilon}_{Nt} dx-2\nu\int_{\Omega}\Phi'(|\vec{P}^{\epsilon}_N|^2)\vec{P}^{\epsilon}_N\cdot\vec{P}^{\epsilon}_{Nt} dx. \tag{5.6.33}$$

From (5.6.29) and (5.6.32) (multiplying (5.6.32) by δ_0, chosen later), it follows that

$$\frac{1}{2}\frac{d}{dt}\int_{\Omega}\left[\left|\nabla\vec{Z}^{\epsilon}_N(\cdot,t)\right|^2+\delta_0\left|\vec{E}^{\epsilon}_N(\cdot,t)+\vec{P}^{\epsilon}_N(\cdot,t)\right|^2+2\delta_0\left|\vec{H}^{\epsilon}_N(\cdot,t)\right|+\delta_0\left|\vec{E}^{\epsilon}_N(\cdot,t)\right|\right]$$
$$+\delta_0\sigma\int_{\Omega}\left|\vec{E}^{\epsilon}_N\right|^2+\alpha_2\left\|\vec{Z}^{\epsilon}_N\times\left(\triangle\vec{Z}^{\epsilon}_N+\vec{H}^{\epsilon}_N\right)\right\|^2_2+(2\beta\delta_0-1)\int_{\Omega}\vec{Z}^{\epsilon}_{Nt}\cdot\vec{H}^{\epsilon}_N dx$$
$$+\delta_0\sigma\int_{\Omega}\left|\vec{E}^{\epsilon}_N+\vec{P}^{\epsilon}_N\right|^2+\delta_0\int_{\Omega}\vec{P}^{\epsilon}_{Nt}\cdot\vec{E}^{\epsilon}_N dx$$
$$=\delta_0\sigma\int_{\Omega}\vec{P}^{\epsilon}_N\cdot\left(\vec{E}^{\epsilon}_N+\vec{P}^{\epsilon}_N\right)+\delta_0\int_{\Omega}\left(\nabla\times\vec{H}^{\epsilon}_N\right)\cdot\vec{P}^{\epsilon}_N. \tag{5.6.34}$$

In order to deal with the term $\int_\Omega \vec{Z}_{Nt}^\epsilon \cdot \vec{H}_N^\epsilon dx$, we multiply (5.6.17) by $(2\beta\delta_0 - 1)\vec{\alpha}_{sN}^\epsilon$ and sum up the product for $s = 1, 2, \ldots, N$ to obtain

$$(2\beta\delta_0 - 1)\int_\Omega \vec{H}_{Nt}^\epsilon \cdot \vec{Z}_N^\epsilon + (2\beta\delta_0 - 1)\int_\Omega \left(\nabla \times \vec{E}_N^\epsilon\right) \cdot \vec{Z}_N^\epsilon = 0. \qquad (5.6.35)$$

Adding (5.6.34) and (5.6.35), one gets

$$\frac{1}{2}\frac{d}{dt}\int_\Omega \left[\left|\nabla \vec{Z}_N^\epsilon(\cdot, t)\right|^2 + \delta_0 \left|\vec{E}_N^\epsilon(\cdot, t) + \vec{P}_N^\epsilon(\cdot, t)\right|^2 + 2\delta_0 \left|\vec{H}_N^\epsilon(\cdot, t)\right| + \delta_0 \left|\vec{E}_N^\epsilon(\cdot, t)\right|\right] dx$$

$$+ \alpha_2 \left\|\vec{Z}_N^\epsilon \times \left(\triangle \vec{Z}_N^\epsilon + \vec{H}_N^\epsilon\right)\right\|_2^2 + (2\beta\delta_0 - 1)\frac{d}{dt}\int_\Omega \vec{Z}_N^\epsilon \cdot \vec{H}_N^\epsilon dx$$

$$= -\delta_0\sigma \int_\Omega \left|\vec{E}_N^\epsilon + \vec{P}_N^\epsilon\right|^2 dx - \delta_0\sigma \int_\Omega \left|\vec{E}_N^\epsilon\right|^2 dx - \delta_0 \int_\Omega \vec{P}_{Nt}^\epsilon \cdot \vec{E}_N^\epsilon dx$$

$$+ \delta_0\sigma \int_\Omega \vec{P}_N^\epsilon \cdot \left(\vec{E}_N^\epsilon + \vec{P}_N^\epsilon\right) dx - (2\beta\delta_0 - 1)\int_\Omega \left(\nabla \times \vec{E}_N^\epsilon\right) \cdot \vec{Z}_N^\epsilon dx$$

$$+ \delta_0 \int_\Omega \left(\nabla \times \vec{H}_N^\epsilon\right) \cdot \vec{P}_N^\epsilon dx. \qquad (5.6.36)$$

Putting (5.6.33) and (5.6.36) together, we have

$$\frac{1}{2}\frac{d}{dt}\int_\Omega \left[\left|\nabla \vec{Z}_N^\epsilon(\cdot, t)\right|^2 + \delta_0 \left|\vec{E}_N^\epsilon(\cdot, t) + \vec{P}_N^\epsilon(\cdot, t)\right|^2 + 2\delta_0 \left|\vec{H}_N^\epsilon(\cdot, t)\right|\right.$$

$$\left. + \delta_0 \left|\vec{E}_N^\epsilon(\cdot, t)\right| + \left|\vec{P}_{Nt}^\epsilon(\cdot, t)\right|^2 + \lambda^2 \left|\nabla \times \vec{P}_N^\epsilon(\cdot, t)\right|^2 + \epsilon \left|\nabla \vec{P}_N^\epsilon(\cdot, t)\right|^2\right]$$

$$+ (2\beta\delta_0 - 1)\frac{d}{dt}\int_\Omega \vec{Z}_N^\epsilon \cdot \vec{H}_N^\epsilon + \alpha_2 \left\|\vec{Z}_N^\epsilon \times \left(\triangle \vec{Z}_N^\epsilon + \vec{H}_N^\epsilon\right)\right\|_2^2$$

$$\leq \left(\mu + 2 + \frac{(\delta_0 - \nu)^2}{4}\right)\left\|\vec{P}_{Nt}^\epsilon\right\|_2^2 + \left(8\nu^2 C_0^2 + \frac{\sigma^2\delta_0^2}{4} + \delta_0\sigma\right)\left\|\vec{P}_N^\epsilon\right\|_2^2$$

$$+ |3 - \sigma\delta_0|\left\|\vec{E}_N^\epsilon\right\|_2^2 + \delta_0\sigma \left\|\vec{E}_N^\epsilon + \vec{P}_N^\epsilon\right\|_2^2 + \left\|\vec{H}_N^\epsilon\right\|_2^2$$

$$+ \frac{(1 - 2\beta\delta_0)^2}{4}\left\|\nabla \vec{Z}_N^\epsilon\right\|_2^2 + \frac{\delta_0^2}{4}\left\|\mathrm{curl}\ \vec{P}_N^\epsilon\right\|_2^2,$$

where C_0 is given by (5.6.5).

Integrating the above inequality with respect to t, we obtain

$$\frac{1}{2}\left\|\nabla \vec{Z}_N^\epsilon(\cdot, t)\right\|_2^2 + \frac{\delta_0}{2}\left\|\vec{E}_N^\epsilon(\cdot, t) + \vec{P}_N^\epsilon(\cdot, t)\right\|_2^2 + \delta_0 \left\|\vec{H}_N^\epsilon(\cdot, t)\right\|_2^2 + \frac{\delta_0}{2}\left\|\vec{E}_N^\epsilon(\cdot, t)\right\|_2^2$$

$$+ \frac{1}{2}\left\|\vec{P}_{Nt}^\epsilon(\cdot, t)\right\|_2^2 + \frac{\lambda^2}{2}\left\|\mathrm{curl}\ \vec{P}_N^\epsilon(\cdot, t)\right\|_2^2 + \frac{\epsilon}{2}\left\|\nabla \vec{P}_N^\epsilon(\cdot, t)\right\|_2^2$$

$$+ (2\beta\delta_0 - 1)\int_\Omega \vec{Z}_N^\epsilon \cdot \vec{H}_N^\epsilon dx + \alpha_2 \int_0^t \left\|\vec{Z}_N^\epsilon \times \left(\triangle \vec{Z}_N^\epsilon + \vec{H}_N^\epsilon\right)\right\|_2^2 dt$$

$$\leq \frac{1}{2}\left\|\nabla \vec{Z}_0\right\|_2^2 + \frac{\delta_0}{2}\left\|\vec{E}_0 + \vec{P}_0\right\|_2^2 + \delta_0 \left\|\vec{H}_0\right\|_2^2 + \frac{\delta_0}{2}\left\|\vec{E}_0\right\|_2^2$$

$$+ \frac{1}{2}\left\|\vec{P}_1\right\|_2^2 + |2\beta\delta_0 - 1|\left\|\vec{H}_0\right\|_2 \left\|\vec{Z}_0\right\|_2$$

$$+ \frac{\lambda^2}{2} \left\| \operatorname{curl} \vec{P}_0 \right\|_2^2 + \frac{\epsilon}{2} \left\| \nabla \vec{P}_0 \right\|_2^2 + \left(\mu + 2 + \frac{(\delta_0 - \nu)^2}{4} \right) \int_0^t \left\| \vec{P}_{Nt}^\epsilon \right\|_2^2 dt$$

$$+ \left(8\nu^2 C_0^2 + \frac{\sigma^2 \delta_0^2}{4} + \delta_0 \sigma \right) \int_0^t \left\| \vec{P}_N^\epsilon \right\|_2^2 dt$$

$$+ |3 - \sigma\delta_0| \int_0^t \left\| \vec{E}_N^\epsilon \right\|_2^2 dt + \delta_0 \sigma \int_0^t \left\| \vec{E}_N^\epsilon + \vec{P}_N^\epsilon \right\|_2^2 dt$$

$$+ \int_0^t \left\| \vec{H}_N^\epsilon \right\|_2^2 dt + \frac{(1 - 2\beta\delta_0)^2}{4} \int_0^t \left\| \nabla \vec{Z}_N^\epsilon \right\|_2^2 dt + \frac{\delta_0^2}{4} \int_0^t \left\| \operatorname{curl} \vec{P}_N^\epsilon \right\|_2^2 dt.$$

Therefore we have

$$\left\| \nabla \vec{Z}_N^\epsilon(\cdot, t) \right\|_2^2 + \delta_0 \left\| \vec{E}_N^\epsilon(\cdot, t) + \vec{P}_N^\epsilon(\cdot, t) \right\|_2^2 + 2\delta_0 \left\| \vec{H}_N^\epsilon(\cdot, t) \right\|_2^2$$

$$+ \delta_0 \left\| \vec{E}_N^\epsilon(\cdot, t) \right\|_2^2 + \left\| \vec{P}_{Nt}^\epsilon(\cdot, t) \right\|_2^2 + \lambda^2 \left\| \operatorname{curl} \vec{P}_N^\epsilon(\cdot, t) \right\|_2^2$$

$$+ \epsilon \left\| \nabla \vec{P}_N^\epsilon(\cdot, t) \right\|_2^2 + 2\alpha_2 \int_0^t \left\| \vec{Z}_N^\epsilon \times \left(\triangle \vec{Z}_N^\epsilon + \vec{H}_N^\epsilon \right) \right\|_2^2 dt$$

$$\leq C_7 + 2 \left(\mu + 2 + \frac{(\delta_0 - \nu)^2}{4} \right) \int_0^t \left\| \vec{P}_{Nt}^\epsilon \right\|_2^2 dt$$

$$+ \left(16\nu^2 C_0^2 + \frac{\sigma^2 \delta_0^2}{2} + 8|\delta_0 \sigma| \right) \int_0^t \left\| \vec{P}_N^\epsilon \right\|_2^2 dt$$

$$+ (|6 - 2\sigma\delta_0| + 4|\delta_0 \sigma|) \int_0^t \left\| \vec{E}_N^\epsilon \right\|_2^2 dt + 2 \int_0^t \left\| \vec{H}_N^\epsilon \right\|_2^2 dt$$

$$+ \frac{(1 - 2\beta\delta_0)^2}{2} \int_0^t \left\| \nabla \vec{Z}_N^\epsilon \right\|_2^2 dt + \frac{\delta_0^2}{2} \int_0^t \left\| \operatorname{curl} \vec{P}_N^\epsilon \right\|_2^2 dt, \qquad (5.6.37)$$

where

$$C_7 = \left\| \nabla \vec{Z}_0 \right\|_2^2 + \delta_0 \left\| \vec{E}_0 + \vec{P}_0 \right\|_2^2 + 2\delta_0 \left\| \vec{H}_0 \right\|_2^2 + \delta_0 \left\| \vec{E}_0 \right\|_2^2 + \left\| \vec{P}_1 \right\|_2^2$$

$$+ 2|2\beta\delta_0 - 1| \left\| \vec{H}_0 \right\|_2 \left\| \vec{Z}_0 \right\|_2 + \lambda^2 \left\| \operatorname{curl} \vec{P}_0 \right\|_2^2$$

$$+ \epsilon \left\| \nabla \vec{P}_0 \right\|_2^2 dx + \frac{(1 - 2\beta\delta_0)^2}{\delta_0} \left\| \vec{Z}_0 \right\|_2^2.$$

On the other hand, we have

$$\left\| \vec{P}_N^\epsilon(\cdot, t) \right\|_2^2 - 2 \left\| \vec{E}_N^\epsilon(\cdot, t) \right\|_2^2 \leq 2 \left\| \vec{P}_N^\epsilon(\cdot, t) + \vec{E}_N^\epsilon(\cdot, t) \right\|_2^2$$

$$\leq 3 \left\| \vec{P}_N^\epsilon(\cdot, t) + \vec{E}_N^\epsilon(\cdot, t) \right\|_2^2. \qquad (5.6.38)$$

Taking $\delta_0 = 3$, we get from (5.6.37) and (5.6.38) that

$$\left\| \nabla \vec{Z}_N^\epsilon(\cdot, t) \right\|_2^2 + \left\| \vec{P}_N^\epsilon(\cdot, t) \right\|_2^2 + 6 \left\| \vec{H}_N^\epsilon(\cdot, t) \right\|_2^2 + \left\| \vec{E}_N^\epsilon(\cdot, t) \right\|_2^2 + \epsilon \left\| \nabla \vec{P}_N^\epsilon(\cdot, t) \right\|_2^2$$

$$+ \left\| \vec{P}_{Nt}^\epsilon(\cdot, t) \right\|_2^2 + \lambda^2 \left\| \operatorname{curl} \vec{P}_N^\epsilon(\cdot, t) \right\|_2^2 + 2\alpha_2 \int_0^t \left\| \vec{Z}_N^\epsilon \times \left(\triangle \vec{Z}_N^\epsilon + \vec{H}_N^\epsilon \right) \right\|_2^2 dt$$

$$\leq C_8 \int_0^t \left[\left\| \nabla \vec{Z}_N^\epsilon \right\|_2^2 + \left\| \vec{P}_N^\epsilon \right\|_2^2 + \left\| \vec{E}_N^\epsilon \right\|_2^2 + \left\| \vec{H}_N^\epsilon \right\|_2^2 + \left\| \vec{P}_{Nt}^\epsilon \right\|_2^2 + \left\| \operatorname{curl} \vec{P}_N^\epsilon \right\|_2^2 \right] dt + C_7,$$

$$(5.6.39)$$

where

$$C_8 = \max\left\{\frac{(1-6\beta)^2}{2}, \left(16\nu^2 C_0^2 + \frac{9\sigma^2}{2} + 24\sigma\right), (|6-6\sigma|+12\sigma), 2[\mu+2+(3-\nu)^2], \frac{9}{2}\right\}.$$

For $\lambda^2 > 0$ (in fact, $\lambda > 0$ denotes the speed of light for the internal field), (5.6.39) combined with Gronwall's inequality yields (5.6.25).

Step 3. By Sobolev imbedding theorem and Hölder inequality, we have (5.6.26). Combining (5.6.36) and (5.6.25), we obtain (5.6.27) if $\alpha_2 > 0$. Lemma 5.6.2 is proved.

Lemma 5.6.3 *Under the condition of Lemma 5.6.2, for the solution $(\vec{Z}_N^\epsilon, \vec{H}_N^\epsilon, \vec{E}_N^\epsilon, \vec{P}_N^\epsilon)$ of the problem (5.6.16)–(5.6.24), there exist $C_9 > 0$ and $C_{10} > 0$, both independent of N, D and ϵ, such that*
(i) *when $\alpha_2 = 0$,*

$$\sup_{0\leq t\leq T}\left[\left\|\vec{Z}_{Nt}^\epsilon\right\|_{H^{-2}(\Omega)} + \left\|\vec{H}_{Nt}^\epsilon\right\|_{H^{-2}(\Omega)} + \left\|\vec{E}_{Nt}^\epsilon\right\|_{H^{-2}(\Omega)} + \left\|\vec{P}_{Ntt}^\epsilon\right\|_{H^{-2}(\Omega)}\right] \leq C_9. \quad (5.6.40)$$

(ii) *when $\alpha_2 > 0$,*

$$\left\|\vec{Z}_{Nt}^\epsilon\right\|_{L^{\frac{3}{2}}(Q_T)} + \left\|\vec{H}_{Nt}^\epsilon\right\|_{L^2(0,T;H^{-1}(\Omega))} + \left\|\vec{E}_{Nt}^\epsilon\right\|_{L^2(0,T;H^{-1}(\Omega))}$$
$$+ \left\|\vec{P}_{Ntt}^\epsilon\right\|_{L^2(0,T;H^{-2}(\Omega))} \leq C_{10}. \quad (5.6.41)$$

Remark 5.6.1 This lemma shows that if $\alpha_2 > 0$, then we may get better estimate like the above lemma.

Proof.
(i) When $\alpha_2 = 0$, for any periodic function $\varphi \in H_0^2(\Omega)$, φ can be represented by

$$\varphi = \varphi_N + \overline{\varphi}_N, \quad \varphi_N = \sum_{s=1}^N \eta_s \omega_s(x), \quad \overline{\varphi}_N = \sum_{s=N+1}^\infty \eta_s \omega_s(x). \quad (5.6.42)$$

For $s \geq N+1$, we have

$$\int_\Omega \vec{Z}_{Nt}^\epsilon \omega_s(x)dx = 0$$

Then, by Lemma 5.6.2, there holds

$$\left|\int_\Omega \vec{Z}_{Nt}^\epsilon \varphi(x)dx\right| = \int_\Omega \vec{Z}_{Nt}^\epsilon \varphi_N(x)dx = \alpha_1 \int_\Omega \vec{Z}_N^\epsilon \times \left(\triangle\vec{Z}_N^\epsilon + \vec{H}_N^\epsilon\right)\varphi_N(x)dx$$

$$\leq |\alpha_1|\left(\|\nabla\vec{Z}_N^\epsilon\|_2\|\vec{Z}_N^\epsilon\|_6 + \|\vec{Z}_N^\epsilon\|_2\|\vec{H}_N^\epsilon\|_2\right)\left(\|\nabla\varphi_N\|_3 + \|\varphi_N\|_\infty\right)$$

$$\leq C_{11}\|\varphi\|_{H^2(\Omega)},$$

where we have used Gagliardo–Nirenberg inequalities

$$\|\varphi_N\|_\infty \leq C\|\varphi_N\|_2^{\frac{1}{4}}\|\triangle\varphi_N\|_2^{\frac{3}{4}}, \qquad \|\nabla\varphi_N\|_3 \leq C\|\nabla\varphi_N\|_2^{\frac{1}{2}}\|\triangle\varphi_N\|_2^{\frac{1}{2}}.$$

In the similar manner, we have

$$\left| \int_\Omega \vec{H}^\epsilon_{Nt} \varphi(x) dx \right| \le C_{12} \|\varphi\|_{H^2(\Omega)},$$

$$\left| \int_\Omega \vec{E}^\epsilon_{Nt} \varphi(x) dx \right| \le C_{13} \|\varphi\|_{H^2(\Omega)},$$

$$\left| \int_\Omega \vec{P}^\epsilon_{Ntt} \varphi dx \right| \le C_{14} \|\varphi\|_{H^2(\Omega)},$$

where C_{11}, C_{12}, C_{13} and C_{14} are independent of N, D and ϵ. (5.6.40) follows.

(ii) Now we assume $\alpha_2 > 0$. For any periodic function $\varphi \in L^3(Q_T)$, we have

$$\left| \iint_{Q_T} \vec{Z}^\epsilon_N \varphi dx dt \right| \le |\alpha_1| \left| \iint_{Q_T} \vec{Z}^\epsilon_N \times \left(\triangle \vec{Z}^\epsilon_N + \vec{H}^\epsilon_N \right) \varphi dx dt \right|$$

$$+ \alpha_2 \left| \iint_{Q_T} \vec{Z}^\epsilon_N \times \left(\vec{Z}^\epsilon_N \times \left(\triangle \vec{Z}^\epsilon_N + \vec{H}^\epsilon_N \right) \right) \varphi dx dt \right|$$

$$\le |\alpha_1| \left\| \vec{Z}^\epsilon_N \times \left(\triangle \vec{Z}^\epsilon_N + \vec{H}^\epsilon_N \right) \right\|_{L^2(Q_T)} \|\varphi\|_{L^2(Q_T)}$$

$$+ \alpha_2 \| \vec{Z}^\epsilon_N \|_{L^6(Q_T)} \left\| \vec{Z}^\epsilon_N \times \left(\triangle \vec{Z}^\epsilon_N + \vec{H}^\epsilon_N \right) \right\|_{L^2(Q_T)} \|\varphi\|_{L^3(Q_T)}$$

$$\le C_{15} \|\varphi\|_{L^3(Q_T)}.$$

Similarly, for any periodic function $\varphi \in L^2(0, T; H^1_0(\Omega))$, using (5.6.42) and Lemma 5.6.2, we get

$$\left| \int_\Omega \vec{Z}^\epsilon_{Nt} \varphi dx \right| \le C_{16} \|\varphi\|_{H^1(\Omega)}$$

$$\left| \iint_{Q_T} \vec{H}^\epsilon_{Nt} \varphi dx dt \right| = \left| - \iint_{Q_T} \left(\nabla \times \vec{E}^\epsilon_N \right) \varphi dx dt - \beta \iint_{Q_T} \vec{Z}^\epsilon_{Nt} \varphi_N dx dt \right|$$

$$\le C_{17} \|\varphi\|_{L^2(0,T;H^1(\Omega))},$$

$$\left| \iint_{Q_T} \vec{E}^\epsilon_{Nt} \varphi dx dt \right| = \left| \iint_{Q_T} \left(\nabla \times \vec{H}^\epsilon_N \right) \varphi dx dt - \iint_{Q_T} \vec{P}_{Nt} \varphi_t dx dt - \sigma \iint_{Q_T} \vec{E}^\epsilon_N \varphi_N dx dt \right|$$

$$\le C_{18} \|\varphi\|_{L^2(0,T;H^1(\Omega))}.$$

For any periodic function $\varphi \in L^2(0, T; H^2(\Omega))$, using (5.6.42) again, we obtain

$$\left| \iint_{Q_T} \vec{P}^\epsilon_{Ntt} \varphi dx dt \right| = \left| \nu \iint_{Q_T} \vec{E}^\epsilon_N \varphi_N dx dt + \epsilon \iint_{Q_T} \triangle \vec{P}^\epsilon_N \varphi_N dx dt \right.$$

$$- 2\nu \iint_{Q_T} \Phi'(|\vec{P}^\epsilon_N|^2) \vec{P}^\epsilon_N \varphi_N dx dt$$

$$\left. - \mu \iint_{Q_T} \vec{P}^\epsilon_{Nt} \varphi_N dx dt - \lambda^2 \iint_{Q_T} \mathrm{curl}\,^2 \vec{P}^\epsilon_N \varphi_N dx dt \right|$$

$$\le C_{19} \|\varphi_N\|_{L^2(0,T;H^1(\Omega))} + \epsilon \int_0^T \|\vec{P}^\epsilon_N\|_2 \|\triangle \varphi_N\|_2 dt$$

$$+ \lambda^2 \int_0^T \|\mathrm{curl}\, \vec{P}^\epsilon_N\|_2 \|\nabla \varphi_N\|_2 dt$$

$$\le C_{20} \|\varphi\|_{L^2(0,T;H^2(\Omega))},$$

where C_{15}–C_{20} are independent of N, D and ϵ. The proof of Lemma 5.6.3 is complete.

Lemma 5.6.4 *Under the condition of Lemma 2.1, for the solution* $(\vec{Z}_N^\epsilon, \vec{H}_N^\epsilon, \vec{E}_N^\epsilon, \vec{P}_N^\epsilon)$ *of the problem* (5.6.16)–(5.6.24), *there exist constants* $C_{21} > 0$, $C_{22} > 0$, $C_{23} > 0$, $C_{24} > 0$ *and* $C_{25} > 0$, *independent of* N, D, *and* ϵ, *such that*

(i) *When* $\alpha_2 = 0$,

$$\|\vec{Z}_N^\epsilon(\cdot, t_1) - \vec{Z}_N^\epsilon(\cdot, t_2)\|_2 \le C_{21}|t_1 - t_2|^{\frac{1}{2}}, \tag{5.6.43}$$

$$\vec{H}_N^\epsilon, \vec{E}_N^\epsilon, \vec{P}_N^\epsilon, \vec{P}_{Nt}^\epsilon \in C([0,T]; H^{-1}(\Omega)). \tag{5.6.44}$$

(ii) *When* $\alpha_2 > 0$,

$$\|\vec{Z}_N^\epsilon(\cdot, t_1) - \vec{Z}_N^\epsilon(\cdot, t_2)\|_3 \le C_{22}|t_1 - t_2|^{\frac{2}{3}}, \tag{5.6.45}$$

$$\|\vec{H}_N^\epsilon(\cdot, t_1) - \vec{H}_N^\epsilon(\cdot, t_2)\|_{H^{-1}(\Omega)} + \|\vec{E}_N^\epsilon(\cdot, t_1) - \vec{E}_N^\epsilon(\cdot, t_2)\|_{H^{-1}(\Omega)} \le C_{23}|t_1 - t_2|^{\frac{1}{2}}, \tag{5.6.46}$$

$$\|\vec{P}_{Nt}^\epsilon(\cdot, t_1) - \vec{P}_{Nt}^\epsilon(\cdot, t_2)\|_{H^{-2}(\Omega)} \le C_{24}|t_1 - t_2|^{\frac{1}{2}}, \tag{5.6.47}$$

$$\|\vec{P}_N^\epsilon(\cdot, t_1) - \vec{P}_N^\epsilon(\cdot, t_2)\|_2 \le C_{25}|t_1 - t_2|^{\frac{1}{2}}. \tag{5.6.48}$$

Proof.   (i) When $\alpha_2 = 0$, by the Sobolev interpolation of negative order, there holds

$$\|\vec{Z}_N^\epsilon(\cdot, t_1) - \vec{Z}_N^\epsilon(\cdot, t_2)\|_2 \le C_{26}\|\vec{Z}_N^\epsilon(\cdot, t_1) - \vec{Z}_N^\epsilon(\cdot, t_2)\|_{H^{-2}}^{\frac{1}{3}}\|\vec{Z}_N^\epsilon(\cdot, t_1) - \vec{Z}_N^\epsilon(\cdot, t_2)\|_{H^1}^{\frac{2}{3}}$$

$$\le C_{27}\left\|\int_{t_1}^{t_2} \frac{\partial \vec{Z}_N^\epsilon}{\partial t} dt\right\|_{H^{-2}}^{\frac{1}{3}} \le C_{22}|t_1 - t_2|^{\frac{1}{3}}.$$

On the other hand, it follows from Lemma 5.6.1 and

$$L^2(\Omega) \hookrightarrow\hookrightarrow H^{-1}(\Omega) \hookrightarrow H^{-2}(\Omega);$$

$$\vec{H}_N^\epsilon \in L^\infty(0,T; L^2(\Omega)) \bigcap \left\{ \Psi : \frac{\partial \Psi}{\partial t} \in L^\infty(0,T; H^{-2}(\Omega)) \right\}$$

that

$$\vec{H}_N^\epsilon \in C([0,T]; H^{-1}(\Omega)).$$

Similarly, we also have

$$\vec{E}_N^\epsilon, \vec{P}_N^\epsilon, \vec{P}_{Nt}^\epsilon \in C([0,T]; H^{-1}(\Omega)).$$

(ii) When $\alpha_2 > 0$, we have

$$\left\|\vec{Z}_N^\epsilon(\cdot, t_1) - \vec{Z}_N^\epsilon(\cdot, t_2)\right\|_3 = \left\|\int_{t_1}^{t_2} \frac{\partial \vec{Z}_N^\epsilon}{\partial t} dt\right\|_3$$

$$\leq |t_1 - t_2|^{\frac{2}{3}} \left(\iint_{Q_T} \left|\frac{\partial \vec{Z}_N^\epsilon}{\partial t}\right| dx dt\right)^{\frac{1}{3}}$$

$$\leq C_{22}|t_1 - t_2|^{\frac{2}{3}},$$

$$\left\|\vec{H}_N^\epsilon(\cdot, t_1) - \vec{H}_N^\epsilon(\cdot, t_2)\right\|_{H^{-1}(\Omega)} = \left\|\int_{t_1}^{t_2} \frac{\partial \vec{H}_N^\epsilon}{\partial t} dt\right\|_{H^{-1}}$$

$$\leq |t_1 - t_2|^{\frac{1}{2}} \left(\int_0^T \left\|\frac{\partial \vec{H}_N^\epsilon}{\partial t}\right\|_{H^{-1}(\Omega)}^2 dt\right)^{\frac{1}{2}}$$

$$\leq C_{23}|t_1 - t_2|^{\frac{1}{2}}.$$

For $|\vec{E}_N^\epsilon(\cdot, t_1) - \vec{E}_N^\epsilon(\cdot, t_2)|$, a similar inequality holds.

At the same time, we have

$$\left\|\vec{P}_{Nt}^\epsilon(\cdot, t_1) - \vec{P}_{Nt}^\epsilon(\cdot, t_2)\right\|_{H^{-2}(\Omega)} = \left\|\int_{t_1}^{t_2} \frac{\partial^2 \vec{P}_N^\epsilon}{\partial t^2} dt\right\|_{H^{-2}(\Omega)}$$

$$\leq |t_1 - t_2|^{\frac{1}{2}} \left(\int_0^T \left\|\frac{\partial^2 \vec{P}_N^\epsilon}{\partial t^2}\right\|_{H^{-2}(\Omega)}^2 dt\right)^{\frac{1}{2}}$$

$$\leq C_{24}|t_1 - t_2|^{\frac{1}{2}},$$

$$\left\|\vec{P}_N^\epsilon(\cdot, t_1) - \vec{P}_N^\epsilon(\cdot, t_2)\right\|_2 = \left\|\int_{t_1}^{t_2} \frac{\partial \vec{P}_N^\epsilon}{\partial t} dt\right\|_2$$

$$\leq |t_1 - t_2|^{\frac{1}{2}} \left(\int_0^T \left\|\frac{\partial \vec{P}_N^\epsilon}{\partial t}\right\|_2^2 dt\right)^{\frac{1}{2}}$$

$$\leq C_{25}|t_1 - t_2|^{\frac{1}{2}}.$$

Lemma 5.6.4 follows.

In fact, it follows from (5.6.25)–(5.6.26) that the solution of ODE (5.6.16)–(5.6.24) does not blow up at any finite time. Hence, from ODE theory, Lemmas 5.6.2–5.6.4, we have the following lemma:

Lemma 5.6.5 *Under the conditions of Lemma 5.6.2, the initial value problem for the system of the ordinary differential equation (5.6.16)–(5.6.24) admits at least one continuously differentiable global solution*

$$\vec{\alpha}_{sN}^\epsilon(t), \quad \vec{\beta}_{sN}^\epsilon(t), \quad \vec{\gamma}_{sN}^\epsilon(t), \quad \vec{\delta}_{sN}^\epsilon(t), \quad (s = 1, 2, \ldots, N; t \in [0, T]).$$

5.6.4　Existence of Weak Solution for the Viscosity Problem

First of all, similar to Definition 5.6.1, we may define the weak solution for the viscosity problem (5.6.1)–(5.6.3), and (5.6.13) and (5.6.8). In the proof of the following theorem, we must use the following lemma which is well known to all.

Lemma 5.6.6 *If $u_n \to u$ strongly in $L^2(Q_T)$ and $v_n \to v$ weakly in $L^2(Q_T)$, then $u_n v_n \to uv$ weakly in $L^1(Q_T)$ and in the sense of distribution.*

Theorem 5.6.1 *Assume the 2D-periodic initial data $(\vec{Z}_0(x), \vec{H}_0(x), \vec{E}_0(x), \vec{P}_0(x), \vec{P}_1(x)) \in (H^1(\Omega), L^2(\Omega), L^2(\Omega), H^1(\Omega), L^2(\Omega))$. Then the periodic initial value problem (5.6.1)–(5.6.3), (5.6.13), (5.6.8) admits at least one global weak solution $\vec{Z}^\epsilon(x,t)$, $\vec{H}^\epsilon(x,t)$, $\vec{E}^\epsilon(x,t)$, $\vec{P}^\epsilon(x,t)$ such that*
(i) *When $\alpha_2 = 0$, there hold*

$$\vec{Z}^\epsilon(x,t) \in L^\infty(0,T;H^1(\Omega)) \bigcap C^{(0,\frac{1}{3})}(0,T;L^2(\Omega)); \tag{5.6.49}$$

$$\vec{H}^\epsilon(x,t) \in L^\infty(0,T;L^2(\Omega)) \bigcap C(0,T;H^{-1}(\Omega)); \tag{5.6.50}$$

$$\vec{E}^\epsilon(x,t) \in L^\infty(0,T;L^2(\Omega)) \bigcap C(0,T;H^{-1}(\Omega)); \tag{5.6.51}$$

$$\vec{P}^\epsilon(x,t) \in L^\infty(0,T;H^1(\Omega)) \bigcap C(0,T;H^{-1}(\Omega)); \tag{5.6.52}$$

$$\vec{P}^\epsilon_t(x,t) \in L^\infty(0,T;L^2(\Omega)) \bigcap C(0,T;H^{-1}(\Omega)). \tag{5.6.53}$$

(ii) *When $\alpha_2 > 0$, we have*

$$\vec{Z}^\epsilon(x,t) \in L^\infty(0,T;H^1(\Omega)) \bigcap C^{(0,\frac{2}{3})}(0,T;L^3(\Omega)); \tag{5.6.54}$$

$$\vec{H}^\epsilon(x,t) \in L^\infty(0,T;L^2(\Omega)) \bigcap C^{(0,\frac{1}{2})}(0,T;H^{-1}(\Omega)); \tag{5.6.55}$$

$$\vec{E}^\epsilon(x,t) \in L^\infty(0,T;L^2(\Omega)) \bigcap C^{(0,\frac{1}{2})}(0,T;H^{-1}(\Omega)); \tag{5.6.56}$$

$$\vec{P}^\epsilon(x,t) \in L^\infty(0,T;H^1(\Omega)) \bigcap C^{(0,\frac{1}{2})}(0,T;L^2(\Omega)); \tag{5.6.57}$$

$$\vec{P}^\epsilon_t(x,t) \in L^\infty(0,T;L^2(\Omega)) \bigcap C^{(0,\frac{1}{2})}(0,T;H^{-2}(\Omega)). \tag{5.6.58}$$

Proof. The uniform estimates for the approximate solution $\vec{Z}^\epsilon_N(x,t)$, $\vec{H}^\epsilon_N(x,t)$, $\vec{E}^\epsilon_N(x,t)$, $\vec{P}^\epsilon_N(x,t)$ in the above section yield that there is a subsequence of $\vec{Z}^\epsilon_N(x,t)$, $\vec{H}^\epsilon_N(x,t)$, $\vec{E}^\epsilon_N(x,t)$, $\vec{P}^\epsilon_N(x,t)$, still denoted by $\vec{Z}^\epsilon_N(x,t)$, $\vec{H}^\epsilon_N(x,t)$, $\vec{E}^\epsilon_N(x,t)$, $\vec{P}^\epsilon_N(x,t)$, and $\vec{Z}^\epsilon(x,t)$, $\vec{H}^\epsilon(x,t)$, $\vec{E}^\epsilon(x,t)$, $\vec{P}^\epsilon(x,t)$, such that

$$\vec{Z}^\epsilon_N(x,t) \to \vec{Z}^\epsilon(x,t), \quad \text{weakly-* in } L^6(Q_T); \tag{5.6.59}$$

$$\vec{Z}^\epsilon_N(x,t) \to \vec{Z}^\epsilon(x,t), \quad \text{strongly in } L^{6-\varrho}(Q_T),\ (\varrho > 0); \tag{5.6.60}$$

$$\vec{Z}^\epsilon_N(x,t) \to \vec{Z}^\epsilon(x,t), \quad \text{weakly-* in } L^\infty(0,T;H^1(\Omega)); \tag{5.6.61}$$

$$\vec{H}^\epsilon_N(x,t) \to \vec{H}^\epsilon(x,t), \quad \text{weakly-* in } L^\infty(0,T;L^2(\Omega)); \tag{5.6.62}$$

$$\vec{E}^\epsilon_N(x,t) \to \vec{E}^\epsilon(x,t), \quad \text{weakly-* in } L^\infty(0,T;L^2(\Omega)); \tag{5.6.63}$$

$$\vec{P}_N^\epsilon(x,t) \to \vec{P}^\epsilon(x,t), \qquad \text{weakly-* in } L^\infty(0,T;H^1(\Omega)); \qquad (5.6.64)$$

$$\vec{P}_{Nt}^\epsilon(x,t) \to \vec{P}_t^\epsilon(x,t), \qquad \text{weakly-* in } L^\infty(0,T;L^2(\Omega)); \qquad (5.6.65)$$

$$\operatorname{curl}\vec{P}_N^\epsilon(x,t) \to \operatorname{curl}\vec{P}^\epsilon(x,t), \quad \text{weakly-* in } L^\infty(0,T;L^2(\Omega)); \qquad (5.6.66)$$

$$\vec{Z}_{Nt}^\epsilon(x,t) \to \vec{Z}_t^\epsilon(x,t), \qquad \text{weakly-* in } L^{\frac{3}{2}}(Q_T),\,(\alpha_2 > 0); \qquad (5.6.67)$$

$$\vec{Z}_{Nt}^\epsilon(x,t) \to \vec{Z}_t^\epsilon(x,t), \qquad \text{weakly-* in } L^\infty(0,T;H^{-2}(\Omega)), \quad (\alpha_2 = 0). \quad (5.6.68)$$

From Lemma 5.6.1(ii), we get that

$$L^\infty(0,T;H^1(\Omega))\bigcap\{\varphi : \frac{\partial\varphi}{\partial t} \in L^\infty(0,T;L^2(\Omega))\}$$
$$\hookrightarrow\hookrightarrow C([0,T];L^2(\Omega)) \subset L^2(0,T;L^2(\Omega)).$$

Since $\vec{P}_N^\epsilon$ is bounded uniformly in $L^\infty(0,T;H^1(\Omega))$ and $\partial_t\vec{P}_N^\epsilon$ is bounded uniformly in $L^\infty(0,T;L^2(\Omega))$, we deduce that there exists a subsequence of $\{\vec{P}_N^\epsilon\}$, still denoted by $\{\vec{P}_N^\epsilon\}$, such that as $N \to \infty$

$$\vec{P}_N^\epsilon(x,t) \to \vec{P}^\epsilon(x,t), \quad \text{strongly in } L^\infty(0,T;L^2(\Omega)). \qquad (5.6.69)$$

For any vector-valued periodic test function $\vec{\Psi}(x,t) \in C^1(\overline{Q}_T), \vec{\Psi}(x,T) = 0$, we define an approximate sequence

$$\vec{\Psi}_N(x,t) = \sum_{s=1}^{N} \vec{\eta}_s(t)\omega_s(x),$$

where $\vec{\eta}_s(t) = \int_\Omega \vec{\Psi}(x,t)\omega_s(x)dx$, then

$$\vec{\Psi}_N(x,t) \to \vec{\Psi}(x,t) \quad \text{in } C^1(\overline{Q_T}) \text{ and in } L^p(Q_T),\, \forall\, p > 1. \qquad (5.6.70)$$

Making the scalar product of $\vec{\eta}_s(t)$ with (5.6.16), (5.6.17) and $e^{\sigma t}\vec{\eta}_s(t)$ with (5.6.18), $\vec{\eta}_s(t)$ with (5.6.19), and summing up the products for $s = 1, 2, \ldots, N$, we get from integration by parts

$$\iint_{Q_T} \vec{Z}_N^\epsilon \cdot \vec{\Psi}_{Nt} - \int_\Omega \vec{Z}_N^\epsilon(x,0) \cdot \vec{\Psi}(x,0)dx + \alpha_1 \iint_{Q_T} (\vec{Z}_N^\epsilon \times \nabla\vec{Z}_N^\epsilon) \cdot \nabla\vec{\Psi}_N$$

$$- \alpha_1 \iint_{Q_T} (\vec{Z}_N^\epsilon \times \vec{H}_N^\epsilon) \cdot \vec{\Psi}_N + \alpha_2 \iint_{Q_T} \left(\vec{Z}_N^\epsilon \times \nabla\vec{Z}_N^\epsilon\right) \cdot \nabla\left(\vec{Z}_N^\epsilon \times \vec{\Psi}_N\right)$$

$$- \alpha_2 \iint_{Q_T} \left(\vec{Z}_N^\epsilon \times \vec{H}_N^\epsilon\right) \cdot \left(\vec{Z}_N^\epsilon \times \vec{\Psi}_N\right) = 0, \qquad (5.6.71)$$

$$\iint_{Q_T} \left(\vec{H}_N^\epsilon + \beta\vec{Z}_N^\epsilon\right) \cdot \vec{\Psi}_{Nt}dxdt - \iint_{Q_T} \left(\nabla \times \vec{\Psi}_N\right) \cdot \vec{E}_N^\epsilon dxdt$$

$$+ \int_\Omega \left(\vec{H}_N^\epsilon(x,0) + \beta\vec{Z}_N^\epsilon(x,0)\right) \cdot \vec{\Psi}_N(x,0)dx = 0, \qquad (5.6.72)$$

$$\iint_{Q_T} (\vec{E}_N^\epsilon + \vec{P}_N^\epsilon) \cdot (e^{\sigma t}\vec{\Psi}_{Nt}) dx dt + \iint_{Q_T} e^{\sigma t} \nabla \times \Psi_N \cdot \vec{H}_N^\epsilon dx dt$$

$$+ \sigma \iint_{Q_T} e^{\sigma t} \vec{P}_N^\epsilon \cdot \vec{\Psi}_N dx dt + \int_\Omega (\vec{E}_N^\epsilon(\cdot,0) + \vec{P}_N^\epsilon(\cdot,0)) \cdot \vec{\Psi}_N(\cdot,0) dx = 0, \quad (5.6.73)$$

$$\iint_{Q_T} \vec{P}_{Nt}^\epsilon \cdot \vec{\Psi}_{Nt} - \lambda^2 \iint_{Q_T} \operatorname{curl} \vec{P}_N^\epsilon \cdot \operatorname{curl} \vec{\Psi}_N - \mu \iint_{Q_T} \vec{P}_{Nt}^\epsilon \cdot \vec{\Psi}_N$$

$$+ \nu \iint_{Q_T} \vec{E}_N^\epsilon \cdot \vec{\Psi}_N - \epsilon \iint_{Q_T} \nabla \vec{P}_N^\epsilon \cdot \nabla \vec{\Psi}_N$$

$$- 2\nu \iint_{Q_T} \Phi'(|\vec{P}_N^\epsilon|^2) \vec{P}_N^\epsilon \cdot \vec{\Psi}_N + \int_\Omega \vec{P}_{Nt}^\epsilon(\cdot,0) \cdot \vec{\Psi}_N(\cdot,0) = 0. \quad (5.6.74)$$

Now we are in the position to prove that $(\vec{Z}^\epsilon(x,t), \vec{H}^\epsilon(x,t), \vec{E}^\epsilon(x,t), \vec{P}^\epsilon(x,t))$ is a weak solution of (5.6.1)–(5.6.3), (5.6.13) and (5.6.8). To this aim, one should set N to ∞ in (5.6.71)–(5.6.74). From (5.6.59)–(5.6.70) and Lemma 5.6.6, it suffices to deal with the nonlinear terms in (5.6.71)–(5.6.74).

First of all, we are able to prove

$$\iint_{Q_T} \Phi'(|\vec{P}_N^\epsilon|^2) \vec{P}_N^\epsilon \cdot \vec{\Psi}_N dx dt \to \iint_{Q_T} \Phi'(|\vec{P}^\epsilon|^2) \vec{P}^\epsilon \cdot \vec{\Psi} dx dt. \quad (5.6.75)$$

In fact, using the Lipschitz condition (5.6.6), we get

$$\left| \iint_{Q_T} \Phi'(|\vec{P}_N^\epsilon|^2) \vec{P}_N^\epsilon \cdot \vec{\Psi}_N - \iint_{Q_T} \Phi'(|\vec{P}^\epsilon|^2) \vec{P}^\epsilon \cdot \vec{\Psi} \right|$$

$$\leq \left| \iint_{Q_T} \left[\Phi'(|\vec{P}_N^\epsilon|^2) \vec{P}_N^\epsilon - \Phi'(|\vec{P}^\epsilon|^2) \vec{P}^\epsilon \right] \cdot \vec{\Psi}_N + \iint_{Q_T} \Phi'(|\vec{P}^\epsilon|^2) \vec{P}^\epsilon \cdot \left(\vec{\Psi}_N - \vec{\Psi} \right) \right|$$

$$\leq C_* \iint_{Q_T} |\vec{P}_N^\epsilon - \vec{P}^\epsilon| |\vec{\Psi}_N| + C \|\vec{P}^\epsilon\|_{L^\infty(0,T;L^2(\Omega))} \int_0^T \|\vec{\Psi}_N - \vec{\Psi}\|_{L^2(\Omega)}$$

$$\leq C_* \|\vec{\Psi}_N\|_{L^2(0,T;L^2)} \|\vec{P}_N^\epsilon - \vec{P}^\epsilon\|_{L^2(0,T;L^2)} + C \|\vec{P}^\epsilon\|_{L^\infty(0,T;L^2)} \int_0^T \|\vec{\Psi}_N - \vec{\Psi}\|_2$$

$$\to 0 \quad (\text{as } N \to +\infty),$$

where we have used (5.6.69).

Secondly, we claim that there exist subsequences of $\vec{Z}_N^\epsilon$, still denoted by $\vec{Z}_N^\epsilon$, such that, as $N \to +\infty$, for $i = 1, 2, 3$,

$$(1) \quad \vec{Z}_N^\epsilon \times \frac{\partial \vec{Z}_N^\epsilon}{\partial x_i} \to \vec{Z}^\epsilon \times \frac{\partial \vec{Z}^\epsilon}{\partial x_i} \qquad \text{weakly-$*$ in } L^\infty(0,T; L^{\frac{3}{2}}(\Omega)), \quad (5.6.76)$$

$$(2) \quad \left(\vec{Z}_N^\epsilon \times \frac{\partial \vec{Z}_N^\epsilon}{\partial x_i} \right)_{x_i} \to \left(\vec{Z}^\epsilon \times \frac{\partial \vec{Z}^\epsilon}{\partial x_i} \right)_{x_i} \qquad \text{weakly in } L^2(Q_T), \quad (\alpha_2 > 0). \quad (5.6.77)$$

In fact, for any periodic test function $\vec{\Psi}(x,t) \in C^1(Q_T)$, we obtain

$$\iint_{Q_T} \left(\vec{Z}_N^\epsilon \times \frac{\partial \vec{Z}_N^\epsilon}{\partial x_i} - \vec{Z}^\epsilon \times \frac{\partial \vec{Z}^\epsilon}{\partial x_i} \right) \cdot \vec{\Psi}\, dx dt$$

$$= \iint_{Q_T} \left[(\vec{Z}_N^\epsilon - \vec{Z}^\epsilon) \times \frac{\partial \vec{Z}_N^\epsilon}{\partial x_i} \right] \cdot \vec{\Psi}\, dx dt$$

$$+ \iint_{Q_T} \left[\vec{Z}^\epsilon \times \left(\frac{\partial \vec{Z}_N^\epsilon}{\partial x_i} - \frac{\partial \vec{Z}^\epsilon}{\partial x_i} \right) \right] \cdot \vec{\Psi}\, dx dt$$

$$\le \|\vec{\Psi}\|_{L^\infty(Q_T)} \|\frac{\partial \vec{Z}_N^\epsilon}{\partial x_i}\|_{L^2(Q_T)} \|\vec{Z}_N^\epsilon - \vec{Z}^\epsilon\|_{L^2(Q_T)}$$

$$+ \iint_{Q_T} \left[\vec{Z}^\epsilon \times \left(\frac{\partial \vec{Z}_N^\epsilon}{\partial x_i} - \frac{\partial \vec{Z}^\epsilon}{\partial x_i} \right) \right] \cdot \vec{\Psi}\, dx dt \to 0, \quad (\text{as } N \to +\infty).$$

Therefore, (5.6.76) is proved.

Now we turn to prove (5.6.77). By Lemma 5.6.2, when $\alpha_2 > 0$, $(\vec{Z}_N^\epsilon \times (\triangle \vec{Z}_N^\epsilon + \vec{H}_N^\epsilon))$ is bounded in $L^2(Q_T)$ uniformly with respect to N. Then there exist a subsequence of $(\vec{Z}_N^\epsilon \times (\triangle \vec{Z}_N^\epsilon + \vec{H}_N^\epsilon))$, still denoted by $(\vec{Z}_N^\epsilon \times (\triangle \vec{Z}_N^\epsilon + \vec{H}_N^\epsilon))$, and a vector $\vec{U}^\epsilon(x,t) \in L^2(Q_T)$, such that for any test function $\vec{\Psi}(x,t) \in C^1(Q_T)$, there holds that as $N \to +\infty$

$$\iint_{Q_T} \vec{Z}_N^\epsilon \times \left(\triangle \vec{Z}_N^\epsilon + \vec{H}_N^\epsilon \right) \cdot \vec{\Psi} \to \iint_{Q_T} \vec{U}^\epsilon \cdot \vec{\Psi}.$$

On the other hand, as $N \to +\infty$,

$$\iint_{Q_T} \vec{Z}_N^\epsilon \times \left(\triangle \vec{Z}_N^\epsilon + \vec{H}_N^\epsilon \right) \cdot \vec{\Psi}$$

$$= -\iint_{Q_T} \left(\vec{Z}_N^\epsilon \times \nabla \vec{Z}_N^\epsilon \right) \cdot \nabla \vec{\Psi} + \iint_{Q_T} \left(\vec{Z}_N^\epsilon \times \vec{H}_N^\epsilon \right) \cdot \vec{\Psi}$$

$$\to -\iint_{Q_T} \left(\vec{Z}^\epsilon \times \nabla \vec{Z}^\epsilon \right) \cdot \nabla \vec{\Psi} + \iint_{Q_T} \left(\vec{Z}^\epsilon \times \vec{H}^\epsilon \right) \cdot \vec{\Psi},$$

where we have used (5.6.76) and the fact that, as $N \to \infty$

$$\left| \iint_{Q_T} \left(\vec{Z}_N^\epsilon \times \vec{H}_N^\epsilon - \vec{Z}^\epsilon \times \vec{H}^\epsilon \right) \cdot \vec{\Psi} \right|$$

$$\le \left| \iint_{Q_T} \left(\vec{Z}_N^\epsilon - \vec{Z}^\epsilon \right) \times \vec{H}_N^\epsilon \cdot \vec{\Psi} \right| + \left| \iint_{Q_T} \vec{Z}^\epsilon \times \left(\vec{H}_N^\epsilon - \vec{H}^\epsilon \right) \cdot \vec{\Psi} \right|$$

$$\le \int_0^T \|\vec{Z}_N^\epsilon - \vec{Z}^\epsilon\|_5 \|\vec{H}_N^\epsilon\|_2 \|\vec{\Psi}\|_{\frac{10}{3}}\, dt + \left| \iint_{Q_T} \vec{Z}^\epsilon \times (\vec{H}_N^\epsilon - \vec{H}^\epsilon) \cdot \vec{\Psi} \right| \to 0. \quad (5.6.78)$$

Then we have

$$\iint_{Q_T} \vec{U}^\epsilon \cdot \vec{\Psi} - \iint_{Q_T} \left(\vec{Z}^\epsilon \times \vec{H}^\epsilon \right) \cdot \vec{\Psi} = -\iint_{Q_T} \left(\vec{Z}^\epsilon \times \nabla \vec{Z}^\epsilon \right) \cdot \nabla \vec{\Psi}.$$

Therefore, one gets in the sense of distribution

$$\vec{Z}^\epsilon \times \triangle \vec{Z}^\epsilon = (\vec{U}^\epsilon - (\vec{Z}^\epsilon \times \vec{H}^\epsilon)) \in L^2(Q_T).$$

So (5.6.77) is proved.

It remains to prove that

$$\iint_{Q_T} \left(\vec{Z}_N^\epsilon \times \nabla \vec{Z}_N^\epsilon\right) \cdot \nabla \left(\vec{Z}_N^\epsilon \times \vec{\Psi}_N\right) - \iint_{Q_T} \left(\vec{Z}_N^\epsilon \times \vec{H}_N^\epsilon\right) \cdot \left(\vec{Z}_N^\epsilon \times \vec{\Psi}_N\right)$$
$$\to \iint_{Q_T} \left(\vec{Z}^\epsilon \times \nabla \vec{Z}^\epsilon\right) \cdot \nabla \left(\vec{Z}^\epsilon \times \vec{\Psi}\right) - \iint_{Q_T} \left(\vec{Z}^\epsilon \times \vec{H}^\epsilon\right) \cdot \left(\vec{Z}^\epsilon \times \vec{\Psi}\right).$$

In fact, we have

$$\iint_{Q_T} \left(\vec{Z}_N^\epsilon \times \nabla \vec{Z}_N^\epsilon\right) \cdot \nabla \left(\vec{Z}_N^\epsilon \times \vec{\Psi}_N\right) - \iint_{Q_T} \left(\vec{Z}_N^\epsilon \times \vec{H}_N^\epsilon\right) \cdot \left(\vec{Z}_N^\epsilon \times \vec{\Psi}_N\right)$$
$$- \iint_{Q_T} \left(\vec{Z}^\epsilon \times \nabla \vec{Z}^\epsilon\right) \cdot \nabla \left(\vec{Z}^\epsilon \times \vec{\Psi}\right) + \iint_{Q_T} \left(\vec{Z}^\epsilon \times \vec{H}^\epsilon\right) \cdot \left(\vec{Z}^\epsilon \times \vec{\Psi}\right)$$
$$= \iint_{Q_T} \vec{Z}_N^\epsilon \times \left(\triangle \vec{Z}_N^\epsilon + \vec{H}_N^\epsilon\right) \cdot \left(\vec{Z}_N^\epsilon \times \vec{\Psi}_N\right)$$
$$- \iint_{Q_T} \vec{Z}^\epsilon \times \left(\triangle \vec{Z}^\epsilon + \vec{H}^\epsilon\right) \cdot \left(\vec{Z}^\epsilon \times \vec{\Psi}\right)$$
$$= \iint_{Q_T} \left[\left(\vec{Z}_N^\epsilon \times \triangle \vec{Z}_N^\epsilon\right) - \left(\vec{Z}^\epsilon \times \triangle \vec{Z}^\epsilon\right)\right] \cdot \left(\vec{Z}_N^\epsilon \times \vec{\Psi}\right)$$
$$+ \iint_{Q_T} \left(\vec{Z}_N^\epsilon \times \triangle \vec{Z}_N^\epsilon\right) \cdot \left[\left(\vec{Z}_N^\epsilon \times \vec{\Psi}_N\right) - \left(\vec{Z}^\epsilon \times \vec{\Psi}\right)\right]$$
$$+ \iint_{Q_T} \left[\left(\vec{Z}_N^\epsilon \times \vec{H}_N^\epsilon\right) - \left(\vec{Z}^\epsilon \times \vec{H}^\epsilon\right)\right] \cdot \left(\vec{Z}_N^\epsilon \times \vec{\Psi}\right)$$
$$+ \iint_{Q_T} \left(\vec{Z}_N^\epsilon \times \vec{H}_N^\epsilon\right) \cdot \left[\left(\vec{Z}_N^\epsilon \times \vec{\Psi}_N\right) - \left(\vec{Z}^\epsilon \times \vec{\Psi}\right)\right]$$
$$\doteq I_N^\epsilon + J_N^\epsilon + K_N^\epsilon + L_N^\epsilon.$$

From (5.6.77), we get $I_N^\epsilon \to 0$ as $N \to +\infty$. At the same time, as $N \to +\infty$, we have

$$|J_N^\epsilon| \leq \|\vec{Z}_N^\epsilon \times \triangle \vec{Z}_N^\epsilon\|_{L^2(Q_T)} \left(\iint_{Q_T} |\vec{Z}_N^\epsilon \times \vec{\Psi}_N - \vec{Z}^\epsilon \times \vec{\Psi}|^2\right)^{\frac{1}{2}}$$
$$\leq C \left(\iint_{Q_T} |\vec{Z}_N^\epsilon \times (\vec{\Psi}_N - \vec{\Psi}) + \left(\vec{Z}_N^\epsilon - \vec{Z}\right) \times \vec{\Psi}|^2\right)^{\frac{1}{2}} \to 0.$$

Similarly, one gets that $K_N^\epsilon \to 0$, as $N \to +\infty$ and

$$|L_N^\epsilon| \leq \|\vec{H}_N^\epsilon\|_{L^2(Q_T)} \|\vec{Z}_N^\epsilon\|_{L^4(Q_T)} \|\vec{Z}_N^\epsilon \times \vec{\Psi}_N - \vec{Z}^\epsilon \times \vec{\Psi}\|_{L^4(Q_T)} \to 0.$$

Finally, from the above arguments, one may take $N \to +\infty$ in (5.6.71)–(5.6.74) to obtain that $(\vec{Z}^\epsilon(x,t), \vec{H}^\epsilon(x,t), \vec{E}^\epsilon(x,t), \vec{P}^\epsilon(x,t))$ is a global weak solution of the viscosity problem (5.6.1)–(5.6.3), (5.6.13) and (5.6.8). This completes the proof.

Note that the above *a priori estimates* are independent of D. By using the diagonal method and letting $D \to +\infty$, we can obtain the global existence of weak solution to the Cauchy problem of system (5.6.1)–(5.6.3) and (5.6.13). For simplicity, we do not state the theorem here.

5.6.5 *A Priori* Estimates Uniform in ε

In Sec. 5, we have obtained a global weak solution for viscosity problem (5.6.1)–(5.6.3), (5.6.13) and (5.6.8) for fixed $\varepsilon > 0$. In this section we will derive the *a priori* estimates uniform in ε for solutions to viscosity problem. These uniform estimates enable us to pass to the limit $\varepsilon \to 0$ and then get the global weak solution to the problem (5.6.1)–(5.6.4) and (5.6.8).

We need the following lemmas

Lemma 5.6.7 *Assume* $\Omega = \{x = (x_1, x_2, x_3); |x_i| < D, i = 1, 2, 3\}$, $\vec{Q} \in X_p(\Omega)$. *Then* $\vec{Q} \in H^1(\Omega)$ *and there holds*

$$\|\vec{Q}\|^2_{H^1(\Omega)} = \|\vec{Q}\|^2_{X_p(\Omega)}.$$

Proof. It follows from the relation

$$\Delta \vec{Q} = \nabla(\nabla \cdot \vec{Q}) - \nabla \times (\nabla \times \vec{Q})$$

that

$$\int_\Omega \vec{Q} \Delta \vec{Q} = \int_\Omega \vec{Q} \nabla(\nabla \cdot \vec{Q}) - \int_\Omega \vec{Q} \nabla \times (\nabla \times \vec{Q}).$$

The periodicity of $\vec{Q}$ implies

$$\int_\Omega |\nabla \vec{Q}|^2 = \int_\Omega |\nabla \cdot \vec{Q}|^2 + \int_\Omega |\nabla \times \vec{Q}|^2,$$

and therefore we obtain the conclusion of the lemma.

From the above estimates and convergence, one easily gets the following.

Lemma 5.6.8 *Assume* $(\vec{Z}_0(x), \vec{H}_0(x), \vec{E}_0(x), \vec{P}_0(x), \vec{P}_1(x)) \in (H^1(\Omega), L^2(\Omega), L^2(\Omega), H^1(\Omega), L^2(\Omega))$. *Then for the solution of the initial value problem* (5.6.1)–(5.6.3), (5.6.13) *and* (5.6.8), *there hold following estimates:*

$$\sup_{0 \le t \le T} [\|\vec{Z}^\epsilon(\cdot, t)\|^2_{H^1(\Omega)} + \|\vec{E}^\epsilon(\cdot, t)\|^2_2 + \|\vec{H}^\epsilon(\cdot, t)\|^2_2$$

$$+ \|\vec{P}^\epsilon(\cdot, t)\|^2_2 + \|\mathrm{curl}\, \vec{P}^\epsilon(\cdot, t)\|^2_2 + \|\vec{P}^\epsilon_t(\cdot, t)\|^2_2] \le M_1, \tag{5.6.79}$$

$$\sup_{0 \le t \le T} \|\vec{Z}^\epsilon(\cdot, t)\|^2_6 \le M_1, \quad \sup_{0 \le t \le T} \|\vec{Z}^\epsilon(\cdot, t) \times \nabla \vec{Z}^\epsilon(\cdot, t)\|_{L^{\frac{3}{2}}(\Omega)} \le M_1, \tag{5.6.80}$$

where the constant M_1 *is independent of* α_2, D, *and* ϵ. *When* $\alpha_2 > 0$, *there is*

$$\|\vec{Z}^\epsilon \times (\Delta \vec{Z}^\epsilon + \vec{H}^\epsilon)\|_{L^2(0,T;L^2(\Omega))} \le M_2, \tag{5.6.81}$$

where the constant M_2 *is independent of* ϵ *and* D.

In the following we will prove that $\nabla \vec{P}^\epsilon$ is uniformly bounded in $L^\infty(0,T;L^2(\Omega))$. We shall consider the compatibility conditions associated with the viscosity problem given by the following set of equations that hold in the sense of distributions

$$\frac{\partial(e^\epsilon + p^\epsilon)}{\partial t} + \sigma e^\epsilon = 0, \tag{5.6.82}$$

$$\frac{\partial(h^\epsilon + \beta\nabla\cdot\vec{Z}^\epsilon)}{\partial t} = 0, \tag{5.6.83}$$

$$\frac{\partial^2 p^\epsilon}{\partial t^2} + \mu\frac{\partial p^\epsilon}{\partial t} - \epsilon\Delta p^\epsilon - \nu e^\epsilon + 2\nu\Phi'(|\vec{P}^\epsilon|^2)p^\epsilon = -4\nu\Phi^{(2)}(|\vec{P}^\epsilon|^2)P_i^\epsilon P_j^\epsilon\frac{\partial P_j^\epsilon}{\partial x_i}, \tag{5.6.84}$$

where

$$h^\epsilon = \operatorname{div}\vec{H}^\epsilon, \quad e^\epsilon = \operatorname{div}\vec{E}^\epsilon, \quad p^\epsilon = \operatorname{div}\vec{P}^\epsilon$$

and P_i^ϵ is the i-th component of $\vec{P}^\epsilon$ and the relation:

$$\operatorname{div}(\vec{P}^\epsilon\Phi'(|\vec{P}^\epsilon|^2)) = \Phi'(|\vec{P}^\epsilon|^2)p^\epsilon + 2\Phi^{(2)}(|\vec{P}^\epsilon|^2)P_i^\epsilon P_j^\epsilon\frac{\partial P_j^\epsilon}{\partial x_i}.$$

In order to obtain the $L^2(\Omega)$ estimate of $\nabla\vec{P}^\epsilon(\cdot,t)$, we shall assume that

$$\operatorname{div}\vec{H}_0, \quad \operatorname{div}\vec{E}_0, \quad \operatorname{div}\vec{P}_0, \quad \operatorname{div}\vec{P}_1 \in L^2(\Omega). \tag{5.6.85}$$

We have the following lemma.

Lemma 5.6.9 *Under the conditions of Lemma 5.6.8 and assuming that the hypotheses (5.6.85) hold, then for the solutions of the viscosity problem, we have*

$$\sup_{0\le t\le T} \|\nabla\vec{P}^\epsilon(\cdot,t)\|_{L^2(\Omega)} \le M_3, \tag{5.6.86}$$

where M_3 is independent of D and ϵ.

Proof. For simplicity we present the proof for the case that $\nabla(\operatorname{div}\vec{P}_0) \in L^2(\Omega)$, since the general case of $\operatorname{div}\vec{P}_0 \in L^2(\Omega)$ can be handled by the modifying technique or the proper approximation of the $\operatorname{div}\vec{P}_0$.

Multiplying (5.6.82) by $3e^\epsilon$ and $2(e^\epsilon + p^\epsilon)$, we have

$$\frac{3}{2}\frac{d}{dt}\int_\Omega |e^\epsilon|^2 dx + 3\int_\Omega e^\epsilon\frac{\partial p^\epsilon}{\partial t}dx + 3\sigma\int_\Omega |e^\epsilon|^2 dx = 0, \tag{5.6.87}$$

$$\frac{d}{dt}\int_\Omega |e^\epsilon + p^\epsilon|^2 dx + 2\sigma\int_\Omega |e^\epsilon + p^\epsilon|^2 dx - 2\sigma\int_\Omega (e^\epsilon + p^\epsilon)p^\epsilon dx = 0. \tag{5.6.88}$$

Multiplying (5.6.84) by $\frac{\partial p^\epsilon}{\partial t}$, one gets

$$\frac{1}{2}\frac{d}{dt}\int_\Omega\left|\frac{\partial p^\epsilon}{\partial t}\right|^2 dx + \mu\int_\Omega\left|\frac{\partial p^\epsilon}{\partial t}\right|^2 dx + \frac{\epsilon}{2}\frac{d}{dt}\int_\Omega |\nabla p^\epsilon|^2 dx - \nu\int_\Omega e^\epsilon\frac{\partial p^\epsilon}{\partial t}dx$$

$$+ 2\nu\int_\Omega \Phi'(|\vec{P}^\epsilon|^2)p^\epsilon\frac{\partial p^\epsilon}{\partial t}dx + 4\nu\int_\Omega \Phi^{(2)}(|\vec{P}^\epsilon|^2)P_i^\epsilon P_j^\epsilon\frac{\partial P_j^\epsilon}{\partial x_i}\frac{\partial p^\epsilon}{\partial t}dx = 0. \tag{5.6.89}$$

Combining (5.6.87)–(5.6.89), we obtain

$$\frac{1}{2}\frac{d}{dt}\int_\Omega\left[2|e^\epsilon + p^\epsilon|^2 + 3|e^\epsilon|^2 + \left|\frac{\partial p^\epsilon}{\partial t}\right|^2 + \epsilon|\nabla p^\epsilon|^2\right] + 2\sigma\int_\Omega|e^\epsilon + p^\epsilon|^2 + \mu\int_\Omega\left|\frac{\partial p^\epsilon}{\partial t}\right|^2$$

$$= -3\sigma\int_\Omega|e^\epsilon|^2 dx + 2\sigma\int_\Omega(p^\epsilon + e^\epsilon)p^\epsilon dx + (\nu - 3)\int_\Omega e^\epsilon\frac{\partial \mu^\epsilon}{\partial t}dx$$

$$- 2\nu\int_\Omega\Phi'(|\vec{P}^\epsilon|^2)p^\epsilon\frac{\partial p^\epsilon}{\partial t}dx - 4\nu\int_\Omega\Phi^{(2)}(|\vec{P}^\epsilon|^2)P_i^\epsilon P_j^\epsilon\frac{\partial P_j^\epsilon}{\partial x_i}\frac{\partial p^\epsilon}{\partial t}dx$$

$$\leq M_3\int_\Omega\left[|e^\epsilon|^2 + |p^\epsilon|^2 + \left|\frac{\partial p^\epsilon}{\partial t}\right|^2\right]dx + M_4 + M_5\int_\Omega|\nabla\vec{P}^\epsilon|^2 dx.$$

Therefore, we get that

$$\frac{1}{2}\frac{d}{dt}\int_\Omega\left[2|e^\epsilon + p^\epsilon|^2 + 3|e^\epsilon|^2 + \left|\frac{\partial p^\epsilon}{\partial t}\right|^2 + \epsilon|\nabla p^\epsilon|^2\right]dx$$

$$\leq M_3\int_\Omega\left[|e^\epsilon|^2 + |p^\epsilon|^2 + \left|\frac{\partial p^\epsilon}{\partial t}\right|^2\right]dx + M_4 + M_5\int_\Omega|\nabla\vec{P}^\epsilon|^2 dx.$$

Integrating with respect to t, we have

$$2\|(e^\epsilon + p^\epsilon)(\cdot, t)\|_2^2 + 3\|e^\epsilon(\cdot, t)\|_2^2 + \left\|\frac{\partial p^\epsilon}{\partial t}(\cdot, t)\right\|_2^2 + \epsilon\|\nabla p^\epsilon(\cdot, t)\|_2^2$$

$$\leq M_6 + 2M_5\int_0^t\int_\Omega|\nabla\vec{P}^\epsilon|^2 + 2M_3\int_0^t\int_\Omega\left[|e^\epsilon|^2 + |p^\epsilon|^2 + \left|\frac{\partial p^\epsilon}{\partial t}\right|^2\right],$$

where $M_6 = 2\|\mathrm{div}\,\vec{E}_0 + \mathrm{div}\,\vec{P}_0\|_2^2 + 3\|\mathrm{div}\,\vec{E}_0\|_2^2 + \|\mathrm{div}\,\vec{P}_1\|_2^2 + \epsilon\|\nabla(\mathrm{div}\,\vec{P}_0)\|_2^2 + 2M_4$ is a constant from hypotheses (5.6.85).

On the other hand,we get

$$\|p^\epsilon(\cdot, t)\|_2^2 = \|p^\epsilon(\cdot, t) + e^\epsilon(\cdot, t) - e^\epsilon(\cdot, t)\|_2^2$$

$$\leq 2\|p^\epsilon(\cdot, t) + e^\epsilon(\cdot, t)\|_2^2 + 2\|e^\epsilon(\cdot, t)\|_2^2.$$

We obtain that

$$\|p^\epsilon(\cdot, t)\|_2^2 + \|e^\epsilon(\cdot, t)\|_2^2 + \left\|\frac{\partial p^\epsilon}{\partial t}(\cdot, t)\right\|_2^2 + \epsilon\|\nabla p^\epsilon(\cdot, t)\|_2^2$$

$$\leq M_6 + 2M_5\int_0^t\int_\Omega|\nabla\vec{P}^\epsilon|^2 + 2M_3\int_0^t\int_\Omega\left[|e^\epsilon|^2 + |p^\epsilon|^2 + \left|\frac{\partial p^\epsilon}{\partial t}\right|^2\right].$$

By Gronwall inequality we get

$$
\|p^\epsilon(\cdot,t)\|_2^2 + \|e^\epsilon(\cdot,t)\|_2^2 + \left\|\frac{\partial p^\epsilon}{\partial t}(\cdot,t)\right\|_2^2
$$

$$
\leq \left(M_6 + M_5 \int_0^t \int_\Omega |\nabla \vec{P}^\epsilon|^2\right)\left(1 + M_3 t e^{M_3 t}\right)
$$

$$
\leq M_7 + M_8 \int_0^t \int_\Omega |\nabla \vec{P}^\epsilon|^2.
$$

Therefore, we obtain that

$$
\|p^\epsilon(\cdot,t)\|_2^2 \leq M_7 + M_8 \int_0^t \int_\Omega |\nabla \vec{P}^\epsilon|^2. \tag{5.6.90}
$$

Using Lemma 5.6.7 for $\vec{P}^\epsilon(x,t)$, we get

$$
\|\nabla \vec{P}^\epsilon(\cdot,t)\|_2^2 \leq M_9(\|\operatorname{curl}\vec{P}^\epsilon(\cdot,t)\|_2^2 + \|\operatorname{div}\vec{P}^\epsilon(\cdot,t)\|_2^2 + \|\vec{P}^\epsilon(\cdot,t)\|_2^2)
$$

$$
\leq M_{10} + M_{11}\int_0^t \int_\Omega |\nabla \vec{P}^\epsilon|^2.
$$

By Gronwall's inequality one gets

$$
\|\nabla \vec{P}^\epsilon(\cdot,t)\|_2^2 \leq M_{12},
$$

where M_{12} is independent of ϵ. Lemma 5.6.9 is proved.

Remark 5.6.2 Lemma 5.6.8 and Lemma 5.6.9 show that $\{\vec{P}^\epsilon\}$ is bounded in $L^\infty(0,T;H^1(\Omega))$.

5.6.6   Global Existence of Weak Solutions

By *a priori* estimates uniform in ε obtained in Sec. 5.6.4 for the viscosity problem and passing to the limit $\varepsilon \to 0$ in Eqs. (5.6.1)–(5.6.3) and (5.6.13), we can get the global weak solution of problem (5.6.1)–(5.6.4) and (5.6.8).

Theorem 5.6.2 *Assume the periodic functions* $(\vec{Z}_0(x), \vec{H}_0(x), \vec{E}_0(x), \vec{P}_0(x), \vec{P}_1(x)) \in (H^1(\Omega), L^2(\Omega), L^2(\Omega), H^1(\Omega), L^2(\Omega))$ *and satisfying* (5.6.85). *Then the periodic initial value problem* (5.6.1)–(5.6.4) *and* (5.6.8) *admits at least one global 2D-periodic weak solution* $\vec{Z}(x,t)$, $\vec{H}(x,t)$, $\vec{E}(x,t)$, $\vec{P}(x,t)$ *such that*
(i) *When* $\alpha_2 = 0$,

$$
\begin{aligned}
\vec{Z}(x,t) &\in L^\infty(0,T;H^1(\Omega))\cap C^{(0,\frac13)}(0,T;L^2(\Omega)); \\
\vec{H}(x,t) &\in L^\infty(0,T;L^2(\Omega))\cap C(0,T;H^{-1}(\Omega)); \\
\vec{E}(x,t) &\in L^\infty(0,T;L^2(\Omega))\cap C(0,T;H^{-1}(\Omega)); \\
\vec{P}(x,t) &\in L^\infty(0,T;H^1(\Omega))\cap C(0,T;H^{-1}(\Omega)); \\
\vec{P}_t(x,t) &\in L^\infty(0,T;L^2(\Omega))\cap C(0,T;H^{-1}(\Omega)).
\end{aligned} \tag{5.6.91}
$$

(ii) *When $\alpha_2 > 0$,*

$$\vec{Z}(x,t) \in L^\infty(0,T; H^1(\Omega)) \cap C^{(0,\frac{2}{3})}(0,T; L^3(\Omega));$$
$$\vec{H}(x,t) \in L^\infty(0,T; L^2(\Omega)) \cap C^{(0,\frac{1}{2})}(0,T; H^{-1}(\Omega));$$
$$\vec{E}(x,t) \in L^\infty(0,T; L^2(\Omega)) \cap C^{(0,\frac{1}{2})}(0,T; H^{-1}(\Omega)); \qquad (5.0.92)$$
$$\vec{P}(x,t) \in L^\infty(0,T; H^1(\Omega)) \cap C^{(0,\frac{1}{2})}(0,T; L^2(\Omega));$$
$$\vec{P}_t(x,t) \in L^\infty(0,T; L^2(\Omega)) \cap C^{(0,\frac{1}{2})}(0,T; H^{-2}(\Omega)).$$

5.7   Bibliography Comments

In this chapter we mainly introduce the readers to the main results of Landau–Lifshitz–Maxwell equations. Guo and Su [80] first proved the existence of global weak solution to three-dimensional L–L–M equations. They also showed the existence of global smooth solution to two-dimensional L–L–M equations [81, 82]. Meanwhile, Guo and Su obtained the existence and uniqueness of global smooth solution to L–L–M equations without Gilbert damping term and small initial data. Guo and Ding [74] gave the existence of global weak solution of Neumann boundary value problem to L–L–M equations. In [84], Guo and Su studied the existence of global weak solution to L–L–M equations on Riemannian manifold.

The Landau–Lifshitz–Maxwell equations coupling with polarizations was first proposed by Ding, Guo, Lin, and Zeng in 2006 which will be published in DCDS 2007.

Chapter 6

Long Time Behavior of Solutions to the System of Ferromagnetic Spin Chain

6.1 Existence and Stability of Steady State Solutions

6.1.1 One-Dimensional Landau–Lifshitz Equations

Consider the existence and stability of a steady state solution to the one-dimensional Landau–Lifshitz equations

$$\frac{\partial \vec{Z}}{\partial t} = \vec{Z} \times \Delta\vec{Z} - \lambda\vec{Z} \times (\vec{Z} \times \Delta\vec{Z}). \tag{6.1.1}$$

The steady state solution solves

$$\vec{Z} \times \Delta\vec{Z} = \lambda\vec{Z} \times (\vec{Z} \times \Delta\vec{Z}), \tag{6.1.2}$$

where $\Delta\vec{Z} = \vec{Z}_{xx}$.

Taking the inner product (6.1.2) by $\vec{Z} \times \Delta\vec{Z}$, we have

$$|\vec{Z} \times \Delta\vec{Z}|^2 = 0. \tag{6.1.3}$$

Thus, it follows that $\vec{Z} = (Z_1, Z_2, Z_3)$ is a steady state solution of (6.1.1) if and only if $\vec{Z}$ is a solution of $\vec{Z} \times \Delta\vec{Z} = 0$ and $\Delta\vec{Z}$ is parallel to $\vec{Z}$. It is clear that $\Delta\vec{Z}$ is parallel to $\vec{Z}$ if and only if there exists a real function $l(r)$ such that

$$\Delta\vec{Z} = l(r)\vec{Z}. \tag{6.1.4}$$

Multiplying the two sides of (6.1.4) by $\vec{Z}$ and noting $|\vec{Z}| = 1$, we get

$$\vec{Z} \cdot \Delta\vec{Z} = l(r). \tag{6.1.5}$$

Since $|\vec{Z}| = 1$, i.e. $\vec{Z} \cdot \vec{Z} = 1$, it follows that $\vec{Z} \cdot \nabla \vec{Z} = 0$. Thus (6.1.5) reads

$$\vec{Z} \cdot \Delta \vec{Z} = -|\nabla \vec{Z}|^2 = l(r). \tag{6.1.6}$$

It is clear that $l(r) \le 0$. Substituting $l(r) = -|\nabla \vec{Z}|^2$ into (6.1.4), we have

$$\Delta \vec{Z} = -|\nabla \vec{Z}|^2 \vec{Z}. \tag{6.1.7}$$

Multiplying (6.1.7) by $\nabla \vec{Z}$, one gets

$$\frac{1}{2} \frac{\partial}{\partial r} |\nabla \vec{Z}|^2 = -|\nabla \vec{Z}|^2 \frac{1}{2} \frac{\partial}{\partial r} |\vec{Z}|^2 = 0.$$

This implies that $l(r) = -|\nabla \vec{Z}|^2 = -l^2 < 0$ independent of r. Thus (6.1.4) implies

$$\Delta \vec{Z} = -l^2 \vec{Z}. \tag{6.1.8}$$

Solving (6.1.8), we know that the steady state solutions can be expressed by

$$\vec{Z}(r) = C_1 \exp(ilr) + C_2 \exp(-ilr), \tag{6.1.9}$$

where C_j $(j = 1, 2)$ are two three-dimensional complex vectors. Because $\vec{Z}$ is a three-dimensional real vector function, $\vec{Z}$ can be rewritten as

$$\vec{Z} = \vec{R}_1 \cos(lr) + \vec{R}_2 \sin(lr).$$

Since $|\vec{Z}| = 1$, we have that $|\vec{R}_1| = |\vec{R}_2| = 1$ and $\vec{R}_1 \perp \vec{R}_2$.

Theorem 6.1.1 $\vec{Z} = (Z_1, Z_2, Z_3)$ *is the steady state solution of* (6.1.1) *if and only if*

$$\vec{Z}_l(r) = \vec{R}_1 \cos(lr) + \vec{R}_2 \sin(lr), \quad \forall \ l \in R, \tag{6.1.10}$$

where $|\vec{R}_1| = |\vec{R}_2| = 1$ *and* $\vec{R}_1 \perp \vec{R}_2$.

Remark 6.1.1 By Theorem 6.1.1, we know that the inhomogeneous steady-state solutions of (6.1.1) are independent of λ, i.e., steady state solutions of (6.1.1) with the damping term are absolutely the same as those of (6.1.1) without the damping term.

6.1.2   Stability of the Steady State Solutions

Now we study the linear stability of the steady state solutions of Landau–Lifshitz equation (6.1.1). We know that Eq. (6.1.1) is equivalent to the following equation in the classical sense:

$$\frac{\partial \vec{Z}}{\partial t} = \vec{Z} \times \Delta \vec{Z} + \lambda \Delta \vec{Z} + \lambda |\nabla \vec{Z}|^2 \vec{Z}. \tag{6.1.11}$$

Linearize (6.1.11) about $\vec{Z}_l = \vec{R}_1 \cos(lr) + \vec{R}_2 \sin(lr)$. Let $\vec{Z} = \vec{Z}_l + \mathbf{V}$, where $\mathbf{V}$ is initially, at least, a small perturbation (i.e., $\mathbf{V}(r,0)$ is arbitrarily small). Then substituting $\vec{Z}$ into (6.1.11), and retaining only linear terms in $\mathbf{V}$, we have

$$\frac{\partial \mathbf{V}}{\partial t} = \vec{Z}_l \times \Delta \mathbf{V} + \mathbf{V} \times \Delta \vec{Z}_l + \lambda \Delta \mathbf{V} + \lambda |\nabla \vec{Z}_l|^2 \mathbf{V} + 2\lambda (\nabla \vec{Z}_l \cdot \nabla \mathbf{V}) \vec{Z}_l, \quad (6.1.12)$$

with the condition $\mathbf{V}(r + \frac{2\pi}{l}) = \mathbf{V}(r)$. Let $\mathbf{V} = \mathbf{d} \exp(ilmr + \Omega(t))$ ($|\mathbf{d}| \ll 1$) in (6.1.12). Then we have

$$\frac{d}{dt} \Omega \mathbf{V} = (1 - m^2) l^2 \vec{Z}_l \times \mathbf{V} + \lambda (1 - m^2) l^2 \mathbf{V} + 2ilm\lambda (\nabla \vec{Z}_l \cdot \mathbf{V}) \vec{Z}_l. \quad (6.1.13)$$

In the following, we calculate the last term on right-hand side of (6.1.13). By the facts $|\vec{Z}| = 1$ and $\vec{Z} = \vec{Z}_l + \mathbf{V}$, we have

$$2(\vec{Z}_l \cdot \mathbf{V}) = -|\mathbf{V}|^2. \quad (6.1.14)$$

On the other hand, $\nabla \vec{Z}_l = -l\vec{Z}_l(r - \frac{\pi}{2l})$. Then one has

$$\nabla \vec{Z}_l(r) \cdot \mathbf{V}(r) = -l\vec{Z}_l\left(r - \frac{\pi}{2l}\right) \mathbf{V}\left(r - \frac{\pi}{2l}\right) \exp\left(im\frac{\pi}{2}\right). \quad (6.1.15)$$

By (6.1.14) and (6.1.15), we have

$$\nabla \vec{Z}_l(r) \cdot \mathbf{V}(r) = -\frac{l}{2}\left|\mathbf{V}\left(r - \frac{2\pi}{l}\right)\right|^2 \exp\left(im\frac{\pi}{2}\right).$$

Thus

$$2ilm\lambda (\nabla \vec{Z}_l \cdot \mathbf{V}) \vec{Z}_l = -il^2 m\lambda |\mathbf{V}|^2 \exp\left(im\frac{\pi}{2}\right) \vec{Z}_l. \quad (6.1.16)$$

Substituting (6.1.16) into (6.1.13) and then multiplying it by $\mathbf{V}$ and noting (6.1.14), we have

$$\frac{d}{dt} \Omega |\mathbf{V}|^2 = \lambda (1 - m^2) l^2 |\mathbf{V}|^2 + \frac{1}{2} i\lambda l^2 m |\mathbf{V}|^4 \exp\left(im\frac{\pi}{2}\right). \quad (6.1.17)$$

Set $\Omega(t) = \Omega_r(t) + i\Omega_i(t)$ and note $|\mathbf{V}| = |\mathbf{d}| \exp(\Omega_r(t))$. (6.1.17) implies

$$\begin{cases} \frac{d}{dt}\Omega_r = \lambda(1 - m^2)l^2 - \frac{1}{2}\lambda l^2 m \sin(\frac{m\pi}{2})|\mathbf{d}|^2 \exp(\Omega_r(t)), \\ \frac{d}{dt}\Omega_i = \frac{1}{2}\lambda l^2 m \cos(\frac{m\pi}{2})|\mathbf{d}|^2 \exp(\Omega_r(t)). \end{cases} \quad (6.1.18)$$

By the first equation of (6.1.18), we have

$$\begin{cases} \frac{d}{dt}\Omega_r = a_m, & m \equiv 0 \bmod(2), \\ \frac{d}{dt}\Omega_r = a_m + b_m \exp(\Omega_r(t)), & m \equiv 1 \bmod(2), \end{cases} \quad (6.1.19)$$

where $a_m = \lambda(1 - m^2)l^2$ and $b_m = -\frac{1}{2}\lambda l^2 m(-1)^{[m/2]}|\mathbf{d}|^2$. Here $[m/2]$ denotes the maximum integer which is less than or equal to $m/2$.

In the following, we solve Eq. (6.1.19). First if $m = 0$, then $a_m = \lambda l^2 > 0$; thus $\vec{Z}_l$ is linearly unstable.

Second, if $m \equiv 0 \mod(2)$ and $m \neq 0$, then $a_m = \lambda(1 - m^2)l^2 < 0$ and $e^{\Omega_r(t)} = e^{a_m t + \Omega_r(0)}$; it is clear that $\lim_{t \to \infty} e^{\Omega_r(t)} = 0$. Thus the steady state solution $\vec{Z}_l$ is linear stable.

Third, if $m \equiv 1 \mod(2)$, by the second equation of (6.1.19), we have that

$$\frac{d\Omega_r(t)}{a_m + b_m e^{2\Omega_r(t)}} = dt;$$

integrating the identity given above implies

$$\left[\frac{t}{a_m} - \frac{1}{2a_m} \ln\left(a_m + b_m e^{2\Omega_r(t)}\right)\right]_0^t = t;$$

solving the above identity yields

$$e^{\Omega_r(t)} = \left\{\frac{1}{b_m}\left[(a_m + b_m e^{2\Omega_r(0)})e^{2(1-a_m)t} - a_m\right]\right\}^{1/2}. \tag{6.1.20}$$

By (6.1.20), it follows that $\lim_{t \to \infty} e^{\Omega_r(t)} = \infty$; thus the steady state solution $\vec{Z}_l$ is linearly unstable.

Theorem 6.1.2 *The inhomogeneous steady state solutions $\vec{Z}_l(r)$, $(l \neq 0)$, is hyperbolic. Moreover, we have*

$$W^u_{\mathrm{loc}}(\vec{Z}_l) = \mathrm{span}\{\mathbf{d}\exp(ilmr + \Omega(t)); \ m \equiv 1 \mod(2) \ and \ m = 0\};$$
$$W^s_{\mathrm{loc}}(\vec{Z}_l) = \mathrm{span}\{\mathbf{d}\exp(ilmr + \Omega(t)); \ m \equiv 0 \mod(2) \ but \ m \neq 0\}.$$

6.2   Asymptotic Behavior of L–L Equations

6.2.1   Estimates for Energy

First, we introduce an energy function

$$E(\vec{Z}(t)) = \frac{1}{2}\int_\Omega |\nabla \vec{Z}(r,t)|^2 dr,$$

where $\Omega = [0, 2\pi]$. In the following we give the properties of the energy function.

Proposition 6.2.1 *If $\vec{Z}(r,t)$ is a solution of (6.1.1), then*
    (i) $\frac{dE(\vec{Z}(t))}{dt} = -\lambda \int_\Omega |\vec{Z} \times \Delta\vec{Z}|^2 dr \leq 0$;
    (ii) $\frac{dE(\vec{Z}(t))}{dt} = 0$ *if and only if $\vec{Z}$ is the steady state $\vec{Z}_l(r)$;*
    (iii) $E(\vec{Z}_l(r)) = \pi l^2$;
    (iv) *the energy function $E(\vec{Z}(t))$ arrives minimum at $\vec{Z} = \mathbf{R}$ ($\mathbf{R}$ is a constant vector with $|\mathbf{R}| = 1$).*

Proof. (i) Multiplying (6.1.1) by $\Delta \vec{Z}$ and integrating it over Ω, we get

$$-\frac{1}{2}\frac{d}{dt}\int_{\Omega}|\nabla\vec{Z}|^2 dr = -\lambda\int_{\Omega}\Delta\vec{Z}\cdot[\vec{Z}\times(\vec{Z}\times\Delta\vec{Z})]dr.$$

Since $\Delta\vec{Z}\cdot[\vec{Z}\times(\vec{Z}\times\Delta\vec{Z})] = (\Delta\vec{Z}\times\vec{Z})\cdot(\vec{Z}\times\Delta\vec{Z}) = -|\vec{Z}\times\Delta\vec{Z}|^2$, (i) holds.

(ii) If $\vec{Z} = \vec{Z}_l(r)$ and (i), then $|\vec{Z}\times\Delta\vec{Z}| = 0$, and $\frac{dE(\vec{Z}(t))}{dt} = 0$. On the other hand, if $\frac{dE(\vec{Z}(t))}{dt} = 0$, we can get $\vec{Z}(r, t_1) = \vec{Z}_{l(t_1)}(r)$ and then by the periodicity of the solution, it follows that $l(t_1)$ is an integer. Thus the item is proved.

(iii) It is clear that $E(\vec{Z}_l(r)) = l^2\pi$. Thus (iv) is also proved.

Lemma 6.2.1 *If $\vec{Z}(r, t)$ is a solution of* (6.1.1), *then*

$$\|\nabla\vec{Z}(\cdot, t)\| \le \|\nabla\vec{Z}(\cdot, 0)\|.$$

Lemma 6.2.2 *If $\vec{Z}(r)$ is a 2π-periodic function, then*

$$\|\nabla\vec{Z}\|^2 \le 2\pi^2\|\Delta\vec{Z}\|^2.$$

Proof. Note the following identity:

$$|\nabla\vec{Z}(x)|^2 + |\nabla\vec{Z}(y)|^2 - 2\nabla\vec{Z}(x)\cdot\nabla\vec{Z}(y) = \left|\int_x^y\Delta\vec{Z}(r)dr\right|^2 \le 2\pi\int_{\Omega}|\Delta\vec{Z}(r)|^2 dr.$$

Integrating twice over Ω, we get

$$4\pi\|\nabla\vec{Z}\|^2 - 2\left(\int_{\Omega}\nabla\vec{Z}(x)dx\right)^2 \le (2\pi)^3\|\Delta\vec{Z}\|^2.$$

By the periodicity of $\vec{Z}(r)$, it follows that $\int_{\Omega}\nabla\vec{Z}(x)dx = 0$; thus the lemma holds.

Lemma 6.2.3 *If $u(r, t) \in C^0(R^+; H^1_{\mathrm{per}}(\Omega))$ and*

$$\lim_{t\to\infty}\int_{\Omega}|\nabla u(r, t)|^2 dr = 0,$$

then $\forall\, r \in \Omega$

$$\lim_{t\to\infty}u(r, t) = V,$$

where V is a constant.

Proof. For any $x, y \in \Omega$, it is clear that

$$u(x, t) = u(y, t) + \int_y^x\nabla u(r, t)dr.$$

Taking the absolute value above, we have

$$|u(x, t)| \le |u(y, t)| + \int_{\Omega}|\nabla u(r, t)|dr \le |u(y, t)| + \sqrt{2\pi}\int_{\Omega}|\nabla u(r, t)|^2 dr.$$

By the hypothesis of the lemma,

$$\lim_{t\to\infty} |u(x,t)| \le \lim_{t\to\infty} |u(y,t)|. \tag{6.2.1}$$

Similar to the proof as above, we have

$$\lim_{t\to\infty} |u(y,t)| \le \lim_{t\to\infty} |u(x,t)|. \tag{6.2.2}$$

Thus by (6.2.1) and (6.2.2) and the continuity of $u(r,\cdot)$ about t, the proof of the lemma is complete.

Theorem 6.2.1 *Assume that the initial data $\vec{Z}(r,0)$ satisfies*

$$\|\nabla \vec{Z}(\cdot,0)\| \le \frac{1-\delta}{K_1 K_2^2 \sqrt{2\pi}}. \tag{6.2.3}$$

Then there exists a constant vector $\mathbf{R}$ such that

$$\lim_{t\to\infty} \vec{Z}(r,t) = \mathbf{R}, \tag{6.2.4}$$

where $\delta \in (0,1)$, $|\mathbf{R}| = 1$ and K_1, K_2 are the constants in the Sobolev inequalities, respectively.

Proof. By (i) of Proposition 6.2.1, we have

$$\frac{1}{2}\frac{d}{dt}\|\nabla\vec{Z}\|^2 = -\lambda\|\vec{Z}\times\Delta\vec{Z}\|^2 = -\lambda\int_\Omega|\Delta\vec{Z}|^2 dr + \lambda\int_\Omega(\vec{Z}\cdot\Delta\vec{Z})^2 dr. \tag{6.2.5}$$

It is clear that $\vec{Z}\cdot\Delta\vec{Z} = -|\nabla\vec{Z}|^2$; thus

$$\lambda\int_\Omega(\vec{Z}\cdot\nabla\vec{Z})^2 dr = \lambda\int_\Omega|\nabla\vec{Z}|^4 dr.$$

By Sobolev's inequality and $|\vec{Z}| = 1$, we have

$$\|\nabla\vec{Z}\|_{L^4}^4 \le K_1\|\Delta\vec{Z}\|\|\nabla\vec{Z}\|^3 \le K_1 K_2^2\|\vec{Z}\|\|\nabla\vec{Z}\|\|\Delta\vec{Z}\|^2 = K_1 K_2^2\sqrt{2\pi}\|\nabla\vec{Z}\|\|\Delta\vec{Z}\|^2. \tag{6.2.6}$$

Substituting (6.2.6) into (6.2.5), we have

$$\frac{1}{2}\frac{d}{dt}\|\nabla\vec{Z}\|^2 + \lambda(1 - K_1 K_2^2\sqrt{2\pi}\|\nabla\vec{Z}\|)\|\Delta\vec{Z}\|^2 \le 0. \tag{6.2.7}$$

By the hypothesis of this theorem and Lemma 6.2.2, it follows that

$$\frac{d}{dt}\|\nabla\vec{Z}\|^2 + 2\lambda\delta\|\Delta\vec{Z}\|^2 \le 0. \tag{6.2.8}$$

By Lemma 6.2.2, it yields

$$\frac{d}{dt}\|\nabla\vec{Z}\|^2 + \frac{\lambda\delta}{\pi^2}\|\nabla\vec{Z}\|^2 \le 0. \tag{6.2.9}$$

Then by Gronwall's inequality, it yields

$$\|\nabla \vec{Z}(t)\|^2 \le \|\nabla \vec{Z}(0)\|^2 \exp(-\lambda \delta t/\pi^2); \tag{6.2.10}$$

thus $\lim_{t\to\infty} \|\nabla \vec{Z}(t)\|^2 = 0$, i.e., $\lim_{t\to\infty} \int_\Omega |\nabla Z_j|^2 dr = 0$ $(j = 1, 2, 3)$. By Lemma 6.2.3, it yields

$$\lim_{t\to\infty} \vec{Z}(r, t) = \mathbf{R}.$$

This completes the proof of the theorem.

Theorem 6.2.2 *Under the hypothesis of Theorem 6.2.1, then*

$$\mathbf{R} = \frac{1}{2\pi} \int_\Omega \vec{Z}(r, 0)dr + \frac{1}{2\pi}\mathbf{r}, \tag{6.2.11}$$

where $\mathbf{R} = \lim_{t\to\infty} \vec{Z}(r, t)$ *and* $\mathbf{r} = \lambda \int_0^\infty \int_\Omega |\nabla \vec{Z}|^2 \vec{Z} dr d\tau$ *satisfies*

$$|\mathbf{r}| \le \frac{(1-\delta)^2}{4\pi K_1^2 K_2^4 \delta}; \tag{6.2.12}$$

here $\delta \in (0, 1)$ *is defined in Theorem 6.2.1.*

Proof. Integrating (6.1.1) over Ω, we have

$$\frac{d}{dt} \int_\Omega \vec{Z}(r, t)dr = \int_\Omega (\vec{Z} \times \Delta \vec{Z})dr + \lambda \int_\Omega \Delta \vec{Z} dr + \lambda \int_\Omega |\nabla \vec{Z}|^2 \vec{Z} dr.$$

Since $\vec{Z} \times \Delta \vec{Z} = \nabla(\vec{Z} \times \nabla \vec{Z})$,

$$\int_\Omega (\vec{Z} \times \Delta \vec{Z})dr = 0, \quad \lambda \int_\Omega \Delta \vec{Z} dr = 0;$$

then

$$\frac{d}{dt} \int_\Omega \vec{Z}(r, t)dr = \lambda \int_\Omega |\nabla \vec{Z}|^2 \vec{Z} dr.$$

It follows that

$$\int_\Omega \vec{Z}(r, t)dr = \int_\Omega \vec{Z}(r, 0)dr + \lambda \int_0^t \int_\Omega |\nabla \vec{Z}|^2 \vec{Z} dr d\tau \tag{6.2.13}$$

and

$$|\mathbf{r}| \le \lambda \lim_{t\to\infty} \int_0^t \int_\Omega |\nabla \vec{Z}|^2 dr d\tau \le \lambda \|\nabla \vec{Z}(0)\|^2 \lim_{t\to\infty} \int_0^t \exp(-2\lambda \delta \tau)d\tau = \frac{1}{2\delta}\|\nabla \vec{Z}(0)\|^2.$$

From the hypothesis of the theorem, we observe that $|\mathbf{r}|$ satisfies the condition. This completes the proof of the theorem.

6.2.2   Bifurcation and Chaos

1. *Magnetization equations, chaos and bifurcation*

Recently, Gibson and Jeffries [69] completed a series of experiments on ferromagnetic resonance in which the signal power exceeds the second order Sullo unstable threshold. When the signal power increases there appear limit cycles, double periodic bifurcations, chaos, and periodic windows. In earlier time, Nahwmuha *et al.* [121] considered the chaos formed from two inhomogeneous spin modes, taking the parallel pump into account. For the transverse pump, if the signal power increases, there are also spin wave double bifurcations and chaos with periodicity 3 and 5 as observed in the GJ experiment.

The magnetization equation is

$$\frac{d\vec{m}}{dt} = -\gamma \vec{m} \times \vec{H} - \alpha \vec{m} \times (\vec{m} \times \vec{H}) \tag{6.2.14}$$

in which the Gilbert damping term has been considered. Field $\vec{H}$ consists of three parts: applied field, diplar field, and exchange field. Writing Eq. (6.2.14) in the back coefficient form and retaining the long terms, we have

$$\begin{cases} \frac{dB_k}{d\tau} = iB_0^2 B_{-k}^* + i\Gamma |B_0|^2 B_k - \eta B_k, \\ \frac{dB_0}{d\tau} = iB_k^2 B_0^* - i\Omega_s - \eta B_0, \end{cases} \tag{6.2.15}$$

where B_0, B_k are the oscillations of the spin wave, η is the damping constant, $d\tau = 0.5\Omega_m dt$, and Γ is a constant proportional to Ω_m. Ω_m is the magnetization, and Ω_s is the strength of the driving field. Suppose $B_k = B_{-k}$, $\eta = 0.005$, and $\Gamma = 0.5$. Equations (6.2.15) and its conjugate equations form 4 ordinary equations whose fixed points are

$$\begin{cases} |B_0| = \Omega_s/\eta, \\ |B_k| = 0. \end{cases} \tag{6.2.16}$$

$$\begin{cases} |B_0| = \eta^{\frac{1}{2}}/(1 - \Gamma^2)^{\frac{1}{4}}, \\ |B_k| = [-\eta(1 - \Gamma^2)^{\frac{1}{2}} + \{\frac{\Omega_s^2}{\eta}(1 - \Gamma^2)^{\frac{1}{2}} - \eta^2\Gamma^2\}^{\frac{1}{2}}]^{\frac{1}{2}}. \end{cases} \tag{6.2.17}$$

When $\Omega_s < \eta_1(\eta/\Omega_m)^{\frac{1}{2}}$ (Sullo threshold), (6.2.17) is stable; when $\Omega_s > \eta_1(\eta/\Omega_m)^{\frac{1}{2}}$ (Sullo threshold), (6.2.11) is unstable, but (6.2.17) is stable. When $|B_k| \ll |B_0|$, the condition of stability for fixed point (6.2.17) is

$$(\lambda + \eta)^3 - (\lambda + \eta)[(1 - \Gamma^2)|B_0|^4 + 4\Gamma|B_k|^2|B_0|^2$$
$$- 8|B_k|^2|B_0|^2 - 4\Gamma^2\eta|B_k|^2|B_0|^2] = 0, \tag{6.2.18}$$

where λ is the Lyapunov index. When $|B_k|^2 \le 0.1|B_0|^2$, λ must be negative. If $|B_k| \sim |B_0|$, the computation is very complicated.

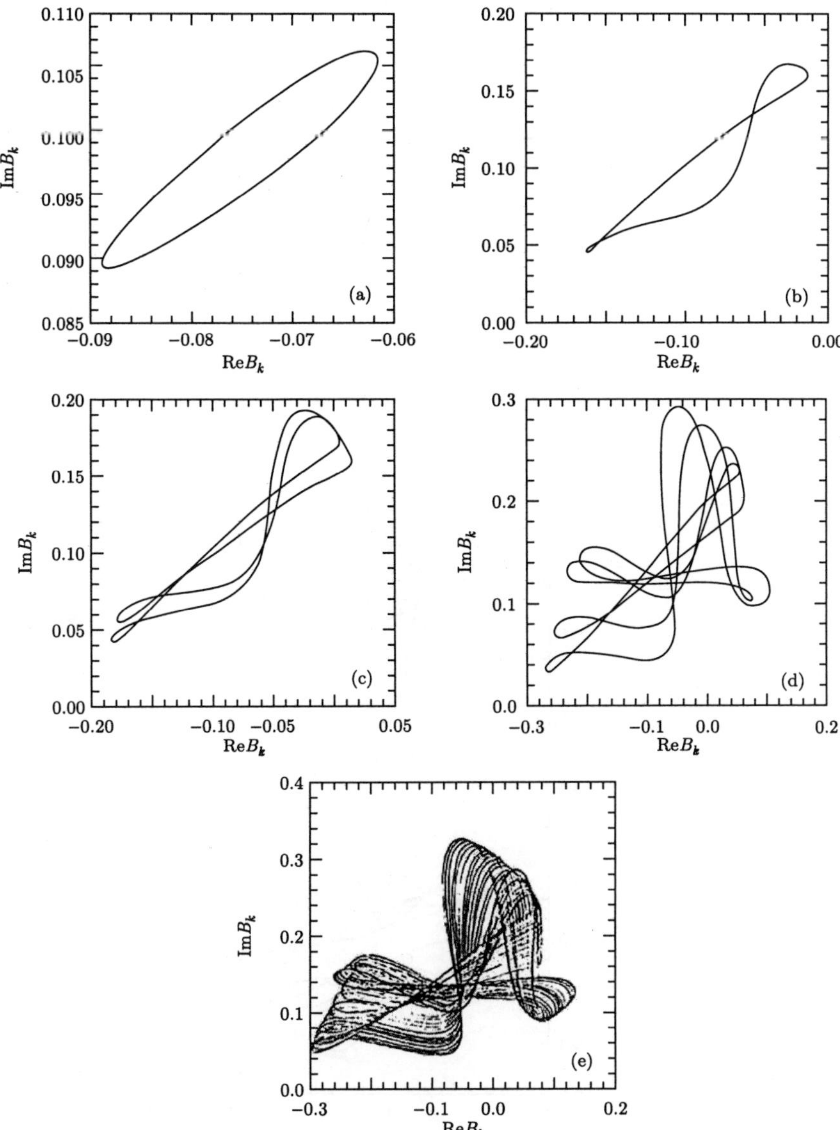

Figure 6.2.1.   (a) *Projection of the first limit cycle on the plan* $\mathrm{Re}\,B_e - \mathrm{Im}\,B_e$, $w_s = 1.518 \times 10^{-3}$, $f = 1.087 \times 10^{-3} w_m$; (b) *twisting and expansion when* $w_s = 0.0022$, $f = 1.470 \times 10^{-3} w_m$; (c) *double period*, $w_s = 0.0045$, $f = 0.018 \times 10^{-3} w_m$; (d) *period 4*, $w_s = 0.0055$, $f = 0.018 \times 10^{-3} w_m$; (e) *chaos*, $w_s = 0.005$.

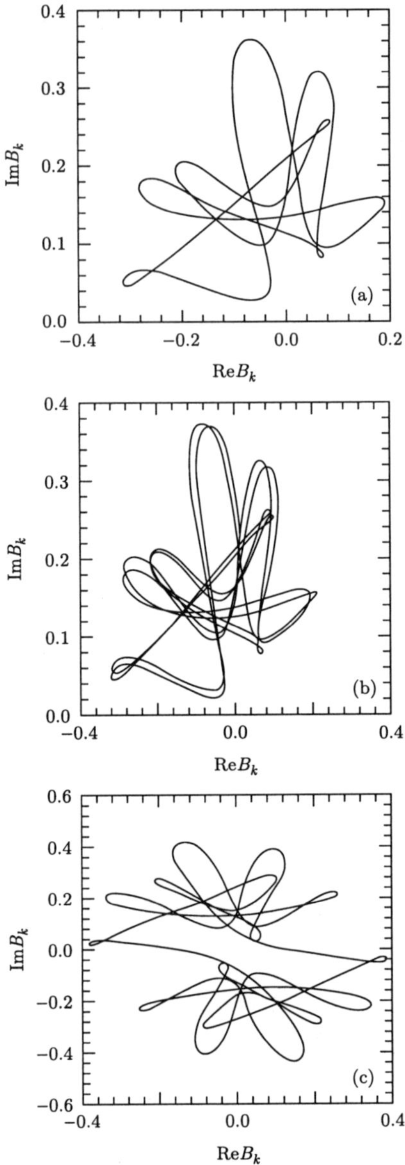

Figure 6.2.2. *Period 3 to period 6 (a) window of period 3, $w_s = 0.00625$, $f = 0.085 \times 10^{-3} w_m$; (b) bifurcation of period 6, $w_s = 0.0065$, $f = 0.442 \times 10^{-3} w_m$; (c) window of period 5, $w_s = 0.0090$, $f = 0.636 \times 10^{-3} w_m$. Ignore the choas between window of period 5 and window of period 6.*

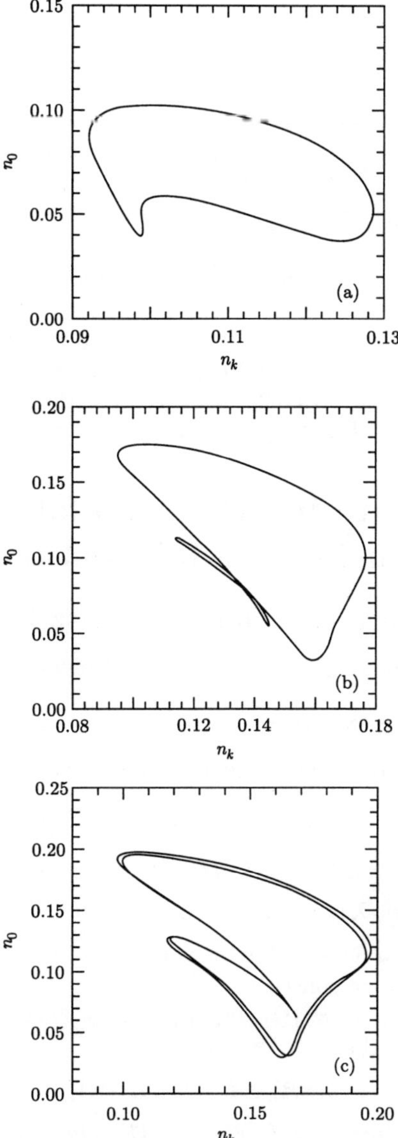

Figure 6.2.3. *Limit cycle (2) (a) projection of limit cycle on plan $n_0 - n_h$, $w_s = 0.00155$, $f = 1.068 \times 10^{-3} w_m$; (b) twisting and expansion when $w_s = 0.0022$, $f = 1.240 \times 10^{-3} w_m$; (c) double period $w_s = 0.0260$.*

2. Number of magnetic waves

It follows from (6.2.14) that

$$\frac{d}{dt}(|B_0|^2 + |B_k|^2) = -\alpha\Omega_s I_m B_0 - 2\eta(|B_0|^2 + |B_k|^2). \tag{6.2.19}$$

$|B_k|^2$ and $|B_0|^2$ can be explained as the number of magnetic wave without damping and driving. The number of magnetic wave is conservative which can be taken as the Hamiltonian system. If (6.2.17) hold, the right-hand side of (6.2.19) becomes zero; If the parcels of the magnetic wave increase (or decrease) such that driving increases (or decrease), the steady state becomes unstable. From this point of view and the fact $R_e B_0 \sim 0$, we see that when $\Omega_s > 4\eta|I_m B_0|$ the fixed point (6.2.17) becomes unstable. Thus we have $\Omega_{s,\text{critical}} = 6.1518 \times 10^{-3}$. At this time, the fixed point is the bifurcation indicated by Jeffries. The frequency of the limit cycle is, from numerical computation, $f = 6.1087 \times 10^{-3}\Omega_m$. In the GJ experiment, $4\pi\Omega_s = 300G$. Hence, $f = 5543 \times 10^6$Hz. Figures 6.2.1–6.2.3 give out the limit cycle, double periodic bifurcation, and periodicities 3, 4, 5, and 6 with different values of f and Ω_s.

6.3   Approximate Inertial Manifold for One-Dimensional L–L Equations

6.3.1   One-Dimensional Landau–Lifshitz Equations

Consider the following one-dimensional Landau–Lifshitz equations:

$$\frac{\partial u}{\partial t} = -\alpha u \times (u \times u_{xx}) + \beta u \times u_{xx}, \quad \Omega \times R^+ \tag{6.3.1}$$

with the initial data and periodic boundary conditions as follows:

$$u(x,0) = u_0(x), \quad |u(x_0)| = 1,$$

$$x \in \Omega = (-D, D), \quad 0 < D < \infty, \tag{6.3.2}$$

$$u(x - D, t) = u(x + D, t), \quad x \in R, \quad t \in R^+. \tag{6.3.3}$$

Here $u = (u_1, u_2, u_3) : R \times R^+ \to R^3$ is the spin vector, α, β are constants, and $\alpha > 0$ denotes the Gilbert damping constant.

As we know, for $u(x_0) \in H^2(\Omega)$, $|u(x_0)| = 1$, problem (6.3.1)–(6.3.3) admits a unique solution $u(t)$ such that

$$u(x,t) \in L^\infty(R^+; H^2(\Omega)).$$

Lemma 6.3.1 *Suppose $|u(x_0)| = 1$, $x \in R$. Then for the solution $u(t)$ to problem* (6.3.1)–(6.3.3);

$$|u(x,t)|^2 = 1, \quad (x,t) \in R \times R^+, \tag{6.3.4}$$

$$\|u_x(t)\|^2 \le \|u_x(0)\|^2, \quad t \in R^+ \tag{6.3.5}$$

hold.

Proof. Multiplying (6.3.1) by u, we have

$$\frac{\partial}{\partial t}|u(x,t)|^2 = 0, (x,t) \in R \times R^+.$$

From this, (6.3.4) follows.

Taking the inner product of (6.3.1) with u_{xx}, we arrive at

$$-\frac{1}{2}\frac{d}{dt}\|u_x\|^2 = -\alpha \int_\Omega (u \times (u \times u_{xx})) \cdot u_{xx} dx$$

$$= -\alpha \int_\Omega (u_{xx} \times u) \cdot (u \times u_{xx}) dx$$

$$= \alpha \int_\Omega |u \times u_{xx}|^2 dx.$$

Hence

$$\frac{d}{dt}\|u_x\|^2 \le 0.$$

And (6.3.5) follows. Thus the proof is complete.

We now introduce subset H_ρ of $H = L_2(\Omega)$

$$H_\rho = \{u \in H : |u(x)| = 1, \|u_x\| \le \rho\}.$$

From Lemma 6.3.1 we assert that H_ρ is an invariant subset of H. Hereafter, we assume that $u_0 \in H_\rho$.

Lemma 6.3.2 *Suppose that the assumptions of Lemma 6.3.1 hold. Then $u(x,t)$ is a classical solution to problem (6.3.1)–(6.3.3) if and only if $u(x,t)$ is a solution to the following problem:*

$$\frac{\partial u}{\partial t} = \alpha u_{xx} + \beta u \times u_{xx} + \alpha |u_x|^2 u, \tag{6.3.6}$$

$$u(x,0) = u_0(x), \quad |u_0(x)| = 1, \quad x \in \Omega, \tag{6.3.7}$$

$$u(x - D, t) = u(x + D, t), \quad x \in R, \quad t \in R^+. \tag{6.3.8}$$

Lemma 6.3.3 *Suppose that $u_0 \in H^2 \cap H_\rho$. Then for solution $u(t)$ to problem (6.3.6)–(6.3.8), the following estimate holds:*

$$\|u_{xx}(t)\| \le C_1, \quad \int_t^{t+1} \|u_{xxx}(t)\|^2 dt \le C_1, t \ge t_1. \tag{6.3.9}$$

Here constant C_1 depends on $(\alpha, \beta, \rho, \Omega)$, and t_1 depends on $(\alpha, \beta, \rho, \Omega)$ and R, $\|u_0\|_{H^2} \le R$.

Proof. Since $|u(x,t)| = 1$, we assert that if $|u_x| \ne 0$, then u, u_x, and $u \times u_x$ form the orthogonal base of R^3. Let $u_{xx} = \alpha_1 u + \alpha_2 u_x + \alpha_3 u \times u_x$. It is easy to see that

$$\alpha_1 = -|u_x|^2, \quad \alpha_2 = \frac{u_x \cdot u_{xx}}{|u_x|^2}, \quad \alpha_3 = \frac{(u \times u_x) \cdot u_{xx}}{|u_x|^2}.$$

Differentiating (6.3.6) with respect to x twice, and taking inner product with u_{xx}, we have

$$\frac{1}{2}\frac{d}{dt}\|u_{xx}\|^2 = \int_\Omega u_{xx} \cdot (\alpha u_{xx} + \beta u \times u_{xx} + \alpha|u_x|^2 u)_{xx} dx$$

$$= -\alpha\|u_{xxx}\|^2 - \beta \int_\Omega (u_x \times u_{xx}) \cdot u_{xxx} dx$$

$$- \alpha \int_\Omega u_{xxx} \cdot (|u_x|^2 u_x + 2u(u_x \cdot u_{xx})) dx. \tag{6.3.10}$$

Differentiate (6.3.4) with respect to x to get

$$u \cdot u_{xxx} = -\frac{3}{2}(|u_x|^2)_x. \tag{6.3.11}$$

Using this we can handle the right-hand side of (6.3.10) as follows:

$$\int_\Omega u_{xxx} \cdot (|u_x|^2 u_x) dx = -\int_\Omega |u_x|^2 |u_{xx}|^2 dx - 2\int_\Omega |u_x \cdot u_{xx}|^2 dx, \tag{6.3.12}$$

$$\int_\Omega u_{xxx} \cdot (u(u_x \cdot u_{xx})) dx = \int_\Omega (u_{xxx} \cdot u)(u_x \cdot u_{xx}) dx$$

$$= -3\int_\Omega |u_x \cdot u_{xx}|^2 dx, \tag{6.3.13}$$

$$\int_\Omega (u_x \times u_{xx}) \cdot u_{xxx} dx = \int_\Omega \left[u_x \times \left(-|u_x|^2 u + \frac{(u \times u_x) \cdot u_{xx}}{|u_x|^2}(u \times u_x) \right) \right] \cdot u_{xxx} dx$$

$$= \int_\Omega |u_x|^2 (u \times u_x) \cdot u_{xxx} dx$$

$$+ \int_\Omega \frac{(u \times u_x) \cdot u_{xx}}{|u_x|^2}(u_x \times (u \times u_x)) \cdot u_{xxx} dx$$

$$= \int_\Omega |u_x|^2 (u \times u_x) \cdot u_{xxx} dx$$

$$- \frac{3}{2}\int_\Omega (|u_x|^2)_x (u \times u_x) \cdot u_{xx} dx$$

$$= \int_\Omega |u_x|^2 (u \times u_x) \cdot u_{xxx} dx$$

$$+ \frac{3}{2}\int_\Omega |u_x|^2 (u \times u_x)_x \cdot u_{xx} dx$$

$$+ \frac{3}{2}\int_\Omega |u_x|^2 (u \times u_x) \cdot u_{xxx} dx$$

$$= \frac{5}{2}\int_\Omega |u_x|^2 (u \times u_x) \cdot u_{xxx} dx. \tag{6.3.14}$$

It follows from (6.3.10) and (6.3.12)–(6.3.14) that

$$\frac{1}{2}\frac{d}{dt}\|u_{xx}\|^2 + \alpha\|u_{xxx}\|^2 = \alpha\int_\Omega |u_x|^2 |u_{xx}|^2 dx + 8\alpha\int_\Omega |u_x \cdot u_{xx}|^2 dx$$

$$- \frac{5}{2}\beta\int_\Omega |u_x|^2 (u \times u_x) \times u_{xxx} dx. \tag{6.3.15}$$

Moreover,

$$\frac{1}{4}\frac{d}{dt}\int_\Omega |u_x|^4 dx = \int_\Omega |u_x|^2 u_x \cdot (\alpha u_{xx} + \beta u \times u_{xx} + \alpha |u_x|^2 u)_x dx$$

$$= \int_\Omega |u_x|^2 u_x \cdot (\alpha u_{xxx} + \beta u_x \times u_{xx} + \beta u \times u_{xxx}$$

$$+ \alpha |u_x|^2 u_x dx + 2\alpha u(u_x \cdot u_{xx}))dx$$

$$= -\alpha \int_\Omega |u_x|^2 |u_{xx}|^2 dx - 2\alpha \int_\Omega |u_x \cdot u_{xx}|^2 dx$$

$$+ \alpha \int_\Omega |u_x|^6 dx - \beta \int_\Omega |u_x|^2 (u \times u_x) \cdot u_{xxx} dx$$

and

$$-\beta \int_\Omega |u_x|^2 (u \times u_x) \cdot u_{xxx} dx = \frac{1}{4}\frac{d}{dt}\int_\Omega |u_x|^4 dx + \alpha \int_\Omega |u_x|^2 |u_{xx}|^2 dx$$

$$+ 2\alpha \int_\Omega |u_x \cdot u_{xx}|^2 - \alpha \int_\Omega |u_x|^6 dx. \qquad (6.3.16)$$

Inserting (6.3.16) into (6.3.15), one has

$$\frac{d}{dt}\|u_{xx}\|^2 - \frac{5}{4}\frac{d}{dt}\int_\Omega |u_x|^4 dx + 2\alpha\|u_{xxx}\|^2 + 5\alpha \int_\Omega |u_x|^6 dx$$

$$= 7\alpha \int_\Omega |u_x|^2 |u_{xx}|^2 dx + 26\alpha \int_\Omega |u_x \cdot u_{xx}|^2 dx. \qquad (6.3.17)$$

Since

$$7\alpha \int_\Omega |u_x|^2 |u_{xx}|^2 dx + 26\alpha \int_\Omega |u_x \cdot u_{xx}|^2 dx$$

$$\leq 33\alpha \int_\Omega |u_x|^2 |u_{xx}|^2 dx$$

$$\leq 5\alpha \int_\Omega |u_x|^6 dx + C \int_\Omega |u_{xx}|^3 dx$$

$$\leq 5\alpha \int_\Omega |u_x|^6 dx + C\|u_{xxx}\|^{\frac{7}{4}}\|u_x\|^{\frac{5}{4}}$$

$$\leq 5\alpha \int_\Omega |u_x|^6 dx + \frac{\alpha}{2}\|u_{xxx}\|^2 + C.$$

We can rewrite (6.3.17) as follows:

$$\frac{d}{dt}\|u_{xx}\|^2 - \frac{5}{4}\frac{d}{dt}\int_\Omega |u_x|^4 dx + \frac{3}{2}\alpha\|u_{xxx}\|^2 \leq C. \qquad (6.3.18)$$

By the Poincaré inequality, we have

$$\frac{\alpha}{2}\|u_{xxx}\| \geq K\|u_{xx}\|^2,$$

since $\int_\Omega u_{xx} dx = 0$. We infer from (6.3.18) that

$$\frac{d}{dt}\left(\|u_{xx}\|^2 - \frac{5}{4}\int_\Omega |u_x|^4 dx\right) + K\left(\|u_{xx}\|^2 - \frac{5}{4}\int_\Omega |u_x|^4 dx\right) + \alpha\|u_{xx}\|^2 \leq C. \quad (6.3.19)$$

Whence,

$$\frac{d}{dt}\left(\|u_{xx}\|^2 - \frac{5}{4}\int_\Omega |u_x|^4 dx\right) + K\left(\|u_{xx}\|^2 - \frac{5}{4}\int_\Omega |u_x|^4 dx\right) \le C.$$

Using the Gronwall inequality,

$$\|u_{xx}(t)\|^2 - \frac{5}{4}\int_\Omega |u_x(t)|^4 dx$$

$$\le \left(\|u_{xx}(0)\|^2 - \frac{5}{4}\int_\Omega |u_x(0)|^4 dx\right)e^{-Kt} + \frac{C}{K}$$

$$\le \|u_{xx}(0)\|^2 e^{-Kt} + C$$

$$\le R^2 e^{-Kt} + C, \quad t \ge 0$$

$$\le 2C, \quad t \ge t_*, \tag{6.3.20}$$

where $t_* = \frac{1}{K}\ln\frac{R^2}{C}$. By the interpolation technique, one has

$$\|u_x\|_4 \le C\|u_x\|^{\frac{3}{4}}\|u_{xx}\|^{\frac{1}{4}}.$$

Hence,

$$\frac{5}{4}\|u_x\|_4^4 \le C\|u_x\|^3\|u_{xx}\|$$

$$\le C\|u_{xx}\|$$

$$\le \frac{1}{2}\|u_{xx}\|^2 + C. \tag{6.3.21}$$

From (6.3.20) and (6.3.21), it follows that

$$\|u_{xx}(t)\| \le C, \quad t \ge t_*. \tag{6.3.22}$$

Using (6.3.22), and integrating (6.3.19) with respect to t over $(t, t+1)$, we arrive at

$$\int_t^{t+1} \|u_{xx}\|^2 dt \le C, \quad \forall\, t \ge t_*.$$

Thus the proof of the lemma is complete.

Lemma 6.3.4 *Assume that $u_0 \in H^{k+1} \cap H_\rho$, $k \ge 1$. Then*

$$\|D_x^{k+1}u(t)\|^2 \le C_k, \quad t \ge t_k,$$

$$\int_t^{t+1} \|D_x^{k+2}u(t)\|^2 dt \le C_k, \quad t \ge t_k,$$

where constant C_k depends only on the initial data and k, and t_k depends only on the initial data, k and R; $\|u_0\|_{H^2} \le R$.

Proof. We apply the induction argument to prove this lemma.

(i) For $k = 1$, invoking Lemma 6.3.3, we assert that the results of Lemma 6.3.4 are valid.

(ii) Assume Lemma 6.3.4 is valid for $k - 1$. We intend to prove that Lemma 6.3.4 is valid for $k \geq 2$. Differentiating (6.3.6) with respect to x and $(k + 1)$ times, taking the inner product with $D_x^{k+1}u$, we have

$$\frac{1}{2}\frac{d}{dt}\|D_x^{k+1}u\|^2 + \alpha\|D_x^{k+2}u\|^2$$

$$= \beta \sum_{i=0}^{k+1} C_{k+1}^i \int_\Omega (D_x^{k+1-i}u \times D_x^i u_{xx}) \cdot D_x^{k+1}u dx$$

$$+ \alpha \sum_{i=0}^{k+1} C_{k+1}^i \int_\Omega (D_x^i|u_x|^2 (D_x^{k+1-i}u \cdot D_x^{k+1}u)dx. \tag{6.3.23}$$

From a simple calculation, one gets

$$\beta \sum_{i=0}^{k+1} C_{k+1}^i \int_\Omega (D_x^{k+1-i}u \times D_x^i u_{xx}) \cdot D_x^{k+1}u dx$$

$$= -\beta C_{k+1}^k \int_\Omega (D_x^k u \times u_{xx}) \cdot D_x^{k+2}u dx$$

$$+ \beta \sum_{i=2}^{k-1} C_{k+1}^i \int_\Omega (D_x^{k+1-i}u \times D_x^i u_{xx}) \cdot D_x^{k+1}u dx$$

$$+ \beta C_{k+1}^k \int_\Omega (u_x \times D_x^{k+2}u) \cdot D_x^{k+2}u dx$$

$$- \beta \int_\Omega (u_x \times D_x^{k+2}u) \cdot D_x^{k+1}u dx$$

$$\leq C\|D_x^k u\|_\infty \int_\Omega |u_{xx}||D_x^{k+2}u|dx$$

$$+ C \sum_{i=2}^{k-1} \|D_x^{k+1-i}u\|_\infty \int_\Omega |D_x^{i+2}u||D_x^{k+1}u|dx$$

$$+ C\|u_x\|_\infty \int_\Omega |D_x^{k+2}u||D_x^{k+1}u|dx$$

$$\leq C\|D_x^k u\|_\infty \|u_{xx}\| \|D_x^{k+2}u\|$$

$$+ C \sum_{i=2}^{k-1} \|D_x^{k+1-i}u\|_\infty \|D_x^{i+2}u\| \|D_x^{k+1}u\|$$

$$+ C\|u_x\|_\infty \|D_x^{k+2}u\| \|D_x^{k+1}u\|$$

$$\leq C\|D_x^{k+1}u\| \|u_{xx}\| \|D_x^{k+2}u\|$$

$$+ C \sum_{i=2}^{k-1} \|D_x^{k+2-i}u\| \|D_x^{i+2}u\| \|D_x^{k+1}u\|$$

$$+ C\|u_{xx}\| \|D_x^{k+2}u\| \|D_x^{k+1}u\|$$

$$\leq C\|D_x^{k+1}u\|\|D_x^{k+2}u\| + C\sum_{i=2}^{k-1}\|D_x^{i+2}u\|\|D_x^{k+2}u\|$$

$$\leq C\|D_x^{k+1}u\|\|D_x^{k+2}u\| + C\|D_x^{k+1}u\|^2$$

$$\leq \frac{\alpha}{4}\|D_x^{k+2}u\|^2 + C\|D_x^{k+1}u\|^2. \tag{6.3.24}$$

By the Leibniz formula, we have

$$D_x^k|u_x|^2 = \sum_{j=0}^{k} C_k^j D_x^{k-j}u_x \cdot D_x^j u_x$$

$$= \sum_{j=1}^{k-1} C_k^j D_x^{k-j+1}u \cdot D_x^{j+1}u_x + 2D_x^{k+1}u \cdot D_x u, \tag{6.3.25}$$

$$D_x^{k+1}|u_x|^2 = \sum_{j=0}^{k+1} C_{k+1}^j D_x^{k+1-j}u_x \cdot D_x^j u_x$$

$$= \sum_{j=2}^{k-1} C_{k+1}^j D_x^{k-j+2}u \cdot D_x^{j+1}u$$

$$+ 2D_x^{k+2} \cdot D_x u + 2C_{k+1}^1 D_x^{k+1}u \cdot D_x^2 u. \tag{6.3.26}$$

Then

$$\alpha \sum_{i=0}^{k+1} C_{k+1}^i \int_\Omega D_x^i|u_x|^2(D_x^{k+1-i}u \cdot D_x^{k+1}u)dx$$

$$= \alpha \int_\Omega |u_x|^2(D_x^{k+1}u \cdot D_x^{k+1}u)dx$$

$$+ 2\alpha C_{k+1}^1 \int_\Omega (u_x \cdot u_{xx})(D_x^k u \cdot D_x^{k+1}u)dx$$

$$+ \alpha \sum_{i=2}^{k-1} C_{k+1}^i \int_\Omega D_x^i|u_x|^2(D_x^{k+1-i}u \cdot D_x^{k+1}u)dx$$

$$+ \alpha C_{k+1}^k \int_\Omega D_x^k|u_x|^2(D_x u \cdot D_x^{k+1}u)dx$$

$$+ \alpha \int_\Omega D_x^{k+1}|u_x|^2(u \cdot D_x^{k+1}u)dx$$

$$\leq C\|u_x\|_\infty^2\|D_x^{k+1}u\|^2$$

$$+ C\|u_x\|_\infty\|D_x^k u\|_\infty\|u_{xx}\|\|D_x^{k+1}u\|$$

$$+ C\sum_{i=2}^{k-1}\|D_x^i|u_x|^2\|\|D_x^{k+1-i}u\|_\infty\|D_x^{k+1}u\|$$

$$+ C\|u_x\|_\infty^2\|D_x^{k+1}u\|^2$$

$$+ C\sum_{j=1}^{k-1}\|u_x\|_\infty\|D_x^{j+1}u\|_\infty\|D_x^{k-j+1}u\|\|D_x^{k+1}u\|$$

$$+ C \sum_{j=2}^{k-1} \|D_x^{j+1}u\|_\infty \|D_x^{k-j+2}u\| \|D_x^{k+1}u\|$$

$$+ C\|D_x u\|_\infty \|D_x^{k+2}u\| \|D_x^{k+1}u\|$$

$$+ C\|D_x^{k+1}u\| \|D_x^2 u\| \|D_x^{k+1}u\|$$

$$\le C\|D_x^2 u\|^2 \|D_x^{k+1}u\|^2$$

$$+ C \sum_{i=2}^{k-1} \sum_{j=0}^{i} C^j \|D_x^{i-j+1}u\|_\infty \|D_x^{j+1}u\| \|D_x^{j+2-i}u\| \|D_x^{k+1}u\|$$

$$+ C \sum_{j=1}^{k-1} \|u_{xx}\| \|D_x^{j+2}u\| \|D_x^{k-j+1}u\| \|D_x^{k+1}u\|$$

$$+ C \sum_{j=2}^{k-1} \|D_x^{j+1}u\| \|D_x^{k-j+2}u\| \|D_x^{k+1}u\|$$

$$+ C\|u_{xx}\| \|D_x^{k+3}u\| \|D_x^{k+1}u\|$$

$$\le \|D_x^{k+1}u\|^2 + C\|D_x^{k+1}u\|^2 \|D_x^k u\|^2 + C\|D_x^{k+1}u\| \|D_x^{k+2}u\|$$

$$\le \frac{\alpha}{4} \|D_x^{k+2}u\|^2 + C\|D_x^{k+1}u\|^2. \tag{6.3.27}$$

It follows from (6.3.23), (6.3.24) and (6.3.27) that

$$\frac{d}{dt} \|D_x^{k+1}u\|^2 + \alpha \|D_x^{k+2}u\|^2 \le C\|D_x^{k+1}u\|^2. \tag{6.3.28}$$

This implies

$$\frac{d}{dt} \|D_x^{k+1}u\|^2 \le C\|D_x^{k+1}u\|^2, \quad t \ge t_*, \tag{6.3.29}$$

where t_* depends on the initial data and k. On the other hand, it follows from the induction assumption that

$$\int_t^{t+1} \|D_x^{k+1}u\|^2 dt \le C, \quad t \ge t_*. \tag{6.3.30}$$

From (6.3.29) and (6.3.30) and the Gronwall inequality, we have

$$\|D_x^{k+1}u\|^2 \le C, \quad t \ge t_* + 1. \tag{6.3.31}$$

For $t \in (t, t+1)$, integrating (6.3.28), we have

$$\int_t^{t+1} \|D_x^{k+2}u\|^2 dt \le C, \quad t \ge t_* + 1. \tag{6.3.32}$$

The conclusion of this lemma follows from (6.3.31) and (6.3.32).

Lemma 6.3.5 *Let $u_0 \in H^{k+2} \cap H_\rho$, $k \ge 0$. There exists a constant C_k depending on the initial data and k such that*

$$\|D_x^k u_t\| \le C_k, \quad t \ge t_k,$$

where t_k depends on the initial data, k, and R with $\|u_0\|_{H^2} \le R$.

Proof. This lemma can be obtained by differentiating (6.3.6) k times with respect to x and applying Lemma 6.3.4.

In order to study the approximate inertial manifold of the Landau–Lifshitz equations, rewrite (6.3.6)–(6.3.8) as the operators on H:

$$\frac{du}{dt} + \alpha Au + B(u, u) + R(u) = 0, \tag{6.3.33}$$

where $A = -\partial_{xx}$ is an unbounded self-conjugate operator defined on

$$D(A) = \{u:\ u \in H^2,\ u \text{ satisfies } (6.3.8)\}.$$

$B(u, 0) = -\beta u \times \nu_{xx}$ is a bilinear form; $R(u) = -\alpha|u_x|^2 u$ is a nonlinear operator: $D(A) \to H$. It is known that H has an orthogonal base consisting of the eigenfunctions of operator A: $\{\Omega_j\}_{j=1}^{\infty}$,

$$A\Omega_j = \lambda_j \Omega_j, \quad 0 = \alpha_1 < \lambda_2 < \cdots < \lambda_j \to \infty, \quad j \to \infty.$$

For a given m, let $P = P_m : H \to \text{Span}\{\Omega_1, \Omega_2, \ldots, \Omega_m\}$ be a project operator. $Q = Q_m = P - P_m$. Acting (6.3.33) by P_m and Q_m, we have

$$\begin{cases} \dfrac{dp}{dt} + \alpha Ap + P_m B(p + q, p + q) + P_m R(p + q) = 0, \\[2mm] \dfrac{dq}{dt} + \alpha Aq + Q_m B(p + q, p + q) + Q_m R(p + q) = 0, \end{cases} \tag{6.3.34}$$

where $p = P_m u$, $q = Q_m u$.

It follows from Lemmas 6.3.1, 6.3.4 and 6.3.5 that

$$\|u\|, \quad \|A^{\frac{3}{2}}u\| \le C, \quad t \ge t_*, \tag{6.3.35}$$

$$\|A^k u\|, \quad \|A^{k+1}u\|, \quad \|A^{k+\frac{3}{2}}u\|, \quad \left\|A^k \frac{\partial u}{\partial t}\right\| \le C_k, \quad t \le t_k, \tag{6.3.36}$$

where t_k, C_k are as before. Note

$$\begin{cases} \|A^{\frac{1}{2}}u\| = \|\frac{\partial u}{\partial x}\|, & u \in H^1, \\[1mm] \|P_m u\| \le \|u\|, \quad \|Q_m u\| \le \|u\|, & u \in H, \\[1mm] \|A^2 u\| \ge \lambda_{m+1}^{\alpha}\|u\|, & \alpha > 0, \quad u \in Q_m D(A^2). \end{cases} \tag{6.3.37}$$

It follows from (6.3.36) that

$$\|q\|, \quad \|Aq\|, \quad \left\|A^{\frac{3}{2}}q\right\|, \quad \left\|\frac{d}{dt}q\right\| \le C_k \lambda_{m+1}^{-k}, \quad t \ge t_k. \tag{6.3.38}$$

Define a map $\Phi : P_m H \to Q_m H$ such that for any $p \in P_m H$, $\Phi(p) = \Psi$ is given by

$$\alpha A\Psi + Q_m B(p, p) + Q_m R(p) = 0. \tag{6.3.39}$$

Let $\Sigma = \text{graph}(\Phi)$. It can be verified that Σ is the approximate inertial manifold of (6.3.6)–(6.3.8). Then

Theorem 6.3.1 *Let $u_0 \in H^{k+2} \cap H_\rho$. Then for any integer k, there exists a constant C_k depending on the initial data and k such that any solution $u(t)$ of (6.3.6)–(6.3.8) is within a H^2-distance no large than $C_k \lambda_{m+1}^{k-1}$ from $\sum$ when $t > t_k$, where t_k depends on the initial data, k and R with $\|u_0\|_{H^2} \leq R$.*

Proof. Subtracting (6.3.34) from (6.3.39), we have

$$\alpha A\Psi - \alpha Aq = \frac{dq}{dt} + Q_m B(p+q, p+q) - Q_m B(p,q)$$
$$+ Q_m R(p+q) - Q_m R(p)$$
$$= \frac{dq}{dt} + Q_m B(q, p+q) + Q_m B(p,q)$$
$$+ Q_m R(p+q) - Q_m R(p). \tag{6.3.40}$$

Since

$$\|B(q, p+q) + B(p,q)\| \leq \beta \|q \times Au\| + \beta \|p \times Aq\|$$
$$\leq \beta \|q\| \|Au\|_\infty + \beta \|p\| \|Aq\|_\infty$$
$$\leq \beta \|q\| \|A^{\frac{3}{2}} u\| + \beta \|p\| \|A^{\frac{3}{2}} q\|$$
$$\leq C_k \lambda_{m+1}^{-k}. \tag{6.3.41}$$

And

$$\|R(p+q) - R(p)\| = \alpha \| |u_x|^2 u - |p_x|^2 p \|$$
$$\leq \alpha \| (|u_x|^2 - |p_x|^2) u \|$$
$$+ \alpha \| (|p_x|^2 (u - p) \|$$
$$\leq \alpha \| (|u_x| + |p_x|)(|u_x| - |p_x|) u \|$$
$$+ \alpha \|p_x\|_\infty^2 \|q\|$$
$$\leq 2\alpha \|Au\| \|Aq\| \|u\| + \alpha \|Au\| \|q\|$$
$$\leq C_k \lambda_{m+1}^{-k}. \tag{6.3.42}$$

It follows from (6.3.40)–(6.3.42) that

$$\|A\Psi - Aq\| \leq C_k \lambda_{m+1}^{-k}, \quad t \geq t_k.$$

Therefore we get

$$\operatorname{dist}_{H^2}\left(u(t), \sum\right) \leq \|u(t) - (p(t) + \Phi(p(t)))\|_{H^2}$$
$$\leq \|\Phi(p(t)) - q(t)\|_{H^2}$$
$$\leq \|A\Psi - Aq\|$$
$$\leq C_k \lambda_{m+1}^{-k}, \quad t \geq t_k.$$

6.4   Attractor of Landau–Lifshitz Equations on Riemannian Manifold

6.4.1   Landau–Lifshitz Equations on Riemannian Manifold

Assume that (M, γ), (N, g) be two compact Riemannian manifolds. M has no boundary, and N is S^2. Consider the following Landau–Lifshitz equations for $u = (u_1, u_2, u_3) : M \to S^2$,

$$\partial_t u = -\alpha_1 u \times (u \times \Delta_M u) + \alpha_2 u \times \Delta_M u, \quad \alpha_1 > 0, \tag{6.4.1}$$

with initial condition

$$u(x, 0) = u_0(x), \tag{6.4.2}$$

where $|u_0(x)| = 1, x = (x_1, x_2, \ldots, x_n) \in M$, Δ_M is the Laplace–Beltrami operator

$$\Delta_M = \frac{1}{\sqrt{\gamma}} \frac{\partial}{\partial x^\beta} \gamma^{\alpha\beta} \sqrt{\gamma} \frac{\partial}{\partial x^\alpha} = \gamma^{\alpha\beta} \frac{\partial^2}{\partial x^\alpha \partial x^\beta} - \Gamma^k_{\alpha\beta} \frac{\partial}{\partial x^k}.$$

In the classical sense, Eq. (6.4.1) is equivalent to

$$\partial_t u = \alpha_1 \frac{1}{\sqrt{\gamma}} \frac{\partial}{\partial x^\beta} \left(\gamma^{\alpha\beta} \sqrt{\gamma} \frac{\partial u}{\partial x^\alpha} \right) + \alpha_2 |\nabla u|^2 u$$

$$+ \alpha_2 u \times \left(\frac{1}{\sqrt{\gamma}} \frac{\partial}{\partial x^\beta} \left(\gamma^{\alpha\beta} \sqrt{\gamma} \frac{\partial u}{\partial x^\alpha} \right) \right). \tag{6.4.3}$$

6.4.2   The *A Priori* Estimates

Our aim is to prove the existence of a global attractor of problem (6.4.1) and (6.4.2) and give the Hausdorff and fractal dimensions. To this aim we first give the *a priori* estimates.

Let ∇ be the Riemannian connection corresponding to metric γ (or covariant derivative):

$$|\nabla u(x)|^2 = \sum_{\alpha,\beta} \sum_i \gamma^{\alpha\beta} \frac{\partial u^i}{\partial x^\alpha} \frac{\partial u^i}{\partial x^\beta}. \tag{6.4.4}$$

For real function $\varphi \in C^k(M)$ $(k \geq 0$ is integer), we define

$$|\nabla^k \varphi|^2 = \nabla^{\alpha_1} \nabla^{\alpha_2} \cdots \nabla^{\alpha_k} \nabla_{\alpha_1} \nabla_{\alpha_2} \cdots \nabla_{\alpha_k} \varphi.$$

In particular, $|\nabla^1 \varphi| = |\nabla \varphi|$, $|\nabla^1 \varphi|^2 = |\nabla \varphi|^2 = \nabla^\nu \varphi \nabla_\nu \varphi$, $\nabla^k \varphi$ is kth order covariant derivative of φ.

Consider the vector space $L^p_k = \{\varphi : \varphi \in C^\infty, |\nabla^l \varphi| \in L^p(M), 0 \leq l \leq k\}$; l, k are integers and $p \geq 1$. Sobolev space $W^k_p(M)$ is the closure of L^p_k in the norm $\|\varphi\|_{W^p_k} = \sum_{i=0}^k \|\nabla^i \varphi\|_{L^p}$. In particular, $W^k_2(M) = H^k(M)$, $\|\cdot\|_2 = \|\cdot\|$.

Lemma 6.4.1 *Let $|u_0(x)|^2 = 1$. Then for the smooth solution of (6.4.1) and (6.4.2) there holds*

$$|u(x,t)|^2 = 1, \quad \forall \ (x,t) \in M \times (0,\infty). \tag{6.4.5}$$

Proof. Multiplying (6.4.1) by u we have

$$u \cdot \partial_t u = 0, \quad \forall \ (x,t) \in M \times (0,\infty);$$

the conclusion follows from (6.4.2) and (6.4.5).

Lemma 6.4.2 *Let $|u_0(x)|^2 = 1$ and $\|\nabla u_0\| < \infty$. Then for the smooth solution of (6.4.1) and (6.4.2),*

$$\begin{cases} \|\nabla u(\cdot,t)\|^2 \leq \|\nabla u_0\|^2, \\ 2\alpha_1 \int_0^t \|u \times \Delta_M u\|^2 dt \leq \|\nabla u_0\|^2, \quad \forall \ 0 \leq t < \infty \end{cases} \tag{6.4.6}$$

hold.

Proof. Multiplying (6.4.1) by $\Delta_M u$ we have

$$\begin{aligned} \Delta_M u \cdot u_t &= -\alpha_1 \Delta_M u \cdot (u \times (u \times \Delta_M u)) \\ &= -\alpha_1 (u \times \Delta_M u) \cdot (\Delta_M u \times u) \\ &= \alpha_1 |u \times \Delta_M u|^2. \end{aligned}$$

On the other hand,

$$\int_M \Delta_M u \cdot u_t dM = -\frac{1}{2}\frac{d}{dt}\|\nabla u(\cdot,t)\|^2.$$

Therefore

$$\frac{d}{dt}\|\nabla u(\cdot,t)\|^2 \leq 0.$$

The lemma follows from this and the assumptions on the initial data u_0.

Lemma 6.4.3 (Sobolev interpolation inequality on Riemannian manifold) *Let M be a compact Riemannian manifold with smooth boundary, q, r be real numbers, $1 \leq q$, $r \leq \infty$, j, m be integers, $0 \leq j < m$. Then there exists a constant C depending only on n, m, j, q, r and α, M such that for any $f \in W_r^m(M) \cap L_q(M)$, there holds*

$$\|\nabla^j f\|_p \leq C\|f\|_{W_r^m}^\alpha \|f\|_{L_q}^{1-\alpha}, \tag{6.4.7}$$

where

$$\frac{1}{p} = \frac{j}{n} + \alpha\left(\frac{1}{r} - \frac{m}{n}\right) + (1-\alpha)\frac{1}{q},$$

for any $\alpha : \frac{j}{m} \leq \alpha \leq 1$ and nonnegative integer p.

Proof. It follows from the Aubin theorem ([8, Theorem 6.3.70]) that

$$\|\nabla^j F\|_p \leq C\|\nabla^m F\|_r^\alpha \|f\|_q^{1-\alpha}, \tag{6.4.8}$$

where

$$F = f - \overline{f}, \quad \overline{f} = \frac{1}{\mathrm{vol(M)}} \int_M f dM,$$

$$\frac{1}{p} = \frac{j}{n} + \alpha\left(\frac{1}{r} - \frac{m}{n}\right) + (1-\alpha)\frac{1}{q}.$$

(i) $i > 0$. It follows from (6.4.8) that

$$\begin{aligned}
\|\nabla^j F\|_p &\leq C\|\nabla^m F\|_r^\alpha (\|f\|_q + \|\overline{f}\|_q)^{1-\alpha} \\
&\leq C_1\|\nabla^m F\|_r^\alpha \|f\|_q^{1-\alpha} \\
&\leq C_1\|F\|_{W_r^m(M)}^\alpha \|f\|_q^{1-\alpha}.
\end{aligned}$$

(ii) $i = 0$, it follows from Hölder inequality that

$$\begin{aligned}
\int_M |f|^p dx &= \int_M |f|^\alpha |f|^\beta dx \\
&\leq \left(\int_M |f|^{\alpha l} dx\right)^{\frac{1}{l}} \left(\int_M |f|^{\beta l'} dx\right)^{\frac{1}{l'}},
\end{aligned}$$

where

$$\alpha + \beta = p, \quad l\alpha = r, \quad l'\beta = q, \quad \frac{1}{l} + \frac{1}{l'} = 1.$$

Then

$$\frac{1}{l} = \left(\frac{q-p}{q-r}\right)\frac{r}{p}, \quad \frac{1}{p'} = 1 - \frac{1}{l}.$$

(6.4.7) follows from (i) and (ii).

Remark. Replacing space $L_k^p(M)$, T. Aubin in [8] introduced space V_k^p which is the closure of space S_k^p in the norm

$$\|\varphi\|_{V_k^p(M)} = \sum_{0 \leq l \leq \frac{k}{2}} \|\Delta_M^l \varphi\|_p + \sum_{0 \leq l \leq \frac{k-1}{2}} \|\nabla \Delta_M^l \varphi\|_p,$$

where S_k^p is a vector space of elements φ : $\varphi \in C^\infty(M)$, $\Delta_M^l \varphi \in L^p(M)$, $0 \leq l \leq \frac{k}{2}$, and $\nabla \Delta_M^l \varphi \in L^p(M)$, $0 \leq l \leq \frac{k-1}{2}$.

Lemma 6.4.4 *Let the conditions of Lemma 6.4.2 hold, and let*

$$\|\nabla u_0\| \leq \lambda, \quad n = 2, \tag{6.4.9}$$

where constant λ is small enough. Then for the smooth solutions of problem (6.4.1) and (6.4.2),

$$\|\Delta_M u(\cdot, t)\| \leq \frac{E_1}{t}, \quad \forall \, x \in M, \quad t > 0, \quad 1 \leq n \leq 2, \tag{6.4.10}$$

holds, where constant E_1 depends only on $\|\nabla u_0\|_{H^1(M)}$ and $0 < t \leq T$.

Proof. Using operator Δ_M on (6.4.7) and taking the inner product with $t\Delta_M \mathbf{u}$, we have

$$\left(t\Delta_M \mathbf{u}, \Delta_M \mathbf{u}_t - \alpha_1 \Delta_M^2 \mathbf{u} - \alpha_1 \Delta_M(|\nabla \mathbf{u}|^2 \mathbf{u}) - \alpha_2 \Delta_M(\mathbf{u} \times \Delta_M \mathbf{u})\right) = 0, \quad (6.4.11)$$

where

$$t(\Delta_M \mathbf{u}, \Delta_M \mathbf{u}_t) = \frac{1}{2}\frac{d}{dt}(t\|\Delta_M \mathbf{u}\|^2) - \frac{1}{2}\|\Delta_M \mathbf{u}\|^2,$$

$$(\Delta_M \mathbf{u}, \alpha_1 \Delta_M^2 \mathbf{u}) = \alpha_1 \int \Delta_M \mathbf{u}. \frac{1}{\sqrt{\gamma}}\frac{\partial}{\partial x^\beta}\left(\gamma^{\alpha\beta}\sqrt{\gamma}\frac{\partial \Delta_M \mathbf{u}}{\partial x^\alpha}\right)\sqrt{\gamma}dx$$

$$= -\alpha_1 \int \gamma^{\alpha\beta}\sqrt{\gamma}\frac{\partial \Delta_M \mathbf{u}}{\partial x^\alpha}\frac{\partial \Delta_M \mathbf{u}}{\partial x^\beta}dx = -\alpha_1 \|\nabla \Delta_M \mathbf{u}\|^2,$$

$$\left|\int \Delta_M \mathbf{u}.\Delta_M(\mathbf{u} \times \Delta_M \mathbf{u})\sqrt{\gamma}dx\right|$$

$$= \left|\int \Delta_M \mathbf{u}.\frac{\partial}{\partial x^\beta}\left(\gamma^{\alpha\beta}\sqrt{\gamma}\frac{\partial}{\partial x^\alpha}(\mathbf{u} \times \Delta_M \mathbf{u})\right)dx\right|$$

$$= \left|\int \gamma^{\alpha\beta}\frac{\partial \Delta_M \mathbf{u}}{\partial x^\beta}\frac{\partial(\mathbf{u} \times \Delta_M \mathbf{u})}{\partial x^\alpha}\sqrt{\gamma}dx\right|$$

$$= \left|\int \gamma^{\alpha\beta}\frac{\partial \Delta_M \mathbf{u}}{\partial x^\beta}\left[\frac{\partial \mathbf{u}}{\partial x^\alpha} \times \Delta_M \mathbf{u} + \mathbf{u} \times \frac{\partial \Delta_M \mathbf{u}}{\partial x^\alpha}\right]\sqrt{\gamma}dx\right|$$

$$\leq C_1 \|\nabla \Delta_M \mathbf{u}\|\|\nabla \mathbf{u}\|_\infty \|\Delta_M \mathbf{u}\|. \quad (6.4.12)$$

In (6.4.12), we have used the fact

$$\int \gamma^{\alpha\beta}\frac{\partial \Delta_M \mathbf{u}}{\partial x^\beta}.\left(\mathbf{u} \times \frac{\partial \Delta_M \mathbf{u}}{\partial x^\alpha}\right)\sqrt{\gamma}dx = 0. \quad (6.4.13)$$

Lemma 6.4.3 implies that

$$\|\nabla \mathbf{u}\|_\infty \leq C_2 \|\nabla^3 \mathbf{u}\|^{\frac{1}{2}}\|\nabla \mathbf{u}\|^{\frac{1}{2}} + C_3 \quad (n = 2),$$

$$\|\Delta_M \mathbf{u}\| \leq C_4 \|\nabla^3 \mathbf{u}\|^{\frac{1}{2}}\|\nabla \mathbf{u}\|^{\frac{1}{2}} + C_5, \quad (6.4.14)$$

where constants C_3 and C_5 depend on $\|\nabla \mathbf{u}_0\|$.

Substituting (6.4.14) into (6.4.12), we get

$$\left|\int_M \Delta_M \mathbf{u}.\Delta_M(\mathbf{u} \times \Delta_M \mathbf{u})dM\right| \leq 2C_1 C_2 C_4 C_6 \|\nabla \mathbf{u}\|\|\nabla^3 \mathbf{u}\|^2 + C_6'. \quad (6.4.15)$$

Here

$$\|\nabla \Delta_M \mathbf{u}\| \leq C_6 \|\nabla^3 \mathbf{u}\|$$

and constant C_6' depends on $\|\nabla \mathbf{u}_0\|$.

Now we are going to estimate term $(\Delta_M(|\nabla u|u), \Delta_M u)$ in (6.4.11).

$$\left|(\Delta_M u, \Delta_M(|\nabla u|^2 u))\right|$$

$$= \left|\int_M \Delta_M u . \frac{\partial}{\partial x^\beta}\gamma^{\alpha\beta}\sqrt{\gamma}\frac{\partial|\nabla u|^2 u}{\partial x^\alpha}dx\right|$$

$$= \left|\int_M \gamma^{\alpha\beta}\sqrt{\gamma}\frac{\partial}{\partial x^\alpha}|\nabla u|^2 u . \frac{\partial}{\partial x^\beta}\Delta_M u dx\right|$$

$$= \left|\int_M \gamma^{\alpha\beta}\sqrt{\gamma}\left(\left(\frac{\partial}{\partial x^\alpha}|\nabla u|^2\right)u + |\nabla u|^2\frac{\partial u}{\partial x^\alpha}\right)\frac{\partial}{\partial x^\beta}\Delta_M u dx\right|$$

$$= \left|\int \gamma^{\alpha\beta}\left[2\gamma^{l\delta}(x)\frac{\partial^2 u}{\partial x^\alpha \partial x^l}\frac{\partial u}{\partial x^\delta}u\right.\right.$$

$$\left.\left.+ \gamma^{l\delta}(x)'\frac{\partial u}{\partial x^l}\frac{\partial u}{\partial x^\delta}u + |\nabla u|^2\frac{\partial u}{\partial x^\alpha}\right]\frac{\partial\Delta_M u}{\partial x^\beta}dM\right|$$

$$\leq C_7\left[\|\nabla u\|_\infty\|u\|_\infty\|\nabla^2 u\|\|\nabla\Delta_M u\| + \|\nabla u\|^2\|u\|_\infty\|\nabla\Delta_M u\|\right.$$

$$\left.+ \|\nabla u\|_\infty^2\|\nabla u\|\|\nabla\Delta_M u\|\right]$$

$$\leq C_7\left[2C_2 C_4\|\nabla u\| + (2C_2^2 + 1)\|\nabla u\|^2\right]\|\nabla^3 u\| + C_8, \tag{6.4.16}$$

where constants C_7 and C_8 depend on $\|\nabla u_0\|$ and $\sup_{x\in M}(|\gamma^{\alpha\beta}(x)|, |(\gamma^{\alpha\beta}(x))'|)$.
Hence from (6.4.11), (6.4.15) and (6.4.16), we have

$$\frac{1}{2}\frac{d}{dt}t\|\Delta_M u\|^2 - \frac{1}{2}\|\Delta_M u\|^2 + \alpha_1 t\|\nabla\Delta_M u\|^2$$

$$\leq 2t\left[|\alpha_2|C_1 C_2 C_4 C_6 + \alpha_1 C_7 C_2 C_4\right.$$

$$\left.+ \alpha_1(C_2^2 + 1)C_7\|\nabla u\|\right]\|\nabla u\|\|\nabla^3 u\| + C_9. \tag{6.4.17}$$

Now we estimate the lower bound for $\|\nabla\Delta_M u\|^2$.
By the Ricci formula

$$\Delta(\nabla^k f) = \nabla^k(\Delta f) + \sum_{i=0}^{k-1} S_{ki}(\nabla^{k-i}f), \tag{6.4.18}$$

where S_{ki} is a linear functional depending on the covariant derivative $\nabla^i R$ of the curvature tensor, we get

$$\|\nabla\Delta_M u\|_2 = \|\Delta_M(\nabla u) - S_{10}(\nabla u)\|$$

$$\geq \|\Delta_M(\nabla u)\| - S_{10}\|\nabla u\|. \tag{6.4.19}$$

Since

$$\|\Delta_M(\nabla \mathbf{u})\|^2 = \int_M \sum_j \left(\sum_i u_{jii}\right)\left(\sum_k u_{jkk}\right) dM$$

$$= \sum_j \int_M \left(\sum_i u_{jii}\right)\left(\sum_k u_{jkk}\right) dM$$

$$= -\sum_j \int_M \left(\sum_i u_{ji}\right)\left(\sum_k u_{jkki}\right) dM$$

$$= -\sum_j \int_M \left(\sum_i u_{ji}\right)\sum_k (u_{jkik} + u_{jl}R^l_{kki} + u_{lk}R^l_{jki}) dM$$

$$= -\sum_j \int_M \left(\sum_i u_{ji}\right)\sum_k (u_{jikk} + u_l R^l_{jkik}$$

$$+ u_{lk}R^l_{jki} + u_{jl}R^l_{kki} + u_{lk}R^i_{jki}) dM$$

$$\geq \int_M \left|\nabla^3 \mathbf{u}\right|^2 dM - C_{11} \int_M |\nabla \mathbf{u}|^2 dM. \tag{6.4.20}$$

Hence from (6.4.17), (6.4.19) and (6.4.20), we have

$$\frac{d}{dt}t\|\Delta_M \mathbf{u}\|^2 - \|\Delta_M \mathbf{u}\|^2 + 2t[\alpha_1 - (2\alpha_2 C_1 C_2 C_4 C_6 + 2\alpha_1 C_7 C_2 C_4$$

$$+ 2\alpha_1(C_2^2 + 1)C_7\|\nabla \mathbf{u}\| + C_{12}\|\nabla \mathbf{u}\|)]\|\nabla \mathbf{u}\|\left\|\nabla^3 \mathbf{u}\right\|^2 \leq C_{13}. \tag{6.4.21}$$

Choose $\|\nabla \mathbf{u}_0\|$ suitably small such that

$$\alpha_1 - (2|\alpha_2|C_1 C_2 C_4 C_6 + 2\alpha_1 C_7 C_2 C_4 + 2\alpha_1(C_2^2 + 1)C_7\|\nabla \mathbf{u}_0\|$$

$$+ C_{12}\|\nabla \mathbf{u}_0\|)\|\nabla \mathbf{u}_0\| > \frac{\alpha}{2} > 0.$$

From (6.4.21), we get

$$t\|\Delta_M \mathbf{u}\|^2 - \int_0^t \|\Delta_M \mathbf{u}\|^2 dt + \frac{1}{2}\alpha \int_0^t t\left\|\Delta_M^3 \mathbf{u}\right\|^2 dt \leq C_{13}t. \tag{6.4.22}$$

In order to estimate term $\int_0^t \|\Delta_M \mathbf{u}\|^2 dt$ on the left-hand side of inequality (6.4.22), we need the following lemma.

Lemma 6.4.5 *Under the conditions of Lemma 6.4.2, we have*

$$\int_0^t \left\|\frac{\partial \mathbf{u}}{\partial t}\right\|^2 dt \leq \frac{(\alpha_1^2 + \alpha_2^2)}{\alpha_1}\|\nabla \mathbf{u}_0\|^2, \quad \forall\ t \in R^+, \tag{6.4.23}$$

$$\int_0^t \|\Delta_M \mathbf{u}\|^2 dt \leq C_{14}, \quad 0 \leq t \leq T, \tag{6.4.24}$$

where constant C_{14} depends on $\|\nabla \mathbf{u}_0\|$ and T.

Proof. Multiplying (6.4.3) by $\partial_t\mathbf{u}$ and integrating with respect to x, $t \in M \times [0, t)$, we have

$$\int_0^t \int_M |\partial_t\mathbf{u}|^2 dMdt + \frac{\alpha_1}{2} \int_0^t \frac{d}{dt}\|\nabla\mathbf{u}\|^2 dt$$

$$- \alpha_2 \int_0^t \int_M \partial_t\mathbf{u}.(\mathbf{u} \times \Delta_M\mathbf{u})dMdt = 0. \tag{6.4.25}$$

By the cross product of Eq. (6.4.3) with $\mathbf{u}$, we get

$$\mathbf{u} \times \partial_t\mathbf{u} = \alpha_1(\mathbf{u} \times (\Delta_M\mathbf{u})) + \alpha_2(\mathbf{u} \times (\mathbf{u} \times \Delta_M\mathbf{u}))$$

$$= \alpha_1(\mathbf{u} \times \Delta_M\mathbf{u}) - \alpha_2\Delta_M\mathbf{u} - \alpha_2|\nabla\mathbf{u}|^2\mathbf{u}.$$

Because

$$-\Delta_M\mathbf{u} - |\nabla\mathbf{u}|^2\mathbf{u} = -\frac{1}{\alpha_1}\partial_t\mathbf{u} + \frac{\alpha_2}{\alpha_1}(\mathbf{u} \times \Delta_M\mathbf{u}).$$

This gives

$$\mathbf{u} \times \partial_t\mathbf{u} + \frac{\alpha_2}{\alpha_1}\partial_t\mathbf{u} = \left(\alpha_1 + \frac{\alpha_2^2}{\alpha_1}\right)(\mathbf{u} \times \Delta_M\mathbf{u}).$$

Multiplying the above equation with $\partial_t\mathbf{u}$, we have

$$\partial_t\mathbf{u}.(\mathbf{u} \times \Delta_M\mathbf{u}) = \alpha_2(\alpha_1^2 + \alpha_2^2)^{-1}|\partial_t\mathbf{u}|^2.$$

Then we have

$$\alpha_2 \int_0^t \int_M \partial_t\mathbf{u}.(\mathbf{u} \times \Delta_M\mathbf{u})dMdt = \frac{\alpha_2^2}{\alpha_1^2 + \alpha_2^2} \int_0^t \int_M |\partial_t\mathbf{u}|^2 dMdt.$$

From (6.4.25), it follows that

$$\frac{\alpha_1^2}{\alpha_1^2 + \alpha_2^2} \int_0^t \int_M |\partial_t\mathbf{u}|^2 dMdt + \frac{\alpha_1}{2}(\|\nabla\mathbf{u}(.,t)\|^2 - \|\nabla\mathbf{u}_0\|^2) = 0,$$

i.e.,

$$\int_0^t \int_M |\partial_t\mathbf{u}|^2 dMdt \leq \frac{\alpha_1^2 + \alpha_2^2}{2\alpha_1}\|\nabla\mathbf{u}_0\|^2, \quad \forall\, t \in R^+. \tag{6.4.26}$$

From (6.4.3), we have

$$\int_0^t \|\Delta_M\mathbf{u}\|^2 dt \leq C_{15}\left(\int_0^t \|\mathbf{u}_t\|^2 dt + \int_0^t \|\mathbf{u} \times \Delta_M\mathbf{u}\|^2 dt + \int_0^t \int_M |\nabla\mathbf{u}|^4 dMdt\right). \tag{6.4.27}$$

By Sobolev's inequality (6.4.7), it follows that

$$\int_M |\nabla\mathbf{u}|^4 dM \leq C_{16}\|\nabla^2\mathbf{u}\|^2\|\nabla\mathbf{u}\|^2 + C_{17}, \quad 0 \leq t \leq T, \tag{6.4.28}$$

where constant C_{17} depends on $\|\nabla\mathbf{u}_0\|$ and $0 \leq t \leq T$.

The definition of Laplace–Beltrami operator implies

$$C_{18}\left\|\nabla^2\mathbf{u}\right\|^2 + C_{19}\|\nabla\mathbf{u}\|^2 \geq \|\Delta_M\mathbf{u}\|^2 \geq C_{18}\left\|\nabla^2\mathbf{u}\right\|^2 - C_{19}\|\nabla\mathbf{u}\|^2, \qquad (6.4.29)$$

where constants C_{18} and C_{19} depend on $\sup_{x\in M}\left|\gamma^{\alpha\beta}(x)\right|$ and $\sup_{x\in M}\left|\Gamma^k_{\alpha\beta}\right|$, respectively. Therefore, from (6.4.27), we get

$$C_{18}\int_0^t\left\|\nabla^2\mathbf{u}\right\|^2 dt - C_{15}C_{16}\int_0^t\left\|\nabla^2\mathbf{u}\right\|^2 dt.\|\nabla\mathbf{u}_0\|^2$$

$$\leq C_{15}\left(\int_0^t\left\|\frac{\partial\mathbf{u}}{\partial t}\right\|^2 dt + \int_0^t\|\mathbf{u}\times\Delta_M\mathbf{u}\|^2 dt + C_{17}\right) + C_{19}\|\nabla\mathbf{u}_0\|^2 t. \qquad (6.4.30)$$

Choose $\|\nabla\mathbf{u}_0\|$ suitably small such that

$$C_{18} - C_{15}C_{16}\|\nabla\mathbf{u}_0\|^2 \geq \frac{C_{18}}{2}. \qquad (6.4.31)$$

From (6.4.27)–(6.4.30), we have

$$\int_0^t\|\Delta_M\mathbf{u}\|^2 dt \leq C_{14}, \quad 0\leq t\leq T. \qquad (6.4.32)$$

The lemma is proved.

From (6.4.22) and (6.4.32), we get

$$\|\Delta_M\mathbf{u}\|^2 \leq \frac{E_1}{t}, \quad \forall x = (x_1, x_2, \ldots, x_n), \quad n\leq 2, \quad t > 0,$$

where constant E_1 depends only on $\|\nabla\mathbf{u}_0\|$ and T.

Now we prove that inequality (6.4.10) holds in the case $n = 1$ without the restriction $\|\nabla\mathbf{u}_0\| \leq \lambda$. In fact, from Lemma 6.4.3, we have

$$\|\nabla\mathbf{u}\|_\infty \leq C_2\left\|\nabla^3\mathbf{u}\right\|^{\frac{1}{4}}\|\nabla\mathbf{u}\|^{\frac{3}{4}} + C_3,$$

$$\|\Delta_M\mathbf{u}\| \leq C_4\left\|\nabla^3\mathbf{u}\right\|^{\frac{1}{2}}\|\nabla\mathbf{u}\|^{\frac{1}{2}} + C_5 \quad (n = 1),$$

$$\int_M |\nabla\mathbf{u}|^4 dM \leq C_{16}\left\|\nabla^2\mathbf{u}\right\|\|\nabla\mathbf{u}\|^3 + C_{17}.$$

The conclusion of inequality (6.4.10) can be easily obtained.

Using the estimates in [77], we have

Theorem 6.4.1 *Let M be a compact manifold without boundary, and suppose that the following conditions are satisfied:*

(1) $\mathbf{u}_0(x) \in H^m(M)$, $m \geq 2$, $|\mathbf{u}_0(x)|^2 = 1$,

$$x = (x_1, \ldots, x_n) \in M, \quad 1\leq n\leq 2,$$

(2) In the case of $n = 2$,

$$\|\nabla\mathbf{u}_0(x)\| \leq \lambda,$$

where constant λ is suitably small. Then there exists a unique global smooth solution $\mathbf{u}(x,t) : M \times [0, \infty) \to S^2$ of the initial value problem (6.4.1) and (6.4.2)

$$\mathbf{u}(x,t) \in L^\infty(0, \infty; H^m(M)).$$

6.4.3   The Global Attractor

Theorem 6.4.2 *Let M be an m-dimensional compact Riemannian manifold without boundary ($n \leq 2$). If conditions*

(i) $\alpha_1 > 0$, $|\mathbf{Z}_0(x)| = 1$, $\mathbf{Z}_0(x) \in H^1(M)$, $x = (x^1, \ldots, x^n) \in M$, $n \leq 2$;

(ii) $\|\nabla \mathbf{Z}_0(x)\| \leq \lambda$, $x \in M$, $n = 2$, *where constant λ is suitably small, are satisfied, then the initial value problem (6.4.1) and (6.4.2) of the Landau–Lifshitz equation has an attractor A which is compact in $H^1(M)$, where*

$$A = \omega(B^1) = \bigcap_{s \geq 0} \bigcap_{t \geq s} \overline{S(t)B_1}. \tag{6.4.33}$$

Here

$$B_1 = \{\mathbf{Z} \in H^1(M),\ |\mathbf{Z}| = 1,\ \|\mathbf{Z}\|_{H^1(M)} \leq \rho_1\} \tag{6.4.34}$$

is a bounded absorbing set for $S(t)$ in $H^1(M)$, and ρ_1 is a positive constant.

Proof. From Theorem 6.4.1, there exists a unique global smooth solution

$$\mathbf{u}(x, t) \in L^\infty(0, \infty; H^m(M))$$

of the initial value problem (6.4.1) and (6.4.2), which means that problem (6.4.1) and (6.4.2) generates a semigroup $S(t) : \mathbf{Z}(x, t) = S(t)\mathbf{Z}_0$.

From Lemma 6.4.1 and Lemma 6.4.2, we know

$$B_1 = \{\mathbf{Z} \in H^1(M),\ |\mathbf{Z}| = 1,\ \|\mathbf{Z}\|_{H^1(M)} \leq \|\nabla \mathbf{Z}_0(x)\| + \mathrm{Vol}\, M = \rho_1\}$$

is a bounded absorbing set for $S(t)$ in $H^1(M)$. From Lemma 6.4.4, we have

$$\|\mathbf{Z}(., t)\|_{H^2(M)} \leq \frac{E_1}{t}, \quad t > 0,$$

where constant E_1 depends only on $\|\nabla \mathbf{Z}_0(x)\|_{H^1(M)}$. This implies that semigroup operator $S(t)$ is a completely continuous operator in $H^1(M)$ for $t > 0$. From Theorem 6.2, the semigroup operator $S(t)$ has a compact attractor

$$A = \bigcap_{s \geq 0} \bigcap_{t \geq s} \overline{S(t)B_1} = \omega(B_1).$$

6.4.4   The Estimates of the Upper Bounds of Hausdorff and Fractal Dimensions for the Attractors

In the following, we give the estimates of the upper bounds of the Hausdorff and fractal dimensions for the attractors of the mentioned problems in the above section.

We consider the linear variation corresponding to problem (6.4.1) and (6.4.2)

$$\mathbf{v}(t) + L(\mathbf{u}(t))\mathbf{v} = \mathbf{0}, \tag{6.4.35}$$

$$\mathbf{v}(0) = \mathbf{v}_0(x), \tag{6.4.36}$$

where

$$L(\mathbf{u}(t))\mathbf{v} = -\alpha_1 \Delta_M \mathbf{v} - \alpha_1 |\nabla \mathbf{u}|^2 \mathbf{v} - 2\alpha_1 \nabla \mathbf{u} \mathbf{u} \nabla \mathbf{v}$$
$$- \alpha_2 A'(\mathbf{u}) \Delta_M \mathbf{u} \mathbf{v} - \alpha_2 A(\mathbf{u}) \Delta_M \mathbf{v}. \qquad (6.4.37)$$

Since solution $\mathbf{u}$ of problem (6.4.1) and (6.4.2) is suitably smooth, we can prove that the linear problem (6.4.35) and (6.4.36) has a global solution $\mathbf{v}(x,t) \in L^\infty(0,\infty; H^2(M))$ as long as the initial data $\mathbf{v}_0(x)$ is mildly smooth.

In fact, for Eq. (6.4.35), the principle part is given by

$$\sum_{\alpha,\beta=1}^{m} D_\alpha(a_{\alpha\beta}(x,t,\mathbf{u})D_\beta \mathbf{u}) = \sum_{\alpha,\beta=1}^{m} D_\alpha \sqrt{\gamma}(\gamma^{\alpha\beta} D_\beta \mathbf{u} + \mathbf{u} \times \gamma^{\alpha\beta} D_\beta \mathbf{u}).$$

The corresponding coefficients $a_{\alpha,\beta}^{ij}$ are given by

$$a_{\alpha,\beta}^{ij} = \sqrt{\gamma}\gamma^{\alpha\beta} g_{ij}.$$

Here

$$g = (g_{ij}) = \begin{pmatrix} \alpha_1 & -\alpha_2 u^3 & \alpha_2 u^2 \\ \alpha_2 u^3 & \alpha_1 & -\alpha_2 u^1 \\ -\alpha_2 u^2 & \alpha_2 u^1 & \alpha_1 \end{pmatrix},$$
$$\mathbf{u} = (u^1, u^2, u^3).$$

Therefore, the Landau–Lifshitz equation (6.4.1) is strongly parabolic by the following relation:

$$\sum_{i,j=1}^{3} \sum_{\alpha,\beta=1}^{m} a_{\alpha,\beta}^{ij}(x,t,\eta)\xi^\alpha\xi^\beta\zeta_i\zeta_j = \alpha_1|\zeta| \sum_{\alpha,\beta=1}^{m} \gamma^{\alpha\beta}\sqrt{\gamma}\xi^\alpha\xi^\beta > 0$$

for all $(x,t,\eta) \in M \times [0,T] \times R^3$, for all $\xi = (\xi^1,\ldots,\xi^m) \in R^m \backslash \{0\}$, and for all $\zeta = (\zeta_1, \zeta_2, \zeta_3) \in R^3 \backslash \{0\}$.

Let G_t be a solution operator of (6.4.1) and (6.4.2) such that $\mathbf{v}(t) = G_t \mathbf{v}_0$.

It can be verified that the semigroup operator $S_t \mathbf{u}_0$ is differentiated in $L_2(M)$, namely the Frechet derivative $S_t \mathbf{u}_0$ exists, and $G_t \mathbf{v}_0 = S_t' \mathbf{u}_0$. In fact, we set

$$W(t) = S_t(\mathbf{u}_0 + \mathbf{v}_0) - S_t(\mathbf{u}_0) - G_t(\mathbf{u}_0)\mathbf{v}_0$$
$$= \mathbf{w}_1(t) - \mathbf{u}(t) - \mathbf{v}(t).$$

Thus we have

$$\partial_t W(t) = L_1(\mathbf{u}_1(t)) - L_1(\mathbf{u}(t)) + L(\mathbf{u}(t))\mathbf{v}(t)$$
$$= L_1(\mathbf{u}(t) + \mathbf{v}(t) + \mathbf{w}(t)) - L_1(\mathbf{u}(t)) + L(\mathbf{u}(t))\mathbf{v}(t), \qquad (6.4.38)$$
$$\mathbf{w}(0) = \mathbf{0},$$

where $\mathbf{u}_t = L_1(\mathbf{u})$ is the operator form of Eq. (6.4.3).

Therefore, (6.4.38) can be rewritten in the form

$$\partial_t \mathbf{w}(t) + L(\mathbf{u}(t))\mathbf{w}(t) = \Lambda_0(\mathbf{u}, \mathbf{v}, \mathbf{w}), \tag{6.4.39}$$

where

$$\Lambda_0(\mathbf{u}, \mathbf{v}, \mathbf{w}) = L_1(\mathbf{u}(t) + \mathbf{v}(t) + \mathbf{w}(t)) - L_1(\mathbf{u}(t)) + L(\mathbf{u}(t))(\mathbf{v} + \mathbf{w}). \tag{6.4.40}$$

By applying the theorem of linear parabolic partial differential equations, we have the following L_2 estimates:

$$\|\mathbf{w}(t)\|^2 \le C\|\mathbf{v}_0\|^2. \tag{6.4.41}$$

This implies that the semigroup operator $S(t)$ is differentiated in $L_2(M)$.

Denote by $V_1(t), V_2(t), \ldots, V_J(t)$ the solution of linear equation (6.4.36) corresponding, respectively, to the initial data $V_1(0) = \xi_1, \ldots, V_J(0) = \xi_J$; here $\xi = (\xi_1, \xi_2, \ldots, \xi_J) \in L_2(M)$, and by the simple computation [15], we can deduce that

$$\frac{d}{dt}\|V_1(t) \wedge V_2(t) \wedge \cdots \wedge V_J(t)\|^2 + 2\operatorname{Tr}(L(\mathbf{u}(t)) \cdot Q_J(t))$$

$$\cdot \|V_1(t) \wedge V_2(t) \wedge \cdots \wedge V_J(t)\|^2 = 0, \tag{6.4.42}$$

where $L(\mathbf{u}(t)) = L(S(t)\mathbf{u}_0)$ is a linear map: $\mathbf{v} \to L(\mathbf{u}(t))\mathbf{v}$, "$\wedge$" denotes the exterior product, "Tr," the trace of operator, and $Q_J(t)$ is the orthogonal projection of space $L_2(M)$ to the spanning subspace generated by $V_1(t), V_2(t), \ldots, V_J(t)$. Therefore, from (6.4.42), we can obtain the change of volume $\wedge_{j=0}^{J}\xi$ of the J-dimensional cube by

$$W_J(t) = \sup_{\mathbf{u}_0 \in A} \ \sup_{\xi_j \in L_2(M), |\xi_j| \le 1} \|V_1(t) \wedge V_2(t) \wedge \cdots \wedge V_J(t)\|^2_{\wedge^j L_2}$$

$$\le \sup_{\mathbf{u}_0 \in A} \exp\left(\int_0^t \inf(\operatorname{Tr}(L(S(\tau)\mathbf{u}_0))(Q_J(\tau))d\tau\right). \tag{6.4.43}$$

Lemma 6.4.6 *Let (M, g) be a Riemannian manifold of dimension n. For every p with*

$$\max\left\{1, \frac{n}{2m}\right\} < p \le 1 + \frac{n}{2m},$$

there exist two positive constants $K(M)$ and $\chi(M)$, such that for every finite family $\{\varphi_1, \ldots, \varphi_m\}$ in $H^m(M)$, which is (sub-orthonormal) orthonormal in $L^2(M)$, we have

$$\left(\int_M \rho^{\frac{p}{p-1}} dM\right)^{\frac{2m(p-1)}{n}} \le k(M) \sum_{j=1}^{N} \int_M |\nabla^m \varphi_j|^2 dM$$

$$+ \chi(M) \int_M \rho \, dM. \tag{6.4.44}$$

Here $\rho = \sum_{j=1}^{N} |\varphi_j(x)|^2$, and constants $k(M)$ and $\chi(M)$ depend on m, n, p and (M, g).

Lemma 6.4.7 *Let M be a compact Riemannian manifold without (or with) boundary, and $\{\lambda_j\}$ is the spectrum of M with respect to Laplace–Beltrami operator. Then the following inequalities hold*

$$\lambda_k \geq \frac{\delta}{e}\left(\frac{k}{\text{Vol}(M)}\right)^{\frac{2}{n}},$$

where $n = \dim M$ and δ is the Sobolev constant of M, that is, for any $\mathbf{u} \in C^\infty(M)$

$$\int_M |\nabla \mathbf{u}|^2 dM \geq \delta \left(\int_M |\mathbf{u}|^{\frac{2n}{n-2}}\right)^{\frac{n-2}{n}}.$$

Theorem 6.4.3 *Under the conditions of Theorem 6.4.6, the Hausdorff and fractal dimensions of the global attractor A of the problem (6.4.1) and (6.4.2) are finite, and*

$$d_H(A) \leq J_0, \quad d_F(A) \leq 2J_0, \tag{6.4.45}$$

where J_0 is the smallest integer, such that

$$J_0 \geq C_0 \alpha_1^{\frac{-4m}{(4-m)(2+m)}}, \quad (1 \leq m \leq 2), \tag{6.4.46}$$

where C_0 is a constant that depends on M and on the norms $\|\mathbf{u}\|_\infty$, $\|\nabla\mathbf{u}\|_\infty$ and $\|\nabla^2\mathbf{u}\|_2$.

Proof. Suppose that $\{\varphi_1(x),\ldots,\varphi_J(x)\}$ is an orthogonal basis of the subspace $Q_J L_2$, we have

$$\text{Trace}(L(\mathbf{u}(t))\cdot Q_J) = \sum_{j=1}^\infty (L(\mathbf{u}(t))\cdot Q_J(\tau)\varphi_j, \varphi_j)$$

$$= \sum_{j=1}^J (L(\mathbf{u}(t))\varphi_j(\tau), \varphi_j(\tau))$$

$$= \sum_{j=1}^J \{-\alpha_1(\Delta_M\varphi_j, \varphi_j) - \alpha_1(|\nabla\mathbf{u}|^2\varphi_j, \varphi_j)$$

$$\quad - 2\alpha_1(\nabla\mathbf{u}\cdot\mathbf{u}\cdot\nabla\varphi_j, \varphi_j) - \alpha_2(A'(\mathbf{u})\nabla\mathbf{u}\varphi_j, \varphi_j)$$

$$\quad - \alpha_2(A(\mathbf{u})\cdot\Delta_M\varphi_j, \varphi_j)\}$$

$$\geq \alpha_1\sum_{j=1}^J \lambda_j - \alpha_1\left\||\nabla\mathbf{u}|^2\right\|_2\|\rho(x)\|_2$$

$$\quad - 2\alpha_1\|\nabla(\nabla\mathbf{u}\cdot\mathbf{u})\|_2\|\rho(x)\|_2 - \alpha_2\|A'(\mathbf{u})\|_\infty\|\nabla\mathbf{u}\|_2\|\rho(x)\|_2$$

$$\geq \alpha_1\sum_{j=1}^J \lambda_j - 2\alpha_1\|\nabla\mathbf{u}\|_\infty\|\nabla\mathbf{u}\|_2\|\rho(x)\|_2$$

$$\quad - 2\alpha_1\|\mathbf{u}\|_\infty\left\|\nabla^2\mathbf{u}\right\|_2\|\rho(x)\|_2 - \alpha_2\|A'(\mathbf{u})\|_\infty\|\nabla\mathbf{u}\|_2\|\rho(x)\|_2$$

$$\geq \alpha_1 \sum_{j=1}^{J} \lambda_j - \left(2\alpha_1 \|\nabla \mathbf{u}\|_\infty \|\nabla \mathbf{u}\|_2 + 2\alpha_1 \|\mathbf{u}\|_\infty \|\nabla^2 \mathbf{u}\|_2 \right.$$

$$\left. + \alpha_2 \|A'(\mathbf{u})\|_\infty \|\nabla \mathbf{u}\|_2 \right) \|\rho(x)\|_2$$

$$\geq \alpha_1 \sum_{j=1}^{J} \lambda_j - H \left(k(M) \sum_{j=1}^{J} \lambda_j + \chi(M) \right)^{\frac{m}{4}}$$

$$\geq \alpha_1 \left(\sum_{j=1}^{J} \lambda_j \right)^{\frac{m}{4}} \cdot \left[\left(\sum_{j=1}^{J} \lambda_j \right)^{1-\frac{m}{4}} - \frac{C_1}{\alpha_1} H \right] - C_2 H$$

$$\geq \alpha_1 \left(\sum_{j=1}^{J} \lambda_j \right)^{\frac{m}{4}} - C_2 H > 0,$$

where

$$H = 2\alpha_1 \|\nabla \mathbf{u}\|_\infty \|\nabla \mathbf{u}\|_\infty \|\nabla \mathbf{u}\|_2 + 2\alpha_1 \|\mathbf{u}\|_\infty \|\nabla^2 \mathbf{u}\|_2 + \alpha_2 \|A'(\mathbf{u})\|_\infty \|\nabla \mathbf{u}\|_2).$$

As

$$J \geq \mathrm{Max}\left\{ \left(C_3 \left(\frac{2}{m}\right) \right)^{\frac{m}{2+m}} \cdot \left[\frac{C_1}{\alpha_1} H + 1 \right]^{\frac{4m}{(4-m)(2+m)}}, \; \left(C_3 \left(\frac{2}{m}\right) \right)^{\frac{m}{2+m}} \left[\frac{C_2}{\alpha_1} H \right]^{2+\frac{4}{m}} \right\}.$$

Here we have used the following inequality [49]:

$$\sum_{j=1}^{J} j^{\frac{2}{m}} \geq C_3 \left(\frac{2}{m}\right) J^{\frac{2}{m}+1},$$

and C_1, C_2 are constants which depend on the manifold M.

Now we consider the initial-boundary value problem for the following Landau–Lifshitz equation

$$\mathbf{Z}_t = -\alpha_1 (\mathbf{Z} \times (\mathbf{Z} \times \Delta_M \mathbf{Z})) + \mathbf{Z} \times J\mathbf{Z}, \tag{6.4.47}$$

$$\mathbf{Z}|_{t=0} = \mathbf{Z}_0(x), \quad x \in M, \tag{6.4.48}$$

$$\frac{\partial \mathbf{Z}}{\partial \nu}\bigg|_{\partial M} = 0, \tag{6.4.49}$$

where $J = \mathrm{diag}(J_1, J_2, 0)$, $\alpha_1 > 0$, and ν is the outer unit normal vector to ∂M.

Theorem 6.4.4 *Assume that A is the global attractor of problem* (6.4.47)–(6.4.49), *and $J_1 J_2 < 0$.*

Then we have

$$\dim A \geq C \alpha_1^{-\frac{n}{2}}, \tag{6.4.50}$$

where $\dim A$ is the Hausdorff or fractal dimension of A and C is a positive constant.

Proof. It is easy to find out that $\mathbf{Z} = (0, 0, 1)$ is a fixed point of semigroup $S(t)$ generated by the problem (6.4.47)–(6.4.49), i.e., $\mathbf{Z}(0, 0, 1)$ satisfies the equation

$$-A_1(\mathbf{Z}) = -\alpha_1(\mathbf{Z} \times (\mathbf{Z} \times \Delta_M \mathbf{Z})) + \mathbf{Z} \times J\mathbf{Z} = 0. \tag{6.4.51}$$

The linear variation equation $A_1'(Z)$ for Eq. (6.4.51) is as follows:

$$-A_1'(Z)\mathbf{v} = \alpha_1 \Delta_M \mathbf{v} + \mathbf{Z} \times J\mathbf{v} = 0. \tag{6.4.52}$$

Let ζ denote the eigenvalues associated with the matrix $B(\mathbf{Z})\mathbf{v} = \mathbf{Z} \times J\mathbf{v} = \zeta \mathbf{v}$, where

$$B(\mathbf{Z}) = \begin{pmatrix} 0 & -J_2 & 0 \\ J_1 & 0 & 0 \\ 0 & 0 & 0 \end{pmatrix}.$$

Therefore,

$$\zeta^2 + J_1 J_2 = 0,$$
$$\zeta_1 = \sqrt{-J_1 J_2} > 0, \quad \zeta_2 = -\sqrt{-J_1 J_2}.$$

Let λ_k, $k \in N$ denote the eigenvalues associated with Neumann problem

$$-\Delta_M \psi_k = \lambda_k \psi_k, \quad \left.\frac{\partial \psi_k}{\partial \nu}\right|_{\partial M} = 0, \tag{6.4.53}$$

$$0 < \alpha_1 \leq \lambda_2 \leq \cdots, \quad \lambda_k \to \infty, \quad k \to \infty.$$

Let μ_k, $k \in N$ denote the sequence of eigenvalues of linearized operator

$$-A'(\mathbf{Z})\mathbf{w_k} = \alpha_1 \Delta_M \mathbf{w_k} + B(\mathbf{Z})\mathbf{w_k} = \mu_k \mathbf{w_k}. \tag{6.4.54}$$

Here $\mathbf{w_k}(x) = \psi_k(x)p_k$, $p_k \in C^3$, that is,

$$(\alpha_1 \lambda_k + B(\mathbf{Z}))p_k = \mu_k p_k. \tag{6.4.55}$$

If μ_k is roots of equation

$$\det(\alpha_1 \lambda_k + B(\mathbf{Z}) - \mu_k I) = 0, \quad \mathrm{Re}\,\mu_k > 0, \tag{6.4.56}$$

then there exists the nonzero solution p_k. Under the assumptions of the theorem, when $\alpha_1 = 0$, there exists at least one root $\zeta_1 = \sqrt{-J_1 J_2} > 0$. Thus we have a root μ_k of Eq. (6.4.56), $\mathrm{Re}\,\mu_k > 0$, as $\alpha_1 \lambda_k < \delta$, where δ is a suitable small constant. From the asymptotic behavior of eigenvalues $\lambda_k \sim Ck^{\frac{2}{n}}$, we get the inequality

$$1 \leq k \leq C_1 \delta^{\frac{n}{2}} \alpha_1^{-\frac{n}{2}} = C_2 \alpha_1^{-\frac{n}{2}}.$$

Therefore, we have

$$\dim A \geq C\alpha_1^{-\frac{n}{2}}.$$

6.5   The Attractors for Landau–Lifshitz–Maxwell Equations

6.5.1   Periodic Initial Value Problem to L–L–M Systems

Consider the following periodic initial value problem for L–L–M system

$$\vec{Z}_t = \alpha_1 \vec{Z} \times (\Delta \vec{Z} + \vec{H}) - \alpha_2 \vec{Z} \times (\vec{Z} \times (\Delta \vec{Z} + \vec{H})), \tag{6.5.1}$$

$$\nabla \times \vec{H} = \frac{\partial \vec{E}}{\partial t} + \sigma \vec{E}, \tag{6.5.2}$$

$$\nabla \times \vec{E} = -\frac{\partial \vec{H}}{\partial t} - \beta \frac{\partial \vec{Z}}{\partial t}, \tag{6.5.3}$$

$$\nabla \cdot \vec{H} + \beta \nabla \cdot \vec{Z} = 0, \quad \nabla \cdot \vec{E} = 0, \tag{6.5.4}$$

$$\vec{Z}(x + 2D\vec{e}_i, t) = \vec{Z}(x, t), \quad \vec{H}(x + 2D\vec{e}_i, t) = \vec{H}(x, t),$$

$$\vec{E}(x + 2D\vec{e}_i, t) = \vec{E}(x, t), \quad (x \in \Omega, t \geq 0, i = 1, 2, \ldots, n), \tag{6.5.5}$$

$$\vec{Z}(x, 0) = \vec{Z}_0(x), \quad \vec{H}(x, 0) = \vec{H}_0(x), \quad \vec{E}(x, 0) = \vec{E}_0(x), \quad (x \in \Omega), \tag{6.5.6}$$

where $\alpha_2 > 0$, $x + 2D\vec{e}_i = (x_1, \ldots, x_{i-1}, x_i + 2D, x_{i+1}, \ldots, x_n)$, $(i = 1, 2, \ldots, n)$, $D > 0$, $\Omega \subset R^n$ is n-dimensional cube with width $2D$.

Our purpose of this section is to study the existence of the attractor.

6.5.2   *A Priori* Estimates

For the sake of simplicity denote $\| \cdot \|_{L_p} = \| \cdot \|_p, p \geq 2$.

Lemma 6.5.1 *Assume* $|\vec{Z}_0(x)| = 1$. *Then for the smooth solution of the periodic initial value problem* (6.5.1)–(6.5.6), *there are*

$$|\vec{Z}(x, t)| = 1, \quad x \in \Omega, \quad t \geq 0. \tag{6.5.7}$$

Proof. Making the scalar product of $\vec{Z}$ with (6.5.1), we get

$$\frac{\partial}{\partial t}|\vec{Z}(x, t)|^2 = 0.$$

Then the conclusion of the lemma is proved.

Lemma 6.5.2 *Assume* $\alpha_2 > 0$, $\beta > 0$, $\sigma \geq 0$, $\nabla \vec{Z}_0(x) \in L_2(\Omega)$, $\vec{E}_0(x) \in L_2(\Omega)$, $\vec{H}_0(x) \in L_2(\Omega)$. *Then for the smooth solution of the periodic initial value problem* (6.5.1)–(6.5.6)

$$\sup_{0 \leq t < \infty} \left[\|\nabla \vec{Z}(\cdot, t)\|_2^2 + \|\vec{E}(\cdot, t)\|_2^2 + \|\vec{H}(\cdot, t)\|_2^2 \right] \leq K_1, \tag{6.5.8}$$

$$\int_0^\infty \|\vec{Z} \times (\Delta \vec{Z} + \vec{H})\|_2^2 dt \leq K_2, \tag{6.5.9}$$

where the constants K_1 and K_2 depend only on $\|\nabla \vec{Z}_0(x)\|_2$, $\|\vec{E}_0(x)\|_2$, and $\|\vec{H}_0(x)\|_2$.

Proof. Making the scalar product of $\vec{E}$ with (6.5.2), and making the scalar product of $-\vec{H}$ with (6.5.3), and then adding these two equalities obtained, we have

$$(\nabla \times \vec{H}) \cdot \vec{E} - (\nabla \times \vec{E}) \cdot \vec{H} = \frac{\partial \vec{E}}{\partial t} \cdot \vec{E} + \sigma |\vec{E}|^2 + \frac{\partial \vec{H}}{\partial t} \cdot \vec{H} + \beta \frac{\partial \vec{Z}}{\partial t} \cdot \vec{H}. \quad (6.5.10)$$

By using the formula

$$(\nabla \times \vec{H}) \cdot \vec{E} - (\nabla \times \vec{E}) \cdot \vec{H} = \nabla \cdot (\vec{H} \times \vec{E}), \quad (6.5.11)$$

and integrating (6.5.10) over $x \in \Omega$, we get

$$\frac{1}{2} \frac{d}{dt} \left(\|\vec{E}\|_2^2 + \|\vec{H}\|_2^2 \right) + \sigma \|\vec{E}\|_2^2 + \beta \int_\Omega \frac{\partial \vec{Z}}{\partial t} \cdot \vec{H} dx = 0. \quad (6.5.12)$$

Making the scalar product of $(\Delta \vec{Z} + \vec{H})$ with (6.5.1), we get

$$(\Delta \vec{Z} + \vec{H}) \cdot \frac{\partial \vec{Z}}{\partial t} = -\varepsilon (\Delta \vec{Z} + \vec{H}) \cdot [\vec{Z} \times (\vec{Z} \times (\Delta \vec{Z} + \vec{H}))], \quad (6.5.13)$$

where

$$-(\Delta \vec{Z} + \vec{H}) \cdot [\vec{Z} \times (\vec{Z} \times (\Delta \vec{Z} + \vec{H}))] = |\vec{Z} \times (\Delta \vec{Z} + \vec{H})|^2. \quad (6.5.14)$$

From integrating (6.5.13) over $x \in \Omega$, it follows that

$$\int_\Omega (\Delta \vec{Z} + \vec{H}) \cdot \frac{\partial \vec{Z}}{\partial t} dx = \varepsilon \int_\Omega |\vec{Z} \times (\Delta \vec{Z} + \vec{H})|^2 dx,$$

and then

$$\int_\Omega \frac{\partial \vec{Z}}{\partial t} \cdot \vec{H} dx = -\int_\Omega \Delta \vec{Z} \cdot \frac{\partial \vec{Z}}{\partial t} dx + \varepsilon \int_\Omega |\vec{Z} \times (\Delta \vec{Z} + \vec{H})|^2 dx$$

$$= \frac{1}{2} \frac{d}{dt} \|\nabla \vec{Z}\|_2^2 + \varepsilon \int_\Omega |\vec{Z} \times (\Delta \vec{Z} + \vec{H})|^2 dx. \quad (6.5.15)$$

By adding (6.5.12) and (6.5.15) we obtain

$$\frac{1}{2} \frac{d}{dt} (\|\vec{E}\|_2^2 + \|\vec{H}\|_2^2) + \sigma \|\vec{E}\|_2^2$$

$$+ \frac{\beta}{2} \frac{d}{dt} \|\nabla \vec{Z}\|_2^2 + \beta \varepsilon \|\vec{Z} \times (\Delta \vec{Z} + \vec{H})\|_2^2 = 0. \quad (6.5.16)$$

Integrating the above equality with respect to $t \in [0, T]$, we have

$$\mathcal{E}(t) \equiv \frac{1}{2} (\|\vec{E}(\cdot, t)\|_2^2 + \|\vec{H}(\cdot, t)\|_2^2) + \sigma \|\vec{E}(\cdot, t)\|_2^2$$

$$+ \frac{\beta}{2} \|\nabla \vec{Z}(\cdot, t)\|_2^2 + \beta \varepsilon \int_0^t \|\vec{Z} \times (\Delta \vec{Z} + \vec{H})\|_2^2 dt$$

$$= \mathcal{E}(0) = \frac{1}{2} (\|\vec{E}_0(x)\|_2^2 + \|\vec{H}_0(x)\|_2^2)$$

$$+ \sigma \|\vec{E}_0(x)\|_2^2 + \frac{\beta}{2} \|\nabla \vec{Z}_0(x)\|_2^2.$$

It follows that the estimates (6.5.8) and (6.5.9) hold. In order to get the uniform estimates in t for the solution $(\vec{Z}, \vec{H}, \vec{E}) \in (H^2(\Omega), H^1(\Omega), H^1(\Omega))$, we first rewrite (6.5.2) and (6.5.3) as the following equivalent second order nonlinear wave equations:

$$\nabla \times (\nabla \times \vec{H}) = \frac{\partial}{\partial t} \nabla \times \vec{E} + \sigma \nabla \times \vec{E},$$

$$\nabla \times (\nabla \times \vec{E}) = -\frac{\partial}{\partial t} \nabla \times \vec{H} - \beta \frac{\partial}{\partial t} \nabla \times \vec{Z}.$$

From the formula

$$\nabla \times (\nabla \times \vec{H}) = \nabla(\nabla \cdot \vec{H}) - \Delta \vec{H}$$
$$= -\beta \nabla(\nabla \cdot \vec{Z}) - \Delta \vec{H},$$
$$\nabla \times (\nabla \times \vec{E}) = \nabla(\nabla \cdot \vec{E}) - \Delta \vec{E} = -\Delta \vec{E},$$

we have

$$-\beta \nabla(\nabla \cdot \vec{Z}) - \Delta \vec{H} = \frac{\partial}{\partial t} \nabla \times \vec{E} + \sigma \nabla \times \vec{E}$$
$$= -\vec{H}_{tt} - \beta \vec{Z}_{tt} - \sigma \vec{H}_t - \sigma \beta \vec{Z}_t,$$

$$-\Delta \vec{E} = -\frac{\partial}{\partial t}(\nabla \times \vec{H}) - \beta \frac{\partial}{\partial t}(\nabla \times \vec{Z})$$
$$= N \vec{E}_{tt} - \sigma \vec{E} - \beta(\nabla \times \vec{Z})_t,$$

$$\vec{H}_{tt} - \Delta \vec{H} + \beta \vec{Z}_{tt} + \sigma \vec{H}_t + \sigma \beta \vec{Z}_t - \beta \nabla(\nabla \cdot \vec{Z}) = 0, \qquad (6.5.17)$$
$$\vec{E}_{tt} - \Delta \vec{E} + \sigma \vec{E}_t + \beta(\nabla \times \vec{Z})_t = 0. \qquad (6.5.18)$$

It is difficult to derive the *a priori* estimate uniform in t for $(\vec{Z}, \vec{E}, \vec{H}) \in (H^2, H^1, H^1)$ from (6.5.11), (6.5.17) and (6.5.18). We prove instead by Lyapunov functional containing small parameter method. We define the Lyapunov functional as follows:

$$e(t) = \frac{1}{2} \left(\|\vec{H}_t(\cdot, t)\|_2^2 + \|\nabla \vec{H}(\cdot, t)\|_2^2 + \|\vec{E}_t(\cdot, t)\|_2^2 + \|\nabla \vec{E}(\cdot, t)\|_2^2 + \|\Delta \vec{Z}(\cdot, t)\|_2^2 \right)$$
$$+ \eta_1(\vec{H}, \vec{H}_t) + \eta_2(\vec{E}, \vec{E}_t), \qquad (6.5.19)$$

where η_1 and η_2 are constants to be determined. We want to prove that $e(t)$ satisfies the following differential inequality:

$$\frac{de(t)}{dt} + ae(t) \leq K, \qquad (6.5.20)$$

where $a > 0$ is a constant, K is independent of t. Then, the *a priori* estimates can be derived.

In fact, it follows from (6.5.19) that

$$\frac{de(t)}{dt} = (\vec{H}_t, \vec{H}_{tt}) + (\nabla \vec{H}, \nabla \vec{H}_t) + (\vec{E}_t, \vec{E}_{tt})$$
$$+ (\nabla \vec{E}_t, \nabla \vec{E}_t) + (\Delta \vec{Z}, \Delta \vec{Z}_t) + \eta_1(\vec{H}_t, \vec{H}_t)$$
$$+ \eta_1(\vec{H}, \vec{H}_{tt}) + \eta_2(\vec{E}_t, \vec{E}_t) + \eta_2(\vec{E}, \vec{E}_{tt}), \qquad (6.5.21)$$

in which

$$(\vec{H}_t, \vec{H}_{tt}) = (\vec{H}_t, \Delta \vec{H} - \beta \vec{Z}_{tt} - \sigma \vec{H}_t - \sigma \beta \vec{Z}_t + \beta \nabla(\nabla \cdot \vec{Z}))$$
$$= (\vec{H}_t, \Delta \vec{H}) - \beta(\vec{H}_t, \vec{H}_{tt}) - \sigma(\vec{H}_t, \vec{H}_t)$$
$$- \sigma\beta(\vec{H}_t, \vec{Z}_t) + \beta(\vec{H}_t, \nabla(\nabla \cdot \vec{Z})),$$

$$(\nabla \vec{H}, \nabla \vec{H}_t) = -(\Delta \vec{H}, \vec{H}_t),$$

$$(\vec{E}_t, \vec{E}_{tt}) = (\vec{E}_t, \Delta \vec{E} - \sigma \vec{E}_t - \beta(\nabla \times \vec{Z})_t)$$
$$= (\vec{E}_t, \Delta \vec{E}) - \sigma(\vec{E}_t, \vec{H}_t) - \beta(\vec{E}_t, (\nabla \times \vec{Z})_t),$$

$$(\nabla \vec{E}, \nabla \vec{E}_t) = -(\Delta \vec{E}, \vec{E}_t),$$

$$(\Delta \vec{Z}, \Delta \vec{Z}_t) = \alpha_1(\Delta(\vec{Z} \times \Delta \vec{Z}), \Delta \vec{Z}) + \alpha_1(\Delta(\vec{Z} \times \vec{H}), \Delta \vec{Z})$$
$$+ \alpha_2(\Delta(|\nabla \vec{Z}|^2 \vec{Z}, \Delta \vec{Z}) + \alpha_2(\Delta^2 \vec{Z}, \Delta \vec{Z})$$
$$+ \alpha_2(\Delta \vec{H}, \Delta \vec{Z}) - \alpha_2(\Delta^2(\vec{Z} \cdot \vec{H})\vec{Z}, \Delta \vec{Z}),$$

$$\eta_1(\vec{H}, \vec{H}_{tt}) = \eta_1(\vec{H}, \Delta \vec{H} - \beta \vec{Z}_{tt} - \sigma \vec{H}_t - \sigma\beta \vec{Z}_t + \beta \nabla(\nabla \cdot \vec{Z}))$$
$$= -\eta_1 \|\nabla \vec{H}\|_2^2 - \beta \eta_1(\vec{H}, \vec{Z}_{tt})$$
$$- \sigma\eta_1(\vec{H}, \vec{H}_t) - \sigma\beta\eta_1(\vec{H}, \vec{Z}_t) + \eta_1\beta(\vec{H}, \nabla(\nabla \cdot \vec{Z})),$$

$$\eta_2(\vec{E}, \vec{E}_{tt}) = \eta_2(\vec{E}, \Delta \vec{E} - \sigma \vec{E}_t - \beta(\nabla \times \vec{Z})_t)$$
$$= \eta_2\|\nabla \vec{E}\|_2^2 - \sigma\eta_2(\vec{E}, \vec{E}_t) - \beta\eta_2(\vec{E}, (\nabla \times \vec{Z})_t).$$

Hence, we have

$$\frac{de(t)}{dt} = -\sigma\|\vec{H}_t\|_2^2 - \sigma\beta(\vec{H}_t, \vec{Z}_t) - \beta(\vec{H}_t, \vec{Z}_{tt})$$
$$- \sigma\|\vec{E}_t\|_2^2 - \beta(\vec{E}_t, (\nabla \times \vec{Z})_t)$$
$$+ \alpha_1(\Delta(\vec{Z} \times \Delta \vec{Z}), \Delta \vec{Z}) + \alpha_1(\Delta(\vec{Z} \times \vec{H}), \Delta \vec{Z})$$
$$+ \alpha_2(\Delta(|\nabla \vec{Z}|^2 \vec{Z}), \Delta \vec{Z}) - \alpha_2\|\nabla \Delta \vec{Z}\|_2^2$$

$$+ \alpha_2(\Delta\vec{Z}, \Delta\vec{H}) - \alpha_2(\Delta(\vec{Z}\cdot\vec{H}), \Delta\vec{Z})$$
$$+ \beta(\vec{H}_t, \nabla(\nabla\cdot\vec{Z})) + \eta_1\beta(\vec{H}, \nabla(\nabla\cdot\vec{Z}))$$
$$+ \eta_1\|\vec{H}_t\|_2^2 - \eta_1\|\nabla\vec{H}\|_2^2$$
$$- \beta\eta_1(\vec{H}, \vec{Z}_{tt}) - \sigma\eta_1(\vec{H}, \vec{H}_t)$$
$$- \sigma\beta\eta_1(\vec{H}, \vec{Z}_t) + \eta_2\|\vec{E}_t\|_2^2$$
$$- \eta_2\|\vec{Z}\|_2^2 - \sigma\eta_2(\vec{Z}, \vec{E}_t) - \eta_2\beta(\vec{E}, (\nabla\times\vec{Z})_t)$$
$$= -(\sigma - \eta_1)\|\vec{H}_t\|_2^2 - (\sigma\eta_2)\|\vec{E}_t\|_2^2$$
$$- \eta_1\|\nabla\vec{H}\|_2^2 - \eta_2\|\nabla\vec{E}\|_2^2 - \alpha_2\|\nabla\Delta\vec{Z}\|_2^2$$
$$- \beta\eta_1(\vec{H}, \vec{Z}_{tt}) - \sigma\eta_1(\vec{H}, \vec{H}_t) - \sigma\beta\eta_1(\vec{H}, \vec{Z}_t)$$
$$- \sigma\eta_2(\vec{E}, \vec{E}_t) - \beta\eta_2(\vec{E}, (\nabla\times\vec{Z})_t)$$
$$+ \beta(\vec{H}_t, \nabla(\nabla\cdot\vec{Z})) + \beta\eta_1(\vec{H}_t, \nabla(\nabla\cdot\vec{Z})) - \beta(\vec{H}_t, \vec{Z}_{tt})$$
$$- \sigma\beta(\vec{H}_t, \vec{Z}_t) - \beta(\vec{E}_t, (\nabla\times\vec{Z})_t)$$
$$+ \alpha_1(\Delta(\vec{Z}\times\Delta\vec{Z}), \Delta\vec{Z}) + \alpha_1(\Delta(\vec{Z}\times\vec{H}), \Delta\vec{Z})$$
$$+ \alpha_2(\Delta(|\nabla\vec{Z}|^2\vec{Z}), \Delta\vec{Z}) + \alpha_2(\Delta\vec{Z}, \Delta\vec{H})$$
$$- \alpha_2(\Delta(\vec{Z}\cdot\vec{H})\vec{Z}, \Delta\vec{Z}), \tag{6.5.22}$$

in which

$$(\vec{H}, \vec{Z}_{tt}) = (\vec{H}, \vec{Z}_t)_t - (\vec{H}_t, \vec{Z}_t),$$
$$(\vec{H}_t, \vec{Z}_{tt}) = (\vec{H}_t, \vec{Z}_t)_t - (\vec{H}_{tt}, \vec{Z}_t)$$
$$= (\vec{H}_t, \vec{Z}_t)_t - (\vec{Z}_t, \Delta\vec{H} - \beta\vec{Z}_{tt} - \sigma\vec{H}_t - \sigma\beta\vec{Z}_t + \beta\nabla(\nabla\times\vec{Z}))$$
$$= (\vec{H}_t, \vec{Z}_t)_t + \frac{\beta}{2}(\vec{Z}_t, \vec{Z}_t)_t - (\vec{Z}_t, \Delta\vec{H}) + \sigma(\vec{Z}_t, \vec{H}_t)$$
$$+ \sigma\beta\|\vec{Z}_t\|_2^2 - \beta(\vec{Z}_t, \nabla(\nabla\cdot\vec{Z})),$$

$$\beta(\vec{Z}_t, (\nabla\cdot\vec{Z})_t) = \beta(\vec{Z}_t, \nabla\times\vec{Z})_t - \beta(\Delta\vec{E} - \sigma\vec{E}_t - \beta(\nabla\times\vec{Z})_t, \nabla\times\vec{Z})$$
$$= \beta(\vec{Z}_t, \nabla\times\vec{Z})_t + \frac{\beta^2}{2}(\nabla\times\vec{Z}, \nabla\times\vec{Z})_t$$
$$+ \sigma\beta(\vec{Z}_t, \nabla\times\vec{Z}) + \beta(\nabla\vec{Z}, \nabla(\nabla\times\vec{Z})),$$

$$\eta_2\beta(\vec{E}, (\nabla\times\vec{Z})_t) = \eta_2\beta(\vec{E}, \nabla\times\vec{Z})_t - \eta_2\beta(\vec{E}_t, \nabla\times\vec{Z}).$$

Set $e_1(t) = \frac{1}{2}G(t) + R(t)$, where

$$G(t) = \|\vec{E}_t\|_2^2 + \|\vec{H}_t\|_2^2 + \|\nabla\vec{E}\|_2^2 + \|\nabla\vec{H}\|_2^2 + \|\Delta\vec{Z}\|_2^2$$
$$= 2e(t) - 2\eta_1(\vec{H}, \vec{H}_t) - 2\eta_2(\vec{E}, \vec{E}_t),$$

$$R(t) = \beta(\vec{Z}_t, \vec{H}_t) + \frac{\beta^2}{2}\|\vec{Z}_t\|_2^2 + \beta(\vec{E}_t, \nabla \times \vec{Z}) + \frac{\beta^2}{2}\|\nabla \times \vec{Z}\|_2^2$$

$$= -\eta_2\beta(\vec{E}, \nabla \times \vec{Z}) + \frac{1}{2}\sigma\eta_1\|\vec{H}\|_2^2 + \frac{1}{2}\sigma\eta_2\|\vec{E}\|_2^2$$

$$+ \eta_2\beta(\vec{E}, \nabla \times \vec{Z}) + \eta_2(\vec{E}, \vec{E}_t) + \eta_1(\vec{H}, \vec{H}_t). \qquad (6.5.23)$$

$$\frac{de_1(t)}{dt} + (\sigma - \eta_1)\|\vec{H}_t\|_2^2 + (\sigma - \eta_2)\|\vec{E}_t\|_2^2$$

$$+ \eta_1\|\nabla\vec{H}\|_2^2 + \eta_2\|\vec{E}\|_2^2 + \alpha_2\|\nabla\Delta\vec{Z}\|_2^2 + \sigma\beta^2\|\vec{Z}_t\|_2^2$$

$$= -(2\sigma\beta - \eta_1\beta)(\vec{Z}_t, \vec{H}_t) + \beta(\vec{Z}_t, \Delta\vec{H}) - (\sigma - \eta_2)\beta(\vec{E}_t, \nabla \times \vec{Z})$$

$$- \beta(\nabla\vec{E}, \nabla(\nabla \times \vec{Z})) + \alpha_1(\Delta(\vec{Z} \times \Delta), \Delta\vec{Z})$$

$$+ \alpha_1(\Delta(\vec{Z} \times \vec{H}), \Delta\vec{Z}) + \alpha_2(\Delta\vec{Z}, \Delta\vec{H})$$

$$+ \alpha_2(\Delta(|\nabla\vec{Z}|^2\vec{Z}), \Delta\vec{Z}) - \alpha_2(\Delta(\vec{Z} \cdot \vec{H})\vec{Z}, \Delta\vec{Z}) - \sigma\beta\eta_1(\vec{H}, \vec{Z}_t)$$

$$+ \beta(\vec{H}_t, \nabla(\nabla \cdot \vec{Z})) + \beta\eta_1(\vec{H}, \nabla(\nabla \cdot \vec{Z})) + \beta^2(\vec{Z}_t, \nabla(\nabla \times \vec{Z})). \qquad (6.5.24)$$

In the sequel we estimate every term in (6.5.24).

(1) Estimate of the first term.

It follows from (6.5.1) that

$$|\vec{Z}_t|^2 = \alpha_1^2|\vec{Z} \times (\Delta\vec{Z} + \vec{H})|^2 + \alpha_2^2|\vec{Z} \times (\vec{Z} \times (\Delta\vec{Z} + \vec{H}))|^2$$

$$\leq (\alpha_1^2 + \alpha_2^2)|\Delta\vec{Z} + \vec{H}|^2 2(\alpha_1^2 + \alpha_2^2)(|\Delta\vec{Z}|^2 + |\vec{H}|^2).$$

Using Sobolev inequality and Hölder inequality we have

$$| - (2\sigma\beta + \eta_1\beta)(\vec{Z}_t, \vec{H}_t)|$$

$$\leq \varepsilon_1\|\vec{H}_t\|_2^2 + C(\varepsilon_1, \sigma, \beta, \eta_1)\|\vec{Z}_t\|_2^2$$

$$\leq \varepsilon_1\|\vec{H}_t\|_2^2 + 2C(\alpha_1^2 + \alpha_2^2)\|\Delta\vec{Z}\|_2^2 + d_1$$

$$\leq \varepsilon_1\|\vec{H}_t\|_2^2 + \frac{\alpha_2}{l}\|\nabla\Delta\vec{Z}\|_2^2 + d_2,$$

where ε_1 and l are constants to be determined.

(2) Estimate of the second term.

Acting ∇ on (6.5.1), we get

$$\nabla\vec{Z}_t = \alpha_1\nabla\vec{Z} \times (\Delta\vec{Z} + \vec{H}) + \alpha_1\vec{Z} \times \nabla\Delta\vec{Z} - \alpha_2\nabla\vec{Z} \times (\vec{Z} \times (\Delta\vec{Z} + \vec{H}))$$

$$- \alpha_2\vec{Z} \times (\nabla\vec{Z} \times (\Delta\vec{Z} + \vec{H})) - \alpha_2\vec{Z} \times (\vec{Z} \times \Delta\vec{Z}) - \alpha_2\vec{Z} \times (\vec{Z} \times \nabla\vec{H}).$$

Hence,

$$
\begin{aligned}
\beta(\vec{Z}_t, \Delta\vec{H}) &= -\beta(\nabla\vec{Z}_t, \nabla\vec{H}) \\
&= -\beta\alpha_1(\nabla\vec{Z} \times (\Delta\vec{Z} + \vec{H}), \nabla\vec{H}) \\
&\quad - \beta\alpha_1(\vec{Z} \times \nabla\Delta\vec{Z}, \nabla\vec{H}) + \beta\alpha_2(\nabla\vec{Z} \times (\vec{Z} \times \Delta\vec{Z}), \nabla\vec{H}) \\
&\quad + \beta\alpha_2(\nabla\vec{Z} \times (\vec{Z} \times \vec{H}), \nabla\vec{H}) + \beta\alpha_2(\vec{Z} \times (\nabla\vec{Z} \times \Delta\vec{Z}), \nabla\vec{H}) \\
&\quad + \beta\alpha_2(\vec{Z} \times (\nabla\vec{Z} \times \vec{H}), \nabla\vec{H}) \\
&\quad + \beta\alpha_2(\vec{Z} \times (\vec{Z} \times \nabla\Delta\vec{Z}), \nabla\vec{H}) - \beta\alpha_2\|\vec{Z} \times \nabla\vec{H}\|_2^2 \\
&\leq -\beta\alpha_2\|\vec{Z} \times \nabla\vec{H}\|_2^2 + (|\alpha_1| + 2\alpha_2)\beta\|\nabla\vec{Z}\|_\infty\|\Delta\vec{Z}\|_2\|\nabla\vec{H}\|_2 \\
&\quad + (|\alpha_1| + 2\alpha_2)\beta\|\nabla\vec{Z}\|_\infty\|\vec{H}\|_2\|\nabla\vec{H}\|_2 \\
&\quad + (|\alpha_1| + \alpha_2)\beta\|\nabla\Delta\vec{Z}\|_2\|\nabla\vec{H}\|_2 \\
&\leq C_1(|\alpha_1| + 2\alpha_2)\beta\|\nabla\vec{Z}\|_2^{\frac{3}{2}-\frac{n}{4}}\|\nabla\Delta\vec{Z}\|_2^{\frac{1}{2}+\frac{n}{4}}\|\nabla\vec{H}\|_2 \\
&\quad + C_2(|\alpha_1| + 2\alpha_2)\beta\|\nabla\Delta\vec{Z}\|_2^{\frac{n}{4}}\|\nabla\vec{H}\|_2\|\nabla\vec{H}\|_2 \\
&\quad + \frac{|\alpha_1| + \alpha_2}{4}\|\nabla\Delta\vec{Z}\|_2^2 + (|\alpha_1| + \alpha_2)\beta^2\|\nabla\vec{H}\|_2^2 \\
&\leq \frac{C_0^2}{\alpha_2}(|\alpha_1| + 2\alpha_2)^2\|\nabla\vec{Z}\|_2^{3-\frac{n}{2}}\|\nabla\Delta\vec{Z}\|_2^{1+\frac{n}{2}} + C_3\|\nabla\Delta\vec{Z}\|_2^{\frac{n}{2}} \\
&\quad + \frac{|\alpha_1| + \alpha_2}{4}\|\nabla\Delta\vec{Z}\|_2^2 + 2\alpha_2(|\alpha_1| + \alpha_2)\beta^2\|\nabla\vec{H}\|_2^2 \\
&\leq
\begin{cases}
\left(\frac{\alpha_2}{l} + \frac{|\alpha_1|+\alpha_2}{4}\right)\|\nabla\Delta\vec{Z}\|_2^2 \\
\quad + 2(|\alpha_1| + \alpha_2)\|\nabla\vec{H}\|_2^2 + C, \quad n = 1; \\
+ \left(\frac{\alpha_2}{l} + \frac{|\alpha_1|+\alpha_2}{4} + \frac{C_0^2}{\alpha_2}(|\alpha_1| + 2\alpha_2)^2\right)\|\nabla\vec{Z}\|_2^2\|\nabla\Delta\vec{Z}\|_2^2 \\
\quad + 2(|\alpha_1| + \alpha_2)\|\nabla\vec{H}\|_2^2 + C, \quad n = 2.
\end{cases}
\end{aligned}
$$

(3) Estimate of the third term of (6.5.24).

$$
\begin{aligned}
\left|-(\sigma + \eta_2)\beta(\vec{E}_t, \nabla \times \vec{Z})\right| &\leq \varepsilon_2\|\vec{E}_t\|_2^2 + C_1\|\nabla\vec{Z}\|_2^2 \\
&\leq \varepsilon_2\|\vec{E}_t\|_2^2 + C_2.
\end{aligned}
$$

(4) Estimate of the fourth term of (6.5.24).

$$
\begin{aligned}
\left|-\beta(\nabla\vec{E}, \nabla(\nabla \times \vec{Z}))\right| &= \left|\beta(\vec{E}, \Delta(\nabla \times \vec{Z}))\right| \\
&\leq \frac{\alpha_2}{l}\|\nabla\Delta\vec{Z}\|_2^2 + C_1\|\vec{E}\|_2^2 \\
&\leq \frac{\alpha_2}{l}\|\nabla\Delta\vec{Z}\|_2^2 + C_2.
\end{aligned}
$$

(5) Estimate of the fifth term of (6.5.24).

$$\left|\alpha_1(\Delta(\vec{Z}\times\Delta\vec{Z}),\Delta\vec{Z})\right| \le |\alpha_1|\left|(\nabla\vec{Z}\times\Delta\vec{Z},\nabla\Delta\vec{Z})\right|$$

$$\le |\alpha_1|\|\nabla\vec{Z}\|_\infty\|\Delta\vec{Z}\|_2\|\nabla\Delta\vec{Z}\|_2^2$$

$$\le C_0'|\alpha_1|\|\nabla\vec{Z}\|_2^{\frac{1}{2}-\frac{n}{4}}\|\nabla\Delta\vec{Z}\|_2^{\frac{3}{2}+\frac{n}{4}}$$

$$\le \begin{cases} \frac{\alpha_2}{l}\|\nabla\Delta\vec{Z}\|_2^2 + C, & n=1; \\ C_0|\alpha_1|\|\nabla\vec{Z}\|_2\|\nabla\Delta\vec{Z}\|_2^2, & n=2. \end{cases}$$

(6) Estimate of the sixth term of (6.5.24).

$$\left|\alpha_1(\Delta(\vec{Z}\times\vec{H}),\Delta\vec{Z})\right| = \left|\alpha_1(\Delta(\vec{Z}\times\vec{H}),\Delta\vec{Z})\right|$$

$$\le |\alpha_1|\|\nabla\vec{Z}\|_\infty\|\vec{H}\|_2\|\nabla\Delta\vec{Z}\|_2 + |\alpha_1|\|\nabla\vec{H}\|_2\|\nabla\Delta\vec{Z}\|_2$$

$$\le \frac{\alpha_2}{4}\|\nabla\Delta\vec{Z}\|_2^2 + \frac{2\alpha_1^2}{\alpha_2}\|\nabla\vec{H}\|_2^2 + C.$$

(7) Estimate of the seventh term of (6.5.24).

$$\left|\alpha_2(\Delta(|\vec{Z}|^2\vec{Z}),\Delta\vec{Z})\right| = \left|-\alpha_2(\nabla(|\vec{Z}|^2\vec{Z}),\nabla\Delta\vec{Z})\right|$$

$$\le |\alpha_2|\|\nabla\vec{Z}\|_6^3\|\nabla\Delta\vec{Z}\|_2 + C|\alpha_2|\|\nabla\vec{Z}\|_\infty\|\Delta\vec{Z}\|_2\|\nabla\Delta\vec{Z}\|_2$$

$$\le C|\alpha_2|\left(\|\nabla\vec{Z}\|_2^{3-\frac{n}{2}}\|\nabla\Delta\vec{Z}\|_2^{1+\frac{n}{2}} + \|\nabla\vec{Z}\|_2^{\frac{3}{2}-\frac{n}{4}}\|\nabla\Delta\vec{Z}\|_2^{\frac{3}{2}+\frac{n}{4}}\right)$$

$$\le \begin{cases} \frac{\alpha_2}{l}\|\nabla\Delta\vec{Z}\|_2^2 + C, & n=1; \\ C\alpha_2(\|\Delta\vec{Z}\|_2^2 + \|\nabla\vec{Z}\|_2^2)\|\nabla\Delta\vec{Z}\|_2^2, & n=2. \end{cases}$$

(8) Estimate of the eighth term of (6.5.24).

$$\left|\alpha_2(\Delta\vec{Z},\Delta\vec{H})\right| \le \frac{\alpha_2}{4}\|\nabla\Delta\vec{Z}\|_2^2 + \alpha_2\|\nabla\vec{H}\|_2^2.$$

(9) Estimate of the ninth term of (6.5.24).

$$\left|\alpha_2(\Delta(\vec{Z}\cdot\vec{H})\vec{Z},\Delta\vec{Z})\right| = \alpha_2\left|(\nabla(\vec{Z}\times\vec{H}),\nabla\Delta\vec{Z})\right|$$

$$\le \alpha_2\left(2\|\vec{H}\|_2\|\nabla\vec{Z}\|_\infty\|\nabla\Delta\vec{Z}\|_2 + \|\nabla\vec{H}\|_2\|\nabla\Delta\vec{Z}\|_2\right)$$

$$\le \frac{\alpha_2}{8}\|\nabla\Delta\vec{Z}\|_2^2 + 3\alpha_2\|\nabla\vec{H}\|_2^2 + C.$$

(10) Estimate of the tenth term of (6.5.24).

$$\left|-\sigma\beta\eta_1(\vec{H},\vec{Z}_t)\right| \le C_1\|\vec{H}\|_2\|\vec{Z}_t\|_2$$

$$\le C_2\|\Delta\vec{Z}\|_2 + d_1$$

$$\le \frac{\alpha_2}{l}\|\nabla\Delta\vec{Z}\|_2^2 + d_2.$$

(11) Estimate of the eleventh term of (6.5.24).

$$\left|\beta(\vec{H}_t, \nabla(\nabla \cdot \vec{Z}))\right| \le \beta\|\vec{H}_t\|_2\|\Delta\vec{Z}\|_2$$

$$\le \varepsilon_3\|\vec{H}_t\|_2^2 + \frac{\alpha_2}{l}\|\nabla\Delta\vec{Z}\|_2^2 + C.$$

(12) Estimate of the twelfth term of (6.5.24).

$$\left|\eta_1\beta(\vec{H}, \nabla(\nabla \cdot \vec{Z}))\right| \le \eta_1\beta\|\vec{H}\|_2\|\Delta\vec{Z}\|_2$$

$$\le \frac{\alpha_2}{l}\|\nabla\Delta\vec{Z}\|_2^2 + C.$$

Combining (1)–(12) and using (6.5.24) we have
(i) If $n = 1$,

$$\frac{de_1(t)}{dt} + (\sigma - \eta_1 - \varepsilon_1 - \varepsilon_2)\|\vec{H}_t\|_2^2$$

$$+ (\sigma - \eta_2 - \varepsilon_2)\|\vec{E}_t\|_2^2 + \eta_2\|\nabla\vec{E}\|_2^2$$

$$+ \left(\eta_1 - \frac{2\alpha_1^2 + 4\alpha_2^2 + 2\beta^2\alpha_2(|\alpha_1| + \alpha_2)}{\alpha_2}\right)\|\nabla\vec{H}\|_2^2$$

$$+ \alpha_2\left(\frac{1}{8} - \frac{7}{l}\right)\|\nabla\Delta\vec{Z}\|_2^2 \le C_1. \tag{6.5.25}$$

(ii) If $n = 2$,

$$\frac{de_1(t)}{dt} + (\sigma - \eta_1 - \varepsilon_1 - \varepsilon_3)\|\vec{H}_t\|_2^2$$

$$+ (\sigma - \eta_2 - \varepsilon_2)\|\vec{E}_t\|_2^2 + \eta_2\|\vec{E}_t\|_2^2$$

$$+ \left(\eta_1 - \frac{2\alpha_1^2 + 4\alpha_2^2 + 2\beta^2\alpha_2(|\alpha_1| + \alpha_2)}{\alpha_2}\right)\|\nabla\vec{H}\|_2^2$$

$$+ \left[\alpha_2\left(\frac{1}{8} - \frac{6}{l}\right) - C(|\alpha_1| + \alpha_2)\right]\|\nabla\vec{Z}\|_2$$

$$- \frac{C_2}{\alpha_2}(\alpha_2^2 + (|\alpha_1| + 2\alpha_2)^2)\|\nabla\vec{Z}\|_2^2]\|\nabla\Delta\vec{Z}\|_2^2 \le C_2, \tag{6.5.26}$$

where C_1, C_2 are independent of t, positive numbers ε_1, ε_2, ε_3, η_1, η_2 and l are to be determined.

In (6.5.25) and (6.5.26) we let

$$\frac{2\alpha_1^2 + 4\alpha_2^2 + 2\beta^2\alpha_2(|\alpha_1| + \alpha_2)}{\alpha_2} < \sigma, \tag{6.5.27}$$

$$\frac{2\alpha_1^2 + 4\alpha_2^2 + 2\beta^2\alpha_2(|\alpha_1| + \alpha_2)}{\alpha_2} < \eta_1 < \sigma, \tag{6.5.28}$$

$$\eta_2 = \varepsilon_2 = \frac{\sigma}{4}, \tag{6.5.29}$$

$$l > 56, \quad n = 1; \quad l = 60, \quad n = 2. \tag{6.5.30}$$

Suppose that $\|\nabla \vec{Z}_0\|_2$, $\|\nabla \vec{E}_0\|_2$, $\|\nabla H_0\|_2$ are small. It follows from Lemma 6.4.2 that

$$C(|\alpha_1| + \alpha_2)\|\nabla \vec{Z}(\cdot, t)\|_2 + C\frac{\alpha_1^2 + (|\alpha_1| + 2\alpha_2)^2}{\alpha_2}\|\nabla \vec{Z}\|_2^2$$

$$\leq C(|\alpha_1| + \alpha_2)\frac{1}{\sqrt{\beta}}\left[\|\vec{E}_0\|_2 + \|\vec{H}_0\|_2 + \sqrt{2}\sigma\|\vec{E}_0\|_2 + \|\nabla \vec{Z}_0\|_2\right]$$

$$+ \frac{C}{\alpha_2}\left\{\alpha_2^2 + (|\alpha_1| + 2\alpha_2)^2\frac{1}{\beta}\left[\left(1 + \frac{\sigma}{\beta}\right)\|\vec{E}_0\|_2 + \|\vec{H}_0\|_2 + \beta\|\nabla \vec{Z}_0\|_2^2\right]\right\} \leq \frac{\alpha_2}{40}. \tag{6.5.31}$$

Therefore, there exists a constant $a > 0$ such that

$$\frac{de_1(t)}{dt} + a\left(\|\vec{H}_t\|_2^2 + \|\vec{E}_t\|_2^2 + \|\nabla \vec{H}\|_2^2 + \|\nabla \vec{E}\|_2^2 + \|\nabla \Delta \vec{Z}\|_2^2\right) \leq C, \tag{6.5.32}$$

where C is independent of t.

Since $\Delta \vec{Z}(x, t)$ is periodic in x, $\int_0^{2D} \Delta \vec{Z} dx = 0$. From Poincaré inequality we have

$$\|\Delta \vec{Z}\|_2^2 \leq \delta \|\nabla \Delta \vec{Z}\|_2^2.$$

Choosing $\delta_0 = \min(a, \frac{a}{\delta_0})$, we have from (6.5.32) that

$$\frac{de_1(t)}{dt} + \delta_0\left(\|\vec{H}_t\|_2^2 + \|\vec{E}_t\|_2^2 + \|\nabla \vec{H}\|_2^2 + \|\nabla \vec{E}\|_2^2 + \|\Delta \vec{Z}\|_2^2\right) \leq C. \tag{6.5.33}$$

Equation (6.5.33) can be rewritten as

$$\frac{de_1(t)}{dt} + 2\delta_0 e_1(t) \leq C + 2\delta_0 R(t) + C + 2\delta_0 \sup_t |R(t)|. \tag{6.5.34}$$

Hence, we have

$$e_1(t)e^{2\delta_0 t} \leq e_1(0) + \left(\frac{C}{2\delta_0} + \sup_t |R(t)|\right)(e^{2\delta_0} - 1),$$

$$e_1(t) \leq e_1(0) + \frac{C}{2\delta_0} + \sup_t |R(t)|,$$

that is

$$G(t) \leq 2C_0 + 2\left(\sup_t |R(t)| - R(t)\right)$$

$$\leq 2C_0 + 4\sup_t |R(t)|.$$

It follows from (6.5.23) that

$$
\begin{aligned}
R(t) \leq{} & \beta \|\vec{Z}_t\|_2 \|\vec{H}_t\|_2 + \frac{\beta^2}{2}\|\vec{Z}_t\|_2^2 \\
& + \beta \|\vec{Z}_t\|_2 \|\nabla \vec{Z}\|_2 + \eta_1 \|\vec{H}\|_2 \|\nabla \vec{E}\|_2 \\
& + \eta_2 \|\vec{E}\|_2 \|\vec{E}_t\|_2 + \frac{\beta^2}{2}\|\nabla \vec{Z}\|_2^2 \\
& + \frac{1}{2}\sigma\eta_1\|\vec{H}\|_2 + \frac{1}{2}\sigma\eta_2\|\vec{E}\|_2^2 + \beta\eta_2\|\vec{E}\|_2\|\nabla\vec{Z}\|_2 \\
\leq{} & \frac{\beta+\beta^2}{2}\|\vec{Z}_t\|_2^2 + \frac{\beta}{2}(\|\vec{H}_t\|_2^2 + \|\vec{E}_t\|_2^2 + \|\nabla\vec{E}\|_2^2) + C_1 \\
\leq{} & (\beta+\beta^2)(\alpha_1^2 + \alpha_2^2)\|\Delta\vec{Z}\|_2^2 \\
& + \frac{\beta}{2}(\|\vec{H}_t\|_2^2 + \|\vec{E}_t\|_2^2 + \|\nabla\vec{E}\|_2^2) + C_2.
\end{aligned}
\tag{6.5.35}
$$

Taking $\beta < \frac{1}{2}$ and $(\beta + \beta^2)(\alpha_1^2 + \alpha_2^2) < \frac{1}{4}$, we have

$$
a_0 = 4\max\left\{\frac{\beta}{2}, (\beta+\beta^2)(\alpha_1^2 + \alpha_2^2)\right\} < 1.
$$

It follows from (6.5.36)

$$
|R(t)| \leq \frac{1}{4}a_0(\|\vec{H}_t\|_2^2 + \|\vec{E}_t\|_2^2 + \|\nabla\vec{E}\|_2^2 + \|\Delta\vec{Z}\|_2^2) \leq \frac{1}{4}a_0 G(t).
\tag{6.5.36}
$$

Substituting (6.5.37) into (6.5.35), we have

$$
G(t) \leq 2C_0 + a_0 \sup_t G(t),
$$

that is

$$
\sup_t G(t) \leq \frac{2C_0}{1 - a_0} = d_0.
$$

This proves the lemma.

Lemma 6.5.3 *Assume that* $\vec{Z}(x,t)$, $\vec{H}(x,t)$, $\vec{E}(x,t)$ *are smooth solutions of* (6.4.1) *and* (6.4.2). $(\vec{Z}_0(x), \vec{H}_0(x), \vec{E}_0(x)) \in (H^2(\Omega), H^1(\Omega), H^1(\Omega))$, $\Omega \subset R^n$, $1 \leq n \leq 2$. $|\vec{Z}_0(x)| = 1$, *and assume*
  (i)
$$
\alpha_2 > 0, \quad \sigma > \frac{4\alpha_2^2 + 2\alpha_1^2 + 2\alpha_2\beta^2(|\alpha_1| + \alpha_2)}{\alpha_2};
$$
  (ii)
$$
0 < \beta < \frac{1}{2}, \quad (\beta + \beta^2)(\alpha_1^2 + 2\alpha_2^2) < \frac{1}{4};
$$
  (iii) *When* $n = 2$,
$$
\|\nabla\vec{Z}_0(x)\|_2^2 + \|\vec{H}_0(x)\|_2^2 + \|\vec{E}_0(x)\|_2^2 \leq \lambda,
$$

where $\lambda = \lambda(\alpha_1, \alpha_2, \beta)$ is a small constant, then

$$\sup_{t \in [0,\infty)} [\|\vec{Z}(\cdot, t)\|^2_{H^2(\Omega)} + \|\vec{H}(\cdot, t)\|^2_{H^1(\Omega)} + \|\vec{E}(\cdot, t)\|^2_{H^1(\Omega)}] \leq K,$$

where the constant K depends on

$$\|\vec{Z}_0(x)\|^2_{H^2(\Omega)} + \|\vec{H}_0(x)\|^2_{H^1(\Omega)} + \|\vec{E}(x)\|^2_{H^1(\Omega)}.$$

Theorem 6.5.1 *Assume $\alpha_2 > 0$, $\beta \geq 0$, $\sigma > 0$ and the initial data $(\vec{Z}_0(x), \vec{H}_0(x), \vec{E}_0(x)) \in (H^k(\Omega), H^{k-1}(\Omega), H^{k-1}(\Omega))$, $k \geq 1 + [\frac{n}{2}]$, $\Omega \subset R^n$, $1 \leq n \leq 2$, $|\vec{Z}_0(x)| = 1$, and $\nabla \cdot \vec{E}_0 = 0$, $\nabla(\vec{H}_0 + \beta \vec{Z}_0) = 0$. When $n = 2$, assume*

$$\|\nabla \vec{Z}_0(x)\|_2 + \|\vec{H}_0(x)\|_2 + \|\vec{E}_0(x)\|_2 < \delta,$$

where δ is a small constant; then the periodic initial value problem (6.5.1)–(6.5.2) admits a unique global smooth solution:

$$|\vec{Z}(x, t)| = 1, \quad x \in \Omega, \quad t \in R^+,$$
$$\vec{Z}(x, t) \in \cap_{s=0}^{[\frac{k}{2}]} W_\infty^s(0, T; H^{k-2s}(\Omega)),$$
$$\vec{H}(x, t) \in \cap_{s=0}^{k-1} W_\infty^s(0, T; H^{k-s-1}(\Omega)),$$
$$\vec{E}(x, t) \in \cap_{s=0}^{k-1} W_\infty^s(0, T; H^{k-s-1}(\Omega)).$$

Theorem 6.5.2 *Assume that the conditions of Theorem 6.5.1 hold. Then the periodic initial value problem (6.5.1)–(6.5.2) admits an attractor $\mathcal{A}$*

(i) $\mathcal{A}$ is weakly compact in $H^2(\Omega) \times H^1(\Omega) \times H^1(\Omega)$;

(ii) $S(t)\mathcal{A} = \mathcal{A}$;

(iii) $\lim_{t \to \infty} \mathrm{dist}(S(t)B, \mathcal{A}) = 0$ for any bounded set $B \in \mathcal{D} \subset H^2(\Omega) \times H^1(\Omega) \times H^1(\Omega)$, where

$$\mathcal{D} = \{(\vec{Z}, \vec{E}, \vec{H}) \in H^2(\Omega) \times H^1(\Omega) \times H^1(\Omega),$$
$$|\vec{Z}(x, t)| = 1, \nabla \cdot \vec{E} = 0, \nabla \cdot (\vec{H} + \beta \vec{Z}) = 0\},$$

$\mathrm{dist}(x, y) = \sup_{x \in X} \inf_{y \in Y} \|x - y\|$. $S(t)(\vec{Z}_0, \vec{H}_0, \vec{E}_0)$ *is the semigroup formed by problem (6.5.1)–(6.5.2).*

Proof. It follows from Theorem 6.5.1 that the periodic initial value problem (6.5.1)–(6.5.2) has a semigroup operator $S(t)(\vec{Z}_0, \vec{H}_0, \vec{E}_0)$ and it is continuous by [70]. Taking the subset $\mathcal{D}$, it follows from Lemma 6.5.2 that the operator $S(t) : E \to E$ is bounded and

$$\overline{A} = \{|\vec{Z}(x, t)| = 1, \nabla \cdot \vec{E} = 0, \nabla \cdot (\vec{H} + \beta \vec{Z}) = 0,$$
$$\vec{Z}(\cdot, t) \in H^2(\Omega), \vec{H}(\cdot, t) \in H^1(\Omega), \vec{E}(\cdot, t) \in H^1(\Omega),$$
$$\|\vec{Z}(\cdot, t)\|^2_{H^2} + \|\vec{H}(\cdot, t)\|^2_{H^1} + \|\vec{E}(\cdot, t)\|^2_{H^1} \leq \varepsilon_0 + \delta_0\}$$

is a bounded absorbing set in $\mathcal{D}$, then $A = \Omega\overline{A}$ is a weakly compact attractor of the periodic initial value problem.

Now we estimate the Hausdorff dimension and Fractal dimension of A. To this aim we consider the linear variational problem of problem (6.5.1)–(6.5.2):

$$z_t = \alpha_2 \Delta z + 2\alpha_2 (\nabla z \cdot \nabla \vec{Z}) \vec{Z} + \alpha_2 |\nabla \vec{Z}|^2 z + \alpha_1 \vec{Z} \times \Delta z$$
$$+ \alpha_1 \vec{Z} \times h - \alpha_1 (\Delta \vec{Z} + \vec{H}) \times z + \alpha_2 h - \alpha_2 (z \cdot \vec{H}) \vec{Z}$$
$$- \alpha_2 (\vec{Z} \cdot h) \vec{Z} - \alpha_2 (\vec{Z} \cdot \vec{H}) z, \tag{6.5.37}$$

$$e_t = \nabla \times h - \sigma e, \tag{6.5.38}$$

$$h_t = -\nabla \times e - \beta z_t, \tag{6.5.39}$$

$$z(0) = z_0, \quad h(0) = h_0, \quad e(0) = e_0. \tag{6.5.40}$$

Simply write (6.5.37)–(6.5.40) as operator form:

$$v_t = -L(u)v, \quad v(0) = v_0, \tag{6.5.41}$$

where $v = (z, e, h)$, $u = (\vec{Z}, \vec{E}, \vec{H})$, $v_0 = (z_0, e_0, h_0)$, z, e, h, $\vec{Z}$, $\vec{E}$, $\vec{H}$ are all three-dimensional vector-valued functions.

Since the periodic initial value problem (6.5.1)–(6.5.2) has smooth solution, the coefficients of the linear system (6.5.37)–(6.5.39) are smooth. If v_0 is properly smooth, then problem (6.5.37)–(6.5.40) admits global smooth solution. Denote its solution operator by $G(t)$, that is $v(t) = G(t)v_0$. Moreover, we can prove that the semigroup operator $S(t)u_0$ is differentiable in $L^2(\Omega)$ and the Frechet differential $S'(t)u_0 = G(t)v_0$. We first prove

Lemma 6.5.4 *The smooth solution* $(\vec{Z}(x,t), \vec{H}(x,t), \vec{E}(x,t))$ *of problem* (6.5.1)–(6.5.2) *is continuously dependent on the initial data.*

Proof. Let $(\vec{Z}_i(x,t), \vec{H}_i(x,t), \vec{E}_i(x,t))$ $(i = 1, 2)$, be smooth solutions of (6.5.1) and (6.5.2) with initial data $\vec{Z}_i(x,0) = \vec{Z}_{0i}(x)$, $\vec{H}_i(x,0) = \vec{H}_{0i}(x)$, $\vec{E}_i(x,0) = \vec{E}_{0i}(x)$ $(i = 1, 2)$. Let $\vec{Z}(x,t) = \vec{Z}_2(x,t) - \vec{Z}_2(x,t)$, $\vec{H}(x,t) = \vec{H}_2(x,t) - \vec{Z}_1(x,t)$, $\vec{E}(x,t) = \vec{E}_2(x,t) - \vec{Z}_1(x,t)$; then $(\vec{Z}(x,t), \vec{H}(x,t), \vec{E}(x,t))$ satisfies

$$\vec{Z}_t = \alpha_2 \vec{Z} \times \Delta \vec{Z}_2 + \alpha_1 \vec{Z}_1 \times \Delta \vec{Z} + \alpha_1 \vec{Z} \times \vec{H}_2$$
$$+ \alpha_1 \vec{Z}_1 \times \vec{H} + \alpha_2 \Delta \vec{Z} + \alpha_2 |\nabla \vec{Z}_2|^2 \vec{Z}$$
$$+ \alpha_2 (\nabla \vec{Z}, \nabla (\vec{Z}_1 + \vec{Z}_2)) \vec{Z}_1 + \alpha_2 \vec{H}$$
$$- \alpha_2 (\vec{Z}_2 \cdot \vec{H}_2) \vec{Z} - \alpha_2 (\vec{Z}_2 \cdot \vec{H} + \vec{H}_1 \cdot \vec{Z}) \vec{Z}_1, \tag{6.5.42}$$

$$\vec{E}_t = \nabla \times \vec{H} - \sigma \vec{E}, \tag{6.5.43}$$

$$\vec{H}_t = -\nabla \times \vec{E} - \beta \vec{Z}_t, \tag{6.5.44}$$

$$\nabla \cdot (\vec{H} + \beta \vec{Z}) = 0, \quad \nabla \cdot \vec{E} = 0, \tag{6.5.45}$$

$$\vec{Z}(x+D,t) = \vec{Z}(x-D,t),$$
$$\vec{H}(x+D,t) = \vec{H}(x-D,t),$$
$$\vec{E}(x+D,t) = \vec{E}(x-D,t),$$

$$\vec{Z}(x,0) = \vec{Z}_0(x), \quad \vec{H}(x,0) = \vec{Z}_0(x), \quad \vec{E}(x,0) = \vec{E}_0(x), \tag{6.5.46}$$

$$|\vec{Z}_0(x)| = 1,$$

$$\nabla \cdot (\vec{H}_0 + \beta\vec{Z}_0) = 0, \quad \nabla \cdot \vec{E}_0 = 0. \tag{6.5.47}$$

We may establish the following inequality

$$\sup_{0 \le t \le T} [\|\nabla\vec{Z}(\cdot,t)\|_{H^1}^2 + \|\vec{H}(\cdot,t)\|_{L^2}^2 + \|\vec{E}(\cdot,t)\|_{L^2}^2]$$

$$\le C(\|\nabla\vec{Z}_0(x)\|_{H^1}^2 + \|\vec{H}_0(x)\|_{L^2}^2 + \|\vec{E}_0(x)\|_{L^2}^2), \tag{6.5.48}$$

where C is an absolute constant. It is clear that if (6.5.48) holds, then the conclusion of Lemma 6.5.6 is proved.

In fact, taking the inner product of (6.5.42) with $\Delta\vec{Z}$, we have

$$\frac{1}{2}\frac{\partial}{\partial t}\int_\Omega |\nabla\vec{Z}|^2 dx + \alpha_2\|\nabla\vec{Z}\|_2^2 \le C_1[\|\nabla\vec{Z}\|_2^2 + \|\vec{H}\|_2^2 + \|\vec{E}\|_2^2]. \tag{6.5.49}$$

Taking the inner product of (6.5.42) with $\Delta\vec{Z}$, we have

$$\begin{aligned}
\frac{1}{2}\frac{\partial}{\partial t}&\int_\Omega |\nabla\vec{Z}|^2 dx + \alpha_2\|\Delta\vec{Z}\|_2^2 \\
&\le -\alpha_1 \int \vec{Z} \times \Delta\vec{Z}_2 \cdot \Delta\vec{Z} dx - \alpha_1 \int \vec{Z} \times \vec{H}_2 \cdot \Delta\vec{Z} dx \\
&\quad - \alpha_1 \int \vec{Z}_1 \times \vec{H} \cdot \Delta\vec{Z} dx - \alpha_2 \int |\nabla\vec{Z}|^2 \vec{Z} \cdot \Delta\vec{Z} dx \\
&\quad - \alpha_2 \int (\nabla\vec{Z} \cdot \nabla(\vec{Z}_1 + \vec{Z}_2))\vec{Z}_1 \cdot \Delta\vec{Z} dx \\
&\quad + \alpha_2 \int \vec{H} \cdot \Delta\vec{Z} dx + \alpha_2 \int (\vec{Z}_2 \cdot \vec{H}_2)\vec{Z} \cdot \Delta\vec{Z} dx \\
&\quad + \alpha_2 \int (\vec{Z}_2 \cdot \vec{H} + \vec{H}_1 \cdot \vec{Z})\vec{Z}_1 \cdot \Delta\vec{Z} dx, \tag{6.5.50}
\end{aligned}$$

in which

$$\begin{aligned}
\left| -\alpha_1 \int \vec{Z} \times \Delta\vec{Z}_2 \cdot \Delta\vec{Z} dx \right| &= \left| \alpha_1 \int \vec{Z} \times \nabla\Delta\vec{Z}_2 \cdot \nabla\vec{Z} dx \right| \\
&\le |\alpha_1| \|\nabla\Delta\vec{Z}_2\|_\infty \|\vec{Z}\|_2 \|\nabla\vec{Z}\|_2 \\
&\le C|\alpha_1|(\|\vec{Z}\|_2^2 + \|\nabla\vec{Z}\|_2^2),
\end{aligned}$$

$$\left| -\alpha_1 \int \vec{Z} \times \vec{H}_2 \cdot \Delta \vec{Z} dx \right| = \left| \alpha_1 \int \vec{Z} \times \nabla \vec{H}_2 \cdot \nabla \vec{Z} dx \right|$$

$$\leq |\alpha_1| \|\nabla \vec{H}_2\|_\infty \|\vec{Z}\|_2 \|\nabla \vec{Z}\|_2$$

$$\leq C|\alpha_1|(\|\vec{Z}\|_2^2 + \|\nabla \vec{Z}\|_2^2),$$

$$\left| -\alpha_2 \int |\nabla \vec{Z}_2|^2 \vec{Z} \cdot \Delta \vec{Z} dx \right| \leq |\alpha_2| \|\nabla \vec{Z}_2\|_\infty^2 \|\vec{Z}\|_2 \|\Delta \vec{Z}\|_2$$

$$\leq \frac{\alpha_2}{K} \|\Delta \vec{Z}\|_2^2 + C(K)\alpha_2 \|\vec{Z}\|_2^2$$

where K is a constant to be determined.

$$\left| -\alpha_2 \int (\nabla \vec{Z} \cdot \nabla(\vec{Z}_1 + \vec{Z}_2))\vec{Z}_1 \cdot \Delta \vec{Z} dx \right|$$

$$\leq \alpha_2 \|\nabla \vec{Z}_1 + \nabla \vec{Z}_2\|_\infty \|\nabla \vec{Z}\|_2 \|\Delta \vec{Z}\|_2$$

$$\leq \frac{\alpha_2}{K} \|\Delta \vec{Z}\|_2^2 + C(K)\alpha_2 \|\vec{Z}\|_2^2,$$

$$\left| \alpha_2 \int \vec{H} \cdot \Delta \vec{Z} dx \right| \leq \alpha_2 \|\vec{H}\|_2 \|\Delta \vec{Z}\|_2$$

$$\leq \frac{\alpha_2}{K} \|\Delta \vec{Z}\|_2^2 + C(K)\alpha_2 \|\vec{H}\|_2^2,$$

$$\left| \alpha_2 \int (\vec{Z}_2 \cdot \vec{H}_2)\vec{Z} \cdot \Delta \vec{Z} dx \right| \leq \alpha_2 \|\vec{H}_2\|_\infty \|\vec{Z}_2\|_\infty \|\vec{Z}\|_2 \|\Delta \vec{Z}\|_2$$

$$\leq \frac{\alpha_2}{K} \|\Delta \vec{Z}\|_2^2 + C(K)\alpha_2 \|\vec{Z}\|_2^2,$$

$$\left| \alpha_2 \int (\vec{Z}_2 \cdot \vec{H} + \vec{H}_1 \cdot \vec{Z})\vec{Z}_1 \cdot \Delta \vec{Z} dx \right|$$

$$\leq \alpha_2 (\|\vec{H}\|_2 \|\Delta \vec{Z}\|_2 + \|\vec{H}_1\|_\infty \|\vec{Z}\|_2 \|\Delta \vec{Z}\|_2)$$

$$\leq \frac{\alpha_2}{K} \|\Delta \vec{Z}\|_2^2 + C(K)\alpha_2 (\|\vec{Z}\|_2^2 + \|\vec{H}\|_2^2).$$

Therefore, it follows from (6.5.50) that

$$\frac{1}{2}\frac{d}{dt} \|\nabla \vec{Z}\|_2^2 + \alpha_2 \|\nabla \vec{Z}\|_2^2 + \alpha_1 \int \vec{Z}_1 \times \vec{H} \cdot \Delta \vec{Z} dx$$

$$\leq 5\frac{\alpha_2}{K} \|\Delta \vec{Z}\|_2^2 + C(K)(\|\nabla \vec{Z}\|_2^2 + \|\vec{Z}\|_2^2 + \|\vec{H}\|_2^2). \qquad (6.5.51)$$

Multiplying (6.5.43) by $\vec{E}$ and multiplying (6.5.44) by $\vec{H}$ and integrating over Ω $(\beta > 0)$, we have

$$\frac{1}{2\beta}\frac{d}{dt} \int_\Omega (|\vec{E}|^2 + |\vec{H}|^2)dx + \frac{\sigma}{\beta}\|\vec{Z}\|_2^2 = -\int_\Omega \vec{Z}_t \cdot \vec{H} dx. \qquad (6.5.52)$$

It follows from (6.5.51) and (6.5.52) that

$$\frac{1}{2}\frac{d}{dt}\int_\Omega(|\nabla\vec{Z}|_2^2+|\vec{Z}\|_2^2+\frac{1}{\beta}(|\vec{Z}|^2+|\vec{H}|^2))dx+\alpha_2\|\Delta\vec{Z}\|_2^2$$

$$\leq-\int_\Omega(\vec{Z}_t\cdot\vec{H}+\alpha_1\vec{Z}_1\times\vec{H}\cdot\Delta\vec{Z})dx+5\frac{\alpha_2}{K}\|\Delta\vec{Z}\|_2^2$$

$$+C(K)(\|\nabla\vec{Z}\|_2^2+\|\vec{Z}\|_2^2+\|\vec{H}\|_2^2). \qquad (6.5.53)$$

Multiplying (6.5.42) by $\vec{H}$, we have

$$\left|\int_\Omega(\vec{Z}_t\cdot\vec{H}+\alpha_1\vec{Z}_1\times\vec{H}\cdot\Delta\vec{Z})dx\right|$$

$$\leq|\alpha_1|\left|\int_\Omega(\vec{Z}\times\Delta\vec{Z}_2+\vec{Z}\times\vec{H}_2)\cdot\vec{H}dx\right|$$

$$+\alpha_2\left|\int_\Omega\Delta\vec{Z}\cdot\vec{H}dx\right|+\alpha_2\left|\int_\Omega|\nabla\vec{Z}|^2\vec{Z}\cdot\vec{H}dx\right|$$

$$+\alpha_2\left|\int_\Omega|\nabla\vec{Z}\cdot(\nabla(\vec{Z}_1+\vec{Z}+2))\vec{Z}\cdot\vec{H}dx\right|$$

$$+\alpha_2\|\vec{H}\|_2^2+\alpha_2\left|\int_\Omega(\vec{Z}_2\cdot\vec{H}_2)\vec{Z}\cdot\vec{H}dx\right|$$

$$+\alpha_2\left|\int_\Omega(\vec{Z}_2\cdot\vec{H})(\vec{Z}_1\cdot\vec{H})dx\right|+\alpha_2\left|\int_\Omega(\vec{H}_1\cdot\vec{Z})(\vec{Z}_1\cdot\vec{H})dx\right|$$

$$\leq\frac{\alpha_2}{K}\|\Delta\vec{Z}\|_2^2+C(K)(\|\nabla\vec{Z}\|_2^2+\|\vec{Z}\|_2^2+\|\vec{H}\|_2^2). \qquad (6.5.54)$$

It follows from (6.5.53) and (6.5.54) that

$$\frac{1}{2}\frac{d}{dt}\int_\Omega\left(|\nabla\vec{Z}|_2^2+|\vec{Z}\|_2^2+\frac{1}{\beta}(|\vec{Z}|^2+|\vec{H}|^2)\right)dx$$

$$\leq C(K,\alpha_1,\alpha_2)[\|\nabla\vec{Z}\|_2^2+\|\vec{Z}\|_2^2+\|\vec{H}\|_2^2].$$

The lemma is proved.

In order to prove that the operator semigroup $S(t)$ is Frechet differentiable, we consider a linear variational problem of (6.5.1)–(6.5.2) as follows:

$$w_t=\alpha_1 w\times(\nabla\vec{Z}_1+\vec{H}_1)+\alpha_1\vec{Z}_1\times(\Delta w+I)$$

$$-\alpha_2 w\times(\vec{Z}_1\times(\vec{Z}_1+\vec{H}_1))$$

$$-\alpha_2\vec{Z}_1\times(w\times(\Delta\vec{Z}_1+\vec{H}_1))$$

$$-\alpha_2\vec{Z}_1\times(\vec{Z}_1\times(\Delta w+I)), \qquad (6.5.55)$$

where

$$\nabla\times I=\vec{F}+\sigma\vec{F}, \qquad (6.5.56)$$

$$\nabla\times F=-I_t-\beta w_t,$$

$$\nabla \cdot (\vec{H} + \beta\vec{Z}) = 0, \quad \nabla \cdot \vec{E} = 0, \tag{6.5.57}$$

$$(w(t), I(t), F(t))|_{t=0} = (\vec{Z}_0, \vec{H}_0, \vec{E}_0), \tag{6.5.58}$$

where $(\vec{Z}_1, \vec{H}_1, \vec{E}_1) = S(t)(\vec{Z}_{01}, \vec{H}_{01}, \vec{E}_{01})$ is a solution of (6.5.1) with initial data $(\vec{Z}_{01}, \vec{H}_{01}, \vec{E}_{01})$. Set

$$\begin{aligned}
(\tilde{Z}, \tilde{H}, \tilde{E}) &= (\vec{Z}, \vec{H}, \vec{E}) - (w, I, F) \\
&= S(t)(\vec{Z}_{01}, \vec{H}_{01}, \vec{E}_{01}) - S(t)(\vec{Z}_0, \vec{H}_0, \vec{E}_0) \\
&\quad - (DS(t))(\vec{Z}_{01}, \vec{H}_{01}, \vec{E}_{01})(\vec{Z}_0, \vec{H}_0, \vec{E}_0).
\end{aligned} \tag{6.5.59}$$

Hence,

$$\begin{aligned}
\tilde{Z}_t &= \alpha_1[\vec{Z} \times (\Delta\vec{Z}_2 + \vec{H}_2) - w \times (\Delta\vec{Z}_1 + \vec{H}_1)] \\
&\quad + \alpha_1[\vec{Z}_1 \times (\Delta\vec{Z} + \vec{H}) - \vec{Z}_1 \times (\Delta w + I)] \\
&\quad - \alpha_2[\vec{Z} \times (\vec{Z}_2 \times (\Delta\vec{Z}_2 + \vec{H}_2)) - \vec{Z} \times (\vec{Z}_1 \times (\Delta\vec{Z}_1 + \vec{H}_1))] \\
&\quad - \alpha_2[\vec{Z}_1 \times (\vec{Z} \times (\Delta\vec{Z}_2 + \vec{H}_2)) - \vec{Z} \times (w \times (\Delta\vec{Z} + \vec{H}))] \\
&\quad - \alpha_2[\vec{Z}_1 \times (\vec{Z}_1 \times (\Delta\vec{Z} + \vec{H})) - \vec{Z} \times (\vec{Z}_1 \times (\Delta w + I))],
\end{aligned} \tag{6.5.60}$$

$$\nabla \times \tilde{H} = \tilde{E}_t + \sigma\tilde{E}, \tag{6.5.61}$$

$$\nabla \times \tilde{E} = -\tilde{H}_t - \beta\tilde{Z}_t, \tag{6.5.62}$$

$$\nabla \cdot (\tilde{H} + \beta\tilde{Z}) = 0, \quad \nabla \cdot \tilde{E} = 0, \tag{6.5.63}$$

$$(\tilde{Z}, \tilde{H}, \tilde{E})|_{t=0} = 0. \tag{6.5.64}$$

Rewrite (6.5.60) as follows:

$$\begin{aligned}
\tilde{Z}_t &= \alpha_1[\tilde{Z} \times (\Delta\vec{Z}_1 + \vec{H}_1) + \vec{Z} \times (\Delta\tilde{Z} + \tilde{H})] \\
&\quad + \alpha_1[\tilde{Z}_1 \times (\Delta\tilde{Z} + \tilde{H})] \\
&\quad - \alpha_2[\vec{Z} \times (\vec{Z} \times (\Delta\vec{Z}_2 + \vec{H}_2)) + \vec{Z} \times (\vec{Z}_1 \times (\Delta\tilde{Z} + \tilde{H})) \\
&\quad + \tilde{Z} \times (\vec{Z}_1 \times (\Delta\vec{Z}_1 + \vec{H}_1))] \\
&\quad - \alpha_2[\vec{Z} \times (\tilde{Z} \times (\Delta\vec{Z}_1 + \vec{H}_1)) + \vec{Z}_1 \times (\vec{Z} \times (\Delta\tilde{Z} + \tilde{H}))] \\
&\quad - \alpha_2[\vec{Z}_1 \times (\vec{Z}_1 \times (\Delta\tilde{Z} + \tilde{H}))].
\end{aligned} \tag{6.5.65}$$

It follows from (6.5.65) that

$$\frac{1}{2}\frac{d}{dt}\|\tilde{Z}\|_2^2 \leq C_1\|\tilde{Z}\|_2^2 + C_2(\|\vec{Z}\|_2^2 + \|\vec{H}\|_2^2 + \|\vec{E}\|_2^2)^4.$$

This implies

$$\begin{aligned}
\|\tilde{Z}(t)\|_2^2 &\leq \|\tilde{Z}(0)\|_2^2 e^{C_1 t} + \int_0^t e^{C_1(t-s)}C_2(\|\vec{Z}(s)\|_2^2 + \|\vec{H}(s)\|_2^2 + \|\vec{E}(s)\|_2^2)^2 ds \\
&= \int_0^t e^{C_1(t-s)}C_2(\|\vec{Z}(s)\|_2 + \|\vec{H}(s)\|_2 + \|\vec{E}(s)\|_2)^4 ds.
\end{aligned}$$

It follows from Lemma 6.5.4 that for $0 \le t \le T$,

$$\|\tilde{Z}(t)\| \le C(T, K)(\|\vec{Z}_0\|_2 + \|\vec{H}_0\|_2 + \|\vec{E}_0\|_2)^2.$$

Similarly, we can estimate $\|\tilde{H}(t)\|_2$ and $\|\tilde{E}(t)\|_2$. We have

Lemma 6.5.5 *If solutions of* (6.5.1)–(6.5.2) *are properly smooth, then* $S(t) : (\vec{Z}_0, \vec{H}_0, \vec{E}_0) \to (\vec{Z}(t), \vec{H}(t), \vec{E}(t))$ *is uniformly differentiable. Its differential at* $(\vec{Z}_0, \vec{H}_0, \vec{E}_0)$ *belongs to* $\mathcal{A}$ *and*

$$DS(t)(\vec{Z}_0, \vec{H}_0, \vec{E}_0) = (w(t), I(t), F(t))$$

is a solution of (6.5.55)–(6.5.58).

6.5.3　Dimension Estimate of Attractor

Now we estimate the Hausdorff dimension and fractal dimension of $\mathcal{A}$. Rewrite (6.5.37)–(6.5.39) as

$$z_t + \vec{f}(\vec{z}, \nabla \vec{z}, \Delta \vec{z}, \vec{H}; \vec{Z}, \nabla \vec{Z}, \Delta \vec{Z}, \vec{H}) = 0, \tag{6.5.66}$$

$$\vec{e}_t + \sigma \vec{e} - \nabla \times \vec{h} = 0, \tag{6.5.67}$$

$$\vec{h}_t + \beta \vec{z}_t + \nabla \vec{e} = 0, \tag{6.5.68}$$

where

$$\begin{aligned}
\vec{f}&(\vec{z}, \nabla \vec{z}, \Delta \vec{z}, \vec{H}; \vec{Z}, \nabla \vec{Z}, \Delta \vec{Z}, \vec{H}) \\
&= -\alpha_2 \Delta \vec{z} - 2\alpha_2 (\nabla \vec{z} \cdot \nabla \vec{Z})\vec{Z} - \alpha_2 |\nabla \vec{Z}|^2 \vec{z} \\
&\quad - \alpha_1 \vec{Z} \times \Delta \vec{z} - \alpha_1 \vec{Z} \times \vec{h} \\
&\quad + \alpha_1 (\Delta \vec{Z} + \vec{H}) \times \vec{z} - \alpha_2 \vec{h} + \alpha_2 (\vec{Z} \cdot \vec{H})\vec{z} \\
&\quad + \alpha_2 (\vec{Z} \cdot \vec{h})\vec{Z} + \alpha_2 (\vec{z} \cdot \vec{H})\vec{Z}.
\end{aligned} \tag{6.5.69}$$

Choosing periodic orthogonal function basis $(\varphi_j(x), e_j(x), h_j(x))$ such that
 (i)
$$\Delta \varphi_j = -\lambda_j^2 \varphi_j;$$

 (ii)
$$\|\varphi_j\|_2 = \|e_j\|_2 = \|h_j\|_2 = 1,$$

we have
$$\|\nabla \varphi_j\|_2 = |\lambda_j|, \quad \|\Delta \varphi_j\|_2 = \lambda_j^2.$$

From the definition we have

$$\begin{aligned}
\mathrm{Trac}\{L(u(t))Q_J(t)\} = \sum_{j=1}^{J} \Big[&(\vec{f}(\varphi_j, \nabla \varphi_j, \Delta \varphi_j, \vec{H}_j; \vec{Z}, \nabla \vec{Z}, \Delta \vec{Z}, \vec{H}), \varphi_j) \\
&+ \sigma(e_j, e_j) - (\nabla \times h_j, e_j) + (\nabla \times e_j, h_j) \\
&- \beta(\vec{f}(\varphi_j, \nabla \varphi_j, \Delta \varphi_j, \vec{H}_j; \vec{Z}, \nabla \vec{Z}, \Delta \vec{Z}, \vec{H}), h_j)\Big].
\end{aligned} \tag{6.5.70}$$

Since

$$-(\nabla \times h_j, e_j) + (\nabla \times e_j, h_j) = \int_\Omega \nabla \cdot (e_j \times h_j)dx = 0,$$

we only have in (6.5.70) the following two terms:

$$(\vec{f}(\varphi_j, \nabla\varphi_j, \Delta\varphi_j, \vec{H}_j; \vec{Z}, \nabla\vec{Z}, \Delta\vec{Z}, \vec{H}), \varphi_j)$$

and

$$(\vec{f}(\varphi_j, \nabla\varphi_j, \Delta\varphi_j, \vec{H}_j; \vec{Z}, \nabla\vec{Z}, \Delta\vec{Z}, \vec{H}), h_j).$$

It follows from (6.5.69) that

$$\begin{aligned}
&(\vec{f}(\varphi_j, \nabla\varphi_j, \Delta\varphi_j, \vec{H}_j; \vec{Z}, \nabla\vec{Z}, \Delta\vec{Z}, \vec{H}), \varphi_j) \\
&= -\alpha_2(\Delta\varphi_j, \varphi_j) - 2\alpha_2((\nabla\vec{Z} \cdot \nabla\varphi_j)\vec{Z}, \varphi_j) \\
&\quad - \alpha_2(|\nabla\vec{Z}|^2\varphi_j, \varphi_j) - \alpha_1(\vec{Z} \times \Delta\varphi_j, \varphi_j) \\
&\quad - \alpha_1(\vec{Z} \times \vec{H}_j, \varphi_j) - \alpha_2(h_j, \varphi_j) + \alpha_2((\varphi_j \cdot \vec{H})\vec{Z}, \varphi_j) \\
&\quad + \alpha_2((\vec{Z} \cdot h_j)\vec{Z}, \varphi_j) + \alpha_2((\vec{Z} \cdot \vec{H})\varphi_j, \varphi_j),
\end{aligned}$$

in which

$$-\alpha_2(\Delta\varphi_j, \varphi_j) = \alpha_2\lambda_j^2,$$

$$\begin{aligned}
\left|-2\alpha_2((\nabla\vec{Z} \cdot \nabla\varphi_j)\vec{Z}, \varphi_j)\right| &\le 2\alpha_2\|\nabla\varphi_j\|_2\|\varphi_j\|_2\|\nabla\vec{Z}\|_\infty \\
&= 2\alpha_2|\lambda_j|\|\nabla\vec{Z}\|_\infty,
\end{aligned}$$

$$\begin{aligned}
\left|-\alpha_2(|\nabla\vec{Z}|^2\varphi_j, \varphi_j)\right| &\le \alpha_2\|\nabla\vec{Z}\|_\infty^2\|\varphi_j\|_2^2 \\
&= \alpha_2\|\nabla\vec{Z}\|_\infty^2,
\end{aligned}$$

$$\begin{aligned}
\left|-\alpha_1(\vec{Z} \times \Delta\varphi_j, \varphi_j)\right| &= \left|\alpha_1(\nabla\varphi_j, \nabla\vec{Z} \times \varphi_j)\right| \\
&\le |\alpha_1|\|\nabla\varphi_j\|_2\|\varphi_j\|_2\|\nabla\vec{Z}\|_\infty \\
&= |-\alpha_1\lambda_j|\|\nabla\vec{Z}\|_\infty,
\end{aligned}$$

$$\left|-\alpha_1(\vec{Z} \times \vec{H}_j, \varphi_j)\right| \le |\alpha_1|\|h_j\|_2\|\varphi_j\|_2 = |\alpha_1|,$$

$$|-\alpha_2(h_j, \varphi_j)| \le \alpha_2,$$

$$\begin{aligned}
\left|\alpha_2((\varphi_j \cdot \vec{H})\vec{Z}, \varphi_j)\right| &\le \alpha_2\|\vec{H}\|_\infty\|\varphi_j\|_2^2 \\
&= \alpha_2\|\vec{H}\|_\infty,
\end{aligned}$$

$$\left|\alpha_2((\vec{Z} \cdot h_j)\vec{Z}, \varphi_j)\right| \le \alpha_2\|h_j\|_2\|\varphi_j\|_2 = \alpha_2,$$

$$\left|\alpha_2((\vec{Z} \cdot \vec{H})\varphi_j, \varphi_j)\right| \le \alpha_1\|\vec{H}\|_\infty.$$

Hence,

$$\begin{aligned}
&(\vec{f}(\varphi_j, \nabla\varphi_j, \Delta\varphi_j, \vec{H}_j; \vec{Z}, \nabla\vec{Z}, \Delta\vec{Z}, \vec{H}), \varphi_j) \\
&\qquad \ge \alpha_2\lambda_j^2 - (2\alpha_2 + |\alpha_1|)\lambda_j\|\nabla\vec{Z}\|_\infty - \alpha_2\|\nabla\vec{Z}\|_\infty^2.
\end{aligned} \tag{6.5.71}$$

Similarly, it follows from (6.5.69) that

$$-\beta(\vec{f}(\varphi_j, \nabla\varphi_j, \Delta\varphi_j, h_j; \vec{Z}, \nabla\vec{Z}, \Delta\vec{Z}, \vec{H}), h_j)$$
$$= \alpha_2\beta(\Delta\varphi_j, h_j) + 2\alpha_2\beta((\nabla\vec{Z} \cdot \nabla\varphi_j)\vec{Z}, , h_j)$$
$$+ \alpha_2\beta(|\nabla\vec{Z}|^2\varphi_j, h_j) + \alpha_1\beta(\vec{Z} \times \Delta\varphi_j, h_j)$$
$$- \alpha_1\beta((\Delta\vec{Z} + \vec{H}) \times \varphi_j, h_j) + \alpha_2\beta(h_j, h_j)$$
$$+ \alpha_2\beta((\varphi_j \cdot \vec{H})\vec{Z}, h_j) - \alpha_2\beta((\vec{Z} \cdot h_j)\vec{Z}, h_j)$$
$$- \alpha_2\beta((\vec{Z} \cdot h_j)\vec{Z}, h_j) - \alpha_2\beta((\vec{Z} \cdot \vec{H})\varphi_j, h_j),$$

in which

$$|\alpha_2\beta(\Delta\varphi_j, h_j)| \leq \alpha_2\beta\|\Delta\varphi_j\|_2\|h_j\|_2 = \alpha_2\lambda_j^2\beta,$$
$$\left|2\alpha_2\beta((\nabla\vec{Z} \cdot \nabla\varphi_j)\vec{Z}, h_j)\right| \leq 2\alpha_2\beta\|\nabla\varphi_j\|_2\|h_j\|_2\|\nabla\vec{Z}\|_\infty$$
$$= 2\alpha_2\beta|\lambda_j|\|\nabla\vec{Z}\|_\infty,$$
$$\left|\alpha_2\beta(|\nabla\vec{Z}|^2\varphi_j, h_j)\right| \leq \alpha_2\beta\|\nabla\vec{Z}\|_\infty^2\|h_j\|_2\|\varphi_j\|_2$$
$$= \alpha_2\beta\|\nabla\vec{Z}\|_\infty^2,$$
$$\left|\alpha_2\beta(\vec{Z} \times \Delta\varphi_j, h_j)\right| \leq \alpha_2\beta\|\Delta\varphi_j\|_2\|h_j\|_2 = \alpha_2\lambda_j^2\beta,$$
$$\left|-\alpha_1\beta((\Delta\vec{Z} + \vec{H}) \times \varphi_j, h_j)\right| \leq |\alpha_1|\beta\|\Delta\vec{Z} + \vec{H}\|_\infty\|\varphi_j\|_2\|h_j\|_2$$
$$= |\alpha_1|\beta\|\Delta\vec{Z} + \vec{H}\|_\infty,$$
$$|\alpha_2\beta(h_j, h_j)| = \alpha_2\beta\|h_j\|_2^2 = \alpha_2\beta,$$
$$\left|-\alpha_2\beta((\varphi_j \cdot \vec{H})\vec{Z}, h_j)\right| \leq \alpha_2\beta\|\vec{H}\|_\infty,$$
$$\left|-\alpha_2\beta((\vec{Z} \cdot h_j)\vec{Z}, h_j)\right| \leq \alpha_2\beta,$$
$$\left|-\alpha_2\beta((\vec{Z} \cdot \vec{H})\varphi_j, h_j)\right| \leq \alpha_2\beta\|\vec{H}\|_\infty.$$

Hence

$$-\beta(\vec{f}(\varphi_j, \nabla\varphi_j, \Delta\varphi_j, \vec{H}_j; \vec{Z}, \nabla\vec{Z}, \Delta\vec{Z}, \vec{H}), h_j)$$
$$\geq -|\alpha_1|\beta\lambda_j^2 - 2\alpha_2\beta\lambda_j^2 - 2\alpha_2\beta|\lambda_j|\|\nabla\vec{Z}\|_\infty$$
$$- \left[\alpha_2\beta\|\nabla\vec{Z}\|_\infty^2 + |\alpha_1|\beta\|\Delta\vec{Z} + \vec{H}\|_\infty + 2\alpha_2\beta\|\vec{H}\|_\infty\right]. \qquad (6.5.72)$$

Substituting (6.5.71) and (6.5.72) into (6.5.70), we have

$$\text{Trac}\{L(u(t))Q_J(t)\} \geq (\alpha_2 - (\alpha_2 + |\alpha_1|\beta)) \sum_{j=1}^{J} \lambda_j^2$$

$$- (2\alpha_2 + 2\alpha_2\beta + |\alpha_1|) \sum_{j=1}^{J} \lambda_j \|\nabla \vec{Z}\|_\infty$$

$$+ \sigma - \alpha_2(1+\beta)\|\nabla \vec{Z}\|_\infty^2 - 2\alpha_2(1+\beta)\|\vec{H}\|_\infty$$

$$- |\alpha_1|\beta\|\Delta \vec{Z} + \vec{H}\|_\infty. \tag{6.5.73}$$

Choosing β small such that $\alpha_2 > (\alpha_2 + |\alpha_1|)\beta$, we have

$$0 < \beta < \frac{\alpha_2}{\alpha_2 + |\alpha_1|}. \tag{6.5.74}$$

Setting

$$\begin{cases} \delta = \alpha_2 - (\alpha_2 + |\alpha_1|)\beta, \\[2mm] X = \left(\sum_{j=1}^{J} \lambda_j^2 \right)^{\frac{1}{2}}, \\[2mm] a = (2\alpha_2 + 2\alpha_2\beta + |\alpha_1|)\|\nabla \vec{Z}\|_\infty, \\[2mm] b = \sigma - \alpha_2(1+\beta)\|\nabla \vec{Z}\|_\infty^2 - 2\alpha_2(1+\beta)\|\vec{H}\|_\infty \\[2mm] \quad - |\alpha_1|\beta\|\Delta \vec{Z} + \vec{H}\|_\infty. \end{cases} \tag{6.5.75}$$

Noting that

$$\sum_{j=1}^{J} |\lambda_j| \leq \left(\sum_{j=1}^{J} \lambda_j^2 \right)^{\frac{1}{2}} J,$$

we can rewrite (6.5.73) as

$$\text{Trac}\{L(u(t))Q_J(t)\} \geq \delta X^2 - aJ^{\frac{1}{2}}X + bJ$$

$$= \delta \left(X - \frac{aJ^{\frac{1}{2}}}{2\delta} \right)^{\frac{1}{2}} + \frac{4\delta b - a^2}{4\delta} J. \tag{6.5.76}$$

It follows from (6.5.76) that for $4\delta b - a^2 \leq 0$

$$\text{Trac}\{L(u(t))Q_J(t)\}$$

$$\geq \delta \left(X - \frac{a + \sqrt{a^2 - 4\delta b}}{2\delta} J^{\frac{1}{2}} \right) \left(X - \frac{a - \sqrt{a^2 - 4\delta b}}{2\delta} J^{\frac{1}{2}} \right).$$

Take J such that

$$X \geq \frac{a + \sqrt{a^2 - 4\delta b}}{2\delta} J^{\frac{1}{2}}, \tag{6.5.77}$$

and estimate λ_j as in [138] as follows:

$$\lambda_j^2 \geq \left[\frac{(j-1)^{\frac{1}{n}}}{2} - 1\right]^2 = \frac{1}{4}(j-1)^{\frac{2}{n}} - (j-1)^{\frac{1}{n}} + 1,$$

that is

$$\lambda_j^2 \geq \begin{cases} \frac{1}{4}(j-1)^2 - j + 2 = \frac{1}{4}j^2 - \frac{3}{2}j + \frac{9}{4}, & n = 1, \\ \frac{1}{4}(j-1) - (j-1)^{\frac{1}{2}} + 1 = \frac{1}{4}j + \frac{3}{4} - (j-1)^{\frac{1}{2}}, & n = 2. \end{cases}$$

(i) $n = 1$.

$$\sum_{j=1}^{J} \lambda_j^2 \geq \frac{1}{4}\sum_{j=1}^{J} \lambda_j^2 - \frac{3}{2}\sum_{j=1}^{J} \lambda_j + \frac{9}{4}J$$

$$= \frac{1}{24}(J+1)(2J+1)J - \frac{3}{4}(J+1)J + \frac{9}{4}J$$

$$= \frac{1}{12}J^3 - \frac{8}{5}J^2 + \frac{37}{24}J.$$

In order to have (6.5.77), choose J_0 such that

$$\frac{1}{12}J_0^3 - \frac{8}{5}J_0^2 + \left[\frac{37}{24} - \frac{(a + \sqrt{a^2 - 4\delta b})^2}{4\delta^2}\right] J_0 > 0,$$

that is

$$2J_0^2 - 15J_0 + 37 - \frac{6(a + \sqrt{a^2 - 4\delta b})^2}{\delta^2} > 0,$$

$$\left(J_0 - \frac{15}{4}\right)^2 + \frac{71}{16} - \frac{3(a + \sqrt{a^2 - 4\delta b})^2}{\delta^2} > 0.$$

When

$$\left(J_0 - \frac{15}{4}\right)^2 + \frac{71}{16} - \frac{3(a + \sqrt{a^2 - 4\delta b})^2}{\delta^2} > 0,$$

that is

$$(a + \sqrt{a^2 - 4\delta b})^2 < 2\delta,$$

we may choose $J_0 = 1$.

When

$$(a + \sqrt{a^2 - 4\delta b})^2 \geq 2\delta,$$

we may choose

$$J_0 > \sqrt{\frac{3\left(a + \sqrt{(a^2 - 4\delta b)}\right)^2}{\delta^2} - \frac{71}{16}} + \frac{15}{4}.$$

(ii) $n = 2$.

$$\sum_{j=1}^{J} \lambda_j^2 \geq \frac{1}{4} \sum_{j=1}^{J} \lambda_j + \frac{3}{4} J - \sum_{j=1}^{J} (j-1)^{\frac{1}{2}}$$

$$= \frac{1}{8}(J+1)J + \frac{3}{4}J - \sum_{j=1}^{J-1} J^{\frac{1}{2}}$$

$$\geq \frac{J^2 + 7J}{8} - \left(\sum_{j=1}^{J-1} j \right)^{\frac{1}{2}} \sqrt{J-1}$$

$$\geq \frac{J^2 + 7J}{8} - \frac{1}{\sqrt{2}} J^{\frac{3}{2}}.$$

In order to have (6.5.77), choose J_0 such that

$$\frac{J_0 + 7}{8} - \frac{1}{\sqrt{2}} J_0^{\frac{1}{2}} > \frac{(a + \sqrt{a^2 - 4\delta b})^2}{4\delta^2},$$

that is

$$(J_0^{\frac{1}{2}} - 2\sqrt{2})^2 - \left(1 + 2 \left(\frac{a + \sqrt{a^2 - 4\delta b}}{\delta} \right)^2 \right) > 0,$$

$$J_0 > \left\{ 2\sqrt{2} + \left[1 + 2 \left(\frac{a + \sqrt{a^2 - 4\delta b}}{\delta} \right)^2 \right]^{\frac{1}{2}} \right\}^2 > 0.$$

We have from the above results the following:

Theorem 6.5.3 *Let* $\Omega \subset R^n (1 \leq n \leq 2)$ *be a bounded set and assume*
(i)

$$\alpha_2 > 0, \quad \sigma > \frac{4\alpha_2^2 + 2\beta^2 \alpha_2(|\alpha_1| + \alpha_2) + 2\alpha_1^2}{\alpha_2}.$$

(ii)

$$0 < \beta < \min \left\{ \frac{1}{2}, \frac{\alpha_2}{\alpha_2 + |\alpha_1|} \right\}.$$

(iii)

$$(\beta + \beta^2)(\alpha_1^2 + \alpha_2^2) < \frac{1}{4}.$$

(iv) *When* $n = 2$

$$\|\nabla \vec{Z}_0\|_2 + \|\vec{H}_0\|_2 + \|\vec{E}_0\|_2 \leq \nu,$$

where $\nu = \nu(\alpha_1, \alpha_2, \beta)$ *is a small constant. Then the periodic initial value problem*
(6.5.1)–(6.5.2) *has a attractor* $\mathcal{A} = \Omega(\overline{A})$ *and*

$$\overline{A} = \{(\vec{Z}, \vec{H}, \vec{E}) \in (H^2(\Omega), H^1(\Omega), H^1(\Omega)) | \|\vec{Z}\|_{H^2} + \|\vec{H}\|_{H^1} + \|\vec{E}\|_{H^1} \leq \varepsilon_0 + \delta_0\}$$

is a bounded absorbing set. The Hausdorff dimension and Fractal dimension of $\mathcal{A}$ are finite and satisfy

(1) If $a^2 - 4\delta b < 0$, then

$$d_H(\mathcal{A}) \leq 1, \quad d_F(\mathcal{A}) \leq 2;$$

(2) If $a^2 - 4\delta b \geq 0$, then

(a) when $n = 1$ and $a + \sqrt{a^2 - 4\delta b} < 2\delta$, then $d_H(\mathcal{A}) \leq 1$, $d_F(\mathcal{A}) \leq 2$; when $n = 1$ and $a + \sqrt{a^2 - 4\delta b} > 2\delta$, then $d_H(\mathcal{A}) \leq J_1$, $d_F(\mathcal{A}) \leq 2J_1$ where J_1 is the smallest integer subject to

$$J_0 > \sqrt{\frac{3\left(a + \sqrt{a^2 - 4\delta b}\right)^2}{\delta^2} - \frac{71}{16}} + \frac{15}{4}.$$

(b) When $n = 2$, then $d_H(\mathcal{A}) \leq J_2$, $d_F(\mathcal{A}) \leq 2J_2$, where J_2 is the smallest integer subject to

$$J_0 > \left\{ 2\sqrt{2} + \left[1 + 2\left(\frac{a + \sqrt{a^2 - 4\delta b}}{\delta}\right)^2 \right] \right\}^2,$$

in which

$$\delta = \alpha_2 - (\alpha_2 + |\alpha_1|)\beta,$$
$$a = (2\alpha_2 + 2\alpha_2\beta + |\alpha_1|)\|\nabla \vec{Z}\|_\infty,$$
$$b = \sigma - \alpha_2(1 + \beta)\|\nabla \vec{Z}\|_\infty^2$$
$$- 2\alpha_2(1 + \beta)\|\vec{H}\|_\infty + |\alpha_1|\beta\|\Delta \vec{Z} + \vec{H}\|_\infty.$$

6.6   Bibliography Comments

In this chapter we have studied the asymptotic properties for the system of ferromagnetic spin chain. Because the systems are weak degenerate, we do not have the expressions of the linearized equation. This makes the studies of properties when $t \to \infty$ very difficult. Guo and Lu [79] studied the stationary solution and its non-stability and the asymptotic behavior as $t \to \infty$ under some assumptions. Guo and Wang [85, 86] got the existence of approximate inertial manifold for the one-dimensional L–L equations. Guo *et al.* [78] proved the existence of attractor and the estimates of upper and lower bounds of Hausdorff and Fractal dimensions for such attractor for the L–L system on Riemannian manifold. Guo and Su [83] proved the existence of attractor of L–L–M system by the method of multi-parameter Lyapunov functional and have given the estimate in three dimensions. It is worth mentioning that in 1957, Suhl [131] proposed the chaotic property for the attractor of L–L system with Gilbert damping term. The numerical results were given in [147].

Bibliography

[1] Aïssa, N. and Hamdache, K., *Energy decay of solutions to Maxwell's equations with conductivity and polarization*, to appear.

[2] Aïssa, N. and Hamdache, K., *Asymptotics of time harmonic solutions to a thin ferroelectric model*, to appear.

[3] Akhiezer, A. I., Yakhtar, V. G. and Peletminskii, S. V., *Spin Waves* (North-Holland, 1968).

[4] Alouges, F. and Soyeur, A., *On global weak solutions for Landau–Lifshitz equations: Existence and nonuniqueness*, Nonlinear Anal. TMA **18** (1992) 1071–1084.

[5] Amann, H., *Quasilinear parabolic systems under nonlinear boundary conditions*, Arch. Rat. Mech. Anal. **92** (1986) 153–192.

[6] Ammari, H. and Hamdache, K., *Global existence and regularity of solutions to a system of nonlinear Maxwell equations*, J. Math. Anal. Appl. **286** (2003) 51–63.

[7] Amrouche, C., Bernardi, C., Dauge, M. and Girault, V., *Vector potentials in three-dimensional non-smooth domains*, Math. Meth. Appl. Sci. **21** (1998) 823–864.

[8] Aubin, T., *Nonlinear Analysis on Manifolds, Monge-Ampere Equations* (Springer-Verlag, 1982).

[9] Aygelman, S. D., *Parabolic System* (1994).

[10] Babin, A. V. and Visik, M. I., *Attractors of partial differential equations and estimate of their dimension*, Vspekhi Mat. Nauk. **38** (1983) 133–187.

[11] Babin, A. V. and Visnik, M. I, *Regular attractors of semi-groups and evolution equations*, J. Math Pures Appl. **62** (1983) 441–491.

[12] Balakrishnan, R., *On the inhomogeneous Heisenberg chain*, Phys. C. Solid State Phys. **15** (1982) L1305–L1308.

[13] Bartels, S., Ko, J. and Prohl, A., *Numerical approximation of the Landau–Lifshitz–Gilbert equation and finite time blow-up of weak solutions*, preprint, 2005.

[14] Bassiouny, E., Ghaleb, A. F. and Maugin, G. A., *Thermodynamical formulation for coupled electromechanical hysteresis effects-I. Basic equations*, Int. J. Engrg. Sci. **26** (1988) 1279–1295.

[15] Bethuel, F., Brezis, H. and Helein, F., *Asymptotics for the minimization of a Ginzburg–Landau functional*, Calc. Vr. P.D.E. **1** (1993) 123–148.

[16] Bethuel, F., Brezis, H. and Helein, F., *Ginzburg-Landau Vortices* (Birkhäuser, 1994).

[17] Böttcher, C. J. F., *Theory of Electric Polarization* (Elsevier, 1952).

[18] Carbou, G., *Regularity for critical points of a nonlocal energy*, Calc. Var. **5** (1997) 409–433.

[19] Carbou, G., *Thin layers in micromagnetism*, Math. Model. Meth. Appl. Sci. **11** (2001) 1529–1546.

[20] Carbou, G. and Fabrie, P., *Regular solutions for Landau–Lifshitz equation in a bounded domain*, Diff. Int. Eqns. **14** (2001) 213–229.

[21] Carbou, G. and Fabrie, P., *Time average in micromagnetism*, J. Diff. Eqns. **147** (1998) 383–409.

[22] Carbou, G., Fabrie, P. and Gués, O., *Couche limite dans un modéle de ferro-magnétisme*, Comm. P.D.E. **27** (2002) 1467–1495.

[23] Carbou, G., Fabrie, P. and Gués, O., *Boundary layers for Landau–Lifshitz equations*, Physica B **343** (2004) 331–336.

[24] Carbou, G. and Labbe, S., *Stability for static walls in ferromagnetic nanowires*, Proc. ENUMATH 2005 (Springer, 2006), pp. 539–546.

[25] Chang, K. C., *Heat flow and boundary value problem for harmonic map*, Anal. Non-Lineaire **6** (1989) 363–395.

[26] Chang, K. C., Ding, W. Y. and Ye, R., *Finite-time blow-up of the heat flow of harmonic maps from surfaces*, J. Diff. Geom. **36** (1992) 507–515.

[27] Chang, N.-H., Shatah, J. and Uhlenbeck, K., *Schrödinger maps*, Comm. Pure Appl. Math. **53**(5) (2000) 590–602.

[28] Chen, Y., *The weak solutions to the evolution problems of harmonic maps*, Math. Z. **201** (1989) 69–74.

[29] Chen, Y., Ding, S. and Guo, B., *Partial regularity for two-dimensional Landau–Lifshitz equations*, Acta Math. Sinica, New Ser. **14** (1998) 423–432.

[30] Chen, Y. and Ding, W. Y., *Blow-up and global existence for heat flows of harmonic maps*, Invent. Math. **99** (1990) 567–578.

[31] Chen, Y. and Guo, B., *Landau–Lifshitz equation from Riemannian surfaces*, J. P.D.E. **9** (1996) 313–322.

[32] Chen, Y., Li, J. and Lin, F. H., *Partial regularity for weak heat flows into spheres*, Comm. Pure Appl. Math. **48** (1995) 429–448.

[33] Chen, Y. and Lin, F. H., *Evolution of harmonic maps with Dirichlet boundary conditions*, Comm. Anal. Geom. **1** (1993) 327–346.

[34] Chen, Y. and Struwe, M., *Existence and partial regularity results for the heat flow of harmonic maps*, Math. Z. **201** (1989) 83–103.

[35] Chen, Y. and Wang, C., *Partial regularity for weak flows into Riemannian homogeneous space*, Comm. P.D.E. **21** (1996) 735–761.

[36] Coifman, R. and Fefferman, R., *Weighted norm inequality for maximal functions and singular integrals*, Studia Math. **51** (1974) 241–250.

[37] Coifman, R., Lions, P. L., Meyer, Y. and Siemmes, S., *Compacite par compensation et espass de Hardy*, C. R. Acad Sci. Paris **30** (1989) 945–949.

[38] Coron, J., *Nonuniqueness for the heat flow of harmonic maps*, Ann. Inst. H. Poincaré, Non-Linéairé **7** (1990) 335–344.

[39] Coron, J. and Ghidaglia, J., *Explosion en temps fini pour le flot des applications harmoniques*, (French) [Finite-time blow-up for the heat flow of harmonic mappings] C. R. Acad. Sci. Paris, I Math. **308** (1989) 339–344.

[40] Constantin, P., Foias, C. and Temam, R., *Attractor representing turbulent flows*, Mem. Am. Math. Soc. **53** (1985) 314–380.

[41] Cooper, J. and Strauss, W., *The initial boundary problem for the Maxwell equations in the presence of a moving boundary*, SIAM J. Math. Anal. **16** (1985) 1165–1179.

[42] Davi, F., *On domain switching in deformable ferroelectrics, seen as continua with microstructure*, Z. Angew. Math. Phys. **52** (2001) 966–989.

[43] DeBye, P., *Polar Molecules* (Dover, 1945).

[44] DeSimone, A., Kohn, R. V., Müller, S. and Otto, F., *A reduced theory for thin film micromagnetics*, Comm. Pure Appl. Math. **55** (2002) 1408–1460.

[45] DeSimone, A., Kohn, R. V., Müller, S. and Otto, F., *Magnetic microstructure-a paradigm of multiscale problem*, Proc. ICIAM 99 (Edinburgh), eds. J. Ball and J. C. R. Hunt (1999), pp. 175–199.

[46] DeSimone, A., Kohn, R. V., Müller, S. and Otto, F., *A compactness result in the gradient theory of phase transitions*, Proc. Roy. Soc. Edinburgh Sect. A **131** (2001) 833–844.

[47] Ding, S. and Guo, B., *Hausdorff measure of the singular set of Landau–Lifshitz equations with a nonlocal term*, Comm. Math. Phys. **250** (2004) 95–117.

[48] Ding, S. and Guo, B., *Initial-boundary value problem for higher dimensional Landau–Lifshitz systems*, Appl. Anal. **83** (2004) 673–697.

[49] Ding, S. and Guo, B., *Existence of Partially regular weak solutions to Landau–Lifshitz–Maxwell equations*, preprint, 2006.

[50] Ding, S., Guo, B. and Su, F., *Measure-valued solutions to the strongly degenerate compressible Heisenberg chain equations*, J. Phys. Math. **40** (1999) 1153–1162.

[51] Ding, S., Guo, B. and Su, F., *Smooth Solution for one-dimensional inhomogeneous Heisenberg chain equations*, Proc. Roy. Soc. Edinburgh **129A** (1999) 1171–1184.

[52] Ding, S. and Wang, C., *Finite time singularity of Landau–Lifshitz–Gilbert equations*, Int. Math. Res. Notices, Vol. 2007(4), 2007.

[53] Donnat, P. and Rauch, J., *Global solvability of the Maxwell–Bloch equations from nonlinear optics*, Arch. Rat. Mech. Anal., **136**(3) (1996) 291–303.

[54] Eidel'man, S. D., *Parabolic Systems* (North-Holland, 1969).

[55] Evans, L. C., *Partial regularity for harmonic maps into spheres*, Arch. Rat. Mech. Anal. **116** (1991) 101–113.

[56] Evans, L. C. and Gariepy, R. F., *Measure Theory and Fine Properties of Functions*, Studies in Advances Math. (CRC Press, 1992).

[57] Fefferman, C. and Stein, E., $\mathcal{H}^p$ *spaces of several variables*, Acta. Math. **129** (1972) 137–193.

[58] Feldman, M., *Partial regularity for harmonic maps of evolution into spheres*, Comm. P.D.E. **19** (1994) 761–790.

[59] Fivez, J., *On the continuum limit of a classical compressible Heisenberg Chain*, Phys. C: Solid State Phys. **15** (1982) L641 L643.

[60] Fogedby, H. C., *Theoretical Aspects of Mainly Low Dimensional Magnetic System*, Lect. Notes. Phys., Vol. 131 (Springer, 1980).

[61] Freire, A., *Uniqueness for the harmonic heat flow in two dimensions*, Calc. Var. P.D.E. **4** (1996) 761–790.

[62] Fröhlich, H., *Theory of Dielectrics* (Oxford University Press, 1949).

[63] Ghidaglia, J. M., Marison, M. and Temam, R., *Generalization of the Sobolev–Lieb–Thirring inequalities and applications to the dimension of attractors*, Diff. Int. Eqns. **1** (1988) 1–21.

[64] Giaquinta, M., *Introduction to Regularity Theory for Nonlinear Elliptic Systems* (Birkhäuser Verlag, 1993).

[65] Gibson, G. and Jeffries, C., Phys. Rev. **429** (1984) 811.

[66] Gill, T. L. and Zachary, W. W., *Existence and Finite Dimension Ality of Attractors for the Landau–Lifshitz Equations, in Differential Equations in Mathematical Physics*, eds. I. Knowles and Y. Saito, Lecture Note in Mathematics (Springer-Verlag, 1985).

[67] Gilbarg, D. and Trudinger, N. S., *Elliptic Partial Differential Equations of Second Order* (Springer-Verlag, 2001).

[68] Gioia, G. and James, R. D., *Micromagnetics of very thin films*, Proc. Roy. Soc. Lond., Ser. A **453** (1997) 213–223.

[69] Greenberg, J. M., MacCamy, R. C. and Coffman, C. V., *On the long-time behavior of ferroelectric systems*, Physica D **134** (1999), 362–383.

[70] Grisvard, P. P., *Equations differentiales abstraites*, Ann. Scient. EC. Norm. Sup. 4 Serie. T. **2** (1969) 311–395.

[71] Guo, B. and Chen, D., *On infinite conservation laws of generalized Heisenberg equation*, Arch. Math. Sci. **13** (1993) 298–302.

[72] Guo, B. and Ding, S., *Initial-boundary value problem for the Landau–Lifshitz system(I)-existence and partial regularity*, Progr. Nat. Sci. **8** (1998) 11–23.

[73] Guo, B. and Ding, S., *Initial-boundary value problem for the Landau–Lifshitz(II)-uniqueness*, Progr. Nat. Sci. **8** (1998) 147–151.

[74] Guo, B. and Ding, S., *Neumann problem for the Landau–Lifshitz–Maxwell system in two dimensions*, Chin. Ann. Math., Ser. B **22** (2001) 529–540.

[75] Guo, B. and Ding, S., *Initial boundary value problem for the unsaturated Landau–Lifshitz system of ferromagnetic spin chain*, Chin. Ann. Math., Ser. B **21** (2000) 389–402.

[76] Guo, B. and Ding, S., *Initial boundary value problem for the Landau–Lifshitz system with applied fields*, Proc. on Nonlinear Partial Differential Equations and Applications (World Scientific, 1998).

[77] Guo, B. and Hong, M., *The Landau–Lifshitz equations of the ferromagnetic spin chain and harmonic maps*, Calc. Var. **1** (1993) 311–334.

[78] Guo, B., Hong, M. C. and Su, F., *The Attractors for the Landau–Lifshitz Equation of the Ferromagnetic Spin Chain on Compact Manifold*, Beijing Math., Vol. 2 (1996), pp. 40–75; AMS/IP Stud. Adv. Math. 3 (Amer. Math. Soc., 1997), pp. 213–227.

[79] Guo, B. and Lu, B., *Stability for steady solution of Landau–Lifshitz equation*, preprint.

[80] Guo, B. and Su, F., *Global weak solution for the Landau–Lifshitz–Maxwell equation in three space dimensions*, J. Math. Anal. Appl. **211** (1997) 326–346.

[81] Guo, B. and Su, F., *The global smooth solution for Landau–Lifshitz–Maxwell equations without dissipation*, J. Partial Diff. Eqns. **11**(3) (1998) 193–208.

[82] Guo, B. and Su, F., *The global solution for Landau–Lifshitz–Maxwell equations*, J. Partial Diff. Eqns. **14**(2) (2001) 133–148.

[83] Guo, B. and Su, F., *The attractors for Landau–Lifshitz–Maxwell equations*, J. Partial Diff. Eqns. **13**(4) (2000) 320–340.

[84] Guo, B. and Su, F., *Global weak solution for Landau–Lifshitz–Maxwell equations on Riemannian manifold*, preprint.

[85] Guo, B. and Wang, B., *Global existence of the solution for the Landau–Lifshitz equation of the ferromagnetic spin chain*, Math. Sin. **17** (1997) 429–436.

[86] Guo, B. and Wang, B., *Approximation to the global attractor for the Landau–Lifshitz equation of the forromagnetic spin chain*, Beijing Math. **1** (1995) 160–175.

[87] Guo, B. and Wang, Y., *Generalized Landau–Lifshitz systems of the ferromagnetic spin chain and harmonic maps*, Science in China, Ser. A **39** (1996) 1242–1287.

[88] Hamdache, K. and Ngningone, I., *Time harmonic solutions to a model of ferroelectric materials*, Appl. Anal. (2005) 1–16.

[89] Harpes, P., *Partial compactness for the 2-D Landau–Lifshitz flow*, Electronic J. Diff. Eqns. **90** (2004) 1–24.

[90] Helein, F., *Regularite des Applications Faiblerment Harmoniques Entre Une Surface et Une Sphere*, Comptes Rendus Acad. Sci., Serie I **T311** (1990) 519–524.

[91] Hungerbuhler, N., private communication.

[92] Iwaniec, T. and Martin, G., *Quasiregular mappings in even dimension*, Acta Math.

[93] Jackson, J. D., *Classical Electrodynamics*, 3rd edn., Chinese version (Higher Education Press, 2004).

[94] Jochmann, F., *Long Time asymptotics of solutions to the anharmonic oscillator model from nonlinear optics*, SIAM J. Math. Anal. **32** (2000) 887–915.

[95] Joly, J. L., Métivier, G. and Rauch, J., *Global solvability of the anharmonic oscillator model from nonlinear optics*, SIAM J. Math. Anal. **27** (1996) 905–913.

[96] Kamlah, M. and Tsakmakis, C., *Phenomenological modeling of the non-linear electro-mechanical coupling in ferroelectrics*, Int. J. Solids Str. **36** (1999) 669–695.

[97] Klaina, S., *Global existence of nonlinear wave equations*, Comm. Pure Appl. Math. **33** (1980) 43–101.

[98] Kohn, R. V. and Slastikov, V. V., *Another thin film limit of micromagnetics*, Arch. Rat. Mech. Anal. **178** (2005) 227–245.

[99] Kurzke, M., Melcher, C. and Moser, R., *Domain walls and vortices in the thin ferromagnetic films*, Physica B **343** (2004) 331–336.

[100] Kurzke, M., Melcher, C. and Moser, R., *Domain walls and vortices in thin ferromagnetic films*, MPI Leipzig preprint, No. 45 (2006).

[101] Ladyzhenskaja, O., Solonnikov, V. A. and Ural'ceva, N. N., *Linear and Quasilinear Equations of Parabolic Type*, AMS Transl. Math. Monogr., Vol. 23 (Amer. Math. Soc., 1968).

[102] Landau, L. D. and Lifshitz, E. M., *On the theory of the dispersion of magnetic permeability in ferromagnetic bodies*, Phys. Z. Sovietunion. **8** (1935) 153–169.

[103] Landau, L. D. and Lifshitz, E. M., *Electrodynamique des milieux continuous*, Cours de Physique Theorique, Vol. VIII (MIR, 1969).

[104] Landau, L. D. and Lifshitz, E. M., *Electrodynamics of Continuous Media* (Pergamon Press, 1960).

[105] Lakshmanan, M., Ruijgrok, T. W. and Thompson, C. J., *On the dynamics of a continuous spin systems*, Phys. A **84A** (1976) 577–590.

[106] Lakshmannan, M. and Nakamura, K., *Landau–Lifshitz equation of ferromagnetism: Exact treatment of the Gilbert damping*, Phys. Rev. Lett. **53** (1984) 2497–2499.

[107] Lakshmanan, M. and Daniel, M., *Soliton damping and energy loss in the classical continuum Heisenberg spin chain*, Phys. Rev. B **24** (1981) 6750–6754.

[108] Lin, J. and Ding, S., *Smooth solution to the one dimensional Inhomogeneous non-automorphic Landau–Lifshitz equation*, Proc. Roy. Soc. Lond., Ser. A **462** (2006) 2397–2413.

[109] Liu, X., *Partial regularity for Landau–Lifshitz system of ferromagnetic spin chain*, Calc. Var. **20** (2004) 153–173.

[110] Magyari, E., *Solitary waves along the compressible Heisenberg chain*, Phys. C: Solid State Phys. **15** (1982) L1159–L1163.

[111] Malek, J., Necas, J., Rokyta, M. and Ruzicka, M., *Weak and Measure-Valued Solutions to Evolutionary PDES* (Chapman & Hall, 1996).

[112] Melcher, C., *Domain wall motion in ferromagnetic films*, Phys. D. **192** (2004) 249–264.

[113] Melcher, C., *Existence of partially regular solutions for Landau–Lifshitz equations in* $\mathbf{R}^3$, Comm. P.D.E. **30** (2005) 567–587.

[114] Michelitsch, T. and Kreher, W. S. , *A simple model for the nonlinear material behavior of ferroelectrics*, Int. Acta Mater. **46** (1998) 5085–5094.

[115] Moser, R., *Partial regularity for the Landau–Lifshitz equation in small dimensions*, MPI (Leipzig) preprint, 2002.

[116] Moser, R., *Ginzburg-Landau vortices in thin ferromagnetic films*, ARMX Appl. Math. Res. Express **1** (2003) 1–32.

[117] Moser, R., *Boundary vortices for thin ferromagnetic films*, Arch. Rat. Mech. Anal. **174** (2004) 267–300.

[118] Moser, R., *Partial regularity for Landau–Lifshitz equations*, Preprint Series, Max-Planck-Institute for Mathematics in the Sciences 26

[119] Moser, R., *Partial Regularity for Harmonic Maps and Related Problems* (World Scientific, 2005).

[120] Moser, R., *Moving boundary vortices for a thin film limit in micromagnetics*, Comm. Pure Appl. Math. **58** (2005) 701–721.

[121] Nahamura, C. K., Ohta, S. and Kawasahi, K., *Chaotic states on ferromagnets in strong parallel pumping fields*, J. Phys. C **15** (1982) 143–148.

[122] Nakamura, K. and Sasada, T., *Soliton and wave trains in ferromagnets*, Phys. Lett. **48A** (1974) 321–322.

[123] Pistella, F. and Valente, V. *Numerical study of the appearance of singularities in ferromagnets*, Adv. Math. Sci. Appl. **12** (2002) 803–816.

[124] Porsezian, K. and Lakshmanan, M., *On the dynamics of the radially symmetric Heisenberg ferromagnetic spin system*, J. Math. Phys. **32** (1991) 2923–2928.

[125] Riviére, T., *Everywhere discontinuous harmonic maps from B^3 into S^2*, C. R. Acad. Sci. Paris **314** (1992) 719–723.

[126] Riviére, T., *Flot des applications harmoniques en dimension deux* (1993).

[127] Schoen, R. and Yau, S. T., *Differential Geom.* (Academic Press, 1998).

[128] Semmes, S., *A primer on Hardy spaces, and some remarks on a theorem of Evans and Müller*, Comm. P.D.E. **19** (1994) 277–319.

[129] Siman, J., *Nonhomogeneous viscous incompressible fluids existence of vecocity, density and pressure*, SIAM J. Math. Anal. **21** (1990) 1039–1117.

[130] Su, F. Q. and Guo, B., *The global smooth solution for Landau–Lifshitz–Maxwell equation without dissipation*, J. P.D.E. **11** (1998) 193–208.

[131] Suhl, H., *The theory of ferromagnetic resonance at high signal powers*, J. Phys. Chem. Solids (1957) 209–227.

[132] Sulem, P. L., Sulem, C. and Bardos, C., *On the continuous limit for a system of classical spin*, Comm. Math. Phys. **107** (1986) 431–454.

[133] Struwe, M., *On the evolution of harmonic maps on Riemannian surface*, Comm. Math. Helv. **60** (1985) 558–581.

[134] Struwe, M., *On the evolution of harmonic maps in higher dimensions*, J. Diff. Geom. **28** (1988) 485–502.

[135] Struwe, M., *Geometric Evolution Problems, in Nonlinear Partial Differential Equations in Differential Geometry* (Park City, UT, 1992), IAS/Park City Math. Ser., 2 (Amer. Math. Soc., 1996), pp. 257–339.

[136] Stein, E. M., *Singular Integrals and Differentiability Properties of Functions* (Princeton University Press, 1970).

[137] Takhtalian, L. A., *Integration of the continuous Heisenberg spin chain through the inverse scattering method*, Phys. Lett. **64A** (1977) 235.

[138] Temam, R., *Infinite Dimensional Dynamical in Mechanics and Physics* (Springer-Verlag, 1998).

[139] Tjon, J. and Wright, J., *Soliton in the continuous Heisenberg chain*, Phys. Rev. B **15** (1977) 3470–3476.

[140] Visintin, A., *On Landau–Lifshitz equations for ferromagnetism*, Jpn. J. Appl. Math. **2** (1985) 69–84.

[141] Visintin, A., *Maxwell equations with vector hysteresis*, Arch. Rat. Mech. Anal., DOI:10.1007/s00205-004-0333-6.

[142] Wang, C. Y., *On moving Ginzburg–Landau vortices*, Comm. Anal. Geom. **12** (2004) 1185–1199.

[143] Wang, C. Y., *On Landau–Lifshitz equation in dimensions at most four*, to appear, Indiana Univ. Math. J.

[144] Wente, H., *An existence theorem for surfaces of constant mean curvature*, J. Math. Anal. Appl. **26** (1969) 318–344.

[145] White, B., *Infima of energy functionals in homotopy classes of mappings*, J. Diff. Geom. **23** (1986) 127–142.

[146] Zakharov, V. E. and Tekhtajan, L. A., *Equivalence of nonlinear Schrödinger equation and Heisenberg ferromagnet*, Theor. Mat. Phys. **38** (1979) 17.

[147] Zhang, X. Y. and Suhl, H., *Spin wave-related period doublings and chaos under transverse pumping*, Phys. Rev. A **32** (1985) 2530–2533.

[148] Zhong W., *Physics of Ferroelectric* (in Chinese) (Science Press, 2000).

[149] Zhou, Y. L., *Finite difference solutions of the nonlinear mutual boundary problems for the ferromagnetic chain systems*, J. Comput. Math. **2** (1984) 263–271.

[150] Zhou, Y., Sun, H. and Guo, B., *Existence of weak solution for boundary problems of systems of ferromagnetic chain*, Sci. Sin. A **27** (1981) 779–811.

[151] Zhou, Y., Sun, H. and Guo, B., *On the solvability of the initial value problem for the quasilinear degenerate parabolic system:* $\vec{Z}_t = \vec{Z} \times \vec{Z}_{xx} + \vec{f}(x, t, \vec{Z})$, Proc. DD-3 Symp. (1982) 713–732.

[152] Zhou, Y., Sun, H. and Guo, B., *Finite difference solutions of the boundary problems for systems of ferromagnetic chain*, J. Comp. Math. **1** (1983) 294–302.

[153] Zhou, Y., Sun, H. and Guo, B., *Existence of weak solution for boundary problems of ferromgnetic chain*, Sci. Sin. A **27** (1984) 799–811.

[154] Zhou, Y., Sun, H. and Guo, B., *The weak solution of homogeneous boundary value problem for the system of ferromagnetic chain with several variables*, Sci. Sin. **4(A)** (1986) 337–349.

[155] Zhou, Y., Sun, H. and Guo, B., *Some boundary problems of the spin system and the system of ferro magnetic chain I: Nonlinear boundary problems*, Acta Math. Sci. **6** (1986) 321–337.

[156] Zhou, Y., Sun, H. and Guo, B., *Some boundary problems of the spin system and the system of ferromagnetic chain II: Mixed problems and others*, Acta Math. Sci. **7** (1987) 121–132.

[157] Zhou, Y., Sun, H. and Guo, B., *Weak solution systems of ferromagnetic chain with several variables*, Science in China **A30** (1987) 1251–1266.

[158] Zhou, Y., Guo, B. and Tan, S., *Existence and uniqueness of smooth solution for system of ferromagnetic chain*, Science in China, Ser. A **34** (1991) 257–266.

[159] Zhou, Y., Sun, H. and Guo, B., *Geometrical extensions for systems of ferromagnetic chain*, Science in China, Ser. A **36** (1993) 927–939; **36** (1993) 1422–1434.